Contents

DISCOVERING GENOMICS, PROTEOMICS, AND BIOINFORMATICS

Second Edition

A. Malcolm Campbell
Davidson College

Laurie J. Heyer
Davidson College

CSHL PRESS

PEARSON
Benjamin Cummings

San Francisco Boston New York
Cape Town Hong Kong London Madrid Mexico City
Montreal Munich Paris Singapore Sydney Tokyo Toronto

Editor: Susan Winslow
Project Editor: Geoffrey Heilpern
Editorial Assistant: Mercedes Grandin
Managing Editor: Michael Early
Production Supervisor: Lori Newman
Production and Composition Services: GGS Book Services
Text Designer: Gary Hespenheide
Cover Designer: Yvo Reizebos
Copyeditor: Brooke Graves
Illustrations: GGS Book Services
Proofreader: Denne Wesolowski
Indexing: Schroeder Indexing Services
Photo Research: Maureen Spuhler
Manufacturing Buyer: Stacy Wong
Marketing Manager: Jeff Hester
Cover Printer: Phoenix Color
Printer: Von Hoffmann

Library of Congress Cataloging-in-Publication Data
Campbell, A. Malcolm
   Discovering genomics, proteomics, and bioinformatics / A. Malcolm Campbell,
Laurie J. Heyer—2nd ed.
   p. cm.
Includes index.
ISBN 0-8053-8219-4
1. Genomics. 2. Proteomics. 3. Bioinformatics. I. Heyer, Laurie, J. II. Title.
QH447 .C35 2002
572.8'6—dc21   2002067456

ISBN 0-8053-8219-4
3 4 5 6 7 8 9 10 — VH — 09

www.aw=bc.com

Unit Three

Foreword

Francis S. Collins
Director
National Human Genome Research Institute
National Institutes of Health

The world of biology has undergone a stunning transformation in recent years. Fueled by the sequencing of the human genome and the creation of a host of databases brimming with information on nucleic acids, proteins, and their properties, biology has at long last joined chemistry and physics as a truly quantitative science.

Yet, many challenges lie ahead. We have just begun to tap into the potential of genomics and proteomics to expand biological understanding and improve human health. For the full benefits of this revolution to be realized, all scientists must be equipped with the knowledge needed to creatively mine these vast troves of biological data in ways that will uncover clues to life's biggest mysteries.

As set forth in the National Human Genome Research Institute's blueprint for the genomic era, efforts to better understand biological pathways, networks, and molecular systems in an integrated fashion will require information from several levels. At the genetic level, one grand challenge facing biologists today is to find ways to simultaneously monitor the expression of all genes in a cell. Even more ambitious is the goal of developing techniques that allow for real-time measurement of the abundance, location, modification, and activity of all proteins in a cell. At the analytical level, computational methods are now intrinsic to modern biological research. As the amount and complexity of genomic and proteomic data increase, and as the questions addressed become more sophisticated, the importance of bioinformatic innovations is destined to grow.

Given the rapid progress being made in genomics, proteomics, and bioinformatics, it may seem a task worthy of Sisyphus to publish a textbook. However, rather than presenting a body of time-sensitive facts, *Discovering Genomics, Proteomics, and Bioinformatics* has wisely chosen to provide students with a solid intellectual framework for exploring these complex disciplines as they move forward in their scientific endeavors. This text clearly sets forth the basic principles of genomics and proteomics, while at the same time emphasizing the integral role of bioinformatics in analyzing the complex datasets generated by these disciplines.

Bringing biology to life through an interactive, problem-based instructional approach, this new edition places increased emphasis on synthetic biology, comparative genomics, and prokaryote genomics. The updated and expanded text also contains many more Math Minutes, which are student exercises that utilize statistical and probability calculations to add quantitative rigor to the interpretation of data sets.

The text's multi-disciplinary focus reflects the increasingly complex dynamics of biological research in the 21st century. As we build upon the foundation laid by the Human Genome Project, our ability to explore uncharted frontiers will hinge upon melding biological know-how with expertise in computer science, physics, math, clinical research, bioethics, and many other disciplines. Such interactions will be needed at the individual level, the collaborative level, and even at the disciplinary level as new disciplines emerge at the interfaces between traditional research boundaries. Consequently, today's biology courses need to cast a wide net to capture the imaginations of students representing many different interests, skills, and viewpoints.

Another forward-looking component of the text is Discovery Questions, which, along with posing standard scientific queries, encourages students to delve further into the ethical, legal, and social issues raised by genomic and proteomic advances. It is increasingly important that every biologist recognizes the responsibility to weigh the ethical, legal, and social implications of his or her research before embarking upon a project or advocating a new technology.

Clearly, a remarkable landscape of opportunity lies before the next generation of biologists. It is impossible to predict where these future explorations will lead or what discoveries will be made. But one thing is certain: a firm understanding of the powerful potential of genomics, proteomics, and bioinformatics will be essential to success in this amazing new world.

Preface

The term "genomics" was derived from the term "genome," which means the complete (haploid) DNA content of an organism; genomics is the field of genome studies. Once genome and genomics became popular terms, a flurry of new terms that ended in "-omes" and "-omics" began to appear in publications. *Discovering Genomics, Proteomics, and Bioinformatics* is more than a "tome of -omes" because the field has expanded beyond a narrow definition of genomics. Genomics, as presented in this book, includes the interaction of molecules inside cells, including DNA, protein, lipids, and carbohydrates. In the spirit of discovery, we will explore the tools and questions behind the revolution that is changing the way biology is studied.

Discovering Genomics, Proteomics, and Bioinformatics is based on two pedagogical principles that have been successful for many teachers; teach in the context of an interesting question and on a need-to-know basis. This is how everyone learns new information, and it is the best way to help students learn so they will be motivated and more likely to retain the information. *Discovering Genomics, Proteomics, and Bioinformatics* is built on "stories" or case studies taken from scientific publications. In answering the many questions raised by these case studies, we use bioinformatics to explore scientific content and process. The content includes all the major areas such as sequences (whole genomes and variations), microarrays, proteomics, and systems biology.

This book is designed as an interactive resource to use when exploring topics in genomics and proteomics. The figures provide real data that you can mine to extract more information than is initially apparent. The online databases engage you in real-time discoveries using the same databases investigators are using for their own research. Discovery Questions focus your attention on critical information and urge you to think for yourself, using the tools and information presented in text and figures. Traditional textbooks supply you with facts and details that you are inclined to memorize for tests. Genomics requires you to analyze, hypothesize, think, and formulate models; this book was designed to develop your critical thinking skills.

As Dr. Francis Collins noted in the Foreword, today's biologists need to think quantitatively and from a multidisciplinary perspective. Therefore, this textbook must also look and "feel" different. You will need to use the computer a lot to access the latest information. To fully understand genomics, proteomics, and bioinformatics, you will read about real and compelling cases that challenge and encourage you to learn. So, immerse yourself in the case studies and discover what genomics is all about.

Writing Style

The text is written in a style that is easy to read and comprehend. It avoids unnecessary jargon, yet new terms are included when essential to help you understand the material. Bold words are defined in the glossary as well as in the text where they first appear.

Discovery Questions

The process of critical thinking is enhanced by Discovery Questions, which are imbedded within the case studies rather than saved for the end of each chapter. Discovery Questions focus your attention on key concepts as well as experimental design, interpretation of data, and the need to support your opinion with data. Analyzing real data reproduced from peer-reviewed publications will allow you to reach your own conclusions and the text will help guide you through the data. To answer some Discovery Questions (see the following example), you will use online public databases, many of which are regularly updated. Discovery Questions that require you to use online resources are indicated by the Genomics Place icon.

DISCOVERY QUESTION

46. Perform an NCBI MeSH (Medical Subject Heading) search for "CAMP Factor." You should see a hit called "CAMP protein, Streptococcus [Substance Name]." On the far right side, click on the "link" link and choose NLM (National Library of Medicine) MeSH Browser. What can CAMP Factors do to our blood cells?

All the Discovery Questions are available via the companion web site, to facilitate your interaction with online resources and permit you to submit your answers to your instructor via email.

Media

The companion web site for Discovering Genomics, Proteomics, and Bioinformatics (www.GeneticsPlace.com) is a tool to enhance your study of genomics. A shortcut <http://aw-bc.com/genomics> can be used to reach the web site directly.

- **Methods:** These web pages explain how molecular and genomic methods are conducted and what type of data they produce. They are intended to supplement the textbook and provide background information if you have not learned the methods in previous courses.
- **Data:** In several Discovery Questions, you will use online bioinformatics tools to analyze protein or DNA sequences. To save you time and the potential problem of typos, all sequences are supplied in web pages that you can copy and paste for analysis. Some non-sequence data are also provided when needed.
- **Structures:** A significant aspect of any protein is its 3D structure. Periodically, web pages with Jmol tutorials have been created to illustrate structural features that are best understood when you interact with them. Jmol is a Java-based method for viewing 3D structures and will replace Chime in the coming years. We are pleased to be the first textbook to adopt the new standard of Jmol.
- **Links:** Two types of links have been collected for each chapter. The first provides direct access to online databases and bioinformatics analysis tools such as the National Center for Biotechnology Information and the Protein Data Bank. The second facilitates easy access to investigators' laboratory web pages when you are particularly interested in a case study or area of research.
- **Math Minutes:** Links to datasets and interactive electronic media are available for you to explore. Some Math Minute Discovery Questions require you to work with these files in order to answer the questions.

Media Menu

Throughout every chapter are Media Menus to alert you to the resources contained on the book's companion web site. From this site, you can read in-depth descriptions of methods, access data, view a 3D structure, explore Math Minute resources, and link to related web sites. These media tools allow you to participate in the interactive process of discovery that is at the core of genomics. The first time a Media Menu key word appears in the text, it is printed in purple to remind readers that online resources can enhance their understanding. You can reach the companion web site through different routes, but we have created a shortcut that may be easier to remember: <http://aw-bc/genomics>.

METHODS
Microarray Animation
STRUCTURES
Cyclooxygenase
DATA
Uncharacterized Protein
LINKS
Conserved Domain
PDB
PREDATOR
MATH MINUTES
dotplot.xls

Math Minutes

Dr. Collins highlighted the importance for biology students to enhance their quantitative skills. Most biologists in the cell/molecular field do not use much math in their work, but genomics, proteomics, and bioinformatics are changing this reality; these fields rely heavily on mathematics. To facilitate appreciation of how the data were analyzed and the role mathematics plays in understanding biology, we have added more Math Minutes as enrichment for those who want to discover the interaction of math and biology. Math Minutes (see page 40 for one example) use the case studies as foundations for concise lessons in statistical analysis, probability, and computational methods. Some Math Minutes will include Discovery Questions so students can discover more through data explorations and manipulations.

Art Program

Detailed and abundant illustrations are reproductions of original data and expand on the basic information provided in the text, as shown in the example below. Your understanding will be enhanced by analysis of the figures from which you can extract additional information.

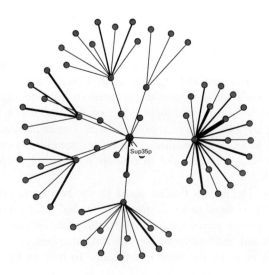

Figures on the Companion Website

All the illustrations in the text appear on the text's companion website (<http://aw-bc/genomics>). Some figures, which are best seen in full color and large format, appear as small thumbnail images in the text with the full color version available on the text's companion website. These illustrations will be noted: Go to www.GeneticsPlace.com to view this figure, though you can use the shortcut <http://aw-bc.com/genomics> to reach the site more directly.

Figure 6.19 Stress-induced expression of 900 genes clustered into the ESR.
Go to www.GeneticsPlace.com to view this figure.

Transition from Genetics to Genomics

Typically, people think of genomics as high-volume genetics. However, a genome is more than the sum of its parts and you need to approach it with a new mindset. The first chapter of each unit is a Case Study chapter to help ease your transition into genomics. Specifically, these three chapters confront the common misconception of "one gene, one protein, one phenotype." The case studies contain real data for you to interpret and discover the interconnectivity of the "cell web." The discovery approach to learning will foster scientific skills of analyzing data and formulating models to explain the data.

For the Instructor

Instructor's Manual ISBN 0-8053-8218-6

The printed version of the Instructor's Manual provides answers to all the Discovery Questions. In addition to the written answers, the electronic version of the Instructor's Manual (<http://aw-bc.com/genomics>) provides color figures to accompany some answers. For example, if students are asked to draw a graph or construct a circuit diagram, your electronic version will illustrate one possible answer. These illustrations, along with all illustrations on the web site, may be used for lecture presentation such as PowerPoint or web pages. Access to these answers is restricted to faculty who have adopted the textbook for course use. Students cannot access this resource. Contact your Benjamin Cummings sales representative for your free password.

Bonus Materials for Adopters

One of the drawbacks of publishing a textbook is it cannot possibly be kept updated as fast as we would like. Therefore, we have created bonus materials for faculty who have adopted the textbook for course use. The bonus materials will include more case studies that can be used to expand your favorite chapters or used as raw material for testing. In addition, new resources will be posted on a regular basis such as laboratory curriculum, electronic files for data exploration, and lessons on data analysis with special emphasis on DNA microarrays and synthetic biology. Contact your Benjamin Cummings sales representative to get your instructor's username and password.

Acknowledgments

It has been an honor to teach genomics to my students at Davidson College. They have taken a leap of faith with me to explore the frontier of genomics. College students have been selected to memorize factoids, but genomics cannot be memorized. My hearty students have taken an irreversible step towards a lifetime of discovery and wonder. However, neither the students nor I could have experienced the thrill of learning through exploration without the support of my colleagues, especially Verna Case, Clark Ross, and Bobby Vagt. They have supported me from the beginning when I set out to create my first genomics course. From that inception, we have created a network of laboratory and computer courses for students to learn by hands-on research and exploration. As with systems biology, the emergent properties of systems education can only be realized when a team of educators share a common vision for improving the curriculum and helping students benefit from the best possible education. And most importantly, Laurie Heyer has been a partner at every step and has enhanced each component of this book and our combined genomics curriculum. Her knowledge of biology and expertise in mathematics makes her the perfect collaborator.

We are continuously inspired by the investigators whose work we use throughout this book. We appreciate their willingness to share information and insights with us. Another source of inspiration has been the growing community of GCAT: the Genome Consortium for Active Teaching <www.bio.davidson.edu/GCAT>. Faculty from many schools have left the security of canned labs with familiar methods to follow their passion and bring genomic methods into their undergraduate laboratory curricula. Since the fall of 2000, faculty and their students have conducted original research using DNA microarrays. We trained ourselves and worked with our students as colleagues and collaborators. The National Science Foundation (NSF) and Howard Hughes Medical Institute (HHMI) have provided funding that allowed us to run workshops and purchase microarrays. GCAT is growing because the enthusiasm of the community is contagious and invigorating. Faculty who teach at institutions with large minority populations are included in the fellowship of scholars as we learn side by side and increase the diversity of talented students eager to learn genomics.

This book was given a second chance in large part through the dedicated work and inspirational vision of Susan Winslow. Susan recognized genomics education as a national need and one that required frequent updates. She combined her years of experience with a youthful energy to bring this second edition to life. Geoffrey Heilpern put his back into the effort and pulled the revisions, new content, and expanded media through the arduous process of producing a book. Mercedes Grandin carried the second edition across the finish line when Geoffrey moved on to a new job. Other critical personnel include: Mark Ong helped design the text and the cover; Yvo Rezebos designed the cover art; Holly Henjum as production editor at GGS was very helpful with the creation of the copy and pages; Lauren Fogel, Mansour Bethoney, Dario Wong, Linda Young, and the entire web development team for overhauling the book's website; Timothy Driscoll created the Jmol web pages, the first ever used with a textbook; and Lori Newman as production editor kept the entire team on track and within the budget. Finally, we thank the many reviewers whose constructive criticism improved the book in substantial ways.

On a personal level, I would never have mustered the energy or confidence without the loving support of my wife Susan, and our daughters Celeste and Paulina. I am blessed with a family who supports my crazy schedule and work habits. But more importantly, they are the source of my greatest happiness. Nothing can beat the shared laughter and "Fun Fridays" when we reinforce the importance of taking time to be together, as a family. My extended family offers a broader base of support, as has Chris Gunn. Without my larger circle of family and friends, I might never have left the floor tile factory.

—A. Malcolm Campbell

Colleagues and administrators at Davidson College have been supportive of every aspect of our genomics and bioinformatics program. In addition to those Malcolm has already thanked, I thank my former and current department chairs Stephen Davis and Rich Neidinger for their encouragement. I appreciate the biology department supporting my efforts to teach and do research with students in bioinformatics. Bill Hatfield, Brian Little, Mur Muchane, Robert Lee, Sarah Hatfield, and Kristen Eshelman have gone the extra mile to provide a reliable and powerful technological infrastructure, without which we could not have created many of the web pages and resources provided with this book. I have learned so much from students who have field-tested Math Minutes, created terrific web pages, and asked all the right questions. I hope all students reading this book will benefit from the insight and inspiration that Davidson students have given me. Coauthoring a book involves a lot of give and take, and my coauthor gives unselfishly and takes graciously. Thanks, Malcolm, for your wisdom and total dedication to this project. Donna Molinek endured thrice-weekly status reports on this project, both times around. Thanks to Mom and Felix, Tom and Kay, and Kristopher and Danielle, for believing in me and being proud of me. I am especially grateful to my husband Bill, who has walked every step of the way with me. His constant love and support are my anchor and my wings.

—Laurie J. Heyer

Reviewers

Daron Barnard
College of the Holy Cross

Shifra Ben-Dor
Weizmann Institute of Science, Israel

Ian Boussy
Loyola University

Steven Brenner
University of California, Berkeley

Michael Buratovich
Spring Arbor University

Scott Cooper
University of Wisconsin, La Crosse

Todd Eckdahl
Missouri Western State College

Irene Evans
Rochester Institute of Technology

Harvey Greenberg
University of Colorado, Denver

Amy Hark
Muhlenberg College

Dan Heruth
William-Jewell College

Laura L. Hoopes
Pomona College

Bob Ivarie
University of Georgia

Mitrick Johns
Northern Illinois University

Elizabeth Ann Joyce
Swarthmore College

David Kass
Eastern Michigan University

Dennis Kibler
University of California, Irvine

Lee Kozar
Stanford University

Carissa Krane
University of Dayton

Ken Kubo
American River College

Mark Lubkowitz
Saint Michael's College

Albert MacKrell
Bradley University

John Merriam
University of California, Los Angeles

Jeffrey Newman
Lycoming College

Desh Ranjan
New Mexico State University, Las Cruces

Peggy A. Redshaw
Austin College

Rebecca Roberts
Ursinus College

Frank Rosenzweig
University of Florida College of Medicine

Carl J. Schmidt
University of Delaware

James A. Shapiro
University of Chicago

Elizabeth Vallen
Swarthmore College

Manuel F. Varela
Eastern New Mexico University

Quinn Vega
Montclair State University

Craig Volker
Cabrini College

Denise Wallack
Muhlenberg State College

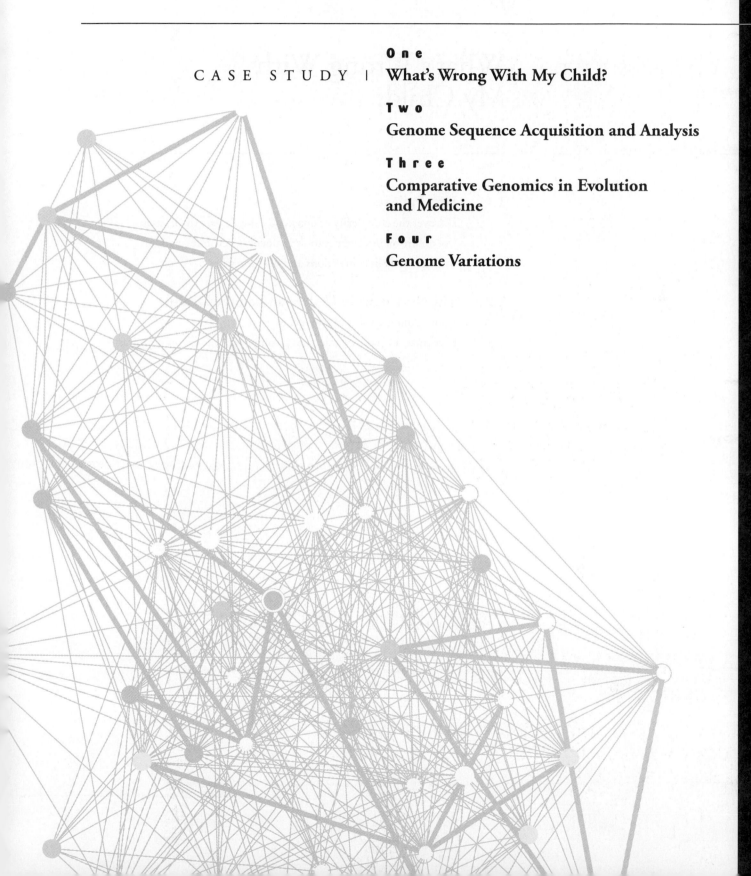

Genome Sequences

Unit | One

CASE STUDY # What's Wrong With My Child?

The purpose of this chapter is to let you analyze genetic data as they are presented to you. Based on the data, formulate rules; then test the rules and look for exceptions that force you to modify them. Use this case study as a way to discover how genetics was used to understand the genomic mechanisms that cause this disease. The case study is written in the style of a short mystery that you as the reader must solve. Put yourself in the place of the physician who tries to help the children in the story.

1.1 First Patients

Section 1.1 is divided into four phases of information. During each phase, you will be given some data and asked to search online sources for conclusions. This process is designed to help you strengthen your analytical skills and provide you with the perspective to appreciate that genomics is not just a collection of methods, but has become an enhanced way of seeing life. Do not read ahead until you have answered the Discovery Questions as they appear in the text.

Phase I: Clinical Presentation

Imagine you are a family physician when a mother and her 6-year-old boy (BB) come in to see you. (Because this case study is based on real patients, only their initials are used here, though a good physician would refer to patients by their names.) On BB's chart, you see that the reason for the office visit is "digestive problems." You have never seen this family before, and the paperwork indicates that they do not have health insurance, even though the mother has a full-time job at a small local company. When you enter the examination room, the mother is sitting in the chair and BB is on the floor playing with some blocks. You greet them and ask BB how he is feeling today, but you get no response. You turn to the mother, assuming that the boy is shy. She says that for a few months now, he has complained of a tummy ache, but he never seems to be really sick.

You ask BB to climb up on the examination table and remove his shirt. He starts to stir a bit, but then the mother gets out of her seat and helps him climb up. You get the sense that BB is unable to stand and walk on his own. You inquire and find out that BB could run and walk normally until he was about three and a half years old. You also notice that his calves seem large, while his thighs and arms are thinner than you would expect for a boy of his size. You test his reflexes and find that they are nearly absent in the knee, diminished in the arms, but normal at the ankles. You listen to his heart and it seems abnormal as well. You ask if he attends school and the mother says that he does but he has to get special help with his writing. Being a parent yourself, you ask if he can write his name and the mother indicates that he has not learned all his letters yet. You know this is not normal for a 6-year-old in first grade.

You decide to get some blood work done and take a small muscle biopsy. A couple of days later, you get these results:

Red blood cells	Normal
White blood cells	Elevated
Glucose levels	Elevated
Creatine kinase	Extremely high
Glucose-6-phosphate dehydrogenase	Normal

BB's muscle fibers did not look normal, with internal nuclei and varying fiber thickness (Figure 1.1). Scattered throughout the fibers were fat cells and connective tissue. Some nuclei are located away from the periphery.

a) Control Muscle Biopsy

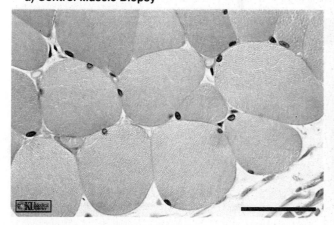

b) Patient's Muscle Biopsy

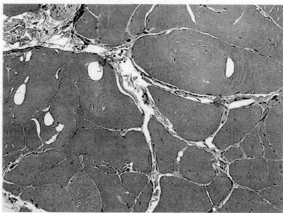

Figure 1.1 Histology of *wt* and patient's skeletal muscle biopsy.
a) Wild-type skeletal muscle histology. Notice that the fibers are regular in size, spacing, and "texture," with peripheral nuclei. Bar is 50 μm long. **b)** Patient's skeletal muscle biopsy. The white spaces show where fat has impregnated the muscle. Nuclei are not peripheral.

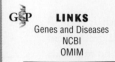

LINKS
Genes and Diseases
NCBI
OMIM

DISCOVERY QUESTIONS

1. Go to Online Mendelian Inheritance in Man (OMIM; OMIM database of human diseases and genes) and see if you can diagnose this disease. Spend only about five to ten minutes at this site.

2. Try the National Center for Biotechnology Information (NCBI), "Genes" Genes and Diseases (different organization from OMIM) and see if you can diagnose this disease. Again, spend only five to ten minutes on this.

Phase II: Family Pedigree

Although you have sensed that this family has financial limitations, you call the mother and ask her to bring BB back. She says she cannot pay, and you suddenly "remember" a new policy that allows you to see him again as part of a free follow-up examination. You know that the mother works, so you suggest that she come by just before 5:00 p.m. when the doors get locked.

When the mother and BB return, you meet them in your office rather than the examination room. You ask some general health questions and then begin to inquire about other family members. BB has two older sisters and one younger brother. The girls exhibit no health problems out of the ordinary; the younger brother is only 2 years old and has just barely begun to walk. BB's older sister, age 24, is married, has one girl (age 2), and is expecting her next child in 7 months. You also ask about the extended family and learn that one of BB's uncles died at age 20 from a crippling disease similar to BB's illness. The mother also had two male cousins on her mother's side who died as young adults. You ask the mother if you may draw some blood from her and other family members.

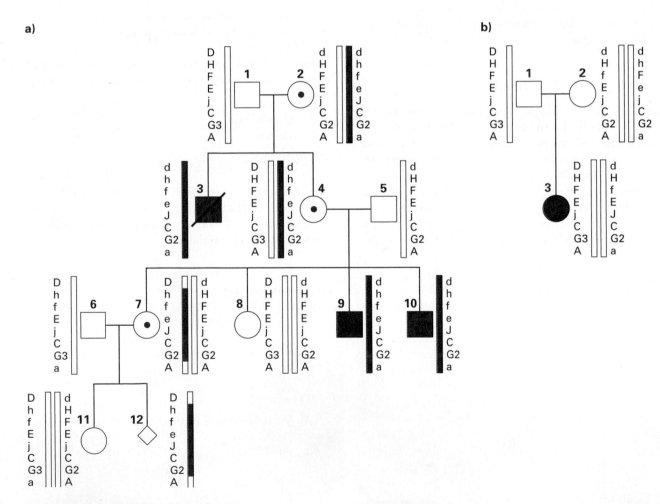

Figure 1.2 Linkage analysis for the families.

a) The original boy (BB) is individual number 9. Dots indicate carrier status, and the bars represent X chromosomes with the RFLP markers as indicated. Purple on the chromosomes indicates regions carrying the disease allele. The capital letters indicate the most frequent markers in the population. **b)** Pedigree and linkage analysis information for the affected girl (GG) and her parents.

DISCOVERY QUESTIONS

3. Go to OMIM again and see if you can diagnose this disease. Spend only about five to ten minutes at this site.
4. Try NCBI Genes and Diseases again and see if you can diagnose this disease. Spend no more than five to ten minutes on this.

Phase III: Karyotyping and Linkage Analysis

During the week between BB's visits, your new policy of free follow-up visits caught the attention of several neighbors of BB's family. Your business is booming with people who have never seen a doctor before, even though all the parents have full-time jobs. In this group, there is one girl (GG) who exhibits the same symptoms as BB. GG is 11 years old, cannot walk on her own, and has enlarged calves; her ankle reflexes are normal, but all other reflexes of the limbs are absent. She appears to have no intellectual deficits, unlike BB. GG has three brothers, all of whom are healthy. You collect pedigree information for GG, but find that no one else in her rather large extended family had any symptoms. You get permission to take blood from her and everyone in her family. You are curious about the karyotypes of both patients and their families. You also ask the lab to perform a restriction fragment length polymorphism (**RFLP**) and linkage analysis with special attention paid to the X chromosome (Figure 1.2).

From BB's pedigree, you see that his disease is recessive and that several of the women in his pedigree are carriers. The next set of data you obtain are photographs of chromosomes taken from your two patients (Figure 1.3). BB appears to have a deletion in his X chromosome: GG has one typical X chromosome, whereas the other has recombined with one copy of chromosome 6. At the top of the der(X) chromosome, the terminus is from chromosome 6 (6qter) and it continues down until band 21 on the long arm of chromosome 6 (6q21). Below this point, the remainder is X chromosome, beginning with band 21 on the short arm of X (Xp21). The der(6) chromosome contains the reciprocal DNA, as indicated in Figure 1.3e.

If this disease is recessive, it does not make sense that GG is suffering from the disease because she has only one mutated X chromosome. She should be a healthy carrier just like BB's mother. What is going on? Luckily, the pathologist who is working with you points out an unusual chromosome (Figure 1.4). Because the pathologist was curious, the karyotype was performed on 56 different cells taken from GG. The wild-type (*wt*) X chromosome appeared to be replicating later than the mutant X chromosome in 55 out of 56 cells examined. In all human female cells, one X chromosome will be inactivated and the other will remain active. In GG, 55 out of 56 times, the mutant X chromosome was the active one and the *wt* X chromosome was the inactive one. For an unknown reason, GG's cells preferentially inactivated the *wt* X chromosome, so in about 98% of her cells, only genes on the mutated X chromosome could be expressed.

DISCOVERY QUESTIONS

5. What can you conclude about individual number 12 in BB's extended family (see Figure 1.2a)? Identify this individual's gender and probable disease status.
6. Go to OMIM and see if you can diagnose this disease with a quick search. Spend only about five to ten minutes at this site.
7. Go to NCBI Genes and Diseases and use the search window to see if you can diagnose BB's disease in five to ten minutes.
8. Does GG represent an exception to the rule for X-linked diseases, or is she consistent with it? Explain your answer.

Phase IV: DNA Sequence Analysis

You decide that a definitive answer can be found only by cycle-sequencing DNA from the two patients. You get permission to use their blood as a source of genomic DNA for sequencing a locus of interest and you obtain the deduced amino acid sequences shown in Table 1.1.

DISCOVERY QUESTIONS

9. Search OMIM to see if you can diagnose this disease. Spend only about five minutes at this site.

10. Go to BLASTp (protein sequence search tool) and see if you can diagnose this disease. To assist you, use the following amino acid sequence from the locus of interest, which was conserved in all family members: **NICKECPIIGFRYRSLKHFNYDICQ SCFF**. Save in electronic form the proteins you find with this search. We will return to these hits.

11. Go to Protein Data Bank (PDB; 3D protein structure database), enter the code "1dxx," and click on "Find a Structure." This file contains four copies of the amino portion of the full protein. Click on "View Structure" and view this fragment using the QuickPDB method (this file is a **homotetramer**). Find the location for the mutations found in GG and BB (see Table 1.1). Where do these mutations lie within the protein? Look at the location of the mutation in the secondary structure.

12. Find and highlight the sequence **DVQKKTFTKW** (amino acids 15–24). What kind of secondary structure does the peptide help form? This answer will be useful later, so note the location of the peptide.

DATA
locus of internet

LINKS
BLASTp
PDB

METHODS
cycle-sequencing
QuickPDB
RFLP

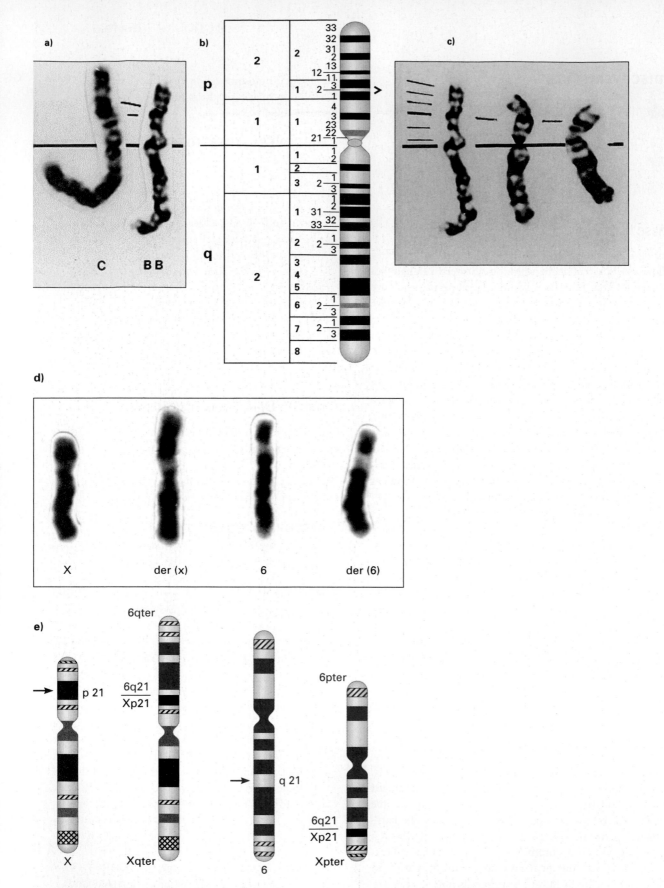

Figure 1.3 Karyotypes for the two patients.
a) Metaphase X chromosome from control (C) and male patient BB. The thick horizontal bars mark the centromeres. The portion deleted from the boy's X chromosome is indicated by two thin lines. **b)** Ideogram of a *wt* X chromosome with the deleted region marked by ">" **c)** The BB chromosome (left) and two additional X chromosomes from the male patient. **d)** From the female patient (GG), *wt* X and 6 chromosomes and two that have recombined. The prefix "der" indicates that the chromosomes are recombinant (*derived*). **e)** Ideogram of the same four chromosomes as in **d)**. Arrows next to the *wt* chromosomes mark the site of recombination.

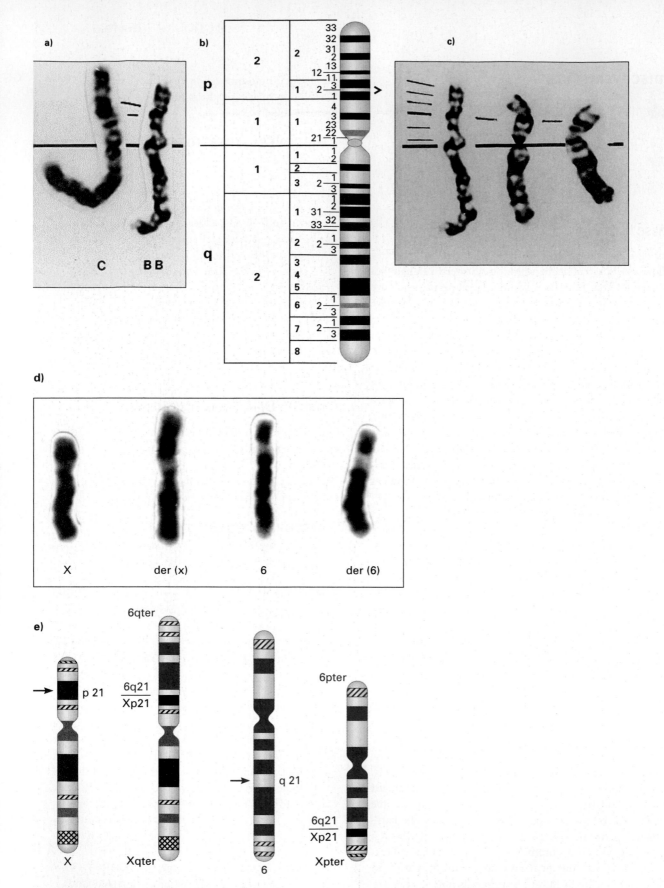

Table 1.1 Deduced amino acid sequences for two patients and family members.

BB:	1	MLWWEEVEDC	YEREDVQKKT	FTKWVNAQFS	KFGKQHIENL	FSDLQDGRRL	LDL**R**EGLTGQ 60
BB's mother:	1	MLWWEEVEDC	YEREDVQKKT	FTKWVNAQFS	KFGKQHIENL	FSDLQDGRRL	LDL**R**EGLTGQ 60
	1	MLWWEEVEDC	YEREDVQKKT	FTKWVNAQFS	KFGKQHIENL	FSDLQDGRRL	LDLLEGLTGQ 60
GG:	1	MLWWEEVEDC	YEREDVQKKT	FTKWVNAQFS	KFGKQHIENL	FSDLQDGRRL	LDLLEGLTGQ 60
	1	MLWWEEVEDC	YEREDVQKKT	FTKWVNAQFS	KFGKQHIENL	FSDLQDGRRL	LDL**R**EGLTGQ 60
GG's mother:	1	MLWWEEVEDC	YEREDVQKKT	FTKWVNAQFS	KFGKQHIENL	FSDLQDGRRL	LDLLEGLTGQ 60
	1	MLWWEEVEDC	YEREDVQKKT	FTKWVNAQFS	KFGKQHIENL	FSDLQDGRRL	LDLLEGLTGQ 60
GG's father:	1	MLWWEEVEDC	YEREDVQKKT	FTKWVNAQFS	KFGKQHIENL	FSDLQDGRRL	LDLLEGLTGQ 60

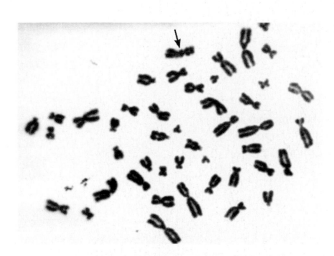

Figure 1.4 Inactivated X chromosome.
A full karyotype that is focused on a *wt* X chromosome of female patient (GG) (arrow). The X chromosome is delayed in its progression through mitosis.

Summary 1.1

You have just progressed through about 40 years of biomedical research. Of course, it is easy to see the linear progress once the final answer has been determined, but imagine what it was like for those who were trying to identify the Duchenne's muscular dystrophy (DMD) gene for the first time. This monumental task required many research groups learning from each other and building on the data. The big break came when an interdisciplinary team of hard-working and creative investigators devised a clever new experiment before anyone had a probe that could be used to clone the full-length cDNA and gene. Once they did identify the gene, you can imagine the collective groan that must have sounded when they realized the mRNA was more than 14,000 nucleotides long and the gene was close to 1 million base pairs (bp) long. At the time of their work, there was no such thing as automated DNA sequencing, so it all had to be done manually by gels. Manually, one person might be able to read 500 bp a day.

Math Minute 1.1   What Is an E-Value?

A BLASTp search returns ***hits***, sequences that produce "significant" alignments to the query sequence. The significance of a hit is measured by its **E-value**, or **expect value**. In the column next to the E-value is a **bit score** (S), a measure of similarity between the hit and the query. The E-value of a hit is the number of alignments that you expect to find by chance (i.e., with no evolutionary relationship) with bit scores at least as large as the bit score of this hit. Biologically significant hits tend to have E-values much less than 1.0. An E-value near 1.0, or substantially larger than 1.0, does not necessarily mean that the corresponding hit is biologically irrelevant; however, the larger the E-value, the greater the chance that the similarity between the hit and the query is mere coincidence. E-values are calculated from the following three factors:

1. *The bit score.* Because a larger bit score is less likely to be obtained by chance than is a smaller bit score, larger bit scores correspond to smaller E-values.
2. *Length of the query.* Because a particular bit score is more easily obtained by chance with a longer query than with a shorter query, longer queries correspond to larger E-values. Remember, the two sequences do not have to match over their entire lengths.

C A S E   S T U D Y

 MATH MINUTES
Reading BLASTp

3. *Size of the database.* Because a larger database makes a particular bit score more easily obtained by chance, a larger database results in larger E-values.

Sequence alignments appear after the table of hits and E-values on the results page. Above each alignment in the BLASTp report is the associated bit score (S) and, in parentheses after the bit score, the raw score (R). Raw scores measure similarity between the query and the hit, but do not account for the three factors in the preceeding list. Math Minute 2.3 explains how raw scores are computed. For now, let's see how bit scores are computed from raw scores, and how E-values are computed from bit scores.

Look at the BLASTp report you produced in Discovery Question 10, and use **Reading BLASTp** reports to guide you. The hits at the top of the list, with E-value 1e-11, each have $S = 70.5$ bits and $R = 171$ (as of spring 2005; the values will differ slightly as the database continues to grow). Hits further down the list, with E-value 2.8, each have $S = 32.7$ and $R = 73$. The bit score is obtained from the raw score using the equation

$$S = \frac{\lambda R - \ln K}{\ln 2}$$

Lambda (λ) and K are normalizing parameters that allow hits to be compared among different BLASTp searches. The values of λ and K are printed at the bottom of the BLASTp report because they change over time. In our example, using $R = 73$, $\lambda = 0.267$, and $K = 0.041$, we find that

$$S = \frac{(0.267)(73) - \ln(0.0410)}{\ln 2} \approx 32.7278$$

The E-value is calculated as $E = mn2^{-s}$, where m is the effective length of the query, and n is the effective length (total number of bases) of the database. (Effective lengths are adjusted from the actual lengths to account for the fact that an alignment cannot start at either end of a sequence—the so-called edge effect.) In our example, $m = 25$ (4 fewer than the 29 amino acids submitted) and $n = 793,175,458$. Plugging these numbers into the E-value equation gives $E = 2.8$, the same value as in the BLASTp report.

MATH MINUTE DISCOVERY QUESTIONS

1. On the BLASTp report from Discovery Question 10, find accession number P11532, click on "P11532", and find the protein sequence at the bottom of this page. You can see your previous query sequence beginning at position 3314. Select amino acids 3241–3420 and submit a new BLASTp query with these 180 letters (BLAST will always ignore numbers and spaces in queries, so it is fine to copy and paste the entire three lines of sequence into the query box.) While you are waiting for the hits, predict how the E-values and number of hits in this report will be different from those in your first search. Explain what happens to E-values corresponding to the exact same bit scores in the two different reports.

2. Perform another BLASTp search, using the same 180 amino acids as in Math Minute Discovery Question 1, but this time restricting the search to the refseq database. Read about this database by clicking the link to "Choose database" on the BLASTp search page. Explain why your E-values changed again in this report. Predict what would happen to your E-value if you performed the exact same search one year later.

Furthermore, both strands should be sequenced, since scientists cannot trust results produced only once; thus, they had to sequence at least 28,000 bases in the cDNA. The full gene proved too large to sequence completely at that time. In fact, a DNA sequence of the entire gene would not become available until years later.

Although the complete gene was not sequenced initially, the work by **Louis Kunkel** and his many collaborators was groundbreaking, establishing the standards and procedures used to identify and clone many other disease-causing genes. Kunkel's research had such a large impact that most other researchers have followed the naming convention established by Kunkel's lab—**dystrophin** is the gene/protein that causes muscular dystrophy. You will find many other *wt* genes that have been named after the mutant phenotypes that led to their discovery.

DISCOVERY QUESTIONS

13. Another form of muscular dystrophy, Becker muscular dystrophy (BMD), affects primarily males. Whereas DMD is diagnosed on average by age 4.6 years, BMD patients do not show signs until they are in their twenties. They develop less severe symptoms and live much longer. Assuming that these two diseases are caused by mutations in dystrophin, hypothesize a mechanism to explain the difference in clinical symptoms.

14. Another phenotype of DMD patients is mental retardation. Hypothesize how a "muscle" protein like dystrophin could cause brain deficiencies. (This issue is addressed again in Section 1.2.)

1.2 The Next Steps in Understanding the Disease

The Need for an Animal Model System

To understand a disease, investigators need a lot of tissue on which to perform experiments. People are understandably reluctant to donate their hearts and skeletal muscle for research purposes, so we have a problem. One solution is to use animal models. For example, it would be helpful if there were a strain of mice that had mutations in dystrophin that we could use to study muscular dystrophy. Fortunately, nature has created many mutations that biologists have utilized for research.

Mutant mice called *mdx*, which develop DMD due to a spontaneous mutation, were isolated by biologists interested in muscle physiology. The mutant phenotype was observed long before dystrophin was cloned and sequenced. Having a mouse that develops muscular dystrophy has been a very helpful research tool. With this model system (the *mdx* mutant strain of mice), investigators can see what happens to a muscle as it

LINKS
BLAST2
Louis Kunkel
mdx
STRUCTURES
amino acids

develops the disease. They can test for the presence of dystrophin, as well as many other muscle-specific proteins, to see how one mutation might alter the expression of other proteins. Finally, investigators can test potential therapies and cures on *mdx* mice before attempting human trials. All these studies would be very difficult or impossible to perform without the mouse model system.

Much of what we know about dystrophin and muscular dystrophy was determined using *mdx* mice. For example, an antibody was made against the mouse dystrophin protein and used to immunologically label *wt* and *mdx* skeletal muscle. Investigators found that dystrophin is localized to the cytoplasmic surface of the plasma membrane of skeletal muscles. They also found dystrophin inside neurons of the brain (which helps to explain the mental retardation in some patients). In *mdx* mice, no dystrophin was detected anywhere. The lack of dystrophin explained why muscular dystrophy is a recessive disease, since it is the loss of function that causes the phenotype and not the creation of a hyperactive or overabundant protein, as is the case with dominant diseases such as Huntington's disease.

What was the Other Protein that Gave Lots of BLASTp Hits?

When you did your sequence search of the dystrophin sequence (Discovery Question 10), you should have gotten a lot of hits for a protein called utrophin. Go to the **BLAST2** page (compares two sequences) and use these accession numbers to compare human dystrophin (P11532) with human utrophin (CAA48829). Be sure to select BLASTp and enter the accession numbers in the box to the right of the prompt "GI."

When you get your BLAST2 results, note the degree of sequence conservation; the statistics below the diagonal line are where you will find the percent amino acid identity and similarity (called positives; see Math Minute 2.3) and the number of gaps needed to maximize alignment. Is this sequence conservation the result of convergent evolution of two unrelated genes, or a pair of duplicated genes (**paralogs**) that have diverged but maintained some common sequence?

Find the first region from the BLAST2 result that is highly conserved between utrophin and dystrophin, DVQKKTFTKW (see Discovery Question 12). The peptide forms part of an alpha helix in dystrophin. To determine what structure this region forms in utrophin, go to PDB and enter the word "utrophin." Choose the file that says "Actin Binding Region Of The Dystrophin Homologue Utrophin." This file contains only the amino portion of the full protein. View this structure using QuickPDB. Find the location for the **amino acids DVQKKTFTKW**. Use the secondary structure view to determine where this portion of utrophin is in relation to the patients' dystrophin mutation.

LINKS
Kevin Campbell
Jim Ervasti

METHODS
immunoprecipitation

Does Utrophin Play a Role in Muscular Dystrophy, Too?

Based on their sequence and structure similarities, you would be insightful if you predicted that dystrophin and utrophin had similar functions. Is utrophin the cause of any diseases similar to DMD? To answer this question, we need to know what the *wt* utrophin protein does.

DISCOVERY QUESTIONS

15. What experiment would you perform to find out where the utrophin protein is normally located? What tissue would you use for this experiment, and why?

16. Would you expect any diseases associated with the loss of utrophin?

Investigators started with the *wt* and *mdx* mice because they were easy sources of tissues. The first thing they wanted to know was the cellular location of utrophin in *wt* mice. Immunological localization images showed utrophin inside skeletal muscles and clustered at the neuromuscular junction (Figure 1.5).

Utrophin is expressed in fetal and regenerating muscle and in many other cell types during early development (utrophin got its name because it is similar to dys*trophin* in structure but is *u*biquitously expressed). Further analysis revealed that the carboxyl terminus of utrophin is required for its localization to neuromuscular junctions of muscle cells. The amino terminus, as with dystrophin, interacts directly with the actin cytoskeleton.

When *mdx* mice were examined, utrophin expression was unaltered compared to *wt* mice. Utrophin was found throughout the muscle plasma membrane in small-caliber skeletal muscles (e.g., muscles that control eye movement) of adult *mdx* mice. Because these muscles show minimal pathologic changes in *mdx* mice, the findings raised the pos-

sibility that induction of utrophin might be a therapeutic approach to muscular dystrophy. To date, no clinical disorders relating to utrophin have been identified.

DISCOVERY QUESTION

17. Why do you think there are no diseases associated with a loss of function in utrophin?

What Does Dystrophin Do?

We have strayed a bit from the original protein, but that's how research goes sometimes; you follow leads as they come. Once antibodies to utrophin and dystrophin were available, a group at the University of Iowa asked a very simple question that had a profound impact on the field. Jim Ervasti and Kevin Campbell led a group that wanted to know if other proteins were physically attached to dystrophin. Up to this point, everyone had focused on dystrophin, which was responsible for the vast majority of cases diagnosed as DMD or BMD. But what did dystrophin do, and why was its loss so crippling? Based on its sequence, some features were known. For example, dystrophin is shaped like a dog bone, with two blobs on either end of a long middle section (Figure 1.6a). The middle section of dystrophin has a sequence **motif** (or characteristic shape) called a coiled-coil. This redundant-sounding name is the same motif seen in myosin, where two identical molecules (a **homodimer**) are wrapped around each other like snakes mating, and each molecule is an alpha helix (Figure 1.6b).

Furthermore, it had been known for a while that the first 240 amino acids folded into a shape that was known to bind actin. This suggested that dystrophin worked as a dimer and was anchored to the cytoskeleton via actin. To answer their question, Ervasti and Campbell used a classic molecular method called immunoprecipitation. The basic idea is to take the cells you are interested in (e.g., skeletal muscles) and create holes in the plasma membrane so that antibody-coated beads can get inside the cell. The antibodies will grab dystrophin and hold on very tightly. When the

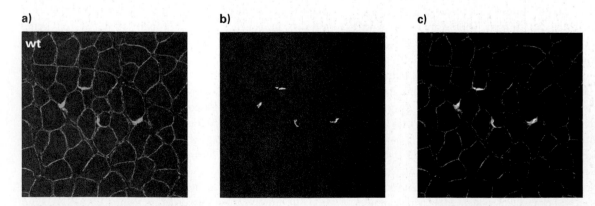

a)   b)   c)

wt

Figure 1.5 Immunofluorescent labeling of dystrophin and utrophin.
Wild-type adult mouse thigh muscle labeled for **a)** dystrophin, **b)** utrophin, and **c)** where they overlap is shown as bright white areas.

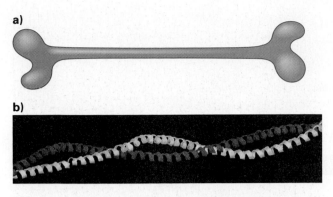

Figure 1.6 Proposed structure of dystrophin.
a) Based on amino acid sequence, dystrophin was believed to be shaped like an elongated bone. **b)** The middle portion of dystrophin was predicted to form a coiled-coil, as shown in this structure view of another protein with a similar motif. (PDB ID# 1C1G.)

mixture is spun gently, the beads will settle to the bottom of the tube and all proteins stuck to the beads will be pulled down with them.

DISCOVERY QUESTIONS

18. Go to the immunoprecipitation animation and study the banding pattern produced for each lane. Explain the banding pattern for each lane.

19. Which lane or lanes in the animation best represents what you would see in an *mdx* mouse or a DMD patient?

Ervasti and Campbell discovered that in addition to actin, dystrophin is attached to a cluster of several more proteins. It took several years and a lot of work by the Iowa team (which included a growing number of investigators), as well as other labs around the world, to identify these proteins. As you can imagine, most of this work was done in *mdx* mice first and later confirmed in humans. Eventually, the picture shown in Figure 1.7 emerged.

Dystrophin plays a critical role in connecting the cytoskeleton to the external skeleton (eventually tendons and bones). You can also see that the carboxyl portion is linked to the dystroglycan complex, which is connected to sarcoglycan complex and laminin. Dystrophin is also linked to a pair of proteins called the syntrophin complex.

There are two ways to look at this portion of the muscle. One is the way most biologists think, in a 2D or 3D representation as seen in Figure 1.7. Equally valid, though, is the way an electrical engineer would see it, as a **circuit diagram** (Figure 1.8). Imagine each protein as a node connected to others by arrows that represent wires. Using a circuit diagram, you can view the model in Figure 1.7 in a slightly different way, which may make some new properties of the system become more obvious.

When you look at this circuit diagram, a few interesting properties emerge. First, it is easier to see the link between the cytoskeleton and the extracellular skeleton. Also, it is more obvious that there are some places where many proteins work in concert to form a single node (e.g., the sarcoglycan complex). In engineering terms, you might describe the sarcoglycan complex as a **redundant** node. In circuit

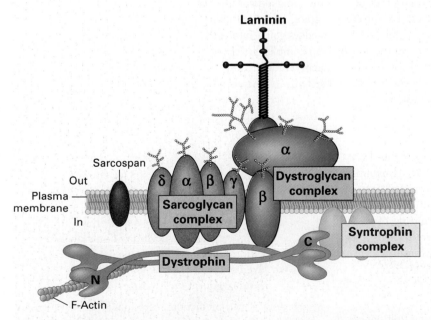

Figure 1.7 Diagram summarizing dystrophin and associated protein interactions.
Notice that some labels do not name each molecule individually, and only the group names are provided. The carboxyl (C) and amino (N) termini of dystrophin have been labeled. The branched beads represent glycosylation of the proteins.

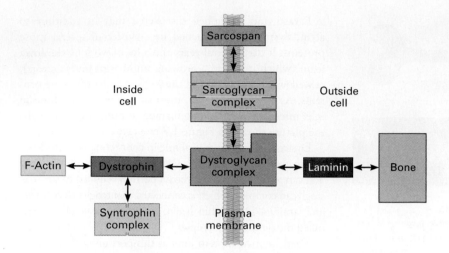

Figure 1.8 Circuit diagram summarizing dystrophin and associated protein interactions. Dystrophin fits into the flow of information between the outside of the cell and the inside. The connection between laminin and bone is simplified here for clarity.

Math Minute 1.2   **What's Special about This Graph?**

Graph theory is the branch of mathematics that studies diagrams called graphs, similar to those in Figure 1.8. In graph theory, the lines between nodes are called arcs or edges. A *directed* graph has arrows on the edges to indicate the direction of information flow between two nodes. In a graph like this one, in which the connections or relationships between nodes are important but there is no directionality, the arrows are normally left off and the graph is called *undirected*.

Graph theorists study properties of graphs such as the degree of connectivity, typical path length from one node to another, and the existence of **cycles** (a path from a node back to itself without visiting any other node more than once). One thing a graph theorist would notice about Figure 1.8 is that there are three edges going into dystrophin and the dystroglycan complex, whereas there are only one or two edges going into each of the remaining nodes. Therefore, dystrophin and the dystroglycan complex are critical nodes: their removal would cause the greatest disruption of the system represented by the graph. We also consider the graph-theoretical view of biological systems in Chapters 10 through 12 and in Chapter 8 when we study proteomics.

---

diagrams, redundancy means that several proteins have similar functions, so that if one is missing, the remaining complex might be sufficient to perform the task. We can also see places in the circuit where only one line connects two proteins in the pathway. Your understanding of electricity must tell you that if a lone wire were broken, the entire circuit would be broken. Therefore, redundancy creates a backup system for critical functions.

DISCOVERY QUESTIONS

20. Use Figures 1.7 and 1.8 to answer these questions.

    a. Predict the phenotype if a person were homozygous for an actin deletion.

    b. Predict the phenotype if a person were homozygous for the deletion of the entire dystroglycan complex.

    c. Predict the phenotype if a person were homozygous for a deletion of only one protein from the dystroglycan complex.

    d. Predict the phenotype if a person were homozygous for a sarcospan deletion.

21. Given your predictions, explain why it makes sense that DMD and BMD were so easily identified as genetic diseases. Why was dystrophin initially identified as the only cause of muscular dystrophy?

22. Predict a molecular mechanism that could produce the differences seen in DMD and BMD.
23. Based on these diagrams and your predictions, do you think there are other human diseases that would present symptoms similar to those of muscular dystrophy?

Why Do DMD Patients' Muscles Deteriorate after the First Three Years?

Now that you have thought about a range of potential genetic causes for muscular dystrophy, you should be wondering how missing dystrophin causes children to develop their symptoms after a few years, and why the muscles gradually deteriorate. It seems that genetic diseases should begin at birth, yet DMD and BMD do not.

First, we should address the differences between BMD and DMD. If you look back to the peptides of the two afflicted children and their families (see Table 1.1), you will see GG and BB had single missense mutations that created one amino acid substitution (**54L → R**; this notation means that amino acid 54 is normally a leucine [L] but in these patients, it is an arginine [R]). This sort of mutation creates a drastic phenotype and DMD. However, mutation 168A → D (alanine to aspartic acid at amino acid 168) causes BMD. The phenotype difference lies in whether the dystrophin circuitry wire is cut completely (54L → R) or only partway (168A → D). In other words, DMD patients lack functional dystrophin. BMD patients have dystrophin that is partially active, and thus their symptoms begin later in life and are less severe.

The remaining question is why are there any symptoms at all. The working hypothesis is that when dystrophin is nonfunctional, the muscle contracts and damages the plasma membrane, because its anchorage to the bone is intact but there is no coupling to the cytoskeleton. In *wt* muscles, the dystrophin connection attenuates and redistributes the physical stress associated with contraction and relaxation. Therefore, the muscle contracts and the tendons pull on the bones, allowing the person to move. However, in DMD the two skeletons (intracellular and extracellular) are connected to each other only by the phospholipid bilayer of the muscle's plasma membrane. This would be analogous to making a bridge out of cement but not including any metal bars to provide the needed durability. Just as a poorly reinforced bridge would crumble, so too do dystrophic muscles. The late onset is due to the fact that children do not move around much during the first few months, so very little damage accumulates.

Although the exact mechanism of damage is uncertain, it seems logical to conclude that if the plasma membranes of muscles are damaged, the cells might die. The most popular idea is that a loss of integrity of the plasma membrane allows calcium to rush into the muscle and activate many proteases

that work only in the presence of high calcium concentrations. As the muscle fibers begin to die, there would be a loss of muscle mass and coordination, which is exactly what we see in DMD. In BMD patients, this damaging process is slowed down, so the symptoms are less extreme.

The damaged membrane hypothesis works well for the muscle symptoms, but the mental retardation is still problematic. Neurons do not contract, so why are there mental deficiencies in some, but not all, patients? There are two parts to this answer. First is the percentage of people who develop the phenotype, which in genetic terms is referred to as **penetrance**. When everyone with a particular mutation gets the same phenotype, the disease is said to have complete penetrance. When few people exhibit the symptoms, the disease is said to have low penetrance.

No one understands the reason for the low penetrance of mental symptoms in DMD. However, it is easy to hypothesize why there might be some functional consequences. All cells, including neurons, have cytoskeletons. Every cell has to be able to bind to, and hold onto, its neighbors. Without intercellular attachment, most cells undergo apoptosis and die. Laminins are a common way to hold onto other cells, and so you might predict that analogs of the molecules in Figures 1.7 and 1.8 are also present in neurons. If the connection between the extracellular and intracellular worlds were broken, there might be some physical consequences for the cells, if not cell death.

Is it Possible to Have DMD and Be Wild-Type for Dystrophin?

About 20 years have passed since BB entered your clinic. Sadly, he died of heart failure; DMD is an incurable disease. It is the Friday of what has been a long week, and you are ready for a break. At 4:55 p.m., a woman enters your clinic holding her son (YY, age 4), and her 12-year-old daughter (DD) follows in a wheelchair. Neither child can walk, and you realize that your generosity continues to attract patients who are often neglected.

During your examination, you notice that the children have hypertrophied calf muscles, but their hips and shoulders look especially small. The torso musculature also looks noticeably reduced. You take a blood sample and tissue biopsies from all three and ask them to come back next Friday for the lab results. You send the samples to a lab, requesting information about the levels of creatine kinase in the blood and asking for immunological detection of dystrophin in the muscle biopsies. Here is what you get back from the lab:

YY and DD:
- elevated creatine kinase serum levels
- normal level and location of dystrophin in the muscle

Mother:
- normal creatine kinase serum levels
- normal level and location of dystrophin in the muscle

CASE STUDY

Next Friday, the family returns and you have to tell them you are not sure what is wrong with the children. You suspect that there is a single point mutation in the dystrophin gene, so you ask for permission to sequence their DNA. They agree and you request that they return for one more free visit. When the sequence data arrive, you are truly baffled. All three individuals have *wt* dystrophin—not a single mutation anywhere. You are certain a mistake has been made, so you ask the lab to sequence a different locus that is highly polymorphic (the MHC locus) to confirm that the samples were contaminated. Surprisingly, the MHC sequence only confirms that the mother really is their mother and that the children had the same father. There is no contamination of DNA, and they really are *wt* for dystrophin.

How Can They Have Muscular Dystrophy if Their Dystrophin Genes Are Normal?

You are completely stumped by this family. You had jumped to conclusions when you first examined the two children because of your previous experiences with DMD. However, your confidence has been shaken. You overlooked the most obvious feature: the mutation did not appear to be X-linked. Luckily, an old college friend of yours does basic research on muscles. You decide to give your friend a call to chat about life in general and this troubling case in particular. The call goes on for almost an hour as the two of you discover that many of your current interests overlap. As a result, you decide to establish a formal collaboration to understand what is going on in this particular family. You contact the family to seek permission for additional research with their tissue samples, explaining what you are doing and why. They are willing to let you use their tissue for research purposes.

Being a PhD researcher in the field of muscle physiology, your friend is more informed about the most recent findings. As a result, you two decide to test the children's muscles with some new monoclonal antibodies that can detect specific subunits of the sarcoglycan complex. Your friend performs the immunofluorescence labeling experiments and sends you the results as an email attachment (Figure 1.9).

DISCOVERY QUESTIONS

24. Do the children have a dominant or a recessive disease? What are the genotypes of the parents?
25. The father was not willing to participate, because of his religious beliefs. Would having access to his tissue change your interpretation of Figure 1.9?
26. Search OMIM and find the chromosomal location for the four genes in Figure 1.9. Spell out the Greek letters when you search. Determine whether the genes are linked or unlinked.

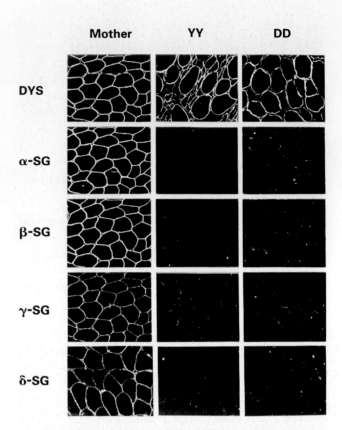

Figure 1.9 Immunofluorescence labeling of biopsies from the mother and her two affected children.
Each panel represents a different section of the thigh muscle labeled with a different antibody. The white area in these photographs marks the presence of the indicated proteins (DYS: dystrophin; SG: sarcoglycan).

27. Given their chromosomal location, do you think that all eight of the alleles were mutated in the two children, or is there another, more likely, answer? Refer to Figures 1.7 and 1.8.

When you look at the immunofluorescence photomicrograph in Figure 1.9, you are stunned by the clarity of the results. Although the father was not willing to participate in the study, you can make some inferences based on the available data. The children are suffering from a recessive disease that is not linked to the X chromosome. It appears that the heterozygous mother is phenotypically normal, though she must carry a mutated allele or alleles. Because each of these four genes is located on a different chromosome, it is highly unlikely that two children would both inherit the same eight mutant alleles to produce the observed phenotype.

Your friend has been following Kevin Campbell's work and knows of some relevant papers. For example, Kathleen Holt in Campbell's lab did some experiments to study how the individual sarcoglycans assemble into a complex and insert into the plasma membrane. She found that if any one

Math Minute 1.3   **What Do You Mean by Highly Unlikely?**

There are two simple explanatory models for the observations in Figure 1.9: (1) both children inherited 8 mutant alleles from 4 different loci, or (2) both children inherited 2 mutant alleles from a single locus, and that one gene affects expression of the sarcoglycan subunits. To help you choose the next step in your research, you calculate the probability that each of these situations occurs by chance, assuming that both parents are carriers.

In the first model, because the 4 genes are on different chromosomes, the 16 inheritance events (8 alleles for each child) are mutually independent. Independent events have a special mathematical property: the probability that all the events occur is the product of the probabilities that the individual events occur. Because each allele is inherited with probability $\frac{1}{2}$ and there are 16 inheritance events (8 alleles and 2 children), the probability that both children inherited 8 mutant alleles from 4 different loci is $\left(\frac{1}{2}\right)^{16}$ (approximately 0.000015).

In the second model, there are 4 independent inheritance events, leading to a probability of $\left(\frac{1}{2}\right)^4 = 0.0625$ that both children inherited 2 mutant alleles from a single locus. Therefore, although the first model remains a possibility and cannot be discarded yet, a single mutant gene is much more likely to be the cause of disease in this family.

---

member is missing, none of the sarcoglycans accumulate in the plasma membrane. Instead, they accumulate in a compartment that, to you, looks like Golgi (Figure 1.10).

In another paper, Rachelle Crosbie and Kevin Campbell led a team that studied γ-sarcoglycan in particular. The team looked at all four sarcoglycan subunits, as well as sarcospan. Although they screened more than 50 families with muscular dystrophies, they found only three had mutations in sarcospan (Figure 1.11), and these mutations did not seem to be the cause of the disease.

The more probable causes were mutations in their γ-sarcoglycan genes. In one particular patient Crosbie

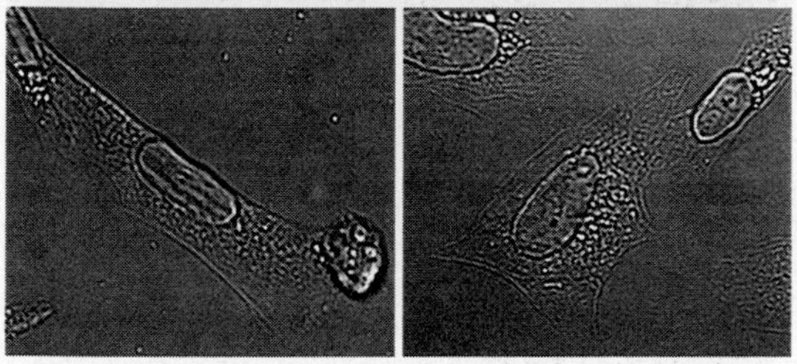

studied, the γ-sarcoglycan contained a frameshift mutation that produced a truncated γ-sarcoglycan protein (Figure 1.12). Interestingly, this patient's complete sarcoglycan complex and sarcospan were located in the plasma membrane, yet this patient still suffered from the disease.

All these data seem like too much detailed information—you are conducting research in addition to maintaining a full patient load. You decide the workload is too much and ask your partners if you may take a one-month sabbatical to research this case more carefully. You are fortunate to have colleagues who think your research is exciting and are willing to work extra hours to help you.

With your new freedom, you rush to the library to read about muscle proteins. You decide that you should sequence γ-sarcoglycan genes from the mother and her two children. The results show that they have point mutations in one codon: **TGT** was changed to **TAT**, producing a 283C → Y amino acid substitution. From your reading, you realize that this family has a very common mutation that may have originated in India 60 to 200 generations ago.

Figure 1.10 Immunofluorescence localization of mutated β-sarcoglycan.
Immunofluorescence labeling of the sarcoglycan complex **a)** in a *wt* cell and **b)** in a cell where the β-sarcoglycan has been mutated. **c)** and **d)** Phase contrast images of the same cells let you see the full extent of the cells.

DISCOVERY QUESTIONS

28. Go to NCBI and select "protein" from the search menu at the top left corner. Enter the word "sarcoglycan" and search the protein database. How many different human sarcoglycan genes are there? You know there are at least four, but can you find any more?

29. What is the name of the disease associated with mutations in sarcoglycans? You might want to click on your results from Discovery Question 28, or search OMIM.

Exon3/ 682 G → A

WT/WT   WT/POLY   POLY/POLY

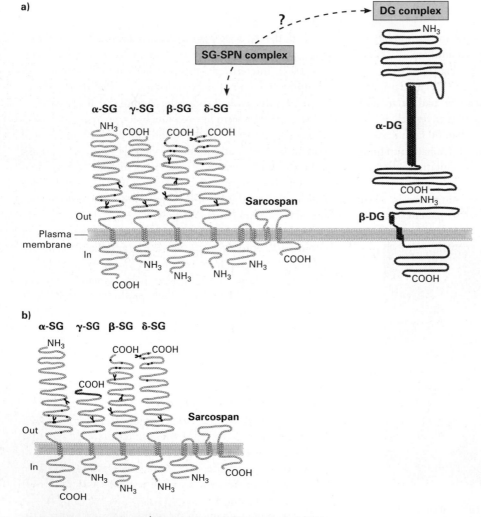

G G G T G T C A   G G G T N T C A   G G G T A T C A

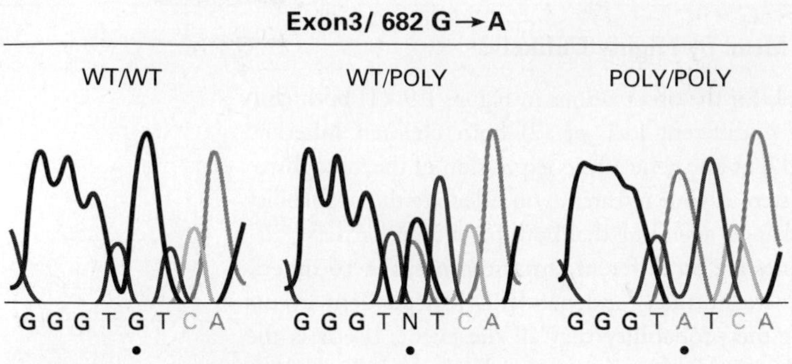

Figure 1.11 Sequence chromatograms **showing homozygous** *wt*, **heterozygous,
and homozygous mutant sequences for sarcospan.**
The mutation is highlighted with a dot below and is called "POLY" because the different
base is a *poly*morphic site.

a)

?

DG complex

SG-SPN complex

α-SG   γ-SG   β-SG   δ-SG

NH₃   COOH   COOH   COOH

NH₃

α-DG

Sarcospan

COOH
NH₃

β-DG

Out

Plasma
membrane

In

NH₃   NH₃   NH₃   COOH

COOH

COOH

b)

α-SG   γ-SG   β-SG   δ-SG

NH₃

COOH   COOH

COOH

Sarcospan

Out

In

COOH

NH₃   NH₃   NH₃   COOH

COOH

Figure 1.12 Sarcoglycan/sarcospan protein-protein interactions.
a) Current model depicting how sarcoglycan/sarcospan complex interacts with the
dystroglycan complex. **b)** Diagram depicting one patient's sarcoglycan and sarcospan
complex in the plasma membrane of dystrophic muscle cells. The small purple
"Y"s represent glycosylation of the proteins (not drawn to scale, for clarity).

Table 1.2 The dystrophin complex proteins, chromosomal location, and associated LGMD classifications.

Disease	Chromosomal Location	Protein
Autosomal Dominant (type 1)		
1A	5q31	myotilin
1B	1q21.2	lamin A/C
1C	3p52	caveolin-3
1D	7q	unknown
Autosomal Recessive (type 2)		
2A	15q15–21	calpain 3
2B	2p13	dysferlin
2C	13q12	γ-sarcoglycan
2D	17q12	α-sarcoglycan
2E	4q12	β-sarcoglycan
2F	5q31	δ-sarcoglycan
2G	7q11–12	telethonin
2H	9q31–34.1	TRIM32
2I	19q13	FKRP

Source: Modified from Betto, et al. 1999. *Italian Journal of Neurological Sciences.* 20(6): 375, Table 2.

Now that you are on sabbatical, you read as many papers on the topic as you can find. From your reading, you realize that there are in fact five sarcoglycan genes in mammals, instead of four. A new member in the family is ε-sarcoglycan, which has almost the same sequence as α-sarcoglycan. Being a student all over again, you draw out the sarcoglycan complex and realize that ε-sarcoglycan may be a redundant subunit that might act as a fail-safe for α-sarcoglycan—an interesting hypothesis you hope to pursue in the future. You also make a table of all the limb-girdle muscular dystrophy (LGMD) diseases and the causes of each form (Table 1.2).

You are not familiar with the proteins dysferlin and dystrobrevin, so you search OMIM to learn more. There are no mammalian orthologs of dysferlin, but its sequence is similar to the spermatogenesis factor *fer-1* of the worm *Caenorhabditis elegans*. People from nine families, all of whom were homozygous for a frameshift mutation in dysferlin, have developed a form of dystrophy. The name "dysferlin" was derived from the role this protein plays in muscular dystrophy and its *C. elegans* ortholog. **Western blots** have demonstrated that dysferlin is 230 kDa in size, and immunofluorescence experiments localized it to the muscle plasma membrane. Dysferlin is expressed at the earliest stages of human development when limbs start to form regional differentiation. The timing and role of dysferlin suggested that it may contribute to the pattern of muscles affected in this form of muscular dystrophy—typically proximal or distal muscles.

OMIM describes the muscle plasma membrane proteins as seen in Figures 1.7 and 1.8, but it describes dystrobrevin localized near syntrophin in skeletal muscle (Figure 1.13). In patients with DMD and some forms of LGMD, dystrobrevin was almost absent from the muscles. Dystrobrevin

amounts and cellular localization were normal in patients with other forms of LGMD in which dystrophin and the rest of the dystrophin-associated protein complex were normally expressed. It appears that dystrobrevin deficiency is a common feature of dystrophies linked to dystrophin and the dystrophin-associated proteins. OMIM states, "This was the first indication that a cytoplasmic component of the dystrophin-associated protein complex may be involved in the pathogenesis of limb-girdle muscular dystrophy." When you finish reading, you appreciate how much has been learned since you were in medical school. The information (the number of different proteins, their roles and subcellular localizations) seems complex and almost overwhelms you.

With all your newfound knowledge, you have another conversation with your collaborating friend. You ask why there was no antibody for ε-sarcoglycan in the analysis of the biopsies, and your friend confesses ignorance of this protein. (You smile with satisfaction for being as knowledgeable as your friend.) You continue to flex your intellectual muscles by discussing the nonframeshift deletion in laminin α2 that produced a mild form of congenital muscular dystrophy, reminiscent of the frameshift deletion in dystrophin that results in BMD.

By now, you and your friend are deep into the causes of muscular dystrophies (note that the disease name is now plural). You begin to focus on the less obvious proteins. Why would proteins that play no apparent structural roles in the linkage between the cytoskeleton and the extracellular skeleton cause a muscular dystrophy? For example, caveolin-3 is a protein involved in endocytosis and not thought of as a skeletal protein in its function (see Figure 1.13b). Plus, LGMD1C (see Table 1.2) is a dominant disease that produces no symptoms until later in life (20s to 40s). Why?

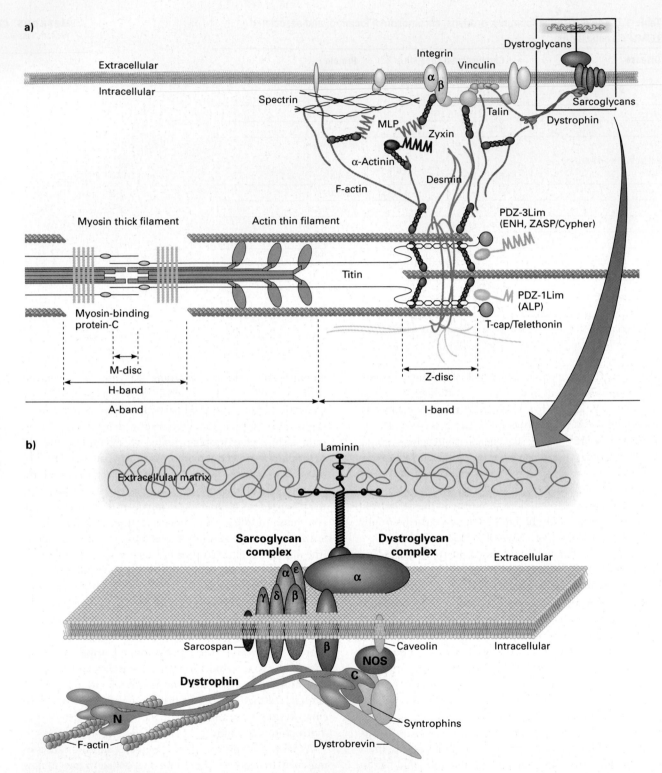

Figure 1.13 Working model of muscle molecules near dystrophin.
a) Schematic diagram of the major proteins involved in striated muscle function.
b) Closer view of the dystrophin complex of proteins. **c)** Circuit diagram showing a different view of panel b. The purple arrows indicate additional connections to other parts of the cell web. Added to this diagram are two large circles that encompass two functional units of the circuitry: the structural unit (circled in purple) and the signaling unit (circled in gray).

c)

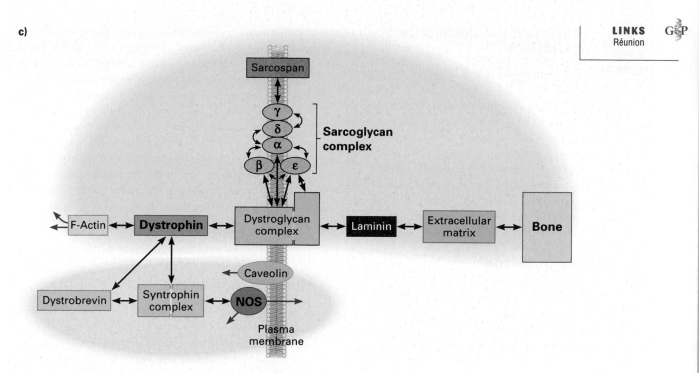

Figure 1.13 *(Continued)*

No one seems to know what dysferlin does (LGMD2B), and the real oddball is calpain 3. Calpains are a family of calcium-activated proteases thought to reside in the cytoplasm. Their job is to cut proteins into smaller pieces to either activate or inactivate them.

Your friend leans back and begins to reminisce about how the mutation linking calpain 3 to LGMD2A was discovered. Your friend was a postdoc in the lab that made this discovery, which focused on a tiny island in the Indian Ocean located several hundred kilometers east of Madagascar. Réunion was initially settled by a few people in the seventeenth century and has a very limited gene pool. In 1642, the island was explored for the first time by Europeans on the French ship *Le Saint-Louis*. In 1644, the governor of Madagascar sent 12 mutineers into exile on the island. Two years later, they were found in excellent health by a delegation that had expected to find them dead. From these genetic founders, the population has remained small, **consanguineous**, and a favorite spot for investigators in search of a good study population for genetic diseases.

Your friend describes how the shallow gene pool on Réunion permitted the team to isolate the genetic cause of LGMD2A, but there was something odd about the findings. Normally, one would expect everyone on this island who suffers from LGMD2A to share the same mutation, as the island has been genetically isolated for many generations—a **founder effect**. Instead of one mutation in calpain 3, the team found six different mutations.

Interestingly, the publication your friend coauthored also discussed the possibility that LGMD2A was **digenic**, or caused by two genes instead of just one gene. The authors also suggested that when mutated, the encoded protein may produce a nonstructural cause of LGMD2A. This hypothesis was radical because muscular dystrophies have "always" been thought to be caused by mutations in structural (i.e., cytoskeleton) proteins.

You sit up abruptly, remembering a paper you read that discussed animal models of muscular dystrophies. In this paper, Volker Straub in Kevin Campbell's lab reported on an investigation of a strain of mice called *dy*, which have a spontaneous mutation in the laminin α2 gene. What was striking was the observation that the plasma membranes in these mice seemed relatively intact. The authors concluded that there may be more than one pathway to produce muscular dystrophies, implying a nonstructural possibility.

DISCOVERY QUESTIONS

30. Is it possible to have two separate pathways (structural and nonstructural) that cause diseases with similar symptoms?

31. Using the circuit diagram in Figure 1.13c, make some predictions about nonstructural causes of muscular dystrophies. Choose any of the proteins discussed so far.

LINKS
Lee Sweeney
PubMed
METHODS
knockout

Where Is the Muscular Dystrophy Field Now?

You and your friend decide to cap off your sabbatical by going to an international meeting on muscular dystrophy. Your good luck continues: This year's meeting is in Paris, where you spent your junior year abroad during college. In addition to the personal joy of returning to France, you are thrilled about the learning opportunity. It is a small meeting with only about 200 investigators, but they are the best in the world. You reserve your ticket and hotel, dust off your French-English dictionary, and do a PubMed search of the latest publications on muscular dystrophies.

DISCOVERY QUESTION

32. Go to PubMed and search for "muscular dys-troph*" (the asterisk is a wild card, so you will get hits for "dystrophy," "dystrophies," and "dystrophin"). Look for papers that suggest nonstructural causes for muscular dystrophies.

Sixth International Conference on Molecular Causes of Muscular Dystrophies

As you fly into Paris, you remind your college friend/collaborator that you went to France for your junior year. Your friend nods politely while silently trying to count the number of times you have said this. You rush off the plane and try out your rusty French as soon as possible, asking, "Where do the ducks get water?" After a puzzled look in response, you see the signs for the toilets, say "*Merci,*" and move on. Next you ask, "Where can I get the cousin in the kitchen?" More blank stares greet you. You begin to doubt your linguistic prowess, and decide to stick to your strength—muscular dystrophies. As you walk to get your luggage, you see several people laughing and asking in English, "Did you find the water for your ducks?" You are thankful that the international meeting will be conducted in English.

The Meeting Begins

The next morning, you head for the first big session, where Lee Sweeney from the University of Pennsylvania School of Medicine will give a brief introduction. In his remarks, you find a few interesting points. He argues that evolution suggests that most parts of dystrophin are important; after all, the worm *C. elegans* contains all the same motifs in its dystrophin protein that humans do. He also discusses a **knockout** mouse in which the gene for syntrophin has been deleted. These mice do not have any pronounced dystrophies, but they do have altered cellular localization of the

enzyme nitric oxide synthase (NOS) that produces nitric oxide (NO), as illustrated in Figure 1.13b. NO, a gas that is produced inside cells, is a part of the **signal transduction** (cellular communication) pathway in muscle cells. Recently, research showed that proper localization of NOS is necessary to increase blood flow during exercise to skeletal muscles in humans. Sweeney speculates that perhaps the mislocalization of NOS could cause dystrophic phenotypes in mice with mutant syntrophin or dystrophin that cannot bind syntrophin. As you listen, you realize that mutations in genes should not be studied under constant laboratory conditions, as is often done. For example, you would like to study syntrophin-deficient mice with and without exercise regimens to see if there is a difference.

Structural Weaknesses

The next speaker is Jim Ervasti, who discovered the glycoprotein complexes (i.e., sarcoglycan and dystroglycan) associated with dystrophin. Ervasti presents some of his work examining the structural link between dystrophin and the underlying cytoskeleton. In his unique approach, Ervasti has peeled the plasma membrane off individual muscles and determined what proteins still adhere to the cytoplasmic side of the removed membrane. He shows a series of beautiful immunofluorescence micrographs of peeled membranes from *wt* and *mdx* mice (Figure 1.14). From

Figure 1.14 Immunofluorescence detection of proteins adhering to plasma membrane.

your reading, you know that α-actinin is part of the cytoskeleton, and that γ-actin can also form polymeric fibers like the more traditional α/β actins do in *wt* muscle actin fibers. Ervasti shows other slides demonstrating that *mdx* mice express γ-actin, but that this protein does not adhere to the plasma membrane.

You realize that this is the first time anyone has directly studied the degree of adhesion between dystrophin and cytoskeletal proteins. All of this is news to you, so you appreciate the circuit model Ervasti proposes to explain his data (Figure 1.15). Ervasti proposes that the lack of dystrophin produces a weak link between the plasma membrane and the cytoskeleton; this weakness is revealed experimentally when the plasma membrane is peeled away. His presentation makes a good case for the structural basis of muscular dystrophies, but you begin to wonder about the other proteins in his circuit diagram. This diagram looks like an integrated circuit, which makes you wonder what roles the other proteins in the circuit might play in this pathology.

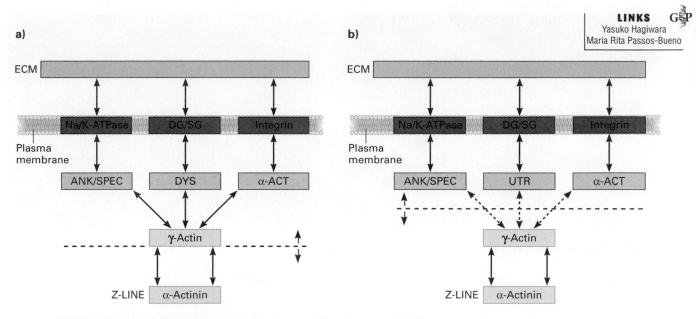

Figure 1.15 Circuit diagrams of proteins adhering to *wt* and *mdx* plasma membranes.
Circuit diagrams of **a)** *wt* and **b)** *mdx* linkage between the plasma membrane and cytoskeleton. The horizontal dashed line marks where the separation occurs in the experiments. Abbreviations: ANK/SPEC: ankyrin/spectrin cytoskeletal proteins; α-ACT: α-actinin cytoskeletal protein; ECM: extracellular matrix; DYS: dystrophin; DG/SG: dystroglycan/sarcoglycan; UTR: utrophin.

DISCOVERY QUESTIONS

33. Using the circuit diagram in Figure 1.15, identify the three functional levels in the model.

34. Using Ervasti's circuit diagrams, predict how there could be a signaling pathway that is also perturbed by the weak linkage between the plasma membrane and γ-actin. Even if you do not know what all these proteins do, use the circuitry to suggest possibilities.

The next talk is presented by **Maria Rita Passos-Bueno**, a respected researcher from Brazil. Passos-Bueno's presentation illustrates how her group discovered telethonin, a recent addition to the growing list of proteins, as the molecular cause of LGMD2G (Figure 1.13a and Table 1.2). The impact of her work is significant for the diagnosis and treatment of patients in São Paulo and all over the world. Her data are clear, but you are still troubled by something. Telethonin is very far away from the plasma membrane, so it is difficult to imagine how mutated telethonin could produce symptoms similar to those seen when a more peripheral protein is mutated—for example, caveolin-3 or one of the sarcoglycans.

As luck would have it, the next speaker is **Yasuko Hagiwara** from Tokyo, presenting his group's work with a knockout mouse that lacks caveolin-3. As you would expect, homozygous mutant mice produce on average 1.8 ± 0.5 **caveolae**/μm^2 (newly forming vesicles used for endocytosis) compared to 14.9 ± 1.6 in *wt* mice. Even the heterozygotes produce less than *wt* mice: 8.0 ± 0.7 (Figure 1.16).

Hagiwara's data are not surprising, but what is hard to understand is the next slide, which shows the muscle pathology of $cav3^{-/-}$ mice (Figure 1.17). During his talk, he says, "Dystrophin and its associated proteins were present at normal levels."

You want to ask Hagiwara some questions, such as, "What proteins did you include when you said 'dystrophin and its associated proteins were present at normal levels'?" but the organizers announce that it is time for a break. You are relieved to stretch your legs when you notice a small stampede to the refreshments. Even your friend has sprinted away and left you alone. You stroll over and discover a lavish setting of *pâté*, *saucisson*, several cheeses, and other yummies you haven't tasted in years. You snap back to the meeting when the lights are lowered and you discover you are the only one still standing at the food table. You have not quite gotten the hang of international meetings yet.

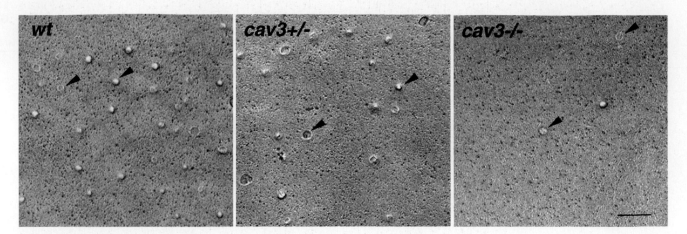

Figure 1.16 Electron micrograph of the cytoplasmic side of muscle plasma membranes from the indicated mice.
The bumps and craters are caveolae caught in the act of forming on the plasma membrane of muscle cells.

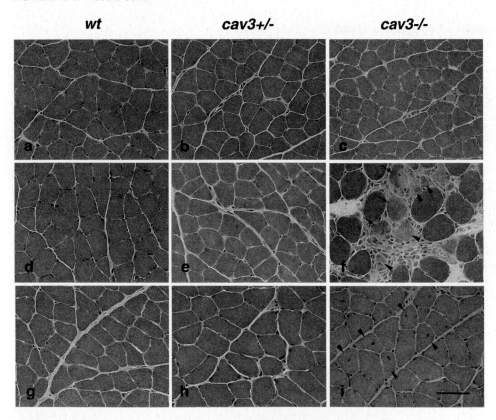

Figure 1.17 Effects of caveolin mutation on muscle development.
Histology of the soleus muscles from **a)** to **c)** 6-week-old mice; **d)** to **f)** 8-week-old mice, and **g)** to **i)** 12-week-old mice from three strains of mice as indicated at the top of each column. Pathology is only apparent at 8 weeks of age and macrophages have invaded the tissue (arrowheads in panel f). By 12 weeks, the muscle has regenerated, but many of the nuclei are abnormally located in the middle of the cells (arrowheads in i). Black bar in the lower right corner of (i) equals 50 μm.

Nonfunctional Mutations

The next speaker is Hiroyuki Sorimachi from the University of Tokyo. He studies the role calpain 3 might play in muscular dystrophies. Sorimachi describes how his lab studied the biochemical properties of nine mutant forms of calpain isolated from LGMD2A patients. Almost all mutant calpains lost their proteolytic activities on a model substrate. Some of them lost their **autocatalytic** capacity (a protein's ability to stimulate or act on itself) or could no longer bind to the structural protein titin (see Figure 1.13a). What was striking about his presentation was that he offered a nonstructural hypothesis. Maybe the enzymatic activity of calpain 3 is responsible for the pathology of muscular dystrophies. But which pathway was disrupted? Why would this lead to a delayed onset of symptoms? How could this lead to muscle atrophy? So many questions, and yet you know Sorimachi's hypothesis represents a step toward a more complete answer, even if it is not clear to you, or anyone, at this time. Science usually moves by slow, sporadic progress, only rarely interrupted by sudden breakthroughs.

The next presentation summarizes a worldwide search for all the different mutations in calpain 3 that can lead to LGMD2A. You were surprised to learn that 97 different pathogenic mutations scattered all over the gene have been discovered (Figure 1.18a). Even more disturbing is the figure that illustrates the age of onset and the rate of disease progression as a function of genotype (Figure 1.18b–d). Once again, severity of the symptoms varies from person to person, which makes you question if a calpain mutation is "the" cause of LGMD2A.

DISCOVERY QUESTIONS

35. Formulate a hypothesis that can explain the data shown in Figure 1.18. Use the data in this chapter to support your hypothesis.

36. If you could obtain tissue from any patient included in Figure 1.18, which ones would you like to study? Why?

New Paradigms: Nonstructural Causes for Muscular Dystrophies

As the next session begins, you are startled by the whispering in the audience. Your friend tells you that the final four presentations are very hot because they are controversial and provocative. Munekazu Shigekawa presents his findings, which suggest a bidirectional communication between the sarcoglycans and integrin. Although you have learned a lot during your sabbatical readings, you are not sure about integrin, though it sounds familiar. Your friend tells you rather tersely that integrins are plasma membrane proteins that allow one cell to hold onto another cell. "It is an adhesion protein!" Then you remember that Ervasti had integrin in his circuit model (see Figure 1.15) that made you think about all the proteins with nonstructural connections.

Shigekawa shows how integrins are immunoprecipitated with dystrophin when an antidystrophin antibody is used. The key point made during his presentation is that when muscle cells adhere to a substrate, α-sarcoglycan and

γ-sarcoglycan become **phosphorylated**. Why? The most popular models do not indicate that sarcoglycans are involved in adhesion. Shigekawa proposes that Ervasti's integrated circuitry (see Figure 1.15) transmits the adhesion status of a cell to other structural proteins such as the dystrophin complex. Why? Does the phosphorylation of sarcoglycans change the ability of the cell to respond to physical stress incurred during muscle contraction and relaxation? When the cytoskeleton circuit is broken, does it create a signaling pathway that is misdirected? Are the structural proteins that normally protect cells less protective when they are not phosphorylated?

LINKS
Mark Grady
Hiroyuki Sorimachi

DISCOVERY QUESTION

37. Modify your earlier hypotheses (Discovery Questions 30 and 31) to explain how muscular dystrophies might be digenic in nature, or at least have two pathways as their molecular causes.

The next speaker is Mark Grady, whose introduction again stimulates the audience to whispers of nervous excitement. Grady begins his talk with a summary slide similar to Figure 1.13b. He plans to discuss the role dystrobrevin plays in muscular dystrophies. He has created a homozygous α-dystrobrevin (*adbn*) knockout mouse strain that he confirms with his next slide (Figure 1.19). He shows how α-dystrobrevin is expressed in skeletal muscle and brain, which reminds you that the mental retardation symptom has yet to be explained by anyone. Could dystrobrevin be the common link between the muscular and mental symptoms?

Grady then shows the pathology in skeletal and cardiac muscle (Figure 1.20), with which you are very familiar at the clinical level; BB died of heart failure. It is clear from Grady's data that the three organs affected by muscular dystrophy are also affected in *adbn*$^{-/-}$ mice. You begin to think that perhaps the dystroglycan/sarcoglycan complexes may be disrupted as well, which would support a model of structural causes rather than signaling causes. The next slide stuns you and the rest of the audience, so there is a collective gasp of amazement (Figure 1.21). You marvel at the clarity of the data and the conclusions they provoke.

DISCOVERY QUESTIONS

38. What effect does a lack of dystrobrevin have on the dystroglycan complex?

39. Where is neural NOS (nNOS) normally located in *wt* muscle cells? Where is nNOS in *adbn*$^{-/-}$ mice? Where is nNOS in *mdx* mice?

40. Given that syntrophin binds to dystrobrevin, how can it be present in *adbn*$^{-/-}$ mice? Refer to Figure 1.13b.

C A S E   S T U D Y

a)

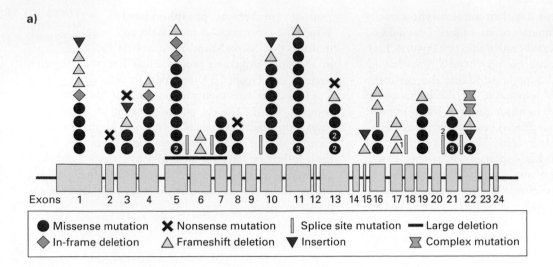

b)

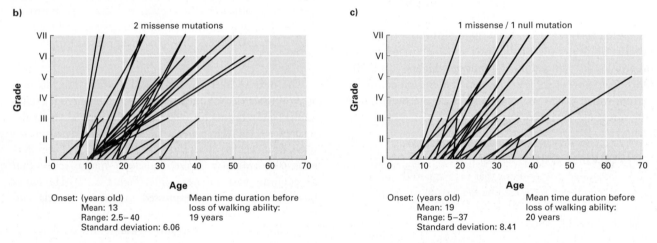

2 missense mutations

Onset: (years old)
Mean: 13
Range: 2.5–40
Standard deviation: 6.06

Mean time duration before
loss of walking ability:
19 years

c)

1 missense / 1 null mutation

Onset: (years old)
Mean: 19
Range: 5–37
Standard deviation: 8.41

Mean time duration before
loss of walking ability:
20 years

d)

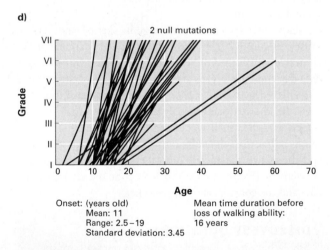

2 null mutations

Onset: (years old)
Mean: 11
Range: 2.5–19
Standard deviation: 3.45

Mean time duration before
loss of walking ability:
16 years

Figure 1.18 Different mutations in calpain and phenotype variations.
a) The 24 exons are indicated by open boxes with exon numbers below. The number of independent mutations is given either inside or above the symbol. **b)** to **d)** Progression-of-disease curves for LGMD2A patients. Functional stages are graded as I–VII. Lines are presented only for patients for whom there are at least two data points of disease progression, whereas the calculation of means and duration takes into account all available data.

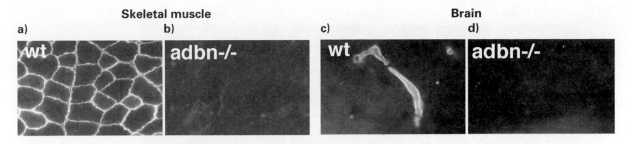

Figure 1.19 Location of α-dystrobrevin in *wt* and mutant mice.
Antibody labeling of *adbn* in skeletal muscle (a and b) and brain tissue (c and d) in:
a) and **c)** *wt*, **b)** and **d)** dystrobrevin knockout mice.

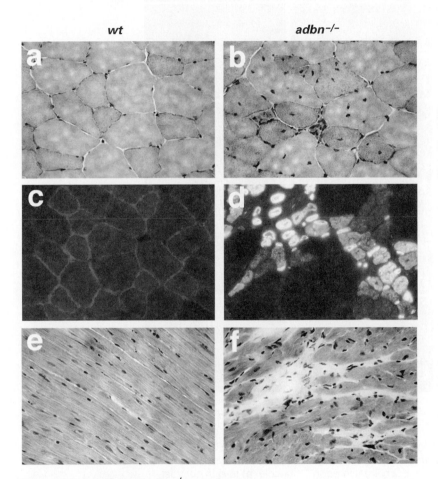

Figure 1.20 Pathology of *adbn*$^{-/-}$ muscle.
a) and **b)** Sections of skeletal muscle from *wt* and *adbn*$^{-/-}$ mice. Small areas of necrosis and centrally nucleated fibers are seen in *adbn*$^{-/-}$ muscle. **c)** and **d)** Sections of skeletal muscle labeled with an antibody to embryonic and fetal myosin heavy chains. The positive fibers in *adbn*$^{-/-}$ muscle indicate actively regenerating muscle fibers.
e) and **f)** Sections of cardiac muscle, showing dystrophic areas in *adbn*$^{-/-}$ tissue.

Your head is swimming, but Grady is not finished. He continues by showing data indicating that the plasma membranes of *adbn*$^{-/-}$ mice are not as weakened as these in *mdx* mice. He also shows data revealing that cytosolic levels of nNOS are normal, so the only perturbation is the plasma membrane localization of nNOS in *adbn*$^{-/-}$ mice. He cannot explain why nNOS fails to localize to the plasma membrane, especially since syntrophin appears to be in its proper location.

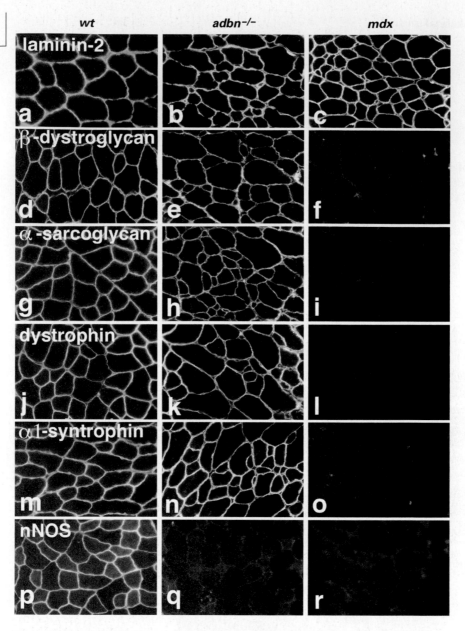

Figure 1.21 Immunofluorescence labeling of skeletal muscle from *wt*, *adbn⁻/⁻*, and *mdx* mice.

a) to **c)** Levels of laminin-α2 were similar in all three genotypes. **d)** to **o)** Levels of the DGC proteins β-dystroglycan, α-sarcoglycan, dystrophin, and α1-syntrophin were markedly reduced in *mdx* muscle but normal in *adbn⁻/⁻* muscle. **p)** to **r)** In contrast, levels of nNOS were greatly reduced in both *mdx* and *adbn⁻/⁻* muscle.

Grady's next slide silences the audience with its implications. He reminds everyone that nNOS plays a signaling role in many cells, and is required to relax the smooth muscles in blood vessels when exercise is vigorous. Nitric oxide stimulates the production of cyclic guanylic acid (cGMP), similar to cAMP, but the base is guanine rather than adenine. In *mdx* mice, nNOS is undetectable near the plasma membrane, cGMP is undetectable, and the blood vessels constrict rather than dilate. Grady wanted to know whether cGMP concentrations could be elevated by stimulation in *adbn⁻/⁻* mice as is the case in *wt* mice (Figure 1.22).

DISCOVERY QUESTIONS

41. What normally happens to cGMP levels in *wt* muscles that have been stimulated to contract? What happens to cGMP levels in *adbn⁻/⁻* mice? What happens to cGMP levels in *nNOS⁻/⁻* mice?

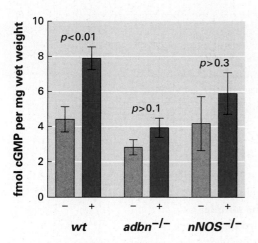

Figure 1.22 cGMP levels in control and mutant muscle. Amounts of cGMP in isolated extensor digitorum muscles from unstimulated (−) or electrically stimulated (+; 30 Hz for 15 s) *wt* (n = 4), *adbn*⁻/⁻ (n = 6), or *nNOS*⁻/⁻ (n = 6) mice. Bar graphs show means +/− s.e.m. The significance of differences between stimulated and unstimulated muscles was assessed by the *t*-test.

> **42.** Do you think that NO is required for the production of cGMP? Support your answer with data from Figure 1.22.
>
> **43.** Can you formulate a model explaining why the loss of nNOS at the plasma membrane of a muscle cell could lead to muscular dystrophy? A correct answer gets you a Nobel Prize nomination!

Grady concludes by stating that he does not know if the dystroglycan complex (DGC) is destabilized even though it is located along the plasma membrane. During the question-and-answer session, Ervasti suggests that his experimental method could answer Grady's question, and they establish a collaboration on the spot. Another audience member asks whether the nNOS might be stretch activated, in which case structural damage could exacerbate the cGMP signaling defect. Grady does not know but says that the question is a good one.

Your head is about to explode from information overload when Kathy Wilson from Johns Hopkins is introduced. Her talk focuses on a form of muscular dystrophy you have never heard of before, Emery-Dreifuss muscular dystrophy. Your brain has neared saturation, and it is hard for you to fully appreciate what she is saying. The general story is about proteins called lamins that interact in the inner nuclear membrane. Somehow, mutations in lamin can lead to sudden death due to irregular rhythms in the heart. Wilson suggests two models. One is more structural and affects the pacemaker cells of the heart. The other suggests that a lack of lamins leads to gene expression irregularities, which you know is another form of signaling. She also notes that during **apoptosis** (genetically determined cell suicide), lamins are cleaved by proteases. You wonder if lamins have anything to do with calpain 3 causing muscular dystrophy, but do not bother to ask because you are not sure you could absorb any answer she might offer.

LINKS GSP
Jacques Beckmann
Kathy Wilson

Final Presentation

The final speaker is Jacques Beckmann from Evry, France. Beckmann gives a wonderful historical summary of the research in muscular dystrophy and LGMD in particular, which was first diagnosed as a unique disease in 1884. He gradually works his way toward the most recent findings and some of the unanswered questions. For example, does the lack of calpain 3 create an enzymatic or a structural problem? It has been shown that calpain 3 binds to titin, and perhaps this is the structural cause of LGMD2A. No one knows for sure. Beckmann wants to know why there is variation in severity of symptoms and suggests that the particular combination of mutations found in a person's two alleles may affect the phenotype. He speculates that other possibilities include linked genes, unlinked genes, or nongenetic factors. This brings him to the original study that identified calpain 3 as the genetic cause of LGMD2A. Beckmann cites your friend's research as seminal in our understanding. Beckmann says the results from Réunion were groundbreaking because nonstructural roles were suggested for the first time, as well as the real possibility of a second genetic locus. Of course, your friend is trying to keep a poker face, but you can see the radiating pride.

As you know, genetically isolated and homogeneous populations make ideal places to find genes that cause diseases. On Réunion, the expectation was to find a single founder allele that had propagated throughout the island and caused LGMD2A. However, the "Réunion Paradox" in this homogeneous population (as well as other isolated populations in the world) is based on the observation that there are many different mutant alleles that seem to cause LGMD2A. Beckmann proposes a simple solution: The founder effect on Réunion is not the calpain 3 gene, but a second genetic locus that has not been discovered yet. The existence of a second locus would explain the Réunion Paradox as well as the wide variation in severity and rate of disease progression seen in LGMD2A, as well as many other forms of muscular dystrophy. Beckmann supports his model with a lot of good logic and unexplained observations. He also cites studies by Richard Rozmahel et al. showing in mice that the very well-documented disease cystic fibrosis can vary in severity due to variations in other genes.

In closing, Beckmann leaves the audience with a proposal in which at least two genes are responsible for muscular dystrophies. Beckmann says, "Thus, the 'one mutation—several diseases' concept could well be explained by a digenic inheritance model." The simplistic view of "one gene—one

---

Math Minute 1.4   **Is cGMP Production Elevated?**

What you want to learn from Figure 1.22 is whether, in each strain of mice, the cGMP levels in unstimulated and stimulated muscles are the same or different. Because both quantities vary from mouse to mouse, the population means (averages of the two quantities over all mice) are compared. You can never know exactly what the population means are (you cannot test every mouse in the world), but their values can be estimated by measuring the two quantities in samples of several individuals. In Figure 1.22, error bars help depict population mean estimates. The following example illustrates how error bars are produced from sample data.

> **Example**: Suppose Grady measured cGMP levels in unstimulated muscles to be 2.7, 3.2, 6.0, and 5.4 in a sample of *wt* mice. The sample size (n) is 4. The **sample mean** ($\bar{y}$) is the average of these measurements (4.3). The sample **standard deviation** (s) is a measure of "spread," or variation from the sample mean. To calculate the sample standard deviation, (1) subtract the sample mean from each measurement, (2) square each difference, (3) sum the squared differences, (4) divide the sum by $n-1$, and (5) take the square root of the quotient. Thus,
>
> $$s = \sqrt{\frac{1}{n-1} \sum_{i=1}^{n} (y_i - \bar{y})^2} = \sqrt{\frac{1}{3}[(2.7 - 4.3)^2 + (3.2 - 4.3)^2 + (6.0 + 4.3)^2 + (5.4 - 4.3)^2]} \approx 1.6$$
>
> where y_i denotes the ith measurement. The *standard error of the mean* (s.e.m.) is estimated by the sample standard deviation divided by the square root of the sample size. The s.e.m. is approximately $1.6/\sqrt{4} = 0.8$, which is how far the error bar extends above and below the sample mean of 4.3 in Figure 1.22.

Notice that the error bar does not tell you anything about the range of the data (2.7–6.0). The error bar extends from 3.5 to 5.1, which does not include some of the observed values. Rather, the sample mean ± s.e.m. error bar marks the boundaries of a **confidence interval**, a range that contains the population mean 60% of the time. In other words, if you repeated the experiment and computed the sample mean ± s.e.m. confidence interval 100 times, the population mean would be contained in approximately 60 of the intervals. The *confidence level* of 60% is obtained from a table of probabilities specifically for $n = 4$.

A rule-of-thumb test for whether two population means are the same or different is based on the error bars for the two samples. If the bars overlap, the means are likely the same. If they do not overlap, the means are likely different. To be more precise, and quantify the likelihood associated with various degrees of overlap, a *t*-**test** is necessary. In this Math Minute, we apply a paired-sample *t*-test because the two samples are not independent; see Math Minute 3.2 for another version of the *t*-test that is used when the two samples are independent.

The paired-sample *t*-test addresses the question of whether the population mean of the difference between cGMP levels in unstimulated and stimulated muscles in a particular genotype is zero. Therefore, you must subtract the unstimulated muscle cGMP level from the stimulated muscle cGMP level for each mouse of this genotype, and find the sample mean and s.e.m. of the resulting numbers. If the population mean of the difference is zero, the sample mean of the difference should be small. How small depends on the sample size and s.e.m.

The result of the *t*-test is a *p*-**value**: the probability of seeing a sample mean of the difference as large or larger than the observed value, under the assumption that the population mean of the difference is zero. The *p*-value is obtained by dividing the sample mean of the difference by the s.e.m. and looking up the result in a table of probabilities called a *t-table*. A small *p*-value (typically 0.1 or less) indicates that the zero-population-mean hypothesis should be rejected; a larger *p*-value indicates a lack of evidence against the zero-mean hypothesis.

The p-values above the bars in Figure 1.22 imply (as expected) that cGMP is elevated by stimulation in wt mice and not in $nNOS^{-/-}$ mice. The borderline p-value in $adbn^{-/-}$ mice leaves room for doubt as to whether cGMP is elevated in this strain: there is a greater than 1 in 10 chance of seeing cGMP levels like those observed in the sample $adbn^{-/-}$ mice even if cGMP is not elevated.

---

function—one phenotype" may well be the exception rather than the rule, even for a single gene. To understand diseases at the genomic level will require understanding of **epistatic** interaction and an improvement in the detection of subtle phenotypes. We need to understand a protein's role in genomic circuits and identify all the proteins with which it interacts. In short, we need to think of cells as complex ecosystems with many interconnected components.

Beckmann continues by citing the many accomplishments of the reductionist approach to biology. However, as we learn more, the clear distinctions between monogenic and multilocus traits may prove to be too simplistic and arbitrary. Beckmann predicts that traits are best described on a continuum, with most traits more accurately described as digenic at least, or conditionally monogenic. He closes his talk with a final caution to those who like to categorize biological traits. "Furthermore, within specific genes, there may be alleles that behave as typical monogenic characters, while the expression of other allelic variants may require the interaction with specific genetic—or even environmental— contexts." In other words, our cells are ecosystems within the larger ecosystem and every component is capable of direct or indirect communication.

Summary 1.2: Your Final Thoughts

On the plane home, you are haunted by Beckmann's remarks. Is any disease monogenic? If a biological "rule" has exceptions, does the rule still hold, or should it be revised? Was Mendel wrong? Are there any traits that can be explained by a single locus? Has all the molecular genetic work performed by studying genes in isolation or artificial settings been a wasted effort? What about all the genomes being sequenced—is there any information in there that will help us understand complex genetics, or is that too a waste of time and money?

On the TV screen in front of you, you watch the silhouette of a plane gradually move over the Atlantic Ocean toward home. The plane seems so directed and certain of its destiny, and you feel so lost.

DISCOVERY QUESTIONS

44. Is muscular dystrophy one disease with many causes? Are muscular dystrophies many different diseases with many related/interconnected causes? Are muscular dystrophies many unrelated diseases with convergent symptoms? Explain your answer.

45. Is the mental retardation seen in some muscular dystrophy patients related to the molecular causes of the muscle pathology, or are they mechanistically unrelated?

46. Perform a PubMed search for the authors "Goyenvalle Vulin Fougerousse" to find a December 3, 2004 paper describing a novel gene therapy approach. If you have access to the paper, examine figure 4. What new genetic strategy did these authors attempt to fix the broken dystrophin gene?

Chapter 1 Conclusions

This DMD case study was designed to illustrate several themes of genomics: (1) nothing is really as simple as it seems; (2) genes are connected in circuits that perform complex functions; (3) research is not a linear process, and it can be frustrating along the way; (4) humans create rules to explain biological processes, but the rules can be wrong. Trust the data when exceptions to established rules are reproducible. One of the major goals of genomics is to build more complex but accurate models and to study proteins in their normal setting—inside functioning cells. Complex interactions inside cells are similar to interactions studied in ecology. The term "food web" refers to the interrelated nature of all the members in an ecosystem. Perhaps we need to think of cells as small ecosystems and the proteins as components of "cell webs." Regardless of the terminology, the complexity of cells is an integral aspect of how proteins function. *Discovering Genomics, Proteomics, and Bioinformatics* will lead you through the challenge Beckmann made: to reconsider cells and their web of interrelated proteins.

References
Phase I

Francke, U., H. D. Ochs, et al. 1985. Minor Xp21 chromosome deletion in a male associated with expression of Duchenne muscular dystrophy, chronic granulomatous disease, retinitis pigmentosa, and McLeod syndrome. *American Journal of Human Genetics.* 37(2): 250–267.

Zatz, M., A. M. Vianna-Morgante, et al. 1981. Translocation (X;6) in a female with Duchenne muscular dystrophy: Implications for the localization of the DMD locus. *Journal of Medical Genetics.* 18(6): 442–447.

C A S E   S T U D Y

Phases II and III

Bakker, E., M. H. Hofker, et al. 1985. Prenatal diagnosis and carrier detection of Duchenne muscular dystrophy with closely linked RFLPs. *Lancet.* 1(8430): 655–658.

Francke, U., H. D. Ochs, et al. 1985. Minor Xp21 chromosome deletion in a male associated with expression of Duchenne muscular dystrophy, chronic granulomatous disease, retinitis pigmentosa, and McLeod syndrome. *American Journal of Human Genetics.* 37(2): 250–267.

Verellen-Dumoulin, C., M. Freund, et al. 1984. Expression of an X-linked muscular dystrophy in a female due to translocation involving Xp21 and non-random inactivation of the normal X chromosome. *Human Genetics.* 67(1): 115–119.

Zatz, M., A. M. Vianna-Morgante, et al. 1981. Translocation (X;6) in a female with Duchenne muscular dystrophy: Implications for the localization of the DMD locus. *Journal of Medical Genetics.* 18(6): 442–447.

Phase IV

Koenig, M., E. P. Hoffman, et al. 1987. Complete cloning of the Duchenne muscular dystrophy (DMD) cDNA and preliminary genomic organization of the DMD gene in normal and affected individuals. *Cell.* 50(3): 509–517.

Kunkel, L. M., et al. 1986. Analysis of deletions in DNA from patients with Becker and Duchenne muscular dystrophy. *Nature.* 322: 73–77.

Monaco, A. P., C. J. Bertelson, et al. 1985. Detection of deletions spanning the Duchenne muscular dystrophy locus using a tightly linked DNA segment. *Nature.* 316: 842–845.

Monaco, A. P., R. L. Neve, et al. 1986. Isolation of candidate cDNAs for portions of the Duchenne muscular dystrophy gene. *Nature.* 323: 646–650.

The Next Steps in Understanding the Disease

Betto, R., D. Biral, & D. Sandona. 1999. Functional roles of dystrophin and of associated proteins. New insights for the sarcoglycans. *Italian Journal of Neurological Sciences.* 20(6): 371–379.

Campbell, K. Molecular Studies of Muscular Dystrophy. http://www.physiology.uiowa.edu/campbell/Netscape%20Site/DGCResearch.htm. Accessed 7 December 2000.

Campbell, K. P. 1995. Three muscular dystrophies: Loss of cytoskeleton-extracellular matrix linkage. *Cell.* 80(5): 675–679.

Crosbie, R. H., C. S. Lebakken, et al. 1999. Membrane targeting and stabilization of sarcospan is mediated by the sarcoglycan subcomplex. *Journal of Cell Biology.* 145: 153–165.

Crosbie, R. H., L. E. Lim, et al. 2000. Molecular and genetic characterization of sarcospan: Insights into sarcoglycan-sarcospan interactions. *Human Molecular Genetics.* 9(13): 2019–2027.

Ervasti, J. Ervasti web site. www.physiology.wisc.edu/faculty/ervasti.html. Accessed 25 March 2005.

Ervasti, J. M., K. Ohlendieck, et al. 1990. Deficiency of a glycoprotein component of the dystrophin complex in dystrophic muscle. *Nature.* 345: 315–319.

Holt, K. H., & K. P. Campbell. 1998. Assembly of the sarcoglycan complex: Insights for muscular dystrophy. *Journal of Biological Chemistry.* 273(52): 34667–34670.

Straub V., J. A. Rafael, J.S. Chamberlain, et al. 1997. Animal models for muscular dystrophy show different patterns of sarcolemmal disruption. *Journal of Cell Biology.* 139(2): 375–385.

Disease Information Web Sites

DMD and BMD: http://www.ncbi.nlm.nih.gov/htbin-post/Omim/dispmim?310200.

Utrophin: http://www.ncbi.nlm.nih.gov/htbin-post/Omim/dispmim?128240.

Where Is the Muscular Dystrophy Field Now?

Beckmann, J. S. 1999. Disease taxonomy—Monogenic muscular dystrophy. *British Medical Bulletin.* 55(2): 340–357.

Beckmann, J. S., I. Richard, et al. 1996. Identification of muscle-specific calpain and β-sarcoglycan genes in progressive autosomal recessive muscular dystrophies. *Neuromuscular Disorders.* 6: 455–462.

Chien, Kenneth R. 2000. Genomic circuits and the integrative biology of cardiac diseases. *Nature.* 407: 227–232.

Exxun.com. 2005. Political map of the world, section 4 of 6. www.exxun.com/exon/sm_world_pol_4.html. Accessed 25 March 2005.

Goyenvalle, A., A. Vulin, et al. 2004. Rescue of dystrophic muscle through U7 snRNA–mediated exon skipping. *Science.* 306: 1796–1799.

Grady, R. M., R. W. Grange, et al. 1999. Role for α-dystrobrevin in the pathogenesis of dystrophin-dependent muscular dystrophies. *Nature Cell Biology.* 1: 215–220.

Hack, A. A., C. T. Ly, et al. 1998. γ-sarcoglycan deficiency leads to muscle membrane defects and apoptosis independent of dystrophin. *Journal of Cell Biology.* 142(5): 1279–1287.

Hack, A. A., L. Cordier, et al. 1999. Muscle degeneration without mechanical injury in sarcoglycan deficiency. *PNAS USA.* 96: 10723–10728.

Hagiwara, Y., T. Sasaoka, et al. 2000. Caveolin-3 deficiency causes muscle degeneration in mice. *Human Molecular Genetics.* 9(20): 3047–3054.

Jung, D., F. Duclos, et al. 1996. Characterization of δ-sarcoglycan, a novel component of the oligomeric sarcoglycan complex involved in limb-girdle muscular dystrophy. *Journal of Biological Chemistry.* 271(50): 32321–32329.

McNally, E. M., C. T. Ly, & L. M. Kunkel. 1998. Human ε-sarcoglycan is highly related to α-sarcoglycan (adhalin), the limb girdle muscular dystrophy 2D gene. *FEBS Letters.* 422: 27–32.

Minetti, C., F. Sotgia, et al. 1998. Mutations in the caveolin-3 gene cause autosomal dominant limb girdle muscular dystrophy. *Nature Genetics.* 18: 365–368.

Moreira, E. S., T. J. Wiltshire, G. Faulkner, et al. 2000. Limb-girdle muscular dystrophy type 2G is caused by mutations in the gene encoding the sarcomeric protein telethonin. *Nature Genetics.* 24: 163–166.

Ono, Y., H. Shimada, et al. 1998. Functional defects of a muscle-specific calpain, p94, caused by mutations associated with limb-girdle muscular dystrophy type 2A. *Journal of Biological Chemistry.* 273: 17073–17078.

Richard, I., O. Broux, et al. 1995. A novel mechanism leading to muscular dystrophy: Mutations in calpain 3 cause limb girdle muscular dystrophy type 2A. *Cell.* 81: 27–40.

Richard, I., C. Roudaut, et al. 1999. Calpainopathy: A survey of mutations and polymorphisms. *American Journal of Human Genetics.* 64(6): 1524–1540.

Rozmahel, R., M. Wilschanski, et al. 1996. Modulation of disease severity in cystic fibrosis transmembrane conductance regulator deficient mice by a secondary genetic factor. *Nature Genetics.* 12(3): 280–287.

Rybakova, I. N., J. R. Patel, & J. M. Ervasti. 2000. The dystrophin complex forms a mechanically strong link between the sarcolemma and costameric actin. *Journal of Cell Biology.* 158(5): 1209–1214.

Sorimachi H., Y. Ono, & K. Suzuki. 2000. Molecular analysis of p94 and its application to diagnosis of limb girdle muscular dystrophy type 2A. *Methods in Molecular Biology.* 144: 75–84.

Sweeney, H. L., & E. R. Barton. 2000. The dystrophin-associated glycoprotein complex: What parts can you do without? *PNAS USA.* 97: 13464–13466.

Wilson, K. L. 2000. The nuclear envelope, muscular dystrophy, and gene expression. *Trends in Cell Biology.* 10: 125–129.

Yoshida, T., Y. Pan, H. Hanada, et al. 1998. Bidirectional signaling between sarcoglycans and the integrin adhesion system in cultured L6 myocytes. *Journal of Biological Chemistry.* 27(3): 1583–1590.

CASE STUDY

Genome Sequence Acquisition and Analysis

2.1 How Are Genomes Sequenced?

Define the field of genomics.

Learn how genomes are sequenced, annotated, and finished.

Use online tools to analyze genome sequences.

Understand the limitations of genome information.

2.2 What Have We Learned from Unicellular Genomes?

Extract meaningful information from genomes.

Gain insights into unexpected interactions.

Appreciate unexplained mysteries revealed through simple genomes.

2.3 What Have We Learned from Metazoan Genomes?

Discover biological capacities and frailties through metabolic reconstructions.

Uncover evolutionary relationships and genome duplications.

Compare finished and draft versions of genomes.

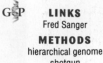

LINKS
Fred Sanger
METHODS
hierarchical genome
shotgun
whole genome
shotgun

With any new field of study, including genomics, come new technology and terminology. The first part of this chapter concentrates on how DNA sequence information is obtained and analyzed. Along the way, we will learn the essential vocabulary. Section 2.2 analyzes genomes of the smallest organisms to reveal secrets we did not know and to clarify long-sought solutions. In Section 2.3, we examine the genomes of model plants and animals to better understand our shared pasts and find ways to improve future research.

2.1 How Are Genomes Sequenced?

This section begins by describing DNA sequencing methodology, so you will understand how data are produced. People often assume that genomics uses only complete genome sequences, but there are rich databases dedicated to short fragments of DNA. Periodically, you will be asked Discovery Questions, which are interactive opportunities to mine public databases. Discovery Questions use specific examples to illustrate general principles of genome data analysis. Because databases are continuously updated, answers and web-page layouts may change over time. You need to learn how to adapt to a field of study that changes rapidly; a gene unknown today may be a newly discovered cancer gene tomorrow.

What Is Genomics?

"Genomics" is an unusual scientific term because its definition varies from person to person. The root word "genome" is universally defined as the total DNA content of a haploid cell or half the DNA content of a diploid cell. You might think the discipline of genomics would be the study of genomes, but this definition is too simplistic. In one sense, all of biology is related to the study of genomes, because every organism is shaped by its genome. However, most biologists would agree that disciplines such as anatomy and zoology should not be lumped into the current usage of the term genomics. How should we define genomics?

For most people, **genomics** involves large data sets (about 3 billion base pairs for the human genome) and **high-throughput** methods (fast methods for gathering data). Genomics includes sequencing DNA, cataloging genome variations within a population, and measuring transcriptional control of genes. Once the terms "genome" and "genomics" gained popularity, a cascade of new terms was initiated; each new area of research became an "-omic," or the subject under investigation was called an "-ome." The best examples are proteome and proteomics; a **proteome** is the complete protein content of a cell

or organism at a given moment. Other terms include transcriptome, metabolome, glycome, and variome. Do these newest fields fit under the larger umbrella of genomics, or are they distinct? It depends whom you ask. Throughout this book, we will define terms as needed and not focus on seldom-used words. In this text, the field of genomics includes a wide range of topics, including proteomics.

One last point about the term *genomics*: Although some people consider genomics to be nothing more than a collection of new methods, we will add one more component to our definition—a new perspective. With new methods come new types of questions and new ways to understand life. For many years, molecular methods were used as **reductionist** tools to dissect cells and understand how the parts work in isolation. In contrast, the field of genomics asks **expansionist** questions to understand how all parts work together. How does a functioning genome respond to environmental changes? What proteins interact with each other? These questions bring new interpretations that have many names, such as networks, systems biology, and circuits. Chapters 10, 11, and 12 consider three organizational levels of circuits: single genes, multiple genes, and whole genomes. Therefore, we use a broad definition of genomics, from DNA sequence analysis to an organism's response to environmental perturbations. The best place to start, however, is at the beginning, so let's learn about sequencing genomes.

How Are Whole Genomes Sequenced?

Genome sequences are collected in three separate phases: (1) preliminary sequencing; (2) **finishing**; (3) **annotating**. Most genomes are subjected to phases 1 and 3, but finishing is often skipped because of cost constraints. The human genome was sequenced more carefully than any other genome because we are especially interested in ourselves. In fact, the human genome was sequenced by two different groups that used slightly different methods: hierarchical genome shotgun (HGS) sequencing and whole genome shotgun (WGS) sequencing. Later, we will compare these methods.

Phase 1: How Is the Preliminary Sequence Collected?
The most popular method of sequencing DNA is the dideoxy method, sometimes referred to as the Sanger method in honor of Fred Sanger, who was awarded a Nobel Prize in 1980. We begin by describing the original radioactive procedure, and then address subsequent modifications that have improved sequencing efficiency.

At the heart of sequencing is DNA replication, which takes place in every dividing cell (Figure 2.1). To sequence DNA, many copies of a double-stranded DNA are denatured into single strands. The investigator mixes denatured

DNA, DNA polymerase, a primer, and deoxyribonucleotide triphosphates (dNTPs) of all four bases (dGTP, dTTP, dATP, and dCTP). One of the dNTPs includes a radioactive atom, for reasons that will become clear. This mixture is aliquotted into four tubes, and to each tube is added a

METHODS G⬡P
X-ray film
STRUCTURES
dNTPs
ddNTPs

small dose of one of the four dideoxyribonucleotide triphosphates (ddNTPs): ddGTP, ddTTP, ddATP, or ddCTP. A ddNTP lacks the 3′ hydroxyl group (-OH) found on normal deoxyribonucleotides. When DNA is being polymerized, the newest dNTP to be added forms a covalent bond onto the protruding 3′ hydroxyl group of the last successfully added dNTP. If a ddNTP was the last one incorporated, the elongation of DNA is terminated because ddNTPs lack a 3′ hydroxyl group. Because each of the four tubes has only one type of ddNTP, all of the terminated DNA strands in a particular tube end with the same base (i.e., in the ddATP tube, all strands terminate with the base adenine). Because only 1% of all adenine-containing nucleotides are ddATP, 99 out of 100 times a normal dATP is incorporated and the strand continues to elongate until a ddATP is incorporated. After a few minutes of elongation, the DNA polymerization reaction is stopped. The contents of each of the four tubes are loaded onto four separate lanes of a gel, and the different length DNA molecules separate according to their sizes, with the smallest molecules migrating the fastest (Figure 2.2). The DNA sequencing gel is exposed to X-ray film, which is developed and read from the bottom of the gel (5′ end) to the top of the gel (3′ end).

a)

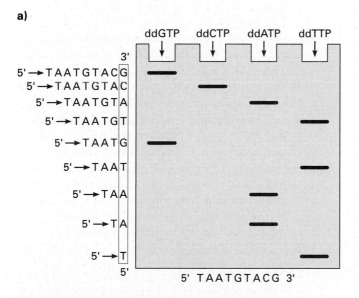

Figure 2.1 DNA replication in three steps.
a) Double-stranded DNA (purple and black interwoven strands) is shown aligned in antiparallel orientation. **b)** The strands are separated and primers (light purple and gray arrows) bind to the 3′ ends along with DNA polymerases (ovals). The dNTPs are available for incorporation into the new second strands. **c)** The final products are two copies of the original DNA with each older strand (dark purple and black) interwoven with the newer strands (gray and light purple).

a)

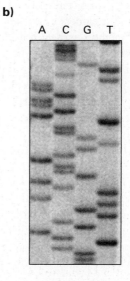

Figure 2.2 How to read a DNA sequencing gel.
a) A DNA sequencing gel where each of the four reactions was added to separate wells of the gel. Charged DNA polymers migrate toward the positive pole at the bottom of the gel, with the smallest fragments moving fastest. On the left side of the gel is the DNA sequence in each band, composed of millions of copies of identical DNA segments. In reality, you cannot see bands as short as shown here for illustration purposes. **b)** Part of an X-ray film is shown; typically DNA gels are loaded at two different times to separate shorter from longer pieces of ddNTP-terminated DNA.

DATA
Chromat 1

LINKS
BLASTn
Norm Dovichi
Genome Sequencing
Center
Leroy Hood
Human Genome
Browser
NCBI
template DNA

METHODS
384-well plate
Arabidopsis
automated sequencing
capillary electrophoresis
chromatogram
colony picking
cycle sequencing
PCR

DISCOVERY QUESTIONS

1. Read the sequence from the real X-ray film in Figure 2.2. Record the sequence for both strands of DNA, with the top strand containing the sequence on the X-ray film. Be sure to keep track of 5′ and 3′ ends for both strands.

Perform a **BLASTn** (nucleotide sequence) search with the top strand of DNA. BLASTn searches allow you to query the constantly updated database of all DNA sequences to find the best matches from the database for your query sequence (see Math Minute 1.1). Read the top "hit" from the BLAST results. What gene did you sequence? Now try a BLASTn search with the bottom strand (remember to enter it 5′ to 3′). Do you retrieve the same gene?

2. To get a complete understanding of the sequencing process, join two students who tour the Genome Sequencing Center at Washington University in St. Louis.

DNA, sequencers produced a four-colored **chromatogram** (**chromat**), which depicted the color of each band (one color for each base), the intensity of the light signal (height of the peaks), and the identity of the bases as determined by automated software (displayed above the peaks).

DISCOVERY QUESTIONS

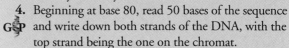

3. Go to the **Chromat 1** web page and examine the entire sequence. Don't bother trying to read the letters yet. Can you tell which end is the 5′ end?

4. Beginning at base 80, read 50 bases of the sequence and write down both strands of the DNA, with the top strand being the one on the chromat.

5. Perform a BLASTn search of the DNA in Chromat 1, but use only the first 30 bases of the 50. What was your best match? Record the **E-value** (measures quality of BLAST hits) presented in the right column. Now BLASTn all 50 bases and compare the new results with the search that used only 30 bases. Explain what happened to the E-value and why. You can read Math Minute 1.1 to understand why the E-value changed for the two BLAST results.

Radioactive sequencing worked well, but it was hazardous and labor intensive. A person could collect only 500 bases of sequence on a gel that took about 24 hours to produce a readable X-ray film. A faster method was needed—ideally, one that did not require radioactivity. Working at Caltech, Leroy Hood and members of his lab developed a nonradioactive dideoxy method with each ddNTP coupled to a different color of fluorescent dye. With their innovation, labs could obtain more sequence information from each reaction and eliminate the time-consuming steps of exposing the gel to X-ray film and manually reading the sequence. Only one microfuge tube was used for a sequencing reaction. Its contents were loaded onto a single lane of a gel in a dark cabinet where a laser scanned the gel as each band migrated down the gel. As each band flashed its color, indicating which ddNTP terminated that segment of DNA, the data were recorded by a computer. Fluorescence-based sequencing began the era of automated sequencing. Without this innovation, sequencing an entire genome, even a small prokaryotic genome, would have been too expensive and slow. Instead of black bars in four lanes for each piece of

With the advent of polymerase chain reaction (PCR), researchers quickly developed cycle sequencing, which combined the improvements of automated sequencing with the power of PCR. PCR works with very small amounts of template, and you don't have to produce a single-stranded template. With the increase in sequencing capacity, there came a need to improve DNA separation. Norm Dovichi and his colleagues helped invent capillary electrophoresis, which uses long, flexible, very thin capillary tubes filled with a grainy matrix that rapidly resolves minuscule amounts of DNA. As cycle sequencing has grown in popularity, companies have created kits that perform 384 reactions simultaneously in a 384-well plate, which can be loaded simultaneously in a 384-capable, high-throughput sequencer. Automated sequencing and capillary electrophoresis pushed the technology far enough that genome sequencing projects were completed ahead of schedule and under budget. In addition, colony picking from libraries and production of template DNA became automated. The number of species and the size of the National Center for Biotechnology Information (NCBI) database demonstrate the success of high-throughput DNA sequencing.

HOW ARE ORGANISMS PICKED FOR GENOME SEQUENCING?

Choosing to sequence the human genome was obvious, but what next? When the Human Genome Project (HGP) began, the "big seven" (*E. coli*, yeast, fly, worm, *Arabidopsis*, mouse, and human) had already been chosen, but it is not obvious why subsequent species were chosen. Why a puffer fish? Of what value is a mosquito genome? Who chose rice? The ultimate choices are made by funding agencies (e.g., U.S.

LINKS   G🧬P
Ensembl
less expensive
near future
sequenced genomes

agencies include the Department of Energy, the National Institutes of Health, and the National Science Foundation), but these agencies respond to arguments put forward by investigators. Thus, investigators ultimately determine which genomes will be sequenced. The primary arguments can be boiled down to four main categories: (1) medical applications; (2) evolutionary significance; (3) environmental impact; (4) food production. Every investigator must make a compelling case that his or her species deserves a portion of the limited pool of research dollars. The number of **sequenced genomes** is probably bigger than you expected (January 2005):

Nonplant eukaryotes	25
Plants	5
Microbes completed	213
Archaea completed	21
Microbes in progress	274
Viruses	1,431
Nonvirus organisms with at least one nucleotide sequence submitted	833

Why have we sequenced so many bacterial genomes and so few animals? Part of the answer is based on the economic reality—it is much less expensive to sequence a prokaryote with one chromosome of 10^6 bp than a plant like the wisk fern with a genome of 10^{11} bp. If every base cost $0.03 to sequence, you could sequence one bacterium genome for $100,000, as opposed to the $10 million it would cost to sequence the wisk fern. Another variable is the amount of **coverage** desired. Fold coverage is calculated by dividing the size of the genome by the total number of bases sequenced (e.g., 25 million bases sequenced for a 2.5 Mb genome represents tenfold or 10X coverage). The first dog genome had 1.5X coverage and produced about 78% of the entire genome and many human **orthologs**, though most of them were not full length. With more genomes for comparison, can we extract sufficient information with less coverage?

An important consideration has been medical relevance: many diseases are caused by prokaryotes. Evolutionary considerations have been another major factor, because prokaryotes (including Archaea) were the first organisms to evolve and thus are important for a more complete understanding of evolution (see Section 3.2). Microbes play a major role in the health of our environment and our ability to produce food. Because the impact of prokaryotes is so great, their abundance in sequencing projects seems justified. What about the puffer fish? It has the most compact genome of any vertebrate and thus is an economical choice for comparison of human genomes. By sequencing the mosquito genome, we may be better able to eradicate malaria. Rice is the staple food for about 2 billion people.

The question that remains to be answered is what will become of the huge genome sequencing centers in Asia, Europe, and the Americas. Some people believe that with improved technology, any individual's genome could be sequenced for $1,000, which would allow for personalized medicine. Who knows, perhaps you will have your genome sequence incorporated into your medical records in the near future, which raises many ethical questions discussed in Chapter 4.

DISCOVERY QUESTIONS

6. Go to Ensembl (European version of NCBI) and click on "Information" G🧬P under the "Docs and downloads" menu on the left side. Click on "Download data files." Are the genome sequences submitted as one single file? What level of organization has been used to post the DNA sequences?

7. Do mammals or amphibians have larger genomes, as revealed on the less G🧬P expensive web site? Why does the answer seem counterintuitive?

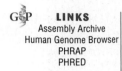

LINKS
Assembly Archive
Human Genome Browser
PHRAP
PHRED

Phase 2: How Are Genome Sequences Finished?

The definition of a finished genome was established by the Human Genome Project, in which investigators wanted to make no more than 1 error in 10,000 bases (99.99% correct). Finishing required several steps once the preliminary sequencing was completed and required the interaction of computers and highly trained technicians called *finishers*. First, a computer program compiled all the preliminary sequences from a segment of DNA and layered the sequences on top of each other (Figure 2.3).

Figure 2.3 Sequences and chromats in assembly archive.
Go to www.GeneticsPlace.com to view this figure.

The software called PHRED produces a quality assessment for each base as measured by the number of times a base was identified by the automated sequencer and the quality of the chromatogram for each base (i.e., peak height and even spacing). If the quality was too low, PHRED automatically notified the sequencers that they needed to resequence this region to resolve the problem, or *finish* this part of the sequence. A second software program called PHRAP took the PHRED output and assembled overlapping segments into larger assemblies of **contiguous DNA**. As the contiguous DNA (**contigs**) grew in size, the finishers began to compile the uninterrupted sequence of whole chromosomes.

Unfortunately, some contigs did not overlap with others, indicating a **gap** in the chromosome sequence. Gaps are more difficult to resolve and require substantial intervention by the finisher. Finishers devise a plan to fill the gap by requesting bench technicians to perform either PCR or new genomic DNA library screens to isolate fragments that span the gap. Once this new fragment of gapped DNA is isolated, it is sequenced and added to the preliminary sequences, then the finishing process is repeated with PHRED and PHRAP.

DISCOVERY QUESTIONS

8. Go to the Assembly Archive (chromat database) to view some chromats from an anthrax sequencing effort. Click on *Bacillus anthracis* str. (strain) Kruger B. You will see a list of many assemblies; click on Contig ID 607. When the new window opens, you will see three frames. The top frame shows the coverage of this 15.3 kb segment of anthrax genome. The middle frame shows the individual sequences used to assemble the 15.3 kb contig. What does the graph in the top frame summarize from the middle frame?

The bottom frame shows you the tiling path of the individual clones that span the entire 15.3 kb contig. The small red dashes indicate marker sequences used to help create the overlapping tiling path of clones. Mouse over the large blue segment that is centered on the 2 kb tick mark with trace ID ti494464459. When you see this trace ID number, click on it. A new window will open to show you the sequence, but click on the box next to "in color" and hit the "Show" button. This will produce a color graph that indicates the quality assessment score produced by PHRED. Where does the quality tend to be best? Scroll to the first two regions with quality scores between 0 and 20.

Now change the menu from "**FASTA**" (plain text format) to "Trace" and hit the "Show" button. You should see the chromat for this sequencing read. Next to the "in color" button should be a new option for the applet size. Change "Normal" to "Big" and hit the "Show" button. Right above the chromat is a "confidence" option; turn that on. On the far left is a scroll bar; move it down and up to see more and less of the chromat, respectively. The confidence is indicated as bar graphs for each base, with higher-confidence bases having longer bars; bar colors match the base colors, not quality assessment values. Find the regions of low quality scores and determine why the scores were so low.

9. Go to the Human Genome Browser and locate section chr19:8,584,715-8,601,616 by typing it into the search window. Click on the large black box in the gap row and read how gaps are depicted. Click the "Back" button once on your browser, and scroll down below the image. Click on "hide all," except modify these individual options: base position = full; chromosome band = dense; gap = pack; Ensembl genes = dense. Be sure to click the "Refresh" button at the top of the display options to implement your modifications; these settings will speed up subsequent navigation.

Click on the "base" button to the right of the 10X zoom-in button. This will show you the consensus sequence where known, and an x where there is no sequence information. Below the DNA sequence, all three reading frames are translated with red boxes marking stop codons and green boxes marking start codons. Zoom out 10X three times. Is this gap near a gene? Do you think this gap affected the nearest neighboring gene annotation? Continue zooming out until you see a second gap. Now hit the >>> button until you find a third gap. Continue to move >>> through the third gap to define the extent of this gap. Which gap is bigger, the first one you looked at or this third one? What chromosomal structure(s) are in the area of the bigger gap?

10. Go to the Finishing web page and determine the order of DNA fragments needed to build the largest possible contig. How many gaps remain after you have created the largest possible contigs?

11. Imagine you're a finisher working on the DNA you assembled in Discovery Question 10. How might you have isolated the gapped DNA if you knew the entire region of DNA was 20 kb long?

When the finished human genome sequence was published in October 2004, only 341 gaps remained, which was a substantial improvement over the 147,821 gaps found in the **draft sequence** of 2001. In the Discovery Questions, you located three gaps in chromosome 19, including the largest remaining gap in the entire finished genome near the centromere. The remaining gaps in the human genome are in areas resistant to current cloning and sequencing methods. Typically, gaps contain highly repetitive DNA that complicates the finishing because repeated DNA pattern (e.g., CTG repeated 1,000 times) is impossible to isolate, or because finishers cannot find unique overlapping sequences for reliable

contig assembly. The chromosomal structures of telomeres and centromeres are especially difficult to sequence due to their highly repetitive DNA.

DATA
Finishing

Finishing is a very expensive process, so most genomes are never finished. Because the human genome was finished, we can evaluate the value of finishing. As taxpayers, we should want to know that the extra expense actually improved the quality of the final human genome sequence. Let's consider chromosome 7 as a case study for the possible improvement between draft and finished sequence (Figure 2.4). If you look at low-resolution alignment, you can see the two sequence versions are very similar, with a diagonal line indicating mutual alignment of DNA. There are a few inversions where the **dot plot** slope is reversed, and a few segments with the correct slope but located off the diagonal due to some insertions or deletions (often collectively referred to as **indels**; see Math Minute 2.1). These misalignments are the macroscopic improvements, but what about the regions that look nearly identical? When we zoom in on a segment, we see that the draft version shows many DNA segments that had not been compiled into a single contig; thus, the 150 kb segment is a jumbled mess that was unscrambled in the finished sequence. If you extrapolate from this small segment to the entire genome (3 billion bp), the true value of finishing a genome sequence becomes apparent.

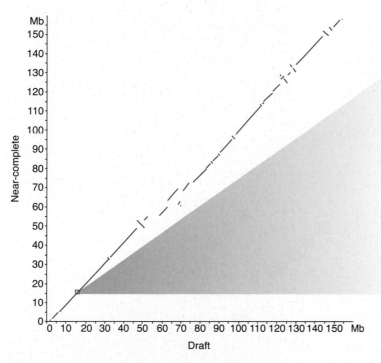
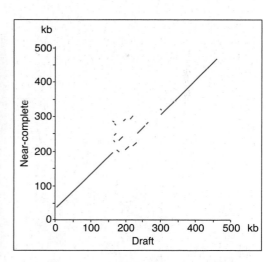

Figure 2.4 Comparing draft and finished versions.
The draft assembly of chromosome 7 was compared with the near-complete Build34 version. Each dot on the diagonal represents sequence agreement in the two versions. Diagonals of the same slope but shifted from the diagonal indicate order disagreements, while diagonals with opposite slope represent inversions. The inset magnifies 500 kb covering 3 BACs at a finer resolution and revealing the central BAC was poorly assembled in the draft version.

Math Minute 2.1   **What Can You Learn from a Dot Plot?**

Humans are very good at identifying patterns visually, so one of the first ways that DNA sequences were compared was with a dot plot (Figure 2.4). In the inset of Figure 2.4, the vertical axis represents 500,000 bases of the finished sequence, listed from bottom to top, and the horizontal axis represents 500,000 bases of the draft sequence, listed from left to right. Think of the rectangular area enclosed by the two axes as a matrix of $500,000^2 = 2.5 \times 10^{11}$ invisible dots, one for each combination of a base from each axis. A dot is defined by its coordinates, and colored black to indicate similarity between the two sequences. To explain how this process works, we will use an interactive example, comparing two DNA sequences that are only 15 bases long: Seq1 = AGGACTGTGATCTGT and Seq2 = CTCTGTGAGTCTTAT.

The simplest way to create a dot plot is to compare one letter at a time, defining similarity as base identity. Using this method, if the seventh base in Seq1 is identical to the fifth base in Seq2, then dot (7,5) is colored black. If they are different bases, the dot (7,5) is left invisible. The resulting dot plot for our example is shown in Figure MM2.1a. You can see that simply finding nucleotide matches does not reveal significant patterns, since every adenine in one sequence will match every adenine in the other sequence, and likewise for the other three bases. Roughly one-fourth of all dots in the matrix would be colored black. Imagine what Figure 2.4 would look like if it had been made this way!

To highlight more meaningful similarity, dot plots use an important concept in bioinformatics: a **sliding window** (see Math Minute 2.2, **GC skew,** GC content [e.g., Figure 2.38a], and Kyte-Doolittle plot for other examples). The sliding window eliminates much of the noise in a dot plot by identifying short strings, or runs, of sequence identity rather than single base identity. To see how this works, suppose we are using a sliding window of size 3. Dot (7,5) will be colored black if bases 6–8 in Seq1 are identical to bases 4–6 in Seq2. If any one of the three bases is different between the two sequences, dot (7,5) is left invisible. The dot plot for our example with a sliding window of size 3 is shown in Figure MM2.1b. Now it is easy to identify two runs of identical bases in the two sequences (bases 5–10 in Seq1 with bases 3–8 in Seq2; bases 11–15 in Seq1 with bases 2–6 in Seq2).

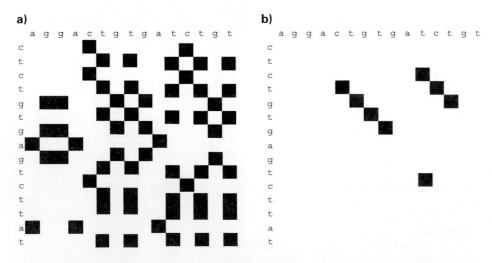

Figure MM2.1 Dot plot example.
Dot plots comparing 15 bp sequences Seq1 and Seq2 using **a)** base identity and **b)** sliding window of size 3. These dot plots were created with Excel file dotplot.xls, which you can download and use to make dot plots to compare other sequences.

When comparing long sequences, as in Figure 2.4, the sliding window size is typically an odd number (to produce a central point with an even number of bases on each side) between 7 and 29. The longer the window, the fewer black dots there tend to be in the plot. Dot plots can indicate various levels of similarity by using a grayscale color scheme, rather than indicating identity with the single color of black, or by allowing a limited number of mismatches within the sliding window. You can experiment with the file **dotplot.xls** and the **Exploring Dot Plots** web page to investigate the simple exercise outlined here, or explore further possibilities to help you answer the following Discovery Questions.

MATH MINUTE DISCOVERY QUESTIONS

1. Suppose that two sequences are identical except that a segment is inverted in one sequence, relative to the other. Explain how such an inversion would be displayed in a dot plot.
2. Suppose that two sequences are identical except that a segment is deleted from the middle of one sequence and not from the other. Explain how such a deletion would be displayed in a dot plot.
3. What would be the value of using a dot plot to compare a sequence to a second sequence, as well as the reverse complement of that second sequence?
4. Dot plots can detect many interesting sequence features by using the exact same sequence on both the horizontal and vertical axes. Sketch the dot plot of a 100 kb sequence in which a 20 kb segment is duplicated.
5. Sketch the dot plot of a 1 kb sequence in which a motif of approximately 50 consecutive bases appears six times in the sequence.
6. What would be the value of using a dot plot to compare a sequence to its own reverse complement?

Phase 3: How Do You Annotate a Genome? The purpose of annotation is to identify the functionally important sections of a sequenced genome. The primary objective is to identify all the genes, but ideally all the noncoding but functionally significant parts of the genome should be identified too. Identifying noncoding parts of the genome is very difficult compared to annotating genes, but identifying a gene is not as easy as it sounds.

Let's imagine you have completed the sequencing phases and compiled the genome into a full-length chromosome or chromosomes with no gaps. Now what? The amount of information you have at this point is overwhelming. You know only a few of the genes and metabolic pathways present. Where do you start?

What Is a Gene? A good place to start is the beginning: What is a gene? Should any piece of DNA that performs a function—even if the DNA is not transcribed—be considered a gene? For example, telomeres and centromeres are critical for chromosome function, but they are not transcribed. Should we consider them to be functional units, and thus genes? Some say that an obvious cellular role is not necessary, and that any piece of DNA that has survived long periods without recombination should be considered a gene, because recombination was selected against and thus its sequence remains intact for some reason.

Most people define a gene as a piece of DNA of which some is transcribed and that includes a promoter, coding sequences, and a signal for the RNA polymerase to stop. For prokaryotes, genes contain open reading frames (**ORFs**; pronounced like the first four letters in the word *orphan*) with no introns. ORFs are segments of DNA ranging from very few to thousands of bases that begin with a start codon and end with a stop codon. Prokaryotic genomes do not contain much **intergenic** sequence (DNA lacking genes), and they have smaller genes. Eukaryotes have more complexity in gene and genome structures.

Promoters often contain specific sequences (e.g., **TATA** [pronounced tah-tah] and **CAAT** [pronounced cat] boxes) that are recognized by RNA polymerase and transcription factors. Further upstream are enhancers with complex sequence patterns that are recognized by transcription factors. The start codon often has a consensus sequence called a Kozak signal (named for Marilyn Kozak, who discovered this pattern), which consists of **GCCRCCATGG**, where the start codon is **ATG** and **R** stands for either purine (**A** or **G**). Eukaryotic genes may contain noncoding introns in addition to the coding exons.

The size of introns can vary, ranging from 70 bp to more than 30,000 bp. The average human gene is 30,000 bp long, but the dystrophin gene is over 2 million bp. In addition to the complexity of real genes, the mammalian genome also contains sections of approximately 225 bp per kb that look like genes but are not transcribed. These 225 bp gene look-alikes are called **pseudogenes**, because a mutation has rendered them nonfunctional. So finding a real gene amidst all this complexity is a very difficult task.

Math Minute 2.2   How Do You Find Motifs?

To locate genes in newly sequenced eukaryotic genomes, we can use the fact that the sequence upstream of a gene contains certain motifs (nucleotide sequence patterns of functional significance). However, we do not know in advance what the sequence patterns are. To complicate matters further, the patterns can vary significantly from gene to gene and organism to organism. Every transcription factor that regulates gene expression has a preferred motif to which it binds—but how can we find these binding sites if we do not know what the transcription factors are, much less which genes they regulate? To get started, bench scientists had to find some binding sites through painstaking molecular biology methods. Once a few binding sites had been identified, a computational analysis using a position weight matrix (PWM) helped find more.

The TATA box is a motif that helps RNA polymerase find the transcription start site in many eukaryotic genes. A similar motif appears in prokaryotic genes as well. We will use the TATA box to explore the PWM method. Although the consecutive letters TATA are the heart of this motif, these four letters also occur randomly in many places that are not immediately upstream of a gene. By discovering some true TATA boxes, investigators began to characterize the sequence pattern that distinguishes true TATA boxes from the random TATA-like sequences. Table MM2.1 summarizes the TATA box sequences for 389 different eukaryotic genes (note that the sum of every column is 389).

The four letters TATA for which the motif is named appear in positions 2–5, but even these four letters are not always TATA. For example, position 2 contains a T only about 80% of the time. Table MM2.1 also shows that position 6 is either A or T, position 7 is almost always A, position 8 is usually A or T, and positions 1 and 10–15 have a slight tendency to be G or C.

Suppose we are examining 15 bp in a potential promoter of a newly sequenced genome. How can we encapsulate all the information in Table MM2.1 to help decide if these 15 bp form a real TATA box? The basic idea is to compare the probability of these 15 letters occurring in a TATA box to the overall probability of these 15 letters occurring anywhere in the genome. For example, if the genome-wide average GC content is 44%, the probability of an A in position 1 is 0.28 (0.56×0.5; probability of not-GC times the probability of A if not GC). However, if these 15 bp form a TATA box, the probability of seeing an A in position 1 is given by the relative frequency of A in position 1, $61/389 \approx 0.1568$.

The TATA probability (e.g., 0.1556) divided by the overall probability in this species (e.g., 0.28) indicates whether a letter is more or less likely to occur at a given position in a TATA box than it is to occur overall. To get the final PWM score, we must multiply the ratios across all 15 positions, but it is easier to work with the logarithm of the ratio, called the **log odds**, and add across all 15 positions. Traditionally, log base 2 is used, because it is easy to detect doublings in the probability.

Table MM2.1 Nucleotide frequencies in 389 known TATA boxes.

Position	1	2	3	4	5	6	7	8	9	10	11	12	13	14	15
A	61	16	352	3	354	268	360	222	155	56	83	82	82	68	77
C	145	46	0	10	0	0	3	2	44	135	147	127	118	107	101
G	152	18	2	2	5	0	10	44	157	150	128	128	128	139	140
T	31	309	35	374	30	121	6	121	33	48	31	52	61	75	71

If the letter is more likely to occur at a given position in a TATA box than it is to occur overall, the ratio of probabilities will be bigger than 1.0 and the log odds will be positive. In contrast, if the letter is less likely to occur at that position of a TATA box than it is to occur overall, the ratio of probabilities will be smaller than 1.0 and the log odds will be negative. In our example of determining the probability of seeing an A in position 1 of a real TATA box, the log odds is $\log_2(0.1568 / 0.28) = -0.84$, which means that an A is not very likely to be in the first position of a 15 bp TATA box. We compute the log odds for each of the other three letters that could be in position 1, and then repeat this process for all 15 positions. The log odds scores for the TATA box motif are given in a position weight matrix in Table MM2.2. Note that we use a large negative number (−99) whenever the log odds is undefined (i.e., ratio of probabilities is 0).

To measure the likelihood that a 15 bp query sequence is a TATA box, we sum the log odds scores for the 15 letters in the sequence. For each position, we use the row that matches the letter in that position. For example, suppose our sequence is **ACATATATAAGCTGG**. The log odds scores to be added are highlighted in Table MM2.3. The sum of all 15 highlighted scores (6.78) is the total PWM score of this sequence. By considering every 15 bp sequence in the genome using a sliding window (see Math Minute 2.1), we can identify sequences that are most likely to be TATA boxes (i.e., with the highest scores). By repeating this scoring process with many different PWMs, representing many known binding-site motifs, we can begin to deduce the location of genes in our newly sequenced genome.

Table MM2.2 Position weight matrix.

A	−0.84	−2.77	1.69	−5.18	1.70	1.30	1.76	1.03	0.51	−0.96	−0.39	−0.41	−0.41	−0.68	−0.50
C	0.76	−0.90	−99.00	−3.10	−99.00	−99.00	−4.80	−5.42	−0.96	0.66	0.78	0.57	0.46	0.32	0.24
G	0.83	−2.25	−5.42	−5.42	−4.10	−99.00	−3.06	−0.96	0.88	0.81	0.58	0.58	0.58	0.70	0.71
T	−1.81	1.50	−1.64	1.78	−1.86	0.15	−4.14	0.15	−1.72	−1.18	−1.81	−1.07	−0.84	−0.54	−0.62

Table MM2.3 PWM score of the 15 bp sequence **ACATATATAAGCTGG**.

	A	C	A	T	A	T	A	T	A	A	G	C	T	G	G
A	−0.84	−2.77	1.69	−5.18	1.70	1.30	1.76	1.03	0.51	−0.96	−0.39	−0.41	−0.41	−0.68	−0.50
C	0.76	−0.90	−99.00	−3.10	−99.00	−99.00	−4.80	−5.42	−0.96	0.66	0.78	0.57	0.46	0.32	0.24
G	0.83	−2.25	−5.42	−5.42	−4.10	−99.00	−3.06	−0.96	0.88	0.81	0.58	0.58	0.58	0.70	0.71
T	−1.81	1.50	−1.64	1.78	−1.86	0.15	−4.14	0.15	−1.72	−1.18	−1.81	−1.07	−0.84	−0.54	−0.62

MATH MINUTE DISCOVERY QUESTIONS

1. Go to JASPAR and select "Browse profiles by class". Scroll down to the TATA box and click on the "View" button in this row. Verify that the values in Table MM2.1 are displayed in this window. Explain how the sequence logo represents the information in Table MM2.1.

2. Return to the JASPAR "Browse profiles by class" page. Find transcription factors with ID numbers MA0040, MA0041, and MA0047. By looking at the sequence logos, explain which of these three transcription factors in rat is most likely to bind to DNA containing the motif **TGTTTA**.

3. Use the spreadsheet pwm.xls to compute the total TATA PWM score of the following three sequences, and determine which one is most likely to be a true TATA box: **ATATATATAGGCTGG**, **CTATATATATGCTGG**, **CTATAAATAGGCCGG**.

4. Use the spreadsheet pwm.xls to compute the total TATA PWM score of **CCGGCCTATTTATAG**. Explain why the score is so high, even though this sequence does not look like a true TATA box. (Hint: How is this sequence related to one of the sequences in Math Minute Discovery Question 3?)

LINKS
JASPAR

MATH MINUTE
pwm.xls

DATA
gDNA1
scramble
TestCode

LINKS
Mark Borodovsky
Entrez
Entrez Gene
GeneCard
GeneMark
GenScan
Glimmer
GlimmerM
ORFinder
Perl

METHODS
cDNA

Obviously, we cannot read 30,000 bases of DNA sequence and visually recognize genes. We need computer algorithms to help us sort through the data. When analyzing entire genomes, there are a few tricks that make gene annotation easier. For organisms that have been studied for years, genetic and cytogenetic markers (chromosome banding patterns) might be mapped on chromosomes to localize the position of a gene. The gene *white*, the first mutant gene of *Drosophila*, was genetically mapped to a small region of the X chromosome, and its cytogenetic location is visible on polytene chromosomes. Or, if a **cDNA** had been sequenced and submitted to a database, recognizing conserved coding sequences and thus identifying particular genes would be easier.

Are There Faster Ways to Annotate Genomes? Prokaryote and eukaryote genomes are significantly different and thus require different tools. Many commercial software packages exist, but luckily some effective free programs are also available. Most of the free programs run on your computer's hard drive, but a few are available online through a web browser. GeneMark was originally created to locate genes in prokaryotes, but Mark Borodovsky and his research group at Georgia Tech have expanded the options to include a significant number of model eukaryotes. MIT hosts another online program to locate genes within a genome. GenScan will accept up to 1 million bp of sequence, but if you want to search multiple sequences, you must download the program and run it on a Unix-based computer. If you want to download and use a **Perl** program (see Math Minute 10.1), you might consider Glimmer or GlimmerM from TIGR; these programs locate genes within whole genomes for prokaryotes and eukaryotes, respectively. There is a GlimmerM web server, but only a few species are available and there is a 200 kb limit. With time, more gene-finding web sites will become more flexible and freely available.

TestCode is a Perl script that finds potential coding regions by analyzing nucleotide triplet frequencies. Once you find a possible coding region, you will probably want to locate the open reading frame (ORF). For many biologists, ORFs are the most interesting part of any genome, but later we will see there are functional units in the noncoding parts of the human genome too. ORFinder will translate the DNA in all six reading frames, with the results shown in graphical format. Remember, DNA is double stranded, so there are three reading frames for the top strand and three more for the bottom strand. ORFs are indicated by the colored boxes, and you are looking for the largest ORF. ORFinder ranks the ORFs from largest to smallest and indicates the reading frame for each ORF.

DISCOVERY QUESTIONS

12. Go to the genomic DNA #1 (gDNA1) web site, where you will see three pieces of DNA sequence. Copy and paste one of the sequences and then click on the "TestCode" button. One at a time, submit the three segments of gDNA to TestCode to find which one harbors the ORF.

13. Copy and paste the real ORF from Discovery Question 12 into the scramble web site to have a Perl script generate a scrambled version of the same DNA. Take the scrambled version and resubmit it to TestCode. Does the randomized version of the coding DNA look like coding DNA? Would you expect it to?

Identifying ORFs or retrieving BLAST **hits** may not tell you much about your sequence, so you might want to search the full NCBI database called **Entrez**. Test Entrez with the phrase "homo cyclooxygenase" to search cyclooxygenases (the protein inactivated by aspirin and similar medications; see Chapter 9 for details) of *Homo sapiens*. The Entrez portal provides you with links to papers, sequences, structures, and many other databases. Click on the number next to PubMed to see a list of all the research and review papers published about "homo" and "cyclooxygenase." Go back and click on the "UniGene" results, then click on the "Mammals" tab at the top. You should see the two human cyclooxygenase genes PTGS1 and PTGS2. Click on the link called "Links" and then the option "gene" from the pop-up menu; then click the "Gene" link to see a complete resource for the gene you have chosen. Each of these searches (e.g., UniGene, Genes, etc.) can be performed directly by choosing the appropriate search menu on the NCBI main page.

For the sake of completeness, you should know there are two other powerful ways to find out about annotated (characterized) genes. GeneCard, created at the Weizmann Institute in Israel, is mirrored all over the world. Enter the term "obese," and click on "Go." You will get at least 150 hits, but you can find leptin receptor by performing a find function with your browser using the word "receptor." Click on the four-letter abbreviation to see all the genomic, proteomic, and phenotypic information. Compare GeneCard with Entrez Gene (NCBI database similar to GeneCard). Type in the word "prion" and you will get a list of loci, an abbreviation for the source species (e.g., Hs for *Homo sapiens*, Mm for *Mus musculus*, Dm for *Drosophila melanogaster*), and a colorful key for the linked databases. You are looking for "prion protein (p27–30) (Creutzfeld-Jakob disease . . .)"; browse until you can find it in humans. If you click on the colored boxes with letters, you can access many different types of information from this one search site.

With all these databases linked to each other, you begin to see the power behind genome projects with free public access. If you are conducting research on a particular gene, phenotype, or disease, you do not have to start from scratch, clone any genes, or do much sequencing. These steps have been done for you and compiled in searchable databases. Now that this grunt work has been completed, we are free to ask interesting and complex questions. However, beware of two common mistakes: (1) Just because you see the same information listed in multiple databases does not mean that there are multiple independent validations of the information. If possible, look at the references or sources of information to see if they all cite an identical source. (2) Do not assume that all these linked databases contain the same information. Behind all the databases are humans who can make mistakes or enter new information manually to one database but not all of them. The cautionary note is to search multiple databases for new information, but not to use multiple listings as independent validation of information. We will see examples of these confounding problems throughout this book.

Sometimes you find two similar genes and you want to compare them directly. BLAST2 allows you to compare two nucleotide or amino acid sequences. Enter the two cyclooxygenase (COX1 and COX2) mRNA accession numbers (NM_000962 and NM_000963) in the small (not the large) boxes, make sure "blastn" is selected near the top, and click on "Align." The genes have three sections that are well conserved at the nucleotide level, but the rest of these two ORFs are much more divergent. To see the bp alignment, click on the small boxes to the left of the diagonal graph.

DISCOVERY QUESTIONS

14. Calculate the average percent nucleotide identity for the three COX gene regions from your BLAST2 alignment.

15. Go back to BLAST2 Sequences and enter the two protein accession numbers for COX2, "NP_000954," and COX1, "NP_000953." Be sure to change the search from BLASTn to BLASTp. Verify that the top blank contains NP_000954, so COX2 will be the query in the resulting page.

  a. What is the overall amino acid identity? Is this higher or lower than the overall nucleotide identity?

  b. Notice that a separate percentage is calculated for similarities (called "Positives"), which takes into account the similar structures of some amino acids. What is the percent similarity?

  c. Which parts of the proteins appear to be poorly conserved? Look at the sequence alignment that uses the single amino acid code and find where one protein has several Xs in a row, to mark areas of low complexity (see Math Minute 2.3).

  d. Use your browser's "Find" function to locate the amino acid sequence **GAPFS**. Serine (S)

is the amino acid modified by aspirin. Is **GAPFS** in a region of high sequence identity or similarity? (See pages 336–337 for details.)

16. Go to HGNC (Human Genome Nomenclature Committee) and perform a gene search for cyclooxygenase to see how many cyclooxygenase genes there are in the human genome. HGNC is a good, quick way to perform a gene search with links to many other databases. However, compare the HGNC results with an OMIM search using the gene name COX3. Do you find any surprises?

LINKS
BLAST2
HGNC
OMIM
METHODS
Arabidopsis
Kyte-Doolittle plot

By now you may be wondering how many genes are in your genome. Based on the number of proteins and a heavy dose of human-centric ego, the most common estimate for the number of genes was 100,000 during the early preliminary sequence acquisition. When the human draft sequence was published in 2001, the number of annotated genes was 35,000. Within a year, the number of genes dropped to 30,000. Now that the finished genome sequence is available, the best estimate is between 20,000 and 25,000 human genes. The number of genes is much smaller than expected, in part because we erroneously think of ourselves as the pinnacle of evolution. (In fact, there was an informal pool to see who could guess the closest; to everyone's surprise, the winner was the absolute lowest guess of 40,000 genes.) However, our genomes are humbled by such species as *Arabidopsis* (about 25,000 genes), the puffer fish (*Tetraodon nigroviridis* with 28,000 genes), and rice, which has only 10% as much DNA as humans but nearly three times as many genes. Actually, our genome has about the same number of genes as the eyelash-sized worm *C. elegans*, whose adult body is composed of about 1,000 (10^3) cells compared to our trillion (10^{12}) cells! Finally, it sounds like an oxymoron that we do not know how many genes there are in the "finished" version of the human genome. Defining a gene is not trivial, and with the recent discovery of very small but functionally important microRNA (see page 107), the size of coding regions is smaller than we had thought.

Can We Predict Protein Functions from DNA Sequence?

Because about 30% of all identified genes have no known function, getting a BLAST match to an unknown gene is not particularly helpful. A good predictive tool is the hydropathy plot (graphical display of a protein's hydrophobic regions), or Kyte-Doolittle plot, which predicts whether a protein might be an integral membrane protein. In a Kyte-Doolittle plot, every peak at 1.8 or higher indicates the potential for a transmembrane domain. Hydropathy scores are computed in a

Math Minute 2.3   **What Are "Positives" and What Do They Have to Do with E-values?**

Because there are many characteristics of amino acids (e.g., charge, hydrophobicity, polarity, size, etc.), it makes sense to develop a more refined comparison between protein sequences than simple identity. In Discovery Question 15, you considered a finer scale of comparison known as "positives." To understand the positive scale of comparison, we need to investigate an even finer scale of comparison called a **substitution matrix**, which measures the evolutionary closeness for each possible pair of all 20 amino acids. A substitution matrix is used by BLASTp to compute the raw scores upon which E-values are based (see Math Minute 1.1).

BLOSUM62 (Table MM2.4) is the default substitution matrix for BLASTp. The score for aligning one amino acid with another is the number in the row labeled with the first amino acid and the column labeled with the second amino acid. For example, the score for aligning proline with proline is 7, proline with cysteine is −3, and proline with threonine is −1. The scores in the BLOSUM62 matrix are based on observed amino acid substitutions in orthologous proteins with aligned sequences. Higher scores correspond to substitutions that are more commonly observed, and thus more highly favored by evolution. For example, the scores listed above indicate that proline tends to be replaced more often by threonine than by cysteine. Of course, the most likely change is no change at all (proline for proline), and the relative degree to which each amino acid resists change is indicated by the different magnitudes of positive scores along the diagonal of the matrix.

The X entry in the BLOSUM62 matrix is used to score low-complexity regions, which are determined by an algorithm called SEG. This algorithm defines a low-complexity region as a segment with periodic repeats of amino acids (e.g. VWVWVWVWVW), or with "biased" composition of amino acids as compared to the rest of the sequence.

Table MM2.4 BLOSUM62 substitution matrix.

	A	R	N	D	C	Q	E	G	H	I	L	K	M	F	P	S	T	W	Y	V	B	Z	X
A	4	−1	−2	−2	0	−1	−1	0	−2	−1	−1	−1	−1	−2	−1	1	0	−3	−2	0	−2	−1	0
R	−1	5	0	−2	−3	1	0	−2	0	−3	−2	2	−1	−3	−2	−1	−1	−3	−2	−3	−1	0	−1
N	−2	0	6	1	−3	0	0	0	1	−3	−3	0	−2	−3	−2	1	0	−4	−2	−3	3	0	−1
D	−2	−2	1	6	−3	0	2	−1	−1	−3	−4	−1	−3	−3	−1	0	−1	−4	−3	−3	4	1	−1
C	0	−3	−3	−3	9	−3	−4	−3	−3	−1	−1	−3	−1	−2	−3	−1	−1	−2	−2	−1	−3	−3	−2
Q	−1	1	0	0	−3	5	2	−2	0	−3	−2	1	0	−3	−1	0	−1	−2	−1	−2	0	3	−1
E	−1	0	0	2	−4	2	5	−2	0	−3	−3	1	−2	−3	−1	0	−1	−3	−2	−2	1	4	−1
G	0	−2	0	−1	−3	−2	−2	6	−2	−4	−4	−2	−3	−3	−2	0	−2	−2	−3	−3	−1	−2	−1
H	−2	0	1	−1	−3	0	0	−2	8	−3	−3	−1	−2	−1	−2	−1	−2	−2	2	−3	0	0	−1
I	−1	−3	−3	−3	−1	−3	−3	−4	−3	4	2	−3	1	0	−3	−2	−1	−3	−1	3	−3	−3	−1
L	−1	−2	−3	−4	−1	−2	−3	−4	−3	2	4	−2	2	0	−3	−2	−1	−2	−1	1	−4	−3	−1
K	−1	2	0	−1	−3	1	1	−2	−1	−3	−2	5	−1	−3	−1	0	−1	−3	−2	−2	0	1	−1
M	−1	−1	−2	−3	−1	0	−2	−3	−2	1	2	−1	5	0	−2	−1	−1	−1	−1	1	−3	−1	−1
F	−2	−3	−3	−3	−2	−3	−3	−3	−1	0	0	−3	0	6	−4	−2	−2	1	3	−1	−3	−3	−1
P	−1	−2	−2	−1	−3	−1	−1	−2	−2	−3	−3	−1	−2	−4	7	−1	−1	−4	−3	−2	−2	−1	−2
S	1	−1	1	0	−1	0	0	0	−1	−2	−2	0	−1	−2	−1	4	1	−3	−2	−2	0	0	0
T	0	−1	0	−1	−1	−1	−1	−2	−2	−1	−1	−1	−1	−2	−1	1	5	−2	−2	0	−1	−1	0
W	−3	−3	−4	−4	−2	−2	−3	−2	−2	−3	−2	−3	−1	1	−4	−3	−2	11	2	−3	−4	−3	−2
Y	−2	−2	−2	−3	−2	−1	−2	−3	2	−1	−1	−2	−1	3	−3	−2	−2	2	7	−1	−3	−2	−1
V	0	−3	−3	−3	−1	−2	−2	−3	−3	3	1	−2	1	−1	−2	−2	0	−3	−1	4	−3	−2	−1
B	−2	−1	3	4	−3	0	1	−1	0	−3	−4	0	−3	−3	−2	0	−1	−4	−3	−3	4	1	−1
Z	−1	0	0	1	−3	3	4	−2	0	−3	−3	1	−1	−3	−1	0	−1	−3	−2	−2	1	4	−1
X	0	−1	−1	−1	−2	−1	−1	−1	−1	−1	−1	−1	−1	−1	−2	0	0	−2	−1	−1	−1	−1	−1

Table MM2.5 Sequence alignment using BLOSUM62.

N	I	C	K	E	C	P	I	I	G	F	R	Y	R	S	L	K	H	F	N	Y	D	I	C	Q	S	C	F	F
+	C	+		+					G	F	R	Y	R		+			F	N	Y		+	C	Q	+	C	F	+
S	F	C	R	S	D	G	M	T	G	F	R	Y	R	C	Q	Q	C	F	N	Y	Q	L	C	Q	N	C	F	W
1	0	9	2	0	-3	-2	1	-1	6	6	5	7	5	-1	-2	1	-3	6	6	7	0	2	9	5	1	9	6	1 = 83

The B entry represents either aspartic acid or asparagine and Z represents either glutamic acid or glutamine.

BLASTp computes the raw score of a protein sequence alignment by adding up the individual amino acid alignment scores from the substitution matrix. For example, the BLOSUM62 score for each pair of aligned amino acids in Table MM2.5 is shown on the bottom row.

The raw score of the alignment is computed by adding up the scores, for a total of 83. Note that this alignment was one of the hits you got in Discovery Question 10 of Chapter 1 (accession number NP_956003); you can see the raw score 83 in parentheses next to the bit score. Math Minute 1.1 explains how to use the raw score of 83 to calculate the bit score of 36.6, and finally the E-value of 0.19 for this hit.

Now we can also see where the percent positive measurement comes from in Discovery Question 15. The middle row of the alignment indicates every position for which the BLOSUM62 score is positive (or identical). There are 20 entries in the middle row, corresponding to the 20 positive numbers below the alignment, and the percent positive is $20/29 \approx 0.6897$. The BLASTp report does not round this decimal, but truncates it to 68%.

BLOSUM62 is a good, general-purpose matrix, which is why it was chosen as the default for BLASTp. Other versions of substitution matrices are BLOSUM45 (for finding more closely related proteins) and BLOSUM80 (for finding more divergent proteins). The numbers 45, 62, and 80 refer to how much weight is given to aligned amino acids that are identical in the related proteins, versus aligned amino acids that differ, when computing the substitution likelihood scores. There is another type of matrix, called PAM, of which there are several variants ranging from PAM1 (for closely related proteins) to PAM250 (for more divergent proteins). In practice, BLOSUM matrices are used more often because they seem to find appropriate hits more often than PAM matrices. BLOSUM62 almost always finds appropriate hits, so it is safe to use for all but the most specialized searches.

BLASTn alignments use a much simpler scoring matrix that measures identity. Instead of simply computing percent identity, a score is assigned to identical nucleotides (matches) and nonidentical nucleotides (mismatches) in an alignment. The default match score is 1; the default mismatch score is −3 in BLASTn, and −2 in BLAST2. You can explore the effects of changing the scoring matrix in the following Math Minute Discovery Questions.

MATH MINUTE DISCOVERY QUESTIONS

1. What are the smallest and largest values in the BLOSUM62 matrix? Explain why the diagonal entries of the table are not all the same, even though they all correspond to matches.

2. Repeat the BLAST2 protein alignment you performed in Discovery Question 15, but with COX1 (NP_000953) as the query this time. By looking at the resulting alignment and the BLOSUM62 matrix, explain the difference in raw scores in this comparison and your original comparison. Focus your search on the low-complexity regions. To verify that the low-complexity regions are the only difference between using COX1 or COX2 as the query, repeat the two comparisons with filtering turned off. The filter check box is just above the box for sequence 1 accession number.

3. Repeat the BLAST2 search you did to compare COX1 and COX2 at the nucleotide level, but this time change "Reward for a match" to 2 and "Penalty for a mismatch" to −3. (Blanks for these numbers are just below where you select the blastn program.) What similarities and differences do you see in your results, as compared to your original search? Now repeat the search with the mismatch score remaining at −3, but the match score returned to the default of 1. Are there any unexpected outcomes? Explain the changes in hits for the three different nucleotide alignments, and how your results relate to the three different BLOSUM matrices described earlier.

4. You can change the nucleotide scoring parameters in a regular BLASTn search, too. Go back to the sequence you read from the chromatogram in Discovery Question 4, and enter the 50 bases as your query sequence in a BLASTn search. Before submitting your search, scroll down to the "Other Advanced" window and type "−r 2" to set the match score to 2. (You can click on the link next to this window to see what other parameters you can change.) Now submit your query. How are your hits different from the hits you got in Discovery Question 5 (with default value of 1), and why? Find the numerical evidence of the change you made at the bottom of the BLAST report.

sliding window (see Math Minute 2.1); a window size of 19 is best for detecting transmembrane domains. Copy the **leptin** amino acid sequence and paste it in the appropriate box on the Kyte-Doolittle Hydropathy plot page. When the plot is visible, print or save it. Repeat this process to produce a plot for the **leptin receptor**. Compare the hydropathy plot predictions with what we know about these two proteins. Leptin is secreted and has no transmembrane domains, but it does contain a signal sequence (first 20 amino acids that target it to be synthesized on the rough endoplasmic reticulum [ER]) that has a peak close to 2. The leptin receptor is an integral membrane protein, so it should have produced one peak of 1.8 or higher. How well did the computer predict the proteins' membrane spans?

The 3D shape of a protein is probably its most important characteristic. If you are lucky, the 3D structure has been determined for your favorite protein, so you know its true shape. For most proteins, however, you will have to make some reasonable guesses about 3D structure based on amino acid sequence similarity to proteins with known structures. Let's look at some methods for predicting 3D structures. Perform a **conserved domain** (CD) accession number search for the protein "AAC83646" (type it in the larger lower box).

DISCOVERY QUESTIONS

17. Mouse over the **domain** boxes to determine the number of different CDs from your search. Don't just count the number of boxes, but determine the types of domains revealed when you mouse over each box. Notice that the E-values are provided when you mouse over each box.

18. Click on the "Show" Domain Relatives button and see what hits you get. At what protein have you been looking?

19. Go back and click on "gnl/CDD/7333," to the left of "smart00291,ZnF_ZZ, . . ." Read the text at the top of your screen. Does this domain have an important function? Explain your answer.

20. Copy and paste this **uncharacterized protein** (as of spring 2005) amino acid sequence into web sites of your choosing to characterize the protein's possible functions. Determine as much as you can about its structure and function.

What Shapes Are the Proteins?

Finally, if you want to know whether the 3D structure of your favorite protein has been determined, go to either Protein Data Bank (**PDB**) or **Entrez Structure** (two protein 3D structure databases) to search by key word. Submit "utrophin" to PDB and follow the links to view the structure (the **QuickPDB** web page contains detailed instructions for 3D views). How many polypeptides do you see, and how do they interact? How does utrophin's structure compare with dystrophin's (see Figure 1.6)?

If you have the good fortune to discover a protein with no known function, you might want to predict its 3D structure. Currently, there are no reliable computer programs to perform this task. However, you can use programs such as **PREDATOR** to predict the secondary structure of your protein. Submit the amino acid sequence for human **COX1**, and compare the PREDATOR prediction to the **real structure**. How closely does the predicted structure match reality? The holy grail of computational structural biology is to predict the 3D structure of a protein based on its primary amino acid sequence. Many mathematicians and computer scientists are working on protein structure prediction, but no one has been successful yet. You can even participate in the structure-prediction effort by **donating** your computer's idle time to a distributed network.

Does Structure Reveal Function?

Throughout this book, we will be looking at the structure of proteins and discussing their functions. However, the term **function** has become a bit outdated, because it is too vague. When we say *function*, do we mean that the protein plays a role in signal transduction, or that it is a kinase, or that it spans the plasma membrane? All three descriptors could be categorized as functions, yet they are not synonymous. As more genomes become sequenced, we need unified terminology to describe the role each gene/protein plays inside the cell. A consortium of investigators who work with different model organisms decided to create a more coherent vocabulary to describe genes and their products. The **Gene Ontology** consortium decided that three hierarchical terms were needed to describe different aspects of every protein: Why? What? Where?

1. **Biological process** is the why—the overall objective to which this protein contributes.
2. **Molecular function** is the what—the biochemical activity the protein accomplishes.
3. **Cellular component** is the where—the location of protein activity.

Let's look at one simple example. The protein aquaporin (see pages 296–297 for more information) is an integral membrane protein in red blood cells, kidney cells, and many others. Aquaporin's cellular component is the plasma membrane. Its molecular function is a channel that permits water to cross the phospholipid bilayer, which otherwise would be impossible since water is polar and cannot pass through the plasma membrane. Maintaining osmotic balance is aquaporin's biological process since it permits osmotic pressure to draw water (but not protons) back and forth across the membrane. Many proteins yield more than one answer to the why, what, and where questions. Nevertheless, Gene Ontology's unification of protein "functions" will help us communicate more effectively as we determine how genomes produce multifunctional cells.

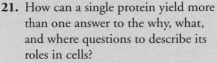

DISCOVERY QUESTIONS

21. How can a single protein yield more than one answer to the why, what, and where questions to describe its roles in cells?
22. Go to WormBase (*C. elegans* gene database), search for the gene "*pmr-1*," and learn its biological process, molecular function, and cellular components (about halfway down, next to Gene Ontology listing). Does *pmr-1* have more than one biological process, molecular function, and cellular component?

DATA COX1
LINKS donating, Gene Ontology, HGP, PREDATOR, prokaryotes, sequenced genomes, WormBase
METHODS cDNA
STRUCTURES real structure

Why Do the Databases Contain So Many Partial Sequences?

Now that we can sequence genomes quickly and cheaply, we need to address the strategies for sequencing a complete genome. When the Human Genome Project (HGP) was proposed, the plan included preliminary steps worth discussing. The HGP was not limited to the human genome; it also targeted yeast, fly, worm, mouse, and the flowering plant *Arabidopsis*. The *E. coli* genome was being sequenced by a single lab at the University of Wisconsin, Madison, and its sequence would be helpful as well (see pages 114–116). By comparing different genomes, we should be able to better understand genomes in general, and the human genome in particular. The current list of **sequenced genomes** includes many species, predominantly **prokaryotes**. To understand the mapping strategy, we will talk about BACs, YACs, STSs, and ESTs. (Molecular biologists love to create acronyms and then use them as if they were real words.)

There is a significant difference between sequencing the genome of a prokaryote with only 2 megabases (Mb; 2 Mb = 2 million bases) of DNA and a mammalian genome with roughly three orders of magnitude more DNA (~$3\cdot10^3$ Mb ≈ 3 billion). Big genomes contain repetitive DNA sequences from old viruses and transposons that have accumulated over the millennia. Piecing together the full genome sequence would be easier with known markers along the way—a type of DNA breadcrumb trail to indicate that we were on the right path. A few genes and cDNAs had been sequenced, but these were too far apart to be useful. The HGP decided to identify short segments of unique DNA sequence along every chromosome, which were called sequence-tagged sites (**STSs**). For the most part, STSs were defined by a pair of PCR primers that amplified only one segment of the genome. In conjunction with STSs, each chromosome was cut into fragments and inserted into vectors that could maintain the human DNA inside bacteria or yeast. Bacterial vectors that could carry large pieces of DNA (about 150 kb) were called bacterial

LINKS
Celera
DOE
Electronic PCR
Human Map Viewer
NIH
TIGR
UniGene
Wellcome Trust
METHODS
Whole genome
shotgun

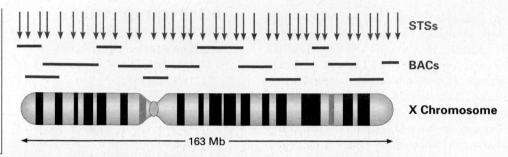

Figure 2.5 Relationships of chromosomes to genome sequencing markers.
Fifteen overlapping clones represent 1,408 BACs needed to span the 163 Mb
X chromosome. Arrows (top) represent STSs scattered throughout the chromosome
and on overlapping BACs, but many more were identified than are shown here.

artificial chromosomes (**BACs**, pronounced "backs") and are replicated in *E. coli*. Yeast artificial chromosomes (**YACs**, pronounced like the plural form of the animal yak) are replicated in yeast and carry DNA inserts from 150 kb to 1.5 Mb. By using restriction maps, investigators could determine which BACs and YACs contained overlapping DNA, and assemble contiguous overlapping segments of DNA, or *contigs*. As contigs were generated, STSs were layered over the chromosomal DNA, and the HGP was able to map the entire human genome. This genomic map would ensure that labs could place their DNA sequences at the correct location along each chromosome (Figure 2.5).

DISCOVERY QUESTIONS

23. Search NCBI's Human Map Viewer using the term "obesity." You will get hits for every locus that has obesity associated with it. How many loci do you see? Are they clustered or distributed throughout the genome?

24. Click on the blue number "10" below the cartoon of chromosome 10. What gene did you identify?

25. Click on the "Maps & Options" button to modify the view. From the new window, you can choose from the list in the left window; your choices are displayed in the right window. Modify the display until only "Gene," "Morbid/Disease," and "Ideogram" are displayed. Click on "Morbid/Disease" and then on the "Make Master" button, followed by "Apply." The ideogram on the far left shows how much of the chromosome you are viewing. You can zoom in or out as needed. This database allows you to search for your favorite disease or condition and track down all this information. You could place an order for this DNA or amplify it yourself using PCR.

26. Go to Electronic PCR and enter the accession number "M18533" in the big open box to determine if there are any STSs in this sequence. What

gene have you located, and how many STS markers are there? Click on one of the blue links and see how much information is there. Do you have all the information you need to amplify this STS? What else did you learn, other than sequences of the primers?

While some hurried to completely sequence the human genome, others continued to map the genome. One mapping technique tried to identify all the coding DNA to distinguish coding DNA from **junk DNA**. To do this, several labs began sequencing very short segments (~500 bases) from either the 5′ or 3′ end of every cDNA they could clone. The short segments of cDNA were called expressed sequence tags (**ESTs**), and the investigators did not care if they could determine what gene the ESTs came from or what the encoded protein did. Their goal was to compile every EST. In addition to identifying genes, ESTs help us detect alternative splicing and tissue-specific transcription. ESTs have been helpful for labs interested in identifying genes. If investigators sequenced a piece of DNA that contained an EST, they would know the DNA was in a coding region and they had sequenced a coding segment. The number of ESTs grew so quickly that a separate database was created for them, although ESTs are also integrated into larger databases such as GenBank and EMBL. To organize all these ESTs, UniGene was created.

By now, you have a sense of the publicly available data, thanks to a combination of your tax dollars (largely through the Department of Energy [DOE] and the National Institutes of Health [NIH]) and charitable contributions from foundations such as Great Britain's Wellcome Trust, the world's largest medical research charity. However, one private company was not satisfied with the speed of the HGP's hierarchical genome shotgun sequencing method, and decided to produce its own human genome sequence using its **whole genome shotgun** (WGS) method. In the mid-1990s, The Institute for Genomics Research (TIGR) was created as a private, nonprofit research institute. In July 1995, TIGR produced the world's first whole-genome sequence (1.8 Mb) from the bacterium *Haemophilus influenzae*. Shortly thereafter, TIGR was divided into two entities and Celera, a for-profit company, was created. Celera was led by TIGR cofounder Craig Venter until early 2002. Interestingly, Celera's name is derived from the Latin word *celere*, which means swift; its goal was to produce genome sequences faster than anyone else. TIGR, and later Celera, had developed a faster method to produce genomic sequencing that relied heavily on computers and new software. Rather than spending a lot of time mapping a genome, they used whole genome shotgun (WGS) sequencing. Shotgun sequencing is a scaled-up version of an old method in which all the DNA is cut into random pieces and all of the pieces are sequenced. The new twist for Celera was the scale. Never before had anyone tried to assemble so much sequence information. To test its ability to assemble whole genomes, Celera started with prokaryote genomes. Later it collaborated with a consortium of academic labs to WGS sequence the *Drosophila* genome; the impressive results are available at Flybase and Berkeley Drosophila Genome Project (BDGP). The shotgun approach has been validated, and the HGP has adopted the method to a limited degree.

DISCOVERY QUESTIONS

27. BLASTn this mystery sequence and select "est_mouse" from the "Choose database" menu. What can you learn based on the hits you obtained? For example, what gene have you identified? Scroll down and see how many tissues are described in this search. Imagine you were studying obesity in mice; how might this help your efforts? (See Chapter 5 for details.)

28. Did the EST database provide you with more than just sequence identification information? With the completion of many genomes, is there any utility for the EST databases? Support your answer with specific examples.

29. Go to the UniGene Statistics page to read the latest information on human ESTs. Based on this information alone, calculate the average number of ESTs for each human gene (assuming 23,000 genes).

WHICH SEQUENCING METHOD WORKED BETTER?

In October 2001, two competing draft versions of the human genome sequence were published—and they were not identical. One version was produced by the publicly funded Human Genome Project (HGP) and generated by academic labs in Asia, Europe, America, and Australia. The other version was produced by one private company, Celera, which initiated its research several years after the HGP began. During the first years of the HGP, the academic labs generated a variety of maps to cover the entire genome, but initially they did not begin large-scale sequencing. The HGP felt it was important to determine markers (segments of DNA that could be used later to verify the sequence assembly) and to organize which segments of DNA should be sequenced. HGP sequencing labs used the mapping markers to choose the minimum number of slightly overlapping fragments that completely spanned each chromosome. These hierarchically optimized fragments were called the "golden tiling path" and were composed of about 45,000 fragments. The mapping and golden tiling path were posted on publicly available web sites. Celera felt that it had sufficient throughput and computer science expertise from sequencing smaller genomes that it did not need to generate genome maps and says it never utilized the HGP maps or golden tiling path, though the information was available. Celera fragmented (like a shotgun blast) whole human genomes into many (27,271,853) small pieces, sequenced all these fragments, and assembled them into contigs that spanned the entire lengths of every chromosome. The HGP argued that because the human genome contains regions of highly repetitive DNA, it would be impossible to assemble a genome full of gene duplications and repetitive DNA without the maps (Figure 2.B1). Who was right?

When the human genome sequence was finished, several investigators compared the finished HGS and Celera's WGS final versions. For most of the human genome, both versions were in close agreement, except when two chromosomal regions had at least 15 kb of more than 97% sequence identity. The WGS method was unable to resolve these two regions and tended to lump them into a single locus. In fact, the WGS version correctly assembled only 8 Mb of DNA with 95% sequence identity, while the

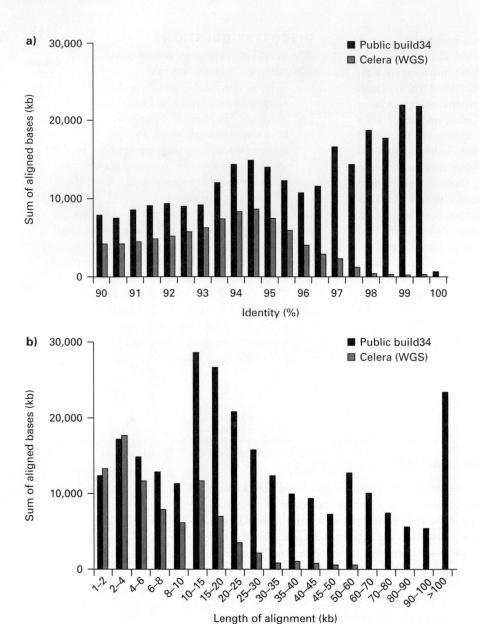

Figure 2.B1 Comparison of HGP and Celera human genome sequences.

a) The number of sequential bases identified for duplicated segments of the genome are compared for WGS and HGS methods. The X-axis indicates the percentage of sequence identity between the duplicated segments, and each method is compared individually. Beginning at 95% identity, the WGS method compares poorly, and after 98% it fails almost completely. **b)** The total number of bases is compared for continuously aligned lengths of duplicated DNA. Note the drop for WGS after the 10–15 kb category.

HGS version assembled 94 MB of DNA with 95% identity. The WGS method was unable to assemble two regions of DNA with 98% or more sequence identity (see Figure 2.B1). The lost Celera DNA contained 103 genes that were either partially or completely absent from the WGS final assembly. The omitted genes included five known disease genes and genes important in your immune system (e.g., *CCL3L1*, an HIV-suppressing chemokine that binds to the HIV coreceptor CCR5). Genes with nearly identical sequences were recently **duplicated** and appear to be mutating at a faster than average rate. Chromosome 16 has the most repetitive DNA and 17% of its length is missing in the WGS assembly (Figure 2.B2).

Given that WGS sequencing fails to recognize large sections of highly repetitive DNA, you could predict that other complicated genomes that were sequenced using the WGS method also failed to identify duplicated portions of those genomes. Perhaps it is not surprising that the WGS-assembled mouse and rat genome sequences exhibited less repetitive DNA than the human genome sequenced by the HGS method. The loss of repetitive DNA is easily detected when we look at a highly repetitive portion of chromosome 16 (Figure 2.B2). The *low-copy-repeat* region called 16a (LCR16a) is 690 kb long and was found in 28 different regions over 60 Mb in the HGS finished sequence, but was compressed down to a single locus of reduced size in the WGS assembly.

Because of LCR16a and other examples, HGP investigators recommend a hybrid solution. In Phase 1, WGS would be used to produce about sixfold coverage for every base in the genome. These short segments would be assembled using Celera's approach and regions of the genome that appeared to have more than sixfold coverage, or higher than normal diversity in DNA sequence, might be duplicated regions with highly conserved or repetitive DNA sequences. Phase 2 would map and produce a mini-golden tiling path of duplicated or repetitive regions for sequencing the entire BAC inserts to resolve potential problems. If this hybrid WGS/HGS method had been used for the human genome, only 3,000 BACs would have been sequenced in their entirety after sixfold WGS coverage; this is only 7% of the total number of BACs actually sequenced using the pure HGS method. Therefore, if we value a finished sequence, using the hybrid method may result in more finished genomes for less cost and fewer errors. As often happens, neither extreme was best, and a compromise of the two methods would have been the best approach, now that we know WGS is prone to certain types of errors.

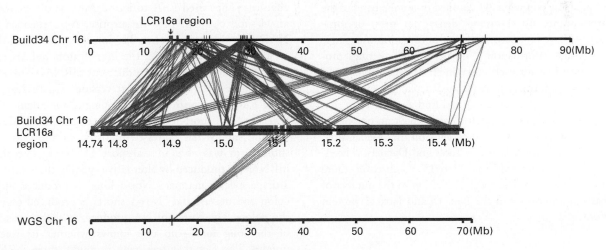

Figure 2.B2 Comparison of WGS and HGS to resolve duplications.
A 690 kb low-copy-repeat (LCR) on Build 34 of chromosome 16 was distributed in 28 locations over about 75 Mb of the chromosome. The WGS assembly of the same region contained only a portion (46 kb total length) of LCR16a at a single site, which contributed to the shorter length estimate for the entire chromosome.

LINKS
Ensembl Genome
Browser
Genome Browser

You now have the background necessary to analyze whole genomes. Imagine what it would be like to annotate millions or even billions of bases. How would you predict where the genes are and what each protein does? How does the entire organism use its genomic information to live, reproduce, and respond to environmental fluctuations? Fully annotating a newly sequenced genome requires many people with different academic backgrounds working together in teams. As you might guess, software development for genome analysis is a very hot research area in computer science, mathematics, engineering, and biology. Few people can master more than one or two of these areas, so collaborations are common, though increasingly biologists are learning more math to improve their research and career options. If you learn both math and biology, you will have many career opportunities ahead of you, but that's another story best told by your advisor.

Annotated Genomes Online

Once a genome has been sequenced and annotated, we can learn a lot about it. Let's look at the X chromosome using the HGP Genome Browser based at the University of California, Santa Cruz (Ensembl Genome Browser in the United Kingdom is an equally powerful site). At the HGP Browser, click on "Genome Browser" at the top left side. Enter "Xq22.1" and hit "Submit." You will get a rather numbing picture that we will modify. Scroll to the bottom to change the settings. Hide all displays except: base position = "dense"; "Known Genes" and "Human mRNAs" = "full"; "Chromosome Band," "STS Markers," and "Human ESTs" = "dense." Hit "Refresh." Now you will have zoomed in on one band of the X chromosome. By choosing "dense" for most of some options, we have compacted a lot of information into single lines of visual information. The three "full" settings produced a new line for each entry in the database and facilitated reading information about a particular gene. Read the list of known genes (blue text), and find *Nox1*. Notice that there are two separate entries for *Nox1*. Click on one of the two blue "*Nox1*" names and then click on the "GeneCards" link under the "Quick Links to Tools and Databases" heading and the GeneCard link "Nox1" to discover more about the gene *Nox1*. Pay special attention to the number of alternative transcripts and the Nox-1S and Nox-1L protein functions.

DISCOVERY QUESTIONS

30. What does GeneCards say about *Nox1*? Is it an NADPH oxidase? Go to the Human MapViewer for the human genome, enter "Nox1," and hit "Find." You will see a red dash next to the X chromosome. Click on the blue "X" under the ideogram and notice the other genes in the area. Next to the gene "NOX1," click on the link "sv." You should see a graphical version of this gene.

  a. How many different mRNAs (listed as **CDS** for *co*d*ing* *se*quences) are produced by this one gene? The color code is just below the expanded view of this gene. Notice that only the first exon is shown in the sequence (as denoted by the red bracket in the cartoon above).

  b. Use the navigation button icon to zoom out by clicking once on the "-" sign. How many genes are in this region of the X chromosome? Do any other genes produce more than one mRNA?

31. Take a few minutes to draw a flowchart illustrating the steps you would take to annotate a newly sequenced genome (define the genes; describe each protein's biological process, molecular function, and cellular component; summarize the major metabolic pathways the organism needs to survive and evolve).

How Many Proteins Can One Gene Make?

A group of investigators in Switzerland and Hungary were interested in ion channels, specifically voltage-gated H^+ ion channels. They used ESTs to locate *Nox1* as the source of an H^+ ion channel. To their surprise, *Nox1* encoded the NADPH oxidase and not an H^+ ion channel. When they searched the EST database, the investigators quickly realized that *Nox1* encodes three different mRNAs. There are two long versions and one short version (Figure 2.6). All three mRNAs appear to encode integral membrane proteins, but the short form does not contain the NADPH binding site. When expressed in tissue culture cells, this short form was able to transport H^+ ions. The three mRNAs are produced by alternative splicing that occurs in a tissue-specific manner: "Nox-1 long" is produced in the colon and uterus, and "Nox-1 short" is produced only in the colon. The third mRNA produces a slight variation of the long form—no one knows whether its role is unique. The fact that one gene produces three mRNAs illustrates the challenge in defining a gene. If one segment of DNA can produce three proteins, should we consider that segment of DNA three genes, or only one? *Nox-1* illustrates the utility of the EST database and how its value will increase as more ESTs are added for different species.

a)

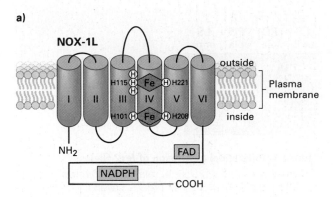

b)

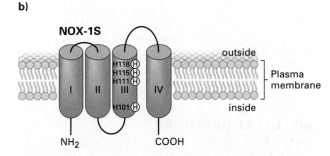

Figure 2.6 Two proteins produced by alternative splicing of RNA.
a) The full-length mRNA was translated to produce the larger protein (NOX-1L) that is an NADPH oxidase.
b) Several exons were excised to produce a smaller protein (NOX-1S) that functions as a H⁺ channel.

DISCOVERY QUESTIONS

32. Analyze the dystrophin gene with the Genome Browser. Enter the name "dystrophin" and click the "Submit" button. You will get a list of hits. Click on the first "dystrophin" option to see a graphic view of the human dystrophin gene. Use the 3X zoom-out button until you can see the entire gene. You may have to modify the view using the options below the display to answer the following questions.

    a. Are there any STS markers in this gene?

    b. Look at the Gap and Coverage lines. Has the public HGP sequenced all of the chromosome in this region?

    c. Change the "Coverage" option to "full" and then count how many BACs span the DMD gene. Gray BACs are draft-quality sequencing and black BACs are finished. Can you determine the minimum number of BACs required to span DMD based on coverage?

    d. How many DMD mRNAs use more than one exon? What can you infer from the number of alternatively spliced mRNAs?

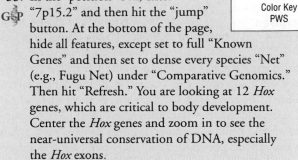

LINKS
AS
Color Key
PWS

33. In the "position" box, enter "7p15.2" and then hit the "jump" button. At the bottom of the page, hide all features, except set to full "Known Genes" and then set to dense every species "Net" (e.g., Fugu Net) under "Comparative Genomics." Then hit "Refresh." You are looking at 12 *Hox* genes, which are critical to body development. Center the *Hox* genes and zoom in to see the near-universal conservation of DNA, especially the *Hox* exons.

    a. Which species highlight the conserved exons the best, closely related species or more distant?

    b. Set repeat masker to "dense" and refresh this diagram. Do *Hox* genes contain a lot of repeats (black indicates repeat sequences) compared to portions outside the *Hox* genes? Do you think this is significant? Explain your answer.

    c. Set the SNPs (single nucleotide polymorphisms, which can be thought of as point mutations) option to dense. Do *Hox* genes have more or fewer SNPs than the surrounding area? (See Color Key.) Do you think the *Hox* frequency of SNPs is significant? Explain your answer. For a comparison, click on the "Move >>>" button.

Can the Genome Alter Gene Expression Without Changing the DNA Sequence?

There are two genetic diseases you probably have not heard of, and yet they have provided us with a valuable insight into the way our genomes work. Prader-Willi Syndrome (PWS) is caused by a megabase-sized deletion in the region of 15q11–q13. A different genetic condition called Angelman Syndrome (AS) is caused by a large deletion in the region of 15q11–q13. As you can see, these two conditions are both caused by the loss of the same section of the genome, and yet the symptoms are different. Whether you have AS or PWS depends on which parent gave you the mutated chromosome. If your father passed on the deletion, you will develop PWS; if your mother passed on the deletion, you will develop AS.

Children with PWS have weakened muscles, fail to suckle, will not grow very tall, have short hands and feet, and will develop obesity and mental retardation as they mature. Children with AS have problems with balance and motor skills, appear hyperactive and overly happy (excessive laughter), and are often more severely retarded than PWS patients. These conditions are not sex linked since they occur in boys and girls with equal frequency, but the sex of the parent who passed on the mutated chromosome always

LINKS
imprinted
Shirley Tilghman
METHODS
RT–PCR

determines whether the child has AS or PWS.

This parental determination of disease status is an example of **imprinting**, a process mammals use to mark a small set of genes during gametogenesis so that only the paternal or maternal copy will be transcribed and the other allele at the same locus will remain silent. As of 2005, 76 human genes are known to be imprinted and therefore differentially transcribed. Silencing one allele depending on parental origin presents an exception to the rule of gene expression. Monotremes and nonmammalian vertebrates do not imprint their genes, which indicates that placental mammals are unique in their requirement of imprinting. Many of the known imprinted genes play a role in embryonic and placental development. When you think about it, the placenta and developing embryos are similar to parasites that drain resources from the mother. In species where the female mates with more than one male, the two parents have opposing interests in the newly formed zygotes. A male will prefer that his offspring extract the maximum amount of nutrients from the mother, even if it hurts the mother's ability to reproduce later with other males. Conversely, a female would prefer to ration resources so that each of her offspring has the potential to survive and she can continue to reproduce with any male she chooses. This opposition of interests creates a genetic tug-of-war, which may have been the selection pressure that led to imprinting. Paternally expressed genes should promote growth, whereas maternally expressed genes should reduce the amount of growth as long as the mother is the sole source of nutrients (pregnancy and nursing).

The tug-of-war hypothesis sounds logically consistent, but we need some data to support it. One of the world leaders in this field was **Shirley Tilghman** from Princeton University, whose lab made significant contributions toward our understanding. One paternally expressed gene, insulin-like growth factor 2 (*Igf2*), promotes the growth of the embryo. Only the paternal allele is expressed in the developing embryo. The Igf2 receptor (*Igf2r*) is expressed only by the maternally inherited allele. By controlling the embryonic expression of the receptor, the mother can maintain control of the paternally driven ligand encoded by *Igf2*.

Tilghman's lab demonstrated imprinting in a series of matings between two closely related strains (abbreviated LS and PO) of mice. Because the strains are closely related, the mice can mate successfully, but their DNA sequences are different enough that each strain's *Igf2* alleles can be distinguished (Figure 2.7). *Igf2* from the PO strain produces three bands, with two of similar molecular weight near the top of the gel. LS *Igf2* produces two major bands, including one of small molecular weight. If the PO *Igf2* is expressed, the top two bands will be detected, whereas LS *Igf2* expression results in a band near the bottom of the gel. When the female LS mated with a male PO, the PO *Igf2* allele was expressed in the embryo and placenta and the maternal LS allele was completely silenced.

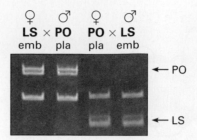

Figure 2.7 Differential expression of *Igf2* alleles. Embryonic (emb) and placental (pla) RNA from two *Peromyscus polionotus* mouse breeds called LS and PO were used for reverse-transcriptase PCR (RT-PCR), digested with a restriction enzyme, and analyzed by agarose gel electrophoresis. Depending on which breed was the father (mating indicated above the emb and pla labels), two different banding patterns are observed due to slight differences in the sequences of the *Igf2* alleles in PO and LS.

DISCOVERY QUESTIONS

In these three questions, we are working with only one strain of mice, not two.

34. If the *Igf2* gene were deleted from the sperm, predict the phenotype in the offspring. What would the phenotype be if the egg's *Igf2* gene were deleted?

35. If the maternally expressed gene *Igf2r* were deleted from eggs, predict the phenotype in the offspring. What would the phenotype be if the sperm lacked an *Igf2r* gene?

36. What would you predict for the offspring if the sperm's *Igf2* gene and the egg's *Igf2r* gene were deleted?

Many more experiments have contradicted the widely held dogma that both alleles are equal. Let's look at one more gene before we try to understand what is known about imprinting. When the promoter of another paternally expressed gene (*Snrpn*) was deleted, *Snrpn* and at least three other genes that are normally transcribed from the paternal alleles were silenced. The *Snrpn* deletion produced symptoms similar to several PWS phenotypes, including failure to suckle. This is consistent with the hypothesis that paternally expressed genes promote the maximum utilization of maternal nutrients and thus promote growth and eating. Surprisingly, the coding region of *Snrpn* was not the critical component—its promoter was. Similarly, the coding portion of the maternally expressed gene *H19* can be deleted with no ill effects, but when its promoter is mutated, normal imprinting is disrupted for both *H19* and the paternally expressed *Igf2*. There are many more hard-to-explain imprinting stories, but let's turn our attention to possible mechanisms.

What Is the Fifth Base in DNA? Methyl-Cytosine

To understand imprinting, we need to discuss **methylated DNA**. The genomes of many species contain a gene called DNA methyltransferase 1 (*Dnmt1*). Dnmt1p is an enzyme that adds a methyl (CH₃) group onto hundreds of thousands—but not all—cytosines throughout the genome. Since the 1960s, methylated cytosines have been associated with genes that are not transcribed. Natural loss of methylation is seen in some cancers and noncancerous aging cells. Regulation of gene expression without alteration of the DNA sequence is called **epigenetic** regulation. In December 1999, an international team created the Human **Epigenome Consortium** to map the human **methylome**, a complete description of the methylation status of the genome. If there are about 400,000 methylated cytosines in any given cell, that alone would take a lot of work to study. However, each cell type has its own unique methylome, and since there are about 100 unique cell types in the human body, the team will need to catalog 40 million methylation events, and perhaps twice that many since gender-specific imprinting utilizes methylation.

The precise mechanism for imprinting is unknown, but there a few models that help us understand how a particular gene could be differentially expressed depending on whether it reached the embryo inside an egg or a sperm.

During gametogenesis, DNA methyltransferase is activated, and it modifies gamete-specific sites that determine which alleles will be expressed in the embryo. Direct methylation may block the transcription of some genes (Figure 2.8a), but this mechanism alone is not sufficient to explain all the data. In some cases, methylation of the promoter allows a distant enhancer sequence to activate an alternative gene on the same chromosome (Figure 2.8b). Still other data indicate that there are insulator regions of chromatin that can block the activity of a downstream enhancer; the insulators are neutralized by methylation (Figure 2.8c). Another model predicts that chromatin structure (e.g., **heterochromatin**) might work over great distances to silence a region; this structure also is regulated by methylation (Figure 2.8d). Tilghman's lab demonstrated that a DNA-binding protein called CTCF recognizes the methylation status of imprint-controlled regions on chromatin, and CTCF only binds to unmethylated DNA. The site of CTCF binding is between imprinted loci and their enhancers, which puts it in the right location to be a key player in Figures 2.8b and 2.8c. The complex regulation of imprinting illustrates that the DNA between genes ("junk DNA") may be more important than we had imagined. In addition, imprinting is even more complex since at least one maternally expressed gene, *Mash*, is unaffected by the

LINKS
Epigenome
Consortium

STRUCTURES
Dnmt1p
methylated DNA

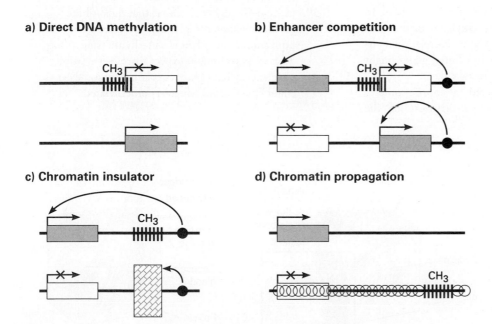

a) Direct DNA methylation

b) Enhancer competition

c) Chromatin insulator

d) Chromatin propagation

Figure 2.8 Four proposed models to explain imprinted gene silencing mechanism.
Panel a–d contains two lines that represent two different copies of the same locus. The active gene is depicted as filled boxes and the silenced genes as open boxes and an X. Methylated DNA is denoted by a series of vertical purple lines and the label CH₃; an enhancer by a small filled circle; a boundary or insulator by a hatched box; closed heterochromatin by overlapping open circles. Arrows emanating from enhancers indicate gene activation.

LINKS
Beckwith-Wiedemann
Albert de la Chapelle
Dnmt1-affected
GenBank
Rudolf Jaenisch
Eric Lander
Yusuke Nakamura
Christopher Wilson

loss of methylation; this indicates that other epigenetic mechanisms may help control imprinted genes.

Imprinting, Methylation, and Cancer

In January 2001, the laboratories of Albert de la Chapelle from Ohio State University and Yusuke Nakamura from the University of Tokyo reported that loss of imprinting of *Igf2* might predispose individuals to colorectal cancer. This team studied the insulator region near *Igf2*, which contains **CpG islands** where CTCF binds. Loss of imprinting of *Igf2* has been associated with a variety of childhood and adult tumors, as well as Beckwith-Wiedemann Syndrome, a congenital overgrowth disorder that predisposes the individual to tumors. With the loss of imprinting, the amount of *Igf2* transcription might be twice as high as normal, and excess growth factor might contribute to Beckwith-Wiedemann Syndrome and tumor formation. Previous research had shown that loss of *Igf2* did affect the growth of intestinal cancers.

Nakamura and de la Chapelle used tumorous and non-tumorous biopsies from cancer patients who had particular DNA repeat units that were unstable and determined the degree of methylation in the *Igf2* insulator region. These samples were compared to others from cancer patients who did not exhibit unstable DNA repeat units. In 20 tumor samples taken from patients with unstable repeat units, 17 were hypermethylated in the CTCF-binding region. In addition, 14 of the 20 were hypermethylated in the non-tumorous colonic mucosa. This positive correlation indicates that the hypermethylation of imprinted genes might predispose some individuals to colorectal cancer.

Laurie Jackson-Grusby conducted a more comprehensive genomic response to a perturbation of the **metabolome** when she led a team from Eric Lander's and Rudolf Jaenisch's labs at MIT and Christopher Wilson from the University of Washington, Seattle. Typically, hypomethylation of the genome is associated with senescence and cell death. Jackson-Grusby wanted to determine how the genome would respond to the deletion of DNA methyltransferase. Using skin cells grown in culture, she measured the change in transcription for about 6,000 mouse genes with and without *Dnmt1*. (See Chapters 6 and 7 to learn how this many genes can be monitored.) Approximately 10% of the genes showed altered expression levels; most of these were induced, but a few were repressed. The *Dnmt1*-affected genes included imprinted genes, cell-cycle control genes, growth factor ligands, and receptors. Jackson-Grusby's results verify that the methylome is a significant component of gene regulation and thus the proteome. To understand a genome, we will also need to understand the methylome and recognize that, functionally, the genome has five bases: G, C, A, T, and methyl-C.

DISCOVERY QUESTIONS

37. Do you think methylation is an ancient mechanism or one limited to vertebrates? How could you answer this question?

38. Go to GenBank and search "methyltransferase." Can you find any DNA methyltransferases in organisms other than vertebrates?

39. Jackson-Grusby et al. found that most of the genes with altered expression increased their expression levels when the methyltransferase was deleted, as you might expect since methylation normally silences genes. Explain how some genes could be repressed when hypomethylated.

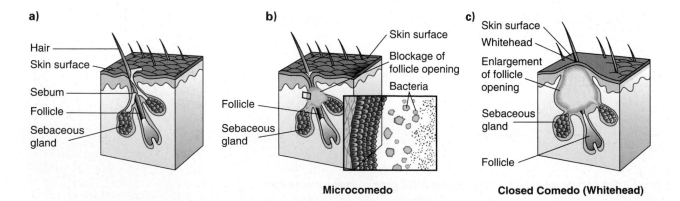

Figure 2.9 Anatomy of a pimple.
a) A pimple is caused by a blocked skin pore through which a hair should emerge.
b) Bacteria in the pore begin to grow and obstruct the secretion of sebum. **c)** When blocked, the associated sebaceous glands continue to secrete sebum and the follicle pore enlarges and becomes inflamed.

Summary 2.1
Genome Sequence and Analysis

In Section 2.1, you learned how genomes are sequenced and how to use public databases that are funded by government and private organizations in the United States, Europe, and Asia. With these public-domain tools, we can use small segments of sequence to retrieve full-length coding regions, deduce protein sequences, and make educated guesses about possible cellular roles for proteins. We can take advantage of evolutionary relationships and compare sequences from different species to examine the degree of conserved domains. You also learned that epigenetic gene changes expand the range of genome regulation. By using dynamic databases, you have become familiar with research tools used by investigators all over the world. In the next section, we explore some of the interesting discoveries made when whole genomes were annotated.

2.2 What Have We Learned from Unicellular Genomes?
Why Do I Get So Many Pimples?

When you entered puberty, you also experienced a phenotype that everyone hates—acne. Every teenager develops pimples that seem to grow and erupt like volcanoes for no apparent reason. Pimples are pus-filled pockets; you know they involve your immune system since pus is a combination of dead white blood cells and whatever microbe they were trying to kill. The microbe your immune system was fighting is called *Propionibacterium acnes*, which had its genome sequenced in July of 2004. *P. acnes* lives on human skin and especially in sebaceous follicles (Figure 2.9). The bacterium feeds on sebum and for an unknown reason its feeding can stimulate an immune response of inflammation. Whenever an area becomes inflamed, white blood cells enter, along with serum from your blood. As the area swells, the follicle opening is sealed shut and the dying cells accumulate and form pus. This process causes the area to enlarge, become tender, and eventually erupt to release the pressure, thereby allowing the follicle to open and expel the mostly dead bacteria and immune cells. What can we learn about *P. acnes* that will help us understand pimples?

Although it would have been fitting for MIT's Whitehead Institute to sequence the genome, *P. acnes* was sequenced by three groups, one in Paris and two in Germany. The labs performed 32,190 sequencing reactions to produce 8.7-fold coverage of the 2,560,265-bp genome, with an estimated error rate of 0.0001 (Figure 2.10). The

Figure 2.10 Map of *P. acnes* genome.
Go to www.GeneticsPlace.com to view this figure.

genome is contained on a single, circular chromosome; it has no additional **plasmids**, which are common in bacteria and function as mini-chromosomes. The groups annotated 2,333 putative genes, which allowed the investigators to reconstruct the metabolism of this pimple-causing microbe. Twelve percent of the genes are believed to encode only RNA products such as rRNA and tRNA. The investigators performed BLAST searches and were able to find orthologs for 1,578 (68%) *P. acnes* genes; 468 others (20%) failed to match any genes. **Ortholog** is the newer and more appropriate term that is replacing the older term **homolog**, though the two are sometimes used interchangeably. Two genes are called orthologs if one gene in a particular species evolved from the other gene in a different species. The term **paralog** describes two genes within a single species where one evolved from the other through gene duplication. The older term homolog does not make the species distinction and thus is less precise than ortholog and paralog.

When a bacterial genome is sequenced, one of the first whole-genome-level evaluations is to identify the origin and terminus for replication. These functional sites are typically located where there is extreme **GC skewing**, a nonuniform distribution of guanine and cytosine bases on the two strands of DNA. It seems logical to predict that the G's and C's would be evenly distributed on the top and bottom strands of the chromosome, but they are unevenly distributed in some portions of the chromosome and evenly distributed in other areas. The origin of replication tends to have the lowest GC skew (evenly distributed G's and C's), while the terminus for replication tends to have higher GC skewing. For example, in *E. coli*, the leading strand has significantly more G's than C's, and therefore the lagging strand has more C's. Although the GC skew trend is widespread, it is not a hard-and-fast rule, so investigators also look for particular genes (such as *DnaA*) that tend to be located near origins. Another indicator for the origin of replication is that genes tend to be transcribed in the direction pointing away from the origin. Collectively, these characteristics help identify where DNA polymerase begins its journey around the chromosome.

Another genome-wide evaluation is to identify genes that may have originated in other species. In this case, we are not referring to orthologs but to genes that appeared in the genome through an unknown mechanism referred to as **horizontal transfer**. To find **alien genes**, you scan the DNA with a **sliding window** for segments that have an abnormal **GC content** (either higher or lower than the species average) and then evaluate the **codon bias** (i.e., which codon is used more often than other codons for a particular amino acid) of these segments. Each species has certain codons that it rarely uses and this bias can help support the hypothesis that a particular gene entered a genome through horizontal transfer. Of course, you would expect BLAST to find similar genes in other species, but since we have not sequenced every genome in the world, the alien gene may not match any in the databases. *P. acnes* has several regions with atypical codon usage. Many of these genes provide metabolic capacity or immunogenic stimulus to our unwanted guest prokaryote. Interestingly, many of these regions are located in or near tRNA genes, which many people believe is an indication that tRNA genes are recombination hot zones.

There is one more GC-related oddity in the *P. acnes* genome that was too small to be detected at the whole-genome level, but was discovered during the quality assurance stage of the sequencing process. The investigators noticed a few genes contained more variation in the number of G's than would be expected due to low-quality chromats. When the finishing reads were performed, the variation was not resolved, which led the investigators to propose that *P. acnes* utilizes the variation of G's to produce **transcriptional phase variation** (Figure 2.11). The hypothesis

Figure 2.11 Poly-cytosine chromats reveal transcription regulatory mechanism. Go to www.GeneticsPlace.com to view this figure.

is that initiation of transcription depends on the number of consecutive guanines on a particular strand at a critical location upstream of the coding region. DNA polymerases tend to have a difficult time accurately replicating regions that have many copies of the same base all in a row. When a region with several G's in a row is replicated, the number of G's may grow or shrink by one or two. This variation would result in a mixed population of bacteria with varying degrees of transcription efficiency and ultimately protein production. By producing a diverse population of bacteria that are clonally derived, *P. acnes* has selected a mechanism to produce diversity that can respond differentially to the changing environment of your sebaceous follicles. In short, we are facing a bug that is optimized to survive any skin treatment you can buy. However, transcriptional phase variation is only one of its tricks, as we will see in the next portion of this story.

DISCOVERY QUESTIONS

40. Perform an NCBI Gene search for PPA1880 (use the pull-down menu on the NCBI main page to select "protein"). Click on the first link, then click on the "CDS" link to see the coding DNA. Copy the DNA sequence into the GC calculator to determine the %GC. *P. acnes* has an average of 60% GC and human has an average of 41%. Which genome does PPA1880 more closely match?

41. Find AAH14236.1 from NCBI's protein pull-down menu. Copy the DNA sequence and determine the %GC. Does this sequence look more human-like or more *P. acnes*-like? What interesting annotation did you uncover? How could this cDNA get into a human cDNA library?

42. Now determine the GC content for one rRNA gene and one tRNA gene. How do they compare to the genome average of 60% GC?

43. Go to the *P. acnes* genome view, enter 1 into the "Start from" box, and hit "Go." Notice that the first gene is *DnaA* (accession number YP_054724). Click the blue arrow pointing to the left. Do you notice anything peculiar about this region upstream of *DnaA*? Compare this region to any other by clicking somewhere on the genome map to see any other region.

Which Genes Cause Pimples?

Examining orthologs of known proteins, the team was able to perform **metabolic reconstruction** for many biochemical pathways. Their first interesting deduction was that *P. acnes* can grow anaerobically as well as in a limited range of aerobic environments. The capacity to survive under limited exposure to oxygen was unexpected. The pathogen has many enzymes able to degrade lipids, esters, and amino acids. These digestive enzymes provide the bacterium with the tools necessary to live happily in your pores (Figure 2.12). It is believed that some of the free fatty acid metabolites from the destruction of sebum enhance bacterial adhesion to your cells.

At this stage, you might be wondering how a bacterium could ingest your cells and sebum in order to feed. The digestive enzymes located on the extracellular wall might be identified by a particular **motif** (conserved sequence of functional importance). Many *P. acnes* digestive enzymes contained the known five amino-acid motif **LPXTG** (leucine, proline, any amino acid, threonine, and glycine) that targets proteins to the extracellular cell wall. Located on the surface of the bacterium, the enzymes chew away on your cells and sebum to produce simplified nutrients. In addition, the cell's exterior is decorated with hyaluronate lyase (gene PPA380), which destroys the extracellular matrix binding your skin cells together and thus facilitates further tissue invasion and digestion.

The ability to digest you for food explains how the cells feed, but it does not explain why we develop pimples. The second half of the puzzle comes from the bacterium's ability to stimulate the immune response. The genome encodes five CAMP (Christie, Atkins, Munch-Peterson) factors, which could be paralogs that evolved within *P. acnes*. **CAMP factors** are secreted proteins that bind to antibodies (*IgG* and *IgM*) and can form pores in eukaryotic cell membranes. Lysis of our

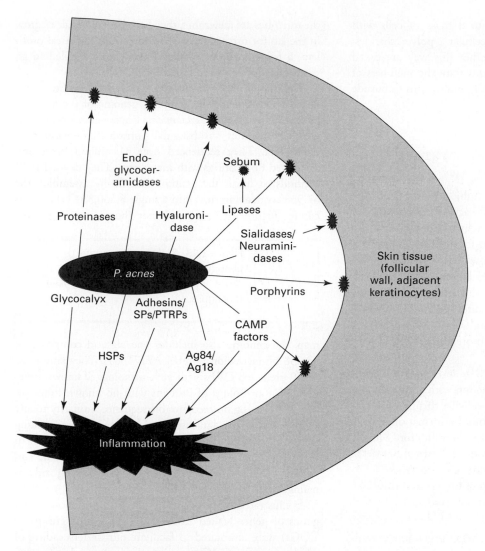

Figure 2.12 Activity of *P. acnes* leading to pimple formation.
Skin keratinocytes lining the pore are digested by enzymes secreted from the centrally
located bacterium. Additional proteins secreted by the bacterium induce an immune
response that contributes to the redness, swelling, and soreness. HSP = heat-shock protein;
SP = surface-associated protein; PTRP = proline-threonine repetitive protein; Ag = antigen.

cells would be one way to trigger an immune response, but this genome has several others too. The dipeptide motif of proline and threonine (**PT**) has been reported for several immune-stimulating antigens of the tuberculosis-causing bacterium *M. tuberculosis*, and is also present in PA1880 and PA2127. Heat shock proteins are also potent stimulators of the human immune response, and *P. acnes* has at least seven genes that encode orthologs of other known heat shock proteins. Finally, *P. acnes* secretes a lot of porphyrins, the same family of ring-shaped molecules present in hemoglobin and chlorophyll. When exposed to increasing amounts of oxygen, the porphyrins produce toxic forms of oxygen that can kill our keratinocytes and further stimulate our immune systems.

If we know all this information, why can't we kill *P. acnes* and get rid of acne? All forms of life, even annoying bacteria, evolve methods to withstand environmental challenges, including acne medication. *P. acnes* is able to withstand antibiotics that would kill most other bacteria. How? One way it resists antibiotics is to sense the environment and signal all the nearby *P. acnes* cells that something has changed. Most bacteria have sensors called **two-component systems**: one component senses the outside world and the other signals the cytoplasm. *P. acnes* has 10 pairs of two-component genes that allow it to sense and signal a wide variety of environmental changes. Furthermore, *P. acnes* cells can detect when the immediate area is crowded with many cells, in a process called **quorum sensing**. This leads to stimulation of the *LuxS* gene, which produces the autoinducer-2 secreted protein that some microbiologists hypothesize may be a universal signal for interspecies communication. Finally, when bacteria need to hunker down and become extra resilient, they transform the population from a collection of free-range

LINKS
COGs
Conserved Domain
Google
Google Scholar
major categories
OMIM
PubMed

individual cells into a mass of cells with specialized extracellular polysaccharides. Cells meshed together this way are called **biofilms**, and in this state the members of the biofilm are nearly impervious to outside attacks.

DISCOVERY QUESTIONS

44. Find the Conserved Domain of **LPXTG**. What does this domain help proteins do?

45. Perform an NCBI Medical Subject Heading (MeSH) search for "CAMP Factor." You should see a hit called "CAMP protein, Streptococcus [Substance Name]." On the far right side, click on the "link" link and choose "NLM [National Library of Medicine] MeSH Browser." What can CAMP factors do to our blood cells?

46. Search Google Scholar for autoinducer-2. Do you see any evidence that autoinducer-2 plays a role in communication? Is this protein expressed in many species?

47. Search for "biofilm" in OMIM and click on the one hit. Perform a find function with your browser for "biofilm" and see what this protein has to do with preventing biofilm formation.

48. Conduct PubMed and Google searches for "Blue Light Acne" and see what you can learn about a novel method to combat acne. Do you think genome sequences can help us understand this method better? Explain your answer.

Bonus Material: A study of the tetanus genome is available on this book's web site.

Are All Bacteria Living in Us Bad for Us?

No one knows for sure how many human cells are in the average person, but some simple math will convince you that more prokaryotes live in our intestines than there are cells in our bodies. An average adult body is composed of about 10 trillion (10^{13}) human cells, give or take a few billion. Every milliliter (mL) of your large intestine's content is estimated to contain 10 billion (10^{10}) microbes, and our intestines contain at least 1,000 mL ($10^{10} \times 10^3 = 10^{13}$ microbes), which means we are a majority microbial and a minority human! Biologists who study the microcosm of the intestines estimate there are 500 to 1,000 different species living in an adult's intestines, which represents approximately 2 to 4 million nonhuman genes inside your gut. You would think we would know a lot about cells composing a substantial portion of our human ecosystem, but we don't. Why? Partly because the location is unappealing to most people, and partly because we had not developed tools to measure what is inside us. Because most of

the microbes are anaerobic, they have been difficult to grow in the lab for further study. To address this technical problem, a few brave and curious investigators decided to go where most of us would rather avoid.

A group from Washington University in St. Louis decided to sequence the complete genome of the gut bacterium *Bacteroides thetaiotaomicron* (theta – iota – omicron), which constitutes a substantial portion of our intestinal microbiota. They sequenced over 31 million bases and assembled 867 contigs with many gaps. They devised PCR methods to finish the genome and finally assembled the 67,938 sequencing runs into a single 6,260,361 bp circular contig (Figure 2.13). Though not shown in the genome

Figure 2.13 *B. thetaiotaomicron* genome map.
Go to www.GeneticsPlace.com to view this figure.

map, the genome also includes one plasmid composed of 33,038 bp (accession number AY171301). They annotated 4,779 predicted ORFs with 58% orthologs of known function, 18% orthologs of proteins with no known function, and 24% having no recognizable sequence similarity to any submissions to the database (as of March 2003). A striking feature of the genome is the extensive number of paralogs in *B. thetaiotaomicron* in categories such as sugar uptake and digestion, cell wall synthesis, environmental sensing and signaling, and mobile DNA elements (transposons).

As illustrated by the chromosome map, functional categories of genes (called clusters of orthologous groups or COGs) were annotated to facilitate our understanding of whole genomes. **COGs** describe phylogenetic classification of the proteins encoded in complete genomes. Each COG includes proteins that are inferred to be orthologs (direct evolutionary counterparts). All of the orthologs are organized into major categories that describe large events such as transcription, energy production, etc. Creating major categories allows us to quickly survey where a particular species invests its genetic resources.

Living in the gut of another species presents physiological challenges for which *B. thetaiotaomicron* has evolved solutions. The first obvious adaptation is the ability to metabolize sugars. Over the millennia, our prokaryotic symbiont has accumulated 170 genes for polysaccharide metabolism; most of these are paralogs of 23 genes typically found once in other species. For comparison, *E. coli* has only eight sugar-metabolizing paralogs. Genes appear to be paralogs and not orthologs when multiple copies are found in a single genome and other species have fewer copies per genome. Of these 170 paralogous proteins, 61% are predicted to be localized outside the cytoplasm. The percentage of extracellular proteins for *B. thetaiotaomicron* is higher than found in other species, which indicates a selection pressure for secreting a substantial number of proteins. Presumably, some of the digested sugars are for the bacterium's own consumption, but other microbial species probably benefit from the extracellular monosaccharides, and it is estimated that 10 to 15% of our daily calories are provided to us by bacteria living in our guts. Humans supply the bacteria with a constant supply of food and the microbial tenants pay rent in the form of simplified sugars for our intestines to transport to the blood. *B. thetaiotaomicron* also has the ability to import polysaccharides into its own cytoplasm; two genes in particular (*SusC* and *SusD*) are represented in the genome by a total of 163 paralogs. Furthermore, the bacterium has many two-component and one-component genes that help it sense its ever-changing biochemical surroundings. Some of these sensors physically interact with transcription factors (called sigma factors) that can migrate to the chromosome when their associated sensors are stimulated.

Microbes have adapted to modern medicines in ways that become apparent when the genomes are sequenced. Another clear advantage contained in this genome are ORFs within some of its 63 transposable elements. One particular type of transposon is known to spread tetracycline and erythromycin resistance between individual cells and between species in the microbiota of the gut. There is another curiosity that may or may not reflect an adaptive state for the microbe: the size of its genes. The **coding capacity** (gene density) for this genome is 89% (i.e., 11% is noncoding DNA), even though the ratio of gene number to genome size is small. This apparent paradox is resolved when you take into account that the average size of a gene is 1,170 bp, the largest for any bacterium sequenced so far (Figure 2.14). Why would a gut bacterium evolve with larger genes than others? Will all gut symbionts share this feature, or will a large gene size be exceptional even among gut microbes?

From this study, we can see that some of our prokaryotic guests are actually beneficial to us. *B. thetaiotaomicron* provides us with predigested sugars that we can use for ourselves. Furthermore, *B. thetaiotaomicron* can stimulate blood vessel formation in the intestine, which improves our ability to take up the monosaccharides provided to us by our microbial guests. Intestinal bacteria are critical to a good immune system in part because they protect us. They crowd out pathogenic bacteria, sequester limited resources such as iron, help stimulate a robust mucosal layer, and promote additional vascularization. Although we had known that bacteria can be our friends, thanks to the genome sequencing of this and future gut microbes, we will better understand how they help us, and perhaps even how we can provide them with an optimum environment to give us as much nutrition and protection as possible.

LINKS   G🧬P
mannose metabolism
Microbial List
Scientific American

DISCOVERY QUESTIONS

49. Go to the Microbial List web page and click on "*Bacteroides thetaiotaomicron* VPI-5482." Then click on the link "4778" to produce a list of all the proteins in the proteome, ordered the way the genes appear on the chromosome. Find "mannosidase" as many times as you can (you can stop when you get tired). This basic search illustrates the high level of polysaccharide utilization enzymes in the proteome—and you only searched for a single sugar.

50. Go back to the list of 4,778 proteins and do a find function for "one-component" (the sensor and signal-transduction proteins are fused). Each time you find one, look to see if a sugar-metabolizing protein is nearby. Perhaps the placement of a sensor gene and a metabolizing gene is not random. Propose a reason why evolution might have selected for these two types of genes to be neighbors.

51. Look at mannose metabolism, then find *Bacterioides* in the pull-down menu, and click on the "Go" button. (The fastest way to do this is to open the menu and start typing *Bacteroides* fairly quickly. The species will be highlighted as you spell out the genus.) Next, click on the oval labeled "Galactose metabolism." Can you verify that our symbiont is well suited to help us digest sugars?

52. Search *Scientific American* for "pylori" to read a surprising proposal published in 2005 that even the ulcer-causing bacterium *H. pylori* might produce beneficial consequences by living in our stomachs. Are we harming ourselves when we take antibiotics unnecessarily?

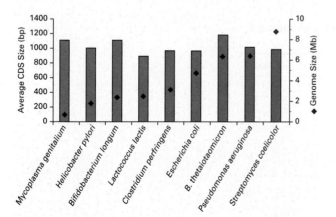

Figure 2.14 *B. thetaiotaomicron* has unusually large ORFs. Average coding sequences (CDS) are displayed as bar graphs for prokaryotes with widely distributed genome sizes (purple diamonds). CDS size does not correlate with genome size; *B. thetaiotaomicron* has the largest average CDS size.

Can Microbial Genomes Become Dependent upon Human Genes?

When we think of pathogenic or symbiotic bacteria, we picture genomes that are able to live completely on their own, without human support. However, some pathogens have become so accustomed to living inside us that they have become host-dependent parasites. There is an old adage that what you don't use, you lose. It appears the same wisdom applies to genomes, because many species have lost genes over the millennia, such as the tuberculosis and leprosy bacteria. One of the most reduced bacterial genomes belongs to the parasite called *Mycoplasma genitalium*, which (as the name suggests) was first isolated from infected cells of the urethra, though it has subsequently been isolated from respiratory tracts. This is the second smallest bacterial genome of a self-replicating species, with only 580,070 bp.

The *M. genitalium* genome was sequenced in 1995 by the same people who sequenced the very first whole genome, and was led by a team from The Institute for Genomic Research (TIGR) in the United States. It took 5 people 8 weeks to assemble 8,472 high-quality sequencing reactions. The overall GC content is 32%, with GC skew revealing the origin of replication located between *DnaA* and *DnaN* genes. Consistent with this interpretation, a majority of genes to the right of the origin of replication on this circular chromosome are transcribed from the plus strand, while those to the left are transcribed from the minus strand (Figure 2.15). As is often the case with low-GC genomes, the tRNA and rRNA genes have elevated GC content of 52% and 44%, respectively.

Figure 2.15 *M. genitalium* genome map.
Go to www.GeneticsPlace.com to view this figure.

Because of the complex and necessary 3D shape of these RNA molecules, the investigators proposed that selection prevents a drift toward AT base pairs for RNA genes.

As you can see from the genome map, the number of ORFs is very small. The investigators annotated 470 ORFs, but of these, 96 had no matches in the database (October 1995). *M. genitalium* has 88% coding capacity, which is similar to other bacterial genomes. Likewise, the average gene is 1,040 bp, which is typical. Therefore, the small genome size is probably due to lost DNA that is no longer advantageous under current growth conditions. By losing DNA, the gene density and lengths were not skewed from common values for other bacteria (Table 2.1). The genome has preferentially lost genes involved in the synthesis of amino acids, cofactors, cell envelope (it lacks a cell wall), and regulatory factors. It has retained proportional numbers of genes for energy metabolism, fatty acid and phospholipid metabolism, nucleotide production, replication, transcription, and protein transport. Interestingly, only one sigma (transcription) factor gene remains in the entire genome, which indicates the species has a limited capacity to adjust its transcription. The only category that is proportionally overrepresented is translation in the form of rRNA and tRNA genes, though in raw numbers, *M. genitalium* has coding capacity for these RNA gene products similar to that of other bacteria. When compared to the only other whole genome available at that time, *Haemophilus influenzae*, the gene order of *M. genitalium* was found to be very similar for two clusters of ribosomal proteins. When a series of genes are conserved in order and orientation between two or more species, the genes are described as being syntenic. Originally, **synteny** described genetic loci located on the same chromosome within a species, even if they were separated by a great distance. With the completion of many genome sequences, the meaning of *synteny* has

Table 2.1 Comparisons of two genomes sequenced very early in the genomic era.
Number of genes listed with percentage of total genes in parentheses.

Biological Role	*M. genitalium*	*H. influenzae*
Amino acid synthesis	1 (0.2)	68 (3.9)
Cell division	4 (0.9)	14 (0.8)
Chaperones	7 (1.5)	6 (0.3)
Detoxification	1 (0.2)	3 (0.2)
Protein secretion	6 (1.3)	15 (0.9)
Transformation	1 (0.2)	8 (0.5)
Intermediate metabolism	3 (0.6)	30 (1.7)
Energy metabolism	31 (6.6)	112 (6.5)
Fatty acid/Phospholipid	6 (1.3)	25 (1.4)
Bases and nucleotides	19 (4.0)	53 (3.1)
Replication	32 (6.8)	87 (5.0)
Transcription	12 (2.6)	27 (1.6)
Translation	101 (21.5)	141 (8.2)
Transport and binding	34 (7.2)	123 (7.1)

shifted to describe multiple genetic loci from different species being located sequentially on a chromosomal region of common evolutionary ancestry.

The metabolic potential for *M. genitalium* is severely limited. *H. influenzae* has 228 genes associated with metabolic pathways, compared to *M. genitalium*'s 44 covering the same categories. Similarly, *M. genitalium* has only 1 amino-acid-synthesizing gene, whereas *H. influenzae* has 68. Any species that cannot make its own amino acids must be dependent upon its environment, which in this case means its human host. *M. genitalium* lacks cytochrome and citric acid cycle genes and cannot produce energy storage molecules of glycogen or butyrate, which indicates its only source of **ATP production** is through substrate-level phosphorylation. Based on the genes that encode sugar uptake proteins, it appears that *M. genitalium* may be limited to glucose metabolism for energy and carbon sources, compared to *H. influenzae*'s ability to utilize six different sugars. The number of integral membrane proteins is estimated to be 140, based on Kyte-Doolittle hydropathy plots and BLAST alignments with previously characterized orthologs.

What Is the Minimum Number of Genes Possible?

The discovery of *M. genitalium*'s tiny genome prompted a number of investigators to determine the fewest genes required by a prokaryote, and they are trying to construct an organism with the fewest possible genes. Craig Venter, who helped establish TIGR, is collaborating with his coauthor and Nobel laureate Hamilton O. Smith, and jointly they have drawn a lot of **attention** to minimum-genome research. They are utilizing a new field called **synthetic biology** (see Section 11.2) to synthesize de novo (from scratch) a functioning genome with as few genes as possible. The purpose of their quest is both basic and applied research. If they can build a simplified life form, they will better understand evolutionary principles and genome **circuitry** (see Chapter 12). Constructing a minimal genome raises many ethical questions that remain to be answered (see Section 4.4).

On a related note, a large group of Japanese and European investigators have tried to identify the essential genes of *Bacillus subtilis*, a very common bacterium. Based on the *M. genitalium* genome, some had predicted the minimum genome would be composed of 260 genes, but the *B. subtilis* group found that only 192 genes were indispensable for *B. subtilis*, and therefore they predict the minimum number of genes is closer to 200 than 260. To help themselves and other investigators, a Chinese team has developed a Database of Essential Genes (DEG) that lists genes required for life in six species. No matter who guesses right or who synthesizes the minimum genome, many people have dreams that these mini-bugs will be able to clean up toxic waste, reduce global warming by fixing carbon, and provide us with biological sources of energy. From these

goals, you can tell that bacteria with tiny genomes will be a hot topic for many years. However, only nature has produced a highly reduced genome and unfortunately, its only significant product is infection.

DATA
ATP production
LINKS
attention
DEG
Firefox
M. genitalium
M. genitalium Genome
NCBI Entrez
Netscape
TIGR CMR

DISCOVERY QUESTIONS

53. Go to the TIGR CMR web site, then choose "Align Whole Genomes." Choose from the two pull-down menus *H. influenzae* and *M. genitalium* with a minimum alignment of 20 nt (this display looks best if viewed in Netscape or Firefox browsers). Do these two genomes look like they evolved from a common ancestor in the recent past? You can increase the sensitivity by changing the minimum alignment to 15 nt; see if this helps.

54. Go to the *M. genitalium* genome page and choose the "Searches" option from the top menu; then choose "Name," search for the genes MG064 and MG101 (see figure 2.15), and follow the "Gen-Bank" link to retrieve protein sequences. Does BLAST provide any insights now?

55. Go to DEG and search for entry number 1169 (DEG10060038 or *clpB*) from *M. genitalium*. Does the name of this gene lead you to believe it might be an essential gene? Copy the ORF sequence from the DEG link and perform a BLASTx search (submit DNA sequence and search for protein matches) search against all DEG entries using DEG's BLASTx program. Then perform a BLASTx search at NCBI. Which search is more informative and why?

56. Go to the *M. genitalium* Genome Browser at Genome.Net. Click on "KEGG" (metabolism database) at the bottom to see an interactive genome map. At the bottom is a genome alignment tool that will show you genes in your query species aligned with the species of your choice (choose *E. coli*, then click on "Exec"[ute]). You will see colored bands for orthologs of the two species and their location in the query genome. Mouse over the genome, and the green bar will show you on which portion it will zoom; click on the section with the most color (red). On the new page, click on the "ORF Color" button to understand the colors, and "View Genome map" to see the position of the conserved genes. What category of genes is in this area? Are the genes clustered or widely distributed?

57. Perform an NCBI Entrez search for *Nanoarchaeum equitans*. Click on the "MeSH" link at the bottom to learn about this species. Go back to the full Entrez results, click on the

DATA
tRNA^His

LINKS
COG functional categories
small genomes

"Genome" link, and follow this until you see the circular map of the smallest cellular genome in the world (as of September 2005). Click on "GenePlot" to align *N. equitans* and *M. genitalium* by changing the default species for the pull-down menu on the right, then click on "compare selected pair." How many protein-coding genes are conserved between these two smallest genomes (number of "bets" [best hits] listed below the 2D alignment maps)? You can navigate around by clicking and changing the zoom button. Identify the gene they have in common that is located nearest to the 0–0 origin of the graph.

58. In Discovery Question 57, you determined that both species have histidyl-tRNA synthetase, but if you studied the whole genome of *N. equitans*, you would discover it lacks a tRNA^His for the codon GUG. Go to the tRNA^His web page and BLASTn each of the three sequences. Note the first hits when you submit each half and then the difference when you submit the full sequence. Do any of these BLAST results indicate they might be a tRNA gene? Why doesn't the full sequence give a better score than just half of the tRNA gene?

Are All Viral Genomes Smaller than All Bacterial Genomes?

So far, we have examined primarily Eubacteria genomes. In Discovery Questions 57 and 58, you learned about an Archaea, *N. equitans*, that has the smallest known genome, about 490 kb. Because viruses infect prokaryotes, we would expect all viruses to be smaller than all prokaryotes. In fact, the typical bacterium size is 1–5 microns (μm), and the largest virus is about 0.4 μm. It would make sense for viral genomes to be small as well, and many well-known viruses have very small genomes (HIV is 9,200 nt; WNV is 10,962 nt; SARS is 29,727 nt; T7 is 39,900 nt; Lambda is 48,502 nt). However, biology is full of surprises, and DNA continues to surprise us the more we study it. Therefore, it pays to be skeptical of absolute statements that contain words such as "all" or "never." A case in point is the largest known virus, the Mimivirus, which infects the tiny eukaryote amoeba.

In 2003, a new virus infecting the amoeba species *Acanthamoeba polyphaga* was isolated from a hospital's air conditioning cooling tower in Bardford, England. The new virus, called Mimivirus, was determined to be the largest virus to date. A group in Marseille, France, sequenced its genome. To everyone's surprise, the double-stranded DNA (dsDNA) genome was 1,181,404 bp long, and was annotated with 1,262 ORFs (Figure 2.16). Each end of the linear

Figure 2.16 Mimivirus genome map.
Go to www.GeneticsPlace.com to view this figure.

chromosome is terminated with inverted repeats, which the investigators interpreted to mean that perhaps the genome is circular at least some of the time, but no one knows for sure. The size of the genome and the large number of predicted ORFs were not the only surprises. The types of proteins encoded by this virus were very unusual for a virus and indicate that some of the rules we have for viruses should be reexamined, if not totally rewritten.

Your first impression of Figure 2.16 might be that the genome looks a lot like a bacterial genome. The genome is only 28% GC, and the investigators predicted the origin of replication is near base 380,000. Notice that the inner strand of DNA encodes more genes to the right of the predicted origin while the outer strand encodes more genes to the left, which is consistent with the origin being located as predicted. Because 380 kb is not centered, it would seem logical that the genome forms a circle at least during replication. The Mimivirus coding capacity (90.5%) is higher than in most prokaryotes. Interestingly, the amino acid isoleucine is used in Mimivirus proteins more than twice as much as in either humans or amoeba proteins. Furthermore, when an amino acid can be encoded by more than one codon, Mimivirus has a very strong bias toward codons lacking G or C. In fact, the most common codon in Mimivirus is the least common codon in amoeba. It appears that the pathogen has been selected to utilize a neglected niche of its host genome.

Most viruses rely heavily, if not exclusively, on their hosts' metabolic capacity to provide replication, transcription, and translational machinery—but not Mimivirus. When annotated with the COG database, 194 of the viral genes fit into 108 distinct COG functional categories, with significant overrepresentation in translation (COG category J; **p-value** $p < 0.001$), posttranslational modification (COG category O; $p < 0.006$), and amino acid transport and metabolism (COG category E; $p < 0.08$). The proteins include DNA repair enzymes (e.g., ATP- and NAD-dependent DNA ligases, DNA mismatch repair, and an endonuclease for ultraviolet damage repair), chaperones (e.g., heat shock protein), translation components (e.g., 6 tRNA genes, 4 tRNA aminoacyl synthases, 2 initiation factors, and 1 termination factor), glycosylation (e.g., mannose-6 phosphate isomerase), amino acid synthesis (e.g., asparagine and glutamine synthases), and a hodgepodge of other proteins (e.g., cholinesterase, ADP-ribosyltransferase). This long list is meant to impress upon you how unorthodox the Mimivirus genome is for a virus; its parts list sounds more like that of a eukaryote than of a lowly virus. In addition, one gene encodes a self-splicing RNA intron, and another

LINKS G⚕P
Genome NCBI

encoded protein contains a self-splicing intein, which is a short piece of peptide that chops itself out of the pre-protein to produce the mature form. Many eukaryotes have topoisomerases to untwist supercoiled DNA that results from transcription and replication, but Mimivirus is the only virus known to encode type IA, type IB, and type II topoisomerases. Interestingly, one of Mimiviruses topoisomerases' BLAST hits with the lowest E-value is an ortholog in *B. thetaiotaomicron*, the human gut symbiont. Mimivirus produces glycosyltransferases that can synthesize polysac-charides and lead to N- and O-linked glycosylation of cap-sid proteins. In fact, the Mimivirus stains positive with the Gram stain that is used to characterize bacteria.

Is Mimivirus Alive?

Mimivirus can synthesize some of its own nucleotides and lipids, as well as glycosylated collagen-like fibers that may explain the virion's microscopic furry appearance. Thirty of its proteins have ankyrin-repeat domains, named for the ankyrin protein in eukaryotic cytoskeletons. All of these oddities led the investigators to ask to which domain of life Mimivirus is most similar. They used many computer algo-rithms and many different gene and protein sequences, but in every test, Mimivirus was found to be most closely asso-ciated with the base of the eukaryote branches (Figure 2.17). Viruses tend to be unstable outside narrow environmental conditions, but Mimivirus was infectious after 1 year of storage at 4°, 25°, and 32°C. One percent of virions sur-vived 90 minutes at 55°C, while these same conditions killed 100% of *E. coli*. Mimivirus also survived 48 hours of desiccation, another treatment that often kills virus particles.

By definition, a virus cannot synthesize its own proteins, but Mimivirus has the capacity to participate in all the major steps of translation: tRNA production, amino acid charging of tRNAs, initiation, elongation, termination, enzyme-enhanced protein folding, and posttranslation modification. Therefore, we are presented with a challenge: is the Mimivirus a life form or a highly modified virus? Per-haps the virus has acquired its unusual genes, but based on GC content and phylogenetic analysis, the authors dis-counted this explanation in favor of a second one. They believe Mimivirus used to contain more genes and over time, it has lost some of its former coding capacity. In para-sitizing a eukaryote, the virus may be more fit with fewer genes that used to participate in translation. In 1957, a pro-posed formal definition of *virus* included a size of less than 0.2 microns and the possession of either DNA or RNA, but not both. Mimivirus breaks both of these rules. How-ever, genome sequence analysis tells us that Mimivirus can-not generate energy from substrates, consistent with the third condition for virus identification. Viruses are unable to grow by binary fission, but we do not yet know if Mimivirus can perform this complex task.

Although the Mimivirus genome is big-ger than some prokaryotes and contains more genes as well, it still lacks some life-defining capabili-ties. However, this bizarre virus has blurred the distinction between prokaryote and virus. It is reminiscent of intracel-lular parasites such as *Buchnera* and *Mycoplasma leprae* (see Section 3.2) which are obligated to remain inside eukary-otic host cells just as Mimivirus does. Perhaps Mimivirus is an ancient relic that resembles some of the earliest forms of complex life that evolved 3 billion years ago. It makes sense that before viruses as we understand them could evolve, host species would have to evolve that could support the viruses' dependent mode of living. Could Mimivirus be a direct descendent of an early form of dsDNA life that was able to replicate independently until it became an intracel-lular parasite and lost some of its original genome and genes? We may never know due to the long time since the beginning of life, but maybe we will discover equally bizarre viruses, and by comparing their genome sequences will begin to piece together how life evolved on earth.

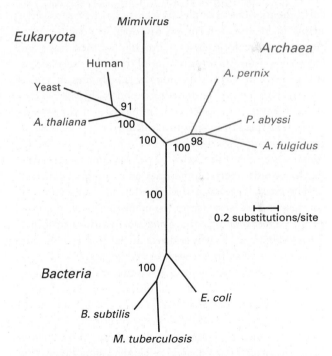

Figure 2.17 Phylogenetic tree of *Mimivirus* with three domains of life.
Seven universally conserved proteins totaling 3,164 amino acids were used to produce this unrooted tree. Bootstrap values are placed at the branch points (see Math Minute 3.3).

DISCOVERY QUESTIONS

59. Perform a Genome NCBI search for "Mimivirus" and click on the link. Change the view of the genome to show only tRNAs and hit "Refresh." How many are in this genome?

60. In the left frame there is a link called "Protein view." Click on it to get a list of Mimivirus protein-coding

genes as they appear on the genome. In the "Start from" field, enter these nucleotide numbers to find three genes and then click on "Go": (1) 234000, R194; (2) 267000, L221; (3) 633000, R480 (L and R refer to genes pointing to the left or right, respectively). Click on each of the three boxes to find out the family of proteins. Which one looks most like a eukaryotic protein, based on the COG sequence similarity results you get with each click of the boxes?

61. If you were going to construct a minimum genome, would you choose a virus or a bacterium? Explain why.

Do Genomes Reflect an Organism's Ecological Niche?

From the previous examples, it seems clear that as a species adapts to its environment, so too does its genome. Symbionts accumulate genes that provide for their hosts and themselves. Parasitic prokaryotes rid themselves of unnecessary genes if their hosts provide the metabolites. It should come as no surprise that the same trend is observed in ecologically important species, too. In the oceans live tiny photosynthetic plankton (**phytoplankton**) that are able to consume more CO_2 than all the rain forests in the world. Perhaps the most productive phytoplankton in the world are the **cyanobacteria**, sometimes misleadingly referred to as blue-green algae even though they are Eubacteria and not algae at all. The two most abundant genera of cyanobacteria are *Prochlorococcus* and *Synechococcus*, two egg-shaped species (Figure 2.18) whose direct ancestors probably converted our ancient anoxic world into the oxygen-rich environment we enjoy today. In the summer of 2003, 3 papers were published in short succession that described 4 genomes of cyanobacteria that literally change our world and that we depend on for the air we breathe.

Three sequencing projects were conducted in France, Germany, Israel, England, and the United States. Because the cells are so small and *Prochlorococcus* was only discovered in the 1990s, we have little experience with or criteria for distinguishing different species. Therefore, individual cells from both genera are referred to by a numbering system to indicate they are different **ecotypes**, meaning a species that occupies different ecological niches. Salinity and temperature stratify ocean water into distinct levels with characteristic amounts of nutrients, light, and life forms (Figure 2.19), and each ecotype was isolated from a different ocean layer. Based on rRNA sequencing ($\geq$97% identical), the three *Prochlorococcus* ecotypes (MED4, MIT9313, and SS120) have been considered a single species. Though we had known for years that each

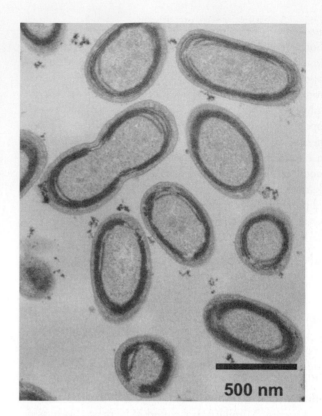

Figure 2.18 Anatomy of *Prochlorococcus*.
Transmission electron micrograph of unicellular prokaryotes with one cell caught in the act of cell division. Note the internal thylakoid membranes that are required for photosynthesis; some prokaryotes do have internal, membrane-bound structures.

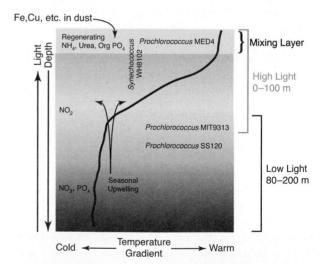

Figure 2.19 Ecology of four distinctive cyanobacteria.
Open ocean ecosystem with vertical gradients of light, nutrients, and temperature, as well as zones of mixing (top) and periodic upwelling of lower water layers (from bottom up). Nutrient labels represent the location of highest concentrations, but the nutrients are not restricted to indicated layers. Location of cyanobacteria names indicates relative depths where each is maximally concentrated.

Table 2.2 Cyanobacteria comparisons.

	Prochlorococcus MED4	*Prochlorococcus* SS120	*Prochlorococcus* MIT9313	*Synechococcus* WH8102
Genome	1,657,990	1,751,080	2,410,873	2,434,428
%GC	30.8	36.4	50.7	60.2
# genes	1,716	1,884	2,275	2,517
% coding	88.0	88.5	82.0	85.6
Features	High light adapted	Very low light adapted	Low light adapted	Wide variety of niches
	Two to three ecotypes, but one species by molecular and classical methods			

ecotype had unique physiology, no one knew if these differences were driven by alleles or genes—until now.

Ecotype MED4 is located in the top layer of the ocean and has the smallest genome of the four (Table 2.2). Down the water column and into the low light level, MIT9313 and SS120 hover and exchange minerals and organic molecules with a different stratum of the ocean than their high-light-adapted MED4 relative. The only oceanic *Synechococcus* isolate to have its genome sequenced is more widely distributed globally, but is less abundant when conditions favor *Prochlorococcus*. *Synechococcus* lives in a larger portion of the water column than MED4, but it too prefers the intense light found near the surface. All four genomes utilize about 85% of their genomes to encode ORFs, but they appear to fall into two genome bins, or categories. The small genome bin with MED4 and SS120 has a very low GC content and fewer than 1,900 ORFs. The large genome bin with MIT9313 and *Synechococcus* has high GC content and more than 2,200 ORFs. It would have been tidier if the two genome bins matched the depth and light-level preferences, but the smallest and the largest genomes prefer bright light near the surface while the two middle genomes prefer deeper water. Why? What has happened to their genomes over time to produce this conundrum?

First, it is helpful to compare whole genomes to get a broad perspective. As you can see in the dot plot of Figure 2.20, MED4 and MIT9313 genomes have undergone many rearrangements. These two ecotypes share 1,352 orthologs, with synteny conserved in a few places as indicated by the short diagonal segments. If the segment has a negative slope, this portion was inverted in one species relative to the other, but the order of genes was not altered. Segments with positive slope but located off the diagonal line indicate chromosomal recombinations from one area to another, and genes remain in the same order and orientation (see Math Minute 2.4). The large number of genes along both axes indicates MED4 has 364 genes not found in MIT9313, and MIT9313 has 923 genes missing from MED4. From the whole-genome perspective, it is clear that these two ecotypes have experienced many indels as well as inversions, which suggests they diverged a long time ago and look more like cousins than siblings.

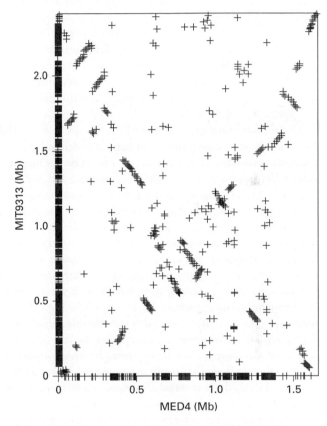

Figure 2.20 Global genome dot plot alignment of MED4 and MIT9313.

Genes present in one genome but not the other are positioned on the appropriate axis. The broken-X pattern indicates numerous inversions, with the intersection located near the origins of replication.

As ecotypes living at different depths, you might expect them to have different genes related to photosynthesis. One particularly striking example is the *pcb* gene family, which encodes the chlorophyll-binding, light-harvesting antenna complex proteins to capture a wider spectrum of light than either chlorophyll a or b alone. In high-light-adapted MED4, only one *pcb* gene is retained; deeper-dwelling MIT9313 has two *pcb* genes (*pcbA* and *pcbB*), and deepest SS120 has eight *pcb* paralogs (*pcbA–pcbH*). MED4's *pcbA*

Math Minute 2.4   **Can You Estimate the Number of Inversions in a Dot Plot?**

Dot plots like Figure 2.20 help us visualize large-scale comparisons of genomes. By coloring the dot (or in this case, drawing a +) at the location of aligned orthologs, we see that large segments of the chromosome were rearranged as MED4 and MIT9313 diverged. If there had been no rearrangements, the dot plot would consist of a diagonal line of plus signs, from the bottom left to the top right of Figure 2.20. (Note that the sequences were written in a different orientation than in Figure 2.4 and Figure MM2.1 in Math Minute 2.1, where the diagonal would have gone from top left to bottom right.)

To help understand the differences between MED4 and MIT9313, and quantify the evolutionary distance between them, we would like to reconstruct the genome rearrangements that might have occurred over time. Visually, reconstruction means moving and/or inverting segments of genes in one genome until total synteny with the other genome is achieved (i.e. until the dot plot is a diagonal line). To explain this reconstruction process mathematically, let's consider a simplified example of 11 genes that appear on both the human and mouse X chromosomes. In this example, we restrict ourselves to rearrangements consisting of chromosomal inversions only. Note that what a biologist calls an inversion, a computer scientist calls a **reversal,** which is the term we will use in this Math Minute.

The rows of numbers in Table MM2.6 show one way the 11 genes (numbered 1 to 11) could be rearranged from their current order and orientation on the mouse X chromosome to match their order and orientation on the human X chromosome. The target gene order/orientation is on the bottom row, and is defined to be positive numbers from 1 to 11. The top row is the gene order/orientation on the mouse X chromosome, where a gene is negative if its orientation in mouse is inverted from its orientation in human. Each of the remaining rows shows the result of reversing the segment highlighted in the row above. Note that the sign of every gene is flipped in a reversal, since the orientation must change for the genes to stay in sequential order.

Following the seven reversal steps in the rearrangement scenario of Table MM2.6 is not too difficult, but starting from scratch with the data in Figure 2.20 and rearranging the segments manually would be an overwhelmingly difficult and tedious task. Fortunately, computer science makes this task feasible. An **algorithm** is a precisely controlled list of steps, like a lab protocol, that allows the computer to do the work and eliminates subjective decisions and human error. If an algorithm could always rearrange one genome to match another using the fewest reversals possible, we could use this number of reversals, called the **reversal distance**, as a numerical measure of differences between two genomes.

Sridhar Hannenhalli, a graduate student in computer science at Penn State University, and his advisor, Pavel Pevzner, found a rearrangement algorithm that nearly always finds the fewest reversals possible. The key to the algorithm is to represent gene order and orientation with a **graph** (see Math Minute 1.2) like that in Figure MM2.2a. This graph contains two nodes for each gene, one for the 5′ end of the gene and one for the

Table MM2.6 Seven reversal steps from mouse X chromosome to human X chromosome.

Query (mouse) genome	1	−7	6	−10	9	−8	2	−11	−3	5	4
↓	1	−7	6	−5	3	11	−2	8	−9	10	4
↓	1	2	−11	−3	5	−6	7	8	−9	10	4
↓	1	2	−11	−3	5	6	7	8	−9	10	4
↓	1	2	−11	−3	5	6	7	8	9	10	4
↓	1	2	−11	−10	−9	−8	−7	−6	−5	3	4
↓	1	2	−11	−10	−9	−8	−7	−6	−5	−4	−3
Target (human) genome	1	2	3	4	5	6	7	8	9	10	11

3′ end. (Adding 5′ and 3′ labels to the nodes would make them hard to read, so we use L to indicate the "left," or 5′ end of the gene, and R to indicate the "right," or 3′ end of the gene.) The Start node (S) is a placeholder at the 5′ end of the segment, upstream of the first gene, and the Finish node (F) is a placeholder at the 3′ end of the segment, downstream of the last gene. The genes are listed clockwise from Start to Finish in the order they appear in mouse, with L before R for positive orientation and R before L for inverted orientation. Before going on to the next paragraph to learn how the edges were drawn in this graph, you should be sure you understand the labels and placement of the nodes in Figure MM2.2a.

Solid edges in the graph connect consecutive genes in mouse. When the R label of one gene is connected with a solid edge to the R label of the next gene, the orientations of these two consecutive genes are different (corresponding to a change in sign from positive to negative in Table MM2.6). In contrast, when the R label of one gene is connected with a solid edge to the L label of the next gene, the orientations of these two consecutive genes are the same. Similarly, L of one gene connected to L of the next indicates opposite orientation (from negative to positive in Table MM2.6), while L connected to R indicates the same orientation. The solid edges are often referred to as *reality edges* because they represent the gene order of the query genome. Again, be sure you understand how the solid edges are placed in the graph before continuing to the next paragraph.

Dashed edges in the graph connect consecutive human genes. Because the target gene order and orientation are defined to be positive numbers from 1 to 11, these edges are always the same (Start → 1L, 1R → 2L, . . . , 11R → Finish). The dashed edges are referred to as *desired edges*, because they represent the gene order and orientation of the target genome. The graph in Figure MM2.2a is called a *reality and desire graph*, because it contains both the query and the target gene orders.

Now we can see how to determine the reversal distance, the minimum number of reversals needed to match the query to the target. Amazingly, the reversal distance is approximately $n + 1 - c$, where n is the number of genes (11) and c is the number of *edge-disjoint alternating cycles* in the reality and desire graph. An *alternating cycle* is a multi-edge path that (1) does not visit any node or traverse any edge more than once, (2) ends at the same node where it started (i.e., forms a cycle), and (3) alternates between reality and desired edges. A set of alternating cycles is *edge-disjoint* if the same

a)   b)

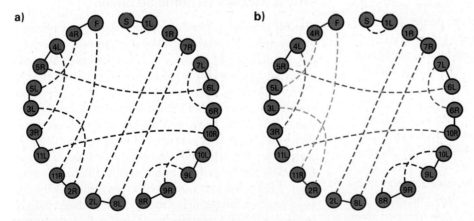

Figure MM2.2 a) The reality and desire graph. **b)** Reality and desire graph decomposed into five edge-disjoint alternating cycles. Cycle 1 (black edges): S → 1L → S. Cycle 2 (dark purple edges): 1R → 2L → 8L → 7R → 1R. Cycle 3 (grey edges): 7L → 6L → 5R → 4L → 3R → 11L → 10R → 6R → 7L. Cycle 4 (black edges): 10L → 9R → 8R→9L→10L. Cycle 5 (light purple edges): 2R → 3L → 5L → 4R → F → 11R → 2R. Note that the direction in which a cycle is traversed is not important, nor is the node selected as the starting node. Therefore, the same five cycles could be listed in many other equivalent orders.

edge is not used in two different alternating cycles. Because of the special way the graph is constructed, with two nodes for each gene and exactly one reality edge and one desired edge touching each node, it can always be decomposed into edge-disjoint alternating cycles, as shown in Figure MM2.2b. In our example, $n = 11$ and $c = 5$, so the reversal distance should be approximately 7 ($11 + 1 - 5$). In Table MM2.6, that is exactly how many reversals were needed. By programming the computer to count edge-disjoint alternating cycles in the reality and desire graph, reversal distances can be determined for much larger sets of genes, such as those in Figure 2.20. Although we have omitted the details of the algorithm that determines the precise reversal steps needed, these steps are also determined from the reality and desire graph.

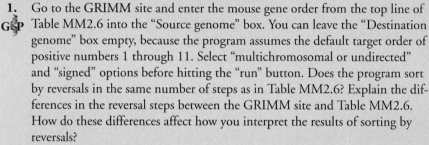

MATH MINUTE DISCOVERY QUESTIONS

1. Go to the GRIMM site and enter the mouse gene order from the top line of Table MM2.6 into the "Source genome" box. You can leave the "Destination genome" box empty, because the program assumes the default target order of positive numbers 1 through 11. Select "multichromosomal or undirected" and "signed" options before hitting the "run" button. Does the program sort by reversals in the same number of steps as in Table MM2.6? Explain the differences in the reversal steps between the GRIMM site and Table MM2.6. How do these differences affect how you interpret the results of sorting by reversals?

2. At the GRIMM site, select "Human Mouse (123 genes)" from the "choose sample data" drop-down menu. Scroll down to see the results. What operations other than reversals have been performed? Why are these additional operations needed?

---

gene is always activated (constitutive) and does not respond to changes in Fe^{+3}, but MIT9313's *pcbB* was induced 7-fold and SS120's *pcbC* 23-fold when exposed to Fe^{+3} in lab growth conditions. Being exposed to more light permits more photosynthesis, but has a disadvantage, too: UV mutagenesis. MED4 has a lower GC content and has lost some DNA repair genes found in the deeper MIT9313, including repair enzymes that normally prevent GC → AT transversions. The negative correlation between light exposure and DNA repair seems maladaptive, but it does help explain the AT-rich genome of MED4. However, as we have seen in previous examples, the tRNA genes of all three ecotypes are nearly identical and near 50% GC, supporting the earlier conclusion that strong selection pressures maintain GC content and 3D structure of tRNAs. By comparison, SS120 has a more complete set of DNA repair enzymes and is devoid of any transposable elements, indicating its genome is more stable than the other two ecotypes. Consequently, the deeper ecotypes are more protected from UV light, but they have less light for photosynthesis. Perhaps to compensate for less effective photosynthesis, the deep-water ecotypes MIT9313 and SS120 have sufficient genes to provide partial heterotrophic energy production, but full pathways were not obvious. SS120 has 640 ORFs of unknown function and MIT9313 has 906. Perhaps some of these uncharacterized genes may

be capable of complementing the missing biochemical pathways to help these cyanobacteria produce energy and incorporate carbon into their metabolomes.

Why Is MED4's Genome So Small?

The MED4 genome is the smallest known for a photoautotroph (photosynthetic organism that can utilize inorganic carbon sources) and may represent the minimum, naturally occurring genome for a photosynthetic organism. Has MED4 lost genes, or have the other two ecotypes gained genes? It appears that MED4 has lost genes, as illustrated in Figure 2.21. If we begin with the largest genome of *Synechococcus* and work down to MED4, it is clear that each ecotype has lost a bit more than the one above it in size. MIT9313 lost several nitrogen metabolizing and uptake genes; MED4 lost some nitrite (NO_2; see Figure 2.19) metabolic capacity. Interestingly, based on codon usage and GC content, MIT9313 apparently acquired a nitrite transport

Figure 2.21 Deletion, acquisition and rearrangement of nitrogen usage genes.
Go to www.GeneticsPlace.com to view this figure.

gene through horizontal transfer. MIT9313 cannot use nitrate (NO_3) and MED4 cannot use either nitrate or nitrite, which is consistent with their vertical positioning in the water column and the nutrients that predominate in their layers. However, it is counterintuitive that SS120's genome requires it to depend upon $NH4^+$ and amino acids for its nitrogen when these constituents are not concentrated in the deepest layer. Perhaps SS120 is able to accumulate other nitrogen sources with some of its ORFs of unknown function. *Synechococcus* is able to utilize nitrate in addition to the other nitrogen sources available to *Prochlorococcus* ecotypes.

The largest genome belongs to the cyanobacterium that has the widest ecological range and is better able to process dissolved ions. *Synechococcus* WH8102 has several transport proteins for importing trace elements and has the ability to extrude excessive ions. For example, WH8102 has two efflux pumps that are thought to be at least partially responsible for its copper toxicity resistance. WH8102 has a surprising capacity for Na^+ ion-coupled transportation (symporters and antiporters) of dissolved compounds, including other ions, amino acids, and polypeptides. Furthermore, WH8102 has adapted its photosynthetic electron transfer pathway to utilize copper, a detoxifying superoxide dismutase to incorporate nickel, and nucleotide reductases to bind cobalt instead of the ocean's growth-limiting mineral of iron. WH8102's generalist capacity for growth in diverse and iron-limited ocean environments has resulted in a larger and more robust genome rather than a leaner genome that is easier to replicate. Once again we see that genomes reflect the tradeoffs inherent in adapting to an ever-changing environment. Because of the ubiquity of cyanobacteria and the paucity of oceanic iron, people are proposing plans to decrease worldwide CO_2, and thus global warming, by fertilizing the ocean with Fe^{+3} to increase the capacity of Fe^{+3}-limited cyanobacteria and diatoms (see Section 4.1). Do we understand these genomes well enough to predict what will happen if we alter the ocean's chemical balance?

How Many Genome Changes Are Required Before a New Species Is Created?

Having sequenced the four cyanobacteria genomes, investigators have created a question for themselves. How do you determine whether an organism is a species or an ecotype? The definition of a eukaryotic species was based on the reproductive failure of the proposed two species, but this definition was already problematic for prokaryotes that reproduce asexually. Microbiologists had used morphology and physiology to define species, but if we applied this same method to dogs, we would think every breed was a different species. With genome sequences in hand, it seems our definition of *species* must be modified—but how much? How many genome indels, inversions, horizontal transfers,

LINKS G&P
Sallie Chisolm
ORNL

etc. are required to classify two ecotypes as distinct species? Do you need population studies to measure variations within ecotypes before you can decide if they are indeed two potential species (see Section 4.1)? Adding to the complexity of cyanobacteria classification are the recently discovered marine cyanophages, viruses that infect oceanic cyanobacteria. At the same time the genome sequences were being published, Sallie Chisholm's group at Woods Hole/MIT surveyed the ability of viruses to infect different ecotypes/species of cyanobacteria. Some of the viruses were restricted to the cyanobacterium from which they were isolated, but others were able to infect all three *Prochlorococcus* ecotypes/species, and *Synechococcus* too. The ability of cyanophage to infect multiple genera and possibly transfer genetic material between the hosts indicates cyanobacteria genomes probably are targets of quick and substantial alterations, which could lead to rapid speciation and evolution of dominant phytoplankton.

By sequencing the genomes of four cyanobacteria, we have seen four different solutions to a common problem: how to survive in a highly competitive, ever-changing environment. Let's imagine 3 billion years ago when cyanobacteria were new residents on earth. As their success and numbers grew, they changed the world environment forever. The abiotic environment changed and so the biotic members had to as well. Some individual cells adapted to the less crowded niche of deeper water, which selected for genomes that harvested light more efficiently and utilized different nitrogen sources. *Synechococcus* took advantage of its large genome to become adept at rapidly adjusting to changing environments by retaining numerous two-component systems for sensing and signaling changes, as well as sigma factors to initiate transcription.

DISCOVERY QUESTIONS

62. Go to the Oak Ridge National Laboratory (ORNL) Microbial Genome web site, choose the *Prochlorococcus marinus* sp. MED4 genome from the "Finished Eubacteria" pull-down menu, and note the %GC. Compare this percentage with the genomes of *P. marinus* MIT9313 and *Synechococcus* sp. WH8102. Do the two *P. marinus* genomes look like two ecotypes of the same species to you?

63. Go back to the *P. marinus* sp MED4 genome and click on the "View genome in Web-Artemis" link. It will take a while to load this Java applet, but it is worth the wait. Warning: Do not close the web page that simply says "loading entry— done" or you will lose the applet. You will see the full, annotated genome in three frames. The top and middle frames are duplicates, but the vertical slider bars allow you to adjust magnification, with the default showing the top frame in medium scale and the middle frame in high-resolution scale. Thin vertical lines in the top frames represent stop codons. The bottom frame is the complete list of every gene and annotated feature (e.g., transmembrane domain, signal peptide, etc.). A "c" in the gene list indicates the "Crick" strand of DNA.

Your browser should display a few new menus that add extensively to the Artemis viewing and analysis. Under the "View" menu, choose "Show CDS Genes and Products" and under the "GoTo" menu, choose "Navigator . . . ." Check the "Goto Feature with This Qualifier Value:" and search for "*cobA*" (see Figure 2.19), telling the navigator to pay attention to case. Double-click on the highlighted gene in the list of "Genes and Products," and you will see the ORF displayed in the main graphic window, complete with DNA and protein sequences. What does *cobA* do, and is it on the Crick or Watson strand?

Now click on the main graphic window to make sure it is the active one (a Java requirement). Under the "Graph" menu, choose "GC content (%)" to see how the GC content shifts with different genes. In the list of features in the bottom frame, scroll down until you see a tRNA gene (light green box). Double-click on the box and look at the %GC. Try a few more genes and describe the pattern you observe. Artemis is very powerful, so feel free to explore and make new discoveries on your own.

64. Go back to the ORNL Microbial Genome page, select *Synechococcus* and note its %GC again. Now launch the Artemis viewer (wait . . .) and view the following regions with attention to the GC graph and how many genes are in these AT-rich sections: (1) 427233; (2) 622199; (3) 912128; (4) 2379775. Did you notice one of the world's longest prokaryote ORFs in one of these sections? Compare this long ORF to the average gene on the *Synechococcus* statistics page. You have just examined four areas with different codon bias and GC content. Propose an explanation for these four apparent anomalies.

In the final two cases of Section 2.2, we will examine two unicellular eukaryote genomes, but you will encounter some questions commonly asked about prokaryotes also. Which genes are redundant and can be deleted? Which paralogs are especially useful and provide a selective advantage? Why are some genomes especially AT-rich? Why do some species have larger or smaller genes and genomes? For example, why does *Synechococcus* have one of the largest

prokaryote genes ever reported? The *swmB* protein (genome annotation number SYNW0953) is 10,791 amino acids long and its gene occupies more than 1% of the entire WH8102 genome. Sometimes evolution favors solutions that seem odd to us, but the solutions may reflect ancestry more than replication efficiency. Consider an analogous situation in your own life. On occasion, all of us have found it easier to copy, paste, and modify a paragraph rather than type a new paragraph from scratch. Evolutionary changes are built from previous solutions, and examples such as *swmB* indicate that more surprises lie within each sequenced genome. To discover more genome surprises, let's turn our attention to one of the world's most lethal genomes: *Plasmodium*.

What Kind of Organism Causes Malaria?

Many diseases are in the news and they deserve attention, but malaria, which is rarely in the news, is a daily threat to nearly 3 billion people in the world who live in tropical and subtropical climates. In 2002, about 500 million people were infected with the malaria-causing eukaryotic parasite of the genus *Plasmodium*, and about 2.7 million people die of malaria each year. Tragically, 90% of all malaria-related deaths occur in children under the age of five, and nearly all of these are sub-Saharan Africans. To provide you with a scale of malaria's devastation, the terrible tsunami in 2004 killed about 300,000 people; malaria kills that many people every six weeks of every year, and the numbers are expected to increase as the world gets more crowded. The cause of malaria has been known for over 100 years, but we cannot control its spread. Can we utilize genomic information to develop new methods for treatment (medication) and prevention (vaccine)?

The most lethal form of malaria is caused by *Plasmodium falciparum*, which is transmitted to humans by the *Anopheles* mosquito. Several *Plasmodium* species have a range of host organisms, including many mammals, birds, and reptiles. The life cycle of *Plasmodium* is divided into two halves (Figure 2.22a). One half begins when an infected mosquito bites someone to draw blood. The parasites leave the mosquito's salivary glands during the process of drawing blood, and move to your liver to infect hepatocytes. In your liver, they mature and hatch out of hepatocytes to infect red blood cells (RBCs) 48 hours after you were bitten. RBC infection is the stage at which malaria infection can be diagnosed microscopically (Figure 2.22b), but *Plasmodium* continues to develop, progressing through a regulated series of biochemical and morphological stages. When new parasite cells are ready to emerge from RBCs, they burst the host cell that has been digested from the inside and release progeny and metabolic waste, which is the cause of the cyclical pattern of fever followed by chills. The RBC stage continues ad infinitum, but occasionally a

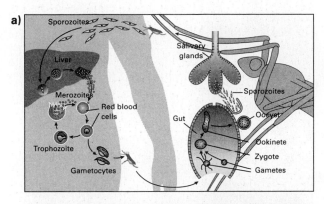

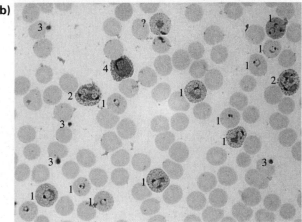

Figure 2.22 Life cycle of _Plasmodium_.

a) When an infected _Anopheles_ mosquito bites a human, the sporozoite form of _Plasmodium_ enters its new host. Sporozoites travel to the liver and mature to merozoites, which infect RBCs; pass through the trophozoite stage; and return to the merozoite stage to begin the cycle again. Some infected RBCs produce gametocytes that enter mosquitoes that ingest new blood; there the gametes fuse to form diploid zygotes that eventually emerge from the insect gut to produce sporozoites and begin the entire life cycle again. **b)** Many human RBCs are visible, some of which have been infected by _Plasmodium_. The stages are: 1 = trophozoite; 2 = schizont; 3 = merozoite; 4 = gametocyte.

few _Plasmodium_ cells differentiate into gametes that move through your blood and can be ingested by a new mosquito to begin the second half of the life cycle. In the mosquito gut, the gametes form zygotes and undergo meiosis to produce haploid cells that migrate back to the salivary glands and begin the full life cycle again.

The most vulnerable time for _Plasmodium_ is during the RBC infection stage when a parasite has to find and bind to one of your cells and then force its way inside your RBC without rupturing any plasma membranes (Figure 2.22a). A human RBC is about 6 μm in diameter and the parasite is about 1.2 μm, so it will be a snug fit. Three structures are of particular importance during the infection stage. First is the extracellular coating of the _Plasmodium_ cells. They have to be sticky to RBCs but invisible to the immune system. Second,

the pointed (apical) end of the infecting cell must be oriented toward the RBC, even though it does not form the first bonds with the host cell. Finally, _Plasmodium_ has a unique organelle called the apicoplast (Figure 2.23). This tiny organelle is the remnant of an internalized alga engulfed millions of years ago. Although the alga symbiont is reduced in size now, it retains a small genome and its function is absolutely essential for _Plasmodium_'s survival and growth. These three aspects of _Plasmodium_ anatomy are of particular importance, so we will revisit them later.

METHODS
pulse-field gel
STRUCTURES
selenocysteine

What Sort of Genome Does _Plasmodium_ Possess?

Plasmodium has three genomes: nuclear, mitochondrial, and apicoplastic, though the latter two are very small. To sequence the nuclear genome, a multinational team utilized a hybrid method of whole chromosome shotgun sequencing. Each of the 14 chromosomes was subjected to **pulse-field gel** electrophoresis to isolate the chromosomes before they were chopped into random fragments for cloning and sequencing. (Chromosomes 6–8 were inseparable by size, so they were treated collectively.)

The _Plasmodium_ nuclear genome contains at least 22,853,764 bp and about 5,268 ORFs. The mitochondrial genome is 5,967 bp and encodes 3 proteins; the apicoplast genome encodes 30 proteins with its 29,422 bp genome carried on two molecules. The nuclear genome is 19.4% GC, which makes it the most AT-rich genome sequenced so far. The coding capacity of the genome is 52.6% and the average gene is 2,283 bp long—longer than those in _B. thetaiotaomicron_, which were longer than most prokaryotic genes. Keep in mind _Plasmodium_ is a eukaryote and 54% of its genes contain one or more introns. The introns have an average of 13.5% GC and the intergenic regions are 13.6% GC, which means the exons have a higher GC content (23.7%) than the rest of the genome. This AT-rich genome proved difficult to finish, and 3 to 37 gaps remained in half of the chromosomes. Annotation of the ORFs was also challenging, but the investigators were able to utilize a substantial EST database to help identify the exons. Nevertheless, approximately 60% of all the predicted ORFs have no known function and no identifiable orthologs—twice the percentage of most genomes. Between the gaps and unknown ORFs, our ability to fully understand the _Plasmodium_ genome is limited, but let's see what we can deduce with the information we have.

Two easily identified genes in any species are the rRNA and tRNA genes. _Plasmodium_ lacks two possible tRNA genes, but they may be located in some of the gaps. Interestingly, the genome contains a tRNA that can carry **selenocysteine** (see Discovery Question 65) to ribosomes which makes _Plasmodium_ capable of using 21 amino acids instead of just 20. The mitochondrial genome does not

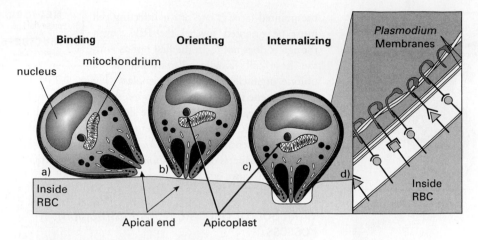

Figure 2.23 Invasion of RBC by *Plasmodium* merozoite.
The eukarotic *Plasmodium* has a nucleus, mitochondria, and a unique but vital organelle called an apicoplast. Infection takes place in three stages. **a)** The parasite binds loosely to the surface of the RBC. **b)** The merozoite orients itself to place the apical surface with specialized binding proteins directly adjacent to the RBC surface. **c)** *Plasmodium* pulls itself inside the RBC by myosin motors **d)** pulling on the merozoite proteins attached to the RBC surface proteins. Note that the internalized parasite does not cross the RBC plasma membrane; it surrounds itself with the host cell membrane.

encode any tRNAs, so it must utilize nuclear genes, but the apicoplast does encode sufficient tRNA genes to meet its own needs during translation. In many species, rRNA genes appear in linear clusters, but not so in *Plasmodium*. Not only does the distribution of the individual rRNA genes vary, but their expression patterns do, too. *Plasmodium*'s translational machinery is host specific—one set of rRNA genes is activated in the human host and the other set is activated in the mosquito host.

The unusual nature of the malarial genome extends to the overall structure of the chromosomes as well. The central portions of each chromosome are similar in length, but the regions near the telomeres vary enormously with several different repeated sequence patterns. The centromeres were fairly easy to identify because they are extremely AT-rich (>97%) and contain short tandem repeats. Telomeres from different chromosomes share these repeated sequences but vary in the number of repeats, which appears to have stimulated chromosomal recombination outside meiosis. The consequence of all the recombination is that some genes located near telomeres are replicated many times, and thus the genome has many paralogs of certain genes. In particular, the surface genes that are responsible both for binding to RBCs and for avoiding the immune system are located in the subtelomere region, which could explain the enormous diversity in these genes (gene families called *var*, *rif*, and *stevor*). All three gene families are polymorphic (59 *var* paralogs, 149 *rif*, and 28 *stevor*) and include both silent pseudogenes and truncated genes that may or may not add to the variation of *Plasmodium*'s extracellular surface. Because so many different proteins can be produced on the

surface, our immune systems cannot learn to recognize the parasite and thus destroy it.

Is the Predicted Proteome Equally Bizarre?

Based on hydropathy plots and other annotation methods, 31% of the encoded polypeptides are predicted to be integral membrane proteins. One percent of the annotated proteins are used for cell–cell adhesion and host cell invasion; another 4% help evade the immune response. These proportions do not sound out of line for an intracellular, obligate parasite. Many odd proteins are found in the apicoplast, a derived chloroplast that is only partially understood. This essential organelle synthesizes fatty acids, isoprenoids, and heme groups. Through a series of elegant experiments, we are learning how to determine by sequence analysis which nuclear-encoded proteins are imported into the apicoplast (Figure 2.24). About 10% of all proteins are destined to target the apicoplast to assist in DNA replication and repair (of the apicoplast genome), transcription, translation, posttranslational glycosylation, protein import, and protein degradation. Only two photosynthetic orthologs remain now that *Plasmodium* lives in relative darkness, and these two proteins probably serve to recycle molecules similar to NADP$^+$. Although a majority of the apicoplast

Figure 2.24 Protein targeting to the apicoplast.
Go to www.GeneticsPlace.com to view this figure.

proteins have unknown functions, many metabolic pathways not found in humans are discernable, which makes them appealing targets for new antimalarial medications.

The metabolic potential of *Plasmodium* is equally mysterious, but we can detect some familiar eukaryotic pathways. About 14% of the ORFs were annotated as enzymes with recognizable GO categories and Enzyme Commission (EC) numbers. EC numbers, a numbering system similar to zip codes, have been established for every known enzyme; enzymes with similar functions get similar numbers. The percentage of the proteome assigned to enzyme GO categories is about 4-fold fewer than is typical for proteomes, so either *Plasmodium* has fewer enzymes (because it is a parasite and relies heavily on its host), or we cannot tell which of the 60% unknown proteins are enzymes. This gap in our understanding is a serious problem for designing drugs to combat the parasite.

In perhaps one of the most ironic twists of fate, the malaria blood parasite utilizes mostly anaerobic glycolysis for energy production. You might have predicted that a life inside RBCs would lend itself to aerobic energy production, but the parasite consumes hemoglobin as its main food source, which should deplete the RBC's oxygen content. The investigators were unable to detect any energy storage pathways for the production of trehalose (storage sugar found in yeast), glycogen (storage sugar found in many animals), or carbohydrates such as starch (used by plants). These metabolic deductions indicate the parasite lives for the moment by extracting little of the potential energy in its food, converting its waste to lactic acid, and storing nothing for lean times. Although a complete citric acid cycle is present, it appears to function mainly to produce the cofactor succinyl-CoA for heme biosynthesis. Another irony is that *Plasmodium* produces its own heme for use by proteins in the mitochondria and apicoplast rather than using human heme from hemoglobin. Although the cells contain mitochondria, their role is not ATP production—at least not in the human half of the life cycle, since the genome lacks crucial components of the ATP synthase.

Being a parasite that digests proteins for food, the pathogen recycles the available amino acids rather than resynthesizing them. No amino-acid-synthesizing genes remain in its genome. Purines for nucleic acids are also recycled, as no synthase genes are discernable, but pyrimidines can be synthesized de novo. Investigators identified genes that enable *Plasmodium* to import sugars, water, and carboxylate while inside RBCs, but not amino acids, which emphasizes the importance of hemoglobin digestion inside the parasitic food vacuole. Although *Plasmodium* has many genes for ATP-ase pumps to transport particular solutes, the RBC parasite also produces a mysterious transporter that can pump a wide range of molecules but the annotators were unable to identify candidate genes as this mysterious, multifunctional transporter. The mystery transporter is another example of a metabolic capacity detected through biochemical analysis, but hidden from sequence annotation by the perplexing nature of the genome.

As you learned in Section 2.1, investigators must justify spending money on a sequencing project. The *Plasmodium* sequencing consortium took 6 years and 45 people in Australia, the United Kingdom, and the United States to determine the genome sequence. Their sincere belief was that with the complete genomes of *Plasmodium*, *Anopheles*, and *Homo sapiens*, we would be able to design effective new antimalarial medications and/or vaccines. While the *P. falciparum* genome was being sequenced, another group sequenced the *P. yoelii yoelii* malaria genome, which infects rodents and is a model system for biomedical research. The two *Plasmodium* genomes are about the same size but *P. y. yoelii* has about 600 additional ORFs. Unfortunately, the *P. y. yoelii* genome was never finished; 5,812 gaps exist in *P. y. yoelii* compared to 93 gaps in *P. falciparum*. Interestingly, the average gene size in *P. y. yoelii* is about half that in *P. falciparum*, but this could be an artifact of poor assembly of the WGS reads with only 5-fold coverage for *P. y. yoelii*, compared to *P. falciparum*'s 14.5-fold coverage. You might expect the two pathogens to have many genes in common, and they do, but only 3,310, which represents 56% of *P. y. yoelii*'s capacity. This raises an uncomfortable question: How good a model system is *P. y. yoelii*, given that it is only 50% similar and half of its life cycle is spent in a rodent?

A few candidate drugs are in clinical trials, but all of these were produced using traditional methods of research and development before the genome was sequenced. Vaccine development against a surface protein that changes frequently seems a distant dream. The best vaccines to date have been irradiated live cells that exist in human cells but fail to produce new cells to perpetuate the life cycle. Injecting viable parasites poses obvious risks, not to mention logistical problems of mass-producing cells that must be grown inside human cells yet are sterile. However, a German and American team genetically modified *Plasmodium* by knocking out a particular gene (*Uis3*) to attenuate the pathogen. They used **reverse genetics** (beginning with a gene sequence and deducing its function) to produce a mutant strain of *Plasmodium* that protects rodents injected with this liver-seeking, attenuated strain. The mice were protected for at least 30 days, which is better than the effectiveness of traditional protein subunit vaccines. The technical issues of mass production and safety involved in producing a human vaccine must still be addressed, but this proof-of-concept discovery gives some investigators hope.

Nevertheless, some people who study malaria are skeptical of a genomic solution. The AIDS epidemic mirrors one of their concerns, in that drugs are available, but funding and public health care infrastructure are lacking in countries with the greatest disease threat. The amount of money being spent on malaria research is very small compared to the amount needed. The Bill and Melinda Gates Foundation

LINKS
Enzyme
Commission
Bill and Melinda Gates

LINKS
KEGG Pathway
PlasmoDB

is one of the world's leading funders of malaria research, but even its millions cannot meet the current and future needs. Furthermore, drug companies do not see a profit motive in malaria pharmaceuticals, since most patients could not pay much for their medication. Of the 1,223 new drugs registered between 1975 and 1996, only 3 were antimalarials. Because of the bioterrorism threat, billions of dollars are being funneled into a field that has not seen many deaths (e.g., anthrax), while the 2.5 million who die each year are not provided with comparable funding.

The *Plasmodium* sequencing consortium worked very hard to post its most recent sequences online, in the hope that investigators around the world, including those where malaria is endemic, would be able to use the free information immediately. However, the research facilities in most sub-Saharan countries are unable to sustain global research. Only South Africa has the resources necessary to launch a bioinformatics institute, but many African investigators have felt excluded from true collaboration with scientists in the developed countries. For example, no African scientists were included in the annotation phase of the genome projects.

The current debate is between helping now with proven technology or investing in a hope that high-technology research will lead to a vaccine. The World Heath Organization knows that the most effective prevention of malaria is insecticide-treated mosquito netting for sleeping under at night. These nets cost only a few pennies and can be produced locally to help the economy as well. Would it be better to spend the same amount of money to protect people today and maybe even break the life cycle of *P. falciparum* by preventing new human infections? Obviously, there is no easy answer to this ethical dilemma, and people will continue to die either way. Because the genomes of both hosts and several species of *Plasmodium* are available, the task at hand is to expedite the translation of basic research to clinical application. Unfortunately, the published genome of *P. falciparum* has not helped us prevent malaria or improve treatment of the 500 million people infected each year.

DISCOVERY QUESTIONS

65. Perform a PubMed search for the term "selenocysteine" and find out what this is. Does it matter functionally whether a protein incorporates a cysteine or a selenocysteine?

66. Search Google for "isoprenoid Wiki" and select the link for Wikipedia to read what isoprenoids are. Explain why loss of the apicoplast would be lethal given it's the source of isoprenoids.

67. Go to the KEGG Pathway web site and click on the "ATP synthesis" link to see a model of ATP synthase. Change the pull-down menu from "Reference pathway" to "Plasmodium falciparum" located near the top of this long list of species, then click on "Go." Does *Plasmodium* have all the parts necessary to synthesize ATP from an H^+-ion gradient? Explain your answer.

68. Return to the KEGG Pathway, choose "glycolysis," and see which enzymes *Plasmodium* has. Follow the pathway from b-D-Glucose to pyruvate and see if any steps are missing. Compare *Plasmodium* to *Saccharomyces cerevisiae* (baker's yeast) to see which has the more robust metabolic capacity. Finally, look at Aminoacyl-tRNA biosynthesis on the list of maps and see if *Plasmodium* is missing any enzymes needed to synthesize tRNAs coupled with their amino acids (aminoacyl-tRNA). Would you predict that a parasite might depend on the host for any of these enzymes? Explain your prediction and then test it by searching the database.

69. Go to PlasmoDB and view bases 400,000–450,000 on chromosome 1. Below the busy graphics, choose to hide all options except "%AT," "BLASTx," "Genefinder," and "Pf Annotation," all of which should be set to "show one line." Hit the "Update" button. Do the BLASTx (DNA query against protein database) hits align with the annotated genes? Did the predictive software Genefinder identify every exon correctly?

Now go back to PlasmoDB and choose to view the mitochondrial genome at the bottom of the page. Change the display so that all RNAs and genes are displayed as "show—expanded." How many genes and how many RNAs are encoded on this organellar genome? Explain why the two numbers are different.

Is There a Model Eukaryote Genome?

You might be surprised to know that of all eukaryotes, baker's yeast, *Saccharomyces cerevisiae*, is the darling of the genomics world (Figure 2.25). *S. cerevisiae* is a unicellular fungus that has been used to understand many aspects of eukaryotes, including cytoskeleton, cell signaling, cell cycle, mitosis, transcription, and, of course, fermentation. This species can be grown as either a diploid or a haploid and has the capacity to reproduce sexually or asexually with very fast generation times. Plus, when you grow a plate of *S. cerevisiae*, it smells much better than a plate of *E. coli*!

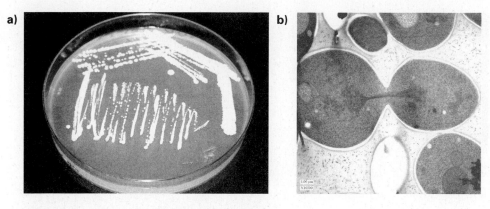

Figure 2.25 Baker's yeast *Saccharomyces cerevisiae*.
a) Yeast is easily grown in the lab by streaking it onto nutrient-containing agar plates; single colonies are easily isolated. The strain in this photo is S288C, which is the same strain used in the genome sequencing project. **b)** Yeast cells grow rapidly through mitosis, as shown in this transmission electron micrograph.

As a model organism, it makes sense that *S. cerevisiae* was the first eukaryote genome to be sequenced. In fact, when the genome sequence was published in October 1996, only *H. influenzae* and *M. genitalium* had been sequenced and made publicly available. To tackle the first large, multichromosomal genome, over 600 scientists in Europe, North America, and Japan worked in a highly coordinated, cooperative effort to sequence the 12,068 kb genome composed of 16 chromosomes. At the time, the investigators annotated 6,275 ORFs, though they also predicted some of these ORFs were "dubious," meaning the ORFs probably did not encode any protein. We will survey the yeast genome as it was annotated in 1996, and later update the information using information available in 2005. The retrospective nature of this case study allows us to see what changes will probably happen over time for all genome sequences, including human. The yeast consortium also made some predictions about the future of genome sequencing, which we will examine to see how well they were able to foresee the future of a new field that changes very rapidly.

Are Eukaryote Genomes Similar to Prokaryotes?

So far, we have seen several Eubacteria genomes and *Plasmodium*'s bizarre eukaryote genome, but we really don't know how eukaryotes compare to prokaryotes. The genome of *S. cerevisiae* is 38.3% GC and the coding capacity is 70.3%. Prokaryotes have wide ranges of GC content, so this statistic is hard to compare overall, but we can say that the GC content for eukaryote coding regions tends to be higher than for noncoding portions of the genome. High GC regions also correlate with regions of increased recombination, similar to the high degree of recombination found

in bacterial tRNA and rRNA genes. Yeast telomeres and centromeres have higher AT content and less recombination, which is different from the telomeres of *Plasmodium*. The coding capacity is much lower than in bacteria, which tend to have very little intergenic DNA. In yeast chromosomes, we find a gene every 2 kb, which is a larger spacing than in bacteria but even in 1996 we knew that eukaryotes had less dense genomes. The genome of *C. elegans* was estimated to contain a gene every 6 kb, and humans every 30 kb or more. Therefore, the first hallmark of a eukaryote is more intergenic (junk) DNA. The term **junk DNA** was coined before genome sequencing became feasible and says more about our ignorance of genome structure than about the DNA's lack of function or content (see Section 2.3).

Another major eukaryote hallmark is gene structure, with enhancers, promoters, and introns adding substantially to the size of a eukaryote gene compared to either Eubacteria or Archaea genes. Only 4% of *S. cerevisiae* protein-coding genes contain short introns, which are located near the 5' end of the gene. The category of genes with the highest proportion of introns is rRNA genes. The four smallest chromosomes (I, III, VI, and IX; Roman numerals are used for yeast chromosomes) have the highest recombination frequencies and some unusual architecture (Figure 2.26). From laboratory work, we knew yeast artificial chromosomes smaller than 150 kb did not fare well during mitosis which led to a hypothesis that shorter chromosomes may have accumulated some DNA to ensure their stability. Furthermore, some investigators believe that small chromosomes preferentially recombine to protect the larger chromosomes from the potential disruption of recombination and still promote at least one "safe" recombination for each meiosis event. The telomeres of

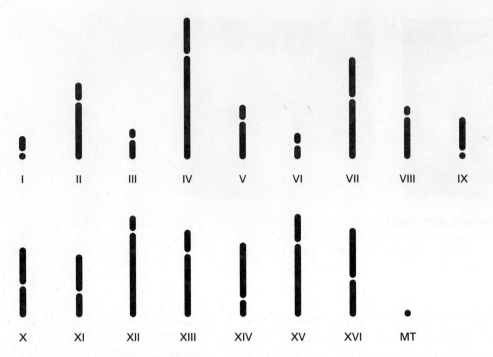

Figure 2.26 *S. cerevisiae* ideogram.
Drawings of the 16 yeast chromosomes numbered with Roman numerals (I—XVI) plus the mitochondrial chromosome (MT). Centromeres are represented by small white gaps in the black chromosomes.

chromosome I have 31 kb of gene-poor DNA with recognizable sequence duplications on the two ends. The telomeres of chromosomes III, V, and XI appear to be intrachromosomal duplicates of each other (paralogs). Furthermore, paralogs of the sugar-fermenting genes *Mal, Suc,* and *Mel* are located in the subtelomeric region of several chromosomes, though many of these appear to be transcriptionally silent. All of these data make it sound as though telomeres are indistinguishable and the right and left sides of a chromosome are identifiable only by the length of DNA from the centromere. However, many of the *Mel* paralogs are found only on one telomere of a given chromosome, which may indicate each chromosome can tell its left from its right better than we can.

What Do All These Genes Do in Yeast? Much of what we learn from studying a genome is based on deductions about the proteome, because much of what a cell does is based on the work performed by proteins (Table 2.3). At the time *S. cerevisiae* was annotated, Gene Ontology (GO) did not exist, and so the investigators were unable to utilized GO's **controlled vocabulary** of functional categories. Nevertheless, based on BLAST sequence similarities, they were able to determine the cellular roles for many of the 5,885 ORFs that were predicted to encode proteins. The sequence conservation between humans and yeast is a reminder of why

S. cerevisiae is a good model organism: orthologs for half of all human disease-causing genes are present in the yeast genome. Knowing GO categories is only part of the goal. We'd like to know which proteins are present at what times, with which proteins each interacts, and the 3D shape of each protein. None of this higher-order information can be deduced from DNA sequence (as of 2005), and the investigators acknowledged that accumulating a gene "parts list" is only the beginning of a full comprehension of yeast biology.

What Is the Evolutionary History of *S. cerevisiae*? Based on a whole-genome sequence analysis, and to explain the

Table 2.3 Percentage of proteome dedicated to major processes.

Metabolism	11%
Transcription	7%
Intracellular trafficking or protein targeting	7%
Translation	6%
Cytoskeleton and other structural roles	4%
Transport proteins	4%
DNA replication, repair, or recombination	3%
Transcription factors	3%
Energy production and storage	3%

substantial number of genes found in duplicate copies (apparent paralogs), the investigators proposed that *S. cerevisiae* had experienced "duplication events." These duplication events were easiest to detect near the centromeres. The investigators found many syntenic regions within the yeast genome, which again indicated large segments of the genome had been duplicated at some point in the species' history. Three examples are paralogous regions of chromosomes V and X, IV and II, and III and XIV. The duplicated region of chromosome III contains four genes that are syntenic on chromosome XIV. One of these genes encodes citrate synthase: the protein encoded by *Cit2* (chromosome III) targets the peroxisome and *Cit1* (chromosome XIV) targets the mitochondrion. *Cit1* and *2* are good examples of what at first appears to be wasteful redundancy of two genes encoding the same metabolic function; however, wet-lab research revealed the two proteins reside in different organelles and their products do not intermingle. Paralogs have been seen as evolutionary experiments where one gene can drift and provide new specialized functional capacity, while the other version remains more consistent with the initial gene's cellular role. Through genome annotation, a third gene was discovered (*Cit3* on chromosome XVI), but its particular cellular role was not known in 1996. Although we do not understand the value of having three *Cit* genes, we do know that yeast evolved under changing environmental conditions, so presumably its genome is well adapted. The challenge now is to design experiments that will help us understand its genome better.

We often treat our first experiences as special and partly for this reason, the *S. cerevisiae* genome will always be special. However, scientists place a lot of value on utility, and for this more practical reason, the stature of yeast as a model organism has increased. We will be encountering yeast biology in many more chapters (Chapters 3, 6, 7, 8, 11, and 12). Many freely available online resources provide a rich educational and research environment that we will explore throughout this book. The epilog section provides a retrospective evaluation of the 1996 yeast genome publication to see how well the original annotation has withstood the test of time, and to see if those investigators were able to foresee the future.

What Did the Investigators Predict for the Future of Genomics?

Because the 1996 paper was a landmark publication, the authors were given some leeway to discuss what they saw as the future of genomics. First, they commented on the impact on yeast biologists. They described plans to delete every gene, one at a time, to produce a collection of knockout mutants and thereby elucidate the function of every gene. However, recognizing the apparent redundancy as exemplified by *Cit1–3*, they proposed that double-deletion mutants—all 18 million of them—might have to be made, not to mention possible triple-deletion mutants.

They addressed the value of making all genome sequence data publicly available. They felt WGS sequencing of genomes over 6 Mb was not feasible and argued that sequencers would need to use the HGS method for eukaryote genomes. They were excited to compare the *S. cerevisiae* genome with other species, including the very distant relative *S. pombe* (brewer's yeast). They also were anxiously awaiting the human genome sequence projected to be ready in 2005 (a draft sequence was published in 2001 and the finished version in 2004). They urged the genome community to focus sequencing efforts on disease-causing microbes such as *Plasmodium*, *Trypanosoma*, and *Leishmania*.

Finally, they commented on their own sequencing consortium, under a heading of "Pride and Productivity." Over half of the genome was sequenced by a large network of 92 different labs working as a cottage industry, while 45% was sequenced at larger, more industrialized sequencing centers. When the quality of data from the two types of sequencing facilities was compared, only 1 to 2 differences per 100 kb of DNA were detected in a 170 kb section of DNA sequenced by both types of facilities. In the finished genome sequence, they estimated an overall error rate of 3 bp per 10 kb of DNA. In the end, they were pleased with the *esprit de corps* established by the multinational effort on their shared sequencing goal. They lamented that future sequencing efforts would probably become the domain of a few large centers and that smaller labs would be called upon for the annotation phase only.

Epilog for Yeast Genome

A lot has changed in the genomic world since 1996. First, many more species have been sequenced, including many yeasts (Table 2.4). From these data, you can see in some ways *S. cerevisiae* is very similar to other yeasts (size and %GC), but in other ways it is different (e.g., coding capacity). But Table 2.4 misses the truly interesting story of evolution (Figure 2.27). The *S. cerevisiae* investigators correctly recognized that the genome was the product of a massive (perhaps a total) duplication, which explains the syntenic paralogs throughout the genome.

With the new genomes came a need to unify the language relating to the cellular roles which produced the multispecies consortium GO. GO permitted a new level of annotation using the controlled vocabulary which produced a better way to compare orthologs and perform computerized database searches. With improvements in computers, search algorithms, and the increased volume of genes in databases, came an increased ability to annotate genomes. Using these new tools and the growing number of genomes for comparison, the yeast community has increased the number of ORFs from 5,885 to 6,672. Many of the new ORFs are small ones, below the 100-codon cutoff used

Table 2.4 Comparison of six yeast species.

Species	Genome Size (Mb)	Average % GC	Codon % GC	Coding Capacity (%)
S. cerevisiae	12.1	38.3	39.6	70.3
S. pombe	13.8	36.0	39.6	57.5
C. glabrata	12.3	38.8	41.0	65.0
K. lactis	10.6	38.7	40.1	71.6
D. hansenii	12.2	36.3	37.5	79.2
Y. lipolytica	20.5	49.0	52.9	46.3

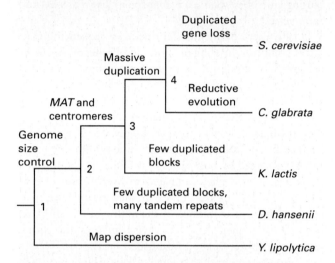

Figure 2.27 Genome evolution in yeast with sequenced genomes.

Schematic diagram of the major genome changes. Map dispersion (branch point 1) indicates that the size of the genome doubled—not in large blocks, but by many small duplication events. Branch 2 separated a genome that produced mostly small blocks of repeated DNA located near the site of origin, while the other three species gained extra copies of the MAT locus that permitted divergent sexual reproduction. Branch 3 separated two species with near whole-genome duplication from *K. lactis*, which is more ancestral than the other two. Branch 4 separated one genome that lost both copies of many paralogs (*C. glabrata*); *S. cerevisiae* often lost only one of the two paralogs.

during annotation of early sequencing projects. With continued bench research on yeast, including ESTs and microarrays, many small genes have been identified and added to the official yeast genome. Some of the new ORFs overlapped with other, larger ORFs and by prearranged rules, smaller genes were not annotated in the original sequencing project.

In addition to finding previously hidden genes, investigators have found 88 sequencing errors. The two most common errors were indels that created frameshift mutations, or base changes that altered stop codons and thus led to incorrectly predicted protein lengths. All of these

changes have been added to the databases, so you will not find the old versions even if you look for them. By their very nature, genome sequence databases are dynamic and frequently updated which means an unknown gene may be annotated and categorized the next time you query the database. Like living genomes, our understanding of a genome sequence changes and evolves over time.

The 1996 authors' observations and predictions about the sequencing process were correct for the most part. Most sequencing projects are being conducted by large, highly efficient sequencing centers. Few genomes are finished and annotation quality is limited by the fold coverage, the number of contigs, and the distribution of gaps in the assembly. However, the authors incorrectly predicted that WGS sequencing could not be performed on genomes longer than 6 Mb. The Human Genome Project adopted a hybrid strategy and the fly genome used the WGS method almost exclusively (see Section 2.3). Another miscalculation was that there would not be enough sequencing capacity to permit non-disease-causing genomes to be sequenced. In fact, the number of genomes being sequenced is growing rapidly. Today, many papers do not present a single genome but two or more because the novelty is wearing off and the interesting biology often comes with comparison, not annotation and analysis of one genome alone.

One final point about the future of sequencing genomes: Who gets scientific credit and who should be doing the work? The 16 authors of the *S. cerevisiae* paper noted that more than 600 scientists (2.5% authorship yield) contributed to the effort, but who were they? Why should a young scientist become involved in a project if they are not given credit in the publication? Could anyone build a research career within a sequencing center? How far down the hierarchy of administration should authorship be shared? If you visit a sequencing center and talk to the people about their work, it sounds a lot like a factory where quality controls, cost per base, and throughput rates are discussed more than biology. Perhaps sequencing centers are ideal places for people who do not want the headaches and pressures of running their own labs, but do want to contribute to the field in a meaningful way within an academic setting. The field of genomics is still young and sequencing methodology is advancing. Some leaders in the field think there will be a quantum leap in sequencing capacity, to the point where you could have your own personal genome sequenced for $1,000. The only certainty at this point is that by 2010, predictions made in 2005 will seem outdated and naïve because of advances in methods and our understanding.

DISCOVERY QUESTIONS

70. Perform a search at the *Saccharomyces* Genome Database (SGD) web site and perform a Quick Search for maltose-metabolizing genes *Mal1, Mal2, Mal3, Mal4, Mal6, Mal10, Mal12,* and *Mal13*. Determine which of these genes are true paralogs or phenotypes with uncertain genomic information. Which gene or genes have the most detailed information?

71. At the bottom of the *Mal13* page, click on the "MIPS" (German genomics database) link to see a different source; click "Protein Info" to see details about Mal13p (p for protein). Are the two databases identical in content, or do they present different information?

72. Go to SGD's Advanced Search to get an up-to-date count on the number of ORFs, ncRNA (nonprotein-coding), pseudogenes, rRNAs, and tRNAs. The search takes a couple of minutes.

73. Compare the yeast gene *Sir2* in all the Model Organisms and determine if this gene is widely conserved. Compare this result with *Mfa1*. Why might you get such different results with two genes?

74. Go to the yeast metabolism map; click on the citric acid cycle, then the "More detail" button until you cannot zoom in any more. Move around the circle until you see the two isocitrate dehydrogenase genes (*Idh1* and *Idh2*). Mouse over the enzymes to see the chemical reaction, then click on the EC number 1.1.1.41. On the resulting page, go down and click on *Idh1* and *Idh2* to find their chromosomal locations. Are these redundant genes located next to each other in the genome?

Summary 2.2
Unicellular Genome Analysis

Section 2.2 used world-class databases to explore the annotation process, discover some unexpected findings, answer long-standing questions, and clarify some biomedical issues that could lead to new disease treatment methods. In Section 2.3, genome sequence analysis will teach us more about plants, other animals, and ourselves.

LINKS
Advanced Search
metabolism map
Model Organisms
Morgan
Nobel Prize
SGD
METHODS
in situ

2.3 What Have We Learned from Metazoan Genomes?
Are Animal Genomes Harder to Finish?

If you were to choose only one non-human animal to sequence, you might select the fruit fly *Drosophila melanogaster*, since it has been studied intensely since T. H. Morgan at Columbia University established the world's first fly lab. The first mutant flies (with white eyes) were isolated in 1910; within three years, Morgan's undergraduate student Alfred Sturtevant produced the first genetic map, placing six genes in a row along the X chromosome. By 1915, these fly geneticists proposed the chromosome theory of inheritance for which Morgan won a Nobel Prize in 1933. A pair of biologists discovered the super-sized polytene chromosomes and within five years, Calvin Bridges published stunning polytene maps for *D. melanogaster* that were so accurate and detailed that they are still used today. Several more Nobel Prizes went to subsequent generations of fly biologists who helped develop the field of molecular biology and ushered in genomics. Using methods developed 100 years ago, fly labs all over the world used phenotypes and genetic crosses to characterize over 2,500 genes. In part because of their genetic successes, many people felt sequencing the fly genome was unnecessary—but not anymore. With most of the genome sequenced and genes identified, fly labs have improved their research and the utility of their favorite model system. Although the animal is diminutive, accomplishing the large fly sequencing project required an unprecedented level of cooperation and resources.

WHAT ARE POLYTENE CHROMOSOMES?

One of your earliest exposures to chromosomes and DNA probably involved a picture of polytene chromosomes—the thick, multicopied chromosome found in the salivary glands of flies. No one knows *why* flies make polytene chromosomes in their salivary glands, but now we have a better understanding of *how* they make them. One of the three genomic libraries made for the sequencing project was a series of overlapping BAC clones (Figure 2.B3). Using these BAC clones as probes for in situ hybridization of the

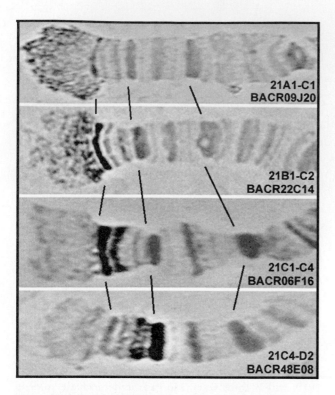

Figure 2.B3 Sequential BACs span the chromosomes.
Part of chromosome 2L is shown hybridized and stained with 4 BAC clones beginning at the telomere (band 21A1) to band 21D2. The BACs had an average insert size of 163 kb and their chromosomal locations were verified by this type of *in situ* hybridization.

Table 2.B1 Polytene bands and estimated sizes not covered by BACs represent gaps in the genome sequence.

Unrepresented euchromatin	Size (kb)
40B4	30
40C3-F7	530
41A1-B3	215
41C3-C7	220
57B4	130
64C5	45
80B3	95
80D3-F9	460
81F1-F6	290
100F5	65
Total	2,080

polytene chromosomes, the investigators estimated the size of each band with an average length of 26.2 kb per band. Using these BACs, some gaps in the assembled genome could be mapped to the polytene chromosome (Table 2.B1).

The X chromosome exhibits a particularly odd banding pattern in section 2B3-8 (Figure 2.B4). Consistent with the original 1938 observations, a part of the chromosome is puffed out and appears to form a V rather than a vertical line. When a particular segment of DNA

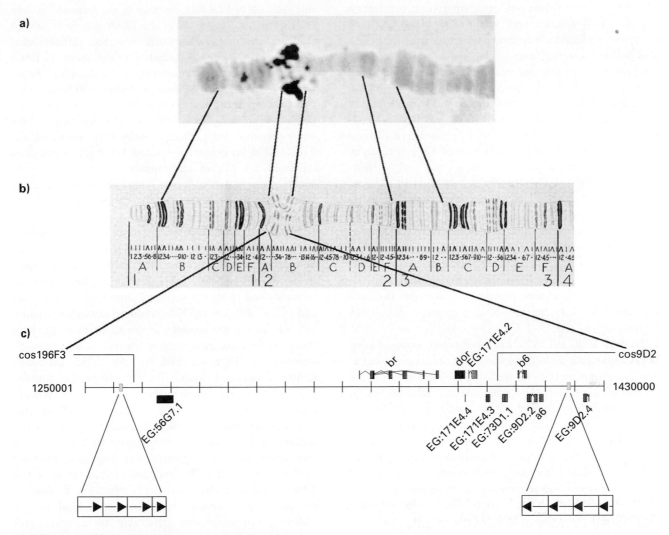

Figure 2.B4 Genomic resolution of chromosomal puff.
Polytene region **a)** 2B puffs out and has horizontal internal bands. **b)** A 40 kb subcloned piece of DNA binds to the peripheral portions of the puff in panel a. **c)** DNA map of 180 kb covers region 2B and reveals paired tandem repeats of 3.5 copies (arrows inside boxes) that flank the puff and presumably contribute to its appearance.

was used to probe this part of the chromosome, the probe labeled horizontally rather than vertically, thus confirming the 1938 drawings. When the probe was sequenced, a series of 3.5 inverted repeats were found to flank the 154 kb region (each repeat was 350 bp and the series totaled 1.2 kb). No other repeated sequences were located in a 1.4 Mb region centered on the puffed-out region of 2B. A small segment (1.2 kb) of repetitive DNA can drastically alter the chromatin structure (154 kb), even when separated by a length of DNA 100 times longer than the repetitive DNA. What remains a mystery is how two tiny inverted repeats can so dramatically alter the chromosome's macro structure. Because four genes are located in the puffed region, it is tempting to speculate that the chromatin structure of the lateral band is functionally significant and controlled at least in part by the two small inverted repeats. If the observations of 1938 had not been so precise and odd, no one would have noticed the tiny repeats that influence transcription or other chromosomal functions.

LINKS
BDGP
EDGP

When you compare a yeast cell (which appears very small even when examined under a microscope) and a fruit fly (which you can see hovering over your bananas from across the kitchen), it seems obvious that the animal's genome would be bigger and probably more complex. Flies probably eat yeast cells while feeding on overripe fruit, so your intuition tells you that the fly, and all animals, should have genomes that dwarf simpler eukaryotes. The fly genome is carried on five types of chromosomes: two copies of #2, two copies of #3, two copies of #4, plus an X and a Y (for males). However, about 80% of the genome is contained in chromosomes 2 and 3, with 4 and Y being very small. The entire DNA content is 180 Mb long, but 60 Mb is a gene-poor, unclonable, sequencing quagmire of repetitive DNA called **heterochromatin**. Gene-rich **euchromatin** constitutes the remaining 120 Mb (Figure 2.28). The investigators working on the public *D. melanogaster* genome project were prepared for a long and slow process, as they began the same way those working on the human genome did—HGS sequencing.

Between 1996 and 1998, a series of papers proposed using WGS sequencing for larger genomes. The idea culminated in a private company (Celera) declaring it would sequence the human genome, with the fly genome as a pilot project. In perhaps the best example of private company/public funding cooperation in the history of genome sequencing, Celera joined forces with the Berkeley *Drosophila* Genome Project (BDGP) and the European *Drosophila* Genome Project (EDGP). Building on the initial efforts of EDGP and BDGP, the new WGS sequencing began in earnest in May of 1999; by March 2000, a coordinated series of papers announced the completion of the fly genome, in record time. The sequencing teams produced 3,156,000 sequencing reactions of about 500 bp each (with a quality control level of at least 98% accuracy) to produce a total length of 1.76 Gb

that resulted in 12.8-fold coverage of the genome. To help span some short segments of the DNA that were hard to sequence, the shotgun approach used three different-sized cloned inserts: 2 kb, 10 kb, and very large inserts in BACs (Bacterial Artificial Chromosomes) and YACs (Yeast Artificial Chromosomes). Each insert had about 500 bp from the two ends sequenced and because the inserts were produced by random shearing of the DNA, the inserts represented the entire genome. When all the contigs were assembled, the 116,117,226 bp genome contained 1,299 gaps totaling an estimated 2.1 Mb of euchromatin (Figure 2.29). Some euchromatin (3.8 Mb) never joined the contigs and the investigators proposed this DNA to be small islands nestled within large heterchromatin oceans.

The fly genomicists were eager to validate their sequencing method and finished genome sequence, so they invited 40 experts from 20 institutions in 5 countries to participate in a 2-week-long annotation jamboree. The mixture of biologists and computer scientists annotated 13,601 genes and 14,113 different mRNAs produced through alternative splicing. To test the accuracy of the annotations, the jamboree participants performed a few in silico (computer) experiments. They searched for the 2,783 previously known genes and found 99.8% within the finished genome. Annotating more mRNAs than genes was possible because of the extensive bench research and EST sequencing that preceded the WGS sequencing. Alternative splicing adds considerable complexity to a genome, so once again the EST sequences were helpful in resolving exon structures. If each gene can produce at least two different mRNAs, the proteome is effectively doubled in complexity and capacity, in a genome of fixed size. In addition, the annotators were able to compare each gene with those from other eukaryotes (Figure 2.30) to integrate many layers of information in diagrams of each chromosome.

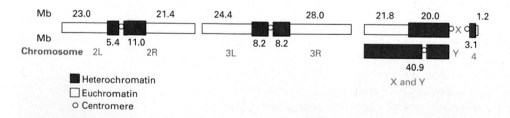

Figure 2.28 Fly euchromatin and heterochromatin distribution.
Euchromatin (open boxes) is the gene-rich region of a chromosome that can be transcribed; heterochromatin (filled boxes) is condensed, gene-poor, and seldom transcribed. Large chromosomes 2 and 3 have arms labeled L (left) and R (right). Chromosome X has only one arm, with a centromere on its far right end. Chromosome 4 is very small and has its centromere on the far left side. The large Y chromosome is nearly 100% heterochromatin.

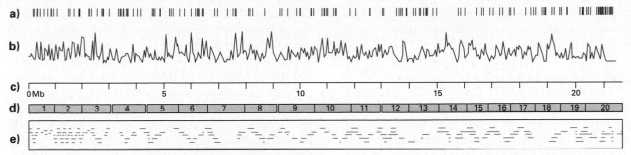

Figure 2.29 Features of X chromosome's sequenced euchromatin.
Major features are indicated: **a)** transposons; **b)** gene density; **c)** length of DNA in Mb; **d)** polytene chromosome bands; **e)** tiling path of individual clones used to sequence the chromosome. All chromosomes can be viewed online.

Figure 2.30 Coding content of *Drosophila* chromosome 4.
Go to www.GeneticsPlace.com to view this figure.

What Makes a Fly Different from Other Eukaryotes?

Fly chromosomes have some features not seen in yeast. First, the telomeres lack the simple repeats found in yeast or vertebrates, and flies lack orthologs to the vertebrate telomerase (the enzyme that polymerizes and maintains telomeres). At the junctions between euchromatin and heterochromatin (see Figure 2.28), the density of transposable elements increases significantly (see line A of Figure 2.29). The average coding capacity in heterochromatin is significantly lower (1 gene per 120 kb) than in euchromatin (1 gene per 9 kb). Compared to the worm *C. elegans*, *D. melanogaster* has about half as many gene duplications (paralogs) that are evenly distributed on the chromosomes.

The authors of the fly genome papers compared the genome sequences of four species (Table 2.5). This simple comparison allowed a number of striking conclusions. First, the size of the organism is not correlated to the size of its genome or the number of genes. Second, the smaller worm has 35% more genes but 62% more paralogs than the fly. As for unique gene families, the worm has only 17% more than the fly. Thirty percent of all fly genes have orthologs in the worm, and 20% of the fly genes are also found in worms and yeast, which probably defines the core eukaryote genes. Interestingly, half of the fly genes have mammalian orthologs while only one-third of worm genes have human orthologs.

You probably have noticed that fruit flies can find a ripe banana as soon as you buy it, and yet flies have only 57 olfactory receptor genes. Comparatively, flies are odor-restricted; fish have about 100 odor receptor genes and mice and worms have about 1,000. Fly odor receptor genes are distributed as clusters of one to three genes instead of the large array of paralogs in the other animals. Flies have about 700 transcription factor genes (4.5% of all genes), compared to 500 in the worm genome (3.5%) and 200 in yeast (also 3.5%), which begs the question, "Why do flies have so many?" There are two ways to produce highly specialized cellular proteomes: either have more genes or have more transcription factors. It appears flies, with their smaller genomes but more numerous cell types, have

Table 2.5 Comparison of diverse genomes.

	H. influenzae	*S. cerevisiae*	*C. elegans*	*D. melanogaster*
Description	Eubacterium	Eukaryote, fungus	Eukaryote, nematode worm	Eukaryote, fruit fly
Size in microns	0.5	5.0	1,000	3,000
Number of predicted genes	1,709	6,241	18,424	13,601
Number of paralogs	284	1,858	8,971	5,536
Number of gene families	1,425	4,383	9,453	8,065

evolved to optimize their cell-specific proteomes by specializing in highly regulated transcription. It will be interesting to see if more complex organisms (number and types of cells) have higher proportions of transcription factor genes.

Is the Fly Still a Good Model Organism?

Having considered what makes flies different, you might wonder if flies are still good model organisms for human biology. Before the completion of the sequencing project, flies had provided us with insights into many human genes. In particular, flies have orthologs to human disease-causing genes in categories such as cancerous, neurological, cardiovascular, congenital, renal, immunological, metabolic, endocrine, and blood-based disorders (Figure 2.31). It is particularly interesting that even though the fly lacks certain anatomical counterparts (e.g., kidney, blood, heart), it has many genes with similar functions and thus remains instructive as a model for genetic and molecular dissection.

Flies are not the perfect model, nor should they be the only model; nevertheless, given its genetics and molecular tools, flies can provide powerful insights into human disease at the **systems** level, revealing how different genes interact in vivo (see Chapters 10–12).

During the annotation phase, several medically important fly orthologs were discovered, including two very important tumor suppressor orthologs (retinoblastoma and p53; see Chapter 11). Interestingly, fly p53 would not have been identified if the annotation had been based solely on BLAST searching since the human and fly orthologs differ substantially, except at functionally critical amino acids. To date, no worm p53 ortholog has been identified. Parkinson's and Alzheimer's diseases may be well suited for study in flies because several key genes are present in the fly genome, including beta amyloid protein precursor, presenilin, tau, and parkin. Related to cancer studies, many of the critical cell cycle genes are present, such as Cyclins A, B, B3, C, D, E, H, K, and T, as well as cyclin-dependent kinases *Cdk1, 2,* and a hybrid *4/6.* Worms have many of these orthologs too,

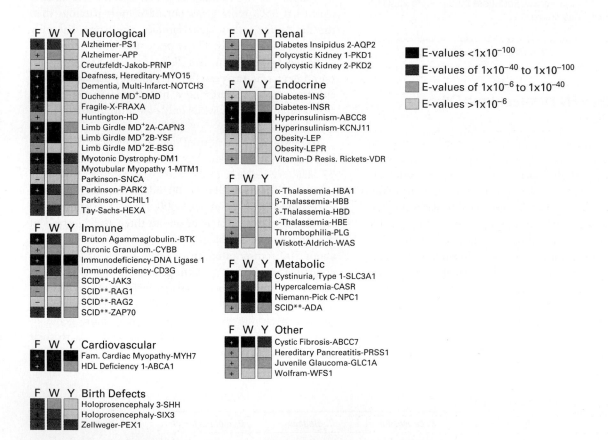

Figure 2.31 Fly remains a good model for many human diseases.
A subset of the fly (F), worm (W), and yeast (Y) genes compared to their human orthologs, with color-coding indicating the relative amount of sequence conservation. Light gray represents weak or no similarity (BLAST E-values $>1 \times 10^{-6}$); dark gray boxes represent E-values of 1×10^{-6} to 1×10^{-40}; light purple boxes represent E-values of 1×10^{-40} to 1×10^{-100}; and dark purple boxes represent the highest degree of sequence conservation (E-values $<1 \times 10^{-100}$). A plus sign (+) indicates the probable functional equivalent to a human gene. Diseases are clustered by tissue type or developmental stage.

but with less sequence conservation to human versions which makes them less appealing for biomedical research.

D. melanogaster has many other interesting genes that you might not expect, such as dystrophin and dystrobrevin (see Chapter 1), molecular motors such as dynein, cell-signaling proteins such as bone morphogenic protein, eight different caspase genes involved in **apoptosis** (worms have 4 and humans have 14), voltage-gated sodium channels (worms have none), over 100 genes for ligand-gated ion channels such as glutamate receptors (fly has 30 and worm has 10). In fact, by reconstructing the metabolic pathways, it is clear flies can produce many neurotransmitters also found in humans, such as dopamine, serotonin, histamine, GABA, and acetylcholine, as well as proteins used to **secrete** these neurotransmitters (e.g. 10 syntaxin genes, 4 SNARE genes, and 3 SNAP-25 genes). The innate half of our immune system is represented in the fly genome, including *Toll* and its receptor and a cytokine-like gene called *Spaetzle*.

The investigators summarized their analysis by stating neither size of the organism nor the genome matter when it comes to complexity. Comparisons of fly and worm genomes in March 2000 surprised many experts, who had expected physiological complexity to be mirrored by gene numbers. They had underestimated two scaling mechanisms: alternative splicing and differential gene regulation. By utilizing these two mechanisms, the fly evolved a smaller genome with fewer genes but a proteome exceeding that of the worm. In part, the fly needs more cell–cell interaction because it has more cells and more cell types than either the worm or yeast. When polyglutamate tracks were engineered in flies, the tiny model organism developed neuronal nuclear inclusions as seen in human cases of Alzheimer's disease and Down syndrome. When the fly chaperone heat shock protein 70 (*Hsp70*) was overexpressed, the nuclear inclusions were completely suppressed, which may provide new insights into human disease therapies.

The authors ended their landmark paper with an open-ended question: Why are 30% of the genes in every sequenced genome unique to that species? Do about a third of our genes evolve more quickly than others? If so, is this rate of genomic change the cause or consequence of speciation?

Fly Genome Epilog

LINKS
BruinFly
FlyBase
secrete
version 4

The fly genome sequence was published in March 2000. Since then, several versions of the genome sequence have been released online (Table 2.6). Some of these updates are due to the ongoing efforts to finish the heterochromatin portions and close the gaps in the euchromatin. As of January 2005, only 23 gaps remained in the fly genome, compared to 1,300 in genome sequence version 1.0. The number of bases has been refined as well, with the **version 4** count at 118,357,599 bp. Interestingly, the authors stated in 2000 that the human genome contained an estimated 80,000 genes which was proven wrong just 11 months later when the human draft sequence was published (see page 99). Their overestimation of human genes reflects a widely held belief that the human genome would be the most complex and the largest. In the remainder of this chapter, we will consider the genomes from a plant and a fish before we finally consider the human genome. However, it should be apparent by now that our assumptions about genomes often are wrong. Only through careful sequencing and annotation can we learn what makes each species different.

DISCOVERY QUESTIONS

75. Enter the BruinFly database created through the original research of many undergraduates at UCLA. First, search for the term "*misshapen.*" Click on the link in the first column and view the eyes. Read the description and then zoom in by clicking on the images. Compare the eye phenotype of *misshapen* to the *patched* eye phenotype. Click on the name "*patched*" to see information about this gene from FlyBase. Go back and click on the "P-element insertion site in the genome" to see where the transposable element landed to cause the *patched* phenotype. Is the insertion in a coding or noncoding portion of this gene? Explain how this insertion could lead to a mutant phenotype. If you look at other genes, notice some of the quirky names used by fly biologists

Table 2.6 Fly genome iterations.

Release Versions	Dates	Number of Genes	Number of mRNAs	Number of Unique Proteins
1.0	March 2000	13,601	14,113	NA
2.0	October 2000	13,474	14,335	13,922
3.0	June 2002	13,659	18,505	16,201
3.1	February 2003	13,659	18,505	16,201
3.2	March 2004	13,792	18,906	19,746
4.0	November 2004	13,792	18,906	19,746

LINKS
BDGP
Ensembl

(e.g., *Ken and Barbie, Sunday driver,* and *deadpan*).

76. Search Entrez for the largest fly gene, called *Kakapo*, then click on the "gene" database option. Notice the polytene band location on the results page, and then click on the link. What name was given to this gene based on its mutant phenotype? How many different mRNAs are produced from this one gene? While on the gene page, click on the "link" link in the top right corner and choose the map viewer option. You should see the gene highlighted with its location shown on the drawing of the polytene chromosome on the left side. Notice how long this gene is. Click on the "hm" (homology) link to the right of the gene name. Is this gene found in many different species?

77. Go to Ensembl and click on the fruit fly button to access the fly database. Enter "P450" in the top text box preceded by the word "with" and click on the top "Look up" button. Only a few hits will be displayed out of a large number possible; click on CG10093, which may be the first one listed. The gene *Cyp313a3* should be displayed with other genes nearby. Do you see any clustered paralogs? Scroll down to determine how many exons are used in the mRNA, as shown in a diagram.

78. Go to BDGP and click on the "Expression Patterns" link to see where a particular mRNA is produced. Search for *bicoid* (abbreviated *bcd*) and follow the links until you see a bar graph and images of the blue-stained mRNA. When and where is *bcd* transcribed? Compare the *bcd* expression pattern to the pattern for *Mkp3* to see how differently some genes are transcribed.

79. Search OMIM for "transferrin" to see what this human protein does. On the page, do a find function for the word "Alzheimer" to see one possible medical role for this gene. Now perform a search on FlyBase for "*Tsf1*" to see if transferrin could be studied in the fly as a possible model for this protein's influence on human Alzheimer's disease.

Do We Need Two Plant Genome Sequences?

Angiosperms (flowering plants) evolved rapidly about 120 million years ago (when dinosaurs ruled). Grasses are relative newcomers, evolving about 70 million years ago (at the time of massive extinction of large animals, including dinosaurs), but now grasses cover about 20% of the earth's land surface. Humans co-evolved with grains, which are a subset of the grasses (Figure 2.32). Humans diverged from apes at the same time the first wheat evolved (3 million years ago). Polyploidy wheat emerged about the same time

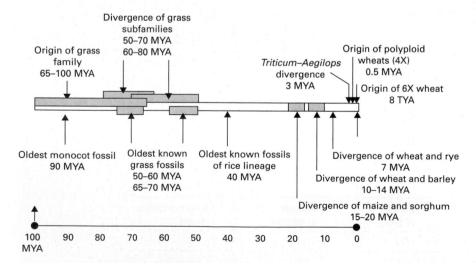

Figure 2.32 History and evolution of grasses.
Time zero represents current date in million-years-ago (MYA) or thousand-years-ago (TYA) units. Some events are displayed with ranges that overlap and show windows of best estimates. Notice the rapid rate of changes beginning 3 MYA when humans diverged from other primates.

LINKS
RGP
IRGSP
BGI

humans migrated from Africa (see Chapter 4), about 200,000 years ago, and wheat was domesticated in the Fertile Crescent (modern-day Iraq, Turkey, Syria, and Iran) when civilization began about 15,000 years ago. Today, cereal grains provide 60% of the calories for a majority of the people in developing countries, with one-third of the world relying on rice (*Oryza sativa*). Therefore, it seems appropriate that the second plant to have its genome sequenced (the model plant *Arabidopsis* was first) would be rice, but the sequencing process itself foreshadowed how complicated plants are.

The International Rice Genome Sequencing Project (IRGSP) took shape in February 1998, though the Japanese Rice Genome Program (RGP) had been slowly building infrastructure and basic tools since 1991. Because Japan grows predominantly variants of the *japonica* subspecies, a particular cultivar (Nipponbare) was used as the source of DNA. In April 2000, Monsanto and the University of Washington produced a crude draft of the rice genome, which was provided to the IRGSP. In May of the same year, China thrust itself into the genomics arena by announcing it would sequence its favorite subspecies, *indica*, at the newly established Beijing Genomics Institute (BGI). Two private companies (Syngenta and Myriad Genetics) announced in January 2001 that they too would sequence a rice genome, and they chose *O. sativa ssp. japonica cv. Nipponbare*, just as the IRGSP had. All of the groups adopted the WGS sequencing method to produce draft versions of the genomes, but the IRGSP was simultaneously developing the tools required for HGS sequencing, which it used for the finishing phase. The race was won by the Chinese team, which released its rice genome sequence to the public in October 2001. By April 2002, BGI and Syngenta had published parallel manuscripts for their respective sequences; the IRGSP released a draft sequence in 2002 but did not publish a paper. Although these competing projects recall the rivalry of the two human genome projects, the plant genomicists were more cooperative and shared their information more readily.

A note about the following sections: Because there are two different draft sequences for two different subspecies, information about both appears in parentheses preceded by either "i" for *indica* or "j" for *japonica*.

Plants Seem Simpler than Animals, but Are Their Genomes?

For years, the rice genome had been estimated to be 450 Mb. Draft sequences were smaller (i361 Mb/ j390 Mb) due to the gaps, though finished sequences were closer to the original predictions (i466 Mb/ j420 Mb). Quality control was calculated for each of the whole genome assemblies, with quantification of coverage presented as an average number (i92% sequenced and i4.2-fold coverage/ j93% sequenced and

j6-fold coverage) since not every base was sequenced the same number of times. It would be difficult to know how much DNA had *not* been sequenced, so to assess the percentage of the genome that was sequenced, the investigators compared their assembled sequences to the known rice genes and ESTs to determine how many of the known genes were present. Rice contains many repeat sequences, especially small repeat units of 2, 3, and 4 bp (i2.1% of the assembly and i42% of all sequenced nucleotides omitted from assembly / j48,351 blocks of 4 to 77 repeat units in a row). In addition, both genomes contained large numbers of transposons or remnants of transposons (i24.9% of the genome), but these mobile elements were intergenic; human transposons tend to be located within introns.

In bacterial genomes, the GC content helped us locate origins of replication, but this simple rule does not apply to eukaryotes. However, GC content is still important to our understanding of the rice genome. Although rice has average GC content for a eukaryote (i38% / j44%), introns have substantially lower GC content than exons (Figure 2.33a, b). You might hypothesize that really long or really short exons are the outliers in Figure 2.33a, with GC closer to 70%, but exon length and GC content are uncorrelated (Figure 2.33c). *Arabidopsis* also has more of its GC content in exons (43% exons, 32% introns), but the distribution of GC in rice is even more complex. The GC content of rice exons depends on the location of the base within the gene (Figure 2.33d and e), with 5' ends enriched for GC. Regardless of the length of the gene, GC content is substantially higher at the 5' end of the gene and decreases to about 50% toward the 3' end of the gene.

Although this unexplainable intragenic GC distribution might have a biological role, it has a clear effect on biologists' ability to annotate the genome. Five software packages are commonly used to predict the location and extent of genes during annotation, but each one differs in its ability to adjust for codon bias. Because rice genes have a gradient codon bias, being GC-heavy at the 5' end and more GC-balanced at the 3' end, most software was ill equipped to detect genes (Figure 2.34) in rice even though these programs performed well in human and other genomes. Using the best available tools, the investigators' estimate of the number of genes in rice yielded surprisingly high numbers (i53,398 to 64,529 / j32,000 to 50,000); in contrast, *Arabidopsis* has only 25,554 annotated genes.

Can We Draw Any Conclusions from Draft Sequences?

Rather than focusing on minutia, let's look at some of the bigger trends and lessons from the rice genome as compared to other eukaryotes. It is unclear if rice exons or introns are different from those of other species (Figure 2.35).

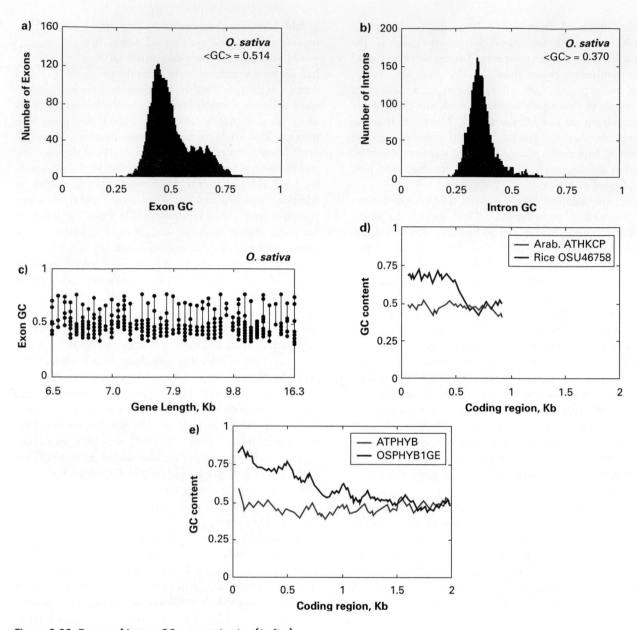

Figure 2.33 Exon and intron GC content in rice (*indica*).
Frequency (Y axis) of GC content (X axis) is displayed as a series of small bar graphs for exons **a)** and introns **b)**. Note the central tendencies and the small shoulders to the right of the major peaks. **c)** Each exon for 41 of the longest genes with full-length cDNAs are plotted as dots and connected to other exons within a single gene. GC content (0.5 = 50% GC) is shown for all the genes. The genes are evenly spaced for clarity, which leads to nonuniform spacing of gene lengths along the X-axis. **d)** and **e)** Sliding 129 bp window GC content from 5′ (left) to 3′ for the coding portions of two rice and *Arabidopsis* genes of different lengths.

The average rice exon (201 bp) is similar to both *Arabidopsis* (214 bp) and human (129 bp), but rice introns (356 bp) are different from *Arabidopsis* (164 bp) and human (3,560 bp) introns. In addition, it appears that the rice genome experienced extensive duplication, such that it carries many paralogs and roughly twice as many genes as *Arabidopsis*. Approximately 58% of the genome (*indica*) is duplicated, with nearly 20% of the duplicated genes in the same order as their paralogous partners. Given this duplication, you could predict that most of the genes in *Arabidopsis* have homologs in rice but not vice versa. This hypothesis is verified by the data: 85% of *Arabidopsis* genes are homologs of *japonica* and 80.6% are *indica* homologs, whereas only i49.4% / j49.5% of rice genes can be paired with homologs

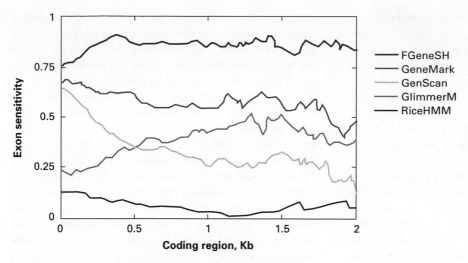

Figure 2.34 Gene prediction programs varied in reliability.
Five gene-predicting programs were compared for their capacity to correctly identify genes within the genome sequence (*indica*). Sensitivity is the probability of finding known exons (1 minus the false negative rate).

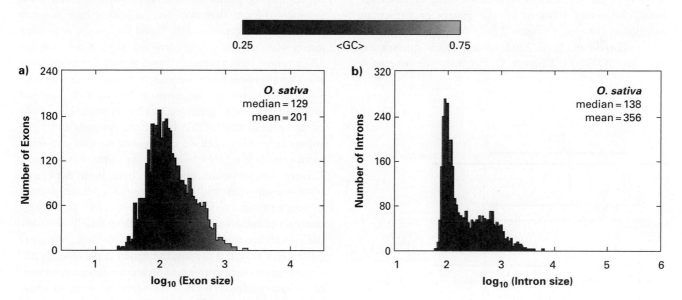

Figure 2.35 Exon and intron size distributions (*indica*).
Frequency of exon and intron sizes with %GC indicated by overlay color scale. X-axis is $\log_{10}$ of base pairs. Note the tendency of GC content to vary for exons but not introns. Compare with data from Figure 33(d) and (e).

in *Arabidopsis*. The authors of both papers deliberately used the term *homolog* instead of *ortholog* because the rice genome duplication made it impossible to know which rice paralog predated the duplication. However, there is a paradox: with a recent genome duplication, you would expect to find high sequence conservation in the paralogous copy, but in fact about half of all rice genes have no known match in GenBank. This lack of sequence conservation in the duplicated rice genes indicates rapid evolution since the

time of **monocot/dicot** evolutionary divergence about 170 million years ago.

Although the only two sequenced plant genomes have many differences, they also share many features and genes, as you might expect of two flowering plants. Because we know a lot about *Arabidopsis* flowering from molecular and genetic experiments, we can determine the probable functions for many rice genes based on sequence conservation. Approximately 8,000 *Arabidopsis* genes are unique to plants. However, about half of these *Arabidopsis* "plant-genes" lack any homology to rice genes, which indicates either some of the *Arabidopsis* genes with unknown function might be incorrectly annotated as genes when they are not, or these 4,000 genes with no homologs in rice represent dicot-specific genes. Interestingly, both plant genomes lack some rather famous genes found in other eukaryotes, such as *Janus kinase* (*Jak*), *Notch*, *Pou*, *p53*, and *hedgehog*, which we know a lot about because of their prominent roles in nonplant species.

Once the grasses evolved, they radiated into many new niches, often duplicating their genomes along the way. Rice may have been one of the earliest grasses to evolve (460 Mb genome); later came sorghum (1,000 Mb), maize (3,000 Mb), barley (5,000 Mb), and finally wheat (16,000 Mb). If substantial portions of genomes were duplicating, then you could predict that many of the rice genes would be represented in multiple loci on the chromosomes from other plant species. Not only are multiple copies of rice genes found in partially mapped plant genomes, they are syntenic as well. Synteny in plant genomes improves the mapping of agriculturally important traits in all the grasses once they are mapped in one species (Figure 2.36). When the rice genome was published in April 2002, about 2,000 **quantitative trait loci** (**QTL**; see Chapter 4) that influence measurable

phenotypes in one or more nonrice cereal species were placed on the rice genome with the assistance of syntenic blocks of genes. For example, a QTL in maize that affects grain yield can be mapped to the syntenic region on rice chromosome 3, which narrows the number of candidate genes to 220. If the QTL gene can be determined in rice, this information can be mapped through synteny back onto maize and the other cereal genomes. This grain-yield QTL represents one of the major desired applications for rice genome sequencing. Rather than using genetic engineering to optimize desired traits, we could use selective breeding with DNA sequence verification to established new cultivars of cereal plants with optimal traits, such as grain production and disease resistance. Currently, we rely on random recombination and phenotype screening, which take more time and are less productive.

What Lessons Have We Learned?

By comparing plants to other eukaryotes, we can draw some conclusions that may need adjusting later as more plants join the genome club. Based on grasses and other plants with extremely large genomes (e.g., Kiwi fruit has 58 chromosomes), it appears plants are subject to different selection pressures when it comes to genome size. Genome duplications give an organism redundancy for critical processes and "spare" genes for adaptive radiation, but replicating twice as much DNA has a cost. Animals and fungi appear to have balanced the cost against the gain and often reduce extraneous DNA faster than do plants. For example, 7 wheat chromosomes appear to have been duplicated twice, leaving wheat with 21 chromosome types and thus 42 chromosomes in every diploid cell. As a result, plants have many more **isozymes** (members of a gene family with very similar cellular roles) than do animals or fungi. We don't really know why, but perhaps the reason is the obvious difference: plants cannot move. If an animal faces a new environmental condition, it can move to another location where its genome is better able to respond. Perhaps plants are selected to accommodate a wider range of environmental conditions, thanks to the high degree of redundant isozymes that have slight differences in kinetics, tolerances, expression levels, etc.

Animals require many different proteins, but our genomes have not expanded the way plant genomes have. To produce more proteins while maintaining a relatively small genome, animals utilize alternative splicing of the exons to produce proteins with a wider assortment of shapes and functions than would be apparent from a simple count of the number of genes. In addition, animals have utilized a wider assortment of transcription factors to produce a greater combination of proteomes that are optimized for the larger number of cell types. Plants have fewer transcription factors but more genes.

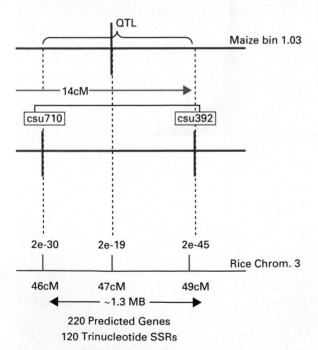

Figure 2.36 Rice QTL localized by homology with maize.
A known QTL (called bin 1.03) for grain yield in maize is syntenic with markers in rice (*japonica*) containing 220 genes and 120 simple sequence repeat markers (SSRs). BLAST E-values for matching sequences are indicated by dotted lines.

Let's consider a basketball analogy to illustrate the apparent solutions offered by plants and animals. Animals are like a basketball team with seven players. The coach sends in one substitute at a time (e.g., changing a defensive player for a long-distance shooter) to change the team dynamics by altering only a small proportion of the players. Plants are like a basketball team with 15 players; the coach changes all 5 players at once to present a different dynamic (switching from defensive to offensive groups). If you have a deep bench (duplicated genome), changing all the players works well (using cohorts of isozymes for different growth conditions). However, if you have a smaller team (nonduplicated genome with fewer isozymes), then you make subtle changes to your team (alternative splicing and more sophisticated transcriptional control). Genomes and basketball teams can succeed with either strategy, so we cannot say one solution is better than the other. However, we can see the strategy for evolutionary success depends on the size of a genome—and plants appear to do well with larger genomes.

Rice Epilog

Since the 2002 publication of the rice genome, what has happened? We still do not have a finished genome sequence, though people are working on this. If you go to the statistical web site, you can monitor the current status of the rice genome sequencing efforts. Finished chromosomes have very few gaps, and we can sift through the repetitive sequences with greater confidence (Figure 2.37).

Figure 2.37 Features of rice chromosome 10.

Go to www.GeneticsPlace.com to view this figure.

We can correlate EST expression data with gene density. Perhaps one of the most surprising traits is the insertion of chloroplast and mitochondria-derived genes into the nuclear genome. For unknown reasons, most of the chloroplast insertions targeted the right side of Figure 2.37, while mitochondrial insertions appear more evenly distributed. One 131 kb block of chloroplast DNA inserted into chromosome 10 and brought with it nearly a complete chloroplast genome. However, these organellar genes are not expected to produce proteins since their promoters are more closely related to those of prokaryotes and thus are not likely to match the transcription factors used for nuclear genes. Furthermore, these relocated organellar genes would be unlikely to produce functional proteins that could enter the organelles of their origin since they lack targeting and transit peptides required of nuclear genes

that encode organellar proteins. The finished chromosome 10 annotation tallied 3,471 genes, many more than in the draft rice genome sequences (i0 / j1,724) due to the criteria for gene annotation and the quality of the assembled DNA sequence. Approximately 96% of *indica*'s chromosome 10 was sequenced, which appeared to be sufficient to permit adequate annotation. However, the large number of small gaps often disrupted genes and led to incomplete understanding of rice genes. For example, the draft sequence predicted the median protein was 232 amino acids long, compared to the finished median of 333 amino acids. This difference in size illustrates that many draft-version gene annotations were artificially truncated, which in turn strengthens the argument that finished-quality genome sequences are worth the added expense and time.

Horizontal gene transfer has been documented for mitochondrial genes in many flowering plants. The ability of genes to move from one species to another without sexual reproduction indicates that plant genomes are more dynamic and unpredictable than we had expected. Their unpredictability supports the cautionary warnings of people who oppose genetically modified food crops (see Chapter 4). Nevertheless, many in science have championed genomics as the solution to global problems as varied as poverty, malnutrition, violent conflicts, and equitable educational opportunities. More than 1 billion people live on less than one U.S. dollar per day. If the world population continues to grow at the current rate through 2025, grain production must increase by 80% to meet the increased need for food. The International Rice Research Institute Genebank has over 100,000 varieties of rice which could contain production-enhancing QTLs that we could uncover through their synteny with *O. sativa* and introduce into commercial varieties. By combining genomic insights with traditional breeding methods, we may be able to offer the benefits of genomics to the world's poorest inhabitants. In November 2003, the wheat research community organized a workshop to address the potential for sequencing this cereal's huge genome. One of nine conclusions adopted at the workshop was to engage all K-12 educational institutions globally in the transfer of technology, workforce training, and promotion of science. Perhaps you will play a role in the realization of this educational effort, since you are in a small minority of people who understand the basic science behind the widely touted genomics revolution.

One final note on the politics and business aspects of genomics: The journal *Science* published both draft versions of the rice genome. The Chinese version was made freely available to all, as is the standard policy of any journal—most of the time. However, for the second time in as many years, *Science* made an exception to its policy when it allowed Syngenta to publish its *japonica* research without releasing the genome sequence into the public

LINKS
notice
Chromosome 8
RAD
RIS
Syngenta

domain. If you read the *Science* editorial, you will see a justification for this policy, with which you may or may not agree. However, if you follow the web links provided in the Syngenta paper, you will see this **notice**: "TMRI has closed its doors, but TMRIs affiliate, SBI, is making the rice genome sequence available to external scientists." Companies merge and liquidate rapidly in today's economy. Syngenta's rice genome sequence was made available to the international sequencing consortium, but today it is difficult to access the data. This fact raises many troubling questions about publication, public access to information, and intellectual property. Scientific discoveries propelled by genomics stimulate unparalleled excitement and understanding, but economic realities and rapidly changing markets for publishers have produced uncertainty in the science community. Perhaps the expression attributed to the Chinese best summarizes the turbulent early years of genomics research: "May you live in interesting times."

DISCOVERY QUESTIONS

80. Go to the IRGSP web page and click on the status tab to see where the project is currently. "PLN" means that the finished sequence has been submitted to the PLaNt database. Click on the finished bar graph for chromosome 5 to see a clone-by-clone list of all the DNA sequenced for this chromosome. Click on the P0668H12 clone link called "INE" (for INtegrated rice genome Explorer) to see an interactive version of the rice genome (this launches a Java applet, so be patient). Move down to about 5.5 cM on the chromosome to locate the marker "S14158"; mouse over this text to see how many different maps it is on, or not on. Click on the text to launch a new window containing information about this marker. How were these data used by the sequencers during the assembly phase of the project?

81. On the clone-by-clone list page, click on "Rice-GAAS:Rice Genome Automated Annotation System" for the same P0668H12 clone you explored by INE. This view gives you a very different insight into the same DNA. For example, does this portion of chromosome 5 contain any tRNA genes? Does the repeat masker identify highly repetitive DNA within genes or outside of genes? Many genes have been predicted by the various computer programs, but how many of these failed to yield BLASTx, cDNA, or EST hits and therefore could not be verified by biological evidence? You can change

the view by altering the default preferences in the bottom frame. If you find a segment you really like, you can select the "MAP Download" link and print out a copy to hang on your wall.

82. Go to the Rice Annotation Database (**RAD**) and click on the link to gene length distribution; then choose, in this order, chromosomes 1, 4, 10, 3. What characteristic changes the most? Does this characteristic correlate with chromosome length? Return to the RAD main page and determine the source of the trend you detected for these four chromosomes by studying the **ideogram** of the chromosomes.

83. Go to the **Chromosome 8** link and you will get basically a blank page. Click on one of the black areas in the ideogram at the top, then scroll to the right and left until you see some content. What area have you landed in, based on what you do not see and the physical location within the chromosome? Explain why you did not see anything when you first explored the chromosome.

84. Go to BGI's Rice Information System (**RIS**), click on the "ComView" tab at the top, and then click on the "Refresh" button on the right side if the settings indicate the base organism is 9331 (*indica*), chromosome 1, the first 1 Mb. Explore the first 5 Mb of chromosomal synteny in jumps of 1,000,000 using the windows at the bottom of the page. Which section has the lowest level of synteny? Between 9311 (*indica*) bases 2,000,000 and 3,000,000, find cDNA OsJRFA 107843 and click on the *japanica* link. How many **SNP**s (single nucleotide polymorphisms) did you find? Click on the Mapviewer link to the right of one SNP's information (you may have to click on the refresh button to see the full display). Can you determine if this SNP alters the encoded protein sequence? Explain what you see and what the limitations are with this visualization of data.

85. Many academics are concerned about free access to genome information. Go to **Syngenta**'s web page and read the second paragraph. Follow the links to see how quickly you can access the data. Compare this effort with the BGI databases in the preceding Discovery Questions and draw your own conclusions about the availability of data.

What Can We Possibly Learn from a Puffer Fish Genome?

At this point, you are probably wondering, "Why a puffer fish?" Rice feeds millions and flies are model genetic organisms, so they make sense. But a puffer fish? As you can imagine, the investigators who argued its genome was worth the money and time for sequencing must have made a compelling case. The **freshwater** puffer fish, *Tetraodon nigroviridis*, has one of the smallest genomes of any vertebrate, which means it would be easier to find the coding sequence (CDS) in its genome than in species with larger genomes. Furthermore, a marine puffer fish (*Takifugu rubripes*) had its genome partially sequenced in 2002, so we would be able to compare two closely related fish species to see what has been conserved and what has diverged over the last 20 million years or so. In October 2004, a team of 61, mostly French, coauthors published the draft genome sequence of *Tetraodon* (German, Spanish, and U.S. labs collaborated, too). The puffer fish genome had been predicted to be about 375 Mb, but the sequencers estimate the finished size will be closer to 342 Mb, even smaller than expected. The true value of this small genome can be appreciated if you consider genes to be needles and genomes to be haystacks. If you can start with a smaller haystack (genome), you increase the chances of finding all the needles (genes). From this little fish, we have learned several major lessons about vertebrate genomes and human origins.

The investigators used WGS sequencing to produce 8.3-fold coverage. The quality assessment indicates that the bases were 99.9% accurate and covered more than 90% of the euchromatin. They were able to annotate 27,918 genes (compared to 20,796 annotated *Takifugu* genes), primarily because of the good coverage and quality of the *Tetraodon* draft sequence. At the time they published, the investigators did not have an EST database to estimate alternative splicing, but they did calculate statistics for an average gene: 7.3 exons; 4,778 bp long; cumulative exon length per CDS = 1,230 bp; and average exon length of 178 bp. The overall GC content is higher in both puffer fish species than in mouse or human (Figure 2.38a), though keep in mind that the two mammalian genome sequences had better coverage than the fish genomes. As with mammals, *Tetraodon*

gene-rich segments have higher GC content, even more so than human genes (Figure 2.38b). Another odd fact about *Tetraodon* is that although the genome has fewer than 4,000 transposons in its genome, they fall into 73 different sequence categories. By comparison, humans have millions of transposons, but only about 20 different types. Those finishing the human genome sequence (see pages 99–109) used the *Tetraodon* genome to lower the estimated number of human genes, and the *Tetraodon* annotators uncovered about 900 new human genes that had been too small to be characterized. Many people are optimistic that human gene regulatory sequences will become apparent with further *Tetraodon*/ human genome comparisons.

Did the Genome Reveal Any Surprises?

Tetraodon has many paralogs (1,078 pairs in *Tetraodon* and 995 in *Takifugu*). The investigators mapped the location of each gene and its paralog and found some interesting trends. First, the paralogs occurred in pairs more than might be expected, instead of triplets or quadruplets as seen in rice. A mere list of the paralogs was not very helpful, so the investigators displayed the locations of all the paralogs and noticed a striking feature: many adjacent genes had their paralogs on the same chromosomes (Figure 2.39). For example, consider the patterns for chromosomes 9 and 11 or 10 and 14, and you will experience the same surprise the investigators got. As often happens in research, knowing the information is not enough; you have to visualize the information in order to gain new insights. In this case, when the investigators created Figure 2.39, they hypothesized the puffer fish genome was the consequence of a whole genome duplication. If their hypothesis were correct, then three predictions would be true: (1) If the duplication occurred after fish diverged from

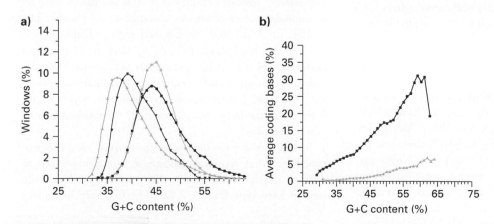

Figure 2.38 GC content in puffer fish genome (*Tetraodon*).

a) Nonoverlapping windows of 5 kb were used to calculate %GC for both puffer fish (*Tetraodon* = purple squares; *Takifugu* = purple circles) while 50 kb windows were used for human (gray triangles) and mouse (black inverted triangles) genomes. **b)** The graph depicts the proportion of nucleotides within exons that have various GC percentages for *Tetraodon* (dark purple squares) and human (gray triangles).

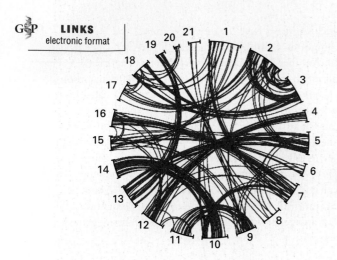

Figure 2.39 Genome duplication in *Tetraodon*.
Display of genome-wide paralogous DNA on the 21 numbered chromosomes, drawn as arcs on the circumference of the circle.

the other vertebrates (about 450 million years ago), then *Takifugu* would have many paralogs located on the same chromosomes. (2) A comparison of a duplicated genome with a nonduplicated but related genome should show orthologous genes occurring in a 2:1 ratio. (3) If humans did not experience the same whole genome duplication, then human genes would have puffer fish orthologs on interleafed chromosomes (i.e., some orthologs on one chromosome and others on the paralogous fish chromosome).

The *Tetraodon* annotators looked for 6,684 syntenic *Tetraodon* orthologs in human or mouse genomes and mapped each one onto human chromosomes (Figure 2.40).

Figure 2.40 Genome duplication revealed in synteny with human genes.
Go to www.GeneticsPlace.com to view this figure.

In this portion of the human genome (the full version can be found in electronic format), you can see how several syntenic human genes on chromosome 16 are located on either *Tetraodon* 13 or 15; similarly, human X chromosome genes appear on *Tetraodon* chromosomes 1 or 7. Notice the interleafed pattern, with blocks of human genes located on alternating fish chromosomes, and in a 2:1 ratio. Therefore, all three predictions have been confirmed by thorough examination of the *Tetraodon* genome. Though it is impossible to prove anything with 100% certainty, the data support the hypothesis that both puffer fish genomes (perhaps those of all bony fish) are the products of a common ancestral genome duplication that has lost many of its original paralogs through attrition over time.

Are There More Big Lessons from *Tetraodon*?

We noted that puffer fish have few copies of many different transposons which turns out to be more than just a curiosity. We have seen before that chromosomes can recombine in areas of highly repetitive DNA such as transposons. However, by not having many copies of the same repetitive DNA, puffer fish chromosomes have undergone more intrachromosomal recombination than interchromosomal recombination. Consistent with this pattern, mammalian chromosomes have experienced more extensive interchromosomal, as opposed to intrachromosomal, recombinations. Again, intrachromosomal recombination may sound like a future college bowl question, but it turns out to be an incredibly fortuitous evolutionary artifact that has provided an insight into our shared ancestral vertebrate genome. Working backward from the *Tetraodon* chromosomal map, the investigators were able to piece together a probable karyotype for an early vertebrate that mammals and fish had as a common ancestor (Figure 2.41). Using Occam's razor to produce the most parsimonious

Figure 2.41 Proposed evolution of puffer fish genome.
Go to www.GeneticsPlace.com to view this figure.

(least complex) primitive karyotype, the investigators proposed that our tetrapod ancestors had 12 chromosomes; they can retrace the chromosomal recombinations, fusions, and fissions, that produced the *Tetraodon* karyotype we see today. For example, consider the ancient chromosome L, which was duplicated. One of the duplicated L chromosomes is nearly intact in *Tetraodon* (chromosome 8), but the other copy of L was split into two smaller fish chromosomes (21 and 6). Conversely, *Tetraodon* chromosome 1 is the fusion product of duplicate copies of ancient chromosomes H and I. The authors predicted that the fish and mammalian common ancestor that lived about 450 million years ago had a dozen chromosomes and fewer genes than we see in *Tetraodon* today. The thrust of their prediction is that mammals would also have fewer genes than modern puffer fish, which is consistent with the latest interpretations of the finished human genome sequence.

Just as they did for *Tetraodon*, the authors used the proposed ancestral tetrapod genome as a starting place to reconstruct the evolutionary changes that produced the human karyotype (Figure 2.42). The first step in human

Figure 2.42 Proposed evolution of human genome.
Go to www.GeneticsPlace.com to view this figure.

genome evolution was the expansion of transposable elements in the chromosomes. As indicated earlier, the existence of multiple copies of highly repetitive DNA not only increases the overall size of the genome, but also permits interchromosomal recombination. Notice that the authors did not propose fusion or fission as much as recombination, though simple math tells us some of the 12 ancestral chromosomes had to split to form the more numerous but shorter human chromosomes. With this visualization of how our chromosomes may have evolved, we can make testable predictions about other vertebrate genomes, such as chicken and chimpanzee.

Perhaps the most surprising lesson from *Tetraodon* is that well-chosen genomes can be more valuable than we might have anticipated. Not only did *Tetraodon* help us annotate human genes, but it also revealed a plausible evolutionary tale of vertebrate and human evolution. At this point, we have genome sequences from vertebrates closely related to humans (rat and mouse), as well as from very distantly related species (puffer fish). It will be interesting to see if any new insights can be gleaned from a vertebrate of intermediate distance (e.g., an amphibian, reptile, or bird). We will consider the genome of a few close human relatives (e.g., dog and chimpanzee) in Chapter 3 when we consider larger-scale, comparative genome analysis.

DISCOVERY QUESTIONS

86. Look at figure 2.39 and consider chromosome 13. Was it duplicated, or was it unlucky enough never to have been duplicated? If 13 was duplicated, describe what happened to the duplicated version. Which chromosome pairs have been duplicated and retained nearly intact?

87. Find the location of the human genes oligophrenin and arrestin, using MapViewer. Now go to the Tetraodon Genome Browser and search for oligophrenin and arrestin. Can you detect the interleaving genes shown in figure 2.40 and the genome duplication in figure 2.39 using these genes? Do they have paralogs near each other in the puffer fish?

88. Go to the Human Genome Browser and search for "sarcospan." Is sarcospan highly conserved in the diverse species shown in the browser? You should see the Tetraodon Net line; click on the *Tetraodon* exons until you get a tabular report showing the gene's summary statistics. What is the size difference between human and fish genes?

89. Go back to Tetraodon Genome Browser and examine the ideograms. Is the genome finished yet? Click on a gap and show the resolution at 50 kb. Change the viewing options so that all are hidden, except turn on "DNA/GC content," "Gap: All Sequence Gap," "Genescan," "Hox

Genes," "*Takifugu* ecotigs," and "*Tetraodon* cDNAs." What effect do gaps have on the number of genes predicted by Genescan? Compare the predicted genes to the number of cDNAs to see if any validating sequences are available. If so, how well did Genescan predict the genes' correct sizes? Now search for *HoxA*, zoom out to 200 kb, and change the settings to highlight mouse, human, *Takifugu*, and *Tetraodon* gene conservation. Which *HoxA* gene is in *Tetraodon* but not the mammals?.

LINKS G⚛P
Human Genome Browser
MapViewer
Tetraodon Genome Browser

Bonus Material: A study of the chicken genome is available on this book's web site.

What Makes Humans Different?

We begin our study of the human genome with the public-domain draft version of the human genome sequence, which was published by the Human Genome Project (HGP) in February 2001. We begin with a list of findings that outline what was learned when 3 billion base pairs were examined for the most obvious traits. Of course, it is impossible to fully summarize the human genome, in the sense that new discoveries will continue for many years. Investigators will need to perform laboratory experiments to glean more information than is available from computer searches. For example, alternative splicing adds a significant degree of complexity in defining the proteome, but we cannot detect it without bench work, although ESTs are helpful. We will briefly survey some of the initial findings and look in depth at others later in this chapter.

Highlights from the human genome draft sequences include:

- Publication of the draft sequence on February 15, 2001, was based on data "frozen" on October 7, 2000. The number of genes was estimated to be 35,000 (20,000–25,000 revised estimate in 2005). Draft sequence means the DNA was sequenced on average 4 times, half the coverage of finished sequence. Gaps, sequencing errors, and contig assembly errors in the draft sequence will be corrected during the finishing process to achieve an error rate of 1 in 10,000 bp.

- For unknown reasons, the mutation rate in males is twice as high as in females. Recombination occurs more frequently nearer the telomeres and on the "petite" (p) arms of chromosomes. Long chromosomes exhibit a recombination rate of 1 cM per Mb, whereas short arms show about twice that rate. The extreme examples are the short (2.6 Mb) pseudoautosomal regions of the petite arms of the sex chromosomes,

Xp and Yp, at about 20 cM per Mb. Crossing-over is believed to be necessary for normal separation of chromosome pairs during anaphase I of meiosis.

- Some human genes initially appeared to have entered our genome via horizontal transfer from bacteria, but this interpretation has been contested by several labs (see page 103 for details).

- Genes have a greater GC concentration than intergenic DNA (Figure 2.43a). CpG dinucleotides form "CpG islands," and the cytosine base is often methylated. When cytosines spontaneously deaminate (lose an amine group), the base is converted into thymidine. Therefore, over time, CpG dinucleotides gradually change to TpG dinucleotides. If the cytosine is not methylated, the deamination produces uracil, which is fixed by the cell's quality control machinery back to cytosine. Some chromosomes would experience more deamination than others (Figure 2.43b).

- There are five classes of repeated sequences: (1) transposon-derived; (2) pseudogenes; (3) simple repeats, sometimes called variable number of tandem repeats (VNTR; e.g., $(CA)_n$); (4) segmental duplications, where one chromosomal region gets copied onto

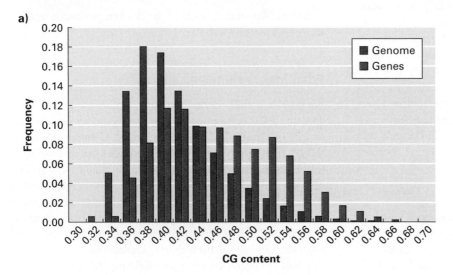

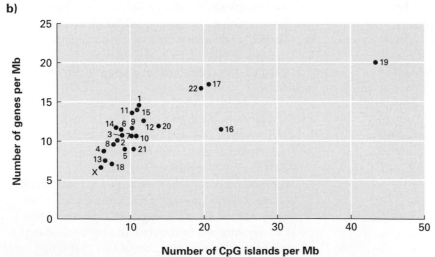

Figure 2.43 GC content of the genome.

a) The ratio of GC content is plotted on the X-axis with frequency of each ratio on the Y-axis. GC content for the whole genome (gray) and individual genes (purple) is not identical. To produce the ratios, gene content was calculated using sections of DNA 20 kb long and centered on the 9,315 known genes; genome content was determined by 20 kb of nonoverlapping adjacent DNA. **b)** For each chromosome, the number of CpG islands per Mb is plotted on the X-axis, genes per Mb on the Y-axis. Most of the chromosomes fit along a diagonal line, but four do not.

Math Minute 2.5 **How Do You Fit a Line to Data?**

Figure 2.43b depicts the relationship between CpG island density and gene density for each chromosome. The investigators report that chromosomes 16, 17, 19, and 22 are outliers because they fail to follow the same linear trend as the remaining chromosomes. To explore how they might have drawn this conclusion from the data in Figure 2.43b, try to draw straight lines on the graph in the figure supporting each of the following statements:

1. Chromosomes 16, 17, 19, and 22 are outliers.
2. Chromosomes 16 and 19 are outliers, but chromosomes 17 and 22 are not.
3. None of the chromosomes are outliers.

Which line do you think fits the data best?

Since you drew these three lines somewhat subjectively, your lines may look different from someone else's lines. However, there is a mathematical approach, called *simple linear regression*, by which everyone can draw the exact same lines, every time. Regression analysis is routinely used in scientific studies to model linear relationships between variables. (The term regression comes from the work of the founder of eugenics, Sir Francis Galton, who described the tendency for offspring to be closer to average size than their parents as "regression towards mediocrity.")

You can quantify how well a particular line fits a set of data points by measuring the vertical distance (i.e., absolute value of the difference in Y-coordinates) from each point to the line. If you seek a line that fits all 23 data points, you might try to find the line that minimizes the sum of these 23 distances. Mathematically, however, it is more convenient to determine the slope and intercept of the line that minimizes the sum of squared distances. Since the quantity to be minimized (sum of squared distances) is a function of two variables (slope and intercept), partial derivatives are key to this process. Squared distances are more convenient because it is easier to work with derivatives of squared quantities than derivatives of absolute values.

For the data in Figure 2.43b, the investigators did not offer a rationale for their choice of line (1) over lines (2) and (3). You can explore one possible explanation in the Math Minute Discovery Questions. Regardless of how the outliers are chosen, the regression line helps us understand the relationship between CpG islands and genes in typical and atypical (i.e., outlier) chromosomes.

MATH MINUTE DISCOVERY QUESTIONS

1. One possible explanation for the choice of line (1) over lines (2) and (3) is that the regression line was constrained to go through the origin. Why would it make sense for the line to pass through the origin?
2. Redraw lines (1), (2), and (3) with the additional restriction that each line must go through the origin. Under this restriction, do you agree with the investigators that line (1) is the best fit?

another chromosome; and (5) regions near telomeres and centromeres and ribosomal genes.

- Transposable elements fall into four major categories: (1) short interspersed elements (SINEs), including Alu, which collectively constitute about 13% of the entire human genome and may confer a functional advantage on us; (2) long interspersed elements (LINEs), which collectively constitute about 21% of the entire genome; (3) long terminal repeat (LTR) retrotransposons, which account for about 8% of the genome; and (4) DNA transposons, which are about 3% of the genome. All together, these mobile elements constiute about 45% of the human genome. One of the great treasures hidden in these repeat elements is evolutionary data. By measuring the degree of sequence divergence within the transposable elements, we can begin to measure time of divergence if care is taken to calibrate the "molecular clock," though some investigators think calibration is impossible.

- The X chromosome has the highest concentration of transposable elements, with a 525 kb section being 89% transposon, of which a 200 kb subset is 98% transposon. Conversely, some developmentally important regions are nearly transposon-free, such as the portion of chromosome 2 containing our *Hox* genes.

- The Y chromosome is the site of greatest LINE transposition, and thus the Y chromosome looks the "youngest" of all the chromosomes due to the high degree of recent DNA insertions. The density of LINEs is probably a consequence of the relative paucity of genes on the Y chromosome, which can tolerate insertions more readily than gene-dense chromosomes.

- One hypothesis has been proposed for a function of SINEs. SINEs are transcribed under conditions of stress, and the resulting RNAs hybridize to a particular protein kinase that normally inhibits translation. Thus, SINEs would enable a cell to produce more proteins when stressed, which is exactly what happens in yeast (see pages 250–252). The hypothesis predicts that SINEs would provide a selective advantage if they enabled cells to survive stress better. Interestingly, Alu sequences are concentrated around genes, which may be selected for as predicted by this hypothesis.

- About 50 human genes were derived from transposons. Included in this list are *RAG1* and *RAG2*, which play a significant role in producing the high degree of diversity in antibody binding capacity.

WHOSE DNA DID WE SEQUENCE?

One area of controversy that arose early in the planning of the HGP was whose DNA would be sequenced. Nine different sources of DNA were used, eight of them identified as male. Interestingly, only three of the nine DNA samples were taken from germ line cells (sperm); the other six were somatic cells, and at least five of these were male.

The issue of gender has raised a number of questions. DNA from a female donor will have two slightly different copies of X and no Y chromosome. Conversely, DNA from a male donor will have only one copy of X, but will also have a Y chromosome. Because some genes (e.g., immunoglobulin and T-cell receptor genes) rearrange in somatic cells, there is an interest in sequencing nonrearranged genes, and thus germ line cells are preferred. However, obtaining eggs from women is much more difficult than obtaining sperm from men. In addition to difficulties in obtaining eggs, the number of sperm cells that can be readily obtained is advantageous as well. For many years, females have been underrepresented in research, both as investigators and as participants in clinical trials. Therefore, there was an appeal to include DNA from women in the Human Genome Project. Of course, every male donor received half of his DNA from his mother; nevertheless, some of the DNA libraries used for sequencing were derived from anonymous female donors.

It would have been easier to obtain DNA from members of the laboratories where the DNA libraries were constructed. However, personal involvement with the research may create difficult ethical issues. A technician might have felt compelled to donate DNA, or confidentiality of the donors' identities could have been breached. How would you feel if a mutation were discovered as a part of the HGP and you knew your DNA was the one being sequenced?

The final consideration was that the HGP might be perceived as elitist if only the top scientists had their DNA sequenced. For these reasons, many donors were recruited from different locations and only a subset of the collected DNA was used for sequencing. The identities of the HGP DNA sources have been hidden to minimize potential ethical dilemmas. You can read more about what steps were taken to choose DNA samples at the Human Genome Funding Statement, Caltech, and the BACPAC Resource Center in Oakland, California.

In April 2002, Craig Venter disclosed that his DNA was used for Celera's sequencing project. This revelation contradicts Celera's public statements made in February 2001. Venter's DNA accounted for three-fifths of the total Celera sequence.

Can We Describe a Typical Human Gene?

Let's look at an average human gene (Table 2.7). Human genes tend to be composed of several exons (the average number is about nine, though some genes have only one). Exons are small (145 bp on average) and are separated by large introns (on average about 3.3 kb). The average 5′ untranslated region (UTR) is 300 bp, and the 3′ UTR averages 770 bp, with a total coding region of 1,340 bp to produce an average protein size of 447 amino acids. For the average gene, the transcribed portion is distributed over 27,000 bp. You can see that the human gene is not a model of efficiency, which is one of the major reasons why deciphering the human genome is difficult (Figure 2.44).

Once the number of human genes was determined, investigators wondered how humans could be more complex at the protein level if we only have twice as many genes as the worm *C. elegans*, which is composed of 1,000 cells. Using a sampling method, the HGP found about 60% of all human genes produced more than one mRNA due to alternative splicing, with an average of 2.6 transcripts per gene. *C. elegans* is not as complex: about 22% of its genes produce more than one mRNA (average of 1.34 transcripts per gene). However, fewer *C. elegans* ESTs are available for analysis, so estimates of alternative splicing may be artificially low, and thus alternative splicing may not be the main source of the greater human proteome complexity.

In addition to the number of genes that produce multiple mRNAs, there is another issue. If each gene can be either on or off, then the combination of active genes in any given cell is 2^x, with x being the number of genes in the genome. Therefore, flies have $2^{18,000}$ possible combinations of on and off genes, while humans have $2^{35,000}$ possible combinations (using gene-number estimates from 2001). Humans have more cells with specialized functions than yeast, flies, or worms, therefore we need to regulate gene expression very carefully. The increased number of special-

LINKS
TIGR investigators

ized cells allows a prediction that humans need more transcription factors than less complex organisms (Figure 2.45). Proteome complexity (subsets of proteins in each cell, tissue, and organism) is regulated by gene expression, and thus worms need more gene regulation than yeast, flies more than worms, and humans more than flies.

What makes humans unique? This question is difficult to answer, but we do have some relevant information (see Figure 2.45). Each species has a unique combination of proteins for the different biological processes, as you would expect. Perhaps the more humbling data are revealed in Figure 2.46, which makes us realize that only 1% of our genes have not been found in other species and that we share one-fifth of our genes with all other organisms, including bacteria.

The field of genomics is a combination of hypothesis testing and discovery science. Discovery science is what Leeuwenhoek performed after inventing the microscope: "I wonder what I'll find if I look at . . . ." While mining data from genome sequences, we can make unexpected discoveries. One such discovery focused on 223 human genes, found only in vertebrates, that appeared to be orthologs of bacterial genes. At least 113 are found in many diverse bacteria, which led some investigators to conclude these genes entered our genome at some point in the vertebrate lineage by horizontal transfer from bacteria. In case you think this small number of genes is not physiologically important, let's consider two genes of questionable origins: two monoamine oxidases (MAOs). MAOs deactivate neurotransmitters such as norepinephrine and serotonin. MAO inhibitory medications are used to treat depression and Parkinson's disease and cannot be taken with painkillers such as Tylenol.

The hypothesis of horizontal transfer of bacterial genes into the human genome is not universally accepted. Contradictory publications proposed that the alleged horizontal transfer of bacterial genes was an illusion created by the limited number of genome sequences. TIGR investigators reported that only about 41 genes might have been

Table 2.7 Average based on 1,804 genes chosen from RefSeq and unambiguously aligned over their entire length with finished sequence.

	Median	Mean	Sample (size)
Internal exon	122 bp	145 bp	RefSeq alignments to draft genome sequence with confirmed intron boundaries (43,317 exons)
Exon number	7	8.8	RefSeq alignments to finished sequence (3,501 genes)
Introns	1,023 bp	3,365 bp	RefSeq alignments to finished sequence (27,238 introns)
3′ UTR	400 bp	770 bp	Confirmed by mRNA or EST on chromosome 22 (689)
5′ UTR	240 bp	300 bp	Confirmed by mRNA or EST on chromosome 22 (463)
Coding Sequence	1,100 bp	1,340 bp	Selected RefSeq entries (1,804)
(CDS)	367 aa	447 aa	
Genomic extent	14 kb	27 kb	Selected RefSeq entries (1,804)

a)

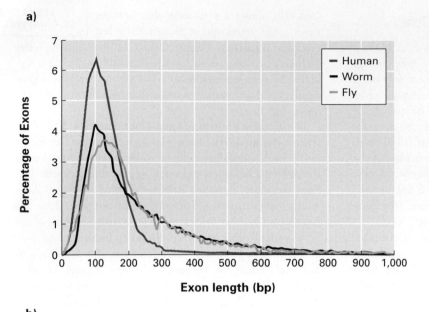

b)

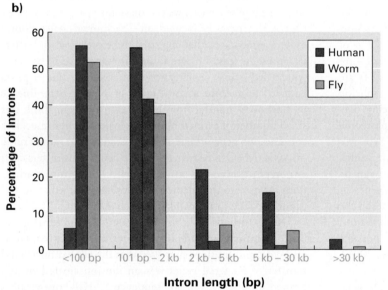

Figure 2.44 Size distribution of exons and introns.
a) Species comparison of exon sizes in human, *C. elegans*, and *Drosophila* genomes.
b) Species comparison of intron sizes in human, *C. elegans*, and *Drosophila* genomes.
Note the inverse correlation between intron and exon sizes.

horizontally transferred, but they also predicted that this number will gradually shrink to zero as more genome sequences are analyzed. The TIGR group proposed that all eukaryotes once had these 223 suspect genes but have lost them over time, thus making the genes appear to have been horizontally transferred from bacteria to mammals. Although the initial idea of horizontal transfer of bacterial genes into the human genome appears to be incorrect, it is a good example of discovery science producing an unexpected hypothesis that was experimentally tested.

DISCOVERY QUESTIONS

90. Let's do some quick estimations about our DNA using these numbers: haploid genome of 3,289,000,000 bp; 23,000 genes; and the numbers from Table 2.7.

    a. What percentage of your genome is spent on genes? Exons? Introns?

    b. What percentage of your genes is spent on exons? Introns?

a)

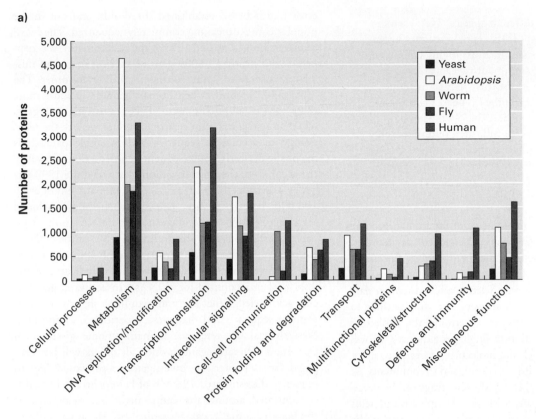

Figure 2.45 Genome-wide comparisons of transcription factor families.
Nine different transcription factor families (X-axis) compared with abundance in each family on the Y-axis for five species.

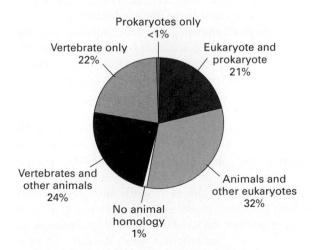

Figure 2.46 Functional categories of proteins.
Distribution of orthologs for the predicted human proteins. These data are based on genome sequences available February 2001.

91. Are any chromosomes missing from Figure 2.43b?

92. Given the information in Figure 2.45, name one aspect that makes humans different from other species.

93. Go to the BLAST2 web page and perform a protein alignment with human sarcoglycan delta (NP_758447) and sarcoglycan gamma (NP_000222) by pasting their protein accession numbers into the smaller accession boxes and choosing "blastp" from the program menu. What are the percent identity and percent similarity between these two proteins? Now align the *C. elegans* sarcoglycan ortholog *sgn-1* (CAA92663) with both of the human sarcoglycan genes and determine which one is more similar (compare identity, similarity, and gaps).

94. Go to UCSC's Genome Browser and search for "syntrophin." How many syntrophin genes do humans have? Click on the link for "(NM_009228) syntrophin, acidic 1." Change the view to hide all except "Known genes" in full view, and all the different species "Net" views in dense, then refresh. Is this gene's structure (combination of exons and introns) highly conserved? Click on the dog alignment twice until you have a nongraphic report for this locus of the dog genome, then click on the "Open Dog browser" link. Set the species "Blat" or "Net" views to full, and "Conservation" to full, then refresh. Is the conservation confined to the exons only? Explain the significance of the conservation in the noncoding regions.

Where will the complete sequence of the human genome lead us? The finished human genome sequence was published in October 2004 with the hope that many current uncertainties would be settled (see Epilog following). Don't expect all the questions to be answered. Science advances when data produce more questions than answers. We will continue to utilize the genome sequence to determine the molecular causes of diseases. In addition to single-gene diseases, we will discover new ways to track the molecular causes of complex phenotypes such as Alzheimer's, cancer, and schizophrenia. As disease-causing genes are identified, new drugs can be developed to modify their cellular roles. For example, two proteins are known to play a role in amyloid plaque formation in the brains of Alzheimer's patients. One is called amyloid precursor protein (APP), which is cleaved by a protease called b-site APP-cleaving enzyme (BACE). BACE was identified in 1999, and is located on chromosome 11. When the investigators searched the human genome, they identified BACE2, which shares a 52% amino acid identity with BACE1. Interestingly, BACE2 is located in the Down syndrome region of chromosome 21. Perhaps the amyloid plaque formation found in individuals with Down syndrome (caused by three copies of chromosome 21) is caused by excessive amounts of BACE2, and both Down syndrome and Alzheimer's might be treatable with a common drug.

Eventually, we want to identify all the regulatory regions for all the genes to better understand what governs transcription (see the Epilog and Chapter 10). We'd like to compare the human genome with other genome sequences (see Chapter 3). Understanding genetic diversity within the human population will help us understand evolution, physiology, disease susceptibility, and drug efficacy (see Chapter 4).

Human Genome Epilog

Three and a half years after the two draft versions of the human genome were published, the public sequencing effort of the HGP published its finished version (Build35 which means the thirty-fifth version). The term *finished* should be used in its sequencing context, which means the error rate is below established thresholds, and not in the popular context meaning completely sequenced. The HGP draft version had missed 10% of the euchromatin, was peppered with nearly 150,000 gaps of varying sizes, and mistakenly assembled some contigs in the wrong order. The 2,851,330,913-bp finished human genome includes more than 99% of all euchromatin. It contains only 341 gaps, which investigators want to sequence, but we lack the technology to accurately sequence and assemble paralogous regions with nearly 100% sequence conservation. However, the accuracy of the finished genome exceeded the 1990 target by a factor of 10, with a final error rate of 1 out of every 100,000 bases. Quality assessment methods included searching Build35 for all of the human cDNAs in the database; investigators were able to find 99.78% of the cDNAs. Keep in mind that "the" human genome is intended as a **reference genome**, not a definitive one (as indicated by the 0.28% of unaccounted-for cDNAs). In addition, genomicists acknowledge that every human has a slightly different genome sequence and that these variations are important. Nevertheless, the reference human genome gives us an important standard against which all others will be compared (see Chapter 4). Investigators also would like to sequence the estimated 198 Mb of heterochromatin that is concentrated around the centromeres and telomeres. If the heterochromatin can be sequenced, the final Build of the reference human genome will total about 3.08 billion bp (Gb).

The HGP improved not only its sequencing capacity, but also its contig assembly. For example, chromosome 7, estimated to be 144.2 Mb in 2001, is now estimated to be 155.6 Mb. Some of the increased length comes from clarification of duplicated regions with high sequence conservation (>97%) that initially had been considered one locus, but with improved coverage were recognized as paralogous loci. One way to visualize the improved assembly is to perform a dot plot comparing the draft sequence with the nearly complete Build35 sequence (see Figure 2.4). At the macroscopic level, you can see about 10 regions of different lengths that were inverted in the draft assembly but have been resolved. If we zoom in on one portion of chromosome 7 that looks well aligned in the two versions, we can see that a large segment had been incorrectly assembled. Of the three BACs shown in detail, the middle BAC was not finished in 2001, although the two flanking BACs were. If you were studying a gene in this region of chromosome 7, you would be very pleased that the incorrect assembly has been corrected. If you extrapolate these 500 kb over the entire genome, the benefits of finished-quality sequencing become more apparent and worth the extra 3.5 years of effort.

With Build35, we have better resolution of the number of human paralogs and other issues that are apparent at the gene level. For example, now we can distinguish the 193 olfactory receptor genes clustered on chromosomes 11, 1,

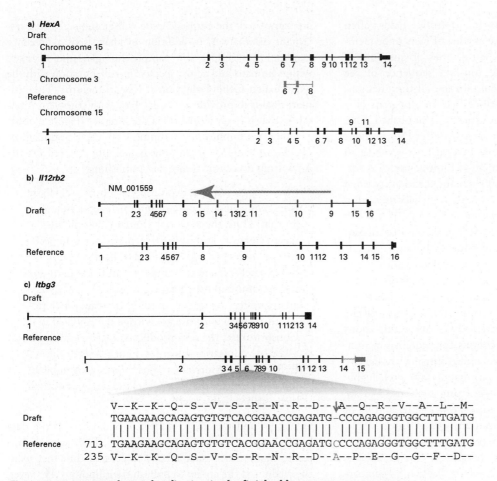

Figure 2.47 Improved gene localization in the finished human genome sequence.
a) Exons 6, 7, and 8 of *hexosaminidase A* (*HexA*) were mistakenly duplicated in chromosome 3, in addition to their correct location on chromosome 15. **b)** *Interleukin 12 beta-2 receptor* (*Il12rb2*) contained an inversion of several internal exons and a duplication of exon 15 in the draft sequence. **c)** A single-base deletion in exon 5 of *integrin beta-3 precursor* (*Itgb3*) mistakenly caused a frameshift at amino acid 246 in the draft sequence; also, the terminal exon was omitted.

and 17, as well as the 14 paralogs of type-2 taste receptor genes clustered on chromosome 12. In the families of duplicated human genes, certain functional categories were overrepresented: olfaction, taste, immune system, and testis-specific proteins. Duplicated regions (≥ 1 kb with $\geq 90\%$ sequence identity, but excluding transposons) account for 5.3% of the entire genome. The Y chromosome was the most difficult to resolve in this regard, because more than 25% of its 26.7 Mb length contains duplicated regions, including two blocks totaling 1.45 Mb with 99.97% sequence identity.

The final number of genes is not a single number, but an estimated range: 20,000–25,000. The HGP cataloged 22,287 genes (with 34,214 cataloged mRNAs), composed of 19,438 known genes and 2,188 predicted genes. The average gene contains 10.4 exons and the average mRNA includes 9.1 exons, which takes into account the previously documented alternative splicing. Build35 fixed sequence errors in 58% of the cataloged genes and corrected 39% of the cataloged genes with at least one omitted exon (Figure 2.47). The 2,188 predicted genes are based on comparisons to other species with highly conserved ORFs. From the outset, genes smaller than 100 codons were not annotated, but with the increasing awareness of microRNAs used in **RNAi** (**RNA interference**), we should see the total number of genes rise again as we annotate smaller genes. For these reasons, the HGP authors preferred a range of genes to an absolute number. Once again, our species is humbled by a downward trend in our number of genes compared to less complex organisms such as fly (13,472), worm (22,246), and *Arabidopsis* (31,222), the only other multicellular organisms to have finished-quality genome sequences as of 2005.

LINKS
Comparative Maps
ENCODE
MGC
PubCrawler

People outside scientific fields often think of science as a list of facts to be memorized rather than a series of questions to be asked. The finished sequence of the human genome is not complete, nor are all the questions answered. In fact, some questions became apparent only when the finished genome sequence was published. One example is unexplainable "ultraconserved" DNA sections preserved from fish to humans. Investigators working in Australia and the U.S. identified 481 ultraconserved sections of at least 200 bp with 100% sequence identity when compared with rats and mice, more than 99% conservation in dogs, and more than 95% conservation in 467 chicken segments. These ultraconserved sections excluded ribosomal genes, but the probability of finding just one ultraconserved section is 10^{-22} if you assume neutral evolution and the slowest mutation rate for any genome of more than 1 Mb. Therefore, investigators wanted to know more about these highly unlikely but highly conserved sections. Of the 481 human ultraconserved sections, 111 are within exons and probably contain critical codons for protein function. Of the remaining 370, 100 are within introns, though we do not know why these 100 introns contain such highly conserved sequences. Of the remaining sections, 140 are more than 10 kb from the nearest known gene, and 88 of them are over 100 kb away from any genes. Interestingly, 156 of the sections located between genes are clustered near genes involved in embryonic development (the probability of this clustering by chance is 2.7×10^{-5}). Perhaps comparative genomics and the identification of ultraconserved sections will help us identify enhancers that are critical for vertebrate development, which opens the door to the next level of genome analysis: noncoding structures.

What Is the Next Goal in Human Genomics?

The HGP has two major spinoff projects: to measure the extent of variation in the human populations (see Chapter 4), and to identify every functional element in the human genome (Figure 2.48a). The ENCyclopedia Of DNA Elements (ENCODE) is a multinational consortium that began a pilot phase, in September 2003, to study only 1% (30 Mb) of DNA. To sample widely, ENCODE has chosen 44 discrete portions of the genome for analysis to figure out the best methods, develop new methods, validate data collection, and identify any new types of data that are currently missing. Once the pilot and methods-development phases

are completed, the entire genome will be subjected to a systematic annotation. In addition to analysis of a reference genome, ENCODE will sample for genomic diversity within humans and across species (in collaboration with the Mammalian Gene Collection (MGC) program). The ultimate goal is to provide a rich database with complete annotation that is freely available via the Internet (Figure 2.48b).

The 2004 publication outlining ENCODE and its goals concluded with an interesting paragraph that reflects the interesting times for science and publishing:

> [We] recognized that deposition in a public database is not equivalent to publication in a peer-reviewed journal. Thus, the National Human Genome Research Institute and ENCODE participants respectfully request that, until ENCODE data are published, users adhere to normal scientific etiquette for the use of unpublished data. Specifically, data users are requested to cite the source of the data (referencing this paper) and to acknowledge the ENCODE Consortium as the data producers. Data users are also asked to voluntarily recognize the interests of the ENCODE Consortium and its members to publish initial reports on the generation and analyses of their data as previously described.

Genomics and the need for free, public access to the data have created a problem for investigators. They are committed to sharing their data as soon as possible, but they want the credit and the ability to publish their findings. However, it is impossible to publish the same story twice, so they are trying to invent a new code of conduct that says "You can see our data and use our data, but don't publish until we have published." You might want to perform a PubMed search or program your PubCrawler search for ENCODE to see who publishes a scientific paper first, the consortium or another lab that utilized their data without permission.

DISCOVERY QUESTIONS

95. Go to the Comparative Maps web page, and click on "Chromosome 9" under the "Rat and Mouse compared to Human" column. Using the "Region Shown" box in the left frame, enter "122M" in the top box and "123M" in the bottom box. Locate the human gene called *PTGS1*, which is the *cyclooxygenase 1* gene and the target of painkillers such as aspirin (see Chapter 9 for details). Use the "Maps & Options" button, activate the "Show Connections" option, and hit "Apply." Do all 3 species have *PTGS1*? Are these regions syntenic, or are the human genes not linearly related to those in the two rodents? What pattern in the alignment of genes is evident just below *PTGS1*? Click on the Rat ortholog and choose the "ev" (evidence) link. Do you think

Figure 2.48 Genomic features being identified by ENCODE.

Go to www.GeneticsPlace.com to view this figure.

there are sufficient data to support the annotation that this is a true ortholog?

96. Go to UCSC's ENCODE web site and choose "Alpha Globin" from the menu in the left frame. Alpha Globin is abbreviated HBA, but you can see several genes that begin with HB. How many can you identify? (You may want to modify the view and zoom in to help you focus on the HB genes.) Considering the conservation in other species, how many of these HB genes are conserved in vertebrates? Do all these genes produce functional protein? Click on "HBA1" to find the answers in text. Click on the "AceView" link in the "Tools and Databases" table to see more information, including graphic depictions of alternative splicing for these genes.

Now go to Ensembl's *Multispecies* ENCODE web site and click on the "MultiSpecies alignment" for the cystic fibrosis gene "CFTR" to display 1 Mb of human, mouse, and chicken genome alignments. On what chromosomes are CFTR orthologs located for each species? Scroll down to examine the alignment and notice the difference for chicken. What must have happened to produce the converging lines seen in the chicken chromosome? Follow the blue line from human through chicken CFTR (the blue line is centered on the whole gene, not the 5′ end). Is CFTR conserved in chicken, or is the gene truncated? You can recenter and zoom in to see the genes in better detail.

97. Perform a Gene search for the human *ITGB3* (integrin beta 3) gene, which is illustrated in Figure 2.47c. Is the correct version of the gene in the database, or is the number of exons incorrect? Click on the KEGG pathway for "regulation of actin cytoskeleton". Where is integrin located in cells (its cellular component, in GO terminology)? Click on the highlighted box labeled "ITG" and use your browser's find function to locate the amino acids "RNRD." Does this database have the old or new acid sequence?

Summary 2.3

Multicellular Genome Analysis

In this section, you learned of common traits between the genomes of unicellular and multicelluar organisms. You learned how to utilize a wider range of databases while uncovering interesting facts and trends about evolution and physiology. Dog, chicken, chimpanzee, and many other model genomes will be sequenced before 2010, and no textbook can stay current with the literature. However, this chapter has given you insights into genome analysis that will enable you to continue genomic research.

LINKS
ENCODE
Multispecies

Chapter 2 Conclusions

In Chapter 2, we learned how genome sequences are produced and annotated, and why some are finished while others remain unfinished. You used a wide assortment of public-domain, online databases that continue to grow in size and in sophistication of the analytical tools provided. Prokaryotes have surprised us by their diversity and metabolic flexibility. Eukaryotes contain a lot of duplicated DNA from earlier and simpler times. From complexity come alternatives and redundancies that provide plants and animals with the potential to evolve rapidly and subtly as selection pressures change over time; our genomes lead the way to our ongoing adaptations. In Chapter 3, we will conduct comparative genomics to learn from a higher order of genomics.

References

Section 2.1

Tools

Adams, M. D., & J. C. Venter. 1996. Should non-peer-reviewed raw DNA sequence data release be forced on the scientific community? *Science.* 274: 534–536.

Bánfi, B., A. Maturana, et al. 2000. A mammalian H+ channel generated through alternative splicing of the NADPH oxidase homolog of NOH-1. *Science.* 287: 138–142.

Bently, D. R. 1996. Genomic sequence information should be released immediately and freely in the public domain. *Science.* 274: 533–534.

Cohen, B. A., R. D. Mitra, et al. 2000. A computational analysis of whole-genome expression data reveals chromosomal domains of gene expression. *Nature Genetics.* 26: 183–186.

Elgin, S. C. R. Genetics/Genomics in K–16 education. http://www.nslc.wustl.edu/elgin/genomics. Accessed 14 February 2005.

Gene Ontology Consortium. 2000. Gene onotology: Tool for the unification of biology. *Nature Genetics.* 25: 25–29.

Gonzalez, E., H. Kulkarni, et al. 2005. The influence of CCL3L1 gene-containing segmental duplications on HIV-1/AIDS susceptibility. *Science.* 307: 1434–1439.

International Human Genome Sequencing Consortium. 2004. Finishing the euchromatic sequence of the human genome. *Nature.* 431: 931–945.

International Human Genome Sequencing Consortium. 2001. Initial sequencing and analysis of the human genome. *Nature.* 409: 860–921.

Jackson-Grusby, L., C. Beard, et al. 2001. Loss of genomic methylation causes p53-dependent apoptosis and epigenetic deregulation. *Nature Genetics.* 27: 31–39.

Kimball, J. W. 2005. Genome sizes. http://users.rcn.com/jkimball.ma.ultranet/BiologyPages/G/GenomeSizes.html. Accessed 12 February 2005.

Kirkness, E. F., V. Bafna, et al. 2003. The dog genome: Survey sequencing and comparative analysis. *Science.* 301: 1898–1903.

Larsen, F., G. Gundersen, et al. 1992. CpG islands as gene markers in the human genome. *Genomics.* 13: 1095–1107.

Nakagawa, H., R. B. Chadwick, et al. 2001. Loss of imprinting of the insulin-like growth factor II gene occurs by biallelic methylation in a core region of H19-associated CTCF-binding sites in colorectal cancer. *PNAS.* 98(2): 591–596.

Natale, D. A., U. T. Shankavaram, et al. 2000. Towards understanding the first genome sequence of a crenarchaeon by genome annotation using clusters of orthologous groups of proteins (COGs). *Genome Biology.* 1(5): research0009.1–0009.19.

Overbeek, R., N. Larsen, et al. 2000. WIT: Integrated system for high-throughput genome sequence analysis and metabolic reconstruction. *Nucleic Acids Research.* 28(1): 123–125.

Pasquinelli, A. E., B. J. Reinhart, et al. 2000. Conservation of the sequence and temporal expression of *let-7* heterochronic regulatory RNA. *Nature.* 408: 86–89.

Pruitt, K. D., & D. R. Maglott. 2001. RefSeq and LocusLink: NCBI gene-centered resources. *Nucleic Acids Research.* 29(1): 137–140.

Salzberg, S. L., D. Church, et al. 2004. The Genome Assembly Archive: A new public resource. *PLoS Biology.* 2(9): 1273–1275.

Schmutz, J., J. Wheeler, et al. 2004. Quality assessment of the human genome sequence. *Nature.* 429: 365–368.

She, X., Z. Jiang, et al. 2004. Shotgun sequence assembly and recent segmental duplications within the human genome. *Nature.* 431: 927–930.

Shoemaker, D. D., E. E. Schadt, et al. 2001. Experimental annotation of the human genome using microarray technology. *Nature.* 409: 922–927.

Tatusov, R. L., E. V. Koonin, & D. J. Lipman. 1997. A genomic perspective on protein families. *Science.* 278: 631–637.

Tilghman, S. M. 1999. The sins of the fathers and mothers: Genomic imprinting in mammalian development. *Cell.* 96: 185–193.

Vrana, P. B., J. A. Fossella, et al. 2000. Genetic and epigenetic incompatibilities underlie hybrid dysgenesis in *Peromyscus. Nature Genetics.* 25: 120–124.

Vrana, P. B., X.-J. Guan, et al. 1998. Genomic imprinting is disrupted in interspecific *Peromyscus* hybrids. *Nature Genetics.* 20: 362–365.

Review and Summary Sources

Aach, J., M. L. Bulyk, et al. 2001. Computational comparison of two draft sequences of the human genome. *Nature.* 409: 856–859.

Adoutte, A. 2000. Small but mighty timekeepers. *Nature.* 408: 37–38.

Birney, E., A. Bateman, et al. 2001. Mining the draft human genome. *Nature.* 409: 827–828.

Feinberg, A. P. 2001. Methylation meets genomics. *Nature Genetics.* 27: 9–10.

Hagmann, M. 2000. Mapping a subtext in our genetic book (from News of the Week). *Science.* 288: 945–946.

Hillis, D. M., & M. T. Holder. 2000. Reconstructing the tree of life: New technologies for the life sciences. *A Trends Guide.* (December): 47–50.

Littlejohn, T. 2000. Phylogenetics—A web of trees. *BioTechniques.* 29(3): 482–483.

Passarge, E., B. Horsthemke, & R. A. Farber. 1999. Incorrect use of the term synteny. *Nature Genetics.* 23: 387.

Pennisi, E. 2004. More genomes, but shallower coverage. *Science.* 304: 1227.

Salzberg, S. L., O. White, et al. 2001. Microbial genes in the human genome: Lateral transfer or gene loss? *Science.* 292: 1848–1850.

Tupler, R., G. Perini, & M. R. Green. 2001. Expressing the human genome. *Nature.* 409: 832–833.

University of California at Santa Cruz. 2001. Human genome browser user guide. http://genome.ucsc.edu/goldenPath/help/hgTracksHelp.html. Accessed 19 March 2002.

Weinberg, R. A., & E. S. Lander. 2000. Genomics: Journey to the center of biology. *Science.* 287: 1777–1782.

Wickware, P. 2000. Next-generation biologists must straddle computation and biology. *Nature.* 404: 683–684.

Wolfsberg, T. G., J. McEntyre, & G. D. Schuler. 2001. Guide to the draft human genome. *Nature.* 409: 824–826.

Section 2.2

Abbott, A. 2004. Gut reaction. *Nature.* 427: 284–286.

Berger, J. D. 2005. *Plasmodium* test field. http://www.zoology.ubc.ca/courses/bio332/Labs/Apicomplexa/plasmodium/tutorial.htm. Accessed 12 February, 2005.

Bibby, T. S., I. Mary, et al. 2003. Low-light-adapted *Prochlorococcus* species possess specific antennae for each photosystem. *Nature.* 424: 1051–1054.

Blaser, M. J. 2005. An endangered species in the stomach. *Scientific American.* February: 38–44.

Brüggemann, H., S. Bäumer, et al. 2003. The genome sequence of *Clostridium tetani,* the causative agent of tetanus disease. *PNAS.* 100(3): 1316–1321.

Brüggemann, H., A. Henne, et al. 2004. The complete genome sequence of *Propionibacterium acnes,* a commensal of human skin. *Science.* 305: 671–673.

Butler, D. 2002. What difference does a genome make? *Nature.* 419: 426–428.

Cowman, A. F., & B. S. Crabb. 2002. The *Plasmodium falciparum* genome—A blueprint for erythrocyte invasion. *Science.* 298: 126–128.

Dalevi, D., & S. Hoope. 1999. GC-skew: Origins of replication. http://www.lcb.uu.se/course/bioinfo_ht99/labs/GCskew.html. Accessed 12 February 2005.

Doolittle, R. F. 2002. The grand assault. *Nature.* 419: 493–494.

Dufresne, A., M. Salanoubat, et al. 2003. Genome sequence of the cyanobacterium *Prochlorococcus marinus* SS120, a nearly minimal oxyphototrophic genome. *PNAS.* 100(17): 10020–10025.

Dujon, B., D. Sherman, et al. 2004. Genome evolution in yeasts. *Nature.* 430: 35–44.

Foth, B. J., S. A. Ralph, 2003. Dissecting apicoplast targeting in the malaria parasite *Plasmodium falciparum. Science.* 299: 705–708.

Fraser, C. M., J. D. Gocayne, et al. 1995. The minimal gene complement of *Mycoplasma genitalium. Science.* 270: 397–403.

Fuhrman, J. 2003. Genome sequences from the sea. *Nature.* 424: 1001–1002.

Gardner, M. J., N. Hall, et al. 2002. Genome sequence of the human malaria parasite *Plasmodium falciparum. Nature.* 419: 498–511.

Goffeau, A., B. G. Barrell, et al. 1996. Life with 6000 genes. *Science.* 274: 546–567.

Hall, N., M. Karras, et al. 2005. A comprehensive survey of the *Plasmodium* life cycle by genomic, transcriptomic, and proteomic analyses. *Science.* 307: 82–86.

Hannenhalli, S., & P. A. Pevzner. 1999. Transforming cabbage into turnip: polynomial algorithm for sorting signed permutations by reversals. *Journal of the Association of Computing Machinery.* 46(1): 1–27.

Hoffman, S. L., G. M. Subramanian, et al. 2002. *Plasmodium,* human and *Anopheles* genomics and malaria. *Nature.* 415: 702–709.

Kellis, M., B. W. Birren, & E. S. Lander. 2004. Proof and evolutionary analysis of ancient genome duplication in the yeast *Saccharomyces cerevisiae. Nature.* 428: 617–624.

Kobayashi, K., S. D. Ehrlich, et al. 2003. Essential bacillus subtilis genes. *PNAS.* 100(8): 4678–4683.

Legionella Genome Project. 2004. Genome replication and transcription. http://legionella.cu-genome.org/annotation/replic.html. Accessed 12 February 2005.

Ménard, R. 2005. Knockout malaria vaccine? *Nature.* 433: 113–114.

Palenik, B., B. Brahamsha, et al. 2003. The genome of a motile marine *Synechococcus. Nature.* 424: 1037–1042.

Pejchal, R., & M. L. Ludwig. 2005. Cobalamin-independent methionine synthase (MetE): A face-to-face double barrel that evolved by gene duplication. *PLoS Biology.* 3(2): 1–12.

Randau, L., R. Münch, et al. 2005. *Nanoarchaeum equitans* creates functional tRNAs from separate genes for their 5'- and 3'-halves. *Nature.* 433: 537–541.

Raoult, D., S. Audic, et al. 2004. The 1.2-megabase genome sequence of *Mimivirus.* 306: 1344–1350.

Rocap, G., F. W. Larimer, et al. 2003. Genome divergence in two *Prochlorococcus* ecotypes reflects oceanic niche differentiation. *Nature.* 424: 1042–1047.

Saccharomyces Genome Database. 2005a. New ORFs: Summary of results. http://www.yeastgenome.org/chromosomeupdates/newORFs.html. Accessed 12 February 2005.

Saccharomyces Genome Database. 2005b. Sequence annotation changes. http://www.yeastgenome.org/chromosomeupdates/annotation_changes.html. Accessed 12 February 2005.

Snow, R. W., C. A. Guerra, et al. 2005. The global distribution of clinical episodes of *Plasmodium falciparum* malaria. *Nature.* 434: 214–217.

Sullivan, M. B., J. B. Waterbury, & S. W. Chisholm. 2003. Cyanophages infecting the oceanic cyanobacterium *Prochlorococcus. Nature.* 424: 1047–1051.

Waters, E., M. J. Hohn, et al. 2003. The genome of *Nanoarchaeum equitans:* Insights into early archaeal evolution and derived parasitism. *PNAS.* 100: 12984–12988.

Wirth, D. F. 2002. Biological revelations. *Nature.* 419: 495–496.

Wood, V., R. Gwilliam, et al. 2002. The genome sequence of *Schizosaccharomyces pombe. Nature.* 415: 871–880.

Xu, J., M. K. Bjursell, et al. 2003. A genomic view of the human–*Bacteroides thetaiotaomicron* symbiosis. *Science.* 299: 2074–2076.

Zhang, R., H.-Y. Ou, & C.-T. Zhang. 2004. DEG: A database of essential genes. *Nucleic Acids Research.* 32: D271–D272.

Section 2.3

Adams, M. D., S. E. Celniker, et al. 2000. The genome sequence of *Drosophila melanogaster. Science.* 287: 2185–2195.

Alonso, J. M., A. N. Stepanova, et al. 2003. Genome-wide insertional mutagenesis of *Arabidopsis thaliana. Science.* 301: 653–656.

Bejerano, G., M. Pheasant, et al. 2004. Ultraconserved elements in the human genome. *Science.* 304: 1321–1325.

Benos, P. V., M. K. Gatt, et al. 2000. From sequence to chromosome: The tip of the *X* chromosome of *D. melanogaster. Science.* 287: 2220–2222.

Bergthorsson, U., K. L. Adams, et al. 2003. Widespread horizontal transfer of mitochondrial genes in flowering plants. *Nature.* 424: 197–201.

Berkeley Drosophila Genome Project. 2005. Release 4.0 notes. http://www.fruitfly.org/annot/release4.html#id. Accessed 14 February 2005.

Chen, J., G. B. Call, et al. 2005. Discovery-based science education: Functional genomics dissection in *Drosophila* by undergraduate researchers. *PLoS Biology.* 3(2): e59 0208–0209.

The ENCODE Project Consortium. 2004. The ENCODE (ENCyclopedia Of DNA Elements) project. *Science.* 306: 636–640.

Gill, B. S., R. Appels, et al. 2004. A workshop report on wheat genome sequencing: International Genome Research on Wheat Consortium. *Genetics.* 168: 1087–1096.

Goff, S. A., D. Ricke, et al. 2002. A draft sequence of the rice genome (*Oryza sativa* L. ssp. *japonica*). *Science.* 296: 92–100.

Hallem, E. A., A. N. Fox, et al. 2004. Mosquito receptor for human-sweat odorant. *Nature.* 427: 212–213.

International Chicken Genome Sequencing Consortium. 2004. Sequence and comparative analysis of the chicken genome provide unique perspectives on vertebrate evolution. *Nature.* 432: 695–722.

International Chicken Polymorphism Map Consortium. 2004. A genetic variation map for chicken with 2.8 million single-nucleotide polymorphisms. *Nature.* 432: 717–722.

International Human Genome Sequencing Consortium. 2004. Finishing the euchromatic sequence of the human genome. *Nature.* 431: 931–945.

Ito, Y., K. Arikawa, et al. 2005. Rice Annotation Database (RAD): A contig-oriented database for map-based rice genomics. *Nucleic Acids Research.* 33: D651–D655.

Jaillon, O., J.-M. Aury, et al. 2004. Genome duplication in the teleost fish *Tetraodon nigroviridis* reveals the early vertebrate protokaryotype. *Nature.* 431: 946–957.

Kornberg, T. B., & M. A. Krasnow. 2000. The *Drosophila* genome sequence: Implications for biology and medicine. *Science.* 287: 2218–2220.

Lercher, M. J., A. O. Urrutia, & L. D. Hurst. 2002. Clustering of housekeeping genes provides a unified model of gene order in the human genome. *Nature Genetics.* 31: 180–183.

Myers, E. W., G. G. Sutton, et al. 2000. A whole-genome assembly of *Drosophila. Science.* 287: 2196–2204.

Nelson, C. R., B. P. Berman, et al. 2000. A BAC-based physical map of the major autosomes of *Drosophila melanogaster. Science.* 287: 2271–2274.

Pejchal, R., & M. L. Ludwig. 2005. Cobalamin-independent methionine synthase (MetE): A face-to-face double barrel that evolved by gene duplication. *PLoS Biology.* 3(2): e31 0001–0012.

Ricchetti, M., F. Tekaia, & B. Dujon. 2004. Continued colonization of the human genome by mitochondrial DNA. *PLoS Biology.* 2(9): 1313–1324.

The Rice Chromosome 10 Sequencing Consortium. 2003. In-depth view of structure, activity, and evolution of rice chromosome 10. *Science.* 300: 1566–1569.

Rubin, G. M., & E. B. Lewis. 2000. A brief history of *Drosophila*'s contributions to genome research. *Science.* 287: 2216–2218.

Rubin, G. M., M. D. Yandell, et al. 2000. Comparative genomics of the eukaryotes. *Science.* 287: 2204–2215.

Skaletsky, H., T. Kuroda-Kawaguchi, et al. 2003. The male-specific region of the human Y chromosome is a mosaic of discrete sequence classes. *Nature.* 423: 825–837.

Spellman, P. T., & G. M. Rubin. 2002. Evidence for large domains of similarly expressed genes in the *Drosophila* genome. *Journal of Biology.* 1(5): 5.1–5.7.

Stark, A., J. Brennecke, et al. 2003. Identification of *Drosophila* microRNA targets. *PLoS Biology.* 1(3): 397–409.

Yu, J., S. Hu, et al. 2002. A draft sequence of the rice genome (*Oryza sativa* L. ssp. *indica*). *Science.* 296: 79–92.

Yu, J., J. Wang, et al. 2005. The genomes of *Oryza sativa*: A history of duplications. *PLoS Biology.* 3(2): e38 0001–0016.

Comparative Genomics in Evolution and Medicine

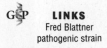

LINKS
Fred Blattner
pathogenic strain

Analyzing genome sequences is more informative when you compare many of them simultaneously. Comparisons allow us to discover trends, new species, and evolutionary legacies. As we learn more, we can detect and identify species from small DNA samples used in forensics. If we can reconstruct a species' full metabolic capacity, we might be able to create medications that kill pathogens and improve health care. If information is power, genome sequences are potential energy for the next century of biological research.

3.1 Comparative Genomics

In Chapter 2, we learned how genomes are sequenced and annotated. Annotation is facilitated by comparing one genome against others. Section 3.1 examines genomes at a variety of scales to teach different types of lessons. What makes a species pathogenic? How many species are alive today? Have genomes finished evolving, or is change still happening and being selected? Can we learn how our chromosomes evolved to their current state? Perhaps the only frustration for comparative genomics is that each week new genomes are sequenced, so conclusions produced today may be outdated next week. If nothing else, this frustration guarantees that every week will offer new opportunities to learn.

How Can *E. coli* Be Lethal and in Our Intestines at the Same Time?

It has been said that every molecular biologist is interested in two organisms, the one he or she studies and *E. coli*. Laboratory strains of *E. coli* enable researchers to clone DNA, express proteins, and isolate genes of special interest; without *E. coli*, labs cannot function. In 1997, the genome of a nonpathogenic laboratory strain of *E. coli* (called K-12) was published. The 4.6 Mb genome was sequenced in the laboratory of Fred Blattner at the University of Wisconsin. This sequencing milestone was heralded as a significant contribution in understanding *E. coli* that could affect how all molecular biology research is performed. In addition, we might be able to make inferences about strains of *E. coli* that are human pathogens. However, genomics taught us one lesson early on: It is risky to use the genome of one organism to make predictions about the genome of another. Blattner knew this lesson as well as anyone, so he sequenced the genome of a pathogenic strain, too.

In January 2001, the Blattner lab produced its second milestone, the complete genome sequence of the **pathogenic** (potential to harm its host) and **virulent** (substantial amount of harm or death to host) *E. coli* strain called **O157:H7**. Every one of us has *E. coli* in our intestinal tracts; this nonpathogenic intestinal strain occupies space and consumes limiting nutrients (e.g. iron) that help us resist infection by pathogenic bacteria. If you consume some undercooked beef that is contaminated with O157:H7, you might have the misfortune of experiencing the differences in genomes. O157:H7 was first recognized as a pathogenic strain in 1983 after an outbreak the previous year that originated from undercooked hamburger meat processed in Michigan. Since then, O157:H7 has been detected worldwide as the cause of occasionally lethal hemorrhagic colitis, of which 75,000 new cases occur annually in the United States. In addition to transmission of O157:H7 via infected beef, pathogenic *E. coli* can be transmitted by raw fruits and vegetables (from fertilizer), person-to-person contact, and contaminated water, which creates the potential for epidemics whenever there is a flood. Therefore, understanding the O157:H7 genome could lead to better diagnosis and treatment.

Blattner's lab used a 1982 isolate of O157:H7 to sequence a 5.5 Mb circular chromosome, of which 4.1 Mb form a "backbone" conserved between O157:H7 and K-12. However, O157:H7 contains about 1.34 Mb, termed "O-islands," not found in K-12; conversely, the K-12 genome contains about 0.53 Mb of "K-islands" not found in O157:H7. Only about half the O-island and K-island genes can be assigned functional roles. The two islands are scattered throughout the genomes, with O-islands containing 1,387 genes (out of 5,416 total genes) not found in K-12, and K-islands containing 528 genes not found in O157:H7. Many of the O157:H7-unique genes are predicted to be virulence genes since they encode toxins, metabolic pathways, transporters, and adhesion molecules. Ironically, K-12 also contains genes that could be predicted to be virulence genes, and yet K-12 is not a human pathogen. The predicted virulence genes might indicate that K-12 could be pathogenic in a nonhuman host, that K-12 is evolving into a human pathogen, or our virulence annotation is wrong. To distinguish between these possibilities will require many years of laboratory research. Deciphering what K-island genes do will help us understand which ones enable O157:H7 to be virulent and K-12 to be a docile lab strain.

One of the most striking features of O-islands and K-islands is their base compositions. The two sets of genomic islands do not have the same ratios of the four bases as found in the backbone. Furthermore, many of the island genes have **orthologs** found in other species and in viruses. Therefore, it appears that large portions of K- and O-islands may be the result of **horizontal transfer** from other organisms. Significant differences in these two genomes begs the question of how we define "species" and "strain." Genome sequences are forcing us to reconsider old definitions that seemed clear just a few years ago (see Section 4.1).

Since about 75% of O157:H7 and K-12 genomes are conserved in the backbone, you might predict that most of

LINKS G&P
genome of HIV

| **Math Minute 3.1** | **How Can You Tell if Base Compositions Are Different?** |

To say that the ratios of the four bases in the O-islands and K-islands are different from the backbone ratio, you must do some statistical analysis. You really want to be able to say they are *significantly* different. The standard tool for testing whether two sets of ratios, or frequencies, are significantly different is the **chi-square test** of homogeneity.

To illustrate how the chi-square test is performed, suppose that you had a 4 kb sequence containing 1,000 bases of each type, and a 3 kb sequence containing 600 A's, 800 C's, 700 G's, and 900 T's (Table MM3.1). For each of the eight cells in the interior of the table, compute the expected frequency of that cell as the cell's row total times the cell's column total divided by the grand total. For example, the expected frequency of A in Sequence 1 is $1,600 \times 4,000/7,000 \approx 914.29$. Subtract the expected frequency from the actual frequency ($1,000 - 914.29 \approx 85.71$), square the result ($85.71^2 \approx 7,346.2$), and divide this number by the expected frequency ($7,346.2/914.29 \approx 8.03$). Repeat this process for each of the eight cells, and sum the results. If this sum is larger than a cut-off value from the chi-square distribution, the frequencies are said to be significantly different.

This is the math behind the scenes of the simple statement that the ratios of the four bases in most O-islands and K-islands are different from the ratios in the backbone. The investigators performed the chi-square test for 108 O-islands longer than 1 kb, and found 101 of them to have a base composition significantly different from that of the backbone.

Table MM3.1 Base frequencies.

Base	Sequence 1	Sequence 2	Total
A	1,000	600	1,600
C	1,000	800	1,800
G	1,000	700	1,700
T	1,000	900	1,900
Total	4,000	3,000	7,000

the genes in the backbone would encode identical proteins. Interestingly, only about 25% (911/3,574) of the backbone genes produce identical proteins, with the remaining proteins varying by at least one amino acid. This may sound like a trivial difference, but remember that wild-type (*wt*) versions of hemoglobin and cystic fibrosis protein (CFTR) differ from disease-causing versions by one amino acid. The difference in pathogenicity between O157:H7 and K-12 may be due to these subtle nucleotide variations.

The genome for O157:H7 had two **gaps** in the sequence database (4 kb and 54 kb) that were filled after publication of the genome sequence. Gap filling is often perceived as uninteresting and not worthy of funding, but genomes can reveal information that can be understood only in the context of a **finished** sequence. Although these gaps are filled, the task is not completed. The genomes of various isolates of O157:H7 have different restriction maps, and thus different genomes. These variations might correlate with disease phenotypes that could help us characterize genes with unknown functions. Nevertheless, the O157:H7 genome sequence is a good starting place for clinical uses such as diagnosis, vaccine production, and treatments. Both *E. coli* genome sequences are in the public domain, so anyone who wants to work in this area of biomedical research may use them.

Now that we have sequenced the genomes of two strains of *E. coli*, we are hopeful that biomedical research will lead to better human health. However, easy fixes are unlikely. For example, the HIV genome was sequenced in 1985, and it contains only nine genes. With such a simple organism and a simplistic genome, you might think genomic methods would lead to a cure and vaccine very quickly. Look at the genome of HIV and you will see that it does not look challenging compared to *E. coli* genomes. However, there is no cure for AIDS, and the only prevention is through behavioral modification. Size can be intimidating, but the complexity of biology does not always correlate with size.

LINKS
CMR
*coli*BASE
Genomes Online
PNAS abstract
sizes range

DISCOVERY QUESTIONS

1. Go to TIGR's CMR and align K-12 and O157:H7 whole genomes. Are the backbones colinear their entire lengths?

2. Since we need some *E. coli* to keep our intestines protected from virulent strains, traditional antibiotics are not ideal therapies. Would it be possible to devise a new class of antibiotics to target only O157:H7 and not K-12? Explain how you might develop this type of drug.

3. Read the PNAS abstract and discover the genomic diversity present in pathogenic *E. coli*. Do you think all pathogenic *E. coli* should be lumped into a single category? What could you do to determine the extent of different *E. coli* pathogens?

4. Go to *coli*BASE and perform a search for the gene *papK*. Follow the link until you see a graphic display of *papK* in the genome (small red arrowhead). Change the default option next to "Colour genes by" to "GC content" and click "Redisplay." Mouse over *papK* to see which gene it is. Do all genes in this area have the same GC contents as *papK*? Change "Colour genes by" to "orthologues" and click "Redisplay." On the new page, you will see many species next to checked boxes; click on the button that says "Uncheck all," click on the button next to "Toggle *Escherichia*" to select this genus, then click on the "Colour by orthologs" button at the very bottom. Does *papK* have orthologs in other *E. coli* strains? Which genes have the highest number of orthologs in this view? Given the percentage and the number of strains, how many orthologs exist?

   Now click on the button next to "View MUMmer alignment with:" and choose strain 0157:H7 (the first strain to be sequenced). How well conserved is this region of strain CFT073's genome in 0157:H7's genome? What challenges do these data present when designing a new drug to kill pathogenic strains but not the beneficial strains?

Two Hundred Genomes: What Can Comparative Genomics Tell Us about Prokaryotes?

Go to Genomes Online to see an impressive list of genome sequences completed and in progress. Now that we have all this information, can we learn anything by comparing the collection of prokaryote genomes? The first thing we discover is that defining a prokaryotic genome is not easy, because the diversity is so great. For example, sizes range from 0.5 Mb to nearly 10 Mb, with GC content ranging from 25% to 75%. Their sizes overlap with the sizes of viruses and eukaryotes (Figure 3.1a). There is a trend within the genome data: Organisms that live in stable niches, such as inside another organism, tend to have the smallest genomes, while those living in volatile environments (e.g., soil, oceans, and legume root nodules) tend to have the largest genomes.

Unlike eukaryotes, the **coding capacity** in prokaryotes is consistent across most genomes, with approximately 1 gene per kb of DNA. As genome sizes increase, prokaryotes have a larger number of gene families and/or more members of each gene family. A particularly interesting trend is that the percentage (not just the number) of genes involved in transcriptional control increases proportionally to the genome size. A different way to see the same trend is to annotate all prokaryote genes by their COG category and then compare the different categories with the size of the genomes. The larger the genome, the more genes a species will have in transcription, as well as signal transduction, cell motility, secondary metabolism, and energy production/conservation. All these categories support the idea that species in dynamic environments need more genes to deal with the changes and disproportionately more genes for transcription regulation.

Consistent with the previous correlation of small genomes and limited metabolic capacity, it is not surprising to see the positive correlation between genome size and GC content (Figure 3.1b). GTP and CTP are more energetically expensive to synthesize, and ATP is more readily available even in the simplest metabolomes. However, this correlation does not explain why even in smaller, AT-rich genomes, intergenic DNA has higher AT content than coding DNA. Promoters tend to be more AT-rich in part because AT enrichment allows DNA to form a rigid but curved shape that is more conducive to transcription. Within a species, the origin of replication is GC-rich, while the terminus is AT-rich. Though widespread, these oddities of GC distribution are still unexplained.

Do All Prokaryotes Have One Circular Chromosome?

This simple question is more complex than it first appears because the term chromosome is controversial in prokaryotic genomes. If we accept the term chromosome, then the original question can be broken into its two parts: single and circular. No doubt you have learned that bacterial genomes are circular, but not all of them are. Three distantly related groups of bacteria have linear chromosomes, some of which have a mixture of circular and linear DNA pieces. Based on their evolutionary distance and the different mechanisms for maintaining the telomeres, these three groups of bacteria evolved linear genomes independently of each other. So, to answer half of the question, no, not all prokaryote chromosomes are circular.

a)

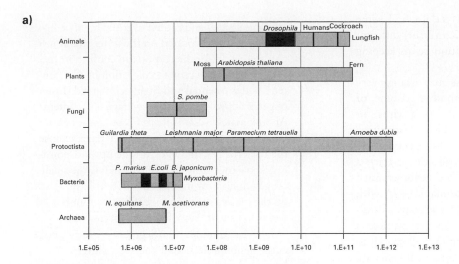

b)

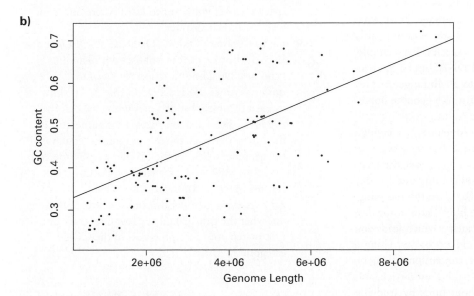

Figure 3.1 Estimates of genome sizes.
a) Size ranges for different classifications of organisms, with some specific examples.
Protoctista includes nucleated microorganisms but excludes fungi, animals, and plants.
b) GC content versus genome length for 146 sequenced prokaryotic genomes.

The number of chromosomes per cell depends on how you define a chromosome. We have already seen a few prokaryotes that have one large chromosome and smaller **plasmids**, similar to eukaryotes having large nuclear genomes and smaller organellar genomes (e.g., mitochondria, chloroplasts, and apicoplasts). When do you call a smaller circular piece of replicating DNA a plasmid rather than a chromosome? Eukaryotes have chromosomes of different sizes, yet we consider all of them to be chromosomes. The distinction becomes even more blurred when you realize some prokaryotes have a "plasmid" nearly as large as the chromosome. If we knew the evolutionary history, perhaps we would know if there had been one giant chromosome that split into two smaller ones or if a small plasmid grew in

size through horizontal transfer. For now, the argument is settled by examining two criteria: (1) Are essential genes on the smaller circular DNA? (2) Are ribosomal genes on the smaller DNA? Yes to either question qualifies the DNA as a chromosome.

Are the Genomes Still Changing?

Most people think of evolution as a slow process of the past, but this is a misconception. Every day, genomes are subjected to selective pressures that can be revealed in comparative genomics. One lesson is that some genes tend to stay linked together over millions of years. Through various recombination events, every gene could have been separated

DATA
80 kb fragment
GS skew

LINKS
coli BASE Browser
DEG
DOGS
GOLD

from its neighbor, and yet we find some gene clusters are **syntenic** in many different species, indicating evolution favors keeping them linked.

The size of a genome may change very rapidly if it fuses with another genome, or accumulates some DNA via a virus, or some other mechanism of horizontal transfer. Acquiring new DNA means acquiring new genes, but genomes do have size limits. Why? Every round of replication extracts a cost for the larger genome, and therefore genomes must balance the expense of replicating redundant DNA with the benefit of having genes that provide a selective advantage only under rare circumstances. Conversely, losing DNA and genes could be advantageous if the cell evolves to fill a new niche, such as inside another species. If genes are no longer advantageous, the DNA can be lost and the more efficient genome provides a selective advantage. From what we can tell so far, genome sizes tend to stay within a fairly narrow size range for a given group of species. For example, K- and O-islands are newly acquired DNA, but all gut bacteria tend to have genomes in the 4 to 5 Mb range. As they acquire new DNA, cells tend to return to a genome homeostasis with an optimal size and gene count.

You might think that 200 sequenced genomes is enough and that we don't need to sequence more, but there is power in numbers for comparative genomics. Consider the analogy of living in a cave all your life and coming out one day and seeing a bluebird and a blue jay. Based on this sampling, you might conclude that all birds are blue. Later in the day, you see a red cardinal and a yellow canary, which leads you to conclude that all birds must be primary colors. Using a small sample size leads to inaccurate conclusions. Imagine your surprise when you see a hummingbird, an ostrich, and a penguin. Just as we learn more about birds by studying their diversity, we learn more about genomes when we have a larger sample size. However, resources are limited so we must choose wisely which genomes we sequence to maximize our ability to learn from them (see Section 2.1).

DISCOVERY QUESTIONS

5. Copy and paste the *E. coli* K-12 80 kb fragment of DNA and perform a GC skew analysis on the DNA. Can you detect the origin of replication?

6. Go to the *M. genitalium* genome picture (Figure 2.15) and use gene orientation to identify the origin of replication. Then find one gene not pointing in the "right" direction and see if you can determine if it is in the Database of Essential Genes (DEG).

7. Search *coli* BASE for the gene *nuoL* and follow the links to *E. coli* K-12 MG1655. Now go to the *coli*BASE Browser and choose K-12 MG1655. Change the coloration to orthologs for all available species and mouse over the highly conserved *nuoL* gene located on the inner strand just past 6 o'clock (see the text box to confirm its location). Change the view to coloration by GC content. Is this essential and highly conserved cluster of genes above or below the average GC content? Approximately what is the GC content for *nuoL*?

8. Go to the DOGS genome size web page, choose the bacterial list, and click on the upward-pointing arrow to sort by size. How small is the smallest genome? How many genes does it have? Click on the species name and see why this number seems too small. Now focus on species with "Main" listed under the "Segment" column to see entire genome sizes. Can you see a pattern between genome size and where small-genome species live? If you want to learn where one lives, read the linked abstract.

9. Go to Genomes OnLine Database (GOLD) and click the button for the published prokaryote genomes. Do a find function for *Bacillus anthracis* Ames 0581. Click on the "MAP" link to see a complete list of genes with *DnaA* at the top of the list. Click on the species name link at the top of this page to see a graphic depiction of the genome. Click on the circular map in the area of GBAA1887. Navigate left or right until you can see genes 1,887 and 1,888. What COG category are these genes? Click on each box to find out what the gene encodes. How does their function match your interpretation of the COG categories?

How Many Genomes Are There?

Determining the number of species in the world has been a challenging problem for many years. Biologists have sampled and counted all over the world, but we tend to count the easiest ones first. For example, half of all described species are insects, including nearly 300,000 beetle species (which prompted naturalist J. B. S. Haldane to remark, "The Creator, if he exists, has a special preference for beetles"). The bigger the organism, the easier it is to count, which means that prokaryotes are the least well-documented organisms. We are uncertain how many Eubacteria and Archaea live on the planet, and we still have trouble defining species-specific characteristics for most prokaryotes. As we saw with cyanobacteria, perhaps genome sequences will provide the best estimate of the number of distinct species in the world.

Craig Venter, the genomicist who pioneered whole genome shotgun (WGS) sequencing, has taken a sabbatical of sorts to sail around the world, collecting DNA samples from all the oceans. He is confident that he can measure

LINKS
Sargasso

species diversity around the world because in 2004 he and 22 coauthors published the first environmental WGS sequencing results from the Sargasso Sea. The Sargasso Sea, located just south of Bermuda in the Atlantic Ocean, is probably the most-studied oceanic region in the world. Venter sampled about 1,500 liters of surface waters 7 times in 4 locations and sorted the species by size to collect organisms bigger than 0.1 μm but smaller than 3.0 μm (i.e., primarily prokaryotes). The cells were lysed and the resulting 7 mixtures of DNA were cut into small pieces and ligated into plasmids. An average of 818 bases from both ends of 990,000 plasmid inserts were sequenced to produce a total of 1.62 billion bp of data. Using a combination of assembly software and hand-curation, the investigators assembled 64,398 **scaffolds** (a collection of contigs lumped together) of 826 bp to 2.1 Mb. The two main questions they wanted to answer were:

1. How many species are there?
2. What is the relative abundance of each species?

Let's look at the number of species first. The investigators identified 1,412 different small subunit rRNA genes or fragments, with 148 of these being new to the database. This indicates that 10% of these ribosomal genes were from species never before sequenced, and illustrates the limitations of previous sampling methods. Most of these species cannot be grown in the lab and therefore we cannot use standard microbiology methods to characterize them. Some investigators have used "universal PCR primers" to amplify all ribosomal genes, but PCR amplification is not uniform and some genes may not bind to these universal primers. Therefore, WGS sequencing identified 148 previously unknown rRNA genes. However, ribosomal genes were sampled randomly with the WGS method, and other genes may give different species counts.

The investigators used 6 additional genes to estimate a range of 341–569 sampled species, or phylotypes (Table 3.1). The term **phylotypes** is the newest effort to clarify the term "species" when classifying prokaryotes. Intended as a functionally equivalent term to "species," phylotypes recognizes that arbitrary distinctions are used to classify a species, since the mating criterion cannot be used with prokaryotes.

The exact number of identifiable phylotypes varies depending on the gene chosen, but the advantage of using ribosomal genes is a huge database of orthologs; rRNA changes very little over time, and every species has ribosomes. The caveat about using ribosomal genes is that two phylotypes with highly conserved ribosomal genes might be collapsed into a single phylotype by the assembly software, causing the loss of one species in our counting. Another bonus of the WGS sampling method was that sequences included dsDNA viruses in the water. Using the virus database as a standard, the investigators identified 71 scaffolds at least 10 kb long, with 50 different viral genes in the scaffolds, and another 150 viral genes in sequence reads that did not assemble into scaffolds.

The authors acknowledged their sampling method was not comprehensive and that they probably failed to sequence DNA from less abundant species. Other species probably were missed due to the random nature of the DNA cloning process and the inability to find an identifiable gene. To address the sampling omissions, Venter's group performed three different calculations to account for missed species. The most conservative estimate was about 1,800 different phylotypes in the water collected. To sequence 95% of all species in their samples, they would need 12 times more sequence data which permits a cost/benefit calculation of expense versus the value of an estimated number of species from a small sample of the world's ocean.

Because the cloned DNA fragments were sequenced randomly, you would expect that more abundant species would have their DNA sequenced more often; thus, the coverage of a particular gene in an abundant species would be greater and a greater number of different genes would be sequenced per species. For example, 53% of all sequenced DNA in water sample #1 were from two genera, *Shewanella* and *Burkholderia*—but this confounds previous knowledge of these species. *Shewanella* is usually found in nutrient-rich water, not nutrient-poor water like the Sargasso Sea. *Burkholderia* is considered a terrestrial species. However, open ocean water contains marine snow, tiny bits of decaying organic matter (including animal feces). If the sampling of water happened in an area with decaying feces, the microbial composition of the sample would be altered.

Table 3.1 Diversity of species defined by six different proteins. Ortholog cutoff refers to the **E-value** used to determine if a sequence was a true ortholog when the *E. coli* gene was queried with BLASTx against the collection of Sargasso DNA.

Protein Name	Sequence ID	Ortholog Cutoff	Observed Phylotypes
AtpD	NTL01EC03653	1×10^{-32}	456
GyrB	NTL01EC03620	1×10^{-11}	569
Hsp70	NT01EC0015	1×10^{-31}	515
RecA	NTL01EC02639	1×10^{-21}	341
RpoB	NTL01EC03885	1×10^{-41}	428
TufA	NTL01EC03262	1×10^{-41}	397

Unfortunately, Venter's group did not view the samples microscopically, so we do not know if their samples contained marine snow. In other water samples, they found abundant cyanobacteria DNA, especially *Prochlorococcus* and *Synechococcus*, whose genomes we studied in Section 2.2. Ninety percent of the cyanobacteria DNA was from *Prochlorococcus*, but again, the sampling methods may have created this bias since *Prochlorococcus* is smaller and thus may have fit into the filters better than the larger *Synechococcus*. Taking advantage of the completed *Prochlorococcus* MED4 genome, the researchers assembled four distinguishable *Prochlorococcus* genomes, indicating the diversity of this phylotype is greater than we had known (Figure 3.2): Find an area where 4 rings of genomes overlap

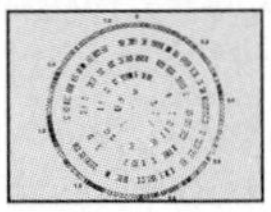

Figure 3.2 Gene conservation among closely related *Prochlorococcus*.
Go to www.GeneticsPlace.com to view this figure.

in the circle (e.g., at ~10 o'clock) and you can see how they arrived at their minimal estimate of *Prochlorococcus* diversity. Notice the large gap (at 8 o'clock) in the 4 inner circles; these genes encode surface polysaccharide synthesizing enzymes and may be either unique to MED4 or highly divergent alleles and thus not recognized as orthologs in the Sargasso DNA. Only by comparing multiple *Prochlorococcus* genomes were we able to identify a diverse cluster of genes in the reference genome sequence.

As you can imagine, sequencing random environmental DNA is bound to uncover some new genes, but you may be surprised by the number of new genes. A total of 1,214,207 genes were identified and added to the databases, including 69,901 novel genes. One class of genes that was very abundant was a recently discovered gene family in marine bacteria called bacteriorhodopsin. Bacteriorhodopsin permits cells to harvest solar energy in the absence of chlorophyll. Previous sampling by PCR had uncovered 67 bacteriorhodopsin homologs, but the Sargasso DNA contained 782 new bacteriorhodopsin homologs—more than 10 times the previous total! The investigators clustered all the bacteriorhodopsin genes and found 13 families from a wider range of phylotypes than we had known before (Figure 3.3).

The purpose of this figure is to impress upon you the degree of our ignorance of the oceans, which influence our global climate (e.g., CO_2 balance; see Section 4.1) and nutrient cycles (nitrogen, phosphorus, and the food chain). Keep in mind, this phylogenetic tree was taken from only 1,500 liters of water in one area, compared to the ocean's estimated volume of 1.37×10^{15} liters. Clearly, we have only begun to sample the oceans' diversity, not to mention the different terrestrial prokaryotes in diverse environments. The survey raises new questions, such as how

nutrient-poor environments can sustain so much genomic diversity (see Section 4.1).

DISCOVERY QUESTIONS

10. Search **ProbeBase**, modify the pull-down menu under the heading "List probes by category" to list the "organisms of medical or hygienical relevance," and click "enter." Do a find function for the probe "Hp16S-2" and view the probe. Copy and paste the sequence into a BLASTn search. Do you get only one species with 100% matching?

11. Based on work published in 1990, oceanographers estimated that 1% to 4% of planktonic bacteria are infected with phage. Based on the number of species in the Sargasso environmental sampling, calculate the number of viruses that should be in that sample. Given their sampling methods, would you expect the number of viruses identified to be higher or lower than the viruses actually present in the sample? Explain your answer.

12. Read a segment from a National Academy of Sciences **report on biodiversity**. Given the bias in sampling, what effect do you think environmental genomics can have on our awareness of the unseen diversity? In 2005, Venter announced an "Air Genome Project" in which his group will sequence DNA sampled from New York City air. Who knows what we're breathing?

13. Perform an **Entrez** search for the accession number AACY01000000. What submission is this? Click on the link and then click on the organism link to see how it is classified. Click on the link "1,986,782" that appears in the bottom right corner of the table. Change the display from "FASTA" to "Trace," click on the color box, then hit the "Show" button. Examine these first few sequences and determine if all the reads were of equal quality (turn on the "Confidence" option). How many bases would you trust from each of the first three reads?

14. Go to NCBI's main page, choose "conserved domain," search for "bacteriorhodopsin," and follow the "pfam" link. Change the view to show as many as are available in the menu; set the color to "identity" and the "Type Selection" to the most diverse members, then hit the "Show Alignment" button. Copy the first stretch of uninterrupted amino acids in the consensus line and perform a BLASTp search. How many good hits did you get? Now try a BLASTp using a region with high conservation as revealed in your modified display. Did you get more hits the second time?

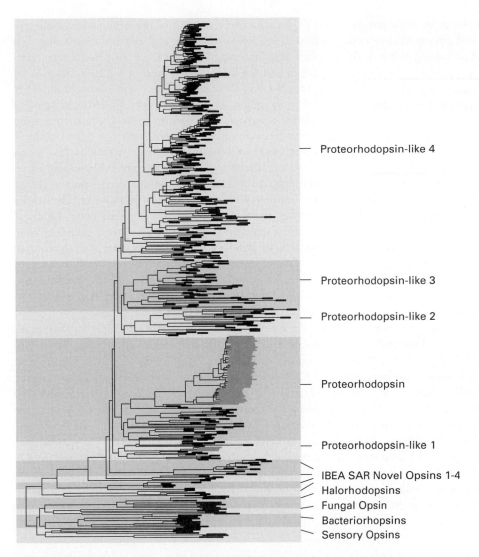

Proteorhodopsin-like 4

Proteorhodopsin-like 3

Proteorhodopsin-like 2

Proteorhodopsin

Proteorhodopsin-like 1

IBEA SAR Novel Opsins 1-4

Halorhodopsins

Fungal Opsin

Bacteriorhopsins

Sensory Opsins

Figure 3.3 Phylogenetic tree of rhodopsinlike genes in the Sargasso Sea data and GenBank.

Sargasso sequences are colored purple, cultured species are black, and other environmental samples are gray. Sequences from uncultured species in the Sargasso Sea. Subfamilies of rhodopsins are indicated on the right. Sequences greater than 75 amino acids long were aligned to each other using CLUSTALw, and a neighbor-joining phylogenetic tree was inferred using Phylip.

What Can We Learn by Comparing Many Whole Genomes?

When we examine a lot of data, what we see depends on what we want to know. If you examine a forest from a distance of 1 meter, you notice the bark. From 20 meters, you notice the shapes of the trees and the foliage; 200 meters reveals species distribution and patchiness; from an airplane you see large patterns of trees combined with other geographical features. Similarly, when we examine many genomes, our perspective determines what we will notice. Zooming in to the amino acid level of the proteome, a group of investigators measured the frequency of amino acid usage in fifteen different species from all three domains of life using three-way comparisons. They identified amino acids in orthologs that were identical in two distantly related species but different in one of two closely related species. By identifying conserved amino acids (in distantly related species) that had drifted during a short evolutionary period (closely related species), they compiled a large number of **amino acid** changes and then charted the frequency for every amino acid, regardless of its position in a protein. It may surprise you to learn that they found a pattern in all species, including humans: cysteine, methionine, histidine, serine, and phenylalanine all showed increased frequency in these variable amino acid positions, while proline, alanine, glutamate, and glycine all decreased (Figure 3.4). The accumulating and decreasing amino acids

Figure 3.4 Color-coded ratio of increasing and decreasing (blue scale) amino acids.

Go to www.GeneticsPlace.com to view this figure.

were fairly consistent across wide taxonomic groups. These two groups of amino acids share interesting features: the decreasing group are proposed to be the first amino acids to be incorporated into the genetic code, while the accumulating group entered the code later. Therefore, all genomes appear to be accumulating more of the newer amino acids at the expense of the older amino acids, in a process that began 3.4 billion years ago and continues today, even in humans.

Zooming out in scale from amino acid to a gene family, a German group led by Svante Pääbo examined the evolution of the family of olfactory receptor (OR) genes in 19 primates plus mouse. In particular, they looked at the number of OR **pseudogenes**. They randomly sampled 100 OR genes from each genome and plotted the percentage of sampled genes that were pseudogenes (Figure 3.5). The nonhuman animals fell into two categories of ≥ 30% or ~18% pseudogenes that correlated with apes and Old World monkeys (30%), as opposed to New World monkeys, a prosimian, and rodent (18%). The New World howler monkey was the lone exception to the correlation since more than 30% of its

OR genes had been inactivated due to mutation. Failing to find a perfect correlation could prove disheartening, but there is a different physiological trait that clarifies the number of pseudogenes. Every species with ~30% OR pseudogenes has full, three-color vision, while males with ~18% OR pseudogenes see only two colors. With the exception of howler monkeys, Old World monkeys have two opsin (color receptor pigment) genes, with one gene on the X chromosome and two alleles (greens or reds) for this locus. Therefore, males can see only two colors (blues and either greens or reds); females can see blue, green, and red if they are heterozygous at the X-linked opsin locus. The howler monkey genome duplicated its X-linked opsin gene and therefore males and females can see all three colors. As often happens in evolution, a gain in one area leads to a loss in another, so those species with full three-color vision cannot sense odors as well. In this case of comparative genomics, we learned how families of genes co-evolve due to selective advantage offered by a second family of genes.

An American team of 72 authors at 8 different institutions zoomed out a bit further to look at a 1.8 Mb segment from human chromosome 7 (including the cystic fibrosis gene *CFTR*) and compared the human sequences to syntenic regions in 11 other species. The entire region encodes 10 genes and the investigators wanted to examine not just the coding regions, but also introns, promoters, and intergenic regions (Figure 3.6). As you would expect, the greatest conservation was in the coding regions; the closer two species were phylogenetically, the higher the DNA conservation.

Surprisingly, rodents are an exception to this trend. Mouse and rat share 10 deletions but their genomes have

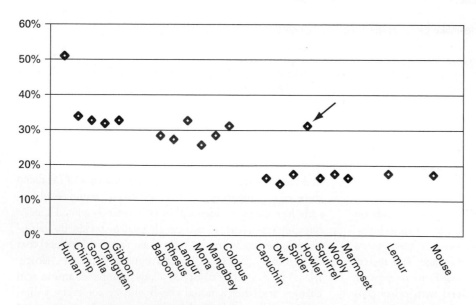

Figure 3.5 Evolution of three-color vision.

Percentage of olfactory receptor pseudogenes form a random sampling of 100 genes from each species. Species are color-coded: humans and apes (purple); Old World monkeys (light purple); New World monkeys, lemur, and mouse (black).

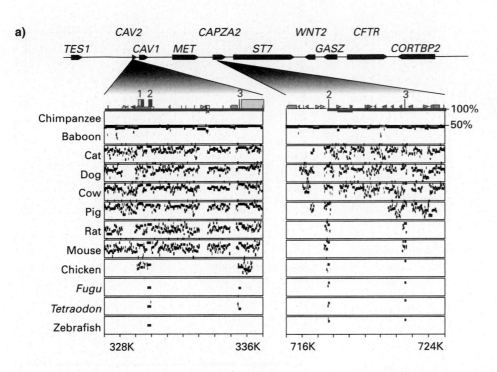

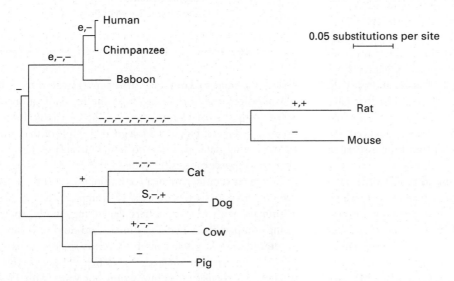

Figure 3.6 Multispecies genome conservation of CFTR region.
a) Pair-wise alignments between the human and 12 vertebrate species. The percentage identity of each gap-free alignment is indicated. Numbered boxes correspond to exons.
b) Deduced phylogenetic tree of indicated mammalian species. Labels on branches reflect differences in exon lengths: +, insertion; −, deletion; e, extension due to alteration of splice site or stop codon; s, early stop codon.

drifted rapidly due to high nucleotide substitution rates, which may explain why they can adapt so quickly to different human environments. However, based on shared deletions, insertions, and transposons, rodents are more closely related to primates than are other mammals. Based on this analysis, rodents are better model animals for studying human physiology than other mammals.

What Can We See at the Chromosomal Perspective?

Evan Eichler at Case Western Reserve University examined human chromosome 7, and his team looked at the entire chromosome for recombination hot spots. By comparing human and mouse synteny, they were able to map locations

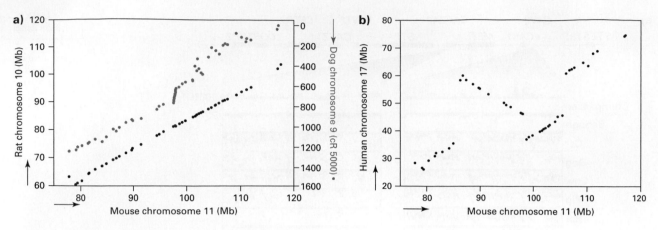

Figure 3.7 Dot plot of gene colinearity compared to mouse.
Arrows point from centromere to telomere. **a)** Rat (black, left Y-axis) genes compared to mouse (X-axis). Dog (gray, right Y-axis) genes compared to mouse (X-axis). **b)** Human genes compared to mouse.

where recombination was more frequent than expected from random recombination events. Chromosome 7 has a total of 27 hot spots for recombination, with 12 on the shorter p arm and 15 on the q arm. Locating recombination hot spots complemented related findings by the international team studying human recombination frequencies that led to common human genotypes (see Chapter 4). A team of 10 investigators from Canada, France, and the U.S. decided to map the exact recombination events that have produced syntenic regions in human, mouse, rat, and dog. Now that all of these genomes have been sequenced, we will examine just one portion of a chromosome associated with a genetic disease in dogs.

Canine tricuspid valve malformation (CTVM) is a genetic disease that leads to thickened heart valves, which stick to the ventricular septum and lead to sudden death. CTVM is especially common in Labrador retrievers, and the CTMV locus was mapped on canine chromosome 9 (dogs have a total of 78 chromosomes). Human chromosome 17 is syntenic to the region containing CTMV and over 50 Mendelian traits are mapped to this region, as well as polygenic traits such as high blood pressure, multiple sclerosis, and diabetes. The investigators looked at whole-chromosome recombination events for CTVM syntenic regions (mouse chromosome 11 and rat number 10; Figure 3.7). As you might expect, the rat and mouse comparison exhibited the fewest chromosomal recombination events. When compared to mouse, dog had one major (4 Mb) recombination event and human experienced two large recombinations (9.2 Mb and 23 Mb).

The dot plots in Figure 3.7 revealed the major recombination events, but the resolution of these figures was limiting. The authors expanded their level of detail, compared all four chromosomes simultaneously, and used lines to help us

Figure 3.8 Mammalian chromosome recombinations.
Go to www.GeneticsPlace.com to view this figure.

follow the major recombination events (Figure 3.8). They also used colored lines to highlight the recombination hot spots, with shaded regions showing the two large human recombined areas. Figure 3.8 is a good example of the power of visualizing data: it is easy to follow inversions (crossing lines) and translocations (lines bending but not crossing).

When we consider smaller recombination events, we can see dog has experienced at least 15, compared to the other mammals, with exact numbers depending on the species being compared (23 between human and dog). The site of recombination is often conserved across species, as is the site of gene loss when it happens. When the perspective is changed to correlate recombination hot spots with DNA sequence, we find highly repetitive DNA, such as Alu sequences, transposons, and RNA genes (including tRNA and rRNA), is often involved. However, the large-scale recombination allows us to create a different type of phylogenetic tree (Figure 3.9). In addition to tracking the number of chromosomal-scale changes between different species, the authors constructed two **most recent common ancestor** (MRCA) chromosomes. The mouse has undergone no new recombination events since it diverged from the rat, but rats have experienced five. Two additional recombinations separate rodents from the human MRCA, as indicated by the two small inversions in the mouse chromosome. Humans have accumulated 10 new inversions or translocations since we diverged from our rodent MRCA.

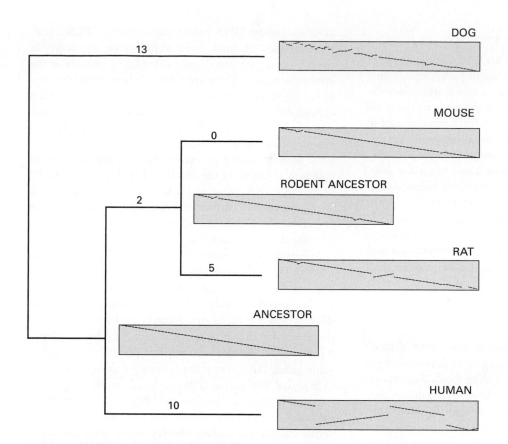

Figure 3.9 Deduced, unrooted phylogenetic tree.
The number of rearrangements that occurred on each edge is shown. Rectangles represent chromosomal arms from Figure 3.8 but are not drawn to scale. A diagonal line provides a visual indication of the gene order and of the positions of rearrangements.

Dogs are placed on the outermost branch of this unrooted tree because its 13 recombinations are unrelated to those shared by the other 3 mammals.

DISCOVERY QUESTIONS

15. Go to the genetic code web page and see if there is a DNA trend based on which amino acids are increasing and which are decreasing.

16. Go to the ECR Browser, choose "Human," and submit this region to be aligned: "chr21:23674384-23707716." What gene have you located? Red bars indicate conserved, noncoding DNA; yellow is conserved *un*translated *r*egion (**UTR**) of exons; blue marks coding exons. Compare the first alignment with the following region: "chr21:26024345–26079899." Is this second region coding or noncoding DNA? How well is this DNA conserved compared to the first alignment?

17. Copy and paste the **Dog v. Mouse** sequences into the dot plot web page. Describe the genomic

changes that have taken place at this region. Look at Figure 3.7a and notice that the rat chromosome is shorter than the mouse one. Look at Figure 3.8 and find a deleted rat region. What visual characteristic in Figure 3.8 makes finding a deletion easier?

18. Use Figure 3.9 to calculate the total recombination distance between each of the species and fill in the following chart:

	Human	Mouse	Rat	Dog
Human	0	—	—	—
Mouse		0	—	—
Rat			0	—
Dog				0

19. Look at the number of human OR pseudogenes in Figure 3.5. Based on these data, what would you predict about human reliance on smell vs. sight for collecting information about our environment?

LINKS
James Lake
tree of life

Summary 3.1

Figure 3.8 illustrates an amazing aspect of genomics research. With all the data in public repositories, you can conduct original genomics research using only the Internet and a computer. Many of the software tools are free, and you can learn to write some Perl scripts yourself to perform analyses never imagined by anyone else. At this time, the only limitations to new comparative genomics discoveries are imagination, time, and some computer skills. We have compared pathogens to beneficial microbes and surveyed environmental microbes to estimate the diversity of prokaryotic life. Looking at vertebrate genomes revealed evolutionary adaptations at many different levels of scale: amino acid, gene families, noncoding DNA, and whole chromosomes. Section 3.2 explicitly addresses the evolution of life.

3.2 Evolution of Genomes

Perhaps the area of biology to benefit most from genome sequences will be our understanding of evolution. Designing nongenomic experiments or finding fossils that provide new insight is extremely difficult. However, organisms alive today carry information in their genomes that have been shaped by evolution. Three fundamental questions have persisted since evolution was accepted as the shaping principle of biology:

1.   What is the origin of life?
2.   How did the nucleus of eukaryotic cells evolve?
3.   What is the origin of human beings?

In this section, we will explore how genomes can be viewed as missing links that enable us to answer these three difficult questions.

What Organism Is the Root of the Tree of Life?

When Darwin described the evolution of life, he used the metaphor of a tree with a trunk giving rise to complex branching as new species evolved. This metaphor has shaped the thinking of biologists since evolution was proposed. The way we visualize information greatly affects the way we think about the data. Since all trees have roots that sprouted from a seed, the image of a tree of life led to the obvious extrapolation: there was a primitive, single-celled organism that can be considered *the* root of the tree of life. Life can be divided into three domains: **Eubacteria** (true bacteria), **Eukarya** (true nuclei), and **Archaea**, the "extremophiles"—lovers of extreme environments. Archaea are distinct from Eubacteria because their cell walls, membrane composition, transcription mechanisms, and rRNA sequences all differ, and Archaea often live where Eubacteria cannot. Archaea live in boiling springs (e.g., *Thermus aquaticus*, whose DNA polymerase is used for **PCR**), consume methane for food, thrive in five-molar salt, or live in the plumes of hot vents under tremendous pressure at the bottom of oceans. These are habitats where textbooks used to say life was impossible. Nature abhors a vacuum, and it appears Archaea can fill all niches. The tree of life metaphor was reinforced by calculating phylogenetic relationships among ribosomal gene sequences of today's living organisms, with branching points representing the most recent common ancestors of the two branches (Figure 3.10a). By analyzing rDNA, three major branches (called **domains**) were apparent with Eubacteria forming one side of the tree and Archea and Eukarya closer to each other evolutionarily forming the other side of the tree.

However, when nonribosomal genes were used to generate a tree of life, the branching pattern was different and placed Archaea closer to Eubacteria than to Eukaryotes. Both trees of life cannot be equally correct. It is precisely this two-tree dilemma that led Maria Rivera and James Lake at UCLA to devise a new way to generate the tree of life. They developed a method that did not look at subtle nucleotide changes in single genes, but rather at the presence or absence of a gene in a digital way (either 0 or 1). Based on recent genome sequence analysis (see Section 2.2), they knew that prokaryotes can transmit genes through two nonsexual means: horizontal transfer and **endosymbiosis**. When genes can move across species that are not evolutionarily close to one another, the tree of life becomes more like a vine with criss-crossed and interwoven branches (Figure 3.10b). Genomic data led to the conclusion that the tree metaphor has outgrown its usefulness and its inherent confinement; there is a need for a new metaphor and visualization of life's origins.

Before we analyze the data, let's consider the implications of visualizing information in a new way (Figure 3.11). Consider a single species (α) that is gradually accumulating mutations and then duplicates its genome, leading to the development of two different species (A is similar to α but very different from B). Over time, A and B continued to drift, with some genes mutating, some being deleted or duplicated, and others evolving with new functions. Some time later, one B cell and one A cell fused genomes such that a new species was formed (species 1), but species A and B also remained intact. When the presence or absence of genes is used to generate a tree (Rivera and Lake's digital method), you can see that the relationships of all four species is somewhat circular, such that any one species is separated from the others by no more than one transition species. Depending on the gene chosen for comparison, species 1 might look more like B or more like A. Therefore, the circular tree in Figure 3.11 allows us to see the dual relationship of genome 1 to A and B.

Using their digital method, Rivera and Lake analyzed a mixture of prokaryote and eukaryote genomes to retrace how different genes had moved over billions of years of evolution. Using different sets of genes, they produced five

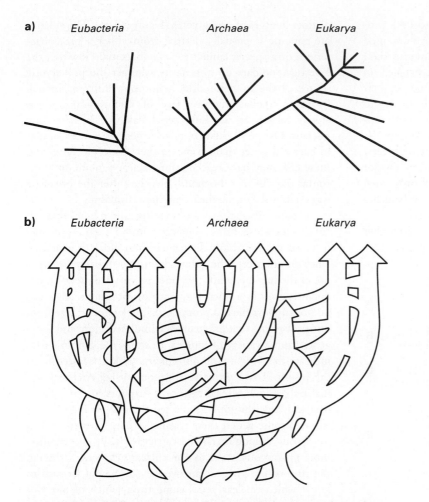

Figure 3.10 Two different schemes displaying phylogenetic relationships.
a) The three-domain proposal based on the ribosomal RNA tree. **b)** The three-domain proposal, with continuous lateral gene transfer among domains.

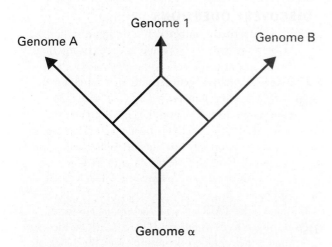

Figure 3.11 Hypothetical example of genome fusion leading to a new domain of life.
Genome α is a progenitor representing the earliest forms of life. Two divergent forms evolved away from each other (A and B). Two representative cells from A and B fused to create the third domain of life (1) as a hybrid of the parental genomes.

different trees that looked different from each other, until they lined up each of the terminal branches in columns for the different species (Figure 3.12). By altering the position of the trees so that similar branches were aligned, they realized these five trees were all cut from a common base. If each of the five trees is drawn as a circle, all five can be seen as variations of a single past. Producing a circular phylogenetic map is very similar to producing a circular map for a plasmid from a series of linear restriction maps. Once the evolutionary relationships were drawn as a circle, the investigators realized the tree terminology was inappropriate, so they called their map a "ring of life." The first ring version had an 87% probability of having the correct branching structure, but if they allowed a bit more ambiguity in their rings, they could increase the probability of having the remaining branches correct.

One of the most striking consequences of this visualization is that the ring of life accommodates all the earlier tree-of-life conclusions about the location of Eukarya relative to Eubacteria and Archaea. As shown in Figure 3.11, when two genomes fuse and produce a new species, the

LINKS
Archaeal genome
TAIR

new species is equally related to both progenitor genomes. The investigators used two very divergent yeast genomes as their Eukarya representatives and these always were perched on top of the ring of prokaryotes. To test the robustness of their digital taxonomy method, they removed each species in their ring of life one at a time to see if the ring would convert to a traditional tree. The only way to convert the ring into a tree with their method is to remove both yeast species at the same time to produce a tree with two major branches (Figure 3.13a). The ring is completed only when genomes that share contents with both of the major branches (Archaea and Eubacteria) are included in the analysis.

The ring of life was constructed from a limited number of species, but they were chosen carefully to represent very

distinct lineages of prokaryotes (Figure 3.13b). The eukaryote genome is partially derived from either a proteobacterium or a species similar to cyanobacteria. However, the investigators' data do not clarify whether the prokaryotic source of the mitochondrial genome (a different branch within proteobacteria; see **Tree of Life** web page) was closely related to the Eubacterial half of the eukaryotic nucleus. Over 680 Eukarya nuclear genes are more similar to bacterial genes than to most other eukaryote genes and these 680 may have reached our nuclear genome by horizontal transfer of mitochondrial genes after the organelle was established inside the eukaryote cytoplasm.

In their discussion, the investigators argue that all eukaryotes arose from a genome fusion event, probably from an endosymbiont living within another prokaryote. They do not accept the old mechanism evoked by Darwin's use of the tree metaphor that one species gave rise to subsequent species in a linear sequence of evolution. They also do not support the old proposal that eukaryotes arose from one prokaryote that gradually accumulated more and more genes from a variety of sources. In short, the genomic data indicate that we, and all our eukaryotic kin, evolved from the fusion of two prokaryote genomes, half Archaeal and half Eubacterial. After the genome fusion event, mitochondria and later chloroplasts entered the cytoplasm and contributed more genes; these organelles have further evolved to provide unique metabolic capacities while contributing more prokaryotic genes to our nuclear genomes. Although the ring of life gives us an overview, it would be nice to know which nuclear genes came from which of our two prokaryote ancestors. We address this question in the next section.

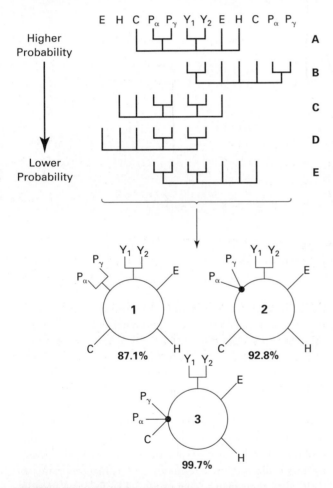

Figure 3.12 Relationships between genomes.
The genomes are from two yeasts (Y_1, *S. pombe*, and Y_2, *S. cerevisiae*), a γ-proteobacterium ($P_γ$, *X. fastidiosa*), an α-proteobacterium ($P_α$, *Brucella melitensis*), a cyanobacterium (C, *Synechocystis*), a halobacterium (H, *Halobacterium* sp.), and an eocyte (E, *S. tokodaii*). The five unrooted trees (A–E) consistent with the ring are shown with leaves positioned to emphasize similar order, with probability decreasing from top to bottom. Circularized unrooted trees show varying degrees of resolution and proportional reliability.

DISCOVERY QUESTIONS

20. Look the rings numbered 1–3 in figure 3.12. Determine where you would have to cut the rings to produce trees A–E in the same figure.

21. View an **Archaeal genome** and do a search for the TATA binding protein by searching with "tbp" in the search field. Click on the gene icon and the "BLink" link on the NCBI gene page. Does this database report any hits outside Archaea? Perform an NCBI Homologene search with "TATA" and see if any other domains use TATA binding proteins

22. Search the **TAIR** web site for *Arabidopsis* for the nitrogen fixation protein NFU5. Click on the one protein link on the results page, then click on "NFU5" listed under "Name," and finally on the "NCBI BLink" to search for orthologs of NFU5. Look at the domain distribution of orthologs at the top of this NCBI BLink result to determine how widespread NFU5 orthologs are. Is NFU5

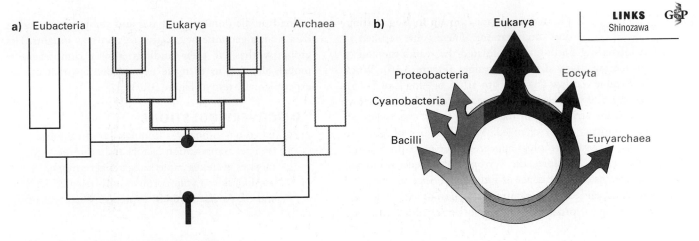

a) Eubacteria   Eukarya   Archaea

b) Eukarya   Proteobacteria   Cyanobacteria   Bacilli   Eocyta   Euryarchaea

Figure 3.13 Two versions of the ring of life.
a) Archaeal (dark purple) and Eubacterial branches (black) with two individual genomes fusing to form the first eukaryote (purple and black). **b)** Stylized version with eukaryotes at the top of the circle and contributing genomes below. Archaea includes euryarchaea and eocyta; Eubacteria includes proteobacteria, cyanobacteria, and bacilli.

found in one, two, or three domains? Propose a mechanism that could explain the distribution of NFU5.

23. Go to the **Tree of Life** web page and click on "Root of the Tree." How accurate does the root look based on what you have read?

What Are the Origins of Our Nuclear Genes?

Most of you learned that the nucleus, chloroplast, and mitochondrion came into existence through endosymbiosis; that is, one cell engulfed another and now the two formerly separate cells rely on each other for survival as a single eukaryotic cell. Although it is difficult to imagine one cell engulfing another, it sounds plausible given that all three organelles contain their own genomes and are encased in double membranes. However, the nuclear genome is very different from prokaryotic, chloroplastic, and mitochondrial genomes. How can the endosymbiotic nuclear genome look so different from the other two?

Four investigators in Takao **Shinozawa**'s lab at Gumma University in Kiryu, Japan, used comparative genomics to determine the origin of nuclear genes. Shinozawa and his colleagues used BLAST alignments to determine the number of orthologous yeast genes in six Archaea and nine Eubacteria genomes. For their BLAST alignments, they used a range of different expect values (**E-values**) and a statistical method called a **t-test** to infer which prokaryote provided the DNA in four categories of yeast genes.

The first group of genes they tested were those involved in yeast "nuclear organization." They compared the number of Eubacterial or Archaeal orthologs (i.e., the hit number) to

each yeast gene (Figure 3.14). Processes of nuclear organization include meiosis, DNA replication, cell cycle control, mitosis, transcription, ribosomal proteins, translation, nuclear structure, and the endoplasmic reticulum (ER), for a total of 2,688 genes. To read Figure 3.14a, compare the number of Archaea orthologs (black lines) to the number of Eubacteria orthologs (purple lines). Archaea genomes contain more nuclear organization orthologs, as denoted by higher hit numbers for the six Archaea lines compared to the Eubacteria lines. The X-axis is a range of stringency or E-values. If the E-value is set to exclude potential orthologs with low sequence similarity (moving right on the X-axis), the hit numbers decline for each genome. As the stringency is lowered to include potential orthologs with lower sequence similarity (moving left on the X-axis), the hit numbers for each genome increase. Therefore, with −log E-values 1 through 95, the greatest number of yeast orthologs were found in Archaea genomes. The t-test (Figure 3.14b) validated the impression that nuclear organization genes originated in Archaea rather than Eubacteria (see Math Minute 3.2 for details).

The next two categories of yeast genes Shinozawa's group analyzed were 601 mitochondria-related **ORFs**, which included tricarboxylic acid cycle (TCA) and electron transport genes (Figure 3.15a and b) and 1,266 genes related to energy/metabolism (Figure 3.15c and d). In both cases, yeast orthologs were more numerous in Eubacteria genomes than Archaea genomes over a wide range of E-values. The Eubacterial origin of mitochondria and energy metabolism was expected.

Shinozawa's team analyzed a fourth group of genes, ones associated with "organization of the cytoplasm," which includes subcategories such as cytoskeleton, Golgi, vacuoles,

DATA
Archaea protein

plasma membrane, protein folding, sorting, and modifications. These genes resulted in an interesting finding that validated the team's method of considering many different E-values (Figure 3.16). When the E-value was low (e.g., 10^{-5} to 10^{-15}, or $-\log$ E of 5–15), the origin of the yeast genes appeared to be Archaea. However, as the stringency of the E-value cutoff was increased, the t-statistic values changed. With E-values of 10^{-50} to 10^{-170}, the yeast orthologs appeared to be more prevalent in Eubacteria. Therefore, the "cytoplasmic" genes were not classified as either Archaea or Eubacteria, since the interpretation depended on an arbitrary selection of an E-value. By examining the prokaryotic origins of yeast genes, Shinozawa

utilized public-domain databases and the power of seeing data from a genomic perspective to gain new insights. His group synthesized their findings into a comprehensive model, allowing us to make new predictions that can be experimentally tested (Figure 3.17).

DISCOVERY QUESTIONS

24. Based on Shinozawa's findings, do you think Archaea transcription factors and DNA polymerases more resemble Eubacterial or human orthologs? Test your prediction by using BLASTp and the Archaea protein sequence. What type of

a)

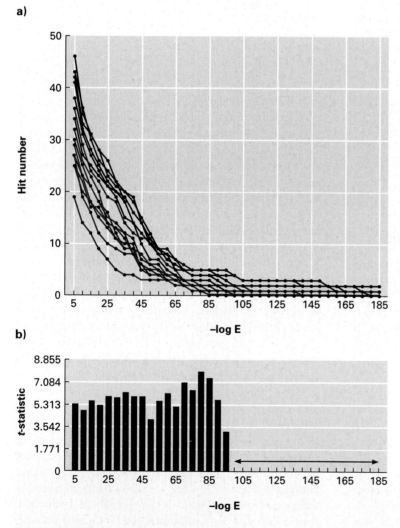

b)

Figure 3.14 Nuclear genes.
Hit numbers **a)** of 2,688 yeast "nuclear organization" genes and t-statistic values **b)** at each E-value threshold. E-values ($-\log$ E-value scale) are shown on the X-axis. Gray lines correspond to Archaea, and purple lines correspond to Eubacteria. t-statistic values are shown in the range of E-values over which the hit number (number of matching sequences in the database) was ≥ 5 in at least one bacterium (bars) or < 5 (double-headed arrow).

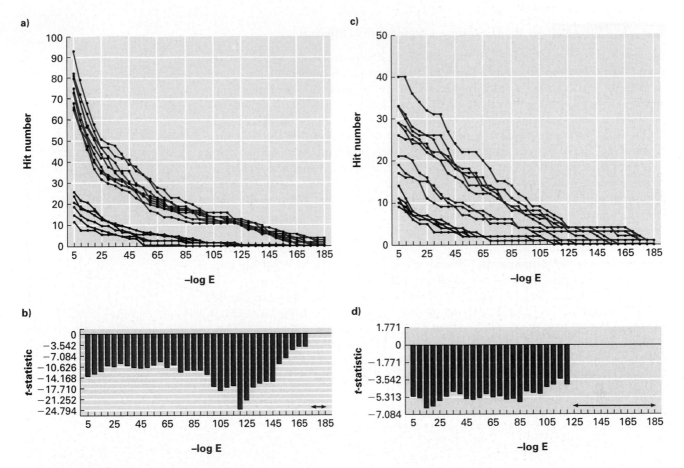

Figure 3.15 Mitochondrial and energy genes.
Hit numbers **a)** of 601 "mitochondrial genes" and *t*-statistic values **b)** at each E-value threshold. Hit numbers **c)** of 1,266 "energy genes" and *t*-statistic values **d)** at each E-value threshold. Gray lines correspond to Archaea, and purple lines correspond to Eubacteria.

protein is this? What species is it from? Are there any human orthologs on the results page (perform a "Find" function with your browser for the word "human")?

25. If you wanted to use an antibiotic to kill an Archaea but not a eukaryote, what cellular processes would make the best targets? Which would be the worst? Explain your answers.

26. Why does it make sense for proteins in the ER and translation machinery to be among those originating in Archaea, rather than proteins in the "cytoplasm" category?

You may have heard some people argue that because no one was present as life first evolved, we cannot be certain about evolution. This argument takes an obvious statement (that no human was alive 3.5 billion years ago) and uses it to imply that science cannot make reasonable conclusions based on vast amounts of data. The life sciences have never claimed to *prove* anything, but we can demonstrate, and support with a large degree of certainty, some conclusions. (For any conclusion you make, someone always can propose alternative explanations that cannot be disproven, and therefore no one can *prove* any given statement. For example, disprove that the reason you are tired today is because last night, you were secretly abducted in your sleep by Martians and brainwashed to forget the entire event.) However, some widely held concepts are as close to the proven state as possible. Gravity is one; DNA as the heritable material is another; evolution is equally well supported. Genomes contain information about an organism's ancestry. As we compare more genomes across wider taxa, we will see with increasing resolution the continuity of life and the overwhelming support of evolution as proposed by Darwin in the mid-1800s. Evolution is not an idea you can choose to *believe* any more than gravity is: it is a scientific

The analysis in Figures 3.14–3.16 is based on BLAST searches for several sets of yeast ORFs at progressively smaller **E-values** (horizontal axes in the figures), limited to six Archaea and nine Eubacteria genomes that are fully sequenced. Since the input sequence length and database size are fixed, BLAST hits must have increasingly larger bit scores as the E-value gets smaller (see Math Minute 1.1). In other words, a lower E-value threshold corresponds to a smaller number of hits. Therefore, every line in Figures 3.14–3.16 either remains flat or decreases as you read from left to right.

What the investigators looked for in these data is a significant separation between the number of Archaea hits and the number of Eubacteria hits. A separation can be detected graphically if the black lines bunch together, the purple lines bunch together, and there is not much overlap between black and purple lines. But how much bunching is necessary? How little overlap? To help answer these questions and determine whether a separation between black and purple lines is statistically significant, the researchers used a two-sample **t-test** (a variation on the paired sample t-test discussed in Math Minute 1.4). Let's walk through the steps of the two-sample t-test on these data.

When the E-value threshold was 10^{-5}, the mitochondria-related yeast genes (Figure 3.15a) returned 12, 15, 19, 21, 24, and 27 hits, respectively, from the six Archaea species, and 65, 66, 68, 73, 75, 80, 81, 82, and 93 hits from the nine Eubacteria species. The first step in the t-test is to describe these two sets of hit numbers with the following summary statistics: sample size of each set (n_1 and n_2), **sample mean** of each set ($\bar{y}_1$ and $\bar{y}_2$), and the sample **standard deviation** of each set (s_1 and s_2; see Math Minute 1.4). In this example, $n_1 = 6$, $\bar{y}_1 \approx 19.67$, $s_1 \approx 5.6$, $n_2 = 9$, $\bar{y}_2 \approx 75.89$, $s_2 \approx 9.09$. Next, the two sample standard deviations are pooled using the following weighted average:

$$s_p = \sqrt{\frac{(n_1-1)s_1^2 + (n_2-1)s_2^2}{n_1+n_2-2}} \approx \sqrt{\frac{5(5.6)^2 + 8(9.09)^2}{6+9-2}} \approx 7.93$$

Finally, the t-statistic for testing whether the Archaea hits and the Eubacteria hits are significantly different is

$$t = \frac{\bar{y}_1-\bar{y}_2}{s_p\sqrt{\frac{1}{n_1}+\frac{1}{n_2}}} \approx \frac{19.67-75.89}{7.93\sqrt{\frac{1}{6}+\frac{1}{9}}} \approx -13.45$$

This is the value shown in Figure 3.15b under the E-value threshold 10^{-5} ($-\log E = 5$). Note that large positive values of t correspond to cases in which $\bar{y}_1$ is significantly larger than $\bar{y}_2$, and large negative values of t correspond to cases in which $\bar{y}_2$ is significantly larger than $\bar{y}_1$.

The final determination of statistical significance is made by comparing each t-statistic to a cutoff value from a table of the t-distribution. For this study, the investigators chose the cutoff value of ± 1.771. That is, t-statistic values greater than 1.771 favor an Archaeal origin, and t-statistic values less than -1.771 favor a Eubacterial origin.

principle that has withstood the test of time and continues to be supported the more we learn. Evolution is a principle you can choose to *accept*, or not. Your personal faith is something you base on beliefs, but religion is not mutually exclusive of evolution. You can be a person of faith and accept evolution as the only logical, scientifically supported explanation of the diversity of life. We discover new evolutionary adaptations every time a new species has its genome sequenced, and some people find their faith strengthened as well.

Is There Evidence of Intermediate Stages in Genomic Evolution?

Data reveal the laws of nature, so errors in science usually arise from poor experimental design, interpretations, or assumptions. With this in mind, you deserve additional data to support the endosymbiotic origin of the nucleus. When we talk about the fossil record and its evidence supporting the principle of evolution, we often think of the "missing link": a hypothetical fossil of a halfway transitional

LINKS
aphids

a)

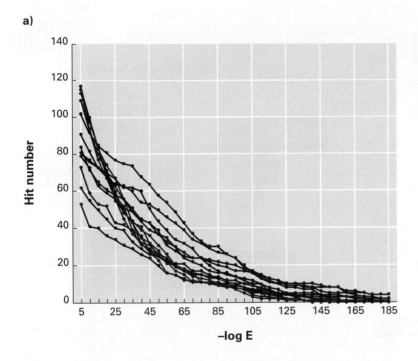

b)

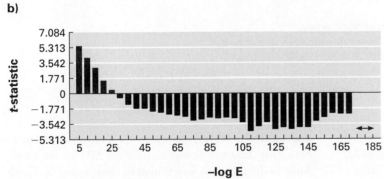

Figure 3.16 Cytoplasmic genes.
Hit numbers **a)** of 2,995 yeast "cytoplasmic genes" and *t*-statistic values **b)** at each E-value threshold. The critical point to note is that the gray (Archaea) and purple (Eubacteria) lines cross at about −log E of 25, and the *t*-statistic value changes with different E-values.

organism that will fill in the missing data point to complete the picture. Wouldn't it be nice if nature provided us with current examples of evolution where prokaryotes are in various stages of symbiosis and genomic flux (Figure 3.18)? Genomes sequenced in 2000 and 2001 provide us with examples of missing links that illustrate what might have happened when the first Archaea was engulfed by a bacterium and eukaryotes were poised to evolve.

Are You Going to Eat That?

This story focuses on a biological curiosity that at first was difficult to explain. Did you know that **aphids**, the tiny insects that draw their food from the phloem of plants, do not excrete nitrogenous waste? This might explain why ants

can drink what comes out of the back end of an aphid, since it lacks toxic nitrogenous waste. However, one rule of biology is that all animals produce nitrogenous waste. Mammals urinate; birds excrete uric acid in their droppings, which is why their feces appear white. We accumulate nitrogenous waste when we break down proteins and amino acids, and we rid ourselves of this waste. Aphids are a curiosity because they are exceptions to this rule. Any time we observe an exception to a rule, we are faced with a few choices: accept these apparent oddities as exceptions that prove the rule, ignore these exceptions, or modify our rules. It seems unwise to say nitrogen waste is either not produced or is not toxic. Therefore, we need to find another reason to keep our rule and still accept the observation that aphids do not excrete nitrogenous waste.

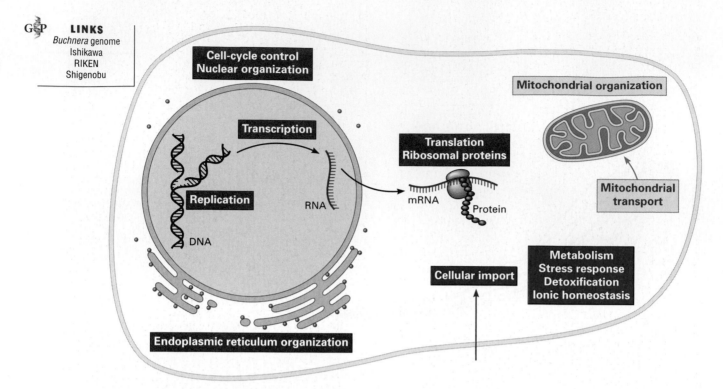

Figure 3.17 Summary of the prokaryotic origins for eukaryotic genes.
Yeast genes of functional groups are colored gray if they originated from Archaea, dark purple if they originated from Eubacteria; mitochondrial genome of bacterial origin is shown in light purple.

Aphids have 60 to 80 large cells in their abdomens called bacteriocytes, which are home to a round bacterium called *Buchnera*. These bacteria are transmitted to the next generation of aphids from the mother, and the mutualism between the insect and the bacterium is so complete (established about 225 million years ago) that neither species can reproduce in the absence of the other. Shuji Shigenobu and colleagues from the University of Tokyo and the **RIKEN** Genome Sciences Centre sequenced the genome of *Buchnera*. The circular chromosome is 640,681 bp, and the genome includes two smaller plasmids of about 8 kb each. This is a small genome (*M. genitalium* has 580,070 bp), which indicates this once free-living bacterium might have lost some of its original DNA.

When Hajime Ishikawa and his collaborators looked for ORFs, they found only 583, with an average gene size of

988 bp. Using BLAST to search the nonredundant database, they assigned biological roles for 500 of the genes. Most of the genes are very similar to those found in *E. coli* (which has 4,288 genes), and many of the *Buchnera* genes are found in the same order as seen in *E. coli*. Seventy-nine of the remaining ORFs are similar to hypothetical genes annotated for other bacterial genomes. Only four genes were found to be unique to the *Buchnera* genome. Based on this, Ishikawa's team concluded that *Buchnera* evolved from a close relative of modern *E. coli*, although *Buchnera* has lost about 75% of its original genome.

What makes the aphid/*Buchnera* symbiosis so interesting is the degree of interdependence. At the creation of eukaryotes, when Archaeal and Eubacterial genomes fused, presumably the new cell retained all genes from both genomes. Current eukaryotic cells contain portions of both genomes, but many genes have been lost over time. The genes found in eukaryotes today are the result of a "compromise," with many of the redundancies between Eubacterial and Archaeal genomes deleted to be more efficient and advantageous. Aphids, like all animals, can synthesize some amino acids, but the remaining essential ones must be supplied through diet. *Buchnera* has lost many genes necessary to synthesize amino acids except the ones aphids cannot synthesize. Therefore, *Buchnera* needs the amino acids aphids produce,

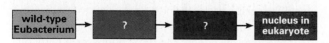

Figure 3.18 Schematic of missing links in the evolution of the eukaryotic nucleus.
We know a lot about bacterial and nuclear genomes, but we would like to discover intermediate steps in this transition.

and aphids need the amino acids *Buchnera* produces. Through this mutualistic relationship, the two are collectively self-sufficient, and they do not need any amino acids to be supplied through diet.

DISCOVERY QUESTIONS

27. Search *Buchnera* for genes in the following list, where you will see an amino acid followed by one gene necessary for its production. Deduce which amino acids are supplied by *Buchnera* and which are supplied by the aphid.

Arginine—ArgA	Histidine—HisA
Valine—IvHI	Tyrosine—TyrA
Isoleucine—IlvC	Proline—ProA
Leucine—LeuA	Serine—SerA
Phenylalanine—PheA	Glycine—GlyA
Tryptophan—TrpAB	Cysteine—CysE
Threonine—ThrA	Glutamate—GdhA
Methionine—MetC	Glutamine—GlnA
Lysine—LysA	Aspartate—AspC
Asparagine—AsnA	Alanine—AlaB

28. Are there any amino acids that are considered nonessential for most animals that *Buchnera* can produce? The nonessential amino acids for most animals are tyrosine, proline, serine, glycine, cysteine, glutamate, glutamine, aspartate, asparagine, and alanine.

29. Based on your findings for Discovery Questions 27 and 28, can you make any predictions about what might happen to the *Buchnera* genome over the next 100 million years?

Given that *Buchnera* has a very accommodating host, you might suspect other genes essential for a free-range life outside aphids might have been lost as well. Indeed, *Buchnera* would not survive for long if it were "set free," since it lacks many DNA repair enzymes and cell-wall synthesizing enzymes, and has no recognizable phospholipid synthesis capacity. It also lacks many signaling pathways most bacteria use to sense and respond to changes in their environment. Therefore, *Buchnera* is dependent on a stable environment where DNA damage is rare and many metabolites are supplied for *Buchnera*'s use. In return, *Buchnera* produces a few amino acids and vitamins as a type of rent to its aphid host. Given that aphids reproduce rapidly and thrive globally, this symbiotic arrangement appears to work very nicely from the host's perspective as well. But no landlord likes to be taken advantage of, and *Buchnera* seems to "know" this rule for subletting space. *Buchnera* retains all the genes necessary to carry out ATP synthesis, including ATP synthase and the electron transport system, although it lacks the genes necessary for fermentation and anaerobic metabolism. Thus, *Buchnera*

LINKS
CMR
Search *Buchnera*

carries its own weight by aerobically producing ATP in addition to a few amino acids and vitamins.

In some ways, *Buchnera* is like a mitochondrion. *Buchnera* produces ATP, contains a small but essential genome, and requires many metabolites produced indirectly by the nuclear genome, including the membrane phospholipids that separate it from the cytoplasm. In this comparison, the aphid supplies the nuclear genome, and *Buchnera* behaves as if it were a newly evolving organelle—"aminoacid-plast." In this way, *Buchnera* looks very much like an intracellular missing link that is making the transition from endosymbiont to organelle, much like the apicoplast in *Plasmodium* (see Figure 2.25). By taking a genome-wide look at *Buchnera*, we developed new connections and insights (Figure 3.19).

Figure 3.19 Whole-cell view of *Buchnera* metabolism as deduced from its genome. Go to www.GeneticsPlace.com to view this figure.

We began this story with the dilemma that either aphids retain their nitrogenous waste or we must revise our rules about animal metabolism. With the full genome sequence of *Buchnera*, the explanation becomes evident, and we realize that aphids are not an exception to the rule. Aphids rarely secrete nitrogenous waste because they recycle the amino groups that all animals produce with protein catalysis. Aphids produce excess glutamine that *Buchnera* uses as a substrate for synthesis of the aphid's essential amino acids. "Waste" is a relative term in biology since every species produces it and inevitably another will consume it. By engulfing *Buchnera*, aphids have streamlined the nutrient cycle and improved their efficiency. Evolution favors those with a selective advantage, and converting waste nitrogen into essential amino acids has to be near the top of the list of adaptive solutions to selection pressures.

DISCOVERY QUESTIONS

30. Aphids can destroy houseplants very quickly. Based on what you have learned, can you design a way to kill aphids but not harm other insects, animals, or plants?

31. Ishikawa and his colleagues propose that *Buchnera* evolved from *E. coli*. Go to TIGR's CMR, where you can analyze whole genomes.

   a. Scroll down and click on the link that says "Align Whole Genomes." Choose *Buchnera* for the Y-axis and *E. coli* K-12 for the X-axis, and then click on "Launch Alignment." You will see a picture that aligns the two genomes,

LINKS
Stewart Cole
M. leprae
M. tuberculosis
PubMed

with a color code described below the graph. Determine the degree of genome alignment for these two species.

b. Go back to the CMR home page and choose "Genome v/s Genome Protein Hits." First, choose *E. coli* K-12 as the reference genome. Second, choose *Buchnera* from the pop-up menu and then click on "Add Molecule." Choose *E. coli* K-12 and "Add Molecule." This will allow you to compare the encoded proteins (as defined by the published annotations). Click on "Generate Display." The results are provided graphically (left side) and in text format (right). If you mouse over the graphic displays of the genomes, you can see where the similarities lie within the reference genome. Knowing that *Buchnera* has 583 ORFs and *E. coli* has 4,288, does this comparison match your expectations?

32. Understanding aphid physiology may sound too esoteric to be of any benefit, but compare this story to one about *Onchocerca volvulus*, the parasitic worm that causes river blindness in Africa and parts of Latin America. Perform a PubMed search using the terms "Immunopathogenesis of Onchocerca volvulus Pearlman." What is the true cause of river blindness?

We have seen our first example of a missing link with *Buchnera* losing some of its genome. But again, you might think gene loss is unsubstantiated speculation since no one can prove a negative. In other words, it is impossible to prove that genes have been lost when they are not present currently. It would be reassuring if we could find a free-living bacterium whose genome is halfway between a typical bacterium and a symbiont that has lost a lot of genes already (Figure 3.20).

A Missing Link of Biblical Proportions

A human pathogen may provide the missing data to complete our understanding of the transitional stages that led to the eukaryotic nucleus. For many centuries, one bacterium was the scourge of the world: *Mycobacterium leprae*.

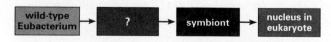

Figure 3.20 Updated schematic of missing links in evolution of eukaryotic nucleus.
Current symbionts (e.g., *Buchnera*) that have lost some of their genome may resemble what happened billions of years ago as an internalized prokaryote became a nucleus.

M. leprae is the causative agent of **leprosy**, which was highly contagious, disfiguring, lethal, and resulted in patients being quarantined in leper colonies. The mode of *M. leprae* transmission is uncertain, although it is believed to enter through the nasal passages. Once inside your body, it lives in the most unlikely places, inside your phagocytic white blood cells (called **macrophages**) that normally engulf and kill pathogenic bacteria. Eventually, it also infects Schwann cells, which provide the fatty insulation for nerves (called **myelin**). Infected Schwann cells lead to the loss of myelin, nerve damage, loss of sensation, and eventually loss of extremities due to reduced blood circulation.

Although we understand leprosy better, we are unable to eliminate it. The World Health Organization reported that 513,798 new cases of leprosy were diagnosed in 2003. The last remaining leper colony in the United States was located in Carville, Louisiana, and closed in the late 1990s. Japan quarantined leprosy patients up to 1996, and even in the 1950s killed some children out of fear, despite the fact that a cure has been available since the 1940s.

M. leprae was first isolated in 1873, but during the last 100 or so years, no one has ever been able to grow it in a culture medium. The best way to grow *M. leprae* is to use the nine-banded armadillo as a surrogate host. In this biological incubator, the bacteria reproduce by fission only once every 14 days (*E. coli* reproduces two to three times an hour). The inability of *M. leprae* to grow in a defined culture medium and its slow in vivo growth rates have baffled the research community, until now. In February 2001, one week after the draft human genome sequence was published, a British and French team led by **Stewart Cole** reported the complete genome sequence of *M. leprae*. Although prokaryotic genomes are easier to decipher than the human genome, it is informative to compare genomes of similar species. Fortunately, in 1998 the complete genome of *Mycobacterium tuberculosis* (the causative agent of tuberculosis) also was published by Cole's team.

The genome of *M. leprae* contains 3,268,203 bp, while *M. tuberculosis* contains 4,411,532 bp. When Cole's lab compared the content of the two related species, the contrast was striking (Table 3.2). Only 50% of the *M. leprae*

Table 3.2 Comparison of *Mycobacterium* genomes.

Feature	M. leprae	M. tuberculosis
Genome size (bp)	3,268,203	4,411,532
GC (%)	57.79	65.61
Protein coding (%)	49.5	90.8
Protein-coding genes	1,604	3,959
Pseudogenes	1,116	6
Gene density (bp per gene)	2,037	1,114
Average gene length (bp)	1,011	1,012
Average unknown gene length (bp)	338	653

Used with permission from Stewart Cole.

genome encodes 1,604 proteins, compared to tuberculosis's 91% coding DNA, which produces 3,959 proteins. The average gene lengths are nearly identical, but *M. leprae* contains over 1,000 **pseudogenes** and *M. tuberculosis* a scant 6. Therefore, it appears that since the time of their evolutionary divergence, the *M. leprae* genome has lost over 2,000 functional genes and over 1 million base pairs of DNA. It is likely that *M. leprae* has evolved as a *Buchnera*-like intracellular recipient of host metabolites.

Let's look at a couple of examples indicating that the leprosy genome has undergone significant changes, including recombinations, rearrangements, and deletions. When the genomes of leprosy and tuberculosis are compared, we find about 65 segments that are syntenic. However, the relative order and orientation of the genes in the regions differ, and some of the genes are missing in *M. leprae* (Figure 3.21). When we look at the gene called *proS*, which encodes prolyl-tRNA synthetase (the enzyme that covalently joins the amino acid proline to its tRNA), we find its sequence is very different from that of *M. tuberculosis*. Instead, *proS* from *M. leprae* resembles the bacterial pathogen that causes Lyme disease, *Borrelia burgdorferi*, as well as *proS* from eukaryotes such as humans, flies, and yeast. *M. leprae* may have acquired a second *proS* ortholog through lateral transfer and later lost its original *proS* gene. Consistent with the recent acquisition of *proS*, the *M. leprae proS* gene is out of order and in the opposite orientation compared to *M. tuberculosis*.

What Else Is *M. leprae* Missing? The leprosy genome is lacking many genes compared to its more vigorous relative *M. tuberculosis* (Figure 3.22). In particular, *M. leprae* is barely able to synthesize its own membrane lipids. Surprisingly, *M. leprae* has one lipid-synthesizing gene not present in *M. tuberculosis*, and this particular lipid is abundant in the *M. leprae* membrane. It has evolved a compensatory mechanism for the loss of more common lipids. *M. leprae* derives most of its energy from lipid catalysis, and its pathway for carbon metabolism is lacking redundant genes found in most other organisms. For example, it lacks malic enzyme, and has only one isocitrate lyase gene, as well as many other genes in the TCA and glycolysis pathways. Interestingly, it is unable to synthesize the amino acid methionine, which is not usually an essential amino acid but appears to be for *M. leprae*.

LINKS
John Fuerst
Taxonomy

Could Nuclei Evolve without Symbiosis?

It is easy to speculate, but much more difficult to support ideas with data. The endosymbiont hypothesis is the favorite model today, but a new discovery may overturn our idea of nuclear origins. **John Fuerst** from the University of Queensland, Australia, has evidence that bacteria can develop nuclei without symbiosis (Figure 3.23). He has stunning electron micrographs that depict a bacterium with a nuclear membrane that contains large openings similar to nuclear pores. Nuclear pores have been an enigma because prokaryote membranes lack them and intermediate steps seem unlikely. Yet without nuclear pores, no information can travel between the eukaryote's cytoplasm and the nucleus. *Gemmata obscuriglobus* may be a different type of missing link. Is it a remnant of a process that long ago led to the formation of the eukaryote nucleus, or is it a newcomer that is in the process of "going eukaryotic"? Did the *G. obscuriglobus* nucleus-like organelle evolve by a mechanism that differs from the eukaryotic one? Initially, unexplained data may seem to weaken support for evolution, but with time and more data, we will find answers that strengthen our understanding of evolution.

DISCOVERY QUESTIONS

33. Explain how the genome sequence could be used to develop a growth medium for *M. leprae*.

34. Given the limited metabolic capacity of *M. leprae*, how could investigators develop new drugs to kill *M. leprae*? Support your answer with data provided in this section.

35. Perform a NCBI Taxonomy search for *Gemmata* to see the full lineage of *G. obscuriglobus*. Now go to the Tree of Life web page and look for the phylum *Planctomycetes*. Does *G. obscuriglobus* look like an ancient species based on its tree location?

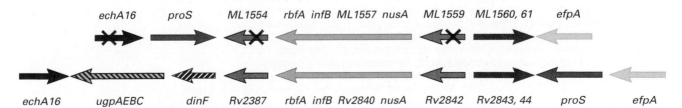

Figure 3.21 Differences between *Mycobacterium* genomes.
Comparison of *proS* genes (solid black arrow) from *M. leprae* (above) and *M. tuberculosis* (below). Genes or operons are depicted as arrows with the point representing the downstream end. X denotes pseudogenes.

LINKS
mitochondrial DNA

With the *M. leprae* genome, we can see another possible step along the way toward creation of nuclei from prokaryotes with reduced genomes (Figure 3.24). Rarely do we have fossil data that represent all the steps in a chain of evolutionary events. Now we are able to examine fossils of a different sort: living bacterial genomes that display different degrees of evolutionary reduction in complexity, size, and ability to live independently. The original eukaryote has had billions of years to adapt to its new genome and niche. Only through genomic analysis were we able to understand how intracellular genomes were formed from Eubacteria and Archaea to produce the third and newest domain of life, Eukarya.

Since Darwin published his historic work outlining evolution, we have wondered how humans evolved. Of course, we did not evolve from present-day monkeys, but we do have common ancestors. We also share common ancestors with flies, worms, yeast, and even Archaea and the first prokaryotic cells, but that was too long ago for most of us to feel any sense of kinship. What did the closest relatives of *Homo sapiens* look like, and where did they live 100,000 years ago? This also leads us to wonder how closely related we are to all the other 6 billion humans alive today.

Are We Related to Rats?

Let's start the human evolution story with the evolution of mammals. A multinational team, including investigators from The Netherlands; the University of California, Riverside; Belfast, Ireland; the University of Cincinnati; the University of Massachusetts, Amherst; and SmithKline Beecham Pharmaceuticals, coordinated efforts to deduce the evolutionary relationships among all taxonomic orders of placental mammals. To accomplish this hairy project, they sequenced some mitochondrial DNA (mtDNA) as well as some nuclear DNA totaling over 8,000 bases for

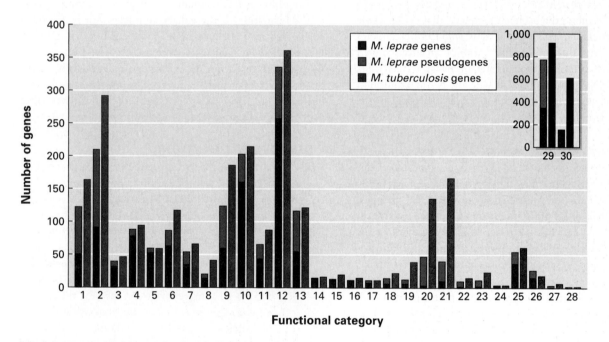

Figure 3.22 Distribution of genes by functional category.

The number of *M. leprae* genes (dark purple), pseudogenes (light purple), and *M. tuberculosis* genes (gray) are listed for each category. Functional categories: 1, small-molecule catabolism; 2, energy metabolism; 3, central intermediary metabolism; 4, amino acid biosynthesis; 5, nucleoside and nucleotide biosynthesis and metabolism; 6, biosynthesis of cofactors, prosthetic groups, and carriers; 7, lipid biosynthesis; 8, polyketide and nonribosomal peptide synthesis; 9, proteins performing regulatory functions; 10, synthesis and modification of macromolecules; 11, degradation of macromolecules; 12, cell-envelope constituents; 13, transport/binding proteins; 14, chaperones/heat-shock proteins; 15, cell division proteins; 16, protein and peptide secretion; 17, adaptations and atypical conditions; 18, detoxification; 19, virulence determinants; 20, IS elements and phage-derived proteins; 21, PE and PPE families; 22, antibiotic production and resistance; 23, cytochrome P450 enzymes; 24, coenzyme F420-dependent enzymes; 25, miscellaneous transferases; 26, miscellaneous phosphatases, lyases, and hydrolases; 27, cyclases; 28, chelatases. Inset: Y-axis shows a different scale for genes in categories 29 (conserved hypothetical proteins) and 30 (hypothetical proteins) that share no significant similarity with any protein in the databases.

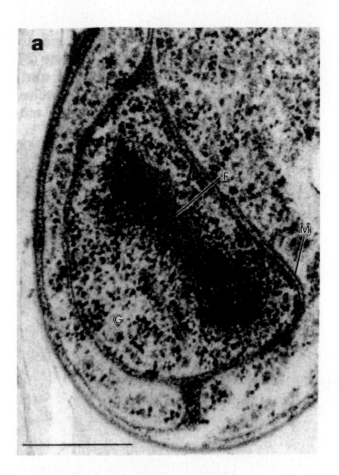

Figure 3.23 Electron micrograph of Eubacteria
***G. obscuriglobus* cell with nuclear body.**
A membrane-bounded nuclear region with central fibrillar area (F) and outer granular area (G). The double-membrane structure of the boundary (M) is apparent. (Bar = 0.5 μm.)

Figure 3.24 Updated schematic of missing links in evolution of eukaryotic nucleus.
Perhaps "genomically weak" prokaryotes engulfed by other prokaryotes sparked a symbiotic relationship that eventually led to the formation of organelles as we recognize them today. This diagram is not intended to suggest that these steps necessarily happened, but it does reveal that intermediate "missing links" are alive today.

each species. When they compared the changes in DNA sequences for these two different genomes, they were able to piece together the mammalian family tree (Figure 3.25). It came as no surprise that our nearest relatives are other primates, specifically the great apes.

In agreement with others, this team concluded that the base from which all mammals arose was an insectivore-like mammal, which is supported by fossil evidence. They also concluded that different insectivore-like lineages have

independently given rise to diverse placental mammalian taxa in different geographical settings.

DISCOVERY QUESTIONS

36. Look at the mammalian tree and find humans and *Galago*. Go to NCBI's Taxonomy home page and find out what a *Galago* is by using the search.

37. Although it is hard to see the exact branching in Figure 3.25, what group of mammals is most closely related to primates?

38. Find mice and rats, the two most common research animals. What is the scientific justification for using rodents as model organisms rather than cats, dogs, or other domesticated farm animals such as pigs?

What Is the Origin of Our Species?

A question related to our evolutionary origin is our evolutionary divergence. How much genetic diversity is present in the human species? When we look at people from different regions of the world, we detect morphologically distinct traits. These traits are determined by our genomes, so we are left with the impression that the human gene pool is extremely diverse. Is the human gene pool more diverse than the gene pools of other species? What about the genetic diversity of our nearest living relatives, the great apes? Do these primates really have less genetic diversity than we do?

To determine the answer, a German group led by Svante Pääbo examined about 10,000 bp of noncoding DNA from the X chromosomes of chimpanzees, lowland and mountain gorillas, and Bornean and Sumatran orangutans. In choosing noncoding DNA, they assumed that this section of the DNA would not be under selective pressure and thus would mutate at a constant rate. They examined nucleotide changes while taking into account the sample size for each species. In conducting their study, the investigators first acknowledged that data from mitochondrial and nuclear DNA do not always agree. For example, mitochondrial DNA indicates each subspecies is distinct, whereas some phylogenetic trees mistakenly place humans closer to gorillas than to chimps, depending on which nuclear DNA is chosen for analysis (see pages 126–129). Given this caveat, they moved forward with their analysis and reached some interesting conclusions.

The time since the most recent common ancestor (MRCA) for all **hominid** species is about 540,000 years ago. Divergence within hominids took place more recently than seen in other great apes. The MRCA for all gorilla species lived about 1.2 million years ago, the chimpanzee's MRCA lived about 1.9 million years ago, and the orangutan's MRCA lived about 2.1 million years ago. In general terms, this agrees with the mammalian tree shown in Figure 3.25. However, the German team concluded that humans are the

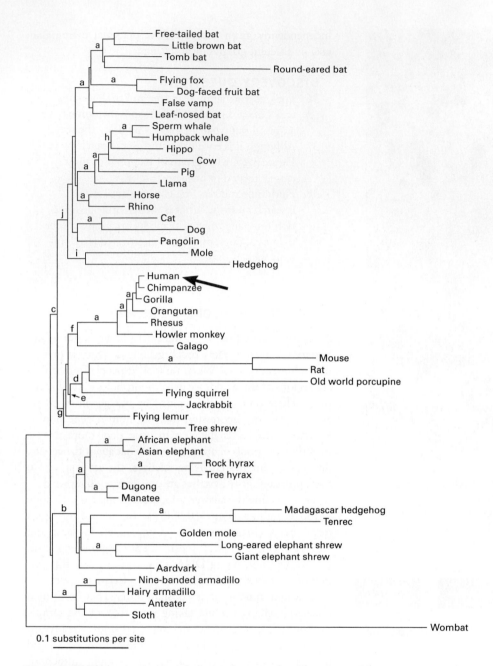

Figure 3.25 Phylogenetic tree of placental mammals with a marsupial root.
This tree used the 2,947 bp nuclear sequences, which were available for a wider range of species than the longer 5,708 bp sequences (mixture of nuclear and mitochondrial sequences). The letters at each branch point indicate a decreasing likelihood score, with "a" being the most likely rating. A purple arrow highlights the location of the human branch.

least diverse of the primates they studied. The great apes have two to three times more diversity (depending on the species) than humans do. They speculated that if the now-extinct hominid Neanderthal were still alive and considered human, then perhaps humans would have as much genetic diversity as the great apes. Since the Neanderthals became extinct about 30,000 years ago, we *H. sapiens* remain as the only surviving hominid. Our gene pool is less diverse than the great apes', even though there are many more humans alive than great apes. Their data indicate that humans

experienced a population expansion approximately 160,000 to 190,000 years ago. At this time, the size of the human population that migrated out of Africa was estimated to be about 3,700. Their estimate for the date of population expansion does not agree with the mitochondrial estimation of about 45,000 years ago. Pääbo's team recognized that their estimate does not include a confidence interval, and therefore their date may be estimating the same population expansion event as the mitochondrial data predict, or perhaps a different and earlier expansion. Nevertheless, the overall finding is striking. Humans, with all our morphological diversity, have less genetic diversity than do the great apes among whom we cannot distinguish individuals. Our limited gene pool is due to a population that was very small and has expanded rapidly over the last 100,000 years.

Are We All of African Descent?

Finally, we are ready to determine where and when modern humans evolved. There are two competing hypotheses to explain the evolution of humans. That *Homo erectus* evolved in Africa about 2 million years ago is widely accepted. The "recent Africa hypothesis" predicts that modern humans originated in Africa 100,000 to 200,000 years ago and a few migrated to cover the rest of the world. The more ancient *H. erectus* became extinct, and there was little or no genetic mixing between the African *H. sapiens* and *H. erectus*. The "multiregional hypothesis" predicts that ancient humans gradually evolved into modern humans at sites around the world. The main evidence for this hypothesis is the fossil evidence, which is updated periodically with new findings. For example, in October 2004, a team of Indonesian and Australian archaeologists discovered several novel, three-foot-tall, adult hominid fossils officially called *Homo floresiensis* but more popularly dubbed "Hobbit." *H. floresiensis* lived in Indonesia about 18,000 years ago. Due to the warm, humid conditions in Indonesia, which are not conducive to preservation of DNA, no mitochondrial DNA has been isolated yet. Were these *H. floresiensis* individuals a different species, or phenotypic outliers (i.e., dwarf) that lived in isolation? Until we can find DNA evidence, we will have to wait for an answer. What is clear though, is the recent-Africa and multiregional hypotheses about human evolution are diametrically opposed to each other; therefore, it should be possible to distinguish which is consistent with the available data.

A pair of labs from Sweden and Germany decided to stop looking at short pieces of genomes. They sequenced the entire mitochondrial genome (~16,500 bp) of 53 people from diverse geographical origins. Mitochondria are transmitted from mother to child and rarely recombine with paternal DNA. The investigators used the chimpanzee mitochondrial genome as the outgroup to root the phylogenetic tree containing the 53 human sequences. Based on genetic and paleontological evidence, chimps and humans diverged about 6 to 7 million years ago. With this as a starting point, they calculated that humans have accumulated 1.7×10^{-8} nucleotide changes per site, per year.

First, the genomic anthropologists looked at mitochondrial genome diversity among current Africans compared to all non-Africans and found that Africans have twice as much diversity among themselves (3.7×10^{-3} nucleotide changes per site) as non-Africans (1.7×10^{-3} nucleotide changes per site). The greater African diversity could be the result of either a larger population size or a significantly longer genetic history. The diversity among Africans is

reflected in the phylogenetic tree (Figure 3.26). Notice how the African half of the tree has many long branches, compared to the non-African half with its short branches. The non-African portion indicates a population bottleneck with a rapid radiation of diversity. Based on the distribution of the nucleotide differences within the mitochondrial genome, the investigators concluded that Africans have experienced a constant population size. Conversely, non-Africans exhibit a more uniform distribution of nucleotide differences that indicate a recent population expansion. The time of expansion was estimated to be 1,925 generations ago. With a generation time of 20 years, they calculated diversification occurred about 38,500 years ago.

Further analysis allowed a prediction for two different MRCAs. If you use their molecular clock (1.7×10^{-8} nucleotide changes per site, per year), it is possible to calculate the time of the MRCA for the two most distantly related human mitochondrial sequences to be 171,500 $\pm$ 50,000 years ago. The MRCA for Africans and non-Africans (arrow in Figure 3.26) is estimated to have lived 52,000 $\pm$ 27,500 years ago. Based on these findings, we are left with only one viable option, the recent-Africa hypothesis. The multiregional hypothesis had predicted that modern humans evolved from ancient hominids who left Africa about 2 million years ago. This hypothesis predicted that *H. neanderthalensis* eventually became modern Europeans and *H. erectus* became modern Asians. Although we know that *H. neanderthalensis* coexisted with modern humans for about 10,000 years, the data indicate that all non-Africans share common ancestors who originated in Africa and left there approximately 130,000 years ago (Figure 3.27). Based on a limited amount of Neanderthal DNA sequence, it has been estimated that the MRCA for all *Homo* species lived about 465,000 years ago and that *H. sapiens* arose approximately 130,000 to 465,000 years ago, with a compromise date of 200,000 years ago. Furthermore, it appears that all 6 billion of us modern humans owe our limited genetic diversity to a small number of breeding individuals (fewer than 10,000) who separated into two main branches. One branch led to the many African populations, while the other branch led to a mixture of African and non-African populations (see Figure 3.26).

DISCOVERY QUESTIONS

39. What two populations are the most distantly related based on mitochondrial sequence? How many mutations accumulated between the two most distantly related human sequences?

40. The Swedish and German team took great pains to demonstrate that the DNA mutation rates in the mitochondrial genome occur at a constant rate. However, they excluded one portion of the mitochondrial genome (called the D loop) because it demonstrates irregular mutation rates. Why is this a critical point for their assumption that a "molecular clock" can be measured?

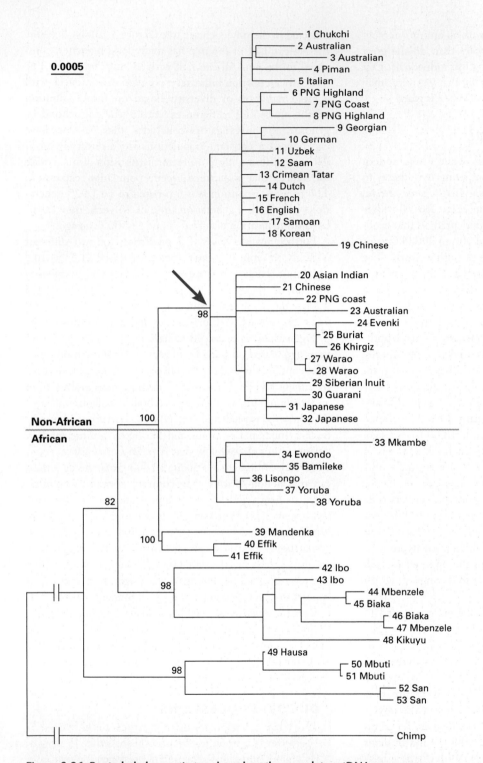

Figure 3.26 Rooted phylogenetic tree based on the complete mtDNA genomes.
Bootstrap values (1,000 replicates) are shown at the branch points, and sources of DNA are listed at the leaves. Arrow highlights the MRCA branch point for branches that contain both African and non-African individuals. The D loop of mitochondrial DNA was excluded due to irregular mutation rates.

Math Minute 3.3	How Do You Know If the Tree Is Correct?

The mitochondrial genome sequences used to form the tree in Figure 3.26 provide evidence of the evolutionary relationships among these individuals. However, several different trees might be consistent with the sequence evidence, depending on how you define "consistent." Researchers typically form phylogenetic trees using one of three standard techniques (distance-based methods, maximum parsimony, or maximum likelihood), each of which implements slightly different criteria for joining individuals (or species, as in Figure 3.25) in the tree. For example, one distance-based method uses the distances between each pair of sequences to form a **hierarchical clustering** of the sequences (see Math Minute 6.3). No matter which method is used, the sum of the horizontal lengths of the tree branches from the root to a leaf is scaled to represent the difference between the root sequence and the leaf sequence.

One way to test the validity of a tree formed using one of these three techniques is to use one of the other techniques and compare the results. Another way to evaluate validity is with **bootstrap analysis**. The idea behind bootstrapping is to repeat the tree-building process several times, each time using different subsequences of the original full-length sequences. If most of the resulting trees agree at a particular branch of the tree, you can place more faith in that branch point than if there is little agreement among the trees.

The bootstrap procedure works as follows. Suppose you have aligned all 54 sequences used to form the tree in Figure 3.26, using a multiple-sequence alignment algorithm. Randomly select columns from the alignment, until you have selected as many columns as were in the original alignment. Column selection is performed "with replacement," which means you may choose the same column more than once and some columns not at all. Take the nucleotides from each sequence of the selected columns, resulting in 54 modified sequences. Build a new tree using these 54 modified sequences. Replicate this selection and tree-building process 1,000 times. Now go back to the tree built with the original sequences and choose a particular branch point, e.g., where the African and non-African peoples first joined (individuals 1–32 join with individuals 33–38). Label this branch point with the percent of the 1,000 trees in which this same branching structure (i.e., separating 1–32 from 33–38) occurs. In Figure 3.26, this particular branching structure occurred 98% of the time in the 1,000 replicates, as shown by the bootstrap value of 98. Branches with bootstrap values of 70 or greater are normally considered reliable. Thus, the tree in Figure 3.26 is a very reliable representation of the relationship among these mitochondrial genome sequences.

---

41. Find the three listings for Australians in Figure 3.26. Do any of these three branches represent the British colonization from the nineteenth century? Propose a hypothesis to explain the separation of Australians on this tree.

Although their research is the most complete study to date (April 2005), other investigators will produce and analyze more data and the recent-Africa hypothesis will be tested further. If it is repeatedly supported, the hypothesis will stay substantially unchanged. If conflicting data are reported, the research community will modify the hypothesis until there is a consensus on our origins. For now, the data are clear that humans alive in the twenty-first century are all cousins to varying degrees.

Have We Stopped Evolving?

Often we think of evolution, especially human evolution, as having already taken place and no further changes are occurring. This perception is demonstrably wrong. Evolution does not stop even though we cannot see all the changes. Natural selection continues to work on genomes, including **mitochondrial genomes**. Have you ever noticed how some people get colder than others, even at the same temperature? Some of this difference may be attributable to our mitochondria— at least, that is the conclusion of a group led by **Douglas Wallace** at the University of California– Irvine.

Wallace and his team sequenced the mitochondrial coding regions of 1,125 people from all over the world. They aligned all the sequences and placed commonly shared mutations closer to the base of the phylogenetic tree and **singleton** mutations at the terminal branches since they

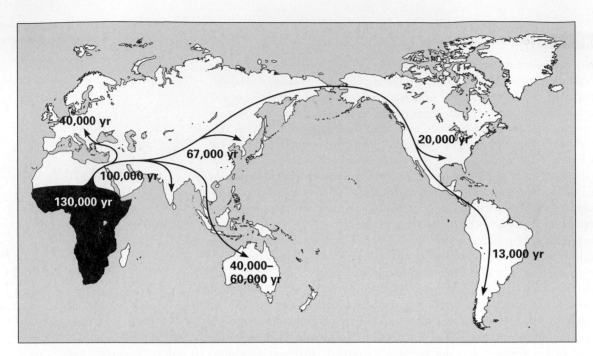

Figure 3.27 Map showing the origin and dispersal of modern humans.
The origin of *H. sapiens* may have been about 200,000 years ago. The earliest known fossil and archaeological evidence on each continent is shown on the map and is consistent with the genome interpretation presented.

were more recent. Natural selection predicts that over time, deleterious mutations will be eliminated from the gene pool and thus older mutations (those closer to the root of the tree and shared by many people) may be advantageous. However, the function of the mitochondria is so vital that it is hard to imagine its efficiency could be improved upon—it has been subject to selection pressures for over 1 billion years. It would be laughable to think that a mere 100,000 years of human evolution could improve mitochondrial performance despite 3 billion years of refinement. However, remember that natural selection pressures are not constant. If human migration out of Africa produced new selection pressures, then perhaps the mitochondrial genome could evolve away from our ancestral mitochondrial genotype.

The investigators compared the nonsynonymous mitochondrial mutations (changed amino acid sequence) of people living near the equator and those living near the Arctic Circle, and found a striking pattern (Figure 3.28a).

Figure 3.28 Modern evolution of OxPhos genes.
Go to www.GeneticsPlace.com to view this figure.

People from colder northern regions had significantly higher mutations in a few genes relating to oxidative phosphorylation (OxPhos) that results in ATP production. OxPhos has two major steps: conversion of chemical energy into a proton gradient, and conversion of the pH gradient into ATP. Another side benefit of OxPhos is heat produced from the inefficiencies inherent in any chemical reaction. If the pH gradient production is tightly coupled to ATP production, less heat is produced and more ATP is produced per unit of food consumed. Evolution has selected for highly efficient ATP production, unless you migrated to Siberia, in which case inefficient ATP production would produce more "wasted" heat that is advantageous in the new environment. Figure 3.28b shows the proteins that were particularly prone to nonsynonymous mutations: ND2 and ATP6 (arctic group A), ND4, and Cytb (Siberian subgroup C).

The ND2, ATP6, ND4, and Cytb mutations that generated more heat have proven advantageous to people who live in cold places. With modern heating, though, wouldn't we expect the mutations to gradually fade away? It turns out that inefficient ATP production has another positive side effect: it appears to protect us from neurodegenerative diseases such as Alzheimer's and Parkinson's, as well as leading to longevity. Reduced ATP production efficiency also results in reduced oxidative damage and its sequela of **apoptosis.**

Therefore, as the average age of human reproduction increases, we may be exerting a new selection pressure that favors inefficient mitochondria that reduce oxidative damage and delay programmed cell death. Our still-evolving, maternally inherited mitochondrial genomes may predispose us to cold adaptation and future disease status. Given these data, it seems clear that even human evolution is an ongoing process.

DISCOVERY QUESTIONS

42. Go to the **KEGG Pathway** database, select "Oxidative phosphorylation," and change the Reference pathway to "Homo sapiens" to view the full pathway with human genes highlighted in green. Find the most commonly mutated genes that lead to extra heat production and determine if their decreased efficiency directly reduces H^+ gradient production or ATP synthesis.

43. Go to **MitoMap** and click on the link to the illustrations. Choose "Mitochondrial energetics" to find the location within the mitochondria for the proteins you investigated in Discovery Question 42. Go back and click on the "Mitochondrial DNA Map" link to find the location of the genes from Discovery Question 42. Finally, go back and click on "World migrations" to see what mtDNA genotype your ancestors probably had.

Summary 3.2

This section has shown how genomic sequences can provide us with a more complete understanding of evolution. As more genomes are completely sequenced, we will fill other gaps in our knowledge. By analyzing complete genomes from different taxa, variations among individuals, the role of parasites and symbionts, and the genomes of organelles (including chloroplasts), we will come to appreciate more fully the gradual population of the earth by organisms with diverse life histories and origins. The unifying principle in biology is evolution, and if genomes could talk, we would be able to hear the story of the creation of life. We are learning how to listen, and soon we will be ready to hear what must be one of the most fantastic stories imaginable. Stay tuned for more details about the pathways of evolution.

3.3 Genomic Identifications

Each species can be defined by its genome, and an individual's DNA sequence can be used to determine its species. However, we cannot easily sequence the entire genome of each species, much less every individual within a species. Therefore, we rely on particular segments of DNA, treating them as molecular bar codes that can be scanned to reveal useful information. In this section, we examine several case studies in which short segments of genomic DNA were used to help identify their origins. The first case study examines how biological weapons can be identified with the aid of genome sequences. Two case studies examine DNA extracted from individuals 1,000 to 250 million years old. Finally, we will see why DNA is the ideal tool for diagnosing emerging diseases and devising new treatments. With these cases, you should discover the diverse applications for genome sequences.

LINKS
Building 470
CDC
KEGG Pathway
MitoMap

How Can We Identify Biological Weapons?

On March 22, 2001, two downtown Toronto office buildings were sealed, with the workers still inside. Simultaneously, two different offices received packages that contained a "gray granular substance" and notes warning that the boxes contained anthrax. The recipients of the boxes called police, who restricted movement in and out of the buildings. After three hours, the contents of the boxes were determined to be harmless and the building occupants were allowed to come and go freely. When anthrax spores were delivered to various people in the United States during the fall of 2001, identification times ranged from a few minutes to several days. Why were there such differences in the amount of time needed to determine whether anthrax spores were present? Is it easier to demonstrate a powder is safe than to verify it is harmful? Can genomes lead to fast and reliable methods for biological weapon identification?

Anthrax is a spore-forming bacterium, *Bacillus anthracis*, that occurs in wild and domestic mammals (cattle, sheep, goats, camels, antelopes, and other herbivores) but can infect humans as well. Anthrax spores can survive for many years and can lead to infection of the skin or lungs. Initial symptoms of respiratory infection resemble a common cold, but within days the symptoms become severe. Symptoms begin with breathing problems and can lead to shock and sometimes death. Fortunately, person-to-person contact is unlikely to spread *B. anthracis*. (For more information, you can consult the Centers for Disease Control and Prevention [CDC].)

The U.S. Army conducted biological weapons research until 1969, when President Nixon outlawed nondefensive research (Figure 3.29). Before the ban, some Americans were intentionally infected with known pathogens to test the feasibility of aerosolized biological weapons; with early intervention of antibiotics, one investigator noted, "We didn't lose any subjects, fortunately." **Building 470** is still located in Fort Detrick, Maryland, where the U.S. Army

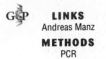

a)

b)

Figure 3.29 American history of anthrax biological weapon development.
a) Building 470 (circa 1952) at Fort Detrick, Maryland, where the U.S. Army began and ended its anthrax weapon development program. Two American employees died from pulmonary anthrax. **b)** Decommissioned aerobic test facility known as the "Eight-Ball" was the largest in the world (1 million liters) and one of six built for "aerobiological studies of agents highly pathogenic to man." It was designed to withstand experiments that used explosives. Its last known use was in 1972 and it became a National Historic Place in 1977 for its "unique function and scientific and technological contributions." U.S. Army photos.

experimented with the production of weaponized anthrax and exposed people to pathogens in a controlled environment where "scientists kept a loaded .45 caliber pistol at their side or on the workbench next to them" as a part of the security procedures. Contrary to recently published accounts, two U.S. military employees died from anthrax contracted in Building 470. William Boyles, a 46-year-old microbiologist, contracted anthrax and died on November 25, 1951. On July 5, 1958, Mr. Joel Willard, age 53, an electrician with the facilities crew, also died of pulmonary anthrax. American experience with anthrax biological weapons proved helpful in the 2001 anthrax attacks in Florida, Washington, D.C., and New York—early and aggressive medical treatment can prevent fatal infections.

How Did They Determine It Was Not Anthrax in Three Hours? Bacterial spores are similar to seeds, so you cannot quickly grow them into rapidly dividing bacteria that can be analyzed by traditional microbiology methods. Even the fastest cells grow too slowly, and you might not know which growth medium is best for unknown samples.

Therefore, the only identifiable material available on short notice is DNA.

Bacterial chromosomes can be isolated in minutes, but raw DNA is not very helpful. We need to sequence the DNA to identify the species. Unfortunately, you would need species-specific primers, sequencing reactions, and sequencing machines, which are not easily transported for use outside the lab. PCR is a robust procedure that performs well in many settings, but it typically takes 2 to 3 hours to amplify the DNA, and then the PCR products have to be analyzed by gel electrophoresis, which takes more time and processing. We need a faster process. The following case studies highlight recent advances in species identification using DNA-based methods.

In 1998, Andreas Manz and his lab published a new and very fast method for PCR, called "chemical amplification continuous-flow PCR on a chip." The key to this technology is that PCR reactions do not need long incubation times; the heat exchange takes place in about 100 milliseconds due to the design of the chip (Figure 3.30). DNA extracted from a biological sample is mixed with PCR reagents; once combined, they rapidly travel through different temperature zones to produce PCR products that are analyzed later. A new capacity was added in 2003 with the option for reverse

LINKS
David Burke
METHODS
RT-PCR

transcriptase PCR (RT-PCR), which can convert RNA into cDNA prior to PCR. RT-PCR allows RNA viruses to be sampled along with DNA viruses and bacteria.

The Manz lab provided a great tool to speed up the PCR process, but detection of the product was still slow and required too much lab equipment for use in remote sites.

What was needed was a self-contained unit that would perform the PCR and gel analysis in one step. David Burke and his colleagues at the University of Michigan have developed an integrated, nanoliter DNA analysis device that is a good solution (Figure 3.31a). The entire chip is

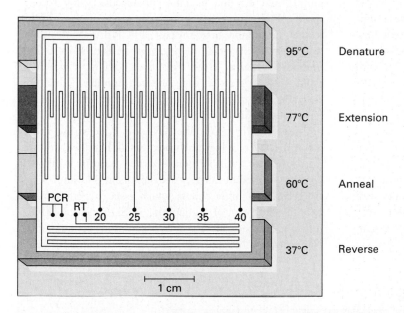

Figure 3.30 Schematic of continuous-flow PCR with reverse transcriptase capacity.
Three well-defined temperature zones are maintained at 60°C for annealing, 77°C for extension, and 95°C for denaturing. Notice that the pathway of the fluid, which is pumped through the channel, keeps the sample at 77°C longer by zigzagging back and forth at this temperature. The duration of time at each temperature is determined by the path length of the channel in each zone. The reagents are mixed just prior to entering the 60°C zone. RT marks the input for reverse transcriptase (37°C) prior to PCR.

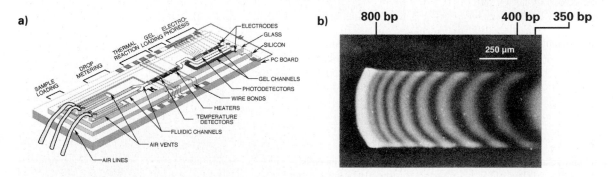

Figure 3.31 Integrated nanoliter DNA analysis.
a) Schematic of integrated device with two liquid samples and electrophoresis gel. The device is able to add set volumes of liquids in the "drop metering" section using a combination of hydrophobic regions and air intakes. **b)** Optical micrograph of a 50 bp DNA ladder in a 0.5 × 1.0 mm polyacrylamide gel. The DNA was loaded onto the gel and electrophoresed for 15 minutes. The band at the far right edge is the 350 bp band. The remaining bands are between 400 and 800 bp.

LINKS
LLNL
METHODS
real-time PCR

produced as a single piece, and the user supplies the standard PCR reagents and template DNA—in our case, unknown bacterial DNA. With this system, the PCR product can be sent directly onto a miniature gel that takes about 30 minutes to resolve the bands (Figure 3.31b). Burke's device offers several advantages. First, the person in the field only needs to load DNA from a biological sample; the chip does everything else, including recording the results onto a computer. The chips are small, portable, require very little electricity, and do not require much training to operate compared to sequencing or agarose gels. This PCR with mini-gel chip is a workable solution, but even faster results would be helpful.

As an alternative to Burke's "lab on a chip," Phillip Belgrader from the Lawrence Livermore National Laboratory (LLNL) helped develop a very different machine that can identify samples in seven minutes. Belgrader's team built a device called **real-time PCR** (often called **qPCR** for quantitative PCR) that performs PCR in microfuge tubes with volumes of 25 µL (Figure 3.32). This device has significantly improved our ability to identify unknown biological samples. qPCR monitors the production of the amplified product at each cycle in the reaction so that, as product accumulates, its production is monitored in "real time." The output is not a gel but a graph that displays the amount of PCR product. The user-friendly output converts a green flat line for no PCR product detected into a red graph to indicate successful identification of the unknown DNA.

Since the specific source of the DNA is unknown, the sample may be one or more pathogens. Many species-specific PCR primer pairs could be tested in a **multiplex** format (using more than one pair of PCR primers in a single reaction tube) in which each species' PCR product emits a unique fluorescent color. As noted, qPCR is extremely rapid, with a detection time of seven minutes (Table 3.3). Belgrader's team can measure the number of cells in the original template and the time it took to detect the PCR product (Figure 3.33). The ANAA is easy to use and is about the size of a big lunch box. Small, handheld versions (HANAA) are available now that can perform 4 different tests for 4 different samples with results in 30 minutes, but the technician must choose which tests to perform. The U.S. Department of Agriculture (USDA) uses HANAA for food inspections to detect common contaminating pathogens. In November of 2002, LLNL demonstrated a

a)

b)

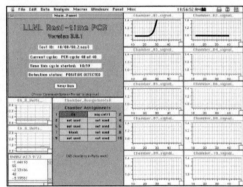

Figure 3.32 The portable Advanced Nucleic Acid Analyzer (ANAA).
a) An ANAA consists of an array of ten reaction modules and a laptop computer.
b) Screen shot of laptop after running a reaction with bacteria (top left trace) and a negative control (top right trace). The detection of a signal produces a warning display.

Table 3.3 Thermal cycling settings and detection times for bacterial cells.
Threshold value is the number of cycles needed to detect a positive signal.

Cycle Time (sec)	Denature Time (sec)	Anneal/Extend Time (sec)	Threshold Value (cycle number)	Detection Time (min)
38	4	19	23	14.6
38	4	10	23	10.7
24	4	5	24	9.6
17	1	1	25	7.0

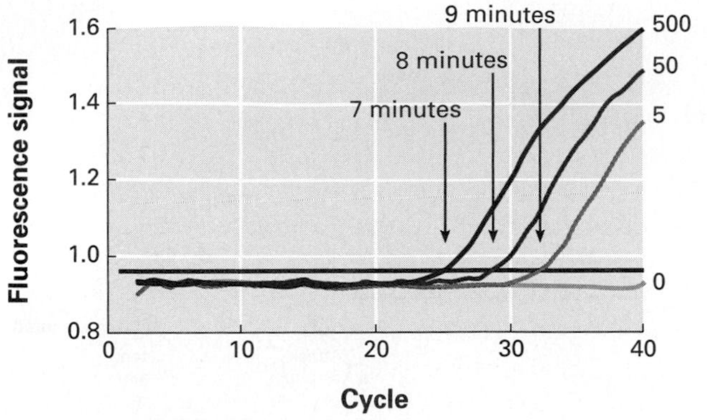

Figure 3.33 Quantitative PCR (qPCR) performed with a 17-second cycle time.

The number of bacteria cells is listed to the right of each trace. The time it took before a positive signal was detected, triggering the warning signal, is shown above the arrows pointing to each trace crossing the threshold value (black line).

combined environmental air sampling device coupled with ANAA to perform 10,000 assays on 1,000 environmental samples in one 8-hour period with only 2 technicians.

Rapid biological detection and identification allowed the workers in Toronto to learn in three hours that they had not been exposed to a lethal biological weapon. They used antibody strips, which are similar to home pregnancy tests, require many spores before detection is possible, and are very good at confirming a negative result (i.e., not anthrax). However, antibody strips can produce false positives due to cross-reactivity of antibodies to similar but harmless bacteria. The Canadian office workers were able to return to their normal lives quickly, and the buildings were reopened for business. For the 2001 cases in the United States, the quickest identifications used antibodies, but the most accurate identification technology is the multi-day, microbiology method of growing cells and performing tests on them. It seems likely that new, high-throughput DNA analysis devices will improve our ability to identify trace amounts of DNA. Since each organism has unique sequences, we will be able to utilize genomic information to identify more species as more genomes are sequenced.

In the case of deliberate release of anthrax, we might find the culprit if we could compare the *B. anthracis* genome to known isolates to identify the source of the spores. For example, TIGR investigators are sequencing whole genomes from many different isolates to generate a phylogenetic tree of divergent isolates (from at least 90 different isolates; Figure 3.34a). The whole genome comparison of different *B. anthracis* isolates has just begun, but the full tree will help us track the source and hopefully the perpetrator (Figure 3.34b) to prevent future attacks. Based on TIGR's sequencing work and earlier efforts by **Paul Keim** at Northern Arizona University, a pioneer in anthrax DNA fingerprinting, it seems clear that the 2001 U.S. anthrax attacks used a single isolate of *B. anthracis*, which indicates a single person or group is responsible in this still-unresolved criminal case.

LINKS GCP
anthrax home
Infectious Diseases
Paul Keim

DISCOVERY QUESTIONS

44. The most important part of any experiment is the design of careful controls. Given that some errors might occur, what controls would be necessary each time a test is performed?
45. The ANAA device can run only ten reactions at a time. How could you address this limitation if you needed to test an unknown sample?
46. Go to the NIH's **Infectious Diseases** web page, find the link to anthrax, and determine how it is toxic. Now go to **anthrax home** page. GCP
If you click on the two other numbers at the top [17680] [17681], you can see the two plasmids (pXO1 and pXO2) that carry the toxin genes. On the home page, click on "TaxPlot" to line up the genomes of three *B. anthracis* genomes (Ames Ancestor, Ames, and A2012), then click on the "compare" button. Zoom out to 5X and then click on the genes that are off the diagonal but along the A2012 genome. Which piece of DNA carries these outlier genes? Scroll down to see the list of genes, with their locations and descriptions. Do you recognize any of the genes there? What does this tell you about the source of differences of these three strains of *B. anthracis*?
Find YP_016503, click on the arrow (⇒) on the right, then on the next page click on the score 4142 for the entry "pox1-107." Find the differences between these two versions of *lef* (lethal factor) near amino acid 300. How might this information be used to track the source of a biological weapon?

How Long Can DNA Survive?

We often think of DNA as a fragile compound that is easily damaged. However, DNA is actually very stable and resists degradation much better than RNA and most proteins. Its ability to withstand extreme conditions has led to the discovery of **ancient DNA** synthesized thousands or even millions of years ago. To anyone familiar with Jurassic Park, this concept is not new, but sometimes fact is more amazing than fiction. Unfortunately, the 1994 *Science* publication claiming to have isolated dinosaur DNA was wrong; the material proved to be human DNA, and the investigators had failed to perform a BLAST search to confirm their findings. Therefore, we must be skeptical when reading about ancient DNA, but that does not imply we should assume all such reports are wrong.

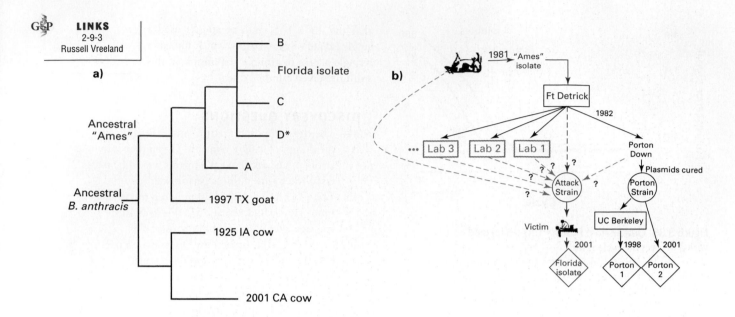

Figure 3.34 Life history of *B. anthracis*.
a) Phylogenetic tree showing the relatedness of strains isolated from farm animals and the Ames strain used in research. Isolates B and C are identical in all markers to the Florida criminal case isolate. Isolate D differs by only one additional adenine, and is marked by an asterisk to indicate this difference. Isolate A has two additional bases (versus the Florida group) and has also lost the pXO1 plasmid. The Texas goat, Iowa (1925), and California (2001) cow isolates are more divergent. **b)** Known distribution (solid arrows) of Ames (Iowa) isolate of *B. anthracis* to Fort Detrick (see figure 3.29) and distribution to nonmilitary labs for basic research purposes. Uncertain distribution is marked by dashed arrows. The toxic plasmid was removed at Porton Down in the 1960s, thus facilitating safer research conditions. Sequenced genomes in panel a) are shown as diamonds.

Bacteria Survives in 250-Million-Year-Old Salt Crystal In 1995, a team of investigators reported the isolation of a previously unknown ancient species of bacteria that had been encapsulated in 25-million-year-old amber (along with an extinct bee species). In October 2000, biologists William Rosenzweig and Russell Vreeland, with geologist Dennis Powers, made a discovery that was ten times as amazing. From deep in the earth, they isolated salt crystals that were dated at about 250 million years old (Figure 3.35). Inside the crystals were small, fluid-filled chambers that occasionally contained bacteria. The team was very careful not to use any crystals that had been penetrated by water or exhibited a crack in the crystal structure. It is possible to identify pristine chambers by their rectangular shape, whereas disturbed chambers have irregular shapes and salt deposits on the back wall of the chamber.

Vreeland's team took great care to sterilize the outside of the crystal and all their tools and equipment; they also worked in sterile environments. They isolated a few micro-

liters of fluid from 66 chambers in 56 different crystals. These small volumes were inoculated into different liquid media and allowed to incubate for three months in sealed tubes. Only 2 of the 56 crystals contained viable bacteria. The growing cells were spread onto solid media where only *Bacillus* colonies were identified. Three additional isolates have been obtained, but only the one they initially called 2-9-3 was characterized.

Chromosomal DNA was isolated from 2-9-3, the complete rDNA region was sequenced and compared with other sequences, and a phylogenetic tree was produced (Figure 3.36). The three species that form a lineage along with 2-9-3 are halophilic (salt-loving) bacteria. Because 2-9-3 was able to form spores, the researchers tested their ability to survive the sterilization techniques used to treat the outside of the crystals and collection tools. When 10^8 cells and 10^8 spores of 2-9-3 were exposed to the sterilizing procedure, none survived, supporting the conclusion that the isolated strain 2-9-3 is not a modern contamination from the outside of the crystals.

LINKS
Ancient Biomolecules

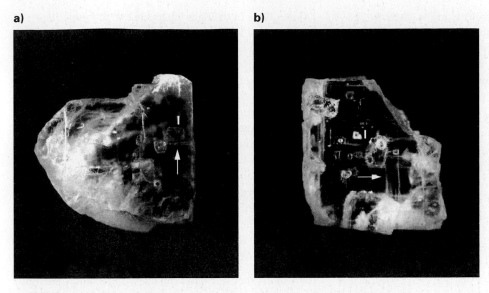

Figure 3.35 Salt crystals from 1,850 feet down an air intake shaft in Carlsbad, New Mexico.

a) This crystal appeared undisturbed due to the clarity and shape of the fluid-filled chamber. This crystal contained the strain 2-9-3. The drill hole used to obtain the sample (shown above the arrow) permitted access to the inclusion (I), or chamber, containing the bacteria. **b)** This crystal was rejected because it contained cracks (arrow points to a large vertical crack) and the inclusion (I) is irregular in shape.

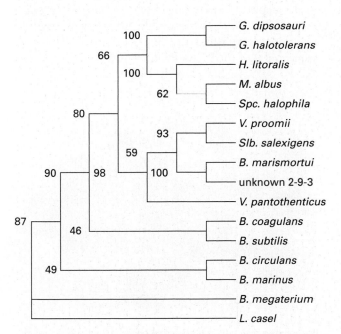

Figure 3.36 Rooted phylogenetic tree using ribosomal DNA sequence from 2-9-3.

Several species were compared with 2-9-3 to produce this tree. Bootstrap values from 100 repetitions are shown at each node.

How could 2-9-3 survive inside salt crystals? Like modern species, 2-9-3 formed spores when cells were exposed to high-salt conditions as water evaporated. The spores became trapped in salt crystals, which protected them from environmental hazards that otherwise might have killed them. If this strain truly is ancient, it is the best example we have of the stability of DNA over long periods of time.

Many scientists think, however, that ancient DNA is really only modern DNA contamination. Given the nature of ancient DNA research, it may be impossible to thoroughly demonstrate the true origin of the 2-9-3 DNA, and thus each of us must reach our own conclusions. Alan Cooper, who directs Oxford University's Ancient Biomolecules Centre, does not accept Vreeland's conclusions. "When we repeated that work with the same [PCR] primers, we were pulling up halobacteria from everywhere. . . . We took some dust from the top of the natural history museum in Oxford, extracted [DNA], used their supposedly halo-specific primers and extracted a whole bunch of sequences, including some that fell within their diversity. . . . I can't see any logic for having 250 million years without any evolution." Vreeland counters these criticisms by stating that Cooper's disbelief is not a scientific objection. "That's like throwing out the baby with the bathwater. If you can show that nothing has penetrated your sample and the DNA is inside, then the age of the DNA has to be equal to the age of the rock." However, we must remember that the crystals are made of salt, and as such they could form and reform over time; thus, they are not as stable as rocks.

DISCOVERY QUESTIONS

47. Go to NCBI's Entrez and search the nucleotide database using the accession number AF166093 (2-9-3 is now called *Bacillus permians*). Perform a BLASTn search with the sequence, and identify which species are returned in your search. Are you able to reproduce results similar to those shown in Figure 3.36?

48. What is the greatest potential weakness in the process of recovering an untainted ancient sample?

49. Based on DNA sequence information from one locus, is it possible to conclude whether something is a prehistoric strain or a modern contamination? What improvements in methodology would you suggest for future studies?

How Did Tuberculosis Reach North America?

If we want to eliminate a disease-causing pathogen, we need to understand how the organism is spread. Does it spread by ticks (Lyme disease), mosquitoes (malaria), worms (elephantiasis), or aerosolized bacteria (leprosy)? Knowing the transmission mechanism enables us to better combat the causative agent. Knowing how the disease was first introduced into humans would provide a more complete understanding of its life history. Unfortunately, most diseases predate modern medicine, so our knowledge is limited. However, with our ability to isolate ancient DNA, can we retroactively diagnose ancient human remains? Can archaeological DNA remains reveal when and where a disease originated and spread?

One increasingly problematic disease is tuberculosis, because drug-resistant forms are becoming more prevalent. A long-standing belief holds that tuberculosis originated in Europe and was brought to North America by European explorers. However, in 1994, **Wilmar Salo** at the University of Minnesota–Duluth reported the identification of *M. tuberculosis* DNA in a Peruvian mummy that predates the arrival of Christopher Columbus. The team located a spontaneously mummified body of a 40- to 45-year-old **Chiribaya** woman. The Chiribaya were an agricultural population that occupied this area of Peru about 1000–1300 A.D. Carbon-dating of the mummy's remains placed the time of death at 934 ± 44 years A.D. The investigators described the tissue sample isolation:

Removal of the chest wall revealed intact, collapsed lungs bilaterally. The left lung and the middle and lower lobes of the right lung were not abnormal. The right upper lobe was diffusely adherent to the chest wall, and a largely calcified, discrete nodule 1.2 cm in diameter was present in a lateral, subpleural position [L in Figure 3.37]. At its point of adhesion to the parietal pleura overlaying the lateral part of the fourth rib, a 1-mm aperture was apparent in the calcified shell, communicating with a central 3-mm diameter cavity within the nodule. In the right hilum, two partly calcified masses, each about $1.5 \times 0.8 \times 0.5$ cm, were found in peribronchial positions [N1 and N2 in Figure 3.37]. No evidence of skeletal change could be identified in the spine, ribs, or elsewhere, and no abnormalities were found in the liver, heart, bowel, skin, trachea, or other soft tissue. No acid-fast bacilli were demonstrable in any of the lesions.

In short, they cut open the chest and found three bony nodules where the upper part of the right lung adhered to the inside lining of the chest wall. They also cut open and examined most of the soft tissue in the abdominal and thoracic cavities. They saw no bone deformities often associated with tuberculosis, nor did they identify any *M. tuberculosis* cells using the classic test of acid-fast staining. They later extracted the DNA from the three nodules and performed PCR on the DNA. The primers were designed to target "a segment of DNA unique to *M. tuberculosis*" called IS6110. They amplified a 97 bp fragment and an overlapping fragment of 232 bp (Figure 3.38). Only tissue sample N2 was "clearly shown to contain [*M. tuberculosis*]'s DNA. The bacteria in the other lesions may have ceased to exist before the individual died." The team sequenced 9 of the 97 bp PCR products and verified that

Figure 3.37 Source of Peruvian DNA.
Accurate reproduction of the published photograph reporting tuberculosis-like lesions on the right lung and hilar lymph region of a Peruvian mummy. The lesions are designated in black letters as N1, L, and N2, referring to the upper lymph node (N1), lung nodule (L), and lower lymph node (N2).

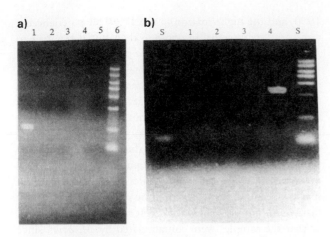

DATA
Egyptian mummy
133 bp
IS6110
LINKS
Kathleen Eisenbach
METHODS
Nested PCR

Figure 3.38 Nested PCR for IS6110.
a) Second-round amplification of the 97 bp target, amplified by two rounds of PCR, in which one microliter of the first PCR product (123 bp long) was used as a template for the second round with nested primers. Lane 1 = DNA extracted from nodule N2; lane 2 = DNA extracted from lung nodule L; lane 3 = control reaction; lane 4 was empty; lane 5 = reaction control; lane 6 = molecular weight standards. From top to bottom: 1,000, 700, 500, 400, 300, 200, 100, and 50 bp. **b)** Nested PCR of the 232 bp target of IS6110. Two rounds of PCR were performed, with the first product (247 bp long) used as the template for the second round of PCR. Lanes labeled S contain molecular weight standards; lane 1 = no first-round PCR template added; lane 2 = control reaction; lane 3 = control extract; lane 4 = N2 DNA extract.

Figure 3.39 Opened thorax of Egyptian mummy.
Black arrow points to pleural adhesions of right lung (left side of photo); destroyed bony elements of lumbar vertebral bodies L4 and L5 are visible at the bottom of the photo.

they had amplified IS6110. Based on these results, the team concluded that *M. tuberculosis* was present in Peruvians prior to the arrival of Columbus.

Three years later, a team led by Andreas Nerlich from the Ludwig-Maximilian University in Munich, Germany, reported finding evidence of tuberculosis in a 3,000-year-old Egyptian mummy. This team's work was published in 1997 as a one-page note in *The Lancet*:

> We were able to examine the torso from a male mummy aged less than 35 years. The head was missing and the mummy had been broken in two parts just above the pelvis. The mummy was loosely wrapped in linen bindings and did not have evidence of evisceration. . . . After careful dissection of the intact anterior body wall, we noted residues of the lung with extensive pleural adhesions to the chest wall in the right thoracic cavity [Figure 3.39]. The left lung was collapsed and macroscopically unremarkable. In addition, there was a severe anterior destruction of two lumbar vertebral bodies, irregular osteolysis, and extensive reactive new bone formation was detectable on radiographs. Tissue samples from both lungs, which we had removed immediately after opening the chest wall under sterile conditions, were subjected to DNA extraction and purification.

Using PCR, they amplified a 133 bp fragment that is not a part of IS6110 but encodes a surface protein expressed in several mycobacterial species, including *M. tuberculosis* and *M. leprae*. They sequenced their PCR product and reported "the sequences showed homology to the DNA of *M. tuberculosis*, thereby confirming that the mycobacterial DNA found in our samples was that of *M. tuberculosis*." They did not publish the sequences from their PCR products or the percent similarity in their BLAST search.

DISCOVERY QUESTIONS

50. Perform a BLASTn search using the IS6110 sequence. Does this piece of DNA match anything besides *M. tuberculosis*?

51. Go to BLAST again and search with the 133 bp fragment sequence amplified from the Egyptian mummy. Does this piece of DNA match anything besides *M. tuberculosis*?

52. When IS6110 was first published, the authors clearly stated IS6110 was identical in all the *M. tuberculosis* complex, which includes three additional species: *M. bovis*, *M. microti*, and *M. africanum*. Does this information affect your interpretations of the two *M. tuberculosis* case studies?

Ancient TB Epilog In 1995, Joe Bates, Kathleen Eisenbach, and their colleagues at the University of Arkansas Medical School published a commentary on the 1994 Salo paper claiming to have found *M. tuberculosis* DNA in a pre-Columbian Peruvian mummy. Bates and Eisenbach were two of the investigators who initially characterized IS6110, and their commentary offered a different interpretation of the Peruvian results. They maintain that IS6110 is unable to distinguish among the four closely related species in the *M. tuberculosis* complex (Figure 3.40). They speculated that the DNA found in the Peruvian mummy was in fact *M. bovis* and not *M. tuberculosis*. They based their conclusions in part on the distribution of host species for *M. bovis*, a population that includes many wild and domesticated animals, such as llamas and seals, with which the Chiribaya people were known to have contact.

Eisenbach's group hypothesized that *M. bovis* predates *M. tuberculosis*, and that as Europeans domesticated animals, they became infected with *M. bovis*, which later mutated into the current *M. tuberculosis* species and was spread by humans as we began to live in more densely populated areas in Europe and the Middle East around 1600 A.D. At this time, virtually 100% of European urban populations became infected with *M. tuberculosis*, and 25% of infected people died from it. High mortality would have created a strong selection pressure on human genotypes that were able to survive the *M. tuberculosis* infection. As humans adapted, tuberculosis changed from a rapidly lethal disease to a chronic, nonlethal, pulmonary infection. Included among the chronically infected would have been European sailors who traveled to North America and came into contact with indigenous peoples. Consistent with this model are modern populations that were first exposed to Europeans in the twentieth century. For example, people in central Africa in 1910, the highlands of New Guinea in 1950, and the high Amazon in the 1990s all experienced a rapid and highly lethal response to their initial exposure to *M. tuberculosis* in the twentieth century.

The hypothesis proposed by Bates et al., that *M. bovis* is more primitive than *M. tuberculosis*, is not universally accepted. However, it is clear that the two instances of ancient *M. tuberculosis* DNA are not as conclusive as they first appeared. The regions of DNA amplified by PCR are not unique to *M. tuberculosis*, and thus alternative explanations are more likely. With the publication of the full *M. tuberculosis* genome, it should be possible to find species-specific DNA that could be diagnostic. It is likely that the mummified individuals were infected with *M. bovis*, or that the samples were contaminated by *M. bovis*, since the mummified remains were exposed to the elements for hundreds or thousands of years. Ancient DNA offers an exciting potential to look into the past, but great care must be taken when interpreting such findings, because we have only partial sequences for some species and they may not be unique to a single species.

In 2000, Cooper and Poinar co-published a paper outlining a list of "authenticity criteria" that should be followed whenever conducting research with ancient DNA, including validation by independent labs. Cooper argued that DNA stored in a permafrost up to 2 million years ago may be legitimate, but he doubts older samples are genuine. For example, DNA samples spanning the last 150,000 years have indicated the North American bison population rapidly declined about 37,000 years ago, which is before human hunting could have been the primary cause of their demise. He feels that working with extinct species is easier because no modern organisms can possibly contaminate the samples—but Cooper assumes you can develop PCR primers that work only on the extinct DNA. However, caution must be used; it is possible that urine from modern bison could seep through dirt and contaminate "ancient" samples, thus confounding the interpretations. In 2003, Cooper's team published findings indicating that very early cultivation of maize by humans (at least 4,400 years ago) selected for alleles that increased crop production.

Cooper has his critics too, who fear his authenticity criteria may become a simple checklist that could lead to a false sense of security and publication of bogus findings, similar to some studies in the 1990s. One critic cited a 2003 study using DNA from Cro-Magnon humans, which adhered to Cooper's criteria, but may in fact have suffered modern human contamination. Keep in mind that when investigators try to amplify trace amounts of ancient DNA, they may accidentally amplify the DNA from a single skin cell from the technician. The criteria may need to be applied differentially to each case, but Cooper feels this will allow the field to revert to the chaos of the 1990s. At this time, perhaps it is best for the casual reader to read a paper's methods section carefully, and be skeptical . . . very skeptical.

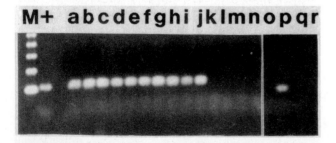

Figure 3.40 Amplification of DNA by PCR using various sources of *Mycobacterium* DNA.
Input DNA was 625 pg of DNA per reaction, with 5 μL of the reaction product loaded on an agarose gel for analysis. Lanes: M = molecular weight marker; 1 = positive control; lanes a–g are different isolates of *M. tuberculosis*; lanes h, i, and j are different strains of *M. bovis*, with j being the BCG vaccine from Glaxo; lanes k–n are *M. kansasii* strains; o = *M. fortuitum*; p = *M. simiae*; q = *M. gordonae*; r = *M. chelonei*.

DATA
ancient DNA
abnormalities
skin lesions
LINKS
extinct species
famciclovir
GOARN
Laura Richman
viral genomes
WHO

DISCOVERY QUESTIONS

53. What additional information would you like to have if you were a peer reviewer for the paper that claimed 3,000-year-old Egyptian mummies had died from *M. tuberculosis*?

54. What controls would you design for similar ancient DNA studies?

55. Go to the ancient DNA sequence web page; copy and paste it into a BLASTn search. What species did you find?

56. Go to the list of extinct species with DNA information. How many extinct prokaryotes are on this list? Any surprises about the list?

How Are Newly Emerging Diseases Identified?

So far, we have focused on genomes inside cells, but this is only a subset of the sequenced genomes. Many viral genomes have been sequenced, enabling a better understanding of viral diseases and perhaps better treatments, too. Initially, however, the greatest benefit has been in the area of diagnosis. Every year, new diseases are identified in humans that have not been characterized before (e.g., SARS, avian flu, monkeypox, etc.). Epidemiologists and pathologists are often the first brave investigators who (willingly) rush into an area of infestation to determine the causative agent. The World Health Organization (WHO) created a group of 40 people who monitor news and web sites for signs of a new infection. WHO also created Global Outbreak Alert and Response Network (GOARN) to coordinate worldwide efforts in the event of a new outbreak. These brave public health workers collect samples from victims, animals, food sources, water, air, etc.

Imagine being the first healthcare worker in a region that has suffered a new outbreak of ebola. You don't know where it hides outside humans. You take samples from a wide assortment of sites and go to the lab with an electron microscope, since many viruses have distinctive shapes that aid in diagnosis. At some point, DNA will be extracted and you will sequence it in hopes of determining the exact cause of the new outbreak. In this section, we will examine in detail two outbreaks, one in North American zoos and the second in New York. Other, more recent, outbreaks will be covered briefly because their status is still evolving as of January, 2006.

Case 1: Elephants Dying in Zoos As a graduate student working with Gary Hayward at Johns Hopkins Medical School, Laura Richman led a team to determine why so many young elephants were dying in captivity. Zoos no longer capture wild elephants, so the only source of new animals is elephants born and raised in captivity. Between 1983 and 1996, 34 Asian elephants were born in North American zoos, of which 7 died with similar symptoms. During the same period, only seven African elephants were born, two of which died prematurely. If this trend continues, most of the adult elephants in zoos will be too old to breed, and there will be too few young adults to sustain the desired populations, especially for African elephants. Therefore, we need to understand the cause of so many premature deaths.

The disease has a sudden onset characterized by fluid retention in the skin of the head and trunk, and internal hemorrhaging. Histological samples revealed abnormalities in the heart, liver, and tongue, with inclusions inside the nuclei of endothelial cells of the smallest blood vessels in these organs (Figure 3.41). Upon electron micrographic analysis, a diagnosis was made of herpesviruses that showed a preference for endothelial cells, which is unusual for herpesviruses. The high mortality rate was due to heart failure from capillary damage and leakage of fluid. Interestingly, some healthy adults had skin lesions that contained the same viral particles, and yet they seemed unharmed by these infections. The investigators were unable to culture the virions on a wide range of cells, so they decided to use DNA sequencing to determine which strain of herpes was killing the elephants.

Based on the diagnosis of herpesvirus, PCR primers were designed to amplify two genes: terminase and DNA polymerase. PCR products of both genes isolated from African and Asian elephants were sequenced to produce unrooted phylogenetic trees (Figure 3.42). The virion sequences isolated from the elephants were highly conserved, but unlike those from the alpha-, beta-, or gamma-herpesviruses. Based on the sequence information, it appeared zoo elephants were infected with a previously unknown type of herpesvirus that was also detected in biopsies taken from noncaptive healthy adult African elephants (samples from healthy adult Asian elephants were unavailable).

One young elephant that became sick was successfully treated with the anti-herpesvirus medication famciclovir immediately after becoming ill. The success of quick treatment for infected elephants indicates that veterinary intervention can save the lives of many newborn elephants. In facilities that have experienced herpes-induced elephant deaths, African and Asian elephants have direct or indirect contact (e.g., from common feeding or bathing areas). Richman and her colleagues speculated that the two elephant species each have evolved resistance to their own variants of this herpesvirus, but not to the virus found in the other species. This is a reasonable hypothesis: in the wild, the two elephant species never come into contact with each other and each harbors its own virus with no ill effects. In captivity, though, the virus has been given the chance to "hop" species and infect a new host whose

DATA
terminase protein

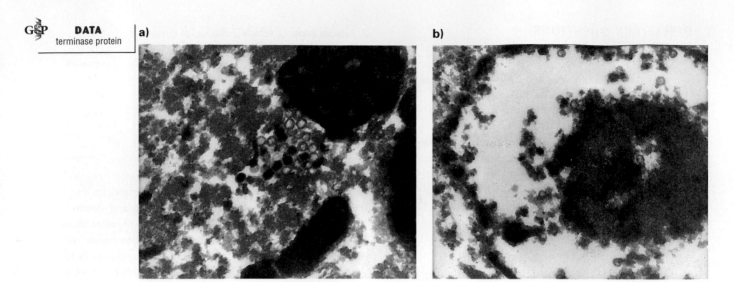

Figure 3.41 Elephant viruses.
a) and **b)** Transmission electron micrographs of herpesvirus "inclusion bodies" inside the nuclei of endothelial cells. These photomicrographs were produced from cardiac tissue obtained in an autopsy of an Asian elephant. See web site for autopsy color photos of the lesions.

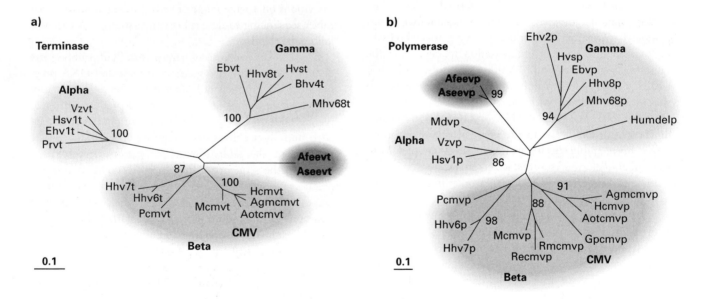

Figure 3.42 New herpesviruses.
Unrooted phylogenetic trees of herpesvirus. **a)** terminase protein sequences and **b)** DNA polymerase protein sequences. The scale bars indicate the distance of a 10% change in amino acid sequence at any given position between any two proteins; numbers on branches indicate bootstrap values.

offspring have no genomic resistance. With the aid of DNA diagnosis, a mystery was solved and an effective treatment devised. This research may have a dramatic impact on the survivability of elephants born in captivity, and may also be a lesson for us about mixing in zoos species that did not evolve in close proximity.

DISCOVERY QUESTIONS

57. Perform a BLASTp search using this region of the terminase protein.
   a. Which virus has the second highest percent identity after the elephants' virus?

b. Which isolate has the highest percent sequence similarity (listed as positives on the results page) but not 100%? You should be able to find some with at least 80% positive.

c. Predict which of these two (answers to a and b) might be the next most likely source of lethal infection for zoo elephants. Explain your answer.

58. Perform a BLASTn search using part of an African viral DNA polymerase gene. Compare these hits with the hits you get from the same region of an Asian viral DNA polymerase gene.

a. What is the source of the hit you received for the Asian elephant virus?

b. Did either of your searches return hits for both types of virus?

c. Given that the authors hypothesized that each species infected the other, why didn't you get hits for both types of viruses?

d. Did you get any hits for any other herpesviruses? Explain why this result is not surprising.

Case 2: New Human Encephalitis Outbreak in Northeastern United States

In August and September 1999, New York (Figure 3.43) experienced an outbreak of human encephalitis (inflammation of the brain) in which two people died. At the same time, there were many cases of both captive and wild birds dying. In a matter of weeks, the cause was diagnosed as a form of West Nile virus (WNV), after a part of the 11,000 bp single-stranded RNA genome was sequenced. This was useful information since it told us that fatalities would be uncommon.

Based on sampling in October 1999, approximately 1,256 people were infected with West Nile virus during the first summer. Approximately 239 people (19%) may have had a mild clinical illness associated with the infection. However, before the summer of 1999, West Nile virus had never been reported in North America. These viruses are spread by insects—mosquitoes in particular. West Nile virus can live in a number of animal hosts, including farm animals, pets, and wild birds that migrate huge distances each year. Where did the 1999 strain originate? How did it arrive? Did migrating birds bring the virus to New York? These questions had to be answered so we could better understand how to prevent the arrival and spread of more West Nile virus.

The complete 11,029 nucleotide genome of the 1999 New York West Nile virus (WN-NY99) was sequenced. The source of the sequenced genome was a dead Chilean flamingo at the Bronx Zoo in New York. To determine the origin of the virus, Robert Lanciotti from the National

DATA
African viral
Asian viral
WNV

a)

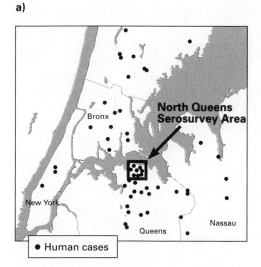

b)

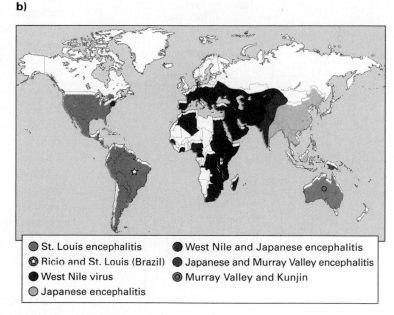

● Human cases

● St. Louis encephalitis
✪ Ricio and St. Louis (Brazil)
● West Nile virus
◐ Japanese encephalitis
● West Nile and Japanese encephalitis
● Japanese and Murray Valley encephalitis
◉ Murray Valley and Kunjin

Figure 3.43 Distribution of West Nile viruses.
a) A map of the New York City area with dots representing the locations where people lived who were confirmed to be infected with WNV during the fall of 1999. **b)** Global distribution of different strains of Nile viruses. Note that before 1999, West Nile virus had not reached North America.

LINKS
NCID
WNV map

Center for Infectious Diseases (NCID) in Fort Collins, Colorado, led an international team to use genome sequences to track it. They compared the genomes of 41 different strains of related viruses and produced a rooted phylogenetic tree (Figure 3.44). The analysis placed WN-NY99 in a cluster of similar viruses, so it did not appear to be new. Hundreds of birds and mosquito samples were collected; all the isolated viruses had 99.8% sequence identity, which indicated a single introduction of the West Nile virus rather than multiple separate introductions. Furthermore, when compared to its nearest neighbor, isolated from a goose that died in Israel in 1998, the two isolates shared greater than 99.8% sequence identity.

Contamination of the 1998 Israeli sample with the WN-NY99 sample could not be the source of sequence similarity since the two genomes were sequenced on differ-

ent continents. Therefore, it seems highly probable that the WN-NY99 virus was brought to North America from Israel. It is impossible to determine if the virus was in a human, an animal such as a pet or an illegally smuggled exotic animal, or an infected mosquito trapped on an airplane. This case illustrates the rapidity with which pathogens can spread between geographically distant locations now that people travel rapidly around the world. The CDC was given several million dollars by the U.S. government to maintain a surveillance program designed to monitor the possible spread of West Nile virus. Samples will be taken from insects, birds, and mammals, which investigators will search for RNA genomes of the West Nile virus. As of summer 2005, West Nile virus had been reported in every U.S. state except Hawaii (only tourists in Alaska have been diagnosed), Canada, Mexico, Central America, and some parts of South America. WNV spread through the continental U.S. at 900 km per year.

Public health officials took the West Nile virus very seriously and canceled summer concerts in Central Park during 1999 because they believed public gatherings might exacerbate the problem. They sprayed extensively to kill mosquitoes, but bad news came when adult mosquitoes that had lived through the winter of 1999–2000 were found to contain the West Nile virus. What precautions would you have advised if you had been in charge of public health?

Alaska's climate will keep it safe, but on September 24, 2004, Hawaiian health officials were notified that a captured bird had tested positive for WNV. The response team sprayed the area where the bird was caught, which everyone knew would be the path of entry: the airport. Based on statistics and probability, they estimate 70 infected mosquitoes arrived by airplane in 2004, with the numbers expected to rise each year. No one knows how many may be riding in the 1,200 container ships that dock in Hawaii each year. Since Hawaii does not have crows, blue jays, or magpies, no one knows which species of local birds will be the sentinel for WNV, so they are conducting tightly controlled experiments with local birds and mosquito species to anticipate what might happen if WNV arrives. How can you defend your borders and remain a tourist destination and commercial center? Perhaps the inevitable is obvious to everyone; the Department of Health is already preparing a rapid response team and improved laboratory testing. Luckily, the initial WNV diagnosis proved to be a false positive once a more thorough but slower test result became available.

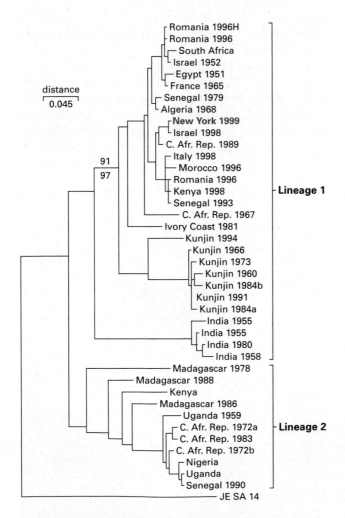

Figure 3.44 Rooted phylogenetic tree of WNV.
Bootstrap value with 500 replicates (above) and confidence probability value (below) based on standard error are located at the branch between West Nile and Kunjin viruses.

DISCOVERY QUESTIONS

59. Go to the WNV map and hypothesize which arrived first: the mosquito vector, infected horses, infected birds, or human patients. Click on a picture and then the text links at the top. You can zoom in on your favorite state by clicking on it. Support your hypothesis with data. Which color is the most common on the map, and what does this tell you about our understanding of WNV?

60. Read the abstract from the March 2005 issue of *Emerging Infectious Disease.* What new information is provided that might alter how WNV should be controlled?

61. Perform a BLAST2 alignment to compare partial coding sequences from an envelope glycoprotein gene of the Israeli 1998 virus (accession number af205882) with the Israeli 1952 virus (accession number af205881). Then compare the Israeli 1998 virus with the New York 1999 virus (accession number af196835). Finally, compare the Israeli 1998 virus with the Central African Republic virus from 1967 (accession number af001566). What hypotheses can you develop from these alignments? To see a very different virus, look at one isolated from the Central African Republic in 1972 (af001563).

What Other Outbreaks Are Coming?

If you are a regular reader of the news, you know that several other "emerging diseases" are poised to change our lives. In the summer of 2003, as many as 100 people in Wisconsin, Illinois, and Indiana developed lesions that looked like large warts and turned out to be monkeypox sores. Fortunately, no one died from the monkeypox, and the only common factor they shared was recent acquisition of a pet from an animal distributor in Milwaukee. Most of the victims had recently been bitten by their new pet prairie dogs. Destroying the animals and treating the people stopped the spread of monkeypox. However, some public health officials were concerned that the response time would have been too slow if the disease had been dangerous. Monkeypox is not transmitted from person to person, but a related and more deadly virus is: smallpox. This monkeypox case illustrates the important roles played by local health officials and primary healthcare workers, who may be unfamiliar with exotic diseases or the appropriate responses.

The virus ebola has been the subject of at least one Hollywood movie. Between 2002 and 2004, 5 outbreaks in Africa infected 313 people and killed 264 of them. In these 5 incidents, at least 11 separate viral strains infected the people. Ebola does not mutate rapidly, indicating that once the outbreak began, more than one person brought in new strains. In every case, the first infected person in the traceable chain of victims was a hunter who had handled dead monkeys or apes. Like many viruses, ebola is not restricted to humans. The large primates in Africa are dying rapidly due to repeated episodes of ebola infections, which not only leads to more human infections, but may also lead to the extinction of gorillas, chimpanzees, and duikers. This may sound like a problem restricted to Africa, but remember how quickly WNV spread across North and South America? Intercontinental travel makes every disease a global threat. As documented in the book *The Hot Zone,* a primate research center in Reston, Virginia experienced an ebola outbreak; as far as we can tell, this version was able to spread as an aerosol as well as by direct contact. The Virginia outbreak was contained and the mutated virus was not as lethal, but the implications are frightening.

Animals are not the only targets of new diseases. In the spring of 2000, a fungus appeared in California that leads to sudden oak death. The cause was quickly identified by DNA sequencing as *Phytophthora ramorum,* a water mold found in northern Europe that had previously infected only rhododendrons. Five years later, the rare infection has become a West Coast epidemic, killing tens of thousands of trees. The fear is that if the mold reaches the East Coast, hardwood forests could be decimated, because the mold has infected a total of 26 species, and may jump to more. The mold spores spread through water splashes. If someone inadvertently transports an infected plant during a move from the West to the East Coast, the disease will have moved more rapidly than WNV. Currently, research is being conducted at Fort Detrick in Maryland to determine the full range of potential hosts for the worst tree disease since the chestnut blight of the early 1900s.

Two respiratory viruses have emerged from southeast Asia, though you may not have heard of both. Severe acute respiratory syndrome (SARS) was caused by a newly identified coronavirus with a genome of 29,751 nucleotides and only 16 ORFs. In November 2002, some vendors and customers at live-animal food markets in Guangdog Province in southern China became ill. The disease started slowly, but soon spread to others living in the area (Figure 3.45). Initially, the Chinese government hid the truth from its people and the world, thinking the virus would go away or they could control its spread. Unfortunately, they were wrong. By February of 2003, nearly 8,000 people in 30 countries (including Canada) on 5 continents had been infected. Once the international community cooperated and pooled expertise, the causative agent was isolated, sequenced, and identified as a new coronavirus unlike any others previously reported. The SARS epidemic was controlled, but a few infections have recurred since then. The most likely natural reservoir is wild animals, civet cats in particular, that are captured from the wild and sold for human consumption. China could choose to close the public markets, but would this prevent all exposure, or just the publicly visible exposure? As the ebola outbreaks have proven, people will eat meat if they can catch the animals.

LINKS
16 ORFs
BLAST2
Ebola
Monkeypox
SARS
March 2005
sudden oak death

LINKS
avian flu
people infected

As of fall 2005, the virus most epidemiologists were watching with fear is the **avian flu** that is persisting at a low level in nine countries in southeast Asia, including China—but this time China cooperated at the early stages. The name stems from the pathogen's original source, chickens, ducks, geese, and other domesticated birds used for food. The strain has a simple name, H5N1, which describes the variation in two proteins on its surface: hemagglutinin (H; #5 out of 15 "alleles") and neuraminidase (N; #1 out of 9 "alleles"). The number of **people infected** has been small, but the virus kills with a high frequency. In September of 2004, investigators found cats (zoo tigers fed tainted birds) infected with the avian flu and discovered cats (including house cats) can spread it to each other directly. As with all flu viruses, the genome is composed of eight RNA molecules; every time two viruses infect an individual, the viral genomes can recombine to produce a new variant combination ($2^8 = 256$ possible combinations). The greatest fear is that, given the opportunity, the virus may mutate and become infectious by human exposure through breathing contaminated air. Since avian flu is an emerging virus, no one has established an immunity to the virus, and thus the entire world is susceptible. In 1918, the world experienced a flu pandemic that killed 40–50 million people. Flu pandemics are episodic and many believe we are due for another one soon.

In early 2005, Europe got dreaded news similar to the report Hawaii got for WNV, but this time the diagnosis was correct and it was avian flu. A man was smuggling endangered mountain hawk eagles for pets, and both of his birds were infected. To reach Brussels, he took two flights: Bangkok to Vienna and Vienna to Brussels. Everyone on board both planes was a potential victim. Luckily, the smuggler was not infected and health officials have not detected any human cases in Europe. The birds were killed in the quarantine area of the airport. That time we were lucky. Illegal bird smuggling may be a small-volume trade, but the global threat is enormous.

As these five cases of viral outbreaks have demonstrated, we need reliable methods for detecting the presence of infectious agents. With improvements in methods and machines, we are getting better at discovering known pathogens using DNA sequences as diagnostic tools. However, as SARS illustrated, knowing about an infection is not the most important component; we need to cooperate and educate on a global scale. If you are interested in this field, unfortunately you probably will find plenty of work. You can monitor the **GOARN** web site to track what is happening in remote places on earth, but keep in mind that modern travel means that each of us is only a few degrees of separation from any person in the world.

DISCOVERY QUESTIONS

62. Visit the WHO web site for **avian flu**. What is the latest confirmed number of people infected, and the number of people killed? In these cases, did the virus spread from human or avian contact?

63. A child has come to your clinic with a black wound that developed after she was bitten by her

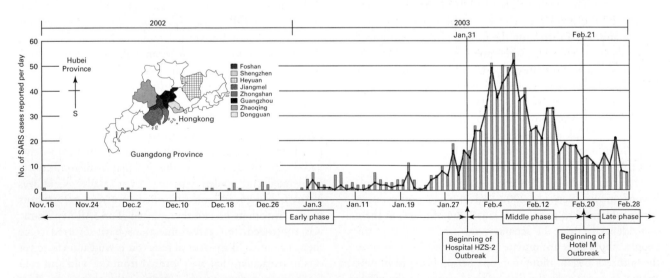

Figure 3.45 The triphasic SARS epidemic in Guangdong Province, China.
The map (inset) shows the geographical distribution of cases belonging to the early phase by administrative districts of Guangdong Province. The daily numbers of SARS cases reported in Guangdong Province (bars) and those in the city of Guangzhou (dots with lines). The early, middle, and late phases of the epidemic are indicated.

pet prairie dog. You take a biopsy and ask a lab to analyze it. The lab returns this **biopsy DNA** sequence. Perform a BLAST search to determine its source.

64. Go to the Sudden Oak Death (SOD) web page and click on the link that says "View/download movie of SOD through time." Watch the spread of SOD. Now go to the **OakMapper** and work with the data. Zoom in a couple of times to the area just north of San Francisco (Sonoma and Napa counties), and turn on the "Symptomatic trees submitted via OakMapper" option. Is SOD spreading north/south or east/west? Does it look like the distribution is following human modes of transportation, or natural barriers? Do you think the mountains will act as an effective barrier to this fungus? Do you predict SOD will spread as fast as WNV? Explain your answer.

65. Go to the **SARS genome** site at NCBI. Zoom out so you can count the number of coding sequences. How many do you see? To zoom in, click on the top blue line over the area where most of the genes are clustered. Navigate until you see sars7b and sars8b in the same view. How can SARS have so many genes in such a small area? What does the annotation reveal about these genes?

66. Visit the **Conservation Medicine** web page and read about their mission. Do you think their holistic perspective is too grandiose to be realistic, or the most appropriate way to think about emerging diseases, their sources, and their cures?

67. Consider the two HIV phylogenetic trees reproduced below. One is from the population at large and the other is from a single patient monitored over time. Although they are not drawn on the same scale, evaluate the challenge of detecting HIV or any new virus by PCR.

DATA
biopsy DNA
OakMapper
SOD

LINKS
Conservation Medicine
SARS genome

Summary 3.3

DNA is a very stable polymer, which allows us to extract it from a wide range of sources. Soon, biological weapon identification will become reliable and quick, which will enable us to respond appropriately to hoaxes as well as real terrorism. At times, it seems that new diseases are evolving faster, but it may be that improving methods help us detect them better. Viral genomes can reveal the cause of diseases and point to potential treatments, but it seems nature has an endless supply of new species to challenge us. Ancient DNA is a perplexing and fascinating issue: we may want to extract DNA from every possible ancient artifact to discover hidden information, but we need to use special caution when working with ancient DNA. Biomedical

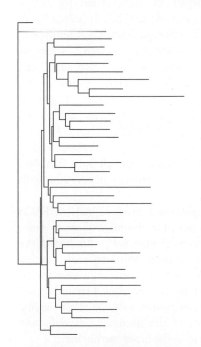

HIV population phylogeny

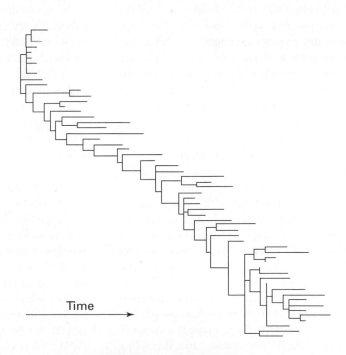

Time

HIV within host phylogeny

LINKS
HGP
translational medicine
METHODS
ELISA
FACS

research will lead to new methods for diagnosis. As the technology improves, look for miniaturized DNA labs in your physician's office or in a wildlife biologist's truck.

3.4 Biomedical Genome Research

"The Human Genome Project (HGP) will change the way we practice medicine." This statement has been repeated in various ways since the HGP first captivated a small number of investigators as a realistic possibility. Now the human genome sequence is available, but where are the changes? Changes in medical treatment will take many years to develop, test, and bring to market (see Chapter 9 for details). In most cases, however, genome sequences of pathogens, not humans, will prove to be easier to convert into medical improvements. This section relates two case studies about using pathogens to develop new treatments. A third case study shows how basic research, molecular methods, and knowledge of genome sequences might lead to one of the first examples of a new type of medication. Finally, we will consider translational medicine, the field in which basic research discoveries are translated into new types of medical treatments.

Can We Use Genomic Sequences to Make New Vaccines?

Meningitis and sepsis involve inflammation of the outer lining of the brain (meningitis) or the blood (sepsis). If left untreated, they can progress rapidly to cause permanent nerve damage or death. The causative agents are often bacterial, so treatment with antibiotics is usually successful unless the pathogen is antibiotic resistant, in which case the infection is often fatal (about 10% of all cases are fatal, and about 25% of survivors have lasting neurological damage). However, the initial symptoms are flu-like, and thus the severity of the infection usually is not appreciated until the immune response is vigorous and harmful. Therefore, vaccines against these types of infections are highly desirable.

A large percentage of meningitis and sepsis cases are caused by *Neisseria meningitidis*, which is a Gram-negative bacterium classified into five major types based on antibody binding: **serotypes** A, B, C, Y, and W135. In the 1960s, vaccines produced for serotypes A, C, Y, and W135 were found to work well in adults, but have been less effective in young children and infants. Modern vaccines currently under development appear to be more effective in younger people, but no vaccine is available for serotype B. Serotype B is responsible for 45% to 80% of the *N. meningitidis* cases, depending

on the country. The outer capsule of serotype B is made of a polysaccharide that is very similar to sugars found on human cells, which means the outer capsule polysaccharide is not a good vaccine candidate. The outer surface of *N. meningitidis* contains a protein known to elicit a strong therapeutic immune response, but this protein is highly variable among different strains of serotype B and thus is also not a good candidate for vaccine production.

To address the need for an *N. meningitidis* vaccine, a group from TIGR and IRIS Chiron in Italy sequenced the 2,272,351 bp genome of *N. meningitidis* strain MC58 (serotype B) isolated from an infected patient. Using standard ORF-finding software and whole genome comparisons with other bacteria, TIGR identified 570 ORFs that potentially encoded novel surface-exposed or secreted proteins (of 2,158 total ORFs). They isolated these ORFs and introduced them into *E. coli* for **overexpression**. Of these 570 ORFs, 350 produced proteins, with 70 predicted to be lipoproteins, 96 periplasmic proteins, 87 inner-membrane proteins, 45 outer-membrane proteins, and 52 with uncertain location. Antibodies were produced that bound to the *E. coli*-produced candidate proteins. Enzyme-linked immunoabsorbent assay (ELISA) and fluorescence-activated cell sorting (FACS) were used to detect surface expression on a panel of diverse serotype B strains. In addition, a bactericidal test of serum was used to determine which proteins elicited an immune response that might kill serotype B. From this series of assays, the group found 85 proteins that were positive in more than one assay. They selected for further testing 7 proteins (termed *genome-derived Neisseria antigens* [GNAs]) that were positive in all assays (representing 1.2% of the original 570 candidate proteins).

To determine the degree of conservation for the 7 GNAs, the investigators sequenced GNA orthologs in 31 *N. meningitidis* strains: 22 serotype B strains, 3 serotype A strains, 2 serotype C strains, and 1 strain each of serotypes X, Y, Z, and W135. In addition, they also sequenced orthologs of the 7 GNAs in 3 different isolates of the closely related species that causes gonorrhoea, *N. gonorrhoeae* (Figure 3.46). From this analysis, they determined that GNA33, GNA1162, GNA1220, GNA1946, and GNA2001 had greater than 99% sequence conservation with the 31 *N. meningitidis* strains, and all but GNA2001 showed greater than 95% sequence conservation with *N. gonorrhoeae*.

Of the 7 GNAs, 2 generated strong signals in FACS analysis when tested against both the original strain and a mixture of serotype B strains (Figure 3.47). Binding of the GNA33 and GNA1946 antisera were especially strong when compared to the prototypical surface protein (called OMV), which works well against its original strain but not against a diverse mixture of strains. GNA33 is an enzyme critical in the formation of the bacterial cell wall, and GNA1946 is a lipoprotein in the outer membrane.

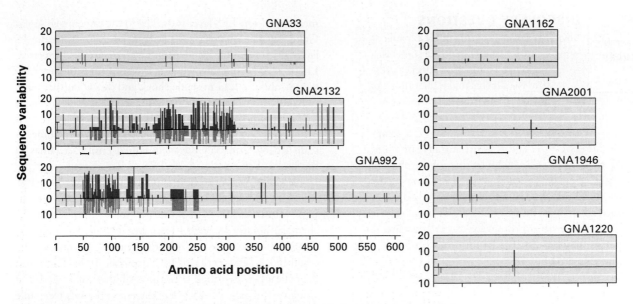

Figure 3.46 Amino acid sequence variation within *N. meningitidis* for the seven potential vaccine antigens.

X-axis is amino acid position and Y-axis is the number of strains analyzed. Line at zero represents the reference genome sequence. Amino acid differences within serotype B are indicated by purple bars above zero. Amino acid differences from the other serotypes are indicated by gray bars below zero. The height of the bars indicates the number of strains with differences. Horizontal bars below GNA2001 and GNA2132 represent DNA segments that are deleted in some strains.

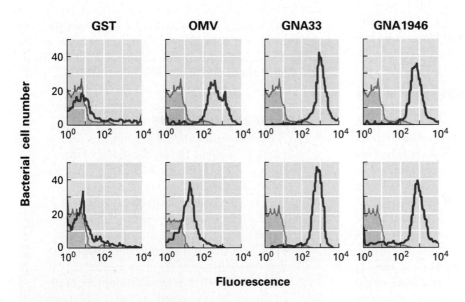

Figure 3.47 Two vaccine candidates.

FACS analysis showing binding of polyclonal OMV, GNA33, and GNA1946 antisera to a uniform antigen (top row) and to a heterogeneous mixture of antigens (bottom row). Gray profiles show binding of negative control antisera; purple profiles show binding with experimental antisera. GST was used as the negative control antigen.

LINKS
Christian Raetz
microbial genome
STRUCTURES
penicillin
transglycosylase

DISCOVERY QUESTIONS

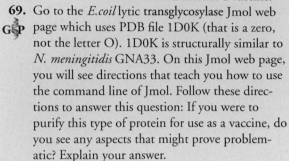

68. The polysaccharide outer capsule of *N. meningitidis* looks like some of the polysaccharides on human cells. Why would this make the outer capsule a bad candidate for a vaccine?

69. Go to the *E.coil* lytic transglycosylase Jmol web page which uses PDB file 1D0K (that is a zero, not the letter O). 1D0K is structurally similar to *N. meningitidis* GNA33. On this Jmol web page, you will see directions that teach you how to use the command line of Jmol. Follow these directions to answer this question: If you were to purify this type of protein for use as a vaccine, do you see any aspects that might prove problematic? Explain your answer.

The investigators started with whole genome sequence information (2,158 annotated ORFs) and selected two highly promising proteins to be used as vaccines against *N. meningitidis* serotype B that had resisted previous attempts. Of course, several years of testing and clinical trials will be required to confirm the utility and safety of the proposed new vaccines, but this work is a milestone. Genome sequence provided the key insight for rational design of new vaccine candidates. As more pathogen genomes are sequenced, this approach will be repeated and refined until new and more protective vaccines are available to the public. However, the production of vaccines is relatively easy compared to the development of new antibiotics, which have additional design constraints, as we will see in the next section.

Can We Make New Types of Antibiotics?

The goal of a genomics approach to new antibiotics is to discover drugs that inactivate new targets. An ideal candidate is outer surface, because the outermost layer helps hide bacteria from the immune system, blocks the entrance of many antibiotics, and is vital for bacterial survival. We need to understand a little about bacterial cell membranes and walls before we discuss the specifics of this case study. We will focus on Gram-negative bacteria, which are the target of one new type of antibiotic.

Gram-negative bacteria are so named because they do not stain with a dye called Gram. Several layers make up the outer structure of Gram-negative bacteria, as shown in Figure 3.48. There are two phospholipid bilayer membranes with a cell wall in between. On the outside of the outer membrane is a lipoprotein layer that contains proteins and lipids coated with different combinations of sugars (Figure 3.48c). Our story focuses on the lipid A portion of a lipopolysaccharide (LPS), which is a component of the outer

membrane. The LPS layer is the key to the impermeability of Gram-negative bacteria; cells with a poorly organized LPS layer are susceptible to immune system attacks. Excessive LPS is toxic to humans when it overstimulates the immune system, and can lead to septic shock and death. Furthermore, mutant bacterial cells that do not make LPS are not viable, so LPS is an ideal target for a new class of antibiotics.

Working with the Gram-negative *E. coli* genome, a team led by **Christian Raetz** identified critical enzymes in the biosynthetic pathway of LPS. Raetz moved from the University of Wisconsin to the pharmaceutical giant Merck, where he continued his efforts to inactivate the pathway for LPS synthesis. Merck has millions of compounds in its chemical library, so Raetz's team began a high-throughput screening procedure to identify candidate drugs that could block LPS synthesis. In November 1996, Raetz and his colleagues identified a compound they call L-161,240 and a variant called L-159,692; these compounds were able to block LPS production and cure mice infected with *E. coli* (Figure 3.49). As predicted, these compounds did not affect Gram-positive bacteria or eukaryotic cells. Unfortunately, their compounds were not effective against two other species of Gram-negative bacteria.

The compound L-161,240 was about as effective as ampicillin, rifampicin, and erythromycin against **wild-type** *E. coli*. L-161,240 blocks a novel pathway and kills *E. coli* cells in a matter of hours, so it looked like the ideal drug. Many bacteria have evolved resistance to commonly used antibiotics and it was hoped L-161,240 might avoid neutralization by the bacteria. Unfortunately, resistance to L-161,240 occurred at a frequency of about 1×10^{-9}, so L-161,240 was not perfect. Cells that were resistant were able to synthesize lipid A, meaning the target might still be viable. The candidate drug will need some additional modifications to improve its ability to avoid resistance.

DISCOVERY QUESTIONS

70. What information would you like to collect to rationally design a variant of L-161,240 that could avoid resistance by *E. coli*? How might you collect this information?

71. Go to NCBI's microbial genome site and click on a name for one of the *E. coli* O157:H7 variants. Use the gene search field to determine if these pathogenic strains synthesize lipid A (the membrane protein targeted by the proposed new class of antibiotics). Here is a list of genes that encode enzymes involved in lipid A synthesis: *envA*, *lpxA*, *lpxB*, *lpxD*, *lpxC*, *firA*, *ssc*, and *omsA*. Could the lipid A antibiotic kill the strain you chose?

72. Look at the structure of penicillin. It has a ring structure called a β-lactam ring consisting of

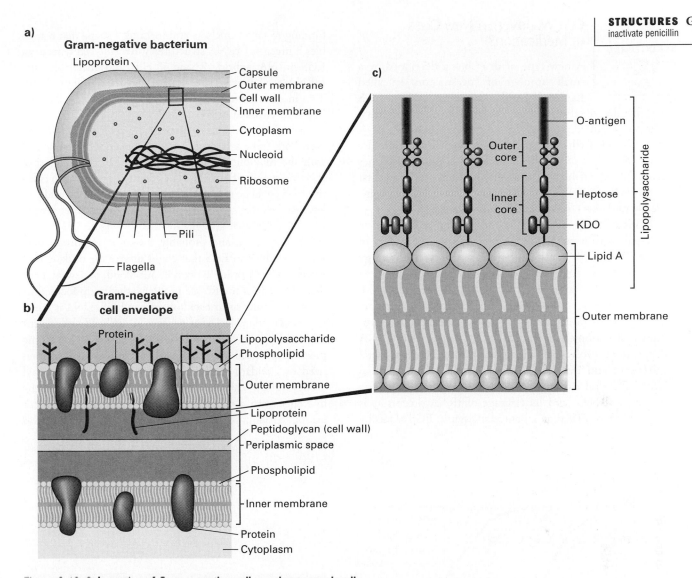

a)

Gram-negative bacterium

- Lipoprotein
- Capsule
- Outer membrane
- Cell wall
- Inner membrane
- Cytoplasm
- Nucleoid
- Ribosome
- Pili
- Flagella

b)

Gram-negative cell envelope

- Protein
- Lipopolysaccharide
- Phospholipid
- Outer membrane
- Lipoprotein
- Peptidoglycan (cell wall)
- Periplasmic space
- Phospholipid
- Inner membrane
- Protein
- Cytoplasm

c)

- O-antigen
- Outer core
- Inner core
- Heptose
- KDO
- Lipid A
- Lipopolysaccharide
- Outer membrane

Figure 3.48 Schematics of Gram-negative cell membranes and wall.
a) Half of a bacterium is shown for orientation. **b)** Magnification of one portion of the cell wall and membranes. **c)** Higher magnification for greater detail of the LPS with the different parts labeled.

three carbons and one nitrogen (click on the button to highlight the ring). Two links from the penicillin page show enzymes that inactivate penicillin. One cleaves the β-lactam ring; the other adds a deactivating group onto penicillin.

a. Why do you think these two bacteria have evolved two different enzymes to inactivate penicillin?

b. The fact that bacteria have evolved at least two different ways to inactivate penicillin tells us something about the selective pressure exerted by penicillin. What does this suggest to you about the possibility of creating an antibiotic that is resistance-proof?

Raetz and his team have made a good start, and no doubt they are testing more drugs that block LPS synthesis and are not neutralized by antibiotic resistance. This case study illustrates how genomic information can stimulate the rational design of new classes of antibiotics. By determining all the genes in the lipid-A synthesis pathway, we should be able to develop a series of antibiotics that block different enzymes in the pathway and produce a cocktail of medications that can kill even the most resistant bacteria.

Can We Invent a New Class of Medication?

A new type of drug being developed by a small number of investigators is called **External Guide Sequence (EGS)**. EGS was first described by Sidney Altman from Yale University, who was awarded a Nobel Prize for his research into RNA enzymes called **ribozymes**. Ribozymes are RNA molecules that can catalyze chemical reactions just as protein enzymes can. EGS, an offshoot of Altman's ribozyme work, was first published in 1990. It has the potential to become a completely different type of pharmaceutical reagent.

EGSs are short oligoribonucleotides that are designed to base-pair with a target mRNA and lead to destruction of the mRNA. EGS is different from **antisense technology** for two reasons. Antisense technology relies upon RNase H to degrade the target mRNA, but EGS uses **RNase P** instead. EGS is more specific in its ability to initiate destruction of only the target mRNA and does not promote destruction of other mRNAs. RNase H-based mRNA degradation evolved in eukaryotes in response to viral infections, and thus its precision is not as important as killing the viral pathogen. RNase P is required to process precursor tRNAs, and its cleavage of tRNA precursors is much more precise than an antiviral assault. EGS works by binding to target mRNAs and forming a shape that resembles a precursor tRNA structure. When the cell detects an EGS-mRNA hybrid, RNase P cuts and inactivates the mRNA as if it were the normal substrate (Figure 3.50). This mechanism allows you to silence any gene by destroying all its encoded mRNA.

Shaji George and his colleagues at Innovir Laboratories in New York (now a part of Sirna Therapeutics) experimented with the minimum size for an effective EGS and possible chemical modifications of it. They demonstrated that an effective EGS could be reduced from the original 68 nucleotides to less than half that size. At 30 nucleotides, their shorter version was equally effective at degrading mRNA (Figure 3.51).

The next step, and probably the more important one, was to produce an EGS that could survive the onslaught of RNases present inside every cell, our blood, our sweat, even dust. To address the RNase problem, the group focused its attention on the sugar backbone of the EGS. DNA is much more stable than RNA, as we saw in our study of ancient DNA. However, EGS made of DNA is ineffective, so the investigators needed a modification that was easily synthesized by standard methods. They chose to add a methyl group to the 2′ carbon of the ribose sugar. By replacing the hydrogen with a CH_3 on all of the EGS ribonucleotides except the loop portion, they produced an EGS that was

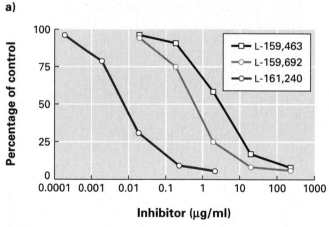

a)

b)

L-161,240   L-159,692   L-159,463

Figure 3.49 Effectiveness and structure of new class of antibiotic.
a) Inhibition of *E. coli* deacetylase in an in vitro assay. Activity of the enzyme is plotted on the Y-axis and concentration of inhibitor on the X-axis. **b)** Three candidate drugs with differences in purple.

a)

b)

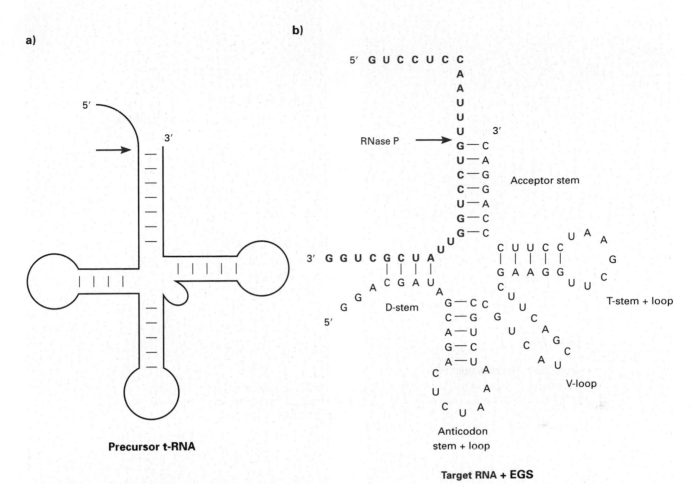

Precursor t-RNA

Target RNA + EGS

Figure 3.50 Two-dimensional representations of tRNA-like EGS structure.
a) Schematic diagram of a precursor tRNA that is cleaved (purple arrow) by RNase P in vivo to produce a mature tRNA. **b)** EGS (black sequence) hybridized to RNA target (purple sequence). The EGS/RNA hybrid structure resembles *E. coli* tRNA^{tyr}. The RNase P cleavage site is indicated by the purple arrow.

stable for 24 hours in 50% human serum. Once they had a stable EGS, they tested its ability to cleave mRNA and found it worked almost as well as the normal RNA version (Figure 3.52). By changing the purines to pyrimidines in the loop, they further increased the stability with no loss of activity. Unlike traditional antisense technology, EGSs are catalytic, which means they are not consumed in the cleavage reaction. The new RNase-resistant EGSs were able to catalyze about three cleavage reactions per molecule in vitro, which is only slightly worse than the normal RNA EGSs. In short, the investigators had created a smaller and more stable EGS that worked almost as well as the original EGS. At that point, they needed to determine if an EGS is capable of targeting specific mRNAs.

In 1998, Deborah Plehn-Dujowich and Sidney Altman wanted to determine if they could block the replication of influenza virus using EGS. They designed EGSs to bind to two different influenza transcripts, one coded for a viral RNA polymerase and the other for a viral nucleocapsid protein. First they tested the two targeted viral mRNAs to see which portions were accessible to RNase digestion. Then they designed several EGSs that would bind to the RNase-sensitive regions of the two viral mRNAs, and tested them in vitro to see which worked best. They chose the best candidates and cloned the corresponding DNA into plasmids that would produce high levels of EGS RNA. These plasmids were incorporated into cultured cells that were later infected with influenza. Plehn-Dujowich and Altman were able to block viral replication by using two EGSs in a single cell. This demonstrated for the first time that EGSs could work on more than one mRNA in a single cell. The drawback was that the EGSs were produced inside the cells—not ideal for use in clinical settings. We would prefer to add EGSs to cells directly from the outside (for

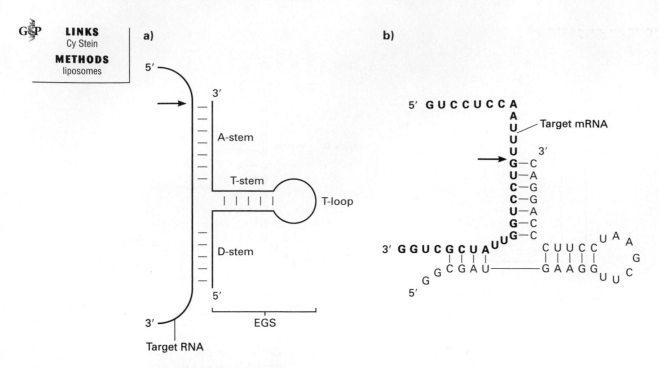

Figure 3.51 The minimum EGS.

a) Schematic showing the key parts in a minimized EGS, only 30 ribonucleotides long.
b) 2D representations of the minimal EGS (black) bound to the same target sequence (purple sequence) as in Figure 3.50, which is cleaved about 61% as efficiently compared to full-sized EGS-induced cleavage.

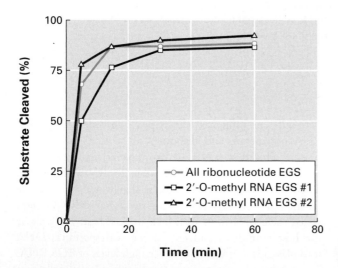

Figure 3.52 A time course assay to measure the effectiveness of modified EGSs.

The all-ribonucleotide EGS (gray line) was compared to two different EGSs that contained 2′ O-methyl modified riboses outside the T-loop. The two experimental EGSs varied by one nucleotide in the loop; dUMP (purple line) and dTMP (black line).

example, as a nasal spray to treat cells lining respiratory passages). Furthermore, we would like to know if more than just viral mRNA can be neutralized by this method.

The therapeutic potential of EGS was assessed when a team attempted to neutralize one specific cellular mRNA but not any other cellular mRNAs with very similar sequences. **Cy Stein** from Columbia University collaborated with Innovir Laboratories to see if they could use EGS technology to block the production of the alpha form of protein kinase C (PKC-α) but not PKC-β or PKC-γ. They designed several candidate EGSs that would bind to the 3′ untranslated region of the PKC-α mRNA. Using **liposomes** to ferry the EGS inside cells, they compared a couple of different EGSs and determined that they had eliminated PKC-α but not PKC-β or PKC-γ.

EGS therapy looks feasible and ideally suited for converting genome sequences into a new type of drug that should be able to combat pathogens as well as human diseases such as cancer. Many more years of research are needed, but EGS technology illustrates how basic research and genomics synergistically may provide new types of medications that could alleviate a lot of suffering.

DISCOVERY QUESTIONS

73. Go to NCBI's Entrez and search the protein database for "RNase P." Has this protein been found in many different species, or just a few?

74. Antibiotics often fail in the long run because pathogens evolve mechanisms to evade them. What do you think about the ability of pathogens to evade EGSs?

75. If you wanted to generate an EGS that was specific to a particular mRNA, what aspect of the gene pool might make this more difficult? How could you address this problem so each person could still benefit from EGS therapy?

Is There an Alternative Way to Inhibit RNAs?

RNA inhibition (**RNAi**) was discovered in the worm *C. elegans* when a negative control failed and the investigators were curious enough to find out why. RNAi is the process whereby small, **double-stranded RNA** (dsRNA) molecules inactivate mRNAs. These little pieces of RNA, only 21 to 22 nucleotides long, are called siRNA (small inhibitory RNA) or miRNA (microRNA). RNAi works by yet another mechanism unrelated to antisense or EGS, and also works in both animals and plants, which indicates it is an old mechanism that evolved long ago. In brief, the siRNAs are produced naturally when dsRNA is cut by the enzyme called dicer. The siRNA is bound by a protein and converted to ssRNA, which basepairs with the target mRNA. The siRNA/mRNA hybrid is cleaved by the RISC complex of enzymes, which can lead to additional siRNA production and further mRNA destruction (see this **RNAi animation**). RNAi is a very efficient and sequence-specific means to silence any gene via mRNA degradation. The question we want answered is whether RNAi can be used in combination with genomic sequence information to generate a new class of medication.

A group of three investigators at the University of Massachusetts Medical School wanted to know if RNAi could inhibit one of the worst RNA molecules of our time: the HIV genome. They produced a series of siRNA molecules that targeted three regions in the HIV genome (Figure 3.53a): the genes *vif* and *nef*, and the long terminal repeat on the 5′ end of the genome. The investigators delivered the siRNA into HIV-infected cells grown in tissue

a)

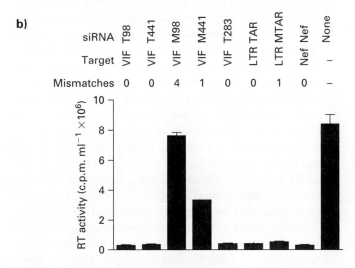

b)

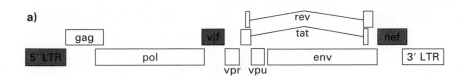

Figure 3.53 siRNAs inhibit HIV replication.
a) HIV genes targeted by siRNAs are highlighted in purple. **b)** Reverse transcriptase activity in HIV cultures targeted by perfectly and imperfectly matched siRNA sequences.

culture and then measured the amount of reverse transcriptase (RT) present in the infected cells (Figure 3.53b). Several different siRNAs were produced for each region of the genome. The data contained good news and bad news. The good news was HIV can be inhibited by RNAi. However, the bad news was some siRNA molecules intended to be negative controls in the experiment also inhibited HIV. Whenever controls fail to give the expected outcomes, interpretation of experimental data becomes very difficult. Therefore, the investigators cannot be certain their specific siRNA is truly specific since even the negative control siRNA reduced HIV activity. Perhaps the siRNA was toxic to the cells and the reduced HIV activity was due simply to its poisoning the host cells more than HIV does.

Over the years, investigators trying to understand how RNAi works have demonstrated that siRNA must match 100% with its target mRNA if it is to silence a gene. The HIV investigators included siRNA molecules with known mismatches to demonstrate the specificity of their reagents, but they also foreshadowed the persistent challenge presented by HIV. HIV mutates about 65 times faster than influenza because of the imprecise replication of its genome by RT. Therefore, all HIV has to do to elude an RNAi therapy is mutate one nucleotide in the 21 nt target region and HIV will become resistant, again. You might think we could simply target a region that is critical for the encoded protein's function, but remember there are multiple codons for nearly every amino acid, so it would be nearly impossible to find the ideal target.

Are There More Stable RNA Genomes We Can Target?

Another group of three investigators from the University of California–San Francisco tried to inhibit polio, which is also an RNA virus. Although polio is not a major problem in most countries that utilize the vaccine, in 2004 the number of countries with polio increased from 6 to 18, with the RNA polio virus infecting about 2,000 children. Therefore, it would be good to have a way to cure those infected with polio. Once again, siRNA worked very well in tissue culture cells, but within four days enough viruses had mutated and become resistant to the particular siRNAs developed by the investigators that the virus infected every cell and flourished. This leads to the obvious question: why bother trying RNAi on viruses if the viruses can mutate so quickly? Let's consider one more virus before we give up.

We have already learned about SARS and its devastating effects on those diagnosed too late. But with rapidly evolving RNA genomes, perhaps we are looking at this problem from the wrong end. We know that codons can change without altering the encoded protein, but do noncoding RNA molecules have any sequence-specific shapes that are critical to their function? If so, could we target these shapes instead of the protein primary structure? A team of biologists and chemists at the University of California–Santa Cruz wanted to see if a highly conserved region of the SARS genome might be required for RNA tertiary structure.

The SARS genome has a 5′ cap and a 3′ poly-A tail, which means it has learned to fool our anti-RNA cellular radar by disguising itself to look like a normal mRNA. SARS is translated directly from its provirus, chromosomally inserted genome, which is transcribed into a large RNA from which several shorter mRNAs are produced ready for translation. Interestingly, every one of the mRNAs has a 32 nt section downstream of the stop codon but upstream of the poly-A tail. These 32 nt are the most highly conserved bases in the entire genome, even in the many variant genomes sequenced during the course of the SARS epidemic. These 32 bases produce a type-two stem loop motif (called s2m) that is composed of dsRNA for the stem and ssRNA for the loop, which makes it look like a lollipop. Why would noncoding RNA be the most highly conserved? In fact, this s2m was highly conserved in many pathogenic viruses, which could indicate a widespread viral Achilles heel for possible medical treatment (Figure 3.54).

The Santa Cruz team crystallized s2m from SARS and determined its three-dimensional (3D) shape. When they resolved its structure, s2m demonstrated a very unusual 90° turn in the stem loop (Figure 3.55). As a part of this unusual turn, four bases exhibit unorthodox base-pairing that helps bend the RNA molecule into the final shape. The investigators wondered how widely this s2m was conserved and found it only in astroviruses and coronaviruses. However, when they allowed some deviation from the exact sequence, they found a similar fold in the 16S rRNA structure. As you know, rRNA is one of the most highly conserved sequences in every genome, and within every one of our ribosomes sits a 90° turn similar to the one in SARS s2m. Furthermore, a literature search revealed that several eukaryotic translation initiation factors bind to this bend in rRNA. What could this mean for SARS?

Having found a connection between s2m, rRNA, and translation initiation factors, the authors speculated that SARS and related viruses utilize the binding of initiation factors to take control of the cell's normal translation process and thus overwhelm the cell with virus production. The s2m sequence is absolutely conserved and presumably serves a function. The hypothesis of hijacking cellular translation via s2m is appealing and testable. More importantly for us, the SARS genome sequence and 3D shape may have revealed the virus's greatest weakness, and thus a potential target for a new class of medication: siRNA.

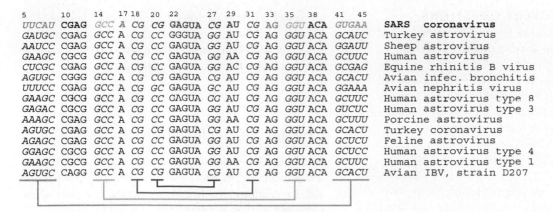

5	10	14	17 18	20	22		27	29	31	33	35	38	41	45	
UUCAU	**CGAG**	*GCC*	A	*CG*	*CG*	**GAGUA**	*CG*	**AU**	*CG*	**AG**	*GGU*	**ACA**	*GUGAA*		**SARS coronavirus**
GAUGC	CGAG	*GCC*	A	*CG*	*CC*	GGGUA	*GG*	AU	*CG*	AG	*GGU*	ACA	*GCAUC*		Turkey astrovirus
AAUCC	CGAG	*GCC*	A	*CG*	*CC*	GAGUA	*GG*	AU	*CG*	AG	*GGU*	ACA	*GGAUU*		Sheep astrovirus
GAAGC	CGCG	*GCC*	A	*CG*	*CC*	GAGUA	*GG*	AA	*CG*	AG	*GGU*	ACA	*GCUUC*		Human astrovirus
CUCGC	CGAG	*GCC*	A	*CG*	*CC*	GAGUA	*GG*	AC	*CG*	AG	*GGU*	ACA	*GCGAG*		Equine rhinitis B virus
AGUGC	CGGG	*GCC*	A	*CG*	*CG*	GAGUA	*CG*	AU	*CG*	AG	*GGU*	ACA	*GCACU*		Avian infec. bronchitis
UUUCC	CGAG	*GCC*	A	*CG*	*GC*	GAGUA	*GC*	AU	*CG*	AG	*GGU*	ACA	*GGAAA*		Avian nephritis virus
GAAGC	CGCG	*GCC*	A	*CG*	*CC*	GAGUA	*GG*	AU	*CG*	AG	*GGU*	ACA	*GCUUC*		Human astrovirus type 8
GAGAC	CGCG	*GCC*	A	*CG*	*CC*	GAGUA	*GG*	AU	*CG*	AG	*GGU*	ACA	*GUCUC*		Human astrovirus type 3
AAAGC	CGCG	*GCC*	A	*CG*	*CC*	GAGUA	*GG*	AA	*CG*	AG	*GGU*	ACA	*GCUUU*		Porcine astrovirus
AGUGC	CGAG	*GCC*	A	*CG*	*CG*	GAGUA	*CG*	AU	*CG*	AG	*GGU*	ACA	*GCACU*		Turkey coronavirus
AGAGC	CGAG	*GCC*	A	*CG*	*CC*	GAGUA	*CG*	AU	*CG*	AG	*GGU*	ACA	*GUCUC*		Feline astrovirus
GGAGC	CGCG	*GCC*	A	*CG*	*CC*	GAGUA	*GG*	AU	*CG*	AG	*GGU*	ACA	*GCUCC*		Human astrovirus type 4
GAAGC	CGCG	*GCC*	A	*CG*	*CC*	GAGUA	*GG*	AA	*CG*	AG	*GGU*	ACA	*GCUUC*		Human astrovirus type 1
AGUGC	CAGG	*GCC*	A	*CG*	*CG*	GAGUA	*CG*	AU	*CG*	AG	*GGU*	ACA	*GCACU*		Avian IBV, strain D207

Figure 3.54 Phylogenetic comparisons of SARS s2m RNA sequences from various coronavirus and astrovirus species.

Conserved sequences are highlighted as bold letters, and covarying sequences involved in conventional RNA helical base-pairing are indicated in italics. Sequence complements are indicated using color-coded brackets.

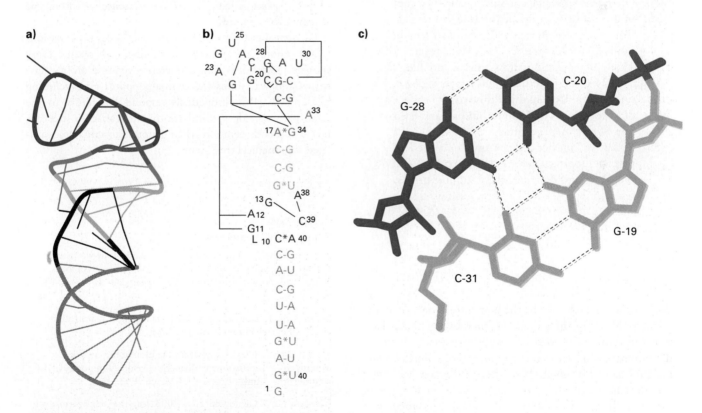

Figure 3.55 3D shape of SARs s2m RNA.

Color-coded 3D **a)** and 2D **b)** shapes are shown with base pairs illustrated as long brackets. The pentaloop structure is shown in darkest purple, Aform RNA helices are shown in light gray and lightest purple, the three-purine asymmetric bulge is in medium purple, and the seven-nucleotide bubble is in dark gray. **c)** Formation of the junction of two perpendicular helices is facilitated by an unusual base quartet composed of two G–C pairs. H-bonds are shown as dashed lines and bases are numbered according to panel b.

LINKS
FirstGlance
Kirchhoff et al.
PLoS Biology
PubMed

DISCOVERY QUESTIONS

76. Refer to Figure 3.53 and then read the abstract by Kirchhoff et al. What do you learn about *nef* that proves both hopeful and problematic for siRNA treatment of HIV? Now refer back to Discovery Question 67 in Section 3.3. Do you foresee additional problems with siRNA treatments?

77. Explain why an RNA genome makes a better target than the proteins encoded by an RNA genome.

78. To make sure your siRNA was specific for only one genome, what type of database search would you perform? Would it be possible to find unexpected surprises in the clinic if you found no matches in the databases? Explain your answer.

79. Imagine you have created the perfect siRNA reagent to cure a particular disease. There is still one major issue that you have to address, though. Explain the problems with delivering siRNA to patients in a clinical setting so that the appropriate cells are reached.

80. Go to PLoS Biology and download a free copy of the January 2005 paper by Robertson et al. Examine figures 3C and 3D to find a literal hole in the SARS genome. Go to FirstGlance and view 1XJR, which is the SARS s2m structure file. View the RNA, and find the tunnel and its exposed nucleotides. Do you think this would be a good place for siRNA to begin its binding and destruction of the SARS genome?

81. Perform a PubMed search using "enhance RNAi potency" to find a February 2005 paper. What improvements described there support further research into RNAi-based medications?

Summary 3.4

Biomedical research may be the best-funded area of all the sciences. Many of the sequenced microbial genomes have direct medical utility. We can use genomes to accelerate the development of new vaccines and antibiotics. Furthermore, insights gained through basic research using molecular methods have led to the development of new types of drugs, such as EGS and RNAi. When combined with genome sequences, we hope EGSs and siRNAs will target a wide range of diseases caused by pathogenic infections and variations in our own genomes. The specificity of RNA-based pharmaceuticals may allow personalized medications to target specific alleles within the human gene pool. As with all research, unexpected discoveries in genome sequences can lead to new questions as well as immediate answers. Through careful reconstruction of metabolic capacity, we can determine which steps are vulnerable to attack and lack redundant alternative paths.

Chapter 3 Conclusions

In addition to new methods and vast amounts of data, whole genome comparisons allow us to detect new patterns and connections that had been impossible to see. Genes come and go from genomes, and some genes may originally have come from evolutionarily distant species. We hope to use sequences to develop better ways to diagnose, prevent, and treat diseases. Learning how to use this information will take a lot of time, money, and effort from people with diverse skills and perspectives. We can use genome sequences to look into the past or make predictions about what roles are performed by different parts of the genome. The completion of a genome sequencing project is no more the end of the story than birth is the end of a person's development. Billions of years of evolution are exposed by completely sequencing a **reference genome** (prototype sequence for a species).

Chapter 4 examines case studies that focus on variations found among the genomes of individuals within every species. Deviations from a reference genome are the raw material for variations we see in any population, including humans. Why do individuals respond differently to the same pathogen or medical treatment? Variation must be taken into consideration when genomic insights are translated into medical treatments.

References

Comparative Genomics

Andelfinger, G., C. Hitte, et al. 2004. Detailed four-way comparative mapping and gene order analysis of the canine ctvm locus reveals evolutionary chromosome rearrangements. *Genomics*. 83: 1053–1062.

Bailey, J. A., R. Baertsch, et al. 2004. Hotspots of mammalian chromosomal evolution. *Genome Biology*. 5(4): 23.1–23.7.

Charity, M. N. 2003. A view from the back of the envelope: Volume m³. <http://www.vendian.org/envelope/TemporaryURL/volume_other.html>. Accessed 11 March 2005.

Clark, A. G., S. Glanowski, et al. 2003. Inferring nonneutral evolution from human-chimp-mouse orthologous gene trios. *Science*. 302: 1960–1963.

Dujon, B., D. Sherman, et al. 2004. Genome evolution in yeasts. *Nature*. 430: 35–44.

Eichler, E. E., & D. Sankoff. 2003. Structural dynamics of eukaryotic chromosome evolution. *Science*. 301: 793–797.

Elert, G. (Ed.) [written by his students]. 2003. Volume of earth's oceans. <http://hypertextbook.com/facts/2001/SyedQadri.shtml>. Accessed 11 March 2005.

Falkowski, P. G., & C. de Vargas. 2004. Shotgun sequencing in the sea: A blast from the past? *Science.* 304: 58–60.

Gilad, Y., V. Wiebe, et al. 2004. Loss of olfactory receptor genes coincides with the acquisition of full trichromatic vision in primates. *PLoS Biology.* 2(1): 0120–0125.

Huang, H., E. E. Winter, et al. 2004. Evolutionary conservation and selection of human disease gene orthologs in the rat and mouse genomes. *Genome Biology.* 5(7): 47.1–47.15.

Huynen, M. A., B. Snel, & V. van Noort. 2004. Comparative genomics for reliable protein-function prediction from genomic data. *Trends in Genetics.* 20(8): 340–344.

Jordan, I. K., F. A. Kondrashov, et al. 2005. A universal trend of amino acid gain and loss in protein evolution. *Nature.* 433: 633–638.

Kent, W. J., & A. M. Zahler. 2000. Conservation, regulation, synteny, and introns in a large-scale *C. briggsae–C. elegans* genomic alignment. *Genome Research.* 10: 1115–1125.

Ovcharenko, I., M. A. Nobrega, et al. 2004. ECR Browser: A tool for visualizing and accessing data from comparisons of multiple vertebrate genomes. *Nucleic Acids Research.* 32: W280–W286.

Thomas, J. W., J. W. Touchman, et al. 2003. Comparative analyses of multi-species sequences from targeted genomic regions. *Nature.* 424: 788–793.

Venter, J. C., K. Remington, et al. 2004. Environmental genome shotgun sequencing of the Sargasso Sea. *Science.* 304: 66–74.

Evolution of Genomes

Adcock, G. J., E. S. Dennis, et al. 2001. Mitochondrial DNA sequences in ancient Australians: Implications for modern human origins. *PNAS.* 98: 537–542.

Baliga, N. S., Y. A. Goo, et al. 2000. Is gene expression in Halobacterium NRC-1 regulated by multiple TBP and TFB transcription factors? *Molecular Microbiology.* 36(5): 1184–1185.

Cole, S. T., K. Eiglmeier, et al. 2001. Massive gene decay in the leprosy bacillus. *Nature.* 409: 1007–1011.

Fuerst, J. A., & R. I. Webb. 1991. Membrane-bounded nucleoid in the eubacterium *Gemmata obscuriglobus. PNAS.* 88: 8184–8188.

Horiike, T., K. Hamada, et al. 2001. Origin of eukaryotic cell nuclei by symbiosis of Archaea in bacteria is revealed by homology-hit analysis. *Nature Cell Biology.* 3: 210–214.

Hutchison, C. A. III, S. N. Peterson, et al. 1999. Global transposon mutagenesis and a minimal Mycoplasma genome. *Science.* 286: 2165–2169.

Ingman, M., H. Kaessmann, et al. 2000. Mitochondrial genome variation and the origin of modern humans. *Nature.* 408: 708–713.

Kaessmann, H., V. Wiebe, et al. 2001. Great ape DNA sequences reveal a reduced diversity and an expansion in humans. *Nature Genetics.* 27: 155–156.

Leakey, M. G., F. Spoor, et al. 2001. New hominid genus from eastern Africa shows diverse middle Pliocene lineages. *Nature.* 410: 433–440.

Madsen, O., M. Scally, et al. 2001. Parallel adaptive radiations in two major clades of placental mammals. *Nature.* 409: 610–614.

Martin, W., & T. M. Embley. 2004. Early evolution comes full circle. *Nature.* 431: 134–137.

Nester, E. W., D. G. Anderson, et al. 2001. Microbiology: A human perspective. Boston: McGraw-Hill.

Ng, W. V., S. P. Kennedy, et al. 2000. Genome sequence of Halobacterium species NRC-1. *PNAS USA.* 97: 12176–12181.

Rivera, M. C., & J. A. Lake. 2004. The ring of life provides evidence for a genome fusion origin of eukaryotes. *Nature.* 431: 152–155.

Ruiz-Pesini, E., D. Mishmar, et al. 2004. Effects of purifying and adaptive selection on regional variation in human mtDNA. *Science.* 303: 223–226.

Shigenobu, S., H. Watanabe, et al. 2000. Genome sequence of the endocellular bacterial symbiont of aphids Buchnera sp. APS. *Nature.* 407: 81–86.

Underhill, P. A., P. Shen, et al. 2000. Y chromosome sequence variation and the history of human populations. *Nature Genetics.* 26: 358–361.

Genomic Identifications

Associated Press. 2001. Canadian office workers quarantined. *New York Times* on the Web. 22 March 2001. <www.nytimes.com/aponline/world/AP-Canada-Biological-Scare.html>.

Belgrader, P., W. Bennett, et al. 1999. PCR detection of bacteria in seven minutes. *Science.* 284: 449–450.

Burns, M. A., B. N. Johnson, et al. 1998. An integrated nanoliter DNA analysis device. *Science.* 282: 484–487.

Chinese SARS Molecular Epidemiology Consortium. 2004. Molecular evolution of the SARS coronavirus during the course of the SARS epidemic in China. *Science.* 303: 1666–1669.

Cole, S. T., T. R. Brosch, et al. 1998. Deciphering the biology of Mycobacterium tuberculosis from the complete genome sequence. *Nature.* 393: 537–544.

Cook, S. M., R. E. Bartos, et al. 1994. Detecting and characterization of atypical Mycobacteria by the polymerase chain reaction. *Diagnostic Molecular Pathology.* 3(1): 53–58.

deMello, A. J. 2003. DNA amplification moves on. *Nature.* 422: 28–29.

Enserink, M. 2001. Biodefense hampered by inadequate tests. *Science.* 294: 1266–1267.

Fort Detrick Standard. 2001. Detrick buildings on National Register of Historic Places. <http://www.dcmilitary.com/army/standard/6_04/local_news/5147-1.html>. Accessed 15 March 2005.

Grenfell, B. T., O. G. Pybus, et al. 2004. Unifying the epidemiological and evolutionary dynamics of pathogens. *Science.* 303: 327–332.

Guan, Y., B. J. Zheng, et al. 2003. Isolation and characterization of viruses related to the SARS coronavirus from animals in southern China. *Science.* 302: 276–278.

Koop, M. U., A. J. de Mello, & A. Manz. 1998. Chemical amplification: Continuous-flow PCR on a chip. *Science.* 280: 1046–1048.

Nerlich, A. G., C. J. Haas, et al. 1997. Molecular evidence for tuberculosis in an ancient Egyptian mummy. *The Lancet.* 350: 1404.

Nicholls, H. 2005. Ancient DNA comes of age. *PLoS Biology.* 3(2): 0192–0196.

Read, T. D., S. L. Salzberg, et al. 2002. Comparative genome sequencing for discovery of novel polymorphisms in *Bacillus anthracis. Science.* 296: 2028–2033.

Salo, W. L., A. C. Aufderheide, et al. 1994. Identification of *Mycobacterium tuberculosis* DNA in a pre-Columbian Peruvian mummy. *PNAS.* 91: 2091–2094.

Stead, W. W., K. D. Eisenbach, et al. 1995. When did *Mycobacterium tuberculosis* infection first occur in the New World? *American Journal of Respiration and Critical Care Medicine.* 151: 1267–1268.

Thierry, D., C. Chureau, et al. 1992. The detection of *Mycobacterium tuberculosis* in uncultured clinical specimens using the polymerase chain reaction and a non-radioactive DNA probe. *Molecular and Cellular Probes.* 6: 181–191.

Vreeland, R. H., W. D. Rosenzweig, & D. W. Powers. 2000. Isolation of a 250-million-year-old halotolerant bacterium from a primary salt crystal. *Nature.* 407: 897–900.

Williams, S. A., M. R. Lizotte-Waniewski, et al. 2000. The filarial genome project: Analysis of the nuclear, mitochondrial and endosymbiont genomes of *Brugia malayi. International Journal for Parasitology.* 30: 411–419.

Biomedical Genome Research

Anderson, J. F., T. G. Andreadis, et al. 1999. Isolation of West Nile Virus from mosquitoes, crows, and a Cooper's hawk in Connecticut. *Science.* 286: 2331–2332.

Blattner, F. R., G. Plunkett III, et al. 1997. The complete genome sequence of *Escherichia coli* K-12. *Science.* 277: 1453–1474.

Cann, A. J. 2001. Malaria. <http://www.wehi.edu.au/MalDB www/intro.html>. Accessed 5 April 2002.

Fraser, C. M., J. Eisen, et al. 2000. Comparative genomics and understanding microbial biology. *Genomics.* 6(5): 505–512.

Glass, J. I., E. J. Lefkowitz, et al. 2000. The complete sequence of the mucosal pathogen *Ureaplasma urealyticum. Nature.* 407: 757–762.

The Institute for Genomic Research. 2001. *The comprehensive microbial resource (CMR).* <http://www.tigr.org/CMR2/NewUsers. shtml>. Accessed 6 April 2002.

Jacque, J.-M., K. Triques, & M. Stevenson. 2002. Modulation of HIV-1 replication by RNA interference. *Nature.* 418: 435–438.

Lanciotti, R. S., J. T. Roehrig, et al. 1999. Origin of the West Nile virus responsible for an outbreak of encephalitis in the northeastern United States. *Science.* 286: 2333–2336.

Ma, M., L. Benimetskaya, et al. 2000. Intracellular mRNA cleavage induced through activation of RNase P by nuclease-resistant external guide sequences. *Nature Biotechnology.* 18: 58–61.

Ma, M. Y.-X., B. Jacob-Samuel, et al. 1998. Nuclease-resistant external guide sequence-induced cleavage of target RNA by human ribonuclease P. *Antisense and Nucleic Acid Drug Development.* 8: 415–426.

Muesing, M. A., D. H. Smith, et al. 1985. Nucleic acid structure and expression of the human AIDS/lymphadenopathy retrovirus. *Nature.* 313: 450–458.

Onishi, H. R., B. A. Pelak, et al. 1996. Antibacterial agents that inhibit lipid A biosynthesis. *Science.* 274: 980–982.

Perna, N. T., G. Plunkett III, et al. 2001. Genome sequence of enterohaemorrhagic *Escherichia coli* O157:H7. *Nature.* 409: 529–533.

Pizza, M., V. Scarlato, et al. 2000. Identification of vaccine candidates against serogroup B Meningococcus by whole-genome sequencing. *Science.* 287: 1816–1820.

Plehn-Dujowich, D., & S. Altman. 1998. Effective inhibition of influenza virus production in cultured cells by external guide sequences and ribonuclease P. *PNAS USA.* 95: 7327–7332.

Richman, L. K., R. J. Montali, et al. 1999. Novel endotheliotropic herpesviruses fatal for Asian and African elephants. *Science.* 283: 1171–1176.

Robertson, M. P., H. Igel, et al. 2005. The structure of a rigorously conserved RNA element within the SARS virus genome. *PLoS Biology.* 3(1): e5 0001–0009.

Salton, M. R. J., & K.-S. Kim. 2001. Structure. Chapter 2 from Medmicro online textbook. <http://gsbs.utmb.edu/microbook/ ch002.htm>. Accessed 5 April 2002.

Tettelin, H., N. J. Saunders, et al. 2000. Complete genome sequence of *Neisseria meningitidis* serogroup B strain MC58. *Science.* 287: 1809–1815.

Weinstock, G. M. 2000. Genomics and bacterial pathogenesis. *Genomics.* 6(5): 496–504.

Werner, M., E. Rosa, et al. 1999. Targeted cleavage of RNA molecules by human RNase P using minimized external guide sequences. *Antisense and Nucleic Acid Drug Development.* 9: 81–88.

Review and Summary Sources

Clarke, T. 2001. Biotechnology: A genetic outcast. *Nature Science Update* (online). <http://www.nature.com/nsu/010301/010301-2. html>. Accessed 27 February 2002.

Eisen, J. 2001. Gastrogenomics. *Nature.* 409: 463–466.

Goodman, B. 2000. Genomic strategies target bacteria. *The Scientist.* May 1, 1, 14.

Hagmann, M. 2000. Leprosy's dying genome. *Science.* 288: 800–801.

Hedges, S. B. 2000. A start for population genomics. *Nature.* 408: 652–653.

Pennisi, E. 1999. Bacterial partners for Filaria [News Focus section]. *Science.* 283: 1105.

Renfrew, C., P. Forster, & M. Hurles. 2000. The past within us. *Nature Genetics.* 26: 253–254.

Rosamond, J., & A. Allsop. 2000. Harnessing the power of the genome in the search for new antibiotics. *Science.* 287: 1973–1976.

Vaara, M. 1996. Lipid A: Target for antibacterial drugs. *Science.* 274: 939–940.

Genomic Variations

LINKS
420,000
David Mann
Ginger Armbrust
METHODS
Microsatellites

In previous chapters, we discussed genomic sequences as if there were one for each species. "The genome sequence" for any species is a reference sequence; the consequences of variations in a species' genome are often overlooked. This chapter focuses on variations within genomes and how they affect species within complex ecosystems as well as an individual's life span. Finally, as we begin to understand the role of genomic variations, it is important to consider the ethical consequences of variations, including those engineered by humans.

4.1 Environmental Case Study

In this section, we examine one of many ecological examples of genomic variations in a species with an unsequenced genome. The case involves some of the ocean's smallest photosynthetic organisms. How can a mitotically dividing population maintain genomic variation, and what effect might this have on the accumulation of global CO_2 levels and global warming? The data that result from exploring these questions offer a cautionary warning about our ability to manage global climate.

Can Genomic Diversity Affect Global Warming?

One type of **phytoplankton** (photosynthetic plankton), the diatom, includes as many as 200,000 different species that can reproduce asexually and sexually. Diatoms are found in fresh and salt waters. David Mann, from the Royal Botanic Garden in Edinburgh, Scotland, estimated that diatoms are

responsible for fixing more CO_2 than all the rain forests combined; others estimate diatoms are responsible for 20–50% of global carbon fixation. Given that CO_2 levels are higher than they have been for at least 420,000 years, it will be important that we understand the full carbon cycle.

Over the last ten years, oceanographers have enriched the oceans with the limiting nutrient iron to test a brash 1985 prediction, "Give me half a tanker of iron and I'll give you an ice age." Phytoplankton respond to added iron with increased density and a dramatic change in species composition. These experiments have spawned serious proposals to use phytoplankton to reduce global levels of CO_2. Since we understand so little of the ocean's complexity, there is a critical need to know more about diatoms and other phytoplankton before we conduct any global experiment. Most people are surprised to learn that genomic tools and genomic insights may help us understand global warming.

For many years, it was assumed that diatoms succeed in a wide range of ecological niches because of their inter- and intraspecific physiological diversity. Individuals isolated from a single species can vary as much as 60-fold in nutrient uptake, photosynthesis, and overall growth rates. We assumed this phenotypic diversity was due to a diversity in genotypes, but the supporting data were lacking. Ecologically important species are rarely considered for whole genome sequencing, so other methods are required to detect genomic diversity. To address this need, Tatiana Rynearson and her thesis advisor, Ginger Armbrust, at the University of Washington's School of Oceanography, used **microsatellite** loci in one species of diatom (*Ditylum brightwellii*) to document genome variation. Microsatellites are genomic regions of DNA variability between individu-

a)

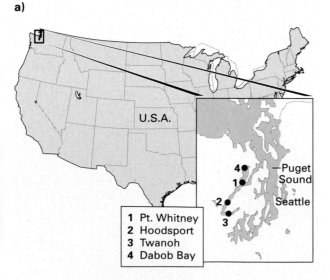

U.S.A.

4 —Puget
1 Sound
2
   Seattle
3

1 Pt. Whitney
2 Hoodsport
3 Twanoh
4 Dabob Bay

b)

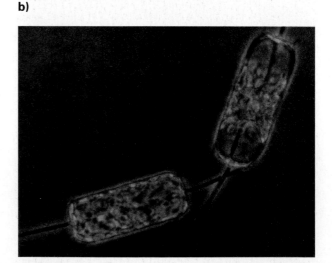

Figure 4.1 Diatom study organism.
a) Map of Washington state and Puget Sound in expanded view. The first three sample collection sites are noted by numbers 1, 2, and 3. The diatom bloom was obtained from site 4. **b)** Light micrographs of two *Ditylum brightwellii* cells.

MATH MINUTES G⌇P
capillary electrophoresis
PCR

als of a species, which can be detected using **polymerase chain reaction** (PCR) and capillary electrophoresis. For example, one individual may have microsatellites of GCGC (i.e., GC$_2$) and GCGCGCGC (GC$_4$), while another individual has GC$_3$ and GC$_6$. The number of repeated units varies in microsatellites, and this variation can be used to measure population diversity.

In their first study, Rynearson and Armbrust collected samples at three places in the Hood Canal near Puget Sound (Figure 4.1). This body of water is relatively stable and therefore should have a minimum of mixing (vertical and horizontal) of different water masses and introduction of new genotypes. All samples were taken from the same depth on November 21, 1997, using a 20 μm mesh net. Individual diatom cells were isolated within 24 hours of the initial collection, maintained in sterile flasks, and grown under constant laboratory conditions for DNA isolation and physiological analysis.

The first step was to define microsatellites for *D. brightwellii*. Nine were cloned and sequenced, with locus Dbr4 used extensively (Figure 4.2). Dbr4 is a compound repeat (a mixture of GA and GT repeats) interrupted by nonrepeat nucleotides. Using one pair of PCR primers, Rynearson and Armbrust defined the genotypes of 23 different clonal isolates from their samples.

Analyzing the PCR products, the investigators identified ten unique genotypes that contained eight different Dbr4 alleles (Table 4.1). Allele 261 was the most abundant, and there were 9 independent isolates that were homozygous 261. As often happens in research, the investigators got some unusual but beneficial results. PCR of Dbr4 often produced three major peaks instead

Table 4.1 10 unique genotypes at Dbr4 locus in *D. brightwellii*.

Genotype*		Number of Isolates
243	243	1
255	259	1
257	257	1
257	259	1
259	259	1
261	261	9
261	263	3
261	267	1
263	263	4
263	271	1

*Allele numbers refer to bp lengths of microsatellite.
Source: Renearson and Armbrust, 2000. *Limnology and Oceanography.* 45(6): 1334.

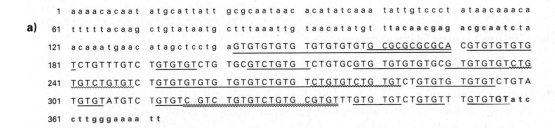

a)

```
  1  aaaacacaat atgcattatt gcgcaataac acatatcaaa tattgtccct ataacaaaca
 61  tttttacaag ctgtataatg ctttaaattg taacatatgt ttacaacgag acgcaatcta
121  acaaatgaac atagctcctg aGTGTGTGTG TGTGTGTGTG CGCGCGCGCA CGTGTGTGTG
181  TCTGTTTGTC TGTGTGTCTG TGCGTCTGTG TCTGTGCGTG TGTGTGTGCG TGTGTGTCTG
241  TGTCTGTGTC TGTGTGTGTG TGTGTCTGTG TCTGTGTCTG TGTCTGTGTG TGTGTCTGTA
301  TGTGTATGTC TGTGTC GTC TGTGTCTGTG CGTGTTTGTG TGTCTGTGTT TGTGTGTatc
361  cttgggaaaa tt
```

b)

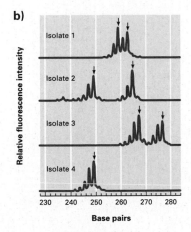

Figure 4.2 Microsatellite Dbr4.
a) Dbr4 sequence; location of PCR primer sequences are in purple; GT dinucleotide is underlined once; GC dinucleotide is underlined twice; (CT(GT)$_2$)$_2$ is underlined as shown; ((CTGTCT(GT)$_2$)$_2$ is underlined as shown. The reference repeat unit is 216 bp long and shown in all capital letters. Accession number AF263004. **b)** Electropherograms of Dbr4 alleles amplified from four different isolates using specific primers. Arrows indicate the allele sizes, with smaller stutter bands also visible.

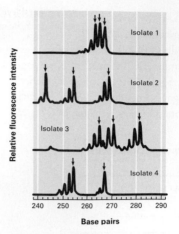

Figure 4.3 Electropherograms of microsatellites.
Alleles from two distinct loci simultaneously were amplified with the first pair of less-specific PCR primers. Arrows indicate the allele size, with smaller stutter bands also visible.

of the expected two (Figure 4.3). The PCR products were cloned and sequenced, revealing that the primers were amplifying two loci simultaneously. By slightly redesigning the primers to be more specific, the investigators were able to distinguish the two distinct loci, increasing their ability to resolve genomic differences among the different isolates.

Using both microsatellite loci, they identified 23 unique genotypes from 24 different isolates and calculated the observed heterozygosity (H_o; number of heterozygotes divided by the total number of samples) and the gene diversity (H_e; reflects the extent of diversity at individual loci). If all individuals in a population were mitotic siblings, H_e would be equal to zero; if all were unrelated, H_e would be very close to one. An examination of only the original Dbr4 locus found that H_o was 0.30 and H_e was 0.69. When the pair of loci were analyzed collectively, H_e was 0.88. The high H_e indicated that the population was very heterogeneous and that most individuals were not mitotic siblings even though the isolates were reproducing by mitosis. The Dbr4 locus H_o and H_e values were significantly different from each other (p-value < 0.0001), indicating that *D. brightwellii* were reproducing asexually, since there were more homozygotes than one would expect if these individuals were reproducing sexually. Consequently, it appears that the 24 diatom isolates were products of mitotic reproduction—and yet the two-locus H_e value indicated most of the isolates were unrelated to each other. The data revealed a surprisingly high degree of genomic variation in a mitotically dividing population, prompting the question, "How can a population of asexual organisms maintain a high degree of genomic diversity?"

Math Minute 4.1   How Do You Measure Genetic Diversity?

Observed heterozygosity (H_o) and goene diversity (H_e; also called expected heterozygosity) are two measures of genetic variation within a population. Rynearson and Armbrust used these measures to infer that the diatom population was reproducing asexually, although most individuals in the population were not mitotic siblings. In this Math Minute, we explore how to compute and interpret H_o and H_e. H_o is easier to calculate, but H_e is more appropriate for inbred or asexually reproducing populations such as *D. brightwellii*.

To calculate H_o for the Dbr4 locus from the data in Table 4.1, divide the number of heterozygous isolates (7) by the total number of isolates (23) to get $H_o = 7/23 \approx 0.30$. This means that 70% of all isolates are homozygotes. (A low value for H_o indicates the cells are reproducing asexually by mitosis. If most individuals are reproducing asexually, you would expect them to be genetically very similar and thus have low allelic diversity [H_e].)

To calculate H_e for the Dbr4 locus, you must first find the frequency of each of the eight unique alleles that appear in Table 4.1. An allele frequency is the number of times the allele occurs in the 23 isolates, divided by the total number of chromosomes among the 23 isolates (46). For example, allele 261 has a frequency of $[(9 \times 2) + (3 \times 1) + (1 \times 1)]/46 = 22/46 \approx 0.48$. To compute H_e, square each of the eight allele frequencies, and sum the resulting eight numbers. Subtract this sum from 1 to get $H_e \approx 0.69$ for the Dbr4 locus.

To interpret our result, we need to understand the relationship between the value of H_e and gene diversity. Let's consider the extreme cases first. At one extreme, if only one allele is represented in the population, its frequency is 1 and $H_e = 1 - 1^2 = 0$. If two alleles appear with equal frequency, $H_e = 1 - \left(\frac{1}{2}\right)^2 - \left(\frac{1}{2}\right)^2 = \frac{1}{2}$. Using calculus (see Math Minute Discovery Questions), you can show that $\frac{1}{2}$ is the largest possible value for H_e when there are two alleles, and therefore H_e is always between 0 and $\frac{1}{2}$ in the two-allele case. More advanced calculus shows that when there are n alleles, the maximum value of H_e occurs when all n alleles have the same frequency (i.e., $1/n$), and the maximum value of H_e is $(n-1)/n$. In our example, $n = 8$, and the largest possible value for H_e is $7/8 \approx 0.875$. Because our value of $H_e \approx 0.69$ is relatively close to the maximum value of 0.875, we conclude that the gene diversity in this population is high. A similar calculation leads to a two-locus (Dbr4 and Dbr9) H_e of 0.88, near the theoretical maximum value for this measure of 0.94.

By calculating H_o (0.3), the investigators determined the cells were reproducing asexually. However, calculating H_e (0.69 for one locus and 0.88 for two loci) revealed that the diatom population was genetically diverse. These two calculations encouraged the investigators to test experimentally how diatoms maintain genomic diversity while reproducing asexually.

MATH MINUTE DISCOVERY QUESTIONS

1. If you know calculus, you can prove for yourself that the largest possible value for H_e is $\frac{1}{2}$ in the two-allele case, and that the largest value occurs when the two alleles occur with equal frequency. To do this, denote the frequency of one of the alleles by p; the frequency of the other allele is $1 - p$. Write H_e as a function of p, and find the maximum value of this function. At what value of p does the maximum occur? How do you know this is a maximum rather than a minimum?

2. Whether you know calculus or not, you can experimentally determine the effects of different allele frequencies, even when there are many alleles. In the Excel file **diatom_he.xls**, there are two scenarios for calculating H_e. The first scenario supposes that the investigators had identified only two unique genotypes, 261/261 in 9 isolates and 261/263 in 3 isolates. Experiment with different numbers of occurrences for each allele, keeping the total number constant at 24, to see the effect on H_e. How many isolates could you collect with each of these two genotypes and produce an H_e of $\frac{1}{2}$?

3. The second scenario in diatom_he.xls represents all the data in Table 4.1.
   a. Change the number of allele occurrences corresponding to the 243/243 isolate being incorrectly genotyped as 255/257. What is the effect of this error on H_e?
   b. Use the Excel file to help you determine which genotyping error of exactly one isolate in scenario two would have the maximum impact on H_e.
   c. Change the number of occurrences for each allele in scenario two to make H_e as large as you can, keeping the total number of alleles at 46. Why is it impossible to reach the theoretical maximum of 0.875?

Trying to understand the genotype variation found in these diatoms brings us back to diatom phenotypic diversity. It has been known for many years that individual cells have different physiological capacities, and now we see the genomic variability is high as well. Rynearson and Armbrust decided to test eight of their isolates for their ability to grow at three different light levels while maintained in otherwise equal growth conditions (Figure 4.4). They chose to test high, medium, and low light intensities since individual cells would experience different light intensities as they drifted up and down the water column. At first glance, the isolates may appear to be very similar, but notice the differences in growth rates and the small error bars for each light treatment.

As you would expect, all the isolates grew faster with more light. However, there are significant differences among them. For example, at 166 µmol, isolates 2 and 8 grew about 20% slower than the others (p-value < 0.01). At 66 µmol, isolate 1 grew the fastest and isolate 2 grew the slowest, with a difference of about 26% (p-value < 0.001). Three different growth rates (p-value < 0.01) were measured at 33 µmol. Isolates 1, 6, and 8 were the fastest; 3, 4, and 7 were intermediate; and 2 and 5 were the slowest, with a 33% slower rate compared to the fastest growers. What was especially interesting was that the growth rate at one light level was not a predictor of growth rate at the other two light levels. Therefore, genomic diversity appears to be the cause of the difference in physiology. The response to light intensity is probably a multilocus trait, since there was no correlation in growth rates among the three light levels. This finding led to the testable prediction that the population must be under changing selection pressures (more than just changes in light intensities), which would explain the genomic variation observed in the 24 isolates.

The power of models is that they allow us to make predictions. Using their laboratory growth rates, Rynearson and Armbrust modeled the growth rates for all eight isolates if an equal number of cells were incubated in the same flask under three different light conditions (Figure 4.5). In each condition, a single isolate quickly dominated the population. The time it took the dominant isolate to represent 50% of all the cells varied, as you could expect since the "losing" isolates had different relative growth rates for each light treatment. If natural selection in the Hood Canal were due to a single and constant factor, you could predict that a diatom population would decrease in genomic diversity and become more homogeneous. Conversely, based on the high genomic diversity of diatoms in Hood Canal, there must be multiple, dynamic selection pressures that maintain genotypic diversity in the population. The 23 different genotypes isolated are each optimized to grow in certain conditions that must occur at specific locations or time periods in the Hood Canal. Otherwise, the genomic diversity would be reduced by the competition for resources. Many questions remain, including the genome composition of a diatom bloom. Blooms are defined by a rapid but localized increase in the number of cells, but no one knows which genotype is responsible.

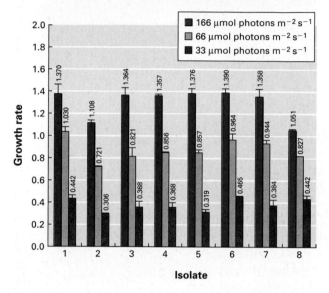

Figure 4.4 Growth rates of genetically distinct isolates.
Cells were grown at 166, 66, or 33 micromoles of photons per square meter per second in culture flasks maintained inside growth chambers. The numbers at the top of the bars are the mean growth rates. Error bars indicate the standard error of the mean growth rate.

DISCOVERY QUESTIONS

1. Would you expect the diatom genomic diversity to be equally high at different locations throughout Puget Sound? Would you expect to find the exact same genotypes, or different ones? Explain your reasoning.
2. The only genomic data we have for diatoms are the nine microsatellite loci. What would you need to know to identify individual alleles that might be responsible for variations in physiological

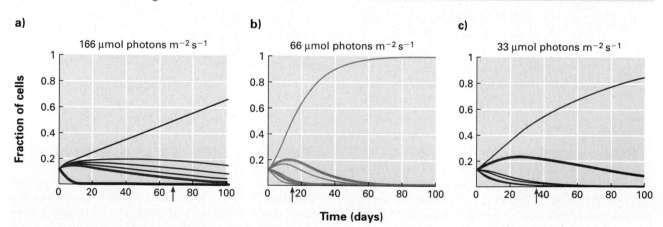

Figure 4.5 Diatom population simulations.
Computer simulations of changes in genomic diversity over time based on the mean growth rates of isolates maintained at 166, 66, and 33 μmol photons m⁻² s⁻¹. Individual curves represent the relative abundance of each isolate. Arrows indicate the times required for the dominant strain to represent 50% of the population. Each light condition produces a different dominant strain and the time to dominance is not correlated with the amount of light.

responses to changing environments? If you could sequence the entire genome of one diatom species, would this reference genome contain all the information you would need to understand diatom blooms? Explain your answer.

3. Submit the 17mer PCR primer
GSP **TACAAAGAGACGCAATC** for Dbr4 to BLASTn using the option "Search for short, nearly exact matches." Do you get any exact matches other than *D. brightwellii*? What aspect of this methodology ensures only this species' DNA is amplified?

4. You might be interested to see how Figure 4.5 GSP was produced. If you go to the modeling activity web page, you can see a Bonus Math Minute and associated Excel spreadsheet to model the growth of the diatom populations.

Over the course of an 11-day sampling period, Rynearson and Armbrust happened to find a diatom bloom (Figure 4.6a) and measured its genomic diversity. Blooms are most common in the spring and fall when nutrients and sunlight are high and mixing of surface water slows. You might predict that a bloom would be composed of one dominant genotype, or a succession of a few dominant genotypes, due to the relative stability of the environment during a bloom, but the results were more complex than anyone expected. The investigators analyzed the microsatellites in 607 isolated cells from a single population (i.e., similar allele frequency to populations previously sampled) and found 496 different

genotypes (82% with unique genotypes) with no indication that they had begun to fully sample the diversity in the bloom (Figure 4.6a). Some clonal siblings were identified, with 69 sampled more than once and 2 sampled 7 times in the 11 days. If you assumed that every genotype was equally abundant, then you would not expect to sample a particular genotype too often (Figure 4.6b). Based on random sampling of a mixed population, you would expect to resample 1.3104 clonal genotypes 4 or more times. The real bloom, however, produced 9 clonal genotypes sampled 4 or more times ($p < 0.05$).

The bloom of cells was ascribed to rapid asexual reproduction, since H_e was high (0.70 ± 0.24, 11-day average and comparable nonbloom values), H_o was low (0.35, Dbr4 11-day average), and microscopic inspection failed to detect any signs of sexual reproduction. With mitotic cell division, we might expect to find more homozygotes and a few dominant genotypes, but the data do not support these assumptions. The genomic heterogeneity was consistently high, even though a windstorm on day 5 did reduce the concentration of cells and alter the prevalence of certain genotypes (Table 4.2), with 52 new genotypes identified on day 11. Two genotypes were resampled more frequently after the storm while 6 other resampled genotypes were equally sampled before and after the storm. Based on assumptions of random sampling and mixing of the cells, the investigators estimated at least 2,400 genotypes contributed to the diatom bloom of a single species. Therefore, diatom blooms appear to be composed of many different genotypes at low frequencies, with a few genotypes more abundant and thus thriving under the bloom conditions. However, the thriving genotypes did not become

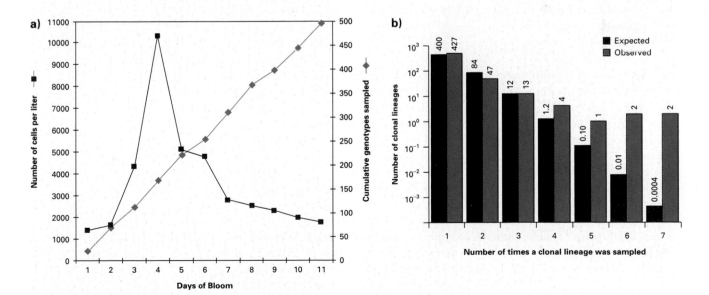

Figure 4.6 Genetic diversity of a diatom bloom.
a) The number of *D. brightwellii* cells from daily samples of Dabob Bay surface waters (left axis) and the cumulative number of genetically distinct clonal lineages identified during each day of sampling. **b)** The expected number (black) and the observed number (purple) of clonally distinct *D. brightwellii* genotypes during the bloom. Numbers above each bar indicate the number of distinct clonal lineages in each category.

Math Minute 4.2 **How Do You Model Population Diversity?**

After 11 days of diatom sampling, Rynearson and Armbrust were able to genotype 607 cells with 496 unique genotypes. What if they had been able to genotype 1,000 times as many cells? Would they have found 1,000 times as many unique genotypes? Probably not. With continued sampling, a growing percentage of cells would have had a previously sampled genotype. Of course, we could never collect and genotype every cell in the diatom population, so we must rely on sampling methods and mathematical tools to model population diversity. In this Math Minute, we explore how to use sample data to estimate the total number of unique genotypes in a population, and to predict the number of genotypes we should see once, twice, three times, and so on, in a sample of 607 cells.

Several different mathematical models are available for predicting the number of unique genotypes in a population based on the number of each genotype in a sample. Rynearson and Armbrust applied two of these models: Chao1 and abundance-based coverage estimators (ACE). These models are adapted from mark-release-recapture statistics used in field biology to estimate animal population sizes. To compute Chao1, square the number of genotypes observed only once, divide by two times the number of genotypes observed exactly twice, and add the number of different genotypes observed. Using the data from Figure 4.6b, we have

$$\text{Chao1} = \frac{427^2}{2 \times 47} + 496 = 2,436.$$

The ACE estimate is similar to Chao1, but includes terms for the number of genotypes observed k times, for k between 1 and 10, and is therefore a more complicated formula. We won't go through the details, but for Figure 4.6b, ACE estimates 2,747 different genotypes. From the Chao1 and ACE estimates, the investigators inferred that at least 2,400 genotypes were in the population, and they used a rough average of the two calculated values (2,550) in their subsequent statistical analyses.

Figure 4.6b shows the expected and observed numbers of genotypes that appear k times in our sample, for k between 1 and 7. The expected numbers are computed under the assumption that all 2,550 genotypes are equally abundant in the population. The equal-abundance assumption is not supported by the data, because the expected numbers are so different from the observed. Can we know how many genotypes we should see exactly once, twice, etc.?

The investigators chose a Poisson distribution to model the number of genotypes seen each number (1–7) of times. As you might have learned in a probability or statistics class, the Poisson distribution is a good model for frequencies, i.e. how many times a single event occurs. For example, if we were just trying to model how many times genotype #42 occurs in the sample of 607 cells, the Poisson distribution would be a good method. But we are trying to model frequencies of frequencies—that is, how many genotypes were sampled each number of times (e.g.,13 genotypes were sampled 3 times)—so there is no single "event" to count. A better model for the expected values in Figure 4.6b can be constructed using more advanced statistical methods. Rather than discuss this theoretical model, we will determine the expected values using a simulation. Simulations are good tools for modeling complex situations like this diatom-sampling data set, and simulations can also be used to validate theoretical models.

A simulation uses random numbers to represent objects or events. In this simulation (download diatom_sim.xls from the web site), each cell type is represented by a random number between 1 and 2,550, the assumed number of genotypes in the entire diatom population. The list of 607 randomly generated numbers represents the sampling and genotyping processes: pull out a cell, genotype it, record the genotype, and repeat this process a total of 607 times. These 607 genotype observations represent one run of the experiment. Once the list of 607 numbers (i.e., genotypes) is generated, you just need to count how many times each genotype occurred, then count how many genotypes occurred once, twice, three times, and so on.

Because each simulated list of 607 cell genotypes is random, you will get slightly different results every time you repeat the experiment. The idea of a simulation is that you can repeat the experiment many times in silico, and average the results to get a good estimate of what to expect in any given run of a nonsimulated, biological experiment. In the Math Minute Discovery Questions, you will use the Excel file diatom_sim.xls to simulate the diatom-sampling experiment and determine the expected numbers of genotypes you should see each number of times. Because the Poisson model used by the investigators was not the best model for the genotype frequency data, your expected values will not agree with those in Figure 4.6b. However, as you will see, the difference between expected and observed values for genotypes occurring four or more times is even greater than under the Poisson model. Therefore, the investigators' conclusion that all 2,550 genotypes are not equally abundant is still supported.

MATH MINUTE DISCOVERY QUESTIONS

1. Run the diatom-sampling simulation in **diatom_sim.xls**, repeating the experiment 10 times, and recording the average number of genotypes occurring k times, for k between 1 and 5. Record your results when you run the simulation 3 times for 10 repetitions each time.

2. Repeat the process from Math Minute Discovery Question 1 another 3 times, but increase the repetitions to 100 for each of the 3 iterations. Compare your answers to those for 10 repetitions. Is it better to use 10 or 100 repetitions? Explain your answer using the data from the simulations.

3. Run the diatom sampling simulation with 500 repetitions. Based on your simulation, how many genotypes do you expect to observe 4 times and 5 times? Compare your results to the expected numbers in Figure 4.6b. How does this difference affect your interpretation of the results of this study?

Table 4.2 Genome diversity during diatom bloom.
G:N ratio is the number of distinct three-locus genotypes (G) divided by the total number of genotypes analyzed (N).

Sample day	Isolates genotyped	G:N (isolates genotyped)
1	30	1.00
2	85	0.95
3	72	0.92
4	78	0.94
5*	83	0.92
6	63	0.92
7	96	0.90
8	95	0.93
9	43	0.97
10	85	0.95
11	90	0.87

*The day of the windstorm.

dominant, nor did they gradually increase in abundance during the 11 days. When the diatom bloom was modeled with 2,400 equally abundant genotypes, it took 60 days before any genotype could be as abundant as the most heavily sampled genotype in the study. Therefore, the model must be too simplistic, indicating the selection pressures and growth rates must be more dynamic and complex than we realized.

Publishing a reference genome sequence is an important starting place, but understanding what might happen if the ocean were enriched with iron would require a greater knowledge of the genomic variation in the oceans. This study of *D. brightwellii* in a small part of a relatively stable body of isolated water found much more genomic diversity than had been expected. A population may exhibit an average phenotype, but this average underestimates the diversity of genomes that remain hidden until changes in the environment favor a different genotype. The interaction of genotype and phenotype diversity is complex and may not respond to human intervention the way we expect it to. In the spring of 2004, a team of 38 authors from New Zealand, Canada, Japan, and the U.S. found that an iron-induced phytoplankton bloom only lasted 18 days, and that much of the sequestered carbon failed to sink into the deep ocean as some had predicted. Much of the carbon was recycled through predation or decomposition by bacteria, which could in fact lead to

LINKS
diatom population
ISMWG

increased CO_2 production. Perhaps we should understand the dynamics of the oceans before conducting an iron-enriching "geoengineering" scheme from which there is no turning back.

DISCOVERY QUESTIONS

5. Given that diatom blooms are not composed of a single dominant genotype, hypothesize how cell numbers can increase but diversity does not decrease. *D. brightwellii* can form dormant spores that germinate later. Do you think these cells could explain the sustained diversity?

6. Is it possible that *D. brightwellii* is so prevalent and diverse that it never produces isolated populations with distinctive physiological traits? Search diatom population abstract.

Summary 4.1

The case study in this section illustrates how genomic methods will affect our understanding of environmental and ecological issues. As often happens in science, answering a simple question can lead to an unexpected discovery. The genomic variation in photosynthetic diatoms is greater than we expected, indicating that the oceans are extremely dynamic environments. Diatoms have evolved diverse genomes to thrive as a species, by mitotically reproducing those genomes that temporarily best suit local conditions. Given this evolutionary response to a complex ecosystem, we should be cautious when devising solutions to complex problems such as global warming. Increasingly, ecologists are turning to genomic methods to measure and understand our environment.

4.2 Human Genomic Variation

How much variation is there in the human genome, and are these variations significant? At the heart of the Human Genome Project is this question of variation in the human genome. You know humans exhibit genome-encoded differences—just look at your classmates. The biomedical field is interested in disease-causing variations. What we often consider "simple" diseases have complex genomic underpinnings (see Chapters 1 and 5). In this section, we examine how genomic variations are used to determine the causes of complex phenotypes and how they influence effective medical interventions.

How Much Variation Is in the Human Genome?

In diatoms, we examined microsatellite variations in genomes, which are commonly used for ecological studies. Similarly, the amount of tandem repeat DNA in the canine

Runx-2 gene affects the length and degree of curve in a dog's snout, while the number of tandem repeats in the canine *Alx-4* gene determines if the dog has 5 or 6 toes on each hind leg (Great Pyrenees are prone to 6 toes). However, another type of polymorphism has garnered more attention in humans: **single nucleotide polymorphisms (SNPs)**. SNPs (pronounced "snips") are single bases at a particular locus that are different in different individuals. SNPs are another form of genomic variation in a population; they may occur anywhere in the genome, and everyone has thousands of SNPs. Let's imagine that 90% of all human chromosomes have the following sequence at a particular location (i.e., at a unique locus):

```
GCATGCATGCATGCAT
||||||||||||||||
CGTACGTACGTACGTA
```

but 10% of all alleles have this slightly different sequence:

```
GCATGCAaGCATGCAT
||||||||||||||||
CGTACGTtCGTACGTA
```

This locus has a single nucleotide polymorphism and thus is one example of a SNP. SNPs differ from microsatellites because microsatellites involve multiple nucleotides and the patterns vary significantly (see Figure 4.2a). Furthermore, microsatellites are prone to continued mutations, because they are caused by "slippage" of the DNA polymerase replicating tandem repeats (facilitating the creation of new dog breeds in short time periods). In contrast, SNPs are more stable, since they differ by single nucleotides and do not induce additional errors by the DNA polymerase.

With the **finished** human genome reference sequence, we are in a position to enumerate all the SNPs. In 2001, the International SNP Map Working Group (ISMWG) identified 1,433,393 SNPs, an average of 1 SNP every 2 kb of genomic sequence (Table 4.3). It had been estimated that SNPs occur about once every 1–2 kb and ISMWG cataloged near the estimated number of SNPs. Based on preliminary data, each ethnic group has its own collection of SNPs. Human SNPs have been classified in one of two ways: major or minor alleles. SNP minor alleles are less common in all humans than other SNPs at the same location. However, within a particular ethnic group, 20% of the group's allelic variation is due to SNP minor alleles. This means that SNPs classified as minor alleles for all humans may be SNP major alleles for an ethnically defined population of humans. Therefore, generalizations for the entire species based on genome-wide definitions of SNP minor alleles should be avoided since these generalizations lead to assumptions for all ethnic groups that may not be correct.

LINKS G&P
Entrez Gene
HGP Glossary
The SNP Consortium

WHAT'S THE DIFFERENCE BETWEEN A MUTATION AND AN ALLELE?

Because nine different people donated their DNA to the Human Genome Project (HGP) for sequencing, there are bound to be some differences in their DNA. This leads us to a philosophical question: Is there such a thing as *the* human genome? No two humans have the exact same DNA sequence. Although identical twins start with the same DNA, by the time they reach sexual maturity, their gametes will not be identical. Thus, it is impossible to say that there is such a thing as *the* human genome. Instead, we use the term "reference genome" to describe a composite genome, but we have recorded all the polymorphisms discovered along the way (about 1.4 million so far).

Comparing individual genomic variation has led to an interesting problem. What is a polymorphism, versus an allele, versus a mutation? The HGP glossary defines the terms as follows:

Allele: Alternative form of a genetic locus; a single allele for each locus is inherited separately from each parent (e.g., at a locus for eye color, the allele might result in blue or brown eyes).

Polymorphism: Difference in DNA sequence among individuals. Genetic variations occurring in more than 1% of a population would be considered useful polymorphisms for genetic linkage analysis.

Mutation: Any heritable change in DNA sequence.

First, you can see that these definitions are fuzzy. An allele is a heritable change in DNA sequence, and it may occur at a frequency above 1% of the population. Alleles and polymorphisms are heritable changes in DNA sequence. What's missing from the list is a definition for "population" and a clarification of "change"—compared to what? If you think these terms are confusing, you're right. Completely sequencing the human genome only made these three terms seem less precise.

DISCOVERY QUESTIONS

7. Which chromosomes have the lowest density of SNPs (see Table 4.2)? Are there any striking patterns? Can you formulate hypotheses to explain this observation?

8. Search The SNP Consortium (**TSC**) for the SNP ID# "19265." Click on "TSC0019265 (a/c)." Click on "View Screening Details," then click on "p1_0k21g02.p1cSCF." Using a Java-ready browser, have it show you the trace using the Retrieve pop-up menu. Turn on "quality score" and then jump to base 530 in the trace window. Do you think base 532 "N" is a real SNP? Go back to the first SNP report and click on "ss2669205" (near the top). What is the frequency of this SNP? Click on "rs1806509" at the top of the ss2669205 report. Use the graphic display to determine where in the gene this "SNP" occurs.

9. You have been examining the human gene *ApoE*. Perform an Entrez Gene search for *ApoE*, and click on the link for *Homo sapiens*. Read the summary and GeneRIFs #3, 7, 8, 22, and 23 (PD is Parkinson's disease and AD is Alzheimer's disease). At the very top right, click on the "Links" text and choose SNP from the pop-up menu. To see how SNPs in this gene were identified, click on the "CGAP-GAI" link in the right column and wait for the Java applet to begin.

10. Search NCBI SNP database for rs11542040. On this SNP page, scroll down to the section heading "GeneView" and click on the very small "[all]" link to see all the SNPs in the *ApoE* gene. Use the color legend to help you understand the significance of the SNPs. Will any of these SNPs alter the protein's primary structure?

11. In the blue, left-hand margin of the rs11542040 SNP report, click the small link called "Detail" under the heading of "Population," which is under the heading of SEARCH. Click the "Search by:" button next to Submitter Population id and enter the population name "MDECODE-3." What population was sampled to find this SNP? Click on the "MDECODE-3" link that

METHODS
Warren Gish

appeared with your search, then use MDECODE-APOE-SNPS to determine the human source of MDECODE1–4. Once you have determined the identity of MDECODE 1–4, go back to the page that has all four MDE-CODE populations listed with the link "detail" under View Frequency. Compare the genotype diversity among the four populations. Do you detect any ethnic differences? Which appears to have more SNPs? How does this SNP frequency compare with the data in figure 3.26?

Why Should We Care about SNPs?

As we saw in Chapter 2, only 2% of our genome encodes protein, so you might think that 98% of all SNPs are not interesting. However, if you define a gene as the coding region plus 10 kb upstream, then 98% of all genes are within 5 kb of a SNP, 93% of all annotated genes contain at least one SNP, 59% of human genes have five or more SNPs, and 39% have ten or more SNPs. It is interesting to note that the sex chromosomes have lower densities of SNPs than other chromosomes, although no one is sure why. Perhaps there is stronger selection on them not to change, but this is speculation.

SNPs may prove useful for several types of research. The first is the study of evolution. For example, you might be familiar with the popular teaching compound phenylthiocarbamide (PTC), which tastes very bitter to some and has no taste to others, depending on which SNPs are present in

Table 4.3 Human SNP distribution by chromosome.

Chromosome	Length (bp)	All SNPs	
		SNPs	**kb per SNP**
1	214,066,000	129,931	1.65
2	222,889,000	103,664	2.15
3	186,938,000	93,140	2.01
4	169,035,000	84,426	2.00
5	170,954,000	117,882	1.45
6	165,022,000	96,317	1.71
7	149,414,000	71,752	2.08
8	125,148,000	57,834	2.16
9	107,440,000	62,013	1.73
10	127,894,000	61,298	2.09
11	129,193,000	84,663	1.53
12	125,198,000	59,245	2.11
13	93,711,000	53,093	1.77
14	89,344,000	44,112	2.03
15	73,467,000	37,814	1.94
16	74,037,000	38,735	1.91
17	73,367,000	34,621	2.12
18	73,078,000	45,135	1.62
19	56,044,000	25,676	2.18
20	63,317,000	29,478	2.15
21	33,824,000	20,916	1.62
22	33,786,000	28,410	1.19
X	131,245,000	34,842	3.77
Y	21,753,000	4,193	5.19
RefSeq	15,696,000	14,534	1.08
Totals	2,710,164,000	1,419,190	1.91

Source: The International SNP Map Working Group. 2001. *Nature.* 409: 929.

Math Minute 4.3 **Are All SNPs Really SNPs?**

A SNP is a position in a genome at which two or more different bases occur in the population, each with a frequency greater than 1%. In general, a SNP can be found by first aligning a set of overlapping DNA sequences and then identifying positions in the alignment at which the same base does not occur in every sequence. For example, the following five sequences appear to have a SNP at position 8. However, is this enough evidence to conclude that position 8 is a SNP?

```
GCATGCAaGCATGCAT
GCATGCAcGCATGCAT
GCATGCAaGCATGCAT
GCATGCAaGCATGCAT
GCATGCAaGCATGCAT
```

An important aspect of the search for SNPs is the assessment of false-positive and false-negative rates, and probability theory is a powerful tool for making these assessments. There are two ways in which a base may be falsely classified as a SNP: (1) inclusion of paralogs in sequence alignments, and (2) errors in sequencing. Let's walk through an approach to SNP finding proposed by investigators in Warren Gish's lab at Washington University in St. Louis, and see how they minimized false-positive SNPs.

The investigators began with approximately 1.3 Mb of finished human genome reference sequence. They found 1,954 hits to the reference sequence in the EST database when they restricted the search to ESTs for which chromatograms were available. By aligning the ESTs and identifying sequence overlaps, they clustered the 1,954 ESTs into 147 contigs that aligned with 80,469 positions in the reference sequence.

Some of the ESTs that aligned with a particular segment of the reference sequence may have come from paralogs. To screen out paralogs, the investigators used the fact that a paralog would have less similarity to the reference sequence than would the same gene containing SNPs. Specifically, the frequency of variation in paralogs is thought to be approximately 1 base in every 50, while the frequency of SNPs is thought to be approximately 1 base in every 1,000. For example, in the following five sequences, EST 2 appears to be a paralog because of the relatively large number of differences between it and the reference sequence. Note that if EST 2 were removed from consideration, there would no longer appear to be a SNP at position 4 or 8.

EST 1	GCATGCAaGCATGCAT
EST 2	GCAgGCAcGCATGCAT
EST 3	GCATGCAaGCATGCAT
EST 4	GCATGCAaGCATGCAT
Reference:	GCATGCAaGCATGCAT

ESTs with significantly more than the expected number of mutations from the reference sequence were classified as paralogs and removed from further consideration. After paralogs were removed, 69,756 positions of the original 80,469 remained to be searched for SNPs.

The second step of the process was designed to screen out positions that appeared to be SNPs due to sequencing errors. To do this, the investigators considered the chromatograms of the 1,954 ESTs and the reference sequence. They determined the probability that a particular position was truly a SNP, considering (1) the bases that were observed at that position, (2) the depth of coverage at that position (i.e., how many ESTs were aligned to that position), and (3) the base quality value at that position in each sequence. For example, in position 8 of the five sequences shown earlier, the bases are {a,c,a,a,a} and the depth of coverage is 4. The base quality values come from processing each chromatogram with the PHRED base-calling program (see page 38). The expected rate of polymorphism (0.001) also entered into their calculations. All positions for which the probability of being a true SNP was estimated to be greater than 0.4 were designated candidate SNPs. There were 59 candidates, with an average probability of 0.78 of being a true SNP.

The candidate SNPs were examined at the corresponding positions in an independent collection of DNA sequences (the validation set). Twenty-three of the 54 candidate SNPs were excluded from validation for technical reasons. Twenty of the remaining 36 candidate SNPs (56%) were found to be polymorphic in the validation set. The fact that the other 16 candidates were not polymorphic in the validation set does not prove that they are not SNPs, but there is no supporting evidence that they are.

SNPs are being added to databases at a rapid pace, but are they all real? The SNP finding process followed by these investigators illustrates that the quality of the data must be taken into account before concluding a SNP is really a SNP.

---

the taste receptor gene *Tas2R*. If your allele has three particular SNPs that encode three particular amino acids (PAV: proline, alanine, valine), then you can taste PTC. If you have two out of the four other alleles in the world (Table 4.4), you probably cannot taste PTC, though at least one

other unknown protein can interact with Tas2Rp to restore taste sensitivity. Investigators tested six nonhuman primates and found that they had only the PAV form of *Tas2R*, indicating that humans evolved these other SNPs after the split from our nearest relative, the chimp. Following SNPs over time will allow us to watch genes as they evolve with changing selection pressures.

SNPs also can be used in DNA fingerprinting for criminal or parental verification. The potential use that is most exciting for biomedical research is genotype-specific medication. Most genes contain at least one SNP, some of which may have functional consequences. SNPs are reliable markers that may allow us to determine which combination of coding alleles is associated with a particular disease.

Before we go further, we must define three terms that are often confusing: linkage, linkage disequilibrium, and haplotype. **Linkage** refers to how close two loci are to each other on a chromosome. If they are near each other, we say the two loci are linked. **Linkage disequilibrium** describes alleles rather than loci. If two alleles (or two SNPs) tend to be inherited together more often than would be predicted, we say the SNPs are in linkage disequilibrium (i.e., inherited together more often than other possible SNP combinations). **Haplotype** refers to a set of linked alleles or SNPs on one particular chromosome. Each person has two haplotypes

in a given region, and one haplotype will be passed on as a complete unit unless recombination occurs to separate this particular set of alleles to form two new haplotypes.

These three terms are used extensively when describing how we deduce which genes are involved in polygenic diseases. For example, let's imagine that a particular population has a higher-than-average incidence of Alzheimer's disease. If we could track SNPs that are in linkage disequilibrium and correlate with the phenotype, we would have SNP markers that form a particular haplotype at several loci potentially involved in causing Alzheimer's. To begin, let's look at a hypothetical example in which two SNPs and one gene are associated with a monogenic disease (Figure 4.7). Notice how the two different populations have different ratios of the three possible genotypes, which is what you would expect to see in a population that experiences a higher-than-average incidence of the disease. Even if you did not know allele *a* was the causative agent, you could evaluate the SNP haplotypes and deduce that SNPs 1′ and 2′ are linked to the disease allele, and that this is a

Table 4.4 Global PTC taster SNP frequencies.
PAV is the only taster allele. Sample size for each population appears in parentheses.

SNP Combinations	European (200)	West Asian (22)	East Asian (54)	African (24)	SW Native American (18)
AVI	0.47	0.67	0.31	0.25	—
AAV	0.03	—	—	0.04	—
AAI	—	—	—	0.17	—
PVI	—	—	—	0.04	—
PAV	0.49	0.33	0.69	0.50	1.00

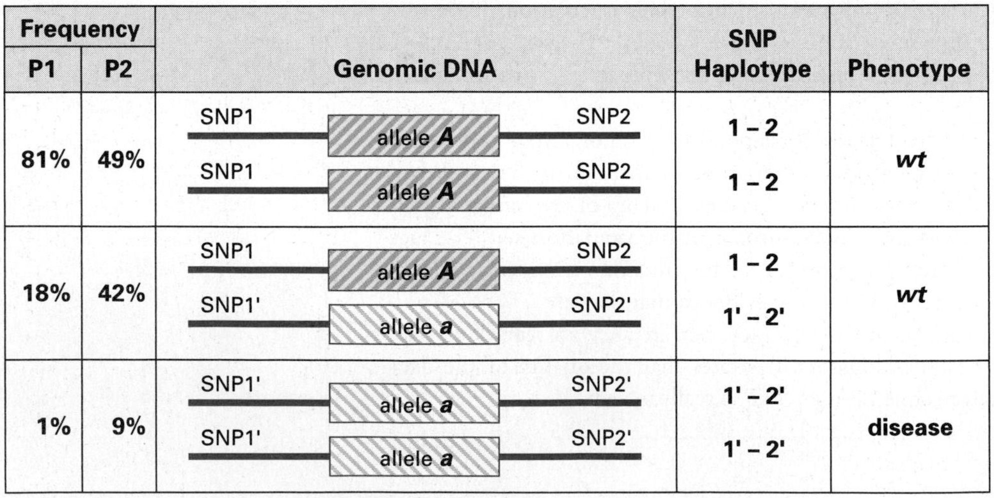

Frequency		Genomic DNA	SNP Haplotype	Phenotype
P1	**P2**			
81%	49%	SNP1 — allele **A** — SNP2 / SNP1 — allele **A** — SNP2	**1 – 2** / **1 – 2**	*wt*
18%	42%	SNP1 — allele **A** — SNP2 / SNP1′ — allele *a* — SNP2′	**1 – 2** / **1′ – 2′**	*wt*
1%	9%	SNP1′ — allele *a* — SNP2′ / SNP1′ — allele *a* — SNP2′	**1′ – 2′** / **1′ – 2′**	**disease**

Figure 4.7 Comparison of SNP data for two populations.
Two populations (P1 and P2) have different frequencies of a single-locus disease.
Included in this figure are diagrams of the genomic DNA showing two SNPs (1′ and 2′)
and a recessive allele (*a*) that are in linkage disequilibrium. An allele and its flanking
SNPs define one locus.

recessive disease because an individual has to be homozygous to contract it.

Traits that are polygenic and can be measured in some quantitative manner can be genetically mapped to **quantitative trait loci** (**QTL**). By studying the inheritance of certain haplotypes that contain identifiable SNPs, we could begin to dissect the genomic causes of quantitative traits such as schizophrenia, high blood pressure, and diabetes. Many investigators want to use similar SNP mapping techniques for genes associated with behaviors such as learning, homosexuality, and daily cycles that determine who needs five hours of sleep each day and who needs nine hours.

DISCOVERY QUESTIONS

12. Look at a more complex trait where three different genes influence the phenotype. In this figure you can see the SNP data, but you cannot see the alleles located near these SNPs that contribute to the trait. In this case, the phenotype could be something similar to cancer or Alzheimer's, where the severity of the disease is similar in all cases but the age of onset varies. (In reality, cancer and Alzheimer's involve more than three genes.) How many different loci contribute to the phenotype of this hypothetical complex trait? Can you tell from these data?

13. If the phenotype in Discovery Question 12 were cancer, which particular SNPs (e.g., 1'-2' or 1-2) would be linked to disease-causing alleles? Which disease alleles are dominant and which are recessive?

14. How many more genotypes would be added to these data if one more locus were included that had only two SNPs at this locus? How much more complexity would be added to these data if the three loci each had a third SNP?

Discovery Questions 12–14 highlight several important points. First, this is a simplistic case, with only 3 genes and 2 alleles for each gene, for a total of 27 different genotype combinations possible. Therefore, mapping polygenic QTL will require large numbers of people if we are to detect linkage disequilibrium. Another point is that some populations will be more informative than others when we try to deduce the causative genes in a genome. For example, Figure 4.7 shows that individuals in P1 have the disease infrequently compared to those in P2. Therefore, P2 would be the better one to study, since you would be more likely to discover informative linkage disequilibrium. Finally, in this simplistic example there are only two haplotypes per locus, while real loci might have many more haplotypes, making it very difficult to determine which SNPs were associated with the trait of interest. With over 1.4 million SNPs, we cannot possible track every possible combination. Luckily, Eric Lander and

his team discovered a simplifying phenomenon in human genome diversity.

In the process of mapping SNPs to identify genes that cause Crohn disease, Lander and his colleagues at the Whitehead Institute mapped the frequency of 103 SNPs in a 500 kb segment of human chromosome 5q31. They surveyed 516 chromosomes from people of European descent and unexpectedly found that humans have less complexity than we had imagined. If you took 103 SNPs (we'll limit it to 2 alleles instead of 4; see Figure 4.8) and permitted random recombination at every possible base in the region, 2^{103} or 10^{31} different combinations would be possible! Instead, the investigators found that many SNPs exhibited linkage disequilibrium and traveled through the generations as haplotypes instead of isolated loci (Figure 4.8). The region can be divided into 11 blocks of DNA that appear to resist recombination and thus carry haplotypes of SNPs from one generation to the next. Even more surprising, each of the 11 blocks only comes in 2 to 4 variations, even the block that is 92 kb long. The haplotypes are not equally abundant in the population studied. For example, block 1 contained 8 SNPs, but there were only 2 major combinations, with the top one occurring in 83% of the chromosomes and the bottom one in 12%. Of the remaining 5%, nearly all differed by only one SNP, so they may be either real SNPs or sequencing errors.

Figure 4.8 Haplotype diversity created in blocks.

Go to www.GeneticsPlace.com to view this figure.

The great news from this study of SNP haplotypes is that if the entire genome has recombinational hot spots as found in 5q31, then it should be possible to map the entire human genome to find the recombination loci. The blocks of SNPs could be treated as alleles and used to find linkage disequilibrium that correlates with disease status. We should be able to sample a subset of SNPs to see which combinations a patient has, and use SNP haplotypes to identify which alleles a person is carrying. This would greatly simplify QTL mapping and eventually enable individualized medical treatment.

DISCOVERY QUESTIONS

15. Which block in figure 4.8 is the least conserved? Hypothesize why and support your answer with data from the figure.

16. Go to HapMap web page, click on "Browse Project Data," and retrieve region chr5: 134530001..134570000, which is 40 kb of 5q31 mapped by Lander's group. What is the name of the SNP that only residents of Utah (CEU) had different from the other groups surveyed? (Note the population code above the ideogram.) Click on this SNP and read the data to determine the frequency of this SNP minor allele in the Utah population.

17. Now search HapMap for the gene *SCN5A*, and click on one of the red triangles to see what gene this is and the variability among the populations. What does this gene encode? At the bottom, you see two alternative splicing cartoons illustrating the two slight variants of this protein that differ by only one amino acid. Can you see the difference in the splicing for one amino acid? Click on the splicing cartoon to read the more complete summary (you cannot see the difference in the cartoon). At the top right of the summary page, click on the "Links" text and choose "Geneview in dbSNP" to see all the SNPs for this gene. Scroll down to find the mutation in codon 1102, which can lead to cardiac arrhythmia and is more prevalent in people of African descent. What is the rs# for this potentially lethal SNP? In which exon is this SNP? (This Discovery Question is designed to show that despite the vast amounts of information available, there are still important gaps.)

18. Search the **Ensembl Human Genome Browser** for *SCN5A*, follow the gene link, choose "Gene variation info." from the menu on the left, and select SNP classification and coding variation from the various options to see Ensembl's method for displaying SNP information. Do you find this display more or less helpful than NCBI's?

19. Search for *SCN5A* using **Genewindow's** display. Click on the line showing the gene below the splicing cartoon and click near the middle of the gene. Navigate until you see exon 18. Can you find the cardiac arrhythmia SNP (converting Ser 1102 to Tyr) in this view?

Do Any SNPs Produce Common Phenotypes?

Increasingly, we can document point mutations that lead to diseases. You probably have already studied at least one case, sickle cell disease, which was the first genetic disease to be understood at the genomic level. Sickle cell disease is caused by a SNP that results in one amino acid difference in the β subunit of hemoglobin. This SNP leads to a different conformation of the protein, which does not function as well as the **wild-type** protein. Examples such as this are not new any more, so we do not need to spend time enumerating multiple cases. Instead, we will study four SNPs that are less well known.

Skin Pigmentation You might think that of all the human phenotypes, surely we understand skin coloration—but we do not. We don't know if there are 50 genes or 500

involved in skin pigmentation. One of the best studies was conducted in 1964, well before any molecular methods were available. The authors concluded that at least 30 to 40 genes were involved and ended their paper with an unusual statement: "The deficiencies in the data in this study are keenly appreciated by the writers, but since there appear at present to be no opportunities for improving the data, it seems justifiable to take the analysis as far as possible."

We certainly know more 40 years later, but not as much as you might expect. We know that melanin is a polymer of two oxidized derivatives of tyrosine called pheomelanin (which appears red-yellow) and eumelanin (which is less water soluble and appears black-brown). You have probably seen red-haired people with pale skin and freckles, all of which can be explained by SNPs that lead to inactive forms of the gene *Mc1R* and the resulting accumulation of pheomelanin and reduction of eumelanin. The linked hair and skin pigmentation phenotypes are familiar, but a team from Louisville, Kentucky, found a surprising correlation with red hair that underscores the complexity of skin pigmentation genomics. People with red hair are significantly less sensitive ($p < 0.05$) to the anesthetic midazolam than non-redheads (blond or dark hair). The clinical investigators did not discern the reason for this drug resistance, but because we don't understand skin pigment genetics, perhaps their shortcoming is understandable.

Malaria Resistance Malaria is a devastating disease that kills millions of people, especially children, every year (see pages 74–78). In East Africa, a team of hematologists studied the frequency of a SNP in the promoter of *Nos2*, which encodes an inducible form of nitric oxide synthase that produces the cell-signaling gas NO. When a particular base was changed from $T \rightarrow C$, children had more NO in their blood and their chances of developing fatal malaria were reduced by about 80%. This discovery may lead to new treatments for malaria if we become able to regulate appropriate NO production through medication. The malaria-protecting SNP was present in 25% of the children studied in Tanzania and Kenya, but we don't know the frequency in other countries.

DISCOVERY QUESTIONS

20. Search OMIM for *Nos2* and click on *NOS2A* for humans. Read the first paragraph and identify two cell types that utilize *Nos2A* and may explain why NO levels may help save children from malarial death. Scroll down to the alleles section and you can find the precise SNP that can save a child's life. At the top right, click on the "Links" text and choose "GeneView in dbSNP" to view all the known SNPs. What weakness do you see in the database, given this particular case study?

21. Search **Entrez Gene** for *Mc1R* and read about the human gene. What feature is unusual about the gene structure? Read the summary and find another phenotype that correlates with accumulation of pheomelanin. At the top again, use "Links" and "GeneView in dbSNP" to view all the known SNPs. This time, what surprising trend do you find in the types of SNPs present in this gene? What does its gene structure have to do with this unusual ratio of SNPs? Search PubMed for SLC24A5. What role does it play in human skin color?

Mitochondrial SNPs Mitochondria produce most of our cells' ATP, and thus their function is critical for our well-being. Most of the 100-odd proteins involved in oxidative phosphorylation are encoded by nuclear genes, but 13 are encoded by mitochondrial genes. Each mitochondrial gene requires the proper function of 22 tRNA genes and 2 rRNA genes that are also encoded in the mitochondrial genome. Phenotypes associated with defective oxidative phosphorylation tend to affect tissues that require the greatest amount of energy. Therefore, cardiac muscle, skeletal muscles, and tissues that have to pump a lot of ions are most often affected. More than 50 different disease-causing mitochondrial SNPs have been identified, and this number is expected to increase as we become more proficient at detecting SNPs.

A panel of 2,000 patients suspected of having diseases due to mitochondrial SNPs was screened for 44 known SNPs. Of these patients, 108 were determined to have a known disease-causing SNP. This research led to several conclusions. First, these 108 patients could be treated more effectively, because the precise nature of their illness was apparent. Second, more mitochondrial SNPs must be tested. Third, maybe some of these 2,000 patients did not have mitochondrial mutations. Finally, there are non-SNP mutations that would have gone undetected in this screen, making it necessary to perform different tests. This study illustrates the need for high-throughput methods to identify as many SNPs as possible through basic research. Eventually, SNP discoveries could be transferred into diagnostic tests for detecting SNPs in clinical settings.

DISCOVERY QUESTIONS

22. If you were to consider the pedigree of a mitochondrial gene mutation (rather than a nuclear gene) that led to a disease, what feature would be diagnostic?

23. Go to OMIM and search for "mitochondrial disease." How many hits did you get? Choose one example from the list.
    a. Is the protein you found involved in ATP production?
    b. Which tissue or tissues are affected the most? Does this make sense, given what you know about the tissue's relative need for ATP?

24. Perform a PubMed search using the terms "mitochondria Alzheimers." Do you see any papers suggesting that mitochondrial mutations may play a role in this disease?

LINKS
Entrez Gene
Adrian R. Krainer
PubMed

Incorrect mRNA Splicing Since high school, you have been aware that mRNA production requires splicing to join the exons and exclude the introns. In Chapter 2, we learned that most human genes use alternative splicing to produce slightly different proteins. In 2001, a team of investigators, led by **Adrian R. Krainer** from the Cold Spring Harbor Laboratory in New York, determined the mechanism responsible for some unsuspected mutations in the breast cancer gene called *BRCA1*. When *BRCA1* encodes a normal protein, breast and ovarian cancers are less likely to occur. However, if the *BRCA1* protein is absent or nonfunctional, many women in the family (and occasionally men) are very likely to develop breast cancer, and women may develop ovarian cancer as well.

We have already considered SNPs that lead to the incorporation of inappropriate amino acids, but there is another form of SNP that can also lead to diseases. When we find a SNP in an exon, we immediately determine whether it alters the amino acid composition of the encoded protein. If the amino acid sequence is unaffected, we tend to think of this SNP as "silent" and call this type of point mutation a "silent mutation." Krainer and his colleagues identified a series of silent mutations in exons of *BRCA1* that lead to alternative splicing. Splicing of RNA to produce a mature mRNA involves the 5′ and 3′ ends of each exon, but internal sequences are required as well. Although the consensus sequences are uncertain, exonic splicing enhancers (ESEs) and exonic splicing silencers (ESSs) are located within exons and are distinct from the terminal splicing junctions. When working properly, ESEs are bound by proteins involved in splicing, and the bound exon will be included in the mRNA. If a SNP occurs in an ESE, an exon that would normally be included in the mRNA will be skipped, which means the encoded protein will not retain its appropriate structure or function. Conversely, ESSs are DNA sequences bound by proteins that exclude an exon that contains a functional ESS. If a SNP disrupts the binding site of an ESS, the exon will be inappropriately included in the mRNA and disrupt normal protein function. Krainer found examples of ESE and ESS mutations in *BRCA1*, which probably explains why some women with silent mutations develop breast and ovarian cancer. This case illustrates that even silent SNPs can have profound influences on phenotypes, including polygenic traits such as cancer.

DISCOVERY QUESTIONS

25. Experimentally, how would you determine if a SNP resulted in alternative splicing?
26. If most genes normally encode alternatively spliced mRNA, why would mutations affecting ESSs and ESEs be problematic?
27. Design an experiment to demonstrate that a splicing mutation in *BRCA1* increases the likelihood of developing breast and ovarian cancer.

Are There Vital SNPs That Can Surprise Me?

The four cases on pages 192–193 illustrate the SNP diversity common in observable phenotypes. Unfortunately, some SNPs do not reveal themselves until it is too late. Here we consider three examples that are a matter of life and death.

Fava Beans and a SNP May Kill You About 2,000 years ago, a Roman named Lucretius Caro described the consequences of this SNP: "What is food to some men may be fierce poison to others." It turns out that some people experience a lysis of their red blood cells from consumption of fava beans (a large white legume). Approximately 10% of people in the world are susceptible to this malady because they fail to make glucose-6-phosphate dehydrogenase (G6PD), a metabolic enzyme found in the cytoplasm of every cell (if you have the right SNP). G6PD produces NADPH, which helps regenerate the enzymes used to neutralize the cellular toxin hydrogen peroxide. The fava bean contains a variety of substances that increase the red cells' sensitivity to oxidants such as hydrogen peroxide. Your red blood cells have only one way to produce NADPH—via G6PD, which is encoded on your X chromosome. Females would need a pair of SNPs, but males only need to inherit one SNP from their mothers to block the production of G6PD in their red blood cells and place them in the 10% of the world's human population lacking G6PD. The SNP 376A → G produces G6PD with normal activity and is found in 20% of African males. A second SNP (202G → A) reduces G6PD activity by about 10% in about 20% of African alleles. Another SNP (563C → T), called "G6PD Mediterranean," produces an enzyme with nearly undetectable activity, and it is found in about 20% of the alleles in Caucasians living around the Mediterranean Sea. Many other G6PD SNPs have been identified in populations around the world.

DISCOVERY QUESTIONS

28. Search OMIM for the word "fava." Click on the "*305900" link; it will be very long, but skim it to determine why a "deleterious" mutation is so prevalent in certain parts of the world. What selection pressures may have contributed to the high proportion of G6PD-deficient individuals?
29. If G6PD is so important, how can these people live without it in any of their cells?

Variations in Medication Responsiveness Many human medications are not administered in their final and active form. The drugs are metabolized in a predictable way, and the enzymatic product is the therapeutic compound. One famous example is AZT, which has to be modified by human enzymes before it can block HIV replication. Cytochrome P450 is a family of enzymes (isozymes) that metabolize a large number of these "pre-drugs." People fall into one of three classifications: typical metabolizers, poor metabolizers, and ultra-rapid metabolizers. Drug studies are performed on large panels of people to determine the optimum dosage for the "average" person. However, any one person may not have the average metabolism, so the ideal dosage for him or her may not be the average dosage. For example, poor metabolizers often fail to convert a pre-drug into its active form, and wind up excreting the inactive version before it can be converted. Conversely, ultra-rapid metabolizers quickly convert nearly all the pre-drug into its active form, which may result in an overdose if the average person normally converts a lower percentage of the pre-drug.

Cytochrome P450 is encoded by two separate genes called *2C19* and *2D6*. The gene *2D6*, which is on chromosome 22, has 9 exons and 8 introns; 12 SNPs have been identified that lead to altered 2D6p activity. The most common mutation is a G → A substitution within exon 4 that alters splicing in mRNA formation and results in no protein being produced. More than 40 different pre-drugs require 2D6p activation, including antiarrhythmics (heart medication), antidepressants, antipsychotics, and painkillers in the opioid family.

The gene *2C19* is known to metabolize mephenytoin, which is used to treat epilepsy. Poor metabolizers do not activate this pre-drug and thus do not receive the same benefit as the average person. Only 2–3% of Caucasians have weak 2C19p activity, but about 23% of Asian populations are poor metabolizers. *2C19* is located on chromosome 10, and the most common mutation is a SNP in exon 5 (681G → A) that creates an aberrant splice site. Therefore, when drugs are administered to different populations, it is important to determine a population-specific recommended dosage.

DISCOVERY QUESTIONS

30. How could you determine if you were a poor or an ultra-rapid metabolizer?
31. Is there any practical way to respond to the information presented in this case study? Who

is a poor metabolizer and who is not? What could a physician do if a patient's genotype were known?

32. Search NCBI's PubChem Substance for "6-mercaptopurine" (also called thiopurine) to see its structure. Now search PubMed for the authors "Coulthard Matheson Hogarth" and read their October 2004 abstract. What are the consequences for a poor metabolizer in this situation?

Food and Drug Interactions So far, have learned that some foods can kill an some drugs can be ineffective if you have Now we examine the interaction of foods w tions. For example, did you know that grapefruit alter your ability to absorb drugs? Typically, pill through your stomach and dissolve in your small intest where the medication is absorbed. One protein involved i absorption is P-glycoprotein, which pumps drugs into intestinal cells. A second protein is a particular allele of the

PATENT LAW AND GENOMICS

No doubt you have heard that it is possible to patent genes, proteins, and even whole organisms. Since these objects are produced by normal biological processes, many people become enraged when they think about this aspect of biotechnology: "It is not right for anyone to patent proteins in my body!"

To understand patents granted for genetic material, we need to understand the principle of patent law. The purpose of a patent is to encourage development of new products and methods by granting "sole right to exclude others to make, use, sell, or import an invention for a limited period of time," currently 20 years. Though this sounds like an unfair monopoly, patent law is designed to reward the investment of time and money needed for research and development required to bring an invention to market. Exclusive rights come with a cost, though; to obtain a patent, you must disclose what the invention is and how it can be used. Since patents are public records, you have just revealed all your secrets to everyone. Human clotting factor made by bacteria would be patentable, but after 20 years, anyone could produce it without any legal restrictions. To see an example, go to the U.S. Patent and Trademark Office and search for patent number 6,180,371. Click on the link and you will see what was patented and why we want to encourage this type of research and development.

Biological patents have been granted for more than 100 years, before Mendel was famous or Watson and Crick were born. Aspirin is a natural product of willow trees, but in its purified form, a form not found in nature, it was a patentable product. Likewise, antibiotics are natural products of fungi, but for medicinal purposes they must be purified, which requires human intervention. Human intervention is the critical distinction between the genes and proteins found inside your body and the DNA and protein sequences that have been patented. Likewise, genetically modified organisms can be patented because they would not exist in nature without human intervention, even though they continue to reproduce naturally.

Patents are granted only to inventions that are novel, nonobvious, and useful. A particular DNA sequence produced by PCR is not a natural product (i.e., it is novel); therefore, it could be patented as long as the inventor could explain its utility. If you were to randomly create a sequence of DNA, it could not be patented unless you could demonstrate a use for it. However, if you tried to file for a patent on a product that has already been published or used by others, it would be refused because this is considered "prior art," which means you cannot have the exclusive rights to this "invention." Likewise, if someone patented a cDNA sequence, you could not patent the mRNA, as that would be considered "obvious."

Therefore, it is possible for a multinational corporation or an undergraduate student to patent a human gene sequence as long as its utility can be demonstrated and it is not obvious or prior art. It is too late, for instance, to file a patent on hemoglobin. However, if you discovered that the hemoglobin gene can be alternatively spliced to produce a different protein with a novel function, you might be able to obtain a patent even if you do not understand the mechanism of this novel function. If you wanted to conduct

patented invention, there is one loophole in the
he invention "merely for philosophical gratifica-
ic phrase is open to interpretation and has been
be careful. If you were working on your thesis
ientific meeting, you probably would be safe. If
ild want to sue you, though they could. If you
tent based on work using someone else's inven-
ion . . . well, you might want to have a lawyer
philosophical enjoyment."

bell, PhD, JD.]

drug metabolizer called cytochrome P450 3A, which converts the potentially beneficial drug into a form that is more
readily excreted. One glass of grapefruit juice can block
P-glycoprotein and cytochrome P450 3A for as long as 24
hours. Cytochrome P450 3A is inactivated by an unknown
component of grapefruit juice, causing the enzyme to be
destroyed. When volunteers were given a cholesterol-lowering
medication along with a typical "breakfast" amount of juice,
the average person showed only slightly elevated levels of the
drug in the blood. When a different panel of volunteers was
given double-strength grapefruit juice, three times a day for
two days, on average the same drug was elevated 12-fold in
the blood. The immunosuppressant cyclosporin, which is
given to organ transplant patients, was shown to be internalized better in the presence of grapefruit juice. Conversely, in
a small trial of 13 individuals, St. John's Wort was found to
enhance the activity of cytochrome P450 3A. The herb St.
John's Wort is popular in health food stores but this study
revealed the herb reduced the effectiveness of drugs used to
control cholesterol, seizures, and transplanted organ rejection. There is very little research in the area of genomic
interactions with drugs and food, even though they affect
human health.

DISCOVERY QUESTIONS

33. Can you imagine any SNPs that occur outside
the coding region of cytochrome P450 3A genes
that might also lead to differences in drug
responses?

34. If grapefruit juice causes a drug to be more concentrated in the blood, which enzyme is affected,
P-glycoprotein or cytochrome P450 3A? Explain
your answer.

Why the SNP Frenzy? Pharmacogenomics!

At this point, you should be keenly aware that genomic variations, ranging from large changes such as aneuploidy to small
SNPs, can have a profound effect on your personal physiology. Your collection of genomic variations makes you a unique
human being and contributes to your potential to learn, your

predisposition to disease and drug addiction, and your
response to pharmaceutical interventions. You should also be
aware that there are variations within as well as between populations. In fact, the variation between individual genomes has
sparked a biotech boon in the area of SNP discovery.

When the relationship between genetic variation and
drug efficacy was first recognized, the term pharmacogenetics was coined. Now the term **pharmacogenomics** is being
applied to genome-wide variations, though conceptually
there is no difference between the terminology of the 1950s
and that of the 21st century. Pharmacogenomics refers to
the complete list of genes that determine overall efficacy,
while pharmacogenetics originally focused on single genes
that metabolized drugs. Pharmacogenomics is trying to
take into account all the genes that include drug metabolism, transporters, receptors, signaling pathways, etc. You
can imagine the high degree of complexity in humans if
each of these components can be polymorphic.

Let's look at a simple example that appeared in a review
written by William Evans and Mary Relling from St. Jude's
Children's Hospital in Memphis, Tennessee (Figure 4.9). In
this example, we consider the efficacy and toxicity of a drug
that requires only two genes, an activator and a binding site,
each with only two alleles. There are nine possible combinations of genotypes, though there are fewer than nine efficacy
phenotypes. Therapeutic effects depend on the genotype of
drug receptors in combination with the amount of active
drug in circulation. From the graphs that overlay efficacy
with toxicity, it becomes apparent that with a homozygous
mutation for drug metabolism, all three drug receptor genotypes suffer from toxicity. This example highlights the complex web of protein interactions that pharmacogenomics
hopes to decipher.

Figure 4.9 illustrates that drug response is polygenic, and
so we will need new technologies to understand the connections between relevant proteins involved in drug
responses. As described earlier in this section, cytochrome
P450 2D6 is involved in metabolizing painkilling medication such as codeine. There are ample clinical data to support the evidence that 2–10% of the population are
homozygous for null alleles and thus cannot use codeine
for pain relief. For most drug-metabolizing enzymes, we do

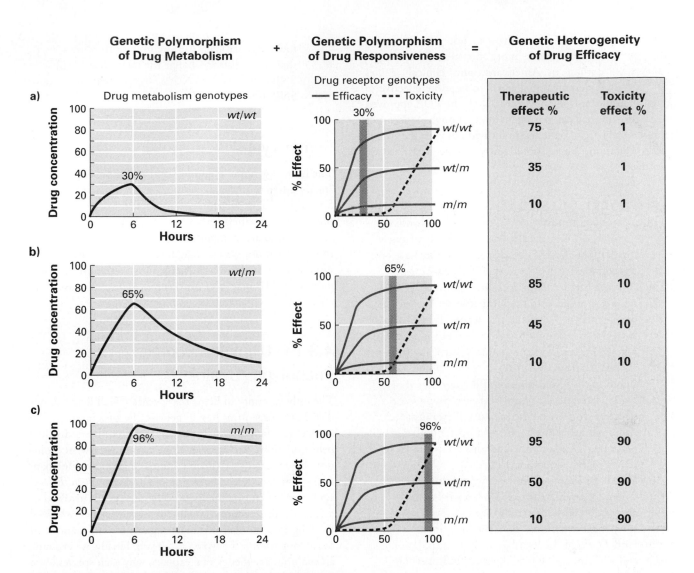

Figure 4.9 Genotype and phenotype interaction for drug efficacy.
Drug metabolism will be one of three genotypes: **a)** homozygous *wt*, **b)** heterozygous, or **c)** homozygous mutant (*m*). For each metabolic genotype, efficacy and toxicity are graphed in the middle column with three lines representing the genotypes of drug receptor locus. The third column summarizes the net effect for each of the nine genotypes.

not know any other phenotypes that alert us to a patient who will not respond appropriately to a particular medication. It has been hypothesized that cytochrome P450 2D6 poor metabolizers are less tolerant of pain because they cannot produce endogenous morphine, but this is not a particularly helpful phenotype for screening patients.

Traditionally, drug development has been aimed at delivering medications that are effective and safe for everyone, but as we have seen, enzyme polymorphisms can have clinically significant consequences. Pharmaceutical companies are spending a lot of money to discover clinically relevant SNPs in order to produce SNP haplotype-specific medications. Industry is still debating whether it makes financial sense to develop genotype-specific drugs when the number of allele

combinations is so large. If genotype-specific medication becomes viable, when you are diagnosed, the physician will need to know your genotype to determine the appropriate medication and dosage for optimal therapy.

Pharmacogenomics may sound too futuristic to be relevant to you, but let's consider one of the first drugs to be proven to work only in some genotypes: Iressa. Every year, 140,000 patients are diagnosed with non–small-cell lung cancer, which is nearly always fatal. During clinical trials of Iressa, clinicians were baffled because about 10% of patients were completely cured (Figure 4.10) but all others died. Why would a drug work so well on some but not at all on others? Two teams of investigators simultaneously found a mutation in one gene (*EGFR*) that determines if

LINKS
GeneSNPs
NPR story
Real Player
Sordella et al.

Iressa will cure you or not. The life-saving mutation is more prevalent in Japanese than in Americans, and clinical studies demonstrated Iressa to be more effective in Japan than in the U.S. In May of 2003, the FDA approved Iressa for lung cancer treatment even though it does not cure the majority of patients. In individuals with the appropriate mutation, Iressa should prove to be nearly 100% effective.

DISCOVERY QUESTIONS

35. Using the simplistic case from Figure 4.9, outline a protocol to test the efficacy and toxicity of a new drug. What effect will pharmacogenomics have on the complexity of clinical trials for new drugs?

36. Imagine this scenario: You and a sibling have been diagnosed with the same disease. You are both genotyped, and the physician tells you there is an effective drug for your sibling but not for you. How would you respond? What if the situation were reversed?

37. Listen to the seven-minute NPR story related to Figure 4.10 and hear how Iressa was found to be effective (you'll need the free Real Player installed). Although Iressa only works in a small number of people today, what hope does this finding offer to the remaining lung cancer patients?

38. Search for Iressa's generic name, gefitinib, in Pub-Chem to see its structure. Now read the 2004 abstract by Sordella et al. to see how even this story is more complex than first imagined.

39. Find *EGFR* in the list of genes at GeneSNPs and click on the "UCSC:hg16:7" link. Let the full image set load. Are any mutations other than SNPs listed? Change the menu from "ALL SNPS" to "NONSYNON." How many missense SNP mutations are listed in this target for Iressa?

Summary 4.2

The human genome might be described more accurately as "many different human genomes." Each of us carries many SNPs, and most of our genes are located near at least one SNP. A lot of effort has been focused on the discovery of SNPs to identify important QTLs. It is hoped that, with this information, we can diagnose and treat more diseases more effectively. However, the genome variation that may identify QTLs may also prove problematic, because drug effectiveness is affected by genomic variations.

4.3 The Ultimate Genomic Phenotype–Death?

The only certainty in life is that eventually all living things die. Each organism has a genetically programmed life expectancy range, which can be affected by individual behavior. However, behavior alone cannot alter what evolution has selected—a predetermined maximum life span. Aging and death are universal phenomena (even *E. coli* ages) and we know a lot about individual genes involved in aging. In this section, we briefly survey genes involved in aging in a number of model organisms and examine in detail longevity in the worm *C. elegans*. Finally, we examine an evolutionary theory that explains why each species has a genomically determined life span.

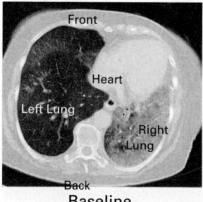
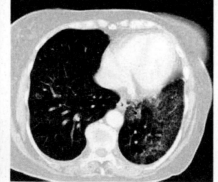
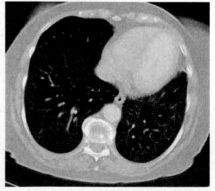

Baseline 3 months 2 years

Figure 4.10 Iressa cures 10% of treated patients.
Three CAT scans for a single patient over a two-year period. At the baseline time point, the patient's right lung is hazy with invasive non–small-cell lung cancer, while the left lung is relatively clear. Within three months, the cancerous lung is clearing; two years later, the lung remains cancer-free.

Why Do We Age?

Bessie and Sadie Delany were sisters who lived for more than 100 years. They were born in Raleigh, North Carolina, in 1891 and 1890, respectively. Their father was an emancipated slave and their mother a mixture of Native American and African American. The Delany sisters never married or had children, but they were highly educated, and each pursued distinguished professional lives. Bessie died on September 25, 1995, at age 104, and Sadie died on January 25, 1999, at the age of 109. What was the secret of their longevity?

Why do some people live longer than others? Did the Delany sisters' genomes permit them to survive while all their age-mates died? How did they live cancer-free for so long? Research in the field of aging and life span has revealed that many genes are involved in the aging process. The model organisms of yeast, *C. elegans*, *Drosophila*, and mice have revealed that the rate at which an individual ages is controlled in part by genomic variations. The list of known genes involved in aging in each species is relatively short, although it grows each year. There are some common themes, though. For example, caloric restriction slows aging. This appears to be a universally conserved evolutionary adaptation to permit individuals to delay reproduction until food is more plentiful. The mechanism is not understood and may vary between species. For example, in yeast, starvation appears to control gene silencing (a form of transcriptional repression controlled by the gene *Sir2*). Why gene silencing prolongs life is still unclear.

Of multicellular animals, the worm *C. elegans* has been especially informative. In addition to caloric restriction, worms may have as many as four different signaling pathways that result in longer life spans. One pathway, regulated by the gene *Daf2*, involves hormonal signaling via a protein similar to insulin. Individuals with certain mutations in this pathway live longer and are more resistant to various forms of stress, including oxidative damage, ultraviolet (UV) light, and heat. Reproductive cells responsible for gamete formation also regulate aging through a second pathway. In the third pathway, mitochondria signal the nucleus to slow the "rate of living" in mutants that live longer. A fourth pathway involves a tyrosine kinase that, when mutated, can prolong life by about 60%, though it does not affect overall development or reproduction the way the other three pathways do.

The famous *Drosophila* mutant allele *methuselah* can prolong the life of individual flies by about 35%. The encoded protein is predicted to be an integral membrane protein in the synapse of nerves that is coupled to G proteins. However, this particular gene has no known orthologs, and we do not know what ligand normally binds to this receptor. Therefore, this pathway may be similar to ones found in worms, or may be a fifth pathway.

"Ames dwarf mice" live longer than *wt* mice, but the molecular mechanism is still a mystery. These mice have altered hormonal levels, which may indicate a mechanism similar to the insulin-like pathway seen in long-lived worms. In *wt* mice, a protein called p66^{shc} helps relay stress signals that initiate cell death. In a different mutant mouse strain, individuals lack a functional version of p66^{shc} and live longer, though no one knows why.

Two general conclusions have emerged from all the research on aging in model systems. First, specific genes selected to promote aging are unlikely to exist. Second, aging is not the function of genes, but results from accumulation of tissue damage, owing to limits on the energy used in tissue maintenance and repair. Longevity is regulated by genes that control the degree of cellular damage and repair.

As you can see, many questions remain to be answered. It is worth noting that the genomes of the model organisms of yeast, worms, and flies have been sequenced fully, and yet we know very little about aging, the most universal of phenotypes. Nevertheless, what has been learned so far has changed the way people think of aging and associated diseases such as cancer. If the aging process can be slowed, might cancer also be postponed? Did the Delany sisters have the right combination of SNPs in their genome that permitted them to live longer than average? Can medicines be developed to prolong life and reduce the incidence of diseases? If humans could experience changes similar to those seen in model organisms, it might be possible to feel 45 at the ripe age of 90. Is this scenario really possible? Does it contain hidden costs that we cannot predict? What ethical issues would be forced upon us in such a long-lived society?

Evolutionarily, it seems counterproductive for individuals to get old and die. If the ultimate selection pressure is to produce as many offspring as possible, then why do we age? What selection pressures have resulted in the progressive loss of functions and fertility, and eventual death? There are two main hypotheses that attempt to explain this apparent evolutionary oxymoron. The **mutation accumulation** hypothesis states that random mutations accumulate gradually and can become fixed by genetic drift in genes that are not subject to selection pressure. Under this model, adaptation to one environment and loss of adaptation to another are caused by distinct and unrelated variations in the genome. Using this model, aging is unrelated to any other traits that may have been beneficial. **Antagonistic pleiotropy** ("antagonistic" denotes opposition and "pleiotropy" indicates more than one function for a single gene) proposes that there are tradeoffs in evolution: A SNP that produces a benefit in one environment will produce a detrimental consequence in another. Under this model, aging is the negative tradeoff for another phenotype that is advantageous. Science progresses through experimental testing of hypotheses such as these.

From the discussion of aging in model organisms, it seems clear that many genes are involved and that aging is a

LINKS

aging
Delany sisters
methuselah

consequence of other processes, rather than the activation of "aging genes." The only area of contention, then, is the mechanism as postulated by these two hypotheses. The existence of a cost for longer life has been documented in many species. Tradeoffs seen in model organisms with longer life spans include lowered fecundity, lower viability, delayed reproductive maturity, and prolonged development. These are well documented, but the genomic mechanisms are still unknown. Let's look at two experiments designed to test which hypothesis is more accurate.

Are There Hidden Costs for a Prolonged Life?

A team of investigators led by Gordon Lithgow from Manchester, England, wanted to test the tradeoff for longevity in the *C. elegans age-1* mutant, which has an altered kinase involved in the insulin-signaling pathway. *age-1* mutants can live as much as 80% longer than *wt* worms. When grown at 20°C, they have similar appearances, developmental rates, activity levels, and total fertility. When *wt* and *age-1* worms were grown together, at three differential ratios (90% *age-1* and 10% *wt*; 50% *age-1* and 50% *wt*; 10% *age-1* and 90% *wt*), the proportion of *age-1* mutants remained essentially unchanged over ten generations (Figure 4.11). These duplicated experiments support the mutation accumulation hypothesis, because the change in aging had no apparent negative consequence.

However, Lithgow recognized that standard laboratory conditions with a constant environment and unlimited food did not resemble life in the natural world where *C. elegans* normally lives. Therefore, his team repeated the experiment, but this time they allowed the worms to consume all their food to mimic temporary starvation for four days. The process was continued for six cycles of feeding and starving. At the conclusion of the experiment, the investigators determined the proportion of surviving *age-1* and *wt* worms (Figure 4.12). They initiated this experiment with a 50:50 mix of *wt* and *age-1* mutant worms, and the experiment was conducted 5 times. Unlike the previous result, this time the ratio of mutant worms in a mixed population decreased dramatically, from 0.5 to 0.06. Upon examination, it was determined that only young *age-1* adults were able to lay eggs during the starvation cycles, which indicated the benefit of prolonged life for *age-1* mutants was lost in reproductive capacity during starvation. These experimental conditions mimicked more realistic growth conditions than unlimited food, and the data support the antagonistic pleiotropy hypothesis. Therefore, Lithgow and his colleagues concluded that genomically determined longevity is regulated by antagonistic pleiotropy and not by mutation accumulation.

DISCOVERY QUESTIONS

40. Do the data support the investigators' claims about *age-1* worms? Can you distinguish between antagonistic pleiotropy and mutation accumulation with this experimental design? Explain your answer.

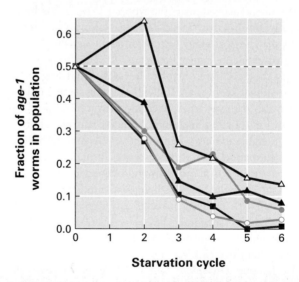

Figure 4.12 Comparison of fitness between food-restricted long-lived *age-1* mutants and *wt* C. elegans.
The worms were maintained at 20°C and were permitted to consume all the available food and then go without food for four days to complete one starvation cycle. The initial ratio of *age-1* worms to *wt* was set at 0.5 and repeated 5 times (each line represents a separate experiment). A reduction of 0.44 allele frequency over 12 generations is estimated to indicate a relative fitness of 0.77 compared to *wt*. The dashed line marks the initial ratio.

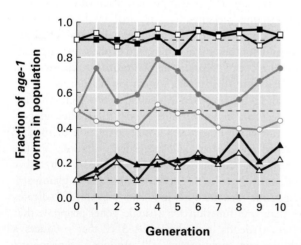

Figure 4.11 Relative fitness of long-lived worms.
Long-lived *age-1* mutants and *wt C. elegans* were maintained at 20°C and given unlimited food. The initial ratios of *age-1* to *wt* were set at 0.9 (black), 0.5 (gray), or 0.1 (purple), and each experiment was duplicated (filled and unfilled shapes). Horizontal dashed lines mark the three initial ratios.

41. Do you know of any drug that does not have a side effect in some individuals? Do you think it would be possible to extend human life span by medication and not experience negative side effects?

Do Bacteria Experience Genomic Tradeoffs Too?

When trying to hypothesize mechanisms for the evolution of aging, you would like those mechanisms to apply to all organisms. The longevity experiments with *C. elegans* supported the antagonistic pleiotropy hypothesis, and now we will look at evolution in the model bacterium *E. coli*. **Richard Lenski** at Michigan State University has been studying real-time evolution in cultures of bacteria that have been growing and evolving continuously for over 3,000 days. Lenski and his graduate student Vaughn Cooper wanted to test the two competing hypotheses on aging and evolution.

They grew 12 different populations of *E. coli* for 20,000 generations in the presence of glucose as the only carbon source. During this time, they removed aliquots and froze them for later analysis. Once they had obtained aliquots from different time points over the 20,000 generations, they measured the ability of each aliquot of bacteria to utilize nonglucose energy sources. They wanted to know if there had been a concomitant tradeoff for evolving optimum metabolism of glucose, as predicted by antagonistic pleiotropy. Or perhaps the cells experienced genetically distinct changes when glucose was the only energy source, so that optimization of glucose metabolism and the loss of other catabolic functions were unrelated, as predicted by mutation accumulation. For each of the aliquots of evolved bacteria, the investigators measured fitness as defined by improved metabolism and growth in the glucose-only environment (Figure 4.13a). The rate of adaptation to glucose was rapid in the first 5,000 generations and then slowed.

In 64 separate experiments, they also measured the ability of each aliquot to grow under new conditions where different sources of energy were available but glucose was not. The ability of each aliquot of cells to catabolize a nonglucose source of food was compared to the original parental strain of bacteria that had not been subjected to glucose-only metabolism. If the loss of nonglucose catabolic potential occurred at the same rate as the increase in fitness, then antagonistic pleiotropy would be supported. If alternative catabolic pathways were lost at a linear rate, then mutation accumulation would be supported (Figure 4.13b).

Over the course of the 20,000-generation experiment, the loss of catabolic potential was rapid initially, and then

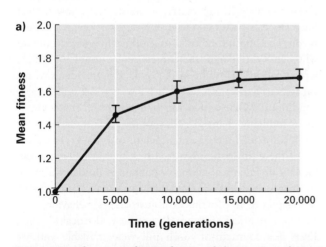

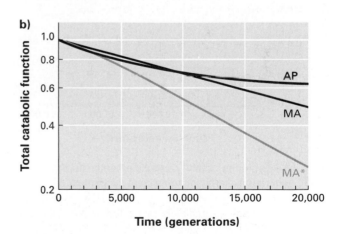

Figure 4.13 Changes in fitness for *E. coli* during an evolution experiment.
a) Measurement of mean fitness during 20,000 generations in minimal glucose medium. Each point is the mean of all 12 populations, with each population tested 5 times. Error bars are the 95% confidence intervals, and the Y-axis is a linear scale. **b)** Computer-simulated trajectories of evolving metabolic specialization: AP (antagonistic pleiotropy), in which functional decay is inversely proportional to gains in fitness; MA (mutation accumulation with original mutation rate), in which loss of catabolic potential occurs at a constant rate and is independent of the pace of adaptation; MA* (mutation accumulation at accelerated mutation frequency) occurs when a population becomes an accelerated mutator strain. Total catabolic function is shown on a logarithmic scale, with mutation accumulation graphed linearly.

slowed (Figure 4.14). During the course of the experiment, 3 of the 12 populations of bacteria developed mutations, so that they accumulated new mutations at a faster rate than the parental strain; these were called "accelerated mutators." The nine normal mutators and the three accelerated mutators lost their ability to catabolize other energy sources at the same rate that they adapted to growth on glucose. The authors concluded that their data support the antagonistic pleiotropy hypothesis for the mechanism of evolution.

We began this section on aging wondering if genomic variations could explain differences in life spans. The growing body of evidence supports a genomic role in aging. We also have abundant data indicating that genomic variations produce variable phenotypes. In the two experiments presented in this section, two teams of investigators concluded that selective advantage in one aspect leads to negative consequences in another, and that aging is a biological certainty that arose as a tradeoff for other adaptations. Reduced fecundity and fitness appear to be the tradeoff for increased longevity. Similar explanations can be proposed for disease alleles that persist in certain populations, such as a reduction in glucose-6-phosphate dehydrogenase and resistance to malaria (see page 194). In February 2001, a genetically engineered mouse that was smarter than normal mice (in learning and memory tests) was shown to be more sensitive to chronic pain. Evolution is a balance between gains and losses, and aging is no exception.

DISCOVERY QUESTIONS

42. In Figure 4.13b, antagonistic pleiotropy (AP) was a curve and mutation accumulation (MA) was a straight line. Do you think it would be easy to distinguish between the lines drawn for AP and MA?

43. Look at the data in Figure 4.14. Given the variation for the replicated experiments, do you think the two curves drawn for the two types of cells accurately summarize the consequences of adapting to glucose-only metabolism?

44. Do we have enough information to determine whether antagonistic pleiotropy or mutation accumulation has been supported by these experiments? Explain your answer.

When a mutation produces a selective advantage early in life, the organism may also accumulate a cost that is exhibited late in life. When model organisms are mutated to prolong life, they lose fitness, meaning the mutation would not spread in a natural population under normal selection pressures. One remaining question about aging is whether we humans have controlled our environment to a similar degree as found in constant laboratory conditions. The Delany sisters did not marry, so we do not know if their longevity came with a loss in reproductive potential. Could medical techniques be used to produce long-lived humans who could thrive in a well-controlled environment? Would a pharmaceutical "fountain-of-youth" pill provide us with a capacity to live longer, just as *age-1* worms did under stable conditions? What would the tradeoffs be if our environment suddenly changed either locally (e.g., restricted diet) or globally (e.g., increased temperatures and CO_2 levels)?

All these futuristic immortality scenarios have a caveat. Genetically identical animals maintained in identical environments do not die at the same time. As will be discussed in Chapter 8, all proteins exhibit stochastic behaviors such that two identical enzymes do not have identical kinetics. Even if a fountain-of-youth pill were available and we cloned humans, there still would be a range of ages when people die. Variation is inherent in the proteome as well as the genome.

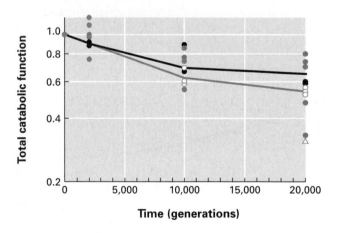

Figure 4.14 Evolution of total catabolic potential during 20,000 generations in minimum medium supplemented with glucose.
Total relative catabolic function is calculated as a weighted average for growth on 64 different substrates, with 1.0 representing the starting point of catabolic potential. Values are shown on a logarithmic scale so a mutation accumulation outcome would have appeared linear. Each point is the mean of three clones from each population. The black line indicates the mean value for the eight populations with original mutation rate, while the gray line indicates the mean for the three accelerated mutators. Closed circles are populations with original mutation rate; open circles are populations with accelerated mutation rate; open triangle is a late accelerated mutator population. An accelerated mutator appeared late in the experiment and was excluded from the mean calculations for both lines.

Summary 4.3

Genomic variation is integral to the survival of any species in a changing environment. From the largest mammals to the smallest phytoplankton, genomes vary. These variations can be used to study population dynamics, including human evolution and diseases. However, a few lessons must be kept in mind when thinking about genomic variation. (1) Not all variations will have an effect on phenotypes. (2) Environmental factors influence the "interpretation" (expression) of genomes. (3) Even genetically identical individuals under identical environments will exhibit phenotypic differences. (4) Evolutionary advances come with tradeoffs. (5) Legitimate ethical concerns must be addressed before technical progress presents us with unanticipated dilemmas. The final portion of this chapter focuses on ethical concerns. With the exception of murder, incest, and other extreme situations, there are few universally accepted ethical conclusions. We will consider case studies on which unanimous opinions are impossible but general agreement may be possible. Then we examine some cases in which ethical objections are obvious, but individuals may proceed nonetheless.

4.4 Ethical Consequences of Genomic Variations

In this section, we examine a few of the ethical issues that surround genomics. The chapter on genomic variations is the best chapter for this topic because if all genomes were identical, there would be no ethical dilemmas to consider. This section highlights subjects that can ignite angry debates, but it is not intended to include every ethical issue. The topics are divided into three general categories: genetically modified organisms, genetic testing, and cloning of humans. Genomics cannot be studied in a moral vacuum and the public deserves a comprehensible explanation of the newest research findings.

Are Genetically Modified Organisms Bad?

Genetically modified organisms (GMOs) are plants, animals, and prokaryotes whose genomes have been deliberately modified through the addition of new genes and/or removal of endogenous genes. This modification is seen as completely evil by some people, and enormously beneficial by others. Let's examine six cases and try to see both sides. As with any new knowledge, there is the potential for great improvements and disastrous consequences. The balance of these two potentials will be determined on a case-by-case basis, and outcomes may depend on your point of view. Humans have been genetically modifying other species for thousands of years through selective breeding. Rice, corn, potatoes, meats, and fruits have all been modified from their original forms to enhance qualities deemed valuable by humans. Now we have the ability to accelerate this process and design traits that would never have been possible through selective breeding.

LINKS
GMOs
Greenpeace
Syngenta

It is worth mentioning one common argument against genetically modified foods—consumers will be forced to eat DNA and proteins from species other than the intended food. Actually, humans have been doing this since we evolved, because bacteria and other microbes live everywhere: in your kitchen, at restaurants, and on your skin. Our digestive and immune systems have evolved to take care of dietary freeloaders. DNA is readily destroyed by the low pH of our stomachs and the DNases we produce. Consuming unintended DNA should not be considered a realistic hazard, but there are other potential hazards that are worth further discussion.

Golden Rice Rice is the staple food for 124 million people worldwide, many of whom suffer from vitamin A deficiency. In Southeast Asia, it is estimated that 250,000 children go blind each year due to vitamin A deficiency. Oral vitamin pills are not feasible due to cost and limitations in infrastructure such as transportation. In January 2000, a team from Switzerland and Germany genetically engineered a strain of rice that can produce β-carotene (normally absent in rice), which humans can convert into vitamin A. To construct this strain, they transformed rice with three genes: two encoded enzymes from daffodil and one encoded a bacterial enzyme. Rice that makes β-carotene appears golden in color, and the name "Golden Rice" is associated with this strain.

In February 2001, the first Golden Rice varieties were shipped to the International Rice Research Station in Los Baüos, Philippines. The research station is conducting safety studies and using traditional breeding techniques to introduce the new phenotype into varieties that perform well in local growth conditions. Similar shipments are underway for research institutes in Africa, China, India, and Latin America. The inventors of Golden Rice transferred all commercial rights to the world's largest agribusiness, called Syngenta, which in turn will give the rice free of charge to all subsistence farmers (earning less than US$10,000 annually). Reaching this agreement was not easy since Golden Rice utilized technology covered in 70 patents held by 32 different companies and universities. In 2005, additional improvements yielded 20 times more β-carotene than the first variety, thus boosting its nutritional value.

It sounds too good to be true—at least that's what groups such as Greenpeace think. They are concerned that this project will enable agribusinesses like Syngenta to introduce more GMOs into developing countries, where organized resistance is limited. The inventors of Golden Rice counter that only varieties consumed in developing countries will be interbred with Golden Rice, so that it does not become a back-door way to grow GMOs in poor countries for consumption in rich countries. Other

LINKS
CAMBIA
Charles Arntzen
Metabolix
Scientific American
U.S. Patent

STRUCTURES
antibiotic resistance
genes

institutions are also using biotechnology to help farmers in developing countries. The Centre for the Application of Molecular Biology to International Agriculture (CAMBIA) in Melbourne, Australia, is one example.

Biological Plastics A large number of labs around the world are busy engineering plants and bacteria that can produce polyhydroxyalkanoate (PHA), a naturally occurring form of polyester. The first PHA was identified by the French microbiologist Maurice Lemoigne in 1925, and currently more than 100 different PHAs have been documented. Rather than having petroleum as the raw material, PHA can be made from renewable resources such as sugars (via bacterial production) or CO_2 and sunlight (via plant production). In addition to consuming less petroleum for synthesis, these plastics are also biodegradable. So, by using GMOs to produce a less toxic and more degradable plastic, we may be able to keep using plastic products and do less harm to the planet at the same time. One company working on this is Metabolix, and you can learn more from the company's perspective at its web site. You can read more in a *Scientific American* article to see if the hype has real potential.

Pharmaceutical Produce In the early 1990s, potatoes and other edible plants were engineered as delivery mechanisms for vaccinations. By engineering plants to produce viral or bacterial peptides, Charles Arntzen and his research group were able to produce the world's first prescription potato. Initially they used raw potatoes, since heat is known to denature most proteins. However, most people (especially children) prefer cooked potatoes. Arntzen's team tested fried potatoes and found that they retained 50% of their potency. You cannot force most children to eat more than one bite of raw potato, but they will eat a bag full of chips, so the loss of potency could be compensated for by increased consumption. New trials are underway to develop prescription bananas instead of potatoes, because they are grown in more places that need access to vaccines and because everyone likes raw bananas.

Sterile Fruit In 1997, a patent was granted for the technology that became known as "Terminator." Terminator biotechnology produces plants that contain sterile seeds. Monsanto announced a merger with one of the two patent holders (the other being the U.S. Department of Agriculture) with the intention of using Terminator technology to "protect the investment companies make in developing genetically improved crops, as well as possibly providing other agronomic benefits." This may not sound too bad, as we don't think much about eating seeds, but to farmers, the consequences are significant. Since humans began farming

thousands of years ago, farmers have saved some seeds to plant next year. If seed companies sold only plants that were sterile, farmers (and thus everyone) would become dependent on seed companies to supply next year's crops. This would put the world's food supply in the hands of a few multinational seed suppliers, a result that does not appeal to farmers or consumers.

In October 1999, Robert Shapiro, CEO of Monsanto, declared that company's intention "not to commercialize sterile seed technologies, such as the one dubbed 'Terminator.'" However, Monsanto does hold patents on technologies for gene protection that inactivate only the transgene introduced by Monsanto. Shapiro indicated that Monsanto continues to develop these technologies for possible future use. You can read the full patent by going to the U.S. Patent and Trade Office and searching for patent number 5,723,765.

Pest-Resistant Plants Many companies have produced plants that contain toxins normally made in bacteria that are capable of killing insect pests. The best-known example is *Bacillus thuringiensis* toxin (Bt). The attraction of Bt was twofold. First, because the toxin evolved through natural selection, it was believed that insect resistance to Bt would be slower to appear. Second, if the plants produced the pesticide, then farmers would not spray their crops as much, thereby saving money and protecting the environment. There were some concerns that Bt corn killed monarch butterflies, but this was shown to be untrue.

We read in Chapter 2 that some genes can jump from one species to another via horizontal transfer. Many species have exchanged genes through nonsexual mechanisms, making pest-resistant crops potentially hazardous because we neither understand nor control the mechanism for horizontal transfer. Plants are pollinated by animals or wind, so we have no control over their gene flow. Therefore, if a pest-resistant gene were transmitted to a weed, we might be unable to control that weed in agricultural fields. Bioengineered genes are often transferred along with **antibiotic resistance genes** in the process of genetic engineering. What if antibiotic resistance were transmitted to a pathogen that we currently control by antibiotics? In addition, insects are becoming resistant to Bt, negating the original assumption that evolution had produced a compound that would never be ineffective on insects.

Xenotransplants Each day in the United States, 13 people die while waiting for an organ transplant. Xenotransplants are organs taken from one species and put into another. For example, in 1984, Dr. Leonard Bailey transplanted an infant baboon's heart into a newborn human identified as Baby Fae, who lived 20 days with the animal heart in her chest while she waited in vain for a human heart donor. Several companies are genetically engineering pigs to be better donors for organ transplantation to humans. PPL Therapeutics engineered pigs that cannot synthesize an extracellular sugar on glycoproteins that hasten transplant rejection in nonpig recipients. Nextran has modified pigs so their plasma membranes contain human proteins; thus, organs from these pigs are not rejected as quickly. How is this research progressing? Here is what Nextran said:

> Nextran recently concluded a Phase I clinical trial that used transgenic pig livers as an ex vivo (outside the body) support system for patients with acute liver failure. The pig liver was used to bridge the gap between organ failure and obtaining an appropriate human liver for transplantation in these patients. The results were encouraging, and our researchers have learned a great deal about the ability of the transgenic organs to withstand the human immune response and about the development of clinical protocol and various assays for the monitoring of patients in this groundbreaking area. With the data gathered from this ex vivo trial, together with information from our ongoing preclinical research, we plan to develop protocol recommendations for a Phase I in vivo (inside the body) clinical trial and to submit those recommendations to the FDA.

Many scientists think xenotransplants could unleash new human pathogens. All pigs contain several copies of porcine endogenous retroviruses (PERVs). In September 2000, pig tissue was transplanted into immunodeficient mice, which became infected with PERVs. Furthermore, PERVs were able to infect human cells grown in culture. This demonstrated that PERVs were able to infect a new species in vivo and human cells in vitro. Because it would be unethical to perform in vivo PERV experiments in humans, this is as close as we will get under controlled conditions to finding out if pig viruses can infect human patients.

It is very unlikely that we have identified every human pathogen; in fact, new ones are discovered each year (see pages 155–161). We know even less about pig pathogens, so it seems reasonable to think there may be porcine pathogens that could jump to humans given increased exposure. If you were the recipient of a pig organ, you would be given very strong immunosuppressants to prevent your immune system from rejecting the new organ. Under these conditions, a pathogen might be able to mutate into a more human-compatible virus, producing a new human pathogen that could spread rapidly to other humans. Family and healthcare workers close to the patient during the recovery period would be the most likely to become infected first. Although the need for organ donors is compelling, the potential for species-wide harm is equally so. One lesson from genomics is that cellular components are interconnected in unexpected ways, so xenotransplants will present us with unexpected consequences.

DISCOVERY QUESTIONS

LINKS G$P
ELSI
PERVs
US Patent

45. Do you think there are any positive uses for GMOs? Do you think there are any inappropriate uses for GMOs? Cite examples to support your answers.

46. Are there more potential uses for GMOs in developing countries, or is this an immoral way to sneak GMOs into the economies of developed countries?

47. If you were a CEO of a biotech company, what criteria would you use to evaluate which products to develop and which not to develop? Remember, you have a legal obligation to maximize profits for your shareholders.

48. Go to the US Patent Office and search for patent G$P number 20050071891. Is the idea behind Terminator seeds still active? What does this invention do?

Is Genetic Testing Good?

From the very beginning of the HGP, it was recognized that ethical issues should be studied and discussed. In 1990, a portion of the HGP budget was dedicated to ethical issues under the heading of **Ethical Legal and Social Issues** (ELSI). ELSI focuses its efforts in four areas:

- Privacy and fairness in the use and interpretation of genetic information. Activities in this area examine the meaning of genetic information and how to prevent its misinterpretation or misuse.
- Clinical integration of new genetic technologies. These activities examine the effect of genetic testing on individuals, families, and society and inform clinical policies related to genetic testing and counseling.
- Issues surrounding genetics research. Activities in this area focus on informed consent and other research ethics related to the design, conduct, and reporting of genetics research.
- Public and professional education. These activities provide education on genetics and related ELSI issues to health professionals, policy makers, and the general public.

In this section, we examine some real-world examples of how genome sequence information can lead to ethical dilemmas. Again, this is not meant to be an exhaustive list; but it does illustrate issues all of us will face.

Every week a new disease-associated gene is identified, which means a new genetic test could be available soon. If you decide to have children, you may be confronted with a long list of possible genetic tests that could be performed on your developing fetus. Even before conception, you and your partner could be tested for the potential to produce

LINKS
acceptable test
Dorothy Nelkin
Innocence Project
Insurance Committee
Peter Neufeld
Steve Hyman

a child with a genetic condition that most people consider unappealing. Since genomics has revealed to us that most medical conditions are not as simple as one gene—one phenotype, there is a growing interest in collecting genome-wide SNPs from diverse populations. Depending on the individual being tested (i.e., family history, ethnicity) and the genetic test being considered, the value of a particular test will vary (Figure 4.15). Genetic testing is expensive and may subject individuals to societal pressures to conform. Is it a new version of eugenics? Is this the best way to spend limited medical budgets? When are genetic tests reasonable and cost-effective? Who has a right to know the results of your tests? Who has a right to obtain some of your DNA for genetic testing? (Is it acceptable for someone to grab a handful of hair from your brush, since you were going to dispose of those dead cells anyway?) The answers to these questions are less obvious than the ones raised in the section on GMOs.

Some ELSI experts were asked about ethical concerns in a New York forum in October 2000. There was a consensus that most people, including journalists, behavioral scientists, and physicians, either were unaware of the new understanding of complexity in the genome or were confused and believe in a new form of **genetic determinism**. Steve Hyman, director of the National Institutes of Mental Health, remembered a newspaper article stating that the "worry gene" had been cloned, but reminded the audience that complex personality traits involve many genes as well as developmental and environmental influences. Dorothy Nelkin from New York University reported her dismay at the large proportion of physicians who "know nothing about genetics" and often believe human behavior and fate are determined by our genes alone. (Her concerns were supported by a paper in the April 2001 edition of the *British Medical Journal*, emphasizing the need for a multifaceted approach to assist physicians in coping with the new genetic tests available.) Peter Neufeld, director of the Innocence Project in New York, soberly stated that

advances in genetics will not overcome racism and classism. He predicted that insurance companies and the judicial system will increasingly use genomic data: "There is no doubt that this body of work will be abused." Let's look at three examples to illustrate some of their concerns.

British Life Insurance As of 2005, life insurance companies in Britain may use data from a Huntington's disease (HD) genetic test, though new regulations are expected as a five-year moratorium draws to a close. The Genetics and Insurance Committee established by the UK Department of Health recommended that insurance companies consider HD as an acceptable test. If the test result is negative, the companies must use rates based on whole-population actuary tables. The committee recommended that the test not be used for life insurance policies of less than £500,000, often required before an individual can secure a home mortgage loan. However, the committee managed to sidestep what happens to individuals who test positive for HD. Chris Smith, an actuary with the insurance company Swiss Re, stated that people who test positive are usually denied insurance. Insurance companies believe that genetic tests are no different from other risk factors, such as smoking and family pedigrees, that are already used universally to calculate insurance rates. Yet the degree of certainty with HD alleles seems to run counter to the very concept that everyone should be able to obtain insurance based on whole-population statistics.

The specifics may be unique to Britain, but the fundamental principles and the desire of insurance companies to test applicants are global. In the United States, applicants for life insurance submit blood to be tested for HIV, but other tests could be performed with the same sample. What tests should be permitted, and how should the results be used? Doesn't it seem fair that those who test negative for a disease trait should pay less than those who test positive? What if you were the one who tested positive and were denied insurance? These issues affect everyone who wants life and medical insurance.

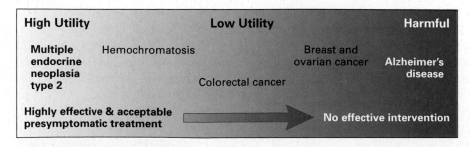

Figure 4.15 Continuum for the utility of a particular genetic test.
Tests for diseases that can be treated easily have a high utility; tests for incurable diseases offer less utility or might even be harmful.

DISCOVERY QUESTIONS

49. If you were asked to write a policy statement for the government, what would you conclude about the use of genetic testing and insurance applications?

50. If you outlawed such tests, would you also outlaw blood tests for HIV? What about family background information? Where would you draw the line?

51. If you permitted such tests, would you include less certain tests such as *BRCA1* and *BRCA2* for breast and ovarian cancer? *ApoE* screening for Alzheimer's? Where would you draw the line? British insurance companies are expected to petition to use *BRCA1* and *2* tests when the moratorium is lifted. Since tests are already in use, would you prohibit denial of insurance but allow the companies to set prices based on the genetic knowledge?

Universal Screening for Cystic Fibrosis Soon, every pregnant woman in America will be informed about the availability of a genetic test for the recessive disease cystic fibrosis (CF). In October 2001, the American College of Obstetricians and Gynecologists and the American College of Medical Genetics recommended that CF screening information be provided to "couples who are in lower-risk racial and ethnic groups and have no known ancestors from high risk groups." This recommendation is made even though the detection rate is only 69% for African Americans, 57% for Hispanics, and 30% for Asian Americans, due to population allele frequencies.

You can be tested for more than 400 genetic diseases or conditions, but none have been implemented nationally until now. CF is the most common genetic disease for Caucasians but not other populations. It is worth noting that the U.S. Census Bureau estimates that by 2100, 60% of the American population will be composed of non-Caucasians, so universal screening may not be appropriate as a national policy. Many studies have been conducted on the cost-benefit ratio of national genetic tests, and almost all recommended against universal screening. However, a physician who does not inform his or her patients about the test may be vulnerable to legal action should a child be born with CF, which happens in the U.S. about once in every 2,500 Caucasian live births and once in every 17,000 African American live births.

For most people, the first time they will learn about CF will be shortly after being told the woman is pregnant. Although CF sounds like a straightforward genetic disease, it is not. Over 900 different CF alleles have been identified, and the consequences are known for about 100 of these alleles. The most common mutation, called ΔF508,

LINKS G&P
American College of Medical Genetics
NIH recommendation

accounts for about 70% of all mutant alleles in Caucasians and another 20 alleles account for an additional 15% of all mutant alleles. Therefore, only 85% of all CF mutations could be detected if all 21 alleles were tested. Physicians would need to explain the test, the relative risk for each partner's ethnic background, the potential outcomes, and the choices facing the couple depending on the outcomes. Furthermore, the long-term prognosis for CF patients has improved in recent years, and childhood death is preventable with aggressive intervention.

The California-based HMO Kaiser Permanente has already conducted a small-scale test under well-controlled conditions. It offered the CF test to all Caucasian women and their partners. If both tested positive, the expecting parents could choose to test the fetus as well. If the fetus proved to be homozygous for CF, the couple had to decide what action to take, if any. Approximately 90% of the fetuses that were homozygous for CF were terminated by abortion. Kaiser screened over 18,000 women during this experiment.

The Kaiser pilot test was well controlled, but would a national program be as uniform or as reliable? A national screening program requires that every physician become fully informed about the possible consequences, and they need to be taught how to communicate this information effectively to a diverse audience. The Kaiser testing was free to patients, but a national program would probably cost patients at least part of the expense, which ranges from $20 to $300. Sixty U.S. laboratories test for different numbers of alleles, from 1 to 87. Different labs will have different error rates, so a single national standard for accuracy seems unlikely.

Although most experts agree that a universal test for CF does not make sense (see National Institute of Health [NIH] recommendation), it is coming anyway. Physicians will be pressured to inform their patients, who may feel pressured to submit to the test. No one knows what will happen as the testing becomes widespread. Physicians will probably rely on well-written booklets to inform patients about CF and genetic testing, but will a booklet be adequate education for everyone? You will become participants in this experiment, and the results will be tabulated as couples face new choices and uncertainties. No doubt you will know people who get tested, or you might be tested. What used to be an abstract hypothetical case will soon be very real and personal. What will you do?

DISCOVERY QUESTIONS

52. Should every pregnant woman be informed about the CF test? Only high-risk Caucasian women? Only those who ask?

53. Currently, U.S. funding is denied to any institution in a developing country that informs women

Math Minute 4.4 **What Does a Positive Test Result Really Mean?**

What does it mean if your test result for a genetic disease (e.g., cystic fibrosis; CF) is positive? What is the probability you actually have the disease, given that the test was positive? You must know the answer to this question before you can respond intelligently to a positive test result. The probability that the positive test result is correct depends on three other probabilities that are commonly associated with disease testing: prevalence, sensitivity, and specificity. To explain these three probabilities and how they are used to determine the probability that your positive result is correct, we consider the case of testing a single Caucasian individual for CF. You can explore other diseases and populations in the Math Minute Discovery Questions.

Prevalence is the overall probability of an individual having the disease, estimated by the proportion of the population with the disease. The prevalence of CF in Caucasians is $P(CF) = 1/2500 = 0.0004$ (assuming no prior knowledge of CF disease alleles in the individual's pedigree).

Sensitivity is the probability that the genetic test will be positive when the individual really has the disease. The sensitivity of the CF genetic test varies with the number of alleles tested. If the 21 common alleles are tested, sensitivity is $P(+ \mid CF) = 0.85$. The vertical bar between "+" and "CF" can be read as "assuming" or "given"; the probability of a positive test result given the person has cystic fibrosis is 85%.

Specificity is the probability that the test will be negative given that the individual does not have the disease. Poor specificity may be caused by technical errors or difficulties in the testing procedures. The specificity of a test can be estimated by studying those who test negative, to see if they later develop the disease. We will suppose that the specificity of the CF test is $P(- \mid no\ CF) = 0.999$. Because a person who does not have CF must test either positive or negative, another way of saying the same thing is $P(+ \mid no\ CF) = 0.001$. In other words, only one out of 1,000 individuals who do not have CF would test positive.

Now we put together prevalence, sensitivity, and specificity to compute $P(CF \mid +)$, the probability of having CF given a positive test result. The formula for calculating $P(CF \mid +)$ is called Bayes' Rule, an important topic in probability and statistics:

$$P(CF \mid +) = \frac{P(+ \mid CF)P(CF)}{P(+ \mid CF)P(CF) + P(+ \mid no\ CF)P(no\ CF)}$$

$$= \frac{0.85 \times 0.0004}{0.85 \times 0.0004 + 0.001 \times 0.9996} \approx 0.2538.$$

Bayes' Rule tells us that if you test positive for CF, there is only about a 25% chance that you actually have the disease. This result is counterintuitive because the test is fairly sensitive (0.85) and very specific (0.999). However, we have proven mathematically that the probability of a correct positive test is much smaller than the sensitivity and specificity might suggest, so we know our calculation is correct.

An alternative way to compute $P(CF \mid +)$ might help you understand Bayes' Rule and see why $P(CF \mid +)$ is lower than you might have expected. Suppose 10 million people are tested for CF. You would expect about 4,000 people in this population to have CF. Of these 4,000, you would expect only 3,400 (4,000 × 0.85) to test positive. In contrast, of the 9,996,000 without CF, you would expect about 9,996 (i.e., 9,996,000 × 0.001) to test positive. Thus, 13,396 people would test positive, but only about 25% (3,400/13,396) of them actually would have CF. Of course, randomness in the population means that none of these numbers are exact, which is why Bayes' Rule uses probabilities instead of specific population counts.

MATH MINUTE DISCOVERY QUESTIONS

1. Who should be more skeptical of a positive CF test, a Caucasian or an African American? Use a calculator to compute the appropriate probabilities

LINKS   **G**♆**P**
deCODE
MATH MINUTES
genetictests.xls

to support your answer. (The prevalence and sensitivity of CF testing in African Americans are given in this section of the text.)

Note: You can use **genetictests.xls** to help you answer Math Minute Discovery Questions 2, 3, and 5.

2. Explore the effects of sensitivity on disease testing by computing $P(CF \mid +)$ when sensitivity is between 0.6 and 0.99, keeping the prevalence 0.0004 and the specificity 0.999.

3. Explore the effect of specificity on disease testing by computing $P(CF \mid +)$ when specificity is between 0.9 and 0.9999, keeping the prevalence 0.0004 and the sensitivity 0.85.

4. Which do you think should be a higher policy priority, finding more CF alleles, or decreasing the false positive rate of CF tests? Explain your answer.

5. Explore the effect of prevalence by computing $P(CF \mid +)$ when prevalence is between 0.0004 and 0.4, keeping the sensitivity 0.85 and the specificity 0.999. For what value of prevalence would $P(CF \mid +)$ be maximized?

---

about abortion options. In the United States, federal money is denied to facilities that perform abortions. What effect might national CF screening have on the supply and demand of abortions in America? Would a similar genetic screen be advisable in developing countries?

54. The CF test may be the first of many national screening programs. What standard should be used to determine which genetic tests are appropriate and which should be limited to high-risk populations?

55. If physicians are unable to adequately educate their patients, due to lack of training or limited time, what steps can be taken to assist them? Is there a national plan to address any anticipated shortcomings?

56. If a woman's insurance company will not pay for a CF test, who should pay for it? If a woman's insurance company does pay for a CF test, who has the right to know the results of her test? The insurance company? The woman's blood relatives? Her partner?

57. Given the guidelines for the utility of genetic testing (see Figure 4.15), what would be your advice about CF screening to a Caucasian couple with no family history of CF? What about a Caucasian man married to an Asian woman when the man had a cousin with CF?

Genomic Diversity Banks and Small Populations

In 1996, Kari Stefansson started a company called deCODE with 20 employees. Its mission was to utilize population genetics to discover new disease-associated genes. This does not sound unique until you learn that the target population is the entire country of Iceland. There are 275,000 living Icelanders; the vast majority are descendants

of a few European explorers who arrived about 1,000 years ago. The population is very homogenous, so finding significant differences that lead to medical conditions will be much easier than in a heterogeneous population. Icelanders have a rich tradition of maintaining family trees, so the whole country can be compiled into one large pedigree. Iceland also has a single healthcare provider, and all medical records are stored in one database. deCODE has purchased the medical records from the government and has correlated family relationships with medical records. Every citizen will give blood to be used to create "genomic fingerprints," unless an individual opts out of the program.

In addition to finding new disease-associated genes, deCODE will be able to track the progress of patients with similar genotypes who are taking newly developed drugs. Roche Pharmaceuticals is collaborating with deCODE to develop data mining software for this purpose. The data collected by deCODE can be sold to other companies that also want to use this island nation to develop new drugs. deCODE employs over 800 people, making it the largest Icelandic corporation. Furthermore, by employing high-tech workers, deCODE reduces Iceland's "brain drain" caused by many of its highly trained citizens seeking employment in other countries.

Although deCODE quickly obtained endorsement from the national government, it does have its critics. Some physicians fear their patients will be less forthcoming with information now that they know their records will be stored in commercial databases. Some feel that patient-physician confidentiality and trust were broken when the records were sold, as the default made informed dissent the only option (5% of the national population chose to opt out). Finally, some think physicians will refuse to comply with a law that requires them to deliver new clinical information to the database, and thus undermine the entire plan.

Stefansson does not agree with his critics, as he explained during an interview with CNN. "Recognize that knowledge is never evil in and of itself. If you run the world by forbidding

LINKS
Estonian Genome Project
Google

new discoveries, you are controlling the world in an unpredictable manner. You are putting yourself in the position of God." However, the Icelandic Supreme Court did not agree with Stefansson when it ruled against the default of inclusion in the database. Citing Iceland's constitution, "Everyone shall enjoy the privacy of his or her life, home, and family," the court ruled in favor of a minor who objected to her dead father's information being included in the database because it would be possible to make inferences about her. The court also ruled that simply removing or encrypting information such as name and address were not sufficient to prevent the identification of individuals in the database. It is not clear where this leaves deCODE legally, though the research continues.

Internationally, the reaction by other investigators has been mixed. In Estonia, formerly a part of the Soviet Union and located in the northwest region of Europe, a group of scientists has formed the Estonian Genome Project (EGP). EGP is targeting the country's 1 million citizens and their centralized health records. Estonian critics of the plan think too much money will be diverted from basic healthcare needs. They also worry that people are not educated enough to make informed decisions and are likely to conform to national policy as they did under Communist rule. Two features distinguish EGP from deCODE: participation is through informed consent only, and EGP will utilize its genomic diversity to find drug targets for genes identified by deCODE as being involved in particular diseases.

DISCOVERY QUESTIONS

58. Did deCODE violate a basic rule of human research, which requires informed consent of the individual rather than informed dissent? Should the people of Iceland share in the profits since their information and DNA are being used to generate profits?

59. deCODE says that DNA from nondissenting donors will be handled under the strictest conditions of anonymity. Do you think this is practical given the small population size and the very complete national pedigree?

60. Everyone has the same health insurance in Iceland, so the motivation to decipher the anonymity might seem minimal. However, there are competing life insurance companies in Iceland. Now can you imagine a motivation to discover the identity of donors?

61. Perform a PubMed search for "Stefansson K," the founder of deCODE, to read his latest research. Is deCODE finding genes associated with complex diseases or conditions?

62. Let's assume you think the deCODE approach is worthwhile, despite the protests of a few fringe physicians and citizens. Does that mean it is also worthwhile for Estonia to launch a similar program, given its differences in economy, health care, and education? Is this a situation where it is acceptable for a developed country but not for a developing country to attempt similar projects?

63. Find the link on the EGP web page that shows you the status of its donor statistics. How many people are in the database? What information is displayed that surprises you?

Who Benefits from Genomic Medicine?

We have all heard of cases in which a health insurance company refused to pay for experimental treatment that might save a patient's life. Genomic tests are expensive and not 100% accurate. Genomics has revealed many secrets, such as biological systems forming complex webs of interconnecting proteins, carbohydrates, lipids, etc. If you and your physician feel a new genomic-based test or treatment might help you, who should pay for it? If gene therapy for a rare disease costs $1 million for each patient, should we all pay higher insurance premiums to cover the treatment of a few patients? How should we spend the limited dollars in the economy that can be used for health care, and who should decide? Are there some medical cases that warrant no treatment even if there is a high probability of a cure? What if there is only a 1% chance of a cure? Or a 0.001% chance? Is it appropriate to allow those who can afford expensive new treatments to be the only recipients? Is it appropriate for only developed countries to benefit from genomic medicine, since developing drugs for poor people will not result in a profit for the pharmaceutical companies?

Medical ethics raises many difficult questions but no easy answers. This section was intended to provoke thought and discussion. Because you know more than most citizens, it is important you think about new discoveries and the consequences of choices. It is important to share your knowledge and facilitate discussions. If you are reading this book, you are in a very small and elite minority. What responsibility do you have, given your place in society?

Are There Simple Applications for Complex Genomes?

In Aldous Huxley's *Brave New World*, human embryos were manipulated to occupy particular niches in society. Today, we are confronted with situations that are eerily similar to Huxley's science-fiction scenario. Here are three cases that are neither fictional nor futuristic: They are all taken from the recent past and deserve our consideration.

New [Fill in the Blank] Gene Discovered! We have all seen headlines touting the discovery of a gene that encodes a complex trait. If you go to Google.com, you can search

SHOULD I GET A GENETIC TEST?

The utility of a genetic test hinges on two main factors: who is asking the question and which test is under consideration. James Evans (an oncologist), Cécile Skrzynia (a genetic counselor), and Wylie Burke (a bioethicist) published a short list to consider when debating the utility of genetic tests.

Increased Utility of Genetic Tests
- High morbidity and mortality of disease.
- Effective but imperfect treatment for the disease.
- High predictive power of the test (i.e., high penetrance for the disease).
- High cost or burdensome nature of nongenetic screening and surveillance methods.
- Preventive measures are expensive or associated with adverse effects.

Decreased Utility of Genetic Tests
- Low morbidity and mortality of disease.
- Highly effective and acceptable treatment.
- Poor predictive power of genetic test (i.e., low penetrance).
- Screening and surveillance methods that are effective, inexpensive, and acceptable.
- Treatment that is inexpensive, effective, and acceptable (e.g., vaccinations).

Most physicians are unprepared to cope with the hundreds of genetic tests that are commercially available. Here are the basic skills needed to integrate successfully genetic testing with primary care:

- Recognition of common inheritance patterns.
- Appreciation of the role ethnicity plays in the assessment of risk for particular diseases.
- Communication and counseling with patients in a nondirective manner.
- Awareness of the inherent limitations in genetic tests and the consequences of testing for insurance coverage.

Unfortunately, most physicians are not prepared to face the challenges of genetic testing, in part because their training was inadequate. It is very difficult to teach medical students the latest information in a field that changes rapidly, and the number of genetic tests available continues to increase. Since keeping track of all this information is daunting, a new model is being evaluated in the UK: the use of community genetic counselors. These counselors act as a liaison between patients and physicians, perform the initial screening to determine when testing is appropriate, and conduct counseling sessions before and after testing. A suggested alternative model—having one physician in a group practice be chosen as a genetics expert—might lead to problems with scheduling and the ability to move from one job to another. More study must be conducted to determine the best way to integrate genetic testing with primary care.

and find reputable web sites that will inform you about a "gay gene," a "smart gene," a "fat gene," a "worry gene," an "Alzheimer's gene," a "cancer gene," even a "fountain-of-youth gene." By this point in the book, you must realize that our genomes are more complex than this. Personality traits and most genetic diseases do not fit neatly into tidy packages such as the "____ gene." As discussed in Chapter 1, even diseases and conditions that are relatively simple, such as muscular dystrophy, are not monogenic. The next time you hear a headline about a new gene, be skeptical, very skeptical.

Maybe you would expect fantastic claims of simple genomic perceptions from tabloids, but they are not alone.

For example, in 1999, ABC News.com ran a story entitled "Where Did the Gay Gene Go?" In this story, ABC traced the history of the "gay gene," first reported in 1993. There is good evidence of a genetic component to homosexuality, but probably only 50% of this complex behavior is genetically based. The original news story fueled arguments on both the left and right sides of the political and ethical spectra and led to confusion, misunderstandings, and misconceptions. In 2000, the Associated Press published a story entitled "Scientists Find Fat Gene," which was different from the other fat genes previously reported (see Chapter 5). CNN's Health Story Page announced "Scientists Discover Cellular 'Fountain-of-Youth.'" CNN was

LINKS
BBC's health page
Leland M. Heller
NBC News
Panos Zavos
Raelians

careful to say in the second paragraph that the finding won't make people any younger or allow them to live forever, but the researchers indicated the gene could keep us healthier longer.

We cannot blame news organizations; their job is to sell a product, and who wants to buy a paper that claims "Investigators Characterize a SNP Associated with Obesity in Mice"? We tend to think physicians would not fall into a similar trap. However, visit the web site of Leland M. Heller, MD and you will find the following medical opinion: "This [serotonin receptor #1] is a genetic worry gene." The sentence does not say *the* worry gene, but to the average reader the difference is unnoticeable. Is oversimplifying biological information just part of a Western tendency to hype science to sell products? The *Indian Express* ran a headline: "Patients of Alzheimer's Take Heart, 'Smart Gene' Is on the Way." It appears that in India, at least, oversimplification is as common as it is in the United States.

We know news organizations are in business to sell products, but at least academics are careful in the way their research is publicized locally, right? In the March/April 1997 issue of *Stanford Today*, you can find the following: "From a vibrating wing's serenade to copulation, the survival of the species depends on one gene," and "Breast Cancer Gene Isolated"—which was neither of the other two "breast cancer genes" called *BRCA1* and *BRCA2*.

The problem of simple headlines seems to arise from time constraints. "Who has enough time to listen to the entire story anymore? Just give me the main point." Genomics is a study of complexity, dynamic interactions, and changing patterns. What we learn does not fit neatly into a 30-second news summary aimed at mass markets. If you take the time to talk with most scientists, they are careful not to hype their findings. On occasion, however, you may hear scientists make claims that sound a bit too bold, too amazing, too . . . simple. The point of these ELSI case studies is to caution you to listen skeptically to news stories, find out more information, and then be prepared to explain any oversimplifications to people who do not understand the technical jargon.

DISCOVERY QUESTIONS

64. Go to any news web site that you prefer (newspapers, TV, magazines, etc.). Search for the word "gene" and discover any genetic hype. Do the stories appear too simplistic? Do scientists share any of the responsibility if a news story distorts their findings?

65. Should we rely on news organizations to be the interpreters of genomic discoveries?

66. Can we blame news organizations for using hyperbole in order to boost sales if we, the consumers, continue to reward such behavior by buying their products?

67. Can you imagine a scenario in which a scientist might intentionally hype his or her findings?

68. Examine the BBC's health page and look for any stories that oversimplify the role genes play in human biology.

Should Humans Be Cloned?

In February 2001, two physicians announced they intended to **clone** a human being by 2003. Panos Zavos from Kentucky and Severino Antinori from Italy discussed the case of a man who had been castrated in an accident but still wanted to have "his own biological children." In an interview with NBC News, Zavos explained his point of view on cloning humans.

Zavos was asked if there would be failed attempts to clone humans, just as there have been for all other mammals. Zavos replied, "Mishaps. And we're ready to face those mishaps. We're going to try to limit those mishaps. We are not perfect, only God is." When asked how he could minimize "mishaps," Zavos said, "Quality control is something we scientists apply all the time. It's Q.C. It's a way of measuring your success. . . . We can find out if that particular embryo carries any hereditary diseases, any genetic deficiencies, and then we decide that embryo should not be implanted." If developmental problems develop after the first trimester, Zavos indicated that parents would be allowed to choose the outcome for these embryos even though it would not reflect well on his team's cloning efforts. "Those are the realities and this is how you learn from any procedure."

Zavos acknowledged there will likely be some miscarriages; when asked how many discarded embryos, miscarriages, or abortions would be allowed for each couple wanting a cloned offspring, his response was, "We hope none. We will be expecting, but obviously we're going to take all precautionary measures to see that this doesn't happen." Zavos sees any potential problems as the price that has to be paid to achieve progress. "It's part of any price that we pay when we develop new technology." In summation, Zavos said, "[Human cloning] is here and it is for the world to harness. We hope to educate the world and tell the world that this technology can be used safely and securely enough without being abused. And that's really what people are afraid of. They don't want monsters. They don't want armies of cloned individuals out there. That's not going to happen."

It is worth mentioning that another group, the religious group called **Raelians**, has announced it too will clone humans. (The Raelians believe that all plants and animals, including humans, are bioengineered products of extraterrestrial beings.) Zavos said that the cloning of a human should not become a race with the Raelians and that his group will follow strict scientific guidelines. His efforts will be carried out in an undisclosed country, by a team of individuals who are not identified, though he suggested the research might be conducted on a boat floating in international waters, because so many countries are outlawing human cloning.

Ian Wilmut, the Scottish principal investigator who cloned the sheep Dolly, and Rudolf Jaenisch (an expert in **epigenetic regulation** of the genome) wrote a strong condemnation of Zavos and Antinori in the March 30, 2001, issue of *Science*. They outlined the number of failed cloning attempts with farm animals and neonatal deaths caused by respiratory and circulatory problems. The few cloned animals that have survived are often oversized, a condition referred to as "large offspring syndrome." Those apparently normal cloned animals may suffer from immune, kidney, or brain malfunctions and may develop abnormally.

Wilmut and Jaenisch proposed that the most likely cause of cloning failure in animals is inadequate genomic reprogramming. This is a series of epigenetic changes (changes that do not alter the DNA sequence) that alter the expression of certain genes (see imprinting in Chapter 2). In fact, expression of normally imprinted genes was significantly altered when cloned mouse or sheep embryos were cultured in vitro prior to implantation. No methods are available to screen embryos for epigenetic alterations that could lead to severe consequences. Therefore, it is impossible to establish substantial quality control for cloned human embryos.

Finally, Wilmut and Jaenisch made the point that attempts to clone a human will repulse the public so much that legitimate efforts to use stem cells for growing new neurons, hearts, and other tissues will suffer. They concluded by saying that attempts to clone humans, while so little is known about animal cloning, would be dangerous and irresponsible.

Many governments, scientific societies, and bioethics communities have uniformly denounced human cloning—and yet the technology is available and inexpensive. The concerns of investigators who first cloned animals is reminiscent of physicists in the 1930s who helped develop atomic energy technology and later urged that atomic weapons not be developed. As Zavos said, the cloning genie is out of the bottle. In addition to technical experts, women must be willing to carry a cloned fetus to term. What thoughts, feelings, and concerns will they have? What effect will the woman's emotional condition have on the hormonal state in which the fetus will grow? If Zavos is successful, will he keep it a secret? Will he publicly disclose his failures too, as one would expect if scientific guidelines were followed? Are there only two groups trying to clone humans, or only two that have gone public? Outside of those immediately involved, it is hard to imagine anyone who supports this enterprise. What childless couple would be willing to attempt this method to produce another human? Nevertheless, Zavos's team has a web site where they claim to be receiving outside support.

It is easy to say you will do something, but much harder to demonstrate success. On their Clonaid.com web site,

in October 2004, the Raelians claimed to have produced 13 cloned children who are alive and well, though no scientific evidence has been offered to support the claim. When asked, Clonaid cited a respect for privacy as the main reason no data will be supplied to answer the questions surrounding these claims. Meanwhile, Zavos reportedly claimed to have cloned a person using a cow's egg and genetic material from three dead people, but the paper was pulled from the scientific journal to which it had been submitted, because of prepublication publicity.

Rather than publishing via press conferences, Korean investigators published their work in a peer-reviewed journal indicating they had indeed cloned human blastocysts and extracted the embryonic stem cells for further research. In May 2005, the Korean team claimed to produce human embryonic stem cells from someone other than the egg donor—an achievement that could open the door for cell or gene therapy that cannot be rejected by the transplant recipient. By the end of 2005, it was clear the Korean data had been fabricated. Search your favorite news source for the latest information on this international story involving research ethics. Perhaps the most promising implication of human embryo cloning may be the therapeutic use of stem cells to cure diseases that today are considered terminal and debilitating. As with xenotransplants, cloning humans is much more complex than simply creating a sphere of dividing cells. The field of systems biology (Chapters 10–12) studies emergent properties that are revealed only when we consider the entire organism.

LINKS G☤P
Clonaid.com
dead people
Dolly
human blastocysts
Ian Wilmut
Rudolf Jaenisch
transplant recipient

DISCOVERY QUESTIONS

69. Do those who developed animal cloning bear any moral responsibility for human cloning? Is there a moral distinction between producing cloned embryonic stem cells and cloned human beings?

70. Is there any situation in which cloning an animal would be beneficial? What if cloning were the only way to raise enough individuals to prevent the extinction of a species?

71. Since cloned animals are not natural products, they can be patented in the U.S. and some other countries. Could a human clone be patented? What if the humans were produced by parthenogenesis?

72. Some think cloning headless humans would be an acceptable way to supply the world's need for organ donors. What do you think?

73. Zavos admits that a cloned human would not be identical to the "parental" donor of the nucleus.

Explain why he is right. Would it be possible to produce cloned twins that would truly be identical?

74. The Raelians claim that the final step to achieving immortality is "personality transfer." The leading Raelian scientist, Dr. Brigitte Boisselier, responded to the UN statement on human cloning, "The real crime against humanity is to deny the right to live forever." Do you think personality transfer to a cloned person is possible?

Summary 4.4

All of us will be affected by genomic research. We will benefit from or suffer the consequences of new genetic tests, GMOs, and those who want to utilize the human genome as objects for simplistic understanding and manipulations. Agronomic companies want to develop better foods to improve the human condition and maximize their profits. Pharmaceutical companies want to create drugs that are genotype-specific. Prolonging life is the objective for much of our medical intervention, and yet longer life may carry unexpected consequences. ELSI questions are important, but the answers are elusive. However, sometimes debating the questions is more important than arriving at the "right" answer.

Chapter 4 Conclusions

Variety is the spice of life, and genomic variation is the engine of evolution. As you saw in Chapters 2 and 3, reference genome sequences provide data that will take decades to decipher. However, many more discoveries lie hidden in the genomic variation found within each species. For example, breeds of dogs and cats exhibit wonderful variations in their phenotypes, not because they have different genes but because they contain allelic variants of the same genes. The variation you see in people is due to genomic variation since we all share the same genes. How can some people who never smoked get lung cancer, while others who smoke heavily stay cancer-free? Why do some people exposed to HIV never develop AIDS? The data are all around us, yet we know very little about the effects of genomic variations.

References

Environmental Case Study

Armbrust, E. V., J. A. Berges, et al. 2004. The genome of the diatom *Thalassiosira pseudonana:* Ecology, evolution, and metabolism. *Science.* 306: 79–86.

Boyd, P. W., C. S. Law, et al. 2004. The decline and fate of an iron-induced subarctic phytoplankton bloom. *Nature.* 428: 549–553.

Jaffe, S. 2004, July 5. Iron seeding just doesn't pay. *Scientist.* 26–27.

López-García, P., F. Rodríguez-Valera, et al. 2001. Unexpected diversity of small eukaryotes in deep-sea Antarctic plankton. *Nature.* 409: 603–607.

Moon-van der Staay, S., R. De Wachter, & D. Vaulot. 2001. Oceanic 18S rDNA sequences from picoplankton reveal unsuspected eukaryotic diversity. *Nature.* 409: 607–610.

Rynearson, T., & E. V. Armbrust. 2000. DNA fingerprinting reveals extensive genetic diversity in a field population of the centric diatom *Ditylum brightwellii. Limnology and Oceanography.* 45(6): 1329–1340.

Rynearson, T., & E. V. Armbrust. 2004. Genetic differentiation among populations of the planktonic marine diatom *Ditylum brightwellii* (Bacillariophyceae). *Journal of Phycology.* 40: 34–43.

Rynearson, T., & E. V. Armbrust. (In press). Maintenance of clonal diversity during a spring bloom of the centric diatom *Ditylum brightwellii. Molecular Ecology.*

Schiermeier, Q. 2004. Iron seeding creates fleeting carbon sink in Southern Ocean. *Nature.* 428: 788.

Human Genomic Variation

Barsh, G. S. 2003. What controls variation in human skin color? *PLoS Biology.* 1(3): 019–022.

Chua, M. V., K. Tsueda, & A. G. Doufas. 2004. Midazolam causes less sedation in volunteers with red hair. *Canadian Journal of Anesthesiology.* 51(1): 25–30.

Daly, M. J., J. D. Rioux, et al. 2001. High-resolution haplotype structure in the human genome. *Nature Genetics.* 29: 229–232.

Dequeker, E., & J.-J. Cassiman. 2000. Genetic testing and quality control in diagnostic laboratories. *Nature Genetics.* 25: 259–260.

Fairbrother, W., D. Holste, et al. 2004. Single nucleotide polymorphism-based validation of exonic splicing enhancers. *PLoS Biology.* 2(9): 1388–1395.

Hill, D. 2002. Misspelled gene tames malaria. *Science.* 298: 1317–1319.

International HapMap Consortium. 2003. The International HapMap Project. *Nature.* 426: 789–796.

International SNP Map Working Group. 2001. A map of human genome sequence variation containing 1.42 million single nucleotide polymorphisms. *Nature.* 409: 928–933.

Kim, U.-K., E. Jorgenson, et al. 2003. Positional cloning of the human quantitative trait locus underlying taste sensitivity to phenylthiocarbamide. *Science.* 299: 1221–1224.

Kruglyak, L. 1999. Genetic isolates: Separate but equal? *PNAS.* 96: 1170–1172.

Liang, M.-H., & L.-J. C. Wong. 1998. Yield of mtDNA mutation analysis in 2,000 patients. *American Journal of Medical Genetics.* 77: 395–400.

Liu, H. X., L. Cartegni, et al. 2001. A mechanism for exon skipping caused by nonsense or missense mutations in BRCA1 and other genes. *Nature Genetics.* 27(1): 55–58.

Maquat, L. 2001. The power of point mutations. *Nature Genetics.* 27: 5–6.

Marshall, E. 2003. Preventing toxicity with a gene test. *Science.* 302: 588–590.

Marth, G. T., I. Korf, et al. 1999. A general approach to single-nucleotide polymorphism discovery. *Nature Genetics.* 23: 452–456.

McLeod, H. L., & W. E. Evans. 2001. Pharmacogenomics: Unlocking the human genome for better drug therapy. *Annual Review of Pharmacology and Toxicology.* 41: 101–121.

Monnens, A. H., L. P. W. J. van den Heuvel, & N. V. A. M. Knoers. 2000. Dominant isolated renal magnesium loss is caused by misrouting of the Na1, K1-ATPase g-subunit. *Nature Genetics.* 26: 265–266.

Morrow, K. J. 2000. Improving patients' responses to drug therapies. *Genetic Engineering News.* 20(10): 18, 75, and 86.

Nakhleh, L. K. 2004. COMP 571 bioinformatics: Sequence analysis. <www.cs.rice.edu/~nakhleh/COMP571/>. Accessed 31 March 2005.

Pennisi, E. 2004. A ruff theory of evolution: Gene stutters drive dog shape. *Science.* 306: 204.

Roberts, L. 2000. SNP mappers confront reality and find it daunting. *Science.* 287: 1898–1899.

Sebat, J., B. Lakshmi, et al. 2004. Large-scale copy number polymorphism in the human genome. *Science.* 305: 525–528.

Sordella, R., D. W. Bell, et al. 2004. Gefitinib-sensitizing EGFR mutations in lung cancer activate anti-apoptotic pathways. *Science.* 305: 1163–1167.

Splawski, I., K. W. Timothy, et al. 2002. Variant of SCN5A sodium channel implicated in risk of cardiac arrhythmia. *Science.* 297: 1333–1336.

Weiss, K. M., & J. D. Terwilliger. 2000. How many diseases does it take to map a gene with SNPs? *Nature Genetics.* 26: 151–157.

Pharmacogenomics

Evans, W. E., & J. A. Johnson. 2001. Pharmacogenomics: The inherited basis for interindividual differences in drug response. *Annual Review of Human Genetics.* 2: 9–39.

Evans, W. E., & M. V. Relling. 2004. Moving towards individualized medicine with pharmacogenomics. *Nature.* 429: 464–468.

Ho, P. C., K. Ghose, et al. 2000. Effect of grapefruit juice on pharmacokinetics and pharmacodynamics of verapamil enantiomers in healthy volunteers. *European Journal of Clinical Pharmacology.* 56: 693–698.

Kanazawa, S., T. Ohkubo, & K. Sugawara. 2001. The effects of grapefruit juice on the pharmacokinetics of erythromycin. *European Journal of Clinical Pharmacology.* 56: 799–803.

Lee, M., D. I. Min, et al. 2001. Effect of grapefruit juice on pharmacokinetics of microemulsion cyclosporine in African American subjects compared with Caucasian subjects: Does ethnic difference matter? *Journal of Clinical Pharmacology.* 41: 317–323.

Lin, Y. 2001. Double-whammy: Food-drug interactions. *Northwest Science and Technology.* Winter: 42–43.

Spahn-Langguth, H., & P. Langguth. 2001. Grapefruit juice enhances intestinal absorption of the P-glycoprotein substrate talinolol. *European Journal of Pharmacological Science.* 12: 361–367.

Aging

Cooper, V. S., & R. E. Lenski. 2000. The population genetics of ecological specialization in evolving *Escherichia coli* populations. *Nature.* 407: 736–739.

Dillin, A., A.-L. Hsu, et al. 2002. Rates of behavior and aging specified by mitochondrial function during development. *Science.* 298: 2398–2401.

Guarente, L. 1997. What makes us tick? *Science.* 275: 943–944.

Guarente, L., & C. Kenyon. 2000. Genetic pathways that regulate aging in model organisms. *Nature.* 408: 255–262.

Kirkwood, T. B., & S. N. Austad. 2000. Why do we age? *Nature.* 408: 233–238.

Kishony, R., & S. Leibler. 2003. Environmental stresses can alleviate the average deleterious effects of mutations. *Journal of Biology.* 2(2): 14.1–14.9.

Lanza, R. P., J. B. Cibelli, et al. 2000. Extension of cell life-span and telomere length in animals cloned from senescent somatic cells. *Science.* 288: 665–668.

Lee, S. S., S. Kennedy, et al. 2003. Daf-16 targets genes that control *C. elegans* lifespan and metabolism. *Science.* 300: 644–647.

Ly, D. H., D. J. Lockhart, et al. 2000. Mitotic misregulation and human aging. *Science.* 287: 2486–2492.

McFadden, K. 2000. The week's famous and infamous women: Centenarian sisters Bessie (1891–1995) & Sadie (1890–1999) Delany. <http://writetools.com/women/stories/delany_bessie.html>. Accessed 11 June 2001.

Michikawa, Y., F. Mazzucchelli, et al. 1999. Aging-dependent large accumulation of point mutations in the human mtDNA control region for replication. *Science.* 286: 774–779.

Pennisi, E. 1998. Single gene controls fruit fly life-span. *Science.* 282: 856.

Rogina, B., R. A. Reenan, et al. 2000. Extended life-span conferred by cotransporter gene mutations in *Drosophila. Science.* 290: 2137–2140.

Stewart, E. J., R. Madden, et al. Aging and death in an organism that reproduces by morphologically symmetric division. *PLoS Biology.* 3(2): 295–300.

Walker, D. W., G. McColl, et al. 2000. Evolution of lifespan in *C. elegans. Nature.* 405: 296–297.

Ethical Consequences of Genomic Variations

Abbot, A. 2004. Icelandic database shelved as court judges privacy in peril. *Nature.* 429: 118.

Beauchamp, T. L., & L. Walters, eds. *Contemporary issues in bioethics.* 6th ed. Australia: Thomson.

Bowen, D., T. A. Rocheleau, et al. 1998. Insecticidal toxins from the bacterium *Photorhabdus luminescens. Science.* 280: 2129–2132.

Breast cancer gene isolated. 1997. *Stanford Today.* March/April: 32.

Celera. 2001. Test case for genetic testing. <http://www.celera.com/genomics/news/articles/03_01/Test_case_CF.cfm>. Accessed 11 June 2001.

Chamberlain, C. 1999. Where did the gay gene go? <http://abcnews.go.com/sections/living/DailyNews/gaygene990422. html>. Accessed 11 June 2001.

Cibelli, J. B., S. L. Stice, et al. 1998. Cloned transgenic calves produced from nonquiescent fetal fibroblasts. *Science.* 280: 1256–1258.

CNN.com. 1998. Scientists discover cellular "fountain of youth." <http://www.cnn.com/HEALTH/9801/13/life.extention/>. Accessed 11 June 2001.

CNN.com. 2000. Iceland sells its medical records, pitting privacy against greater good. <http://www.cnn.com/2000/WORLD/europe/03/03/iceland.genes/index.html> Accessed 11 June 2001.

deCODE Genetics. 2.1. Company web site. <http://www.decode.com/company/history/>. Accessed 11 June 2001.

Eastman, D. 2004. Huge Icelandic genealogy database receives legal setback. <http://eogn.typepad.com/eastmans_online_genealogy/2004/07/huge_icelandic_.html>. Accessed 5 April 2005.

Editorial. 2000. Testing times. *Nature Genetics.* 26: 251–252.

Emery, J., & S. Hayflick. 2001. The challenge of integrating genetic medicine into primary care. *British Medical Journal.* 322: 1027–1030.

Environment News Service. 1999. Monsanto sacks sterile seeds. <http://www.wired.com/news/lycos/0,1306,31786,oo.html>. Accessed 11 June 2001.

Evans, J. P., C. Skrzynia, & W. Burke. 2001. The complexities of predictive genetic testing. *British Medical Journal.* 322: 1052–1056.

Finkel, E. 1999. Australian center develops tools for developing world. *Science.* 285: 1481–1483.

Frank, L. 1999. Storm brews over gene bank of Estonian population. *Science.* 286: 1262–1263.

Gewin, V. 2003. Genetically modified corn—environmental benefits and risks. *PLoS Biology.* 1(1): 15–18.

Heller, L. M. 2000. Is BuSpar really safe? <http://www.biologicalunhappiness.com/AskDoc/IsBuSpar.htm>. Accessed 11 June 2001.

Hwang, W., S. I. Roh, et al. 2005. Patient-specific embryonic stem cells derived from human SCNT blastocysts. *Science.* 308: 1777–1783.

The Indian Express. Patients of Alzheimer's take heart, "smart gene" is on the way. 2000. <http://www.expressindia.com/ie/daily/20000621/iin21008.html>. Accessed 11 June 2001.

Jaenisch, R., & I. Wilmut. 2001. Don't clone humans! *Science.* 291: 2552.

Kahn, P. 1996. Coming to grips with genes and risk. *Science.* 274: 496–498.

Kono, T., Y. Obata, et al. 2004. Birth of parthenogenetic mice that can develop to adulthood. *Nature.* 428: 860–863.

NBC Dateline. 2001. Brave new world? <http://www.msnbc.com/news/525661.asp?cp1=1>. Accessed 11 June 2001.

Nowlan, W. 2002. A rational view of insurance and genetic discrimination. *Science.* 297: 195–196.

Ochman, H., J. G. Lawrence, & E. A. Groisman. 2000. Lateral gene transfer and the nature of bacterial innovation. *Nature.* 405: 299–304.

Paine, J., C. A. Shipton, et al. 2005. Improving the nutritional value of golden rice through increased pro-vitamin A content. *Nature Biotechnology.* 23: 482–487.

Phelps, C. J., C. Koike, et al. 2003. Production of α1,3-galactosyl-transferase-deficient pigs. *Science.* 299: 411–414.

Polejaeva, I. A., S.-H. Chen, et al. 2000. Cloned pigs produced by nuclear transfer from adult somatic cells. *Nature.* 407: 86–90.

Ritter, M. 2000. Scientists find fat gene. <http://abcnews.go.com/sections/science/DailyNews/fatgene000327.html>. Accessed 11 June 2001.

Schiermeier, Q. 2001. Designer rice to combat diet deficiencies makes its debut. *Nature.* 409: 551.

Sex and the single gene. 1997. *Stanford Today.* March/April: 31–32.

Steinberg, D. 2000. N.Y. panel explores genomics issues. *Scientist.* 14(20): 9.

Tian, X. C., J. Xu, & X. Yang. 2000. Normal telomere lengths found in cloned cattle. *Nature Genetics.* 26: 272–273.

Uehling, M. D. 2003. Decoding Estonia. <http://www.bio-itworld.com/archive/021003/decoding.html>. Accessed 5 April 2005.

van der Laan, L. J. W., C. Lockey, et al. 2000. Infection by porcine endogenous retrovirus after islet xenotransplantation in SCID mice. 2000. *Nature.* 407: 90–94.

Wikcken, B. 2003. Ethical issues in newborn screening and the impact of new technologies. *European Journal of Pediatrics.* 162: S62–S66.

Wu, C. 1997. Weight control for bacterial plastic. *Science News.* 151: 23.

Wuethrich, Bernice. 2003. Chasing the fickle swine flu. *Science.* 299: 1505–1520.

Ye, X., S. al-Babili, et al. 2000. Engineering the provitamin A (β-carotene) biosynthetic pathway into (carotenoid-free) rice endosperm. *Science.* 287: 303–305.

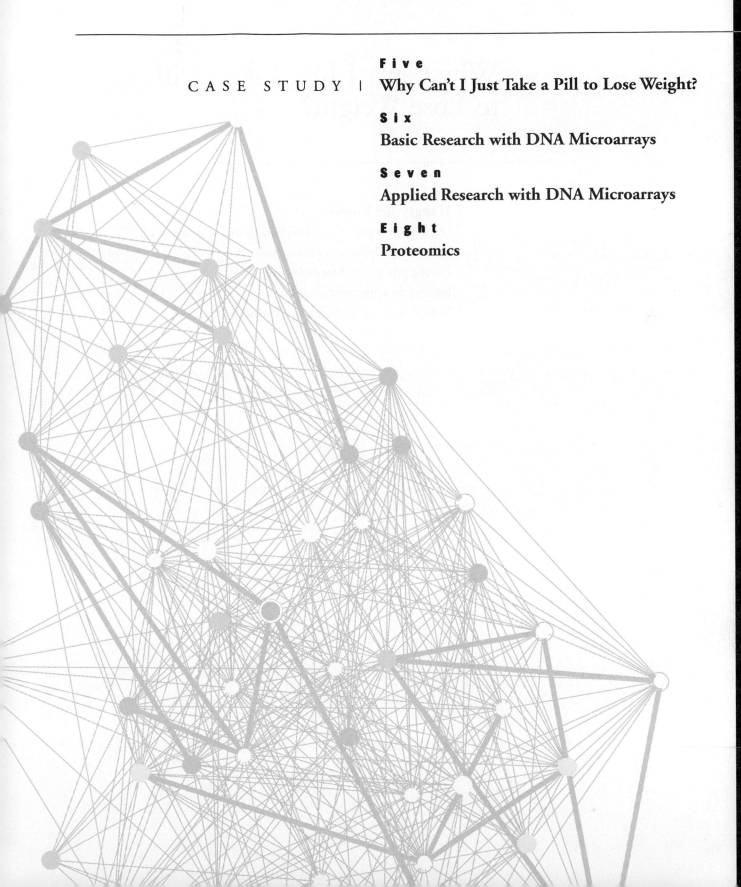

Genome Expression

Unit Two

CASE STUDY

Why Can't I Just Take a Pill to Lose Weight?

Hungry for Knowledge

Discover the complexity of a supposedly simple genetic trait.

Extract useful information from databases.

Perceive proteins in three dimensions.

Improve data interpretation skills.

Analyze circuit diagrams for new insights.

In this chapter, we use a combination of a case study and a short story to immerse you in a genomic perspective. Most students assume that each gene encodes a single protein and that each protein has one function that affects one phenotype. This Mendelian outlook works well when you are learning the fundamental rules of inheritance, but is too simplistic to fully explain more complex traits such as body weight. Each person has both a genetic and a behavioral component to his or her weight. This chapter examines the genetic and behavioral control of weight and what you can and cannot do to regulate it. While you read the story, put yourself in the lead role, and keep an open mind as the data are presented. Chapters 6, 7, and 8 will build on the genomic perspective developed by considering the complex phenotype of body weight.

Hungry for Knowledge
Saturday, 21 October. 7:30 a.m.

It's a cool fall Saturday morning, and you want to sleep late. You love this time of the week when school pressures are temporarily lifted and you can do what you want. Just as you adjust your pillow for a bit more sleep, the phone rings. You answer and hear the chipper voice of your grandma Pauline. She lived with you when you were a child and is like a second mother. Pauline came to America from Germany between the two world wars. She arrived at Ellis Island in 1924 speaking no English, but was well schooled and very smart. You answer the phone with a groggy voice that sounds very sleepy and somewhat angry at being disturbed. Your grandma responds with one of her classic phrases and remnants of her German heritage, "Only vimps sleep in!" This makes you smile and clear your throat to sound wide awake, though she knows better.

Grandma is calling because of an article she just read in the Science section of the *New York Times* describing a protein that can make mice lose weight. She wants to know what protein this is so she can add it to her diet. Since you are a biology major, Grandma likes to ask you about all kinds of things. "Why are my plants turning yellow?" was last month's question. Today, she asks a question that you cannot answer right away (certainly not before breakfast), so you promise to investigate the matter and call her back.

Library Opens at 8:30 a.m. on Saturdays

You don't often visit the library on Saturday morning, but for Grandma Pauline you make an exception—you would do anything for her, including Saturday morning library research. You go to PubMed and do a search for "obesity gene" and get a lot of hits. You scan the titles and quickly find a word that reappears often: **leptin**.

You decide to start at the beginning and find the first paper that talks about leptin. It is a 1994 paper written by Jeffrey Friedman's lab at Rockefeller University. You know this is a big-splash paper since it made the cover of the journal *Nature*. You begin your reading with a news summary in the same issue about the leptin discovery. You learn:

LINKS
Jackson Labs
Jeffrey Friedman
ob
PubMed

1. A mutant strain of mice isolated in the 1950s was called *ob* for obese. The obese phenotype is observed only in homozygous mice. These mice are sterile and often get sick. (To find out what it would take to purchase a pair of these mice for research purposes, browse Jackson Labs in Maine and search for "leptin.")
2. Localized damage to the brain region called the hypothalamus also causes an obese phenotype (Figure 5.1).
3. Surgical removal of fat from an individual produces increased food consumption and fat storage to replace what was removed. This led to the coining of the term **lipostat**, meaning a lipid (or fat) regulator that works similarly to the way a thermostat regulates temperature.
4. If fat is removed from one wild-type (*wt*) animal and put into another *wt* individual, the recipient will lose the weight and revert to its original size.
5. If a homozygous *ob* mouse (*ob/ob*) is surgically joined to a *wt* mouse so that they co-circulate blood, the *ob* mouse will eat less and lose weight. The *wt* mouse is unaffected.
6. Proteins can be extracted from fat in *wt* mice and injected into an *ob* mouse. The injected *ob* mouse will eat less and lose weight.
7. There is another mouse strain called *db* (for diabetic) that has the obese phenotype when homozygous for the *db* mutation.

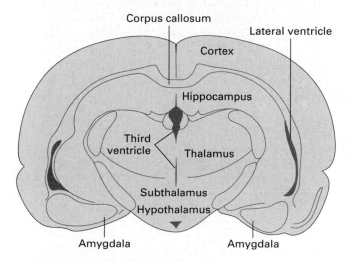

Figure 5.1 Cross-sectional view of a rat brain.
The hypothalamus is located near the bottom of the brain, in about the middle from front to back.

LINKS
Comparative Map
METHODS
chromosomal walk
Northern blot

8. When an *ob/ob* mouse is surgically joined to a *db/db* mouse, the *ob* mouse loses weight but the *db* mouse does not.
9. When a *db* mouse is joined to a *wt* mouse, the *wt* mouse will refuse to eat and starve to death.
10. If you inject a *db* mouse with the *wt* fat protein extract, it does not lose any weight.

DISCOVERY QUESTION

1. What conclusions can you make from the ten preceding observations?

Building a Model for Weight Homeostasis

After reading these facts in the library, you list your conclusions. You then construct a very simple model so you can keep track of what's going on (Figure 5.2). You know Grandma will want to know everything, and she likes pictures.

Preliminary Conclusions:

1. *ob* and *db* are recessive mutations, so these mice fail to make something produced in *wt* mice.
2. The brain has a role in maintaining weight.
3. An individual has an "ideal fat content;" this level is maintained even if fat cells are removed or added surgically. Lipostats explain why most people who go on diets or have liposuction return to their original weight. For example, Grandma has tried every fad diet, but always returns to her previous plump size.
4. Because joining an *ob* mouse with a *wt* mouse **complements** the mutation (makes the *ob* mouse normal size), there must be something in *wt* blood that can cause an animal to lose weight.
5. *db* and *ob* are not alleles of one gene, because when surgically joined to a *wt* mouse, the *ob* mouse loses weight but the *db* mouse does not.

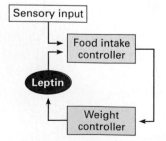

Figure 5.2 Your initial model for weight control.
Sensory input includes both environmental and physical stimuli.

6. *db* mice produce something (the same as *wt*?) that causes weight loss in *ob* mice. However, the blood of *ob* mice does not contain anything that can complement the *db* mutation.
7. There is a protein in fat cells that causes an individual to lose weight. Is this the same protein produced by *db* but lacking in *ob* mice?
8. If there is a protein that controls weight, it does not work in *db* mice. Therefore, the *db* mutation might lack a receptor for this protein.
9. If *db* mice can cause *wt* mice to starve, then the *db* mouse produces an excess amount of this weight-controlling protein, and it works "too well" in *wt* mice.

Cloning the Leptin Gene

Since the phenotype for *ob* has been known for a long time, Friedman's group started with some classical breeding experiments to map the genetic locus for *ob*. It is located on mouse chromosome 6, between loci *Pax4* and *Ptn*. The investigators used probes to *Pax4* and started a chromosomal walk toward *Ptn*.

DISCOVERY QUESTION

2. Go to the Chromosome Comparative Map, which aligns the rat, mouse, and human chromosomes, so you can choose one species' chromosome and find out where the orthologs are in the other species. Under the column heading "Rat and Human compared to Mouse," choose chromosome "6" to align these three chromosomes.
   a. Locate "Ptn" by typing the gene abbreviation into the search box—but do not hit "return"; click on the button that says "Find in This View." Enter the values "27M" and "38M" in the two text boxes of the far left frame to view these 11 Mb of DNA. Did you see the leptin gene (called *Lep*) in between the flanking genes *Ptn* and *Pax4*? On which chromosome is the human leptin gene located? Is *Lep* closer to *Ptn* or *Pax4*? Are all the genes evenly spaced in this region?
   b. Click on "MGI" to see a summary of *Lep*. On this summary page, click on "Mammalian Orthology" to see a multispecies comparison. Which species is the most similar to mouse leptin? Do you think this site is complete, or are some data missing? Explain your answer.

The Friedman lab isolated a portion of genomic DNA, which they called 2G7, that was used as a probe for a Northern blot using mRNA from many tissues in *wt*

LINKS   G&P
Genbank
PDB
METHODS
QuickPDB

C A S E   S T U D Y

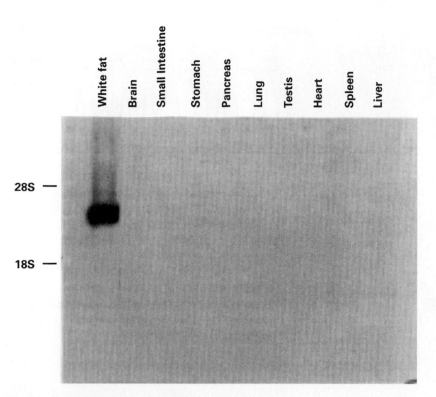

Figure 5.3 Northern blot using 2G7 DNA as probe.
RNA was isolated from *wt* mouse tissues as indicated. 28S and 18S are ribosomal RNA
bands used as molecular weight markers.

mice (Figure 5.3). The investigators also probed the same
blot with a probe specific for actin, and a band was seen in
each lane.

DISCOVERY QUESTIONS

3. What would be the purpose of an actin probe in
   Figure 5.3?
4. What can you conclude about 2G7 DNA? Does
   2G7 encode the protein extracted from fat that
   can cause mice to lose weight? Support your conclusion with data.

    There are two strains of *ob* mice, so Friedman's group
wanted to see if 2G7 was expressed in either strain (Figure
5.4). Given that *ob* is a recessive mutation, they did not
expect to see the leptin mRNA in these mice. Fat from several strains were tested: SM/Ckc-+Dac+/+, which is a
lean mouse, as indicated by the homozygous *ob* +/+ genotype; SM/Ckc-+Dac *ob*²ᴶ/*ob*²ᴶ, which is an obese mouse
homozygous for the *ob* allele called 2J; C57BL/6J +/+,
which is a lean mouse and the *wt* parental stock for the
original *ob* strain found in the 1950s; and C57BL/6J *ob/ob*,
which is the original *ob* mutant strain.

DISCOVERY QUESTIONS

5. Which strains transcribe 2G7 DNA? Do you
   think 2G7 is the *ob* gene? Support your conclusion with data.
6. Figure 5.4 is a Northern blot that measures transcripts only. What kind of data are needed to see
   if functional leptin protein is produced in these
   mice?
7. Search Genbank with the accession number
   NM_008493. How long is the mRNA, and how
   long is the protein? Save the protein sequence for
   Discovery Question 9.
8. What kind of posttranslational processing must
   happen to the leptin protein for it to be secreted
   into the serum?
9. Go to PDB and enter leptin's ID (1AX8). Click
   on "View Structure" and then select
   "QuickPDB." What is the first amino acid in the
   protein structure, as compared to the sequence
   from Discovery Question 7? Rotate the protein
   and notice it has two separate parts. Click on 3I
   and 24I. This should highlight the two ends of
   the separate alpha helix. Begin clicking on amino

METHODS
DNA sequence
Southern blot
STRUCTURES
leptin structure

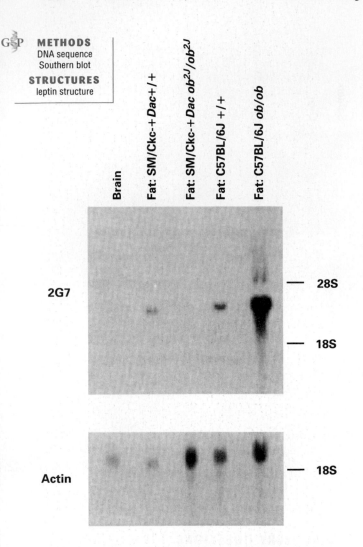

Figure 5.4 Northern blot of fat RNA from various strains of mice.

The probes were either 2G7 (see Figure 5.3) or actin cDNA. 28S and 18S are ribosomal RNA bands used as molecular weights. RNA was extracted from fat or brain as indicated. The source of the brain mRNA was a *wt* mouse.

acids to the right of number 24 until you highlight more of the structure. How many amino acids are missing, and how can this be?

10. Go to the leptin structure page to see more of leptin's structure. This web site adds visual information that may provide insights for the preceding questions.

When the investigators compared the DNA sequence of *wt* mice with *ob* mice, they found a significant point mutation (Figure 5.5). Friedman's group understood the mechanism for two different alleles of *ob*. SM/Ckc-+Dac *ob²ᴶ/ob²ᴶ* does not make any mRNA (see Figure 5.4) and C57BL/6J *ob/ob* makes truncated versions (see Figure 5.5).

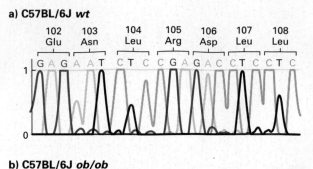

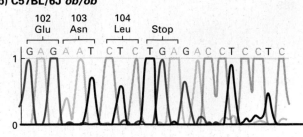

Figure 5.5 Genetic cause of obese phenotype.
Sequencing chromatograms of **a)** *wt* C57BL/6J and **b)** *ob/ob* C57BL/6J mouse leptin coding regions.

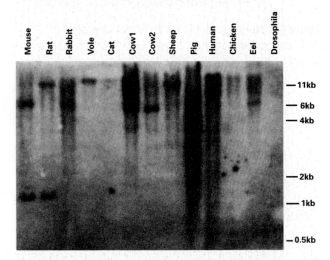

Figure 5.6 Conservation of leptin gene in animals.
Southern blot using genomic DNA from a wide range of species, as indicated. The probe was made from *ob* cDNA.

The investigators wanted to determine if the leptin gene was conserved in other species. They performed a form of Southern blot that is jokingly called a zooblot, because the DNA from many different animals is probed on the same blot (Figure 5.6).

DISCOVERY QUESTIONS

11. Describe the mutation found in C57BL/6J mice.
12. How does this mutation help you understand the Northern blot data from Figure 5.4? What can you conclude about the *ob²ᴶ* mice?

13. If you made an antibody against leptin that bound upstream of leucine number 104, would you see any protein in C57BL/6J *ob* serum? Explain your answer.
14. How widespread is the leptin gene? Can you conclude whether the cat or *Drosophila* genomes contain leptin genes? Explain your answer.
15. Has Friedman demonstrated that 2G7 encodes the leptin protein?

Functional Tests for Leptin

As you read these papers in the library, you are uncertain whether Friedman's group cloned the leptin gene or a different gene by mistake. The definitive proof must always be a functional test. If they could make the 2G7-encoded protein in bacteria, inject this into *ob* mice, and see a weight loss, that would be a good test. That's what they did (Figure 5.7).

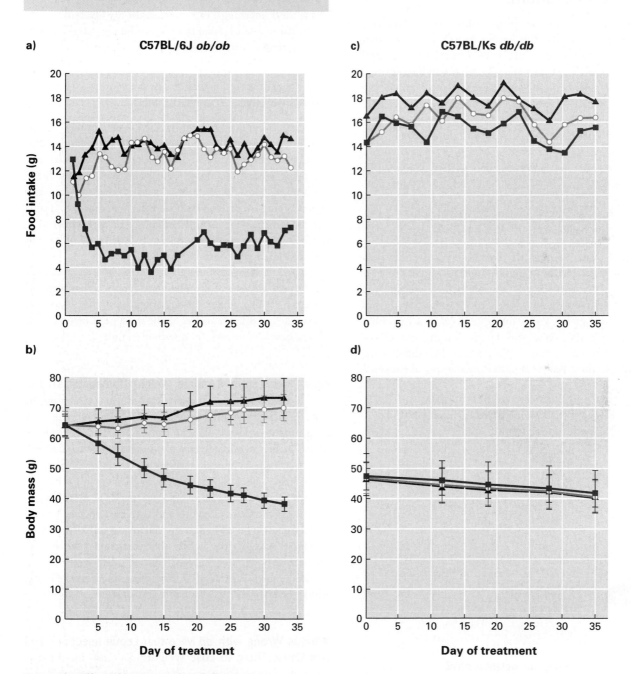

Figure 5.7 Effect of leptin on *ob* and *db* mice.
a) and **b)** Ten homozygous *ob* mice, and **c)** and **d)** 10 homozygous *db* mice received daily intraperitoneal injections of 2G7-encoded protein (filled squares) or saline (open circles), or no injections (filled triangles).

DISCOVERY QUESTIONS

16. Does 2G7 encode leptin?

17. What can you conclude about the *db* mice?

18. Predict what would happen to *ob* mice if they were taken off the leptin treatment.

Time to Visit Grandma

You have spent Saturday morning reading about leptin and how it controls weight. You have a couple of figures that you photocopied for Grandma so she can see the data, and your revised model that incorporates what you have learned this morning (Figure 5.8). You are excited about going over to her house, for two reasons. First, you are happy to help her understand some science, because she helped raise you. Second, you know she is going to make lunch for you and she is a great cook.

As you walk over to her house, you see this headline in a local paper:

Fat Gene Found: Amgen Pays $20 Million
for Rights to Protein to Cure Obesity

You chuckle rather confidently as you buy a copy. Now you have the definitive answer for Grandma: All she has to do is take a leptin pill and she will be able to lose weight. You explain the data and your models to Grandma, and from her tone, you can tell she is impressed. All you can think about now are the wonderful smells coming out of the kitchen. You stand up to get ready to eat, but she says she has a few questions that should be easy for you to answer.

1. How does leptin regulate weight homeostasis?
2. What is wrong with *db* mice that injected leptin did not cause them to lose weight?

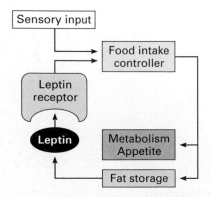

Figure 5.8 Revised model for weight control.
Notice that the leptin receptor has been added. In addition, the controller now produces two separate signals: one for metabolism/appetite and the other for weight gained through fat production.

3. Are all obese people either *ob* or *db* mutants?
4. Why are *ob* mice sterile? Why do they get sick?
5. What is the link between obesity and diabetes? Does leptin control this, too?
6. Can leptin pills or injections cure all cases of obesity?
7. What foods are rich in leptin, and can I eat these to control my weight?
8. What role does the brain play in weight homeostasis?
9. Is there anything I can do to adjust my lipostat?
10. What would you like to have for dinner tonight when you return with all these answers?

Grandma Gives You Homework!

The lunch Grandma fed you was delicious, though for the first time you were consciously aware of its high fat content. All this research into the circuitry of weight homeostasis has made you think more about your own weight. Since Grandma is rather large, and you realize that obesity has a genetic component, you wonder what role diet has, if any. But as you eat your second helping of dessert, you realize that she is asking too many questions and you will have to spend more of your Saturday in the library.

You made a tactical mistake. Rather than searching out the most recent information, you stopped after reading the original leptin cloning paper. Although it gave you the foundation you needed, it did not answer all the questions Grandma has. So you decide to answer her questions in order.

How Does Leptin Regulate Weight Homeostasis? The term leptin comes from the Greek word *leptos*, which means thin. Associating leptin with the word "thin" will help you remember how leptin works. Each person has a lipostat set for his or her own ideal percentage of body fat (Figure 5.9). Leptin is made by fat cells; the more fat you have, the more leptin you make. When a person gains weight through increased fat storage, he or she will produce more leptin, resulting in lowered appetite and increased metabolism so that less fat will be created. If the genetic lipostat is unaltered, temporary weight loss is just that: temporary. When the amount of body fat is reduced (through dietary restriction, surgical removal, or starvation), leptin production is reduced, which results in increased appetite and fat storage.

What Is Wrong with *db* Mice that Leptin Injections Did Not Cause Them to Lose Weight? *db* stands for the diabetes phenotype, and these obese mice have been used as a model for diabetes research for many years. The molecular cause of the *db* mutation was identified in 1995 by a group from **Millennium Pharmaceuticals** Inc. Because it had been

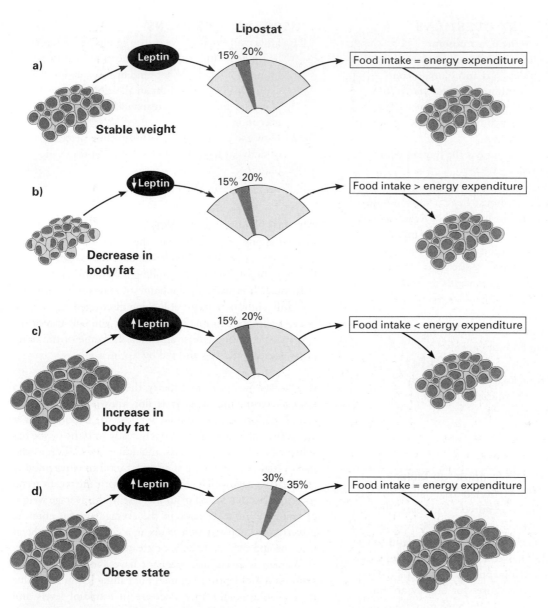

Figure 5.9 Lipostat controls body fat content.

a) Leptin affects the lipostat (scale) in a feedback loop regulating fat mass. At an individual's stable weight (shown as 15–20% fat for a nonobese subject), the amount of circulating leptin elicits a state in which food intake equals energy expenditure. Lipostats respond **b)** to a loss in body fat, and **c)** an increase of fat. **d)** A lipostat in an obese person has a higher set point for body fat (30–35% fat).

documented that *ob* and *db* mice had almost identical phenotypes, it was believed that *db* was probably the leptin receptor. In 1996, the same investigators confirmed that *db* mice have defective leptin receptors. This explains why *db* mice do not lose weight when injected with leptin.

The leptin receptor is produced in three forms that are all derived from a single gene: a long, a short, and a very short form. The long form has a long cytoplasmic domain that allows it to initiate signal transduction when leptin binds. The short form lacks the cytoplasmic domain but is still an integral membrane protein. This form is expressed mostly in the choroid plexus of the brain and is believed to act as a transporter protein that allows leptin to cross the blood-brain barrier. This would explain how leptin can be made in fat cells and found in the cerebrospinal fluid. The very short form lacks the **transmembrane domain** and therefore is secreted into the blood. The very short form is produced by pregnant women, allowing them to increase their fat stores to help ensure a full-term pregnancy. By secreting a **soluble receptor**, the woman's cells sense less leptin and thus increase food intake and store more fat (more about this later).

DISCOVERY QUESTIONS

19. Go to the Chromosome Comparative Map, select chromosome 4 under "Rat and Human compared to Mouse Chromosome," and search for the leptin receptor (*Lepr*). If *Lepr* is not visible, enter "Lepr" in the search field and click on the "Find in this View" button.

a. On which chromosome is the human leptin receptor located?

b. Perform a BLAST2 (choose the program "blastp") with the mouse *Lepr* (accession number P48356) and the human *LEPR* (accession number P48357) by entering these numbers into boxes for accession numbers. What percentage of the receptor sequence is shared between human and mouse? Is the intracellular or extracellular domain larger? How many times does the receptor span the plasma membrane? Do you think human or mouse leptin could work in both organisms?

20. How could three different forms of leptin receptor be derived from a single gene? Use the BLAST2 alignment annotation to help you determine this answer.

Are All Obese People Either *ob* or *db* Mutants?

One set of human twins was identified that failed to make leptin and thus had the *ob* genotype. However, they did not suffer from hypothermia or diabetes the way *ob* mice do. These twins were successfully treated with leptin injections. From 1995 to 2000, over a dozen key components in the neurocircuitry underlying appetite control were discovered. Therefore, the genetic causes for obesity are numerous. Five percent of obese children have mutations in the melanocortin circuit, making it the most common set of genetic defects associated with obesity. (The brain circuitry is discussed later.)

Most obese people are resistant to leptin to varying degrees. In addition, serum leptin levels vary among people who have the same percentages of body fat. These data indicate that a variety of factors, both genetic and environmental, probably contribute to people's sensitivity to leptin. For example, it is known that a high-fat diet can cause mice to become insensitive to leptin and therefore obese. It has also been shown that the set point of a lipostat can be lowered by exercise. Exercise and a low-fat diet are the two most important environmental factors that can alter a lipostat's set point. Leptin levels change during the day, and in most people there is a significant rise in serum leptin at midnight and 4 a.m. When the leptin cycle is abnormal, it may contribute to a behavior pattern that leads to obesity due to an inadequate suppression of nighttime appetite.

DISCOVERY QUESTIONS

21. How could twins have the *ob* genotype (mutant leptin gene) but not have all the same symptoms found in *ob* mice? What kind of information about this particular human allele of leptin would you want in order to understand this apparent exception to the rule?

22. How can obese people with the same percentage of body fat have different levels of leptin in the blood?

Why Are *ob* Mice Sterile? Why Do They Get Sick?

These two questions are specific, but they also ask a larger and more general question: Does leptin have more than just one function, beyond simply controlling the amount of fat in an individual? It is probably the hardest to answer because leptin's different functions are still being discovered (e.g., leptin has a role in bone formation). However, you will answer the two specific questions first and then list some of the other physiological processes affected by leptin.

Why Sterility?

Just prior to puberty, the amount of leptin in a person's serum increases. It is not known if this plays a causal role or not, but it does support the hypothesis that an individual is not reproductively ready until he or she has achieved a certain body mass, including fat. Children who have a higher proportion of body fat tend to enter puberty at an earlier age, which also correlates with the recent trend (over the last 200 years) of an increase in the average weight of children and a decrease in the average age for entering puberty. For menstruation to occur, a girl's percent body fat must be approximately 24% or greater.

Women who are not pregnant have about 40% higher serum levels of leptin than men do. Falling leptin levels due to starvation results in a decrease in estradiol levels and amenorrhea (loss of menstrual cycle). Amenorrhea is commonly seen in women with anorexia nervosa or in female athletes who exercise strenuously over many days. A requirement for body fat is a part of the evolutionary selection pressure on women to have enough energy stores to be able to carry a pregnancy to term, and may also explain why, on average, women have more subcutaneous fat than men. Male and female *ob* mice are sterile, but *ob* mice treated with leptin become fertile. Leptin allowed the maturation of the hypothalamic-pituitary-gonad circuit, which corrected their sterile phenotype (Figure 5.10).

DISCOVERY QUESTION

23. If *ob* mice are sterile, how could researchers breed *ob* mice starting in the 1950s, before leptin was discovered? (Hint: They did not give the breeding mice any special diet or injections.)

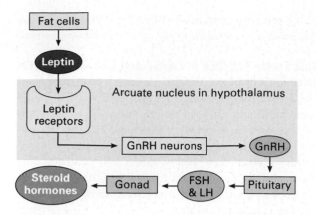

Figure 5.10 Circuit diagram illustrating the role fat accumulation plays in sexual maturity.
Abbreviations: GnRH is gonadotropin-releasing hormone; FSH is follicle-stimulating hormone; LH is luteinizing hormone.

In mice, leptin is not required for any stage of reproduction in females after mating (i.e., it is needed only for sexual maturation). Surprisingly, leptin treatment of *ob* females during pregnancy did not reduce food intake. This apparent contradiction has not been resolved fully, but we do have some suggestive evidence.

Pregnant women secrete a variant form of the leptin receptor, one that has no transmembrane domain. This tailless form of the leptin receptor is called a soluble receptor. By secreting a soluble leptin receptor into the serum, a pregnant woman **chelates** leptin (binds to and thus causes the ligand to be removed from circulation). By having less leptin in her circulation, her body responds by eating more and converting her energy into fat. This explains why so many women gain weight during pregnancy, which would have been selected for during times when a steady supply of food was not assured. It has been hypothesized that women who have difficulty losing weight after giving birth may also have difficulty turning off the production of the soluble leptin receptor.

DISCOVERY QUESTION

24. How could anyone produce a soluble receptor if there is only one gene for the receptor?

Why Sick? This question is not easy to answer, but here are some relevant facts.

- Leptin's structure is similar to that of interleukin-6 (a well-known cytokine secreted by white blood cells), which stimulates cell division and blocks apoptosis in T cells, leukemia cells, and hematopoietic (blood-cell formation) progenitors. There is also a relationship between serum leptin levels and inflammatory cytokines such as tumor necrosis factor-α in patients with chronic heart failure.
- There are leptin receptors on a subset of your white blood cells.
- People with infections and inflammations produce more leptin.

It is easy to imagine that insensitivity to leptin or abnormal production of leptin might lead to a compromised immune system. Hyperleptinemia (high levels of leptin in a person's serum) has been observed in 90% of obese individuals; the remaining 10% have normal leptin levels.

Does Leptin Have Other Functions? Again, you cannot answer this with any certainty, but it appears that leptin is a very busy protein.

1. Leptin receptors are present on the pancreatic islet cells, liver, kidney, lung, vascular endothelial cells, and skeletal muscles.
2. Leptin plays a role in angiogenesis (blood-vessel formation).
3. *ob* and *db* mice show increases in bone formation, though leptin does not stimulate signaling in osteoblasts (bone-forming cells).
4. Leptin increases sympathetic nerve activity to tissues not normally thought of as thermogenic, including the kidney, hind limbs, and adrenal glands. Chronic administration of leptin to the central nervous system increases arterial pressure and heart rate. Therefore, leptin probably plays a role in maintenance of arterial pressure and may be involved in hypertension (high blood pressure), which can lead to problems such as heart attacks and strokes.
5. Obesity is a major risk factor for obstructive sleep apnea syndrome. Subjects treated with nasal continuous positive airway pressure (blowing air up the nostrils all night to prevent snoring and apnea) for six months demonstrated significant reduction in serum leptin levels, which one would expect to increase fat production.

What Is the Link Between Obesity and Diabetes? Pancreatic islet cells die when they accumulate excess fat in the cytoplasm. This occurs frequently in obese individuals, which explains part of the link between obesity and diabetes. There might be other, more direct roles, but this was the only clear one you found during your day of reading.

Can Leptin Pills or Injections Cure All Cases of Obesity? Leptin has been used to cure a few people who fail to produce it, but only a small percentage of those who are obese. In a News Feature in *Nature*, you read, "In October [1999], came the news that [leptin] had performed poorly in its first clinical trial." But the October 1999 *JAMA* publication cited by *Nature* stated, "A dose-response relationship with weight and fat loss was observed with subcutaneous recombinant leptin injections both in lean and obese subjects" (Figure 5.11).

DISCOVERY QUESTIONS

25. Do the data support the authors' claim or the News Feature in *Nature*? Explain your decision with data from Figure 5.11.

26. In the *JAMA* paper that claimed the positive results for leptin, it is clearly indicated that 5 of the 11 authors are employees of Amgen. Remember, Amgen paid $20 million for the commercial rights to leptin. How does this affect your view of the different perspectives (*Nature* vs. *JAMA*) of the therapeutic potential for leptin?

27. What additional information would you like to know about the leptin study and the individuals who participated in the trial?

Five critical points in the fat homeostatic circuitry are available for medical intervention to alter body weight (Table 5.1). Intervention to control weight must fight millions of years of evolutionary selection that have resulted in a lipostat (remember mammals, birds, and eels all have leptin genes; see Figure 5.6). Drugs that control only intake or only metabolism are doomed to failure, because obesity is controlled by a feedback mechanism (see Chapter 8). For example, altering food intake will reduce fat, lower leptin, and result in increased fat storage and lowered metabolism.

What Foods Are Rich in Leptin, and Can Grandma Eat These to Control Her Weight? Leptin is a protein, and our digestive tracts are designed to break down proteins. Eating leptin would not cause anyone to lose weight. You know high-fat meats contain a lot of leptin and high-fat diets can lead to leptin resistance, though the mechanism is uncertain. It is known that "westernization" has created obesity in many nonwestern peoples. It is believed that a change in diet (increased fat) and reduced activity levels are key factors in westernization obesity.

What Role Does the Brain Play in Weight Homeostasis? Deciding whether to eat is a consequence of a complex biochemical mental process. There must be integrated circuitry that can process the genetic (e.g., leptin circuitry), environmental (e.g., smells, time of day, room location), emotional (e.g., distress or worry makes some people want to eat), and physical input (e.g., full or empty stomach). Understanding this circuitry will provide insights into many other behavioral circuitries as well.

Leptin is a link between the nervous system and adipose tissue. It is the central player in the control of **satiety** and energy metabolism. Leptin also plays a key role in sexual maturity, as you have already outlined. Although leptin has an effect on both energy and reproduction, the pathways for each are distinct. For example, if the MC4 (melanocortin-4) receptor is knocked out in mice, or transgenic mice are created that **overexpress** agouti-related protein (AgRP), the mice gain weight but are not sterile. This

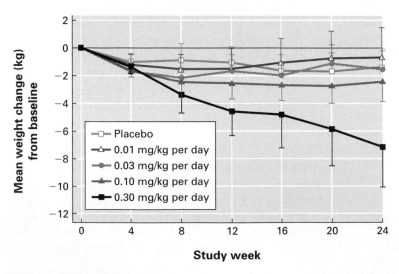

Figure 5.11 Pattern of weight change over 24 weeks in obese people who received recombinant human leptin.

The amount of leptin given to people depended on their weight as indicated in the figure key. The number of people was not constant over the course of the study. Error bars indicate s.e.m.; thin horizontal line indicates no change.

Table 5.1 Five critical points in the fat homeostatic circuitry.

METHODS
knockout

Category	Action	Negative Side Effects
1. Reduce food intake by reducing appetite.	• *Sibutramine* reduces serotonin and dopamine reuptake from synaptic junctions. • *Desensitization* of serotonin and dopamine receptors by high level of agonist (ligand that activates receptor). • *Stimulation* of dopamine receptors reduces food intake; also influences preference for food and the enjoyment factor associated with it.	Dry mouth, insomnia, asthenia, increased blood pressure and heart rate. Leads to increase in seizures in mice.
2. Block nutrient absorption in the gut.	• *Orlistat* inhibits pancreatic lipase and thus lipid uptake.	Fecal fat loss that can lead to loose stools and diarrhea, decreased absorption of fat-soluble vitamins such as E and β-carotene.
3. Increase thermogenesis.	• *Ephedrine* and *caffeine* increase oxygen consumption (i.e., metabolism and heat production). • *Dinitrophenol* uncouples oxidative phosphorylation from ATP production.	Increased heart rate and sense of palpitations. Cataracts, neuropathy.
4. Block adipose synthesis.	• Not possible currently.	—
5. Modulate lipostat set point.	• *Exercise* on a regular basis.	None.

indicates that the same ligand (leptin) binds to sites in the brain that initiate at least two separate circuits.

In January 2001, a new hormone called resistin was discovered. It is produced by fat cells and appears to be involved in the link between obesity and diabetes, though fat homeostasis will not be understood completely for many years. The following are three partial lists of neurological players in this very complex integrated circuitry controlled by leptin.

Appetite Suppression

- Pro-opiomelanocortin (POMC) is the precursor for α-melanocyte-stimulating hormone (α-MSH).
- α-MSH binds to MC4 receptor located in the hypothalamus. α-MSH belongs to the melanocortin family of peptides that regulate pigment formation, food intake, fat storage, immune function, and nervous system function. Initial interest by pharmaceutical companies in α-MSH was for cosmetic tanning.
- Ciliary neurotrophic factor (CNTF) binds to receptors in the hypothalamus and produces a leptin-like weight response when injected into mice.
- MC4: knockout mice eat more; binds the ligand α-MSH.
- Stimulation of opioid receptors μ and κ reduces appetite; this helps explain why drug addicts are often very skinny.
- Serotonin (also called 5-HT) binding to its receptor blocks appetite, but may increase seizure susceptibility.

Appetite Stimulation

- Agouti, which is normally expressed only in hair follicle cells, stimulates pigment formation (dark colors).

Certain mutations lead to agouti production in every cell. Agouti is an antagonist to MC4, which means it binds to MC4 but blocks its signaling ability; thus, agouti causes an increase in appetite.
- AgRP is an antagonist to melanocortin-3 (MC3) and MC4 receptors, which leads to reduced activity, increased fat production (MC3), and increased appetite (MC4).
- Neuropeptide Y (NPY) is overproduced in the hypothalamus of *ob* mice. *ob* mice with knocked-out NPY are less obese because of reduced appetite and increased energy expenditure.

Regulation of Energy Metabolism

- Perosixome proliferator-activated receptor-γ (PPAR-γ).
- PPAR-γ co-activator-1 (PGC-1).
- MC3: Loss of MC3 reduces activity level and increases fat production.

The complete neurological pathways are not known, but we can summarize what happens when a person rapidly loses fat or accumulates fat (Figure 5.12).

Is There Anything I Can Do to Adjust My Lipostat? There are only two ways a lipostat can be adjusted. As noted earlier, a high-fat diet can override the lipostat and result in an obese phenotype even if the individual has a lean lipostat genotype. The only other thing that has been shown to reduce the set point of a lipostat is exercise. For unknown reasons, exercise reduces the set point for lipostats and can result in the conversion of an obese phenotype to a lean

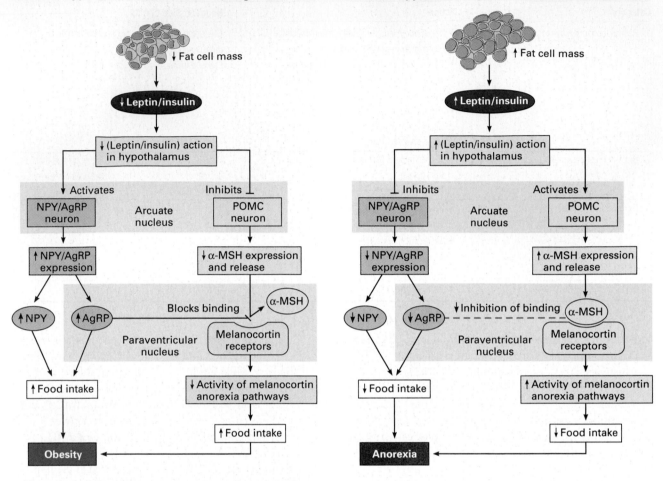

a) What happens when you lose fat through "fad" diet:

b) What happens when you eat too much at one meal:

Figure 5.12 Lipostat response to rapid changes in body fat.
These two circuits are not complete, but they illustrate the complexity inherent in biological pathways. **a)** When fat is lost through surgery or "fad" diet, the body responds by producing less leptin. The outcome of reduced fat is increased food intake and fat storage. **b)** When you eat too much at a holiday meal, you accumulate body fat, which produces more leptin. The result of increased fat storage is reduced food intake, technically called "anorexia" but not to be confused with the pathological condition anorexia nervosa. Eventually, the body fat will be reduced to its preholiday level.

phenotype. The mechanism is not understood, but it is more than simply burning more energy. Sustained aerobic exercise alters the food intake and fat storage ratios so that the net result is a loss in body fat.

In 2005, a team from the Mayo Clinic in Rochester, New York, collected some surprising and encouraging data. They divided a group of self-professed "couch potatoes" and measured their daily activity for 10 days. The participants were either lean or mildly obese prior to the study. The investigators found that the lean couch potatoes were standing or walking, on average, 526 minutes each day, whereas the mildly obese individuals were standing or walking only 373 minutes. The investigators called this form of activity NEAT: nonexercise activity thermogenesis.

They proposed that an increase in daily NEAT is sufficient to cause a person to lose weight. It is surprising that 2.5 hours of additional standing or walking are sufficient to distinguish 2 different lipostat set points even when the 2 groups are fed equal amounts of food. The encouraging news is that with even modest changes in daily routine, obese individuals (25% of all Americans) can lose weight without surgery or fad diets.

What Would You Like to Have for Dinner Tonight When You Come Back with All These Answers? You know that no one can make *lenza und spetzla* like Grandma does, so you request this treat. She smiles and is happy to spend her afternoon making dinner for you. She tells you about an article she found online in the *New York Times Magazine* (December 24, 2000). Lisa Belkin had written a provocative story about why girls enter puberty sooner than they used to (the average age is 1–2 years lower than textbooks say). Theories include obesity, sexual images in the media, pheromones, diet, and the relatedness of males living in the same house. Grandma thinks back to her childhood when children were not exposed to such variables and cannot imagine the answer will be simple or singular. "Humans are the most complex people on earth," she proclaims. (She loves to invent sayings that she hopes will catch on, but you're pretty sure this one won't leave the room.)

As you kiss her goodbye and give her a big hug, you remember one of your childhood wishes. When you were six years old, you wished aloud that someday you could put your arms all the way around Grandma. Now that you are full grown, your hands still cannot touch when you hug her. You realize that her lipostat is set rather high and/or her diet has probably made her insensitive to leptin. It makes you a little sad as you walk away. But then you hear her phoning her best friend, bragging about her new insights and why she can never lose weight. She brags about you, and you realize this was a great way to spend your Saturday after all.

Chapter 5 Conclusions

The obesity case study presents you with increasingly complex information and is still an open research question without any easy solutions. What started as a simple question quickly evolved into an intricate genomic circuit in which components interacted with multiple partners and produced varied outcomes. Gregor Mendel was right: for each locus you have two alleles, and you can make predictions about the inheritance patterns of any gene. Mendel chose his traits carefully because they were simple and not parts of complex circuits that produced confounding outcomes. Simple genetics cannot explain all physical traits, and the growing field of genomics is demonstrating how few traits are simple.

Perhaps one of the most fundamental traits for any animal is the ability to utilize energy sources efficiently. When mammals evolved, there must have been a big advantage in the ability to survive periods of starvation. Accumulation of fat allows us to save energy for later, which could be tied to reproductive maturity and immune capacity. Genomes can evolve over time but they rarely create new genes from scratch. Typically we see gene duplication and/or the utilization of one protein for multiple functions. In your study of genomics, you will need to keep an open mind for unexpected outcomes. Data come in many forms and can accumulate rapidly. As a genomicist, you will need to evaluate the validity and meaning of large data sets. The next three chapters will enable you to use bioinformatics tools to analyze DNA microarray and proteomic data that must be pieced together in order to understand the complete story.

Don't settle for simple answers if the data indicate a more complex model is more accurate.

References

Batterham, R., M. A. Cowley, et al. 2002. Gut hormone PYY_{3-36} physiologically inhibits food intake. *Nature.* 418: 650–654.

Boston, B. A., K. M. Blaydon, et al. 1997. Independent and additive effects of central POMC and leptin pathways on murine obesity. *Science.* 278: 1641–1644.

Boutin, P., C. Dina, et al. 2003. GAD2 on chromosome 10p12 is a candidate gene for human obesity. *PLoS Biology.* 1(3): 361–371.

Bray, G. A., & L. A. Tartaglia. 2000. Medicinal strategies in the treatment of obesity. *Nature.* 404: 672–677.

Campfield, L. A., F. J. Smith, et al. 1995. Recombinant mouse *ob* protein: Evidence for a peripheral signal linking adiposity and central neural networks. *Science.* 269: 546–549.

Chehab, F. F. 2000. Leptin as a regulator of adipose mass and reproduction. *Trends in Pharmacological Sciences.* 21: 309–314.

Chen, A. S., D. J. Marsh, et al. 2000. Inactivation of the mouse melanocortin-3 receptor results in increased fat mass and reduced lean body mass. *Nature Genetics.* 26: 97–102.

Chen, H., O. Charlat, et al. 1996. Evidence that the diabetes gene encodes the leptin receptor: Identification of a mutation in the leptin receptor gene in *db/db* mice. *Cell.* 84: 491–495.

Chicurel, M. 2000. Whatever happened to leptin? *Nature.* 404: 538–540.

Chua, S. C., I. K. Koutras, et al. 1997. Fine structure of the murine leptin receptor gene: Splice site suppression is required to form two alternatively spliced transcripts. *Genomics.* 45: 264–270.

Doehner, W., & S. D. Anker. 2000. The significance of leptin in human: Do we know it yet? *International Journal of Cardiology.* 76: 122–124.

Elmquist, J. K., & J. S. Flier. 2004. The fat-brain axis enters a new dimension. *Science.* 304: 63–64.

Fantuzzi, G., & R. Faggioni. 2000. Leptin in the regulation of immunity, inflammation, and hematopoiesis. *Journal of Leukocyte Biology.* 68: 437–446.

Fried, S. K., M. R. Ricci, et al. 2000. Regulation of leptin production in humans. *Journal of Nutrition.* 130(12 Suppl.): 3127S–3131S.

Friedman, J. M. 2000. Obesity in the new millennium. *Nature.* 404: 632–634.

Fruebis, J., T.-S. Tsao, et al. 2001. Proteolytic cleavage product of 30-kDa adipocyte complement-related protein increases fatty acid oxidation in muscle and causes weight loss in mice. *PNAS.* 98(4): 2005–2010.

Halaas, J. L., K. S. Gajiwala, et al. 1995. Weight-reducing effects of the plasma protein encoded by the obese gene. *Science.* 269: 543–546.

Haynes, W. G. 2000. Interaction between leptin and sympathetic nervous system in hypertension. *Current Hypertension Reports.* 2(3): 311–318.

Heymsfield, S. B., A. S. Greenberg, et al. 1999. Recombinant leptin for weight loss in obese and lean adults: A randomized, controlled, dose-escalation trial. *JAMA.* 282: 1568–1575.

Mantzoros, C. S. 2000. Role of leptin in reproduction. *Annals of the New York Academy of Sciences.* 900: 174–183.

Marik, P. E. 2000. Leptin, obesity, and obstructive sleep apnea. *Chest.* 118: 569–571.

Morgan, H. D., G. E. Heidi, et al. 1999. Epigenetic inheritance at the agouti locus in the mouse. *Nature Genetics.* 23: 314–318.

Mykytyn, K., D. Y. Nishimura, et al. 2002. Identification of the gene (BBS1) most commonly involved in Bardet-Biedl syndrome, a complex human obesity syndrome. *Nature Genetics.* 31: 435–438.

Ollmann, M. M., B. D. Wilson, et al. 1997. Antagonism of central melanocortin receptors in vitro and in vivo by agouti-related protein. *Science.* 278: 135–138.

OMIM. Leptin. <http://www.ncbi.nlm.nih.gov:80/entrez/dispomim.cgi?id=164160>. Accessed 21 December 2000.

Pelleymounter, M. A., M. J. Cullen, et al. 1995. Effects of the obese gene product on body weight regulation in *ob/ob* mice. *Science.* 269: 540–543.

Rink, T. J. 1994. News and views: In search of a satiety factor. *Nature.* 372: 406–407.

Schwartz, M. W., S. C. Woods, et al. 2000. Central nervous system control of food intake. *Nature.* 404: 661–671.

Stephenson, J. 1999. Knockout science: Chubby mice provide new insights into obesity. *JAMA.* 282: 1507–1508.

Steppan, C. M., S. T. Balley, et al. 2001. The hormone resistin links obesity to diabetes. *Nature.* 409: 307–312.

Zhang, Y., R. Proenca, et al. 1994. Positional cloning of the mouse obese gene and its human homolog. *Nature.* 372: 425–432.

Zimmermann, R., J. G. Strauss, et al. 2004. Fat mobilization in adipose tissue is promoted by adipose triglyceride lipase. *Science.* 306: 1383–1386.

Basic Research with DNA Microarrays

LINKS
GCAT
Joseph DeRisi
Pat Brown
Vishwanath Iyer

Chapter 5 presented the genomic complexity of one phenotype—body weight. To understand its true complexity, we will need new tools such as **DNA microarrays**. Perhaps more than any other method, DNA microarrays (often referred to as **DNA chips**) have transformed genomics from a discipline restricted to large sequencing labs to a cottage industry practiced by labs of all sizes. DNA microarrays allow investigators to measure simultaneously the level of transcription for every gene in a genome. With DNA chips, we can collect data from an entire genome as it responds to its environment. Although DNA microarrays were invented elsewhere, their popularity grew in response to publications written by the graduate students and postdocs working with Pat Brown and David Botstein at Stanford University; therefore, many of the case studies in this chapter are taken from their publications, which use genes from the first eukaryote to have its genome sequenced, the yeast *Saccharomyces cerevisiae*. The last case study uses DNA microarrays to improve the annotation of the human genome draft sequence.

6.1 Introduction to Microarrays

This section uses a fictitious story to guide you through one case study: A student trying to make home brew in the dorm has a problem with the biochemistry of fermentation. In this story, you will learn how DNA microarrays are created, experiments are performed, and data analyzed. You will see how DNA microarrays provide insights into the dynamics of a living genome. The Discovery Questions will lead you through public-domain databases that contain raw data you can mine. Since the data sets are immense, much of the data remain unanalyzed. By posting the data online, the original investigators have made it possible for you to make your own original discoveries.

What Happened to My Home Brew?

When you were in third grade, your grandfather gave you a microscope. The next year he gave you a junior chemistry set. Since then, you have not been able to stop playing with science toys. You're the type of person who still likes to put an avocado pit into water so you can watch it sprout roots. Now you are in college, and you are taking real chemistry courses. You cannot believe they actually let you design experiments and give you academic credit for it! Still, you could not give up your hobby of dabbling in science, so you decided to brew your own beer. It was your understanding that it was illegal to buy alcohol, not make it (what you know about legal issues could fit in a shot glass). You are also taking cell biology, so it was just a short hop from cell cycle mutants to yeast biochemistry. You had

intended to take biochemistry, but it did not fit into your schedule, and you considered brewing beer a form of "distance learning" in metabolic pathways. All of this made sense at the time, but now . . . .

It was about 3:00 a.m. on a Tuesday when it happened. You were sleeping soundly when the first one went—bam! It sounded like an explosion, followed by the sound of suds running down your walls. Then another—Bam! Soon, you couldn't count them fast enough—BAM! BANG! CRASH! It sounded like the Fourth of July. Your first batch of home-brewed beer was gone in a 21-gun salute of sorts. You decided there was nothing left to do but roll over and clean up in the morning.

It was a terrible sight to see all your home brew pooled on the floor. You got a shop-vac from the custodian and cleaned up. You didn't feel like eating breakfast, so you went straight to class—cell biology. As classmates asked you what happened, you confirmed some rumors and put others to rest. The instructor entered the class and wrote on the board:

Beer vs. Bread: yeast metabolism

You cannot believe the cruel joke fate has played on you. The topic for today is how yeast cells can switch from fermentation to oxidative phosphorylation. Inside your notebook sits the unread paper that would have saved your beer: Joseph DeRisi, Vishwanath Iyer, and Pat Brown's "Exploring the Metabolic and Genetic Control of Gene Expression on a Genomic Scale." If only you had read it before starting your brew . . . . Oh well, no use crying over spilled beer. But you can't help wondering what went wrong.

The paper you should have read examines the genomic shift in metabolism, but you have no idea what is in this paper. Your look of confusion catches the instructor's eye, and she calls on you to start today's discussion. You hem and haw as you try to think of something witty to say, but she interrupts: "Too busy cleaning the suds off your wall to read, huh?"

You can't believe she knows, too. She calls on a student next to you, who explains, "DNA chips are tools used to measure simultaneously the transcription level of every gene in a cell (in this case yeast). By comparing an experimental **transcriptome** (a complete collection of all transcripts in a cell at any given time point) with a reference transcriptome, you can detect changes in transcription levels for every gene in the genome." You never would have come up with an answer like that. The student continues by saying that this past summer she conducted research using yeast DNA chips (microarrays) as a part of the **Genome Consortium for Active Teaching** (GCAT), a consortium of undergraduates and their faculty mentors. (Now you don't feel so badly that her answer was perfect.)

Your instructor begins to explain how microarrays are produced. The yeast genome has been sequenced and about 6,200 genes have been annotated. All 6,200 genes are amplified by **polymerase chain reaction** (**PCR**), using 6,200 pairs of primers that are commercially available. The PCR products are verified, purified, and spotted onto an ordinary glass microscope slide by a robot. The robot spots all 6,200 PCR products onto 144 slides in about 12 hours. The spotted DNA is denatured and covalently linked to the glass slide. After being spotted, the microscope slide does not look any different, but it has an invisible gridlike pattern (an array) of spots, each containing many amplified copies of a single gene (Figure 6.1). If you fog a microarray with your breath, you can see tiny dots about the size of the period at the end of this sentence. The microarray is covered (all 6,200 spots will fit under a 22-mm coverslip), and stored dry at room temperature.

Another student asks why the DNA doesn't degrade at room temperature. This time you know the answer and say, "Polymerized DNA is a very stable molecule; that's why it can be found intact in ancient samples. So a few days sitting on the shelf should be fine." Answering this question makes you feel a little better about not reading the paper. You get another boost when your instructor says that the Brown lab has provided all the necessary directions to make your own one-armed microarray robot. Since you love to build gadgets, this sounds like a great idea for an independent study project.

LINKS G&P
commercially available
METHODS
cDNA
directions
microscope slide
PCR
robot
Southern blots

DNA microarrays are the functional equivalents of 6,200 Southern blots, in that single-stranded DNA has been immobilized onto a solid support. However, on a Southern blot, the immobilized DNA is usually a mixture of unknown DNAs, such as total genomic DNA. A probe of a known sequence is labeled and allowed to hybridize to the immobilized DNA. For this reason, your instructor explains, there is some debate in the field whether to call the immobilized DNA on a microarray "probes" (because the sequences are known) or "targets" (because the DNA is immobilized). Luckily, your instructor tells you that this nuance is insignificant, so you don't worry about it. (Sometimes she can be kind.)

Where's the Probe?

Now that the yeast DNA chip has been made, you begin to wonder what will be used to probe the equivalent of 6,200 Southern blots. You cannot imagine making that many probes individually, but you're not about to ask since you didn't do the reading. The instructor continues her lecture with a description of experiments that are performed with a yeast DNA chip. Cells are grown in two different conditions, such as in the presence and absence of oxygen. The two populations of mRNA are harvested from each population of cells and separately converted into **cDNA**s. The nucleotides used to make the cDNA include either a green

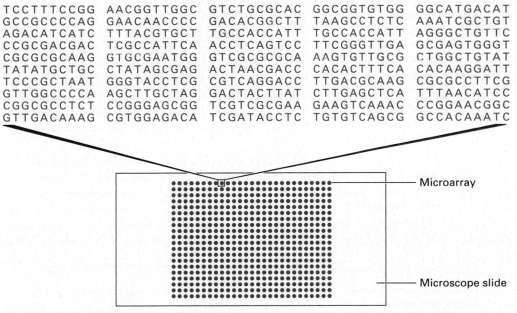

Sequence of one gene

```
TCCTTTCCGG  AACGGTTGGC  GTCTGCGCAC  GGCGGTGTGG  GGCATGACAT
GCCGCCCCAG  GAACAACCCC  GACACGGCTT  TAAGCCTCTC  AAATCGCTGT
AGACATCATC  TTTACGTGCT  TGCCACCATT  TGCCACCATT  AGGGCTGTTC
CCGCGACGAC  TCGCCATTCA  ACCTCAGTCC  TTCGGGTTGA  GCGAGTGGGT
CGCGCGCAAG  GTGCGAATGG  GTCGCGCGCA  AAGTGTTGCG  CTGGCTGTAT
TATATGCTGC  CTATAGCGAG  ACTAACGACC  CACACTTTCA  CACAAGGATT
TCCCGCTAAT  GGGTACCTCG  CGTCAGGACC  TTGACGCAAG  CGCGCCTTCG
GTTGGCCCCA  AGCTTGCTAG  GACTACTTAT  CTTGAGCTCA  TTTAACATCC
CGGCGCCTCT  CCGGGAGCGG  TCGTCGCGAA  GAAGTCAAAC  CCGGAACGGC
GTTGACAAAG  CGTGGAGACA  TCGATACCTC  TGTGTCAGCG  GCCACAAATC
```

Microarray

Microscope slide

Figure 6.1 DNA microarray.
Each purple spot indicates the location of a PCR product on the glass slide. One particular spot has been chosen to illustrate the presence of one gene's sequence. On a real microarray, each spot is about 100 µm in diameter.

METHODS
microarray animation

dye called **Cy3** or a red dye called **Cy5**. Therefore, the two populations of cDNAs are colored either green or red, each color representing the transcriptome from one population of cells (Figure 6.2a).

Figure 6.2 Production of cDNA probes for a DNA chip.
Go to www.geneticsplace.com to view this figure.

The two populations of cDNAs (green and red) are mixed and incubated with the DNA chip. In theory, the colored cDNAs will bind to one strand of one PCR product—the PCR product that encodes the cDNA. During the incubation, there is a competition of sorts (like musical chairs), with all the different cDNAs trying to find places to bind, except in this competition, there are more than enough places to land (Figure 6.2b). After a long incubation (typically overnight), the cDNAs that did not bind to any spots are washed off and the chip is allowed to dry in the dark (light bleaches the fluorescent dyes).

When the microarray is dry, it is put into a scanner that uses light (to excite the dyes) and sensors (to detect the dyes) to record the location and two-color intensities for each spot. Scanning one chip takes about 20 minutes to obtain a high-resolution image. When completed, the two color images (one green and one red) are stored in a computer for image analysis. The computer also generates a new merged image, with yellow spots indicating which **open reading frames** (**ORFs**) are transcribed in both transcriptomes (Figure 6.3). Your instructor turns down the

Figure 6.3 Results from a single DNA chip.
Go to www.geneticsplace.com to view this figure.

lights and shows a DNA microarray animation that summarizes the procedure. Yellow spots are a visual way of depicting a red-to-green ratio of 1:1, but it is unusual for a ratio to be exactly 1:1. More typically, the merged image will be a bit more green or a bit more red. Figure 6.4 illustrates examples of real data.

Figure 6.4 Measuring fluorescence on a DNA chip.
Go to www.geneticsplace.com to view this figure.

DISCOVERY QUESTIONS

1. Why is there a dark center in the middle of each spot? (See Figure 6.4.)
2. How is cDNA made? Why does the DNA on the array have to be denatured?
3. What differences and similarities are there between a DNA chip and a Southern blot?
4. Do you foresee any problems spotting the entire ORF on the chip, instead of only part of it?

From the colored spots, the data are converted to numbers that represent the light intensity of red dye, green dye, and the ratios of red to green (Table 6.1). A long table of numbers is difficult to read and understand. Humans have evolved as visual creatures, and we process visual data much faster than numerical data. Therefore, the group at Stanford converted the numerical ratios of red to green into a visually comprehensible system. They created a color scale to represent the numerical ratios of red to green, but unfortunately they chose a red-to-green scale that can lead to some confusion (Figure 6.5).

Figure 6.5 Red-green color scale for changes in transcription.
Go to www.geneticsplace.com to view this figure.

If there was no change in a gene's transcription between control and experimental growth conditions (i.e., an mRNA ratio of 1:1), the gene is colored black for that time point. If a gene was **induced** (stimulated to produce more mRNA) in the experimental condition, it is given a red color for that time point: the greater the induction, the brighter the red color. If a gene was **repressed** (transcribed less) in the experimental condition, it is given a green color for that time point: the greater the repression, the brighter the green color. This color scale makes sense if you follow two simple rules:

1. The cDNA produced from the control population of cells is green and cDNA from cells grown in the experimental condition (e.g., absence of oxygen) is red.
2. Always place the numerical value for the red dye (from the experimental cells) in the numerator and the green value (control cells) in the denominator when calculating the ratio.

red experimental mRNA ÷ green control mRNA

Following rule 1, mRNA isolated from cells grown in the presence of oxygen is converted into green cDNA, and mRNA isolated from cells grown in the absence of oxygen is converted into red cDNA (i.e., green = with O_2 and

Table 6.1 Summary table of data for one yeast DNA microarray.
The fluorescence intensity for each color was determined, and the ratio of red signal to green signal was calculated. This table shows data for only 14 of the 6,200 genes on the full microarray. Also contained in the data is the location for each spot. When a microarray is fabricated, it is important to know where each gene is located on it.

Block	Column	Row	Gene Name	Red	Green	Red : Green Ratio
1	1	1	*Tub1*	2,345	2,467	0.95
1	1	2	*Tub2*	3,589	2,158	1.66
1	1	3	*Sec1*	4,109	1,469	2.80
1	1	4	*Sec2*	1,500	3,589	0.42
1	1	5	*Sec3*	1,246	1,258	0.99
1	1	6	*Act1*	1,937	2,104	0.92
1	1	7	*Act2*	2,561	1,562	1.64
1	1	8	*Fus1*	2,962	3,012	0.98
1	1	9	*Idp2*	3,585	1,209	2.97
1	1	10	*Idp1*	2,796	1,005	2.78
1	1	11	*Idh1*	2,170	4,245	0.51
1	1	12	*Idh2*	1,896	2,996	0.63
1	1	13	*Erd1*	1,023	3,354	0.31
1	1	14	*Erd2*	1,698	2,896	0.59

red = without O_2). Following rule 2, red number ÷ green number will result in a ratio that represents the change in transcription for one gene at one time point. Let's look at two particular cases:

1. Gene A is expressed in control cells (1,200 light units of green dye bound) but is repressed in the absence of oxygen (only 400 light units of red dye bound). Numerically, expression of gene A would look like this: 400 ÷ 1200 = 0.33, a 3X (X stands for fold) repression of gene A. Using the ratio conversion color scale (see Figure 6.5), we see that 3X on the repression scale would be converted to a dark green color.
2. Gene B is expressed in control cells (180 light units of green dye bound) but is induced in the absence of oxygen (1,800 light units of red dye bound). Numerically, expression of gene B would look like this: 1,800 ÷ 180 = 10, a 10X induction of gene B. Using the ratio conversion color scale, we see that 10X on the induction scale would be converted to a medium red color.

Therefore, we are interested in numerical ratios (converted to colors) for each spot on the microarray. The color scale is meant to be easier to read when there are thousands of data points to examine, as you will see later.

Microarray Data Look Good, But Are They Real?

One problem with inventing a new method is that it is difficult to be sure whether it produces meaningful data. Your colleagues will be skeptical and require evidence to convince them. Recognizing scientific skepticism, the Brown group compared their microarray data with more traditional data—Northern blots (Figure 6.6). Northern blots measure the amount of mRNA produced one gene at a time, and have been used for many years; researchers are very comfortable with them.

Figure 6.6 Comparison of Northern blots with DNA microrray data.
Go to www.geneticsplace.com to view this figure.

DISCOVERY QUESTIONS

5. Choose two genes from Figure 6.6 and draw a GᎦP graph to represent the change in transcription over time.
6. Look at Figure 6.7, which depicts the loss of oxygen over time and the transcriptional response of GᎦP three genes. These data are the ratios of transcription for genes X, Y, and Z during the depletion of oxygen.
    a. What is the fold change for each gene at each time point?
    b. Using the color scale from Figure 6.6, determine the color for each ratio in Figure 6.7b.
7. Were any of the genes in Figure 6.7b transcribed similarly?

How Do You Analyze These Data?

As you can see, microarray data are not difficult to obtain. You grow cells in two different conditions, make fluorescent cDNA of two colors, mix them, and allow them to **hybridize** (form base pairs with complementary strands) to the appropriate spots, wash, dry, scan the chip, and have a computer produce ratios and the appropriate color scheme. When all this is done to one chip, you have 6,200 spots, each with its own numerical ratio that is converted to a color.

However, biologists tend to be more interested in the process of adaptation, not just its end point. We'd like to measure the cellular response every 2 hours to a depletion of oxygen over a 10-hour period. At the end of the experiment, we would have produced 31,000 gene expression ratios. No one can comprehend a spreadsheet of this magnitude, nor 31,000 colored spots on microarrays, each one representing a gene expression ratio. Similarly, it would be difficult to evaluate 6,200 different graphs, each with 5 data points. How would you know whether any genes responded in similar ways to the depletion of oxygen? To answer this fundamental question, you must organize the data into meaningful patterns.

To facilitate the visualization and comprehension of huge data sets, and to look for genes that respond in similar ways, the Stanford microarray team devised a new computer program. Their software looks at each gene's response over time and tries to find other genes with similar responses. Let's look at a simple example of how this can be done. Table 6.2 contains 12 genes organized by name, which makes it difficult to discern gene expression patterns.

When the genes are reorganized, or clustered, according to the similarity of their expression ratios, it is easier to detect genes with similar activity (Table 6.3). If these ratios are then converted to colors, we can quickly understand patterns of gene activity (Figure 6.8).

Table 6.2 Data showing fold change (experimental ÷ control) in mRNA production for 12 hypothetical genes (C − N).

Name	0 hours	2 hours	4 hours	6 hours	8 hours	10 hours
gene C	1	8	12	16	12	8
gene D	1	3	4	4	3	2
gene E	1	4	8	8	8	8
gene F	1	1	1	0.25	0.25	0.1
gene G	1	2	3	4	3	2
gene H	1	0.5	0.33	0.25	0.33	0.5
gene I	1	4	8	4	1	0.5
gene J	1	2	1	2	1	2
gene K	1	1	1	1	3	3
gene L	1	2	3	4	3	2
gene M	1	0.33	0.25	0.25	0.33	0.5
gene N	1	0.125	0.0833	0.0625	0.0833	0.125

Table 6.3 Reorganization, or clustering (see Math Minute 6.3) of gene order from Table 6.2 based on similarity of expression patterns or profiles.

Name	0 hours	2 hours	4 hours	6 hours	8 hours	10 hours
gene M	1	0.33	0.25	0.25	0.33	0.5
gene N	1	0.125	0.0833	0.0625	0.0833	0.125
gene H	1	0.5	0.33	0.25	0.33	0.5
gene K	1	1	1	1	3	3
gene J	1	2	1	2	1	2
gene E	1	4	8	8	8	8
gene C	1	8	12	16	12	8
gene L	1	2	3	4	3	2
gene G	1	2	3	4	3	2
gene D	1	3	4	4	3	2
gene I	1	4	8	4	1	0.5
gene F	1	1	1	0.25	0.25	0.1

a)

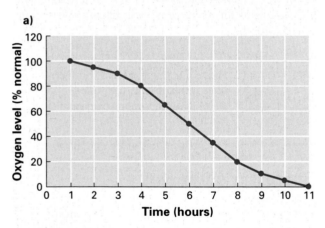

b)

	1 hour	3 hour	5 hour	9 hour
gene X	1.0	2.2	1.0	0.15
gene Y	1.0	4.5	0.95	0.05
gene Z	1.0	1.5	2.0	2.0

Figure 6.7 Transcriptional response of three genes to the gradual loss of oxygen.

a) Graph of oxygen consumption over time by yeast growing in a closed container. **b)** DNA microarray data given in the form of ratios. To calculate a ratio, one gene's activity in cells gradually consuming all the available oxygen was divided by its activity in control cells with unlimited oxygen.

Figure 6.8 Clustered gene expression data.

Go to www.geneticsplace.com to view this figure.

| Math Minute 6.1 | **Why Should You Log-Transform Microarray Data?** |

You have seen how red-to-green intensity ratios are computed from the intensity levels determined in the scanning process, and how they represent repression or induction of the genes. For example, a four-fold repression in gene expression results in a ratio of approximately 0.25. Similarly, a 16-fold repression results in a ratio of 0.0625, a 4-fold induction 4.0, and a 16-fold induction 16.0. But there is a much greater numerical difference between 4.0 and 16.0 than there is between 0.25 and 0.0625. Graph the expression patterns for genes C and N from Table 6.1 on a single set of axes, and notice how the induction stands out much more clearly than the repression, even though the repression is of the same magnitude. In addition to biasing our visual interpretation of gene expression patterns, the compression of ratios between 0 and 1 causes problems with mathematical techniques for analyzing and comparing gene expression patterns (Math Minute 6.2). Furthermore, to interpret a ratio less than 1 as a fold repression, you must take its reciprocal (e.g., 1/0.0625 = 16.0), an operation that most people find difficult to do quickly and accurately in their heads.

To avoid these problems, investigators often use a **log transformation** of the ratio data. Log transformation is illustrated in the following example. Suppose the ratio is 0.0625. Take the base 2 logarithm of this number:

$$\log_2(0.0625) = \log_2(1/16) = \log_2(1) - \log_2(16) = -\log_2(16) = -4.$$

Because the $\log_2$ of 1/16 is the negative of the $\log_2$ of 16, a 16-fold induction and a 16-fold repression have the same magnitude (one positive and one negative) in the $\log_2$-transformed data. The magnitude of the number is the power of 2 needed to get the induction/repression number (e.g., $16 = 2^4$). The $\log_2$ transformations of the ratios in Table 6.1 are given in Table MM6.1.

Sometimes $\log_{10}$ is used instead of $\log_2$; in this case, the magnitude of the transformed data is the power of 10 needed to get the induction/repression number. Transforming the same ratios as above leads to the following:

$$\log_{10}(4) \approx 0.6 \qquad \log_{10}(.25) \approx -0.6,$$
$$\log_{10}(16) \approx 1.2 \qquad \log_{10}(0.0625) \approx -1.2.$$

In $\log_2$ transformed data, a value of 2 corresponds to a ratio of 4; however, you would be surprised to see a value as large as 2 in $\log_{10}$ transformed data, since 2 corresponds to a ratio of 100. In general, a $\log_2$ transformation helps you easily identify doublings or halvings in ratios, while a $\log_{10}$ transformation helps you see order-of-magnitude changes. The key attribute of log-transformed expression data is that equally sized induction and repression receive equal treatment visually and mathematically.

Table MM6.1 Log_2 transformation of gene expression data in Table 6.2.

Name	0 hours	2 hours	4 hours	6 hours	8 hours	10 hours
gene C	0	3	3.58	4	3.58	3
gene D	0	1.58	2	2	1.58	1
gene E	0	2	3	3	3	3
gene F	0	0	0	−2	−2	−3.32
gene G	0	1	1.58	2	1.58	1
gene H	0	−1	−1.60	−2	−1.60	−1
gene I	0	2	3	2	0	−1
gene J	0	1	0	1	0	1
gene K	0	0	0	0	1.58	1.58
gene L	0	1	1.58	2	1.58	1
gene M	0	−1.60	−2	−2	−1.60	−1
gene N	0	−3	−3.59	−4	−3.59	−3

When gene expression ratios are clustered and converted to colors, trends are easier to see than when the data were unorganized ratios (see Table 6.2). With the need to convert large amounts of data from unorganized ratios to clustered colors, a wonderful collaboration was born between biologists and computer scientists/applied mathematicians, resulting in software that automatically produces the colored ratios and clusters genes according to their expression patterns. Collaborations of this sort have created a new interdisciplinary field known as **bioinformatics**. In the preceding examples, each row represents a particular gene, and each column represents a time point. When all 6,200 genes are clustered this way, the boxes are reduced to thin lines (Figure 6.9).

Figure 6.9 Gene expression clusters of several thousand genes.
Go to **www.geneticsplace.com** to view this figure.

Now you have a pretty good understanding of DNA chips, but you have not figured out why your beer bottles exploded. You had hoped the DeRisi paper would offer you some practical advice, but class is over, and no one has told you the bottom line. You decide you'll have to read it yourself.

Math Minute 6.2   **How Do You Measure Similarity between Expression Patterns?**

In Table 6.1, the transcriptional responses of genes G and L are clearly similar, since they have the same ratios at each time point. But what about gene D? How similar is its response to that of genes G and L? A common way to measure the similarity between gene expression patterns like those in Table 6.1 is with the Pearson **correlation coefficient**, abbreviated with the letter r.

Correlation quantifies the extent to which the expression patterns of two genes go up together and down together over several time points or experimental conditions, even if the numbers are not of the same magnitude. A correlation coefficient of 1.0 between two genes means that their expression patterns track each other perfectly. A correlation coefficient of –1.0 between two genes means that their expression patterns track perfectly, but in opposition to one another (i.e., one goes up while the other goes down). A correlation coefficient near zero means that the expression patterns of the two genes do not track each other at all. You can experiment with the **correlation guide** web page and the Excel file **correl_explore.xls** to gain an intuitive understanding of correlation.

In the following example, we will work with the expression values given in Table 6.1 to further illustrate the disadvantages of analyzing raw (untransformed) expression ratios, as discussed in Math Minute 6.1. To find the correlation between genes D and L, denoted r_{DL}, first compute the sample mean and population standard deviation of the expression values for each gene (i.e., each row):

$$\bar{X}_D \approx 2.83 \quad \bar{X}_L = 2.5 \qquad \sigma_D \approx 1.067 \quad \sigma_L \approx 0.957.$$

Subtract $\bar{X}_D$ from each value in the D row and divide each result by σ_D. The result is a row of **normalized** values in the D row:

$$D_{norm} = -1.715, 0.1593, 1.097, 1.097, 0.1593, -0.7779.$$

Do the same in the L row, this time subtracting $\bar{X}_L$ and dividing by σ_L, to produce the following normalized row:

$$L_{norm} = -1.567, -0.5225, 0.5225, 1.567, 0.5225, -0.5225.$$

Now multiply the first number in D_{norm} by the first number in L_{norm}, the second number in D_{norm} by the second number in L_{norm}, and so on, keeping a running sum of these products. (You might recognize this operation as the dot product of the two vectors D_{norm} and L_{norm}.) Finally, divide this sum (5.386) by the number of elements in each row (6) to get the correlation coefficient $r_{DL} \approx 0.897$.

Table MM6.2 Correlation coefficient between each pair of genes, based on log$_2$-transformed gene expression data in Table MM6.1.

	gene C	gene D	gene E	gene F	gene G	gene H	gene I	gene J	gene K	gene L	gene M	gene N
gene C	1	0.94	0.96	−0.40	0.95	−0.95	0.41	0.36	0.23	0.95	−0.94	−1
gene D	0.94	1	0.84	−0.10	0.94	−0.94	0.68	0.24	−0.07	0.94	−1	−0.94
gene E	0.96	0.84	1	−0.57	0.89	−0.89	0.21	0.30	0.43	0.89	−0.84	−0.96
gene F	−0.40	−0.10	−0.57	1	−0.35	0.35	0.60	−0.43	−0.79	−0.35	0.10	0.40
gene G	0.95	0.94	0.89	−0.35	1	−1	0.48	0.22	0.11	1	−0.94	−0.95
gene H	−0.95	−0.94	−0.89	0.35	−1	1	−0.48	−0.21	−0.11	−1	0.94	0.95
gene I	0.41	0.68	0.21	0.60	0.48	−0.48	1	0	−0.75	0.48	−0.68	−0.41
gene J	0.36	0.24	0.30	−0.43	0.22	−0.21	0	1	0	0.22	−0.24	−0.36
gene K	0.23	−0.07	0.43	−0.79	0.11	−0.11	−0.75	0	1	0.11	0.07	−0.23
gene L	0.95	0.94	0.89	−0.35	1	−1	0.48	0.22	0.11	1	−0.94	−0.95
gene M	−0.94	−1	−0.84	0.10	−0.94	0.94	−0.68	−0.24	0.07	−0.94	1	0.94
gene N	−1	−0.94	−0.96	0.40	−0.95	0.95	−0.41	−0.36	−0.23	−0.95	0.94	1

Notice that the expression ratios for genes H and M are the reciprocals of the ratios for genes L and D, respectively. In other words, gene H is repressed to exactly the same extent that gene L is induced, and gene M is repressed to the same extent as D is induced. We would thus expect r_{DL}, which compares the patterns of induction, to be the same as r_{HM}, which compares the patterns of repression. However, r_{HM} is 0.97, which is quite a bit larger than r_{DL}. This strange behavior occurs because correlation is sensitive to the relative magnitudes of the patterns, and can be prevented by first log-transforming the data (see Math Minute 6.1). The correlation coefficients of each pair of genes in Table MM6.1, computed with the log-transformed data, are shown in Table MM6.2. Note that r_{HM} is now the same as r_{DL}.

Because the correlation between genes D and L is close to 1, we conclude that gene D is highly similar to gene L (and thus also highly similar to G). This conclusion can lead to hypotheses about the function of gene D if genes G and L are well understood, but gene D is not. In Math Minute 6.3, we will see how correlations can be used to group highly similar genes so that these hypotheses can be made genome-wide.

MATH MINUTE DISCOVERY QUESTIONS

1. The diagonal entries of Table MM6.2 need not be calculated, because $r_{AA} = 1$ for any gene A. Furthermore, the entries above (or below) the diagonal entries need not be calculated, because correlation is a symmetric function (i.e., $r_{AB} = r_{BA}$, for any two genes A and B). How many entries in Table MM6.2 must be calculated?

2. Find a simple formula that represents the number of entries in Table MM6.2 that must be calculated (i.e., your answer to question 1) as a function of the number of genes being compared.

Math Minute 6.3   **How Do You Cluster Genes?**

Figure 6.9 shows the results of clustering several thousand genes, just as Table 6.3 and Figure 6.8 illustrated the results of clustering the 12 genes in Table 6.2. The purpose of cluster analysis is to organize the genes into groups whose members' expression patterns are all similar to one another according to a particular similarity measure (e.g., Pearson correlation coefficient; see Math Minute 6.2). In Figure 6.8, you could arrange the 12 gene expression patterns by hand so that similar patterns were adjacent to one another as much as possible. However, computer algorithms are required to achieve the

MATH MINUTES
clustering
software

genome-wide clustering shown in Figure 6.9. There are many methods for clustering genes; the one used in our case studies is **hierarchical clustering**.

Hierarchical clustering works as follows. First, find the two most similar genes in the entire set of genes. Join these together into a cluster. Now join the next two most similar objects (an object can be a gene or a cluster), forming a new cluster. Add the new cluster to the list of available objects, and remove the two objects used to form the new cluster. Continue this process, joining objects in the order of their similarity to one another, until there is only one object on the list—a single cluster containing all genes.

To find the two most similar objects, we need a way to measure similarity when one or both objects being compared are clusters of genes. One way is to average the log-transformed expression patterns of the genes in a cluster, forming an average expression pattern that represents that cluster. The pattern is then treated as though it were a single gene, meaning that we must compute its correlation with the pattern of every other currently available object. Let's walk through the process to cluster the genes in Table 6.2, using the similarities in Table MM6.2. First, find the two most similar genes in the entire set of genes. Genes L and G are the most similar, because $r_{LG} = 1$. Join these together into a cluster, denoted [LG]. Cluster [LG] is added to the list of available objects, and the single genes L and G are removed from the list. Now join the next two most similar objects, using the procedure described earlier. (Note that in this case, the average of L and G is equal to both L and G, so we are saved the job of computing new correlations.) The most similar gene to the cluster [LG] is gene C, with $r_{CG} = r_{CL} = 0.95$. However, gene C and cluster [LG] are not the two most similar objects; rather, genes C and E are, with $r_{CE} = 0.96$. Thus, we join genes E and C to form cluster [EC].

At the next iteration, we need to know the correlation of each object with the average log-transformed expression patterns of genes E and C: 0, 2.5, 3.29, 3.5, 3.29, 3. The correlations of all available objects with this pattern representing [EC] are in Table MM6.3.

From Table MM6.3, we see that the object most similar to [EC] is cluster [LG], with a correlation of 0.93. Gene D is even more similar to [LG], since $r_{DG} = 0.94$. However, the two most similar objects now are genes N and H, with $r_{NH} = 0.95$. Therefore, we join genes N and H to form cluster [NH]. We have now completed 3 iterations of the hierarchical clustering algorithm. The entire clustering process for these 12 genes takes 11 iterations; the steps are summarized in Table MM6.4. Note that the final object created is the clustering of all 12 genes shown in Figure 6.8.

The hierarchical clustering process can also be summarized in a **dendrogram** similar to those discussed on pages 139–145. Figure MM6.1 shows the dendrogram for the hierarchical clustering detailed in Table MM6.4 and Figure 6.8. Notice that genes L and G are consolidated into a single node in the tree. The depth (from right to left) at which a node connects two objects represents the similarity between them. At any node that joins two branches, the top and bottom branches can be exchanged without changing the interpretation of the tree. Therefore, many different orderings of the leaves are consistent with the branching structure of a particular dendrogram. Dendrograms are used extensively in Chapter 7 to represent clusters.

Hierarchical clustering is the most popular method for finding trends in gene expression data, but there are several others. Another common method is the k-means cluster algorithm, which tries to find the best partition of the entire set of genes into precisely k groups. Several software programs for clustering gene expression data with hierarchical, k-means, and other methods are freely available for academic use. You can also experiment with clustering microarray data online. Each cluster algorithm may result in

Table MM6.3 Correlations between [EC] and all other objects.

D	F	H	I	J	K	M	N	[LG]
0.90	−0.48	−0.93	0.32	0.33	0.32	−0.90	−0.99	0.93

Table MM6.4 Summary of the hierarchical clustering algorithm applied to the 12 genes in Table 6.2.

| Iteration | Two Most Similar Objects | | Correlation | New Object |
	Object 1	Object 2		
1	L	G	1.00	[LG]
2	E	C	0.96	[EC]
3	N	H	0.95	[NH]
4	M	[NH]	0.95	[MNH]
5	[LG]	D	0.94	[LGD]
6	[EC]	[LGD]	0.94	[ECLGD]
7	I	F	0.60	[IF]
8	J	[ECLGD]	0.29	[JECLGD]
9	K	[JECLGD]	0.19	[KJECLGD]
10	[KJECLGD]	[IF]	−0.12	[KJECLGDIF]
11	[MNH]	[KJECLGDIF]	−0.96	[MNHKJECLGDIF]

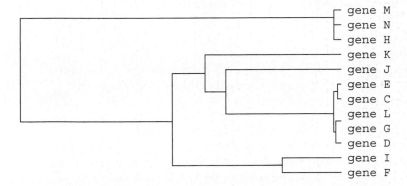

Figure MM6.1 Dendrogram of clustered genes from Table MM6.3 and Figure 6.8.

a different overall clustering of the data. Although there are mathematical methods for evaluating the extent to which clusters agree with the input similarity measurements, the last word in cluster evaluation belongs to the investigators who form and test hypotheses based on the clusters.

MATH MINUTE DISCOVERY QUESTIONS

1. Compute the correlations between cluster [NH] and all other objects, forming a table similar to Table MM6.3.
2. Explain why iteration 4 of the hierarchical clustering algorithm joins gene M with cluster [NH].
3. What new correlations must be computed in iteration 5 of the hierarchical clustering algorithm?
4. How many correlations must be computed to perform the first iteration of hierarchical clustering in the DeRisi diauxic shift data?

You begin reading DeRisi's paper and are impressed that it summarizes more than 43,000 expression ratios. Surely you can find the answer you're looking for in this much data. In this experiment, DeRisi's group grew two populations of yeast. The experimental cells gradually depleted the glucose, while the control cells had ample glucose in their flask (Figure 6.10). As indicated in the figure, aliquots (small volumes) of cells were sampled from the glucose-limited population of cells over the course of the experiment. The resulting cDNA was labeled red and compared to green cDNA produced from control cells with unlimited glucose. You recognize the paper's first figure as a "get acquainted" figure that explains what microarrays are and how to read them (Figure 6.11).

G❄P
DATA
metabolism of glucose
TCA cycle
LINKS
Function Junction

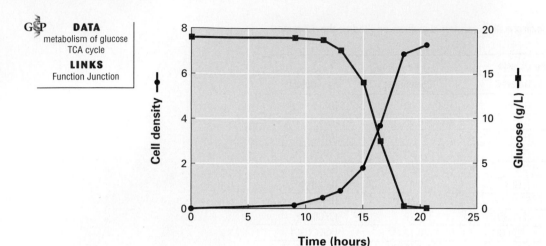

Figure 6.10 Growth of cells as glucose is consumed.
The graph shows the loss of glucose and the concurrent rise in cell density (OD_{600}) during the time course of the experiment by DeRisi et al.

Figure 6.11 Example of real DNA microarray data.
Go to www.geneticsplace.com to view this figure.

Figure 6.12, a close-up view of the white-boxed area in Figure 6.11, illustrates gene regulation over time. Some genes are induced or repressed earlier than others.

Figure 6.12 Time course of gene regulation.
Go to www.geneticsplace.com to view this figure.

DISCOVERY QUESTIONS

8. Go to Function Junction and choose two genes labeled in Figure 6.11, one repressed and one constitutive, or always active, as represented by a ratio of 1:1. Find out what each gene does. To do this, leave all the boxes selected to search all the different databases, and enter the abbreviated name of the first gene you have chosen. Use the metabolism of glucose web page to help you determine why each gene would be repressed or constitutive.

9. Why would most spots be yellow at the earliest time point?

10. Go to Function Junction, search for *Tef4*, and you will see it is involved in translation. Look at the time point labeled OD 3.7 in Figure 6.12, and find the *Tef4* spot. Over the course of this experiment, was *Tef4* induced or repressed? Hypothesize why *Tef4*'s gene regulation was a

part of the cell's response to a reduction in available glucose (i.e., the only available food).

About half the genes that change their expression patterns have no known function yet, which is not too surprising since about 30% of all genes in a genome have no known function. In the DeRisi paper, only a few genes are highlighted for comment because it would be impossible to analyze and discuss all the data in one publication.

The first genes discussed in the paper are *Ald2* and *Acs1*, which function together to convert the products of alcohol dehydrogenase into acetyl-CoA, which enters the tricarboxylic acid (TCA) cycle. Simultaneously, the gene encoding pyruvate decarboxylase is repressed while pyruvate carboxylase is induced, which shifts pyruvate away from acetaldehyde and toward oxalacetate and into the TCA cycle. Furthermore, phosphoenolpyruvate carboxykinase and fructose 1,6-biphosphatase are both induced. These two enzymes catalyze the two irreversible steps in glycolysis, which in turn reverses the overall flow of carbon toward glucose-6-phosphate. Trehalose synthase and glycogen synthase are also induced; these enzymes convert glucose-6-phosphate into storage sugars.

Over the course of the experiment, the yeast cells responded as if they sensed the glucose concentration was decreasing and thus they might soon starve. As a result, genes in entire pathways were induced or repressed in concert (Figure 6.13). Genes involved in protein synthesis were coordinately repressed.

Microarray data have great potential for revealing new insights. First, they can uncover entire pathways that respond in concert to changing conditions. Second, genes that currently have no known function can cluster with genes of known function. When an unknown gene and a known gene are regulated the same way, you can make a good guess that the unknown gene has a function related to

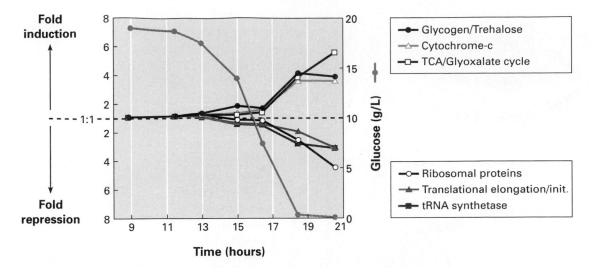

Figure 6.13 Coordinated regulation of gene groups encoding enzymes that work together.
This graph shows that as the glucose was consumed, some genes changed their expression levels as a group. Three groups of energy metabolism genes were induced and three protein production gene groups were repressed.

the known gene's. This **guilt by association** method can lead to predictions of possible gene functions and eventually experimental testing of these predictions (Figure 6.14). In Figure 6.14, a transcription factor is induced twofold and genes 1 and 2 are induced sixfold. Genes 1 and 2 are known to be induced by the transcription factor. A third gene with unknown function is also induced sixfold, which raises two possibilities. First, gene 3 may be regulated by the transcription factor; second, gene 3 may be involved in the same cellular process as genes 1 and 2. These two possibilities can be tested experimentally, and the microarray data provided the information needed to make testable predictions. Real data support the use of guilt by association to propose functions for unknown genes (Figure 6.14b and c).

DISCOVERY QUESTIONS

11. Why would TCA cycle genes be induced if the glucose supply were running out?
12. What mechanism could the genome use to ensure that genes for enzymes in a common pathway are induced or repressed simultaneously?

Can Chips Reveal Regulatory Sequences?

The genome contains a lot more information than just the coding sequences, such as regulatory regions that control gene expression. By clustering genes that exhibit similar expression patterns, we can predict which genes are regulated in similar ways. Because the entire yeast genome has been sequenced, we can examine the promoter regions of clustered genes and look for conserved sequences. By searching for conserved sequences (see Math Minute 2.2), we can identify transcription factor binding sites responsible for regulating clustered genes.

A common regulatory sequence was identified in one particular cluster of genes with a striking expression profile (Figure 6.15). This cluster of seven genes is induced at least tenfold between the 18- and 21-hour time points. All seven genes were known to be repressed in the presence of glucose, and five of them were known to share a common promoter sequence. The gene *Acr1* had not been reported to contain this promoter sequence, but when examined, it did indeed contain the same promoter sequence as the other five. However, *Idp2*, which encodes the cytosolic form of $NADP^+$-dependent **isocitrate dehydrogenase** (**IDH**), does not contain the common promoter motif. Thus, there must be an additional mechanism to control transcription of *Idp2*. Furthermore, the yeast genome contains 4 genes not included in this cluster of 7 that contain the shared promoter sequence, and yet their expression differs from the 7 genes clustered in Figure 6.15. Guilt by association follows two rules: genes with similar expression profiles have similar promoters, and genes with similar promoters have similar expression profiles. In this cluster of seven genes, we found two exceptions to the guilt-by-association rules. Do we keep the rules and live with these exceptions? Ignore these data? Modify the rules?

Can We Formulate Testable Predictions with These Data?

When you think you have a system figured out, it is time to make a prediction that is testable. The data above indicated that, based on expression patterns, we can identify shared transcription factor binding sites. If transcription factors regulate gene expression, then changing the level of a transcription factor should alter the expression pattern of associated genes. What would happen if we delete a transcription factor gene, or overexpress a transcription factor?

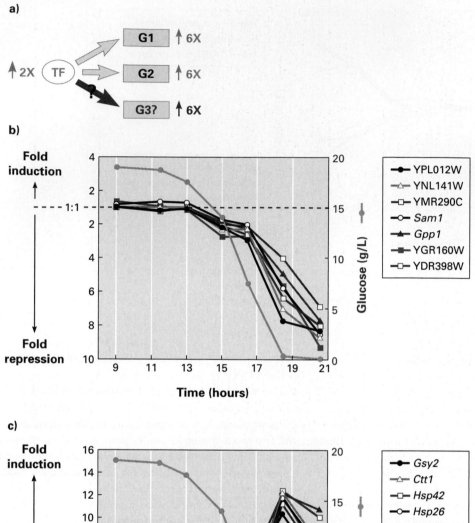

Figure 6.14 Guilt by association.

a) The transcription factor (TF) is induced twofold and is known to activate genes 1 (G1) and 2 (G2), which are both induced sixfold. Gene 3 (G3) is also induced sixfold. Is it regulated by TF? Does it participate in the same circuit as G1 and G2? **b)** and **c)** Clustering of genes suggests functions for unknown genes. In (b), known genes cluster with unknown genes, all of which are co-repressed in response to glucose consumption. In (c), known and unknown genes are induced before the glucose is completely consumed, but the level of induction drops when the glucose is gone.

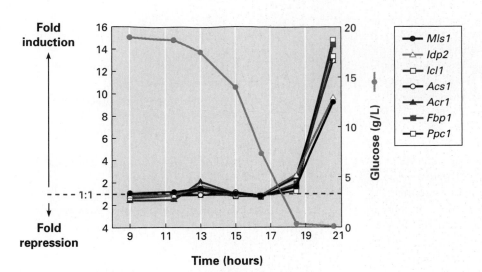

Figure 6.15 Cluster of seven genes with similar expression patterns.
The light purple line with filled circles indicates the level of glucose present; the level of induction for the seven genes is indicated by the remaining lines as noted in the legend.

These questions are not hypothesis driven as much as they are "discovery science." Cell and molecular biology have been powered by hypothesis-driven research for many years, but with the advent of genomic methods such as microarrays, people are asking different types of questions: "What if . . . ?"

The Brown lab asked similar questions: What would happen if we deleted the gene that encodes the repressor **Tup1p**? (p indicates that we are talking about the protein and not the gene.) Tup1p is one protein responsible for the repression of glucose-repressed genes. Of course, the sugar glucose cannot block the transcription of a gene directly, so several proteins are needed, including Tup1p. What would happen if we overexpressed the transcription factor Yap1p? **Yap1p** is a transcription factor known to confer resistance to environmental stresses such as hydrogen peroxide, heavy metals, and osmotic shock.

DISCOVERY QUESTIONS

13. Given rule 1 on page 236, what color spots on a DNA microarray would be more common when cells had the repressor gene *Tup1* deleted?

14. What color spots would be more common when the transcription factor Yap1p is overexpressed?

15. Could the loss of a repressor or the overexpression of a transcription factor result in the repression of a particular gene?

16. What types of control spots would you like to see when altering the expression of a transcription factor? How could you verify that you had truly deleted or overexpressed a particular gene?

Figure 6.16 Transcriptional effects in mutant strains.
Go to www.geneticsplace.com to view this figure.

When the analysis was performed on the two mutant strains, as predicted, there were many changes in gene expression. A large number of red spots are seen in the *Tup1* deletion (Δ*Tup1*) strain. Likewise, overexpression of the transcription factor Yap1p (YAP1+++) resulted in many red spots (Figure 6.16).

DISCOVERY QUESTIONS

The following series of directions and questions shows you how to mine the expression databases for the *Tup1* deletion and *Yap1* overexpression experiments. You will be directed to find particular genes and consider the data obtained for each one.

17. Go to the Tup1 database and determine the change in *Tup1* expression in the Δ*Tup1* mutant strain. You must enter "Tup1" in the gene name box; then hit "start search." Describe what you find. Is it what you expected? Notice the amount of variation in the replicated experiments (this will be discussed later). *Technical note:* The form you will see is delivered by a web server at Stanford. However, the database is delivered from a nonstandard server at the University of California, San Francisco. The IP address for the database server is 128.218.121.63, but it connects via a nonstandard port called 591.

DATA
Yap1 database
LINKS
PubMed

Most web servers connect via port 80, and many institutions have default firewalls that block any server using non-standard ports. Thus, if you do not get results from the Tup1 database, contact your IT department to see if your firewall is blocking access to it.

18. Go back to the Tup1 database to find genes in the Δ*Tup1* strains that were repressed rather than induced. To perform this type of search, delete "Tup1" from the gene box. Go down to the "Average fold induction in Tup1 deletion" box, choose "less than" from the menu, and enter the number "1." Don't sort the results for now. More than 3,000 genes are in this list, but only 25 appear on the first screen. From those first 25, find the gene with the greatest average repression. Is the variation in the two trials high or low?

19. From the list of 3,000-plus hits, go to record number 52 (click on "next page" a couple of times). Look at its average repression, then look at the variation in the two trials. Is its variation high or low? What is the function of gene number 52 from this list?

20. Go back to search the Tup1 database and find the gene with the greatest repression. To do this, have it sort by "average fold induction." What gene was repressed the most by deleting a repressor? What gene was second?

21. Search the Tup1 database again to find the gene that was induced the most. This time change the menu to "greater than," sort by "average fold induction," and choose "descending" for the order of display. Identify a gene family shared by many of the most-induced genes. Which gene was induced the most? (Hint: You might want to have two browser windows open for the next series of questions, one for Function Junction and one to explore the Tup1 database.)

22. Go to Function Junction and enter the name of the shared gene family that describes the trend for Discovery Question 21 (but don't add the "p," which stands for protein). Describe the role this family plays in the cell.

23. Based on your answer from Discovery Question 22, is there a significant effect on the cell when one repressor is deleted?

24. Go through a similar process for the *Yap1* overexpression set in the Yap1 database.
    a. Which gene was induced the most? Was there variation in the two trials?
    b. Go to Function Junction and find out what this most-induced protein does. Does it affect any other genes, or is it more of a dead end?
    c. Which gene was repressed the most? Was there variation in the two trials?
    d. Go to Function Junction and find out what this most-repressed protein does. Does it affect any other genes, or is it more of a dead end?
    e. How much was *Tup1* induced or repressed?
    f. How much was *Yap1* induced?
    g. How many genes were induced when *Yap1* was overexpressed? How many genes were repressed when *Yap1* was overexpressed?

In Discovery Question 24, you should have discovered that YNL331C is the most-induced gene, which by sequence similarity appears to be an aryl-alcohol reductase, although there are no experimental data on YNL331C's function. However, if you followed the Function Junction links long enough, you would have stumbled on a 1999 reference that had not been published when DeRisi et al. published their paper in 1997. Search PubMed for the phrase "AAD gene set yeast." Click on the December 1999 reference and read the abstract by Delneri et al.

Based on the Delneri data, the aryl-alcohol reductase story has not been completely told. What is so amazing and exciting is that you have followed up only one of the 6,200-plus stories buried in data that contain many gems still waiting to be mined. The cell is a complex web with many interconnections that are unknown, even in one species of the "simple" yeast. When first analyzing huge data sets, the best stories may be missed by the original investigators. For example, what are the odds that a randomly chosen book in your library would be interesting to you? Like genes, books are clustered in libraries based on similarities, but clustering does not guarantee you will select the most interesting stories.

Microarrays Seem Too Good to Be True—Are They?

The DeRisi paper was a tipping point in the scientific community, and stimulated many labs to perform microarray experiments. However, the world of microarrays has been in such a hurry to produce results and make new discoveries that some shortcuts have been taken, and must be addressed. Recall your answers to Discovery Questions 17–24 about the variation in the data for the *Yap1+++* and Δ*Tup1* mutant strains. Some genes produced very "tight" data, but others did not. DeRisi repeated each experiment only once. How many times have you repeated experiments once and gotten different answers? Does this mean your entire methodology is wrong? Does it affect your level of confidence? Does it mean you simply need to repeat each experiment a few more times? Although microarray experiments have been presented here to make them look simple, they are actually tricky and expensive. Repeating experiments once when microarrays were being developed must have seemed sufficient. As chips, protocols, and reagents have become more available, people have become more concerned about the reproducibility of their data.

The DeRisi paper contained a short discussion about the variation of the data. The correlation coefficient for the duplicated experiment was 0.87. Was this variation due to variation in experimental procedure, or to "cellular indecision" for a subset of genes (i.e., stochastic genetic circuits that can yield more than one outcome)? DeRisi and his colleagues also stated that their data did not always agree with previous publications. Do contradictory data indicate the method was flawed, or something more biologically significant? We don't know, and won't know, until someone does the experiments necessary to find out. That's why many call this "discovery science."

Let's look at Table 6.4, which appeared in the DeRisi paper. It lists genes that are induced when Yap1p is overexpressed. Topping the list, as you know, is the putative aryl-alcohol reductase. The fold increase is 12.9, but the standard error is not reported. Interestingly, footnote number 50 associated with this table says, "In addition to the 17 genes shown in [Table 6.4,] three additional genes were induced by an average of more than threefold in the duplicate experiments, but in one of the two experiments, the induction was less than twofold (range 1.6- to 1.9-fold)." This footnote tells us that 3 out of 20 genes (about 15%)

were induced an average of threefold but showed induction of less than twofold in one of the two trials. These three genes were all induced, but the magnitude varied. Why? Does this indicate a need for more replications and/or published standard errors for future papers?

DISCOVERY QUESTIONS

25. Now that you have a more complete understanding of how microarray data are obtained and the potential problems in reporting them, here is a challenging question. If full-length PCR products are used for spotting, does this create a problem? If one gene is 300 bp and another is 3,000 bp long, you would expect the shorter cDNA to have only 0.1 as much dye as the longer cDNA. Does the length of the PCR product present a problem when looking at ratios? Clustering? Comparing the expression levels of two genes? Explain your answers.

26. Let's imagine the green-dye labeling of the cDNA did not work as well as it normally does. To compensate, you adjust the scanner so that it brightens the signal from the green channel. Does this present any problems with data interpretation? Explain your answer.

Why Did the Beer Blow?

It is approaching dinnertime, and you are ready to head back to your room. You have only one figure left to read (Figure 6.17), and you are hoping the answer will be clear

Table 6.4 Genes induced by Yap1p overproduction.
This list includes all the genes that were induced over twofold in both trials of the same experiment, and the average is over threefold.

ORF	Gene	Description	Fold Increase
YNL331C		Putative aryl-alcohol reductase	12.9
YKL071W		Similarity to bacterial csgA protein	10.4
YML007W	Yap1	Transcriptional activator involved in oxidative stress response	9.8
YFL056C		Homology to aryl-alcohol dehydrogenases	9.0
YLL060C		Putative glutathione transferase	7.4
YOL165C		Putative aryl-alcohol dehydrogenase (NADP$^+$)	7.0
YCR107W		Putative aryl-alcohol reductase	6.5
YML116W	Atr1	Aminotriazole and 4-nitroquinoline resistance protein	6.5
YBR008C		Homology to benomyl/methotrexate resistance protein	6.1
YCLX08C		Hypothetical protein	6.1
YJR155W		Putative aryl-alcohol reductase	6.0
YPL171C	Oye3	NAPDH dehydrogenase (old yellow enzyme), isoform 3	5.8
YLR460C		Homology to hypothetical proteins YCR102C and YNL134C	4.7
YKR076W		Homology to hypothetical protein YMR251W	4.5
YHR179W	Oye2	NAPDH oxidoreductase (old yellow enzyme), isoform 1	4.1
YML131W		Similarity to *A. thaliana* zeta-crystallin homolog	3.7
YOL126C	Mdh2	Malate dehydrogenase	3.3

to you so you won't waste all that good beer next time. You know your beer shattered the bottles because the pressure was too great. If you had used thicker glass bottles, it might have worked. The downside is that the beer would have spewed the way champagne does when opened. But the question is, what increased the pressure more than normal? After studying Figure 6.17, you realize . . .

Figure 6.17 Yeast cells utilize reversible pathways to metabolize sugar quickly.
Go to www.geneticsplace.com to view this figure.

DISCOVERY QUESTIONS

27. How did the metabolism of glucose generate the increased pressure inside the bottle?

28. There are two reasons why the bottles may have blown up. The more common mistake is that there was too much sugar left when the beer was bottled, but Figure 6.17 offers another possible explanation. What is it?

You know that the increased bottle pressure was produced by metabolic waste product of CO_2. You also know that CO_2 is a byproduct of fermentation and of the TCA cycle. Fermentation cannot continue in the absence of glucose, but could the TCA cycle continue even if all the glucose were gone? The microarray data revealed that the genome sensed the loss of glucose and the presence of alcohol, which the yeast produced by fermentation. Later, the cells switched from fermentation to utilization of the ethanol. Oxygen must have been present in your bottles, so the yeast was able to convert the alcohol into glycogen and trehalose. The genes needed to metabolize ethanol were induced as glucose was depleted. You suspect your bottles were airtight and trapped the CO_2 as it was produced by the metabolism of ethanol to produce acetyl-CoA, which entered the TCA cycle. The continuation of metabolism caused your bottles to explode.

Why would yeast waste its time making alcohol first only to consume it later? Yeast cells are not multicellular, so each cell is doing its best to survive by out-competing its genetically identical siblings. Cells that accumulate the most energy (in the temporary storage form of alcohol) have the resources necessary to produce more offspring. Once the glucose is gone, the cells convert the somewhat toxic alcohol to safer storage molecules such as glycogen.

You put the paper down and wonder, "Wow, how did yeast get to be so smart?" Of course, they have had a few hundred million years to optimize their genomes to adjust their metabolism as needed. You hope it will not take you as long to learn how to brew beer.

What Can We Learn from Stressed-Out Yeast?

By now it must be clear that yeast has a well-deserved reputation as a model organism. The Human Genome Project sequenced several model organisms as well as the human genome. Knowing the nonhuman genome sequences has already proven useful in the case of yeast. The *Drosophila* and *C. elegans* genomes have begun to influence medical research, too. With the full human and mouse genomes, we will see many new uses of microarrays in biomedical research and diagnosis—which brings up another story worth telling.

It has been said that all yeast can do is reproduce and excrete, but this is an oversimplification. Yeast can do so much more than you might expect. Think about what a yeast cell is: a complete eukaryotic organism with about 6,200 genes. It can live as either a diploid or a haploid organism and can switch from aerobic to anaerobic metabolism when it needs to. Yeast can reproduce through sexual and asexual methods, and its metabolism has given us bread and ethanol. That is an impressive assortment of behaviors for a simple fungus with about half as many genes as a fly.

The most amazing thing about unicellular organisms is how they sense their surroundings and respond appropriately. They contain most of the same cell-signaling pathways humans possess, without all that excess baggage of cellular specialization and tissue formation. Yeast has become a model for cancer research because mitosis is almost identical in all species. Our understanding of the human cell cycle is due in large part to yeast biologists, because yeast cells and human cells have many genes in common. That is why another paper by Pat Brown's lab is of such interest to us at this point.

The Brown lab has teamed up with many others, and some members of his lab have moved on to start their own labs. Of all Brown's collaborations, though, perhaps none has proven more successful than his close interactions with David Botstein. Botstein and Brown have become a "dynamic duo" by rapidly publishing many groundbreaking microarray papers. In December 2000, they published a rather weighty paper that examined the genomic response of 6,200 genes under 142 different experimental conditions, which required analysis of more than 880,000 gene expression ratios. This tour de force contained more information than they described, but they did highlight a few interesting trends. Botstein and Brown have led the way in keeping scientific publications and raw data sets in the public domain. Due to their concern for free access to scientific information, you are able to analyze their work and continue to mine some of the richest microarray data sets available. The Yap1 and Tup1 databases were two examples; we will explore others in upcoming Discovery Questions.

Do Fungi Feel Stress?

In the Gasch et al. (2000) paper, the combined labs of Botstein and Brown performed three types of experiments. First, they measured the time course of gene expression in cells exposed to changes in the environment. Second, they looked at the genome responses in cells subjected to a gradient of heat shocks. Third, they examined gene expression profiles in cells that had already adapted to growth in these stressful conditions. Here is a list of the environmental challenges used in their study: heat shock, hydrogen peroxide, exposure to a superoxide-generating drug called menadione, a sulfhydryl oxidizing agent called diamide, a disulfide reducing agent called dithiothreitol (DTT), hyperosmotic shock, amino acid starvation, and nitrogen starvation. The severity of the treatments was designed so that about 80% of the cells would survive. To live, survivors had to change their protein makeup by altering gene expression. The final stress condition was growth into **stationary phase**, which means that the cells had exhausted the nutrients and were neither dead nor dividing.

From this collection of data, the investigators observed four classes of phenomena: (1) It became possible to hypothesize the functions of genes with previously unknown functions when they clustered repeatedly with genes of known function. (2) Genes that clustered together often contained conserved promoter sequence motifs. (3) Clustered gene expressions revealed the components of genomic circuits that work in concert to perform a single "task." (4) Individual genomic circuits could be observed to work in conjunction with other individual circuits to reveal integrated circuits working as larger units to accomplish complex "tasks." The multiple layers of understanding gained, from single genes to integrated circuitry, serve as an excellent example why genomics is more than a new set of tools. Genomic data permit us to envision more complete and complex models of how cells function as dynamic and evolving systems.

Figure 6.18 Expression of genes in response to 142 environmental perturbations.
Go to www.geneticsplace.com to view this figure.

By now you are probably beginning to wonder what all the data look like (Figure 6.18). Even at low magnification, you can observe one trend. Two large clusters of genes mirror one another: the repressed cluster F is bright green when the induced cluster P is bright red. Environmental stress has already killed 20% of the survivors' genetically identical siblings, so it's time to adapt or die. The integrated genomic circuits are working collectively to save the

remaining cells from death. The collective effort of the 900 genes in clusters F and P was dubbed the **environmental stress response (ESR)**. What is so surprising is that these 900 genes altered their expression patterns in a coordinated fashion and in response to many, though not all, of the environmental perturbations (Figure 6.19). How do they do this? Is yeast smarter than we are? Could you consciously initiate 900 different responses to changes in your environment? Suddenly the tiny fungus doesn't look so simple.

MATH MINUTES GP
clustering

Figure 6.19 Stress-induced expression of 900 genes clustered into the ESR.
Go to www.geneticsplace.com to view this figure.

What Goes Up?

Although it is difficult to notice, an interesting phenomenon lies buried in these data. With the exception of starvation conditions, all of the ESR gene expression profiles were transient (Figure 6.20). You can see from the heat shock experiments that both mild and harsh temperature shifts initiated the ESR (gene repression and induction), but for only a short time before the level of transcription returned toward previous levels.

Figure 6.20 Transient ESR response.
Go to www.geneticsplace.com to view this figure.

DISCOVERY QUESTIONS

29. What is the selective advantage of a transient ESR, if the stressor stimulus is continuous? Why is the ESR substantially reduced once the cells have reached steady-state growth at the elevated temperatures?

30. What differences do you see between the mild and harsh heat shock ESR profiles (Figure 6.20)? Propose a mechanism to explain these differences.

31. Go to the online clustering web site where you can analyze conditions and genes from the environmental stress response experiment. Select both "choose all" boxes from the two "Heat Shock" experiments. Examine the list of genes and their color-coded gene profiles. Identify three genes that would be clustered together when the

DATA
IDH isozyme

LINKS
Expression Connection
GenomeNet

threshold is set for 0.6, but one of these three is not clustered when the threshold is set for 0.9.

Once again we see how adaptable the yeast genome is. It would be wasteful (and thus not adaptive) to constantly respond to an environmental shock when it is no longer shocking. For example, in winter you get used to colder temperatures. If you flew to the other side of the world where it was summer, you would be shocked by how hot the day felt. However, after a few days that same temperature would not be shocking to you because you would have adapted to the hotter temperatures. Your body would not respond to a hot day the same way if it occurred in the winter rather than the summer. Once the appropriate physiological adjustments are made, there is no need for cells to transcribe the genomic equivalent of 911.

Why Are There So Many Copies of Some Genes But Not Others?

Like all organisms, the yeast genome contains genes that encode for **isozymes**, or enzymes that share a lot of sequence identity and appear to perform redundant functions. In the Delneri abstract (page 248), we saw that aryl-alcohol reductase is encoded by seven genes that produce seven isozymes. Why do some genes appear multiple times in a genome with only slight differences in their coding regions? Yeast seems to understand isozymes better than we do (Figure 6.21). Although isozyme coding sequences are very

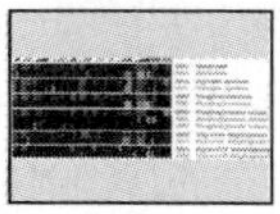

Figure 6.21 Differentially regulated isozyme genes in the ESR.
Go to www.geneticsplace.com to view this figure.

similar and isozymes catalyze the same biochemical reaction, the yeast genome treats isozyme-encoding genes differently. Some are repressed while their functional equivalents are induced. Why?

DISCOVERY QUESTIONS

32. Why would an organism want to regulate differentially genes that encode isozymes? What does the yeast genome know that we don't?

33. Look at the IDH isozyme web page. Examine the expression profiles of these five apparent isozymes that all encode isocitrate dehydrogenase.

a. Are any of them differentially regulated?

b. Should they be differentially regulated? How would your interpretations change if you knew the following:

- Idp1 uses $NADP^+$ as a coenzyme and is located in mitochondria.
- Idp2 uses $NADP^+$ as a coenzyme and is located in the cytosol.
- Idp3 uses $NADP^+$ as a coenzyme and is located in peroxisomes.
- Idh1 uses NAD^+ as a coenzyme, is located in mitochondria, and is one subunit from a heterodimer with Idh2.
- Idh2 uses NAD^+ as a coenzyme, is located in mitochondria, and is one subunit from a heterodimer with Idh1.

c. The ESR database has all five isozymes listed as participating in the TCA cycle. Is this the complete story? Go to GenomeNet, select "PATHWAY" from the search menu, and one at a time find these three pathways: sce00020, sce00720, and sce00480. Click on the pathway link and you will see a metabolic pathway map. IDH has the E.C. number 1.1.1.42, and it will be highlighted. Is IDH restricted to the TCA cycle?

34. Go to Expression Connection, perform a "Search I" using a gene from Figure 6.21 (e.g., *Hxk1*). Enter the name in the blank under "Step 1" and choose "Expression during the diauxic shift" and "Expression in response to environmental changes" from the list of available experiments in "Step 2," then hit "Submit" from "Step 3." You will see a graphic display of your gene's expression ratios in all the available experiments above the results for the two experiments you selected. Click on a red dot under the heading "Similarly expressed genes." Can you find any isozymes that were clustered with your gene? If you click on any of the colored boxes, you will retrieve as many as 20 genes that cluster with the new gene you selected. If you click on one isozyme's colored box, do both still cluster together?

There are three reasons why a smart genome like yeast's might want to differentially regulate isozyme genes. Perhaps what appears to us as enzymes that do the same job may do different jobs, given their subcellular localization. Or perhaps these isozymes catalyze the same reactions in vitro, but act quite differently in vivo. Or maybe they catalyze the same biochemical reaction, but the cell needs complex regulation of gene expression. For example, when the cell is stressed, it may need isozyme #1 for a global response; for a more subtle change in the environment, it may need a more specific, non-ESR, response by isozyme #2. If IDH performs only one reaction, but there are four different situations that require IDH, evolution may have selected four separate promoters (with four coding regions) rather than one promoter with four sets of transcription factors.

DISCOVERY QUESTION

35. Look at Yeast Stress Figure 4 and you will see six subsets of genes that are part of the ESR expression profile and have related functions.

    a. Design one common promoter for one of the six subsets of genes (just draw a box with labels; don't worry about the DNA sequence) that would allow all these genes to be up-regulated at one time.

    b. Add to your design from Discovery Question 35a so that when needed, each of the six genes could be regulated separately from the ESR.

    c. How many different transcription factor genes do you imagine are needed to provide collective and individual control over gene expression?

How Well Do Promoters Control Gene Expression?

In Discovery Question 35, you designed different combinations of transcription factors and promoters to elicit the adaptive responses present in the yeast genome. Some transcription factors have been the subject of intensive investigation, so we know a lot about how they work. Three transcription factors that have been characterized in detail are Msn2p, Msn4p, and Yap1p. DNA microarray data for 15 genes that contain binding sites for Msn2p/Msn4p and Yap1p were analyzed for their responses to heat and hydrogen peroxide when *Msn2/Msn4* and *Yap1* were deleted (Figure 6.22). Three strains of yeast were used for this study: *wt*, an *msn2/msn4* double-deletion strain, and a *yap1* deletion strain.

Figure 6.22 Dependence of ESR gene regulation on transcription factors. Go to www.geneticsplace.com to view this figure.

DISCOVERY QUESTIONS

36. What role does the Yap1p transcription factor play in hydrogen peroxide and heat shock responses? What role do the Msn2p and Msn4p

DATA
Stress Figure 4

transcription factors play in hydrogen peroxide and heat shock responses?

37. How can gene regulation be used to provide the flexibility a genome needs when responding to environmental stress, as opposed to the typical needs of a cell? It appears that all 3 transcription factors can activate these 15 genes. Why does yeast need two types of transcription factors to do the same job?

38. When you think of cells, you probably imagine a functional unit that is very efficient and does not waste its resources. That's certainly what you would expect after a billion years of evolutionary refinement, but the ESR appears to be an exception to the rule of efficiency. Genes that do reciprocal jobs are induced simultaneously as a part of the ESR. For example, transcription was stimulated for the genes encoding the positive effectors of protein kinase A and the negative regulators of protein kinase A. Hypothesize the evolutionary advantage of this apparent extravagance.

39. An old question revisited: Are there any problems associated with spotting PCR products from the entire gene rather than smaller portions of a gene?

The yeast genome has been stressed and shocked for a billion years. It is so accustomed to shocks that it no longer panics every time the red-alert warning is sounded. Genomes sense a change in the environment that could prove lethal if not responded to quickly and appropriately. The 900 ESR genes are induced or repressed according to standard emergency procedure. This includes genes that govern activation and repression of a common critical controlling factor, such as protein kinase A. Producing an activator and a repressor at the same time seems wasteful until you realize the environment can change quickly. Let's imagine that two yeast cells are riding on the back of a deer that has stopped to feed in the sunlight. It gets very hot in the summer sun, so both yeast cells initiate the heat-stimulated ESR set of 900 genes, with a few exceptions. Yeast number one senses the heat and, rather than producing both an activator and a repressor, it "knows" that heat shock only requires the repressor for the cell to survive. Therefore, yeast number one begins the specialized "I'm too hot" response and makes no allowances for alternative responses. Yeast number two responds in the traditional way by producing both an activator and a repressor.

These two yeast cells represent a classic example of genetic variation where two cells express slightly different proteins and thus phenotypes. If the deer stays in the sunlight longer, yeast number one has a slight head start, but yeast number two will eventually respond in the same way. However, if the deer decides to lick its back and a very basic saliva coats our two yeast, there may be a different outcome. Yeast number one decided too soon that this was an "I'm too hot" emergency and committed full speed in one direction. Yeast number two decided to wait a little longer and collect more information before committing. In this instance, yeast number two has both of its regulatory proteins ready to respond either way and is capable of initiating very quickly a response to the new "It's too basic!" emergency. In other words, after a billion years, it appears that the yeast genome has been selected to respond in stages. First is the red alert and the full 900-gene ESR, followed by the more finely tuned and specific response. The yeast genome appears to live by the motto: "Always be prepared." In the lab, when the investigator is applying only one environmental stress at a time, the yeast ESR seems wasteful. In the real world where yeast evolved, though, it may be advantageous to respond by simultaneously producing activators and repressors of a few critical genomic "circuit switches" so that the cell can respond appropriately as it collects more information.

Do Promoters Work in Reverse?

Most of what we have covered has been under the heading of "What happens if . . . ?" As we learn more about how cells respond, however, we are able to build models that are testable. Because these data sets are so rich, the data to test the models are already available. For example, if evolution has produced promoters that work well in response to heat shock, they should be able to respond in the opposite direction when the cell is subjected to cold shock. It would be a disaster for a cell to respond the same way for both heat and cold shock. Imagine being able to sense the weather (hot and cold), but being hardwired so that your only response was to add more layers of clothing, even in the summer. Luckily, you and the yeast genome can sense these two different stimuli and respond in opposite ways (Figure 6.23). When the cells are shifted from cold to hot (Figure 6.23a; solid ramp), about 600 genes are repressed and 300 are induced. Conversely, when cells are shifted from hot to cold (open ramp), the same 600 genes are induced and the same 300 genes are repressed. A similar pattern was seen when sorbitol was used to produce an osmotic shock (Figure 6.23b).

Figure 6.23 Reciprocal gene expression.
Go to www.geneticsplace.com to view this figure.

DISCOVERY QUESTIONS

40. How can one promoter respond differently to hot and cold?

41. The response for every gene is not 100% reciprocal in Figure 6.23. How could a cell control which gene to treat reciprocally and which not?

42. Gray bars indicate no data available. Why do you think there are so many in the steady-state columns?

43. Go to Expression Connection and search for *Trx2*, listed in Figure 6.22. Did all the genes in this figure share expression patterns when exposed to environmental stress other than H_2O_2 or heat?

44. Notice *Sod1* is clustered just below *Trx2*. Go to the MIPS database, search for *Sod1* (use the ORF name YJR104C in the upper-left corner), and read the "FUNCTIONAL CATEGORIES" for *Sod1*. Do you think *Sod1* should have been included in Figure 6.22? Test your prediction by searching the Yeast Stress Figure 4 image for *Sod1*.

Summary 6.1

If you perform a PubMed search with the word "microarray," you will begin to see the effect this method is having on basic and biomedical research. The case studies in Section 6.1 focused on basic research intended to help investigators understand fundamental biology shared by all eukaryotes (equally powerful studies have been conducted in prokaryotes). We can learn much about our own biology when we understand how yeast genomes respond to environmental perturbations. The Nobel Prize for Medicine and Physiology for 2001 was awarded to three yeast geneticists who opened the way for us to understand the cell cycle and human cancer. Similarly, the yeast DNA microarray case studies in this section have opened new areas of basic and applied research. Section 6.2 addresses some remaining questions about the method and gives two examples of how microarrays were used to reinterpret old information, including the paper by DeRisi et al.

6.2 Alternative Uses of DNA Microarrays

Now that you understand how DNA chips work, you might wonder whether there are alternative ways to utilize these data. Can old data be mined a second time to make new discoveries? Section 6.2 addresses these questions—the same questions being discussed at research institutions around the world. Academic and industry labs are trying to maximize the potential of DNA chips, but no one has perfected them yet. You will read how one group reanalyzed some public-domain data (discussed in Section 6.1) to make a new discovery. The final case study illustrates how one company used DNA microarrays to verify annotated genes in the human genome, and even discover genes that had been overlooked or incorrectly defined.

Why Do So Many Unrelated Genes Share the Same Expression Profile?

DNA microarrays are very powerful tools that can detect changes in mRNA levels. However, they are like caller ID on a telephone: they can tell you who's there, but they can't tell you why. It is always important to know what data your method can and cannot provide. When performing microarray experiments, the amount of data can be overwhelming, so careful analysis is always crucial for extracting good conclusions. The investigators at Rosetta Inpharmatics have applied rigorous statistical analysis to their data, enabling them to cluster genes and present the data in a variety of useful ways. Let's begin this section by answering a few questions to set the stage.

DISCOVERY QUESTIONS

45. If a DNA chip detects an increased level of mRNA for several genes, we use guilt by association and hypothesize that they have related functions. Go to Function Junction and determine what these four genes have in common: YFL067W (that's a zero before the 6), *Cos4*, *Sno3*, and *Aad6*. Don't spend more than ten minutes looking.

46. Let's imagine that all four of these genes were induced under many experimental conditions. What conclusions could you make?

47. Let's assume their promoters have no sequence motifs in common. How could they all be induced under all experimental conditions?

48. Go to MIPS chromosome listing, and find the genes YFL067w, Cos4, Sno3, and Aad6. What do they have in common?

These four genes have nothing in common except their location—near a telomere on chromosome 6. With this information, maybe their unified induction is a consequence of duplication on chromosome 6, so there are twice as many copies of these four genes and thus an increased level of mRNAs from each gene.

Timothy Hughes and his colleagues at Rosetta obtained some microarray data that did not make sense. Why would a collection of genes with no shared function or promoter

motifs be co-regulated? It was a puzzle until someone noticed they were all located on a single chromosome, chromosome 7. They were examining the expression profiles of two mutant strains: *erg4* and *ecm18*. Erg4p is an enzyme used to produce lipids, and Ecm18p is a part of the cell wall structure. When Hughes and his colleagues performed statistical analyses of the gene expression in these two mutants, they calculated a correlation coefficient of 0.63 (Figure 6.24).

LINKS MIPS chromosome listing Rosetta Timothy Hughes

Figure 6.24 Two mutants with similar gene expression.
Go to www.geneticsplace.com to view this figure.

When they excluded all genes on chromosome 7, the correlation dropped to near zero. As a control, they excluded genes on chromosome 4 and got a correlation of 0.63. The genes responsible for producing the high correlation were all located on chromosome 7. When Hughes's team examined the mean expression ratios for all genes on each chromosome, they found that transcription for genes on chromosome 7 in these two mutant strains (as compared to *wt*) was higher than the genes on all the other chromosomes. Expression ratios from chromosome 7 were 58% higher in *erg4* and 35% higher in *ecm18* compared with parental *wt* control strains. Genomic DNA ratios from chromosome 7 were 66% higher in *erg4* and 41% higher in *ecm18* compared with parental *wt* strains. The higher level of mRNA correlated proportionately with the higher amount of genomic DNA on chromosome 7 (see Figure 6.24).

DISCOVERY QUESTIONS

49. From Figure 6.24a–c, what can you conclude about the coincidence of genes induced in both mutants?

50. From Figure 6.24d and e, what can you conclude about these two mutant strains?

51. Why did the expression ratios of these two mutant strains have a higher correlation coefficient than you would have expected given the different functions of *Erg4* and *Ecm18*?

With Hughes's results, we now have one more type of information we can extract using microarrays: aneuploidy. **Aneuploidy** is the existence of an abnormal amount of DNA in a particular cell. In this case, both mutants had duplicated some of chromosome 7. It is worth pointing out that these duplication events occurred separately in separate

Is It Useful to Compare the Columns of a Gene Expression Matrix?

The correlation coefficients reported in Figure 6.24 represent a new way of looking at expression profiles. Rather than comparing genes (rows in an expression matrix), the investigators are comparing two experimental conditions (columns): the *erg4* and *ecm18* mutant strains. The Pearson correlation coefficient (see Math Minute 6.2) can be used to measure similarity in both cases. A high correlation coefficient between genes leads researchers to associate their functions, but correlation coefficients between conditions are interpreted differently.

In Figure 6.24a–c, each graphed point represents a gene. The X-coordinate of each point is the expression ratio of that gene in *erg4*, compared to *wt*. The Y-coordinate of each point is the expression ratio of that gene in *ecm18*, compared to *wt*. The correlation coefficient measures the closeness of the points to the line of best fit (see Math Minute 2.5). If the points tend to lie along the line of best fit, and the line has a positive slope, then the correlation is near 1. If the points tend to lie along the line of best fit, and the line has a negative slope, then the correlation is near −1. But if the points do not tend to lie along a line, the correlation is near 0. The linear tendency of the points in Figure 6.24a illustrates a high correlation between the X- and Y-axes, that is, between the expression ratios of the two mutants relative to *wt*, across all genes. In contrast, the "cloud" of points near the origin in Figure 6.24b signifies a low correlation between the expression ratios of the two mutants relative to *wt*.

In some experimental designs, you might expect two distinct columns of an expression matrix to track one another. For example, the two columns may represent the same phase of the cell cycle, one from the first cycle and another from the second. Alternatively, as we will see in Chapter 7, two columns may represent two different samples of the same tissue type. However, in the experiment shown in Figure 6.24, the strong similarity between the two mutants is surprising. Ratios that were significantly elevated in both mutants (almost exclusively genes on chromosome 7) were the primary cause of the high correlation. Thus, identifying an unusually high correlation between columns led immediately to the hypothesis of aneuploidy.

MATH MINUTE DISCOVERY QUESTIONS

1. Hypothesize a situation in which two mutant strains exhibit aneuploidy, but the correlation between the two columns of expression ratios is still near 0.
2. Describe how you might adapt the methods used in Math Minute 3.2 to test for aneuploidy in the data shown in Figure 6.24d–e.

strains that were isolated in different labs. It points to a new possibility that whenever a mutant strain of yeast is made, there is a chance for aneuploidy to occur.

Once you observe a new phenomenon, you want to know how widespread it is. Hughes and his colleagues examined 290 mutant strains and found 22 (about 8%) exhibited aneuploidy! That is a much higher number than anyone expected. How can aneuploidy be so common in yeast that abnormal DNA duplication happens in 8% of all deletion mutant strains? Through a careful analysis of the mutant strains, the team developed a reasonable hypothesis (Figure 6.25). The growth rates for deletion mutants compared to their parental strains indicated the deletion mutants often grow slower.

DISCOVERY QUESTIONS

52. On which chromosome(s) would you guess *Rnr1* and *Rps24a* (Figure 6.25) are located?
53. Go to the MIPS database and search for the two GP deleted genes: *Rnr1* and *Rps24a*. What do they have in common?
54. Hypothesize why chromosome 9 would be dupli-GP cated in *Rnr1* and *Rps24a* mutant strains.

55. Looking at Figure 6.25a, why do you think there is a high percentage of aneuploidy in deletion strains maintained in labs? Explain your answer in evolutionary terms.

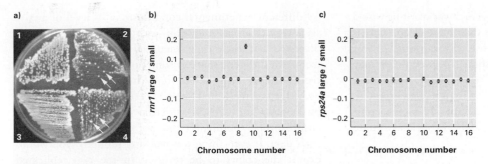

Figure 6.25 Development of aneuploidy in deletion strains.
a) Two different mutant strains (2 is *rnr1* and 4 is *rps24a*) were compared to their parental strains (quadrants 1 and 3) for their ability to grow in standard lab conditions. Two different-sized colonies (small and large) are denoted by white arrows. **b)** and **c)** Relative chromosomal content of deletion strain was calculated and the log$_{10}$ of the ratio was plotted on the Y-axis; chromosome number is plotted on the X-axis. Error bars represent error of the mean log$_{10}$ (ratio).

Rnr1 and *Rps24a* have nothing in common except their chromosomal locations. Surprisingly, they are both located on chromosome 5, not chromosome 9 as you might have guessed. If they had been on chromosome 9, though, the duplication of a deleted gene would not have helped the cells at all. Why would both mutants have duplicated chromosome 9? Both of these genes have isozymes encoded on chromosome 9. *Rps24a* encodes a protein 97% identical to *Rps24b*. Rnr1p has an isozyme (Rnr3p) with 80% amino acid identity. So the only reason both deletion strains duplicated chromosome 9 is that each had functionally redundant genes on chromosome 9. The fact that both strains duplicated the same chromosome does raise the possibility that certain chromosomes may duplicate more readily, or be less toxic than others when duplicated. The possibility of chromosome bias for duplication will require further analysis of many more deletion/aneuploidy mutants. By increasing the amount of mRNA produced from isozymes of the deleted genes, these strains evolved under lab conditions to grow faster, which is what every researcher subconsciously wants because faster-growing strains produce results faster. However, we now appreciate the consequences when deletion strains have agendas that differ from those of investigators.

It would be interesting to determine whether aneuploidy is common. Luckily for Hughes, public data sets from Brown and Botstein were available that could be mined further to determine if other labs had similar strains with aneuploidy. If you refer back to Figure 6.16, you will see that one of the strains DeRisi used had deleted the repressor gene called *Tup1* (*ΔTup1*), which is located on chromosome 7. When Hughes and his team analyzed DeRisi's expression ratios, they found a surprising result (Figure 6.26).

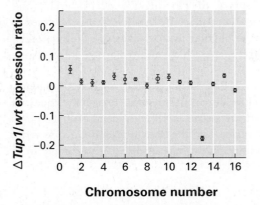

Figure 6.26 Reanalysis of data produced by DeRisi et al.
(Pages 247–248.) Spots indicate the average gene expression ratio (log$_{10}$ on the Y-axis) in *ΔTup1* compared to the parental *wt* strain for each chromosome (X-axis). Error bars indicate the error of the mean log$_{10}$ of the published expression ratios.

DISCOVERY QUESTIONS

56. What do you conclude about the *ΔTup1* strain used by DeRisi et al.?

57. What additional information would you want to have before saying you are certain that aneuploidy exists in the *ΔTup1* genome?

58. Propose a mechanism to explain why the *ΔTup1* strain has a chromosomal deletion rather than a duplication.

59. What does the discovery of frequent aneuploidy suggest should become standard analysis in all microarray experiments?

From the data, it appears that DeRisi was working with a deletion strain that had lost more than he realized. *Tup1* is on chromosome 7, and in response, chromosome 13 lost some of its DNA. The chromosome 13 aneuploidy may have accelerated the growth of the *ΔTup1* strain, and DeRisi never knew his strain was less than he had expected. Based on this analysis by Hughes, aneuploidy is more common than we expected and should be investigated more rigorously.

Can Cells Verify Their Own Genes?

Although the exact definition of a gene is debated, most people agree that an RNA molecule should be produced by a real gene. If the gene produces mRNA, DNA microarrays should be able to detect it. Rosetta Inpharmatics, near Seattle, specializes in performing DNA chip experiments. They have helped develop an "inkjet" printer that allows them to synthesize DNA **oligonucleotides** (short polymers of single-stranded DNA) for each spot. Instead of ink, the printer sprays **dNTPs**: dGTP, dCTP, dATP, and dTTP. Rosetta can produce DNA microarrays composed of 25,000 different oligonucleotides, each one 60 nucleotides long (called **60-mers**, which means polymer of 60 bases; Figure 6.27).

For this work, Rosetta investigators focused on chromosome 22, the first human chromosome to be fully sequenced and annotated (published in 1999). A total of 8,183 exons had been annotated, and the Rosetta team decided to verify these predictions by using labeled cDNA pairs from 69 different experimental conditions. This level of analysis is very rigorous: each exon was tested by two different 60-mers. For each experiment, cDNAs were produced from two different sources; sometimes they were biopsied tissue samples and other times human cell lines. The investigators replicated their experiments (Figure 6.28a) by labeling the cDNAs with the opposite-color dye to ensure probe production did not introduce a bias into their data. If everything was working properly, you would expect the colors to be reversed in the two replicated experiments, as shown in Figure 6.28b and 6.28c. These duplications were performed 69 times on a total of 138 DNA chips.

Figure 6.28 Chromosome 22 DNA microarray data.
Go to www.geneticsplace.com to view this figure.

DISCOVERY QUESTIONS

60. How many different data points would this experiment generate if you count each color for a given spot as a different data point?

61. What would you expect to see in Figure 6.28 if a gene contained 9 exons and the spots were designed so that each 60-mer pair is side by side and each 60-mer pair is placed in the same order in which they are arranged on the chromosome?

62. What would you expect to see if a gene was turned on in only a subset of the 69 experimental conditions?

63. What would you expect to see if a predicted gene is not a real gene and thus was never transcribed?

64. What would you expect to see if a predicted gene is real but produced very little mRNA, or was transcribed in a condition not included in the 69 used for this experiment?

65. What would you expect to see if a gene contained only one exon?

66. What would you expect to see if a gene was alternatively spliced?

67. How could you distinguish a real gene (Discovery Question 64) from a pseudogene (Discovery Question 63)?

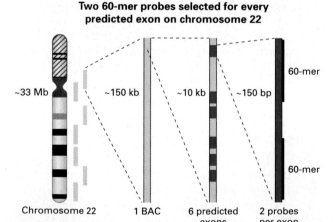

Two 60-mer probes selected for every predicted exon on chromosome 22

~33 Mb · ~150 kb · ~10 kb · ~150 bp · 60-mer · 60-mer

Chromosome 22 · 1 BAC · 6 predicted exons · 2 probes per exon

Figure 6.27 Designing DNA microarrays for determining exon boundaries.
From left to right, increased magnification of chromosome 22, one BAC clone, 10 kb of one BAC, a 150 bp exon, and two 60-mer oligonucleotides. Two 60-mers for each annotated exon were spotted on the chip to create a comprehensive chromosome 22 DNA microarray.

Once all these data points were collected, they had to be displayed in a comprehensible manner (Figure 6.29a). The average expression ratio (control vs. experimental) was represented as a color-coded line with the 69 experiments listed on the Y-axis and 16,366 60-mers listed on the X-axis. Each exon was represented by a pair of 60-mers, so we should look for pairs of data points with identical color patterns.

For example, when a gene with five exons was transcribed, we should see ten spots with similar color patterns.

Figure 6.29 DNA microarray validation of predicted exons on chromosome 22. Go to www.geneticsplace.com to view this figure.

Which Predicted Genes Are Real and Which Ones Aren't?

If a particular gene was transcribed only under certain experimental conditions, we would be hard-pressed to summarize that information, because Figure 6.29a is overwhelming. For single-gene analysis, investigators can reorder the data to place next to each other experiments that affect particular genes in the same way (Figure 6.29b–e). In Figure 6.29b, we see a known gene induced in 39 experimental conditions (red boxes) and repressed in 8 experimental conditions (green boxes). In c), we see the results for another known gene, but one predicted exon (denoted by arrow) does not appear to be included in the mRNA from this gene. In d), we see two different EST sequences transcribed under similar conditions, which indicates they might be encoded by the same gene. Finally, e) shows one set of exons that is transcribed at high levels only in experiments 21 and 68. Experiment 21 compared mRNA from a collection of cell lines to mRNA from testes, while experiment 68 compared testes to uterine mRNAs, Only these two experimental conditions examined testicular mRNA, and only they have clear color patterns for these 60-mers. The data indicate these predicted exons are expressed only in the testes and appear to constitute a real gene.

DISCOVERY QUESTIONS

This series of questions relates to Figure 6.29. For these questions, assume there are no errors due to experimental technique.

68. In panel b), why do some individual spots appear black even when they occur in the middle of a red or green patch? Why might *pairs* of spots appear black even when they occur in the middle of a red or green patch?

69. One interpretation for c) is that this exon is not a true one. Propose an alternative interpretation.

70. In panel d), one pair of spots appears black most of the time. What does this indicate?

71. In panel e), the exons had not been annotated as a gene, but this experiment indicates that they form a real gene. How many exons would be in this gene?

Can Microarrays Improve Annotations?

LINKS G⚬P
Sanger Center

Computer programs are very good at rapidly performing tasks we tell them to do, but the programs have no more insight into the truth than the people who program them. Unfortunately, no one fully understands what defines the beginning or ending of an exon. We know that most exons are preceded by the dinucleotide **AG** and followed by the dinucleotide **GT**, but these dinucleotides also occur by chance where there are no exons. We rely on other sequence patterns with vague rules to help us locate exons, but we make mistakes. When the Rosetta team identified a new gene based on predicted exons, they wanted to determine experimentally the exact size of the six exons. The Sanger Center (one of the biggest genome labs in the world) had defined the length of each potential exon with their software. The Rosetta team made a new set of DNA chips, but this time they produced a new 60-mer every 10 bp over the entire length of a 113,000-bp piece of DNA that contained the newly discovered testes-specific gene (Figure 6.30). After creating this microarray of overlapping 60-mers, they probed the 60-mers with cDNAs that contained the transcript of interest and determined which spots bound the labeled cDNAs. By using a systematic approach, the team was able to confirm some exon predictions (exons 1, 2, 4, and 5) and correct the mistakenly short predictions for exons 3 and 6.

DISCOVERY QUESTIONS

72. How many different 10 bp overlapping 60-mers did Rosetta have to produce to cover both strands of the 113,000 bp fragment of chromosome 22?

73. Why did they produce 60-mers for *both* strands of the DNA?

74. Would the DNA chip data alone be sufficient to determine exactly which specific nucleotide began or ended an exon? Explain your answer.

Could a Microarray Validate Annotation of an Entire Genome?

After the Rosetta group produced a list of microarray-confirmed exons for chromosome 22, they decided to survey the complete human genome. Using the **frozen genome** from June 15, 2000, Rosetta produced 50 DNA microarrays that contained 1,090,408 60-mers to cover all 442,785 predicted exons (Figure 6.31a). The investigators probed these DNA chips using colored cDNA produced from two cell lines: a human colon cancer and a human lymphoma. With these two cDNA populations, they were able to detect expression from 58% of the confirmed exons and about 123,000 additional exons out of 364,299 predicted

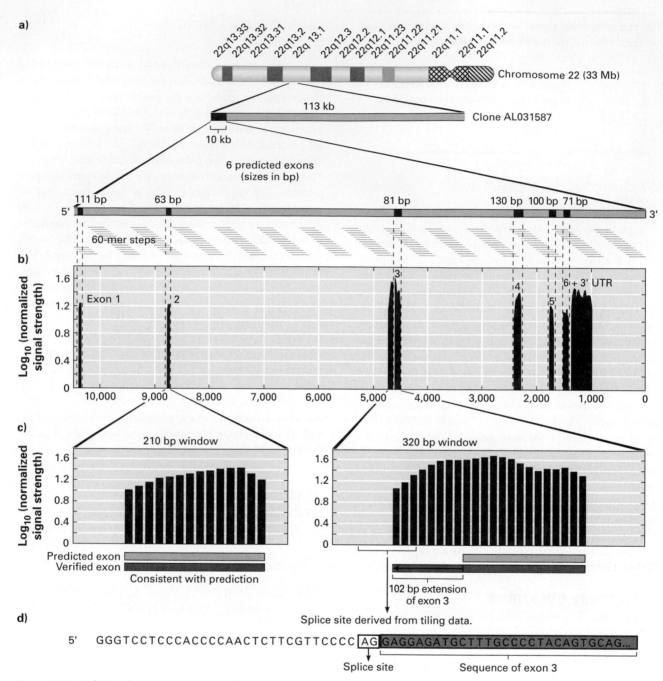

Figure 6.30 Defining the exons of a newly discovered gene.
a) Exons for the testes-specific gene (Figure 6.29e) were defined by 60-mers every 10 bp over a 113,000 bp piece of chromosome 22. **b)** Six exons were identified over a 10 kb region as shown. **c)** Two exons are shown at higher magnification, with each 60-mer represented as a separate bar graph. Below the graph are comparisons of exon predictions (light purple) using current software technology and the microarray exon detection (dark purple). **d)** Sequence from the 5' end of exon 3 showing the consensus splice site that is consistent with the microarray prediction for the 5' boundary of exon 3.

but previously unconfirmed exons. They estimated a 5% false positive rate, which may reduce the number of confirmed exons, but their pioneering work is very impressive (Figure 6.31b–d). By devising these experiments, they introduced a method for large-scale and experimental annotation of the human genome.

DISCOVERY QUESTIONS

75. Only 60% of the "confirmed" exons were detected. Is this method not very accurate, or are the "confirmed exons" not really exons? Explain your answer.

Figure 6.31 Whole genome DNA microarray to validate every predicted exon.
Go to www.geneticsplace.com to view this figure.

76. The investigators detected mRNA from some predicted exons but not all of them. Interpret these data.

77. DeRisi pioneered another use of DNA microarrays. Search the PLoS Biology web site with the phrase "DeRisi SARS." Read the Wang et al. abstract and study figure 2. What innovation helped DeRisi's lab identify the SARS virus?

Summary 6.2

One yeast DNA chip produces as much data as about 12,000 Northern or Southern blots. With this much information, you can appreciate why data analysis and visualization are fundamental to our ability to understand the data. We often use guilt by association to propose functions for nonannotated genes, but variations in clustering software can produce different associations between genes. The way we display this information may limit our perceptions or open new views of the data that lead to new interpretations. With carefully designed microarray experiments, we saw how genomes can lose or gain parts of their chromosomes through aneuploidy. As the resolution and production of DNA microarrays improve, we can use these high-throughput methods to better understand genomic sequences, as was demonstrated by the first use of chips on the human genome sequence.

Chapter 6 Conclusions

DNA microarrays are powerful tools to analyze genomes in vivo. We can see how a genome inside a nucleus responds to an ever-changing environment: the cell. When one component of the extracellular environment changes, the genome adapts by altering its cellular environment. Genomic research is repainting the picture we see inside cells. Traditionally, we have altered single genes and then measured the cellular phenotype. DNA microarrays have shown us that genomes are capable of altering themselves through aneuploidy, so we unknowingly study substantially different cells controlled by dynamic genomes, ones that can alter their expression and composition. When applied to the human genome, we can discover new genes, or refine gene annotation. The success of DNA microarrays depends in large part on how we analyze and visualize the data. Mathematicians, computer scientists, and biologists are collaborating

to create bioinformatics tools that will improve our understanding of genome-wide behaviors. In Chapter 7, several examples illustrate how basic research methods can be applied to better understand cancer, how medicines work, which medications should be used for individual patients, and how leptin modulates obesity and many other genome circuits.

References

Barry, C. E. III, & B. G. Schroeder. 2000. DNA microarrays: Translational tools for understanding the biology of *Mycobacterium tuberculosis*. *Trends in Microbiology.* 8(5): 209–210.

Berns, A. 2000. Gene expression in diagnosis. *Nature.* 403: 491–492.

Brazma, A., & J. Vilo. 2000. Gene expression data analysis. *FEBS Letters.* 480: 17–26.

Brent, R. 1999. Functional genomics: Learning to think about gene expression data. *Current Biology.* 9: R338–R341.

Chee, M., R. Yang, et al. 1996. Accessing genetic information with high-density DNA arrays. *Science.* 274: 610–614.

Chu, S., J. DeRisi, et al. 1998. The transcriptional program of sporulation in budding yeast. *Science.* 282: 699–705.

Conklin, B. 2001. GenMapp: Gene MicroArray Pathway Profiler. <http://gladstone-genome.ucsf.edu/introduction.asp>. Accessed 8 February 2002.

Dalton, R. 2000. DIY (Do it yourself) microarrays promise DNA chips with everything. *Nature.* 403: 236.

DeRisi, J. L., V. R. Iyer, & P. O. Brown. 1997. Exploring the metabolic and genetic control of gene expression on a genomic scale. *Science.* 278: 680–686.

Eisen, M. B., P. T. Spellman, et al. 1998. Cluster analysis and display of genome-wide expression patterns. *PNAS.* 95: 14863–14868.

Gasch, A. P., P. T. Spellman, et al. 2000. Genomic expression programs in the response of yeast cells to environmental changes. *Molecular Biology of the Cell.* 11: 4241–4257.

GCAT: The Genome Consortium for Active Teaching. 2001. <http://www.bio.davidson.edu/GCAT>. Accessed 8 February 2002.

Geschwind, D. H. 2000. Mice, microarrays, and the genetic diversity of the brain. *PNAS.* 97: 10676–10678.

Getz, G., E. Levine, & E. Domany. 2000. Coupled two-way clustering analysis of gene microarray data. *PNAS.* 97: 12079–12086.

Gilbert, D. R., M. Schroeder, & J. van Helden. 2000. Interactive visualization and exploration of relationships between biological objects. *Trends in Biotechnology.* 18: 487–496.

Harris, T. M., A. Massimi, & G. Childs. 2000. Injecting new ideas into microarray printing. *Nature Biotechnology.* 18: 384–385.

Heyer, L. J., S. Kruglyak, & S. Yooseph. 1999. Exploring expression data: Identification and analysis of coexpressed genes. *Genome Research.* 9: 1106–1115.

Hughes, T. R., C. J. Roberts, et al. 2000. Widespread aneuploidy revealed by DNA microarray expression profiling. *Nature Genetics.* 25: 333–337.

Kyoto Encyclopedia of Genes and Genomes (KEGG). 2001. <http://www.genome.ad.jp/kegg/>. Accessed 8 February 2002.

Lee, M.-L. T., F. C. Kuo, et al. 2000. Importance of replication in microarray gene expression studies: Statistical methods and evidence from repetitive cDNA hybridizations. *PNAS.* 97(18): 9834–9839.

Lemkin, P. F., G. C. Thornwall, et al. 2000. The microarray explorer tool for data mining of cDNA microarrays: Application for the mammary gland. *Nucleic Acids Research.* 28(22): 4452–4459. <http://www.lecb.ncifcrf. gov/MAExplorer/>.

Liao, B., W. Hale, et al. 2000. MAD: A suite of tools for microarray data management and processing. *Bioinformatics.* 16(10): 946–947. <http://pompous.swmed.edu/>.

Lockhart, D. J. 1998. Mutant yeast on drugs. *Nature Medicine.* 4: 1235–1236.

Lockhart, D. J., & C. Barlow. 2001. Expressing what's on your mind: DNA arrays and the brain. *Nature Reviews.* 2: 63–68.

Madden, S. L., C. J. Wang, & G. Landes. 2000. Serial analysis of gene expression: From gene discovery to target identification. *Drug Discovery Today.* 5(9): 415–425.

Manduchi, E., G. R. Grant, et al. 2000. Generation of patterns from gene expression data by assigning confidence to differentially expressed genes. *Bioinformatics.* 16(8): 685–698.

National Center for Genome Resources. 2001. GeneX: A collaborative Internet database and toolset for gene expression data. <http://www.ncgr.org/research/genex/>. Accessed 8 February 2002.

Pinkel, D. 2000. Cancer cell, chemotherapy, and gene clusters. *Nature Genetics.* 24: 208–209.

Schaffer, J. L., M. Pérez-Amador, & E. Wisman. 2000. Monitoring genome-wide expression in plants. *Current Opinion in Biotechnology.* 11: 162–167.

Sherlock, G., T. Hernandez-Boussard, et al. 2001. The Stanford Microarray Database. *Nucleic Acids Research.* 29(1): 152–155.

Shoemaker, D. D., E. E. Schadt, et al. 2001. Experimental annotation of the human genome using microarray technology. *Nature.* 409: 922–927.

Velculescu, V. E. 1999. Tantalizing transcriptomes: SAGE and its use in global gene expression analysis. *Science.* 286: 1491–1492. <http://www.sagenet.org/>.

Walker, G. M. 1998. Yeast physiology and biotechnology. New York: John Wiley & Sons.

Young, D. B., & B. D. Robertson. 1999. TB vaccines: Global solutions for global problems. *Science.* 284: 1479–1480.

Additional Readings of Interest

Arimura, G.-I., K. Tashiro, et al. 2000. Gene responsiveness in bean leaves induced by herbivory and by herbivore-induced volatiles. *Biochemical and Biophysical Research Communications.* 277: 305–310.

Brown, P. O., et al. 2001. The Mguide. <http://cmgm.stanford. edu/pbrown/mguide/>. Accessed 14 July 2005.

Cho, R. J., M. Fromont-Racine, et al. 1998. Parallel analysis of genetic selections using whole genome oligonucleotide arrays. *PNAS.* 95: 3752–3757.

Cummings, C. A., & D. Relman. 2000. Using DNA microarrays to study host-microbe interactions. *Genomics.* 6(5): 513–525.

Kane, M. D., T. A. Jatkoe, et al. 2000. Assessment of the sensitivity and specificity of oligonucleotide (50mer) microarrays. *Nucleic Acids Research.* 28(22): 4552–4557.

Khodursky, A. B., B. J. Peter, et al. 2000. Analysis of topoisomerase function in bacterial replication fork movement: Use of DNA microarrays. *PNAS.* 97: 9419–9426.

Khodursky, A. B., B. J. Peter, et al. 2000. DNA microarray analysis of gene expression in response to physiological and genetic changes that affect tryptophan metabolism in *Escherichia coli. PNAS.* 97(22): 12170–12175.

Kumar, A., & Z. Liang. 2001. Chemical nanoprinting: A novel method for fabricating DNA microchips. *Nucleic Acids Research.* 29(2): e2.

Okamoto, T., T. Suzuki, & N. Yamamoto. 2000. Microarray fabrication with covalent attachment of DNA using bubble jet technology. *Nature Biotechnology.* 18: 438–441.

Selinger, D. W., K. J. Cheung, et al. 2000. RNA expression analysis using a 30 base pair resolution *Escherichia coli* genome array. *Nature Biotechnology.* 18: 1262–1268.

Talaat, A. M., P. Hunter, & S. A. Johnston. 2000. Genome-directed primers for selective labeling of bacterial transcripts for DNA microarray analysis. *Nature Biotechnology.* 18: 679–682.

Ye, R. W., W. Tao, et al. 2000. Global gene expression profiles of *Bacillus subtilis* grown under anaerobic conditions. *Journal of Bacteriology.* 182(16): 4458–4465.

Applied Research with DNA Microarrays

7.1   **Cancer and Genomic Microarrays**

Improve diagnosis and treatment of cancer patients.

Understand the complexity of cancer biology.

Discover genome dynamics in cancerous cells.

7.2   **Improving Health Care with DNA Microarrays**

Compare the effectiveness of a drug treatment for different types of cancers.

Explore genomic responses to leptin treatments and fat accumulation.

LINKS
Ash Alizadeh
David Botstein
Louis Staudt
Mike Eisen
Pat Brown

METHODS
cDNA
DNA microarrays

Chapter 6 introduced DNA microarrays and how they can be used for basic research. Chapter 7 discusses biomedical applications for DNA chips. The first half of the chapter focuses on cancer. You will see how DNA microarrays were used to better understand two forms of cancer, a lymphoma and breast cancer. Ideally, microarrays would allow us to diagnose and treat patients more successfully. In addition, we would like to understand more fully the genomic circuitry underlying the diversity and complexity found in clinical settings. In the second half of the chapter, you will examine case studies in which microarrays were used to better understand how medications work and how we can improve their effectiveness. The final case study explores the complexity of the fat-regulating hormone leptin (see Chapter 5). In each of the case studies, basic and applied research are interwoven as we try to understand the basic biology in order to develop more effective treatments.

7.1 Cancer and Genomic Microarrays

As life expectancy increases, there appears to be a tradeoff: increased incidence of diseases. You probably know someone who has been diagnosed with cancer. Many times the patients are elderly, but not always. Since some cancers run in families, we know that an important component of cancer formation is determined by the genome. Can we use the dynamic activity of a person's genome to cure him or her? Are some cancers more amenable to DNA microarray diagnosis than others? When a cell becomes cancerous, can we use microarrays to search for chromosomal instability? The three case studies in this section illustrate the potential for DNA microarrays, and their limitations.

Are There Better Ways to Diagnose Cancer?

Microarrays are being touted as the next great tool for diagnosis of disease. Pat Brown's and David Botstein's laboratories teamed up with several other labs, including Louis Staudt at the National Cancer Institute in Frederick, Maryland, to determine if DNA microarrays could be used to diagnose diffuse large B-cell lymphoma (DLBCL). The research involved several investigators, such as Ash Alizadeh, who did much of the microarray work, and Mike Eisen, a former postdoc at Stanford who now runs his own lab at the Lawrence Berkeley National Laboratory. DLBCL is a very aggressive malignancy of B-cells. Each year in the United States, 25,000 new cases are reported, accounting for about 40% of all non-Hodgkin's lymphoma cases. Sadly, only about 40% of all DLBCL patients are treated

successfully with chemotherapy. No one is sure why some respond and others do not. DLBCL is diagnosed by a loose combination of morphological and molecular characteristics. For example, the **immunoglobulin** (antibody) genes in non-Hodgkin's lymphomas show signs of **somatic hypermutation**, which is a process of increased changes in the DNA sequence of the variable regions of antibody genes. B-cells undergo somatic hypermutation to improve the ability of antibodies to bind to their antigens. Somatic hypermutation occurs in the germinal centers of secondary lymphoid tissue, such as lymph nodes.

To conduct this work, the team constructed human DNA microarrays consisting of 17,856 spots each. They produced 128 microarrays and converted mRNA samples into cDNA probes from 96 normal and malignant lymphocytes to collect about 1.8 million ratios (Figure 7.1). Although the data look like a mess, Eisen's software has extracted some interesting trends. In these experiments, the 96 human samples were used individually to produce 96 different red cDNAs and compared to common green control cDNA produced from 9 different lymphoma cell lines, which were pooled and used as a consistent source of cDNA.

Figure 7.1 DLBCL gene expression analysis.
Go to www.GeneticsPlace.com to view this figure.

The **dendrogram** shown in Figure 7.1 was produced by first clustering the genes according to expression pattern similarities and then clustering the 96 samples based on their similarity of gene expression profiles. The result produced **signature genes**, indicated by colored bars on the right side of the figure. Signature genes can be used to describe a particular cell type. For example, activated B-cells are known to induce and repress particular genes; these genes constitute the activated B-cell signature genes.

The first notable feature is that a subset of genes could be used to classify the samples based on cell type (e.g., germinal center B-cell) or cell process (e.g., proliferation). It is striking to see the DLBCL samples clustered into the left major branch of samples. Of the genes within the six sets of signature genes, some had no known function (Figure 7.2) but their expression profiles were similar to those of known genes within the signature set.

Figure 7.2 Biologically distinct DLBCL gene expression signatures.
Go to www.GeneticsPlace.com to view this figure.

DISCOVERY QUESTIONS

1. Which three types of samples have the highest overall expression of proliferating genes (see Figure 7.2)?

2. Are all the DLBCL samples clustered together? Hypothesize why.

3. Which of the signature genes might be the best to subdivide the DLBCL into subcategories?

4. Go to **Explore Lymphoma** Figure 1 web site, an interactive version of the complete data set from the DLBCL paper. In the left frame is a "radar" view that is zoomed out. In the right frame is a "zoom" window that gives you a closer look at 100 genes at a time. The top frame allows you to change views. Perform the following steps:
   a. In the top frame, "Change zoom X factor to:," enter the number 3, and click on "submit." This will zoom out the view on the right.
   b. In the "radar" frame, scroll down and click on the orange bar to view all the GC-like genes.
   c. Search for a gene/protein that, based on its expression profile, might be a good target for killing DLBCL but not normal B- or T-cells. Explain why you chose this particular target.

All but three samples from the DLBCL collection were clustered into one large group. The three outliers were clustered apart from the others because a majority of the cells in these three biopsies may have contained larger numbers of T cells or less lymph tissue than the other DLBCL samples. DLBCL cells tend to express a lot of proliferation genes, which is consistent with DLBCL being an aggressive malignancy. The germinal center B-cell signature genes appear to be the most heterogeneous, and thus might prove useful for reclustering DLBCL into new subcategories. Furthermore, the germinal center signature genes might help reveal different stages of B-cell development that occurred in the original cancerous cell.

The DLBCL samples and two germinal center B-cell lines were reclustered using only the germinal center B-cell signature genes (Figure 7.3a); this reclustering produced two subgroups. About half of the clustered samples (colored orange) have induced germinal center (GC) genes, while the other half (colored blue) have repressed GC genes. The two normal GC B-cells (colored black) clustered with the DLBCL half that had induced GC genes. Using this organization of samples, the investigators returned to the full collection of genes and reclustered the expression profiles of all 17,856 genes based on the orange/blue subclassification of DLBCL (Figure 7.3b). The 2,984 genes that were inversely repressed (marked by blue

side bar) or induced (marked by orange side bar) in the DLBCL samples were reclustered while maintaining the DLBCL dendrogram, as produced in Figure 7.3a. Based on the gene expression profiles of DLBCL samples, about half of the lymphomas looked like GC B-cells, and the other half looked like activated B-cells (Figure 7.3c). Some heterogeneity still existed in these two halves of DLBCLs, which indicated additional subdivision might be possible.

LINKS
Explore Lymphoma

Figure 7.3 Discovery of DLBCL subtypes. Go to www.GeneticsPlace.com to view this figure.

DISCOVERY QUESTIONS

5. Summarize the steps taken in the three parts of Figure 7.3.

6. Why did the investigators bother doing steps b and c if they just wound up with the same organization of the DLBCL biopsy samples?

The microarray data supported clinical observations: Not all DLBCLs are alike. Although interesting, this was not clinically useful information. The team analyzed the accompanying clinical data to see how the different patients fared after standard chemotherapy treatments. Forty patients had been treated over the course of many years, and 22 (55%) of them had died, which is close to the national mortality rate for DLBCL. When the patients were matched with their samples and charted, there was a startling outcome (Figure 7.4). Those whose cells appeared to be GC-like had a much higher survival rate than those who had activated B-cell-like lymphomas. However, classifying patients based on gene expression patterns is experimental, so it is important to compare microarray-based classification with a more familiar method. Therefore, the patients were regrouped according to traditional clinical criteria (International Prognostic Index [IPI]) and charted (Figure 7.4b). The patients were divided into high- and low-risk categories. The IPI was also good at distinguishing the types of patients, and it correlated well with the new experimental method.

DISCOVERY QUESTIONS

7. Were all of the human genes included on the microarray? Might larger numbers of genes improve the resolution of this method?

8. Other than predicting survival rates, could microarray data be used in other clinically useful ways?

LINKS
Lymphoma Figure 3
UniGene

| Math Minute 7.1 | **What Are Signature Genes, and How Do You Use Them?**

In Math Minute 6.3, we saw how genes can be grouped using a hierarchical clustering algorithm. With the experimental design of the DLBCL study, genes expressed in particular types of cells or processes are grouped together; the investigators called them "signature genes." You can identify signature genes by searching for a single gene whose role and/or subcellular location matches certain criteria, and including genes clustered nearby. However, because all genes are eventually clustered into one dendrogram, you need a systematic way to identify nearby genes.

One definition of a gene's neighbors is the set of genes that are joined to the original gene at a similarity level above an arbitrary threshold. For example, in Figure MM6.1 and Table MM6.3, gene C has only gene E as a neighbor at the 0.96 threshold, but it has genes E, L, G, and D at the 0.94 threshold. Defining clusters in this way is sometimes referred to as "cutting the tree." A vertical line drawn in Figure MM6.1 at a predefined similarity level would cut the tree into disjoint subtrees, each one a cluster formed at the predefined similarity level or above. As shown in Figure 7.1, hierarchical clustering was also applied to the columns of the gene expression matrix, resulting in the clustering of the 96 samples. Samples that are clustered together have similar expression profiles across all 17,856 genes. The clusters labeled A and G in the dendrogram of Figure 7.1a were obtained by cutting this tree.

To find genes useful for classifying the clinical prognosis of DLBCL samples, you want to find gene clusters (signature genes) that have patterns of red in some DLBCLs and green in others. (Note that this cannot be done without first clustering the genes.) It appears that the investigators identified the signature genes by carefully studying the patterns in Figure 7.1b, but the same job could be done mathematically as follows. Create a fictitious gene with an expression pattern consisting of 2X repression in each of the first 6 columns (i.e., samples marked as cluster A), 2X induction in each of the next 10 columns (i.e., samples marked as cluster G), and no change in expression in the remaining 80 columns. The germinal center B-cell signature genes have a large positive correlation with the fictitious gene. The activated B-cell signature genes have a large negative correlation with the fictitious gene. The correlations of the remaining signature genes with the fictitious gene are closer to zero.

Once signature genes have been identified, they can be used to recluster the samples. For example, Figure 7.3a shows the results of a new clustering of the samples, formed by using only the germinal center B-cell signature genes to assess similarity between samples. Note that a strong linear relationship (i.e., high correlation) exists between the expression profiles of samples DLCL-0051 and DLCL-0033 when the comparison is restricted to germinal center B-cell signature genes. Accordingly, DLCL-0051 and DLCL-0033 are clustered together in Figure 7.3d. However, as most other genes did not exhibit a distinctive expression pattern in the DLBCL samples, including them in the comparison diluted the correlation. The dilution effect on correlation may explain why DLCL-0051 and DLCL-0033 were not clustered tightly together when all 17,856 genes were used to assess similarity between samples (see Figure 7.1a).

9. Go to Lymphoma Figure 3 to see a list of genes determined to be the best at distinguishing the two subclasses of DLBCL. Scan the expression profiles and see if any gene/protein might be a good target for chemotherapy treatment of both classes of DLBCL.

10. One gene you might have noticed from Discovery Question 9 was *Ttg-2*, which is highly induced in more than half of the DLBCL samples. Go to UniGene and search for *Ttg-2*. Based on its tissue distribution, do you think this would be a DLBCL-specific target? Support your answer with the UniGene data.

At this point, the study might seem to be over, but the team makes an intriguing suggestion. What would happen

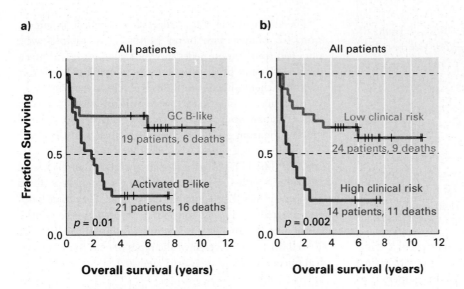

Figure 7.4 Clinical distinctions of DLBCL.
Kaplan-Meier plot of 40 patients being treated for DLBCL. The graph moves down a step each time a patient dies. The tick marks indicate when a patient was surveyed by a clinician. **a)** When the patients were separated into GC-like or activated B-cell-like categories, the survival rates were noticeably different, as indicated. **b)** When more traditional IPI indices were used, the patients again segregated into high-risk and low-risk categories, as indicated. The X-axis is the number of years survival postdiagnosis, and the Y-axis is the fraction of surviving patients.

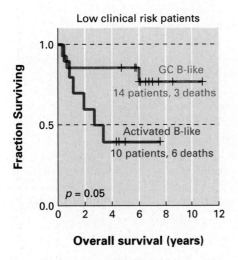

Figure 7.5 Clinical distinction of IPI low-risk DLBCL patients.
The graph moves down a step each time a patient dies. The tick marks indicate when a patient was surveyed by a clinician. The 24 low-risk patients from Figure 7.4b were reanalyzed using the GC signature gene analysis and segregated into two categories as indicated.

if the 24 IPI-rated low-risk patients were reanalyzed using the GC B-cell signature gene criteria? Would it be possible to characterize people with better accuracy than the IPI method alone? Using their clustering definition, the investigators were better able to predict who would benefit from standard chemotherapy and who might have benefited from bone marrow transplantation (Figure 7.5).

Table 7.1 Probability of long-term survival when genes induced in DLBCL.

High	Low
LMO2	BCL2
BCL6	SCYA3
FN1	CCND2

Since this first paper appeared, several other papers have presented their own lists of optimal genes for predicting the outcome of a DLBCL diagnosis. In 2004, Izidore Lossos and colleagues at Stanford University conducted a meta-study of 36 genes from 66 patients using real-time PCR as an independent method to validate the microarray findings. From the list of 36, 6 genes were found to be best at predicting the survival rate of patients with DLBCL (Table 7.1). The reliability of prediction was independent of the IPI rating. The authors tested these six genes in the other published data sets and found that they were also predictive in other studies, regardless of the particular method used to collect the microarray data.

DISCOVERY QUESTIONS

11. Summarize the data illustrated in Figure 7.5.
12. What can you conclude about the potential for using DNA microarrays to diagnose DLBCL?

DATA
Alizadeh et al. database

LINKS
Charles Perou
GEO Profiles
Molecular Portraits
Histology
Therese Sørlie

13. Search for the gene "BCL-6" in the Alizadeh et al. database. Do these online data support the findings by Lossos et al.?

14. Search using "LMO2 DLBCL" in GEO Profiles. Does this microarray database at NCBI support Lossos et al.? Search for *CCND2* and *SCYA3*. Are these two genes repressed in all DLBCL samples? Would you expect them to be? Are they repressed or induced in other cancers? Explain why your findings make sense given the clinical information presented.

We would like to use such information to predict which patients would benefit from chemotherapy and which would not. There is no point in subjecting a person to chemotherapy if he or she will not respond favorably. Furthermore, we might be able to utilize this information to develop more effective drugs for cancer therapy (see pages 276–277). Discovery Question 14 highlights why DNA microarrays may be very helpful for improving diagnoses in the future.

Can Breast Cancer Be Categorized with Microarrays?

Botstein and Brown teamed up again to examine whether breast cancer treatment might benefit from microarray data, as appears to be possible for DLBCL. The two lead authors, Charles Perou (a postdoc in Botstein's lab) and Therese Sørlie of Oslo, Norway, knew that unlike DLBCL, breast cancer does not exhibit two distinct clinical categories, so the challenges in categorizing breast cancer would be greater than for DLBCL. With the discovery of genes that play a role in suppressing breast cancers (*BRCA1* and *BRCA2*), many people outside the research community think the battle is nearly won. This impression is fostered by simplistic headlines and a lack of understanding of the complexity and diversity of breast cancer.

DISCOVERY QUESTIONS

15. Go to the Molecular Portraits Histology site and see the tissue biopsies used in this study. Click on at least three of the samples and look at the diversity of cell types. Be sure to view at least one of each of the grade levels of tumors (1 is the least and 3 is the most progressed). What problems do you foresee if tissue biopsies are used as the sources for sample mRNA?

16. Experimental mRNAs have to be compared to a control source of mRNA. What would you choose as your control mRNA?

The international team collected 65 surgical specimens from 42 women with breast tumors. Two samples were obtained from lymph nodes to which the cancer had metastasized. Twenty of the tumors were sampled twice, before and after chemotherapy with doxorubicin, a standard treatment. Human microarrays were printed that contained 8,102 human **open reading frames** (**ORFs**). For control cDNA, the team produced green **Cy3**-labeled cDNAs from 11 different cell lines as a reliable source of reference mRNA. The cDNAs produced from biopsy samples were labeled red with **Cy5** dye. The investigators chose to focus on only 1,753 ORFs whose expression in at least three biopsies differed by at least fourfold compared to the control median expression (Figure 7.6). In addition to the biopsies, 17 cell lines that represented the types of cells found in typical breast cancer biopsies were analyzed.

Figure 7.6 Variations in expression of 1,753 human ORFs.
Go to www.GeneticsPlace.com to view this figure.

DISCOVERY QUESTIONS

17. Are there any obvious trends in the gene profiles from the breast cancer biopsies?

18. What pattern do you notice for most of the paired tissue samples? Are there any exceptions to it?

19. Find the normal breast tissue cluster. Where did these samples cluster relative to the tumors? What does this indicate?

20. Return to the Molecular Portraits Histology images. Examine the two available biopsies of the samples nearest to normal breast tissue (Norway 61 and Norway 101). Are they grade 1, as you might predict based on the clustering?

The data lack any obvious patterns in the hierarchical clustering of the biopsies. There are a couple of aspects of particular interest. First, the tissue samples exhibit a high degree of heterogeneity in gene expression profiles. The variation in gene profiles appears to correlate with the variation in tumor growth rates and signaling pathways activated in each tumor (e.g., estrogen, interferon, etc.). Second, the before and after biopsies clustered together in all but three instances. The normal tissues clustered right in the middle of the tumors and not by themselves, indicating that the tumors are very much like normal breast tissue.

The primary objective of this study was to determine which genes might prove useful in identifying subclasses of breast tumors and thus be clinically useful. This study was especially well designed because it takes advantage of breast tumor biopsies before and after chemotherapy. The investigators looked for genes that were regulated in similar ways within a patient but not in other patients. The team identified 496 genes that met the criteria for distinguishing one person's paired biopsies from biopsies taken from other patients. The expression of this **intrinsic gene subset** of 496 genes varied more when tumors from different patients were compared than when paired biopsies from one patient were compared (Figure 7.7).

Figure 7.7 Cluster analysis using the intrinsic gene subset.
Go to www.GeneticsPlace.com to view this figure.

DISCOVERY QUESTIONS

21. Go to the web version of breast cancer GeneXplorer, and click on Figure 2 to explore the database. Set the radar for 400%, and then click on regions in the left frame where you would like to see the names of the genes. Do the genes you selected have anything in common?

22. Do you think that all the categories of breast cancer have been identified by the four subclasses in Figure 7.7?

23. In GeneXplorer Figure 2, type "ERB" in the blank box at the top and click on "GO." You will see a list of four *ERBB2* genes. *ERBB2* is on chromosome number 17. Click on one of the brightest red boxes in an *ERBB2* row. You will retrieve genes that were co-expressed with *ERBB2*. Notice the expression pattern for a gene called *FLOT2*, and click on the blue FLOT2 name to learn what this gene does. Does it have any functions that indicate it might lead to cancer if mutated?

24. Go to Entrez Gene. Enter "ERBB2" and notice the chromosomal location for the human gene. Perform a new search for "FLOT2," and note its chromosomal location.

   a. What genomic event might explain several linked genes that all have elevated expression profiles?

   b. Other genes nearby include an apoptosis antagonist called *MAP kinase kinase kinase*, which is needed for inducing mitosis, and *BRCA1*, the "breast cancer gene." This region is often amplified in breast cancers. Do you think chromosomal amplification (i.e., aneuploidy) is a cause, an effect, or unrelated? How could you determine which?

LINKS
Entrez Gene
GeneXplorer

It appears that the easiest way to classify the tumors was by their expression of the estrogen receptor (ER), with ER-positive cells constituting about half of the tumors. One class of tumors appears to be related to a genomic amplification of a portion of chromosome 17 that includes many important genes for breast cancer. Unfortunately, the sample size was too small to definitively identify most classes of breast cancer.

Interestingly, most tumors look more like themselves than another tumor sample within a cluster. Almost all of the biopsy pairs look more like individual tumor signatures than class signatures. Of course, the intrinsic gene set was designed to maximize this effect, but even before this last round of clustering, biopsy pairs predominantly clustered near each other. The weeks of chemotherapy did not significantly alter the gene expression profile in 85% of the tumor pairs.

In 2005, Liat Ein-Dor and colleagues at Wiezmann Institute in Israel reconsidered three different studies of breast cancer microarray data. The authors were struck by the small number of genes identified as predictors in all three public data sets: only 2 out of over 200 genes. Although each study ranked about 70 genes as predictors of clinical outcome, the meta-analysis found that many different collections of 70 genes were equally predictive of clinical outcome based on correlation of expression. However, predictive power based on correlation does not indicate a gene's importance in the cause or cure of the cancer. Ein-Dor et al. concluded that the full list of all correlated genes (not just the top 70 from a single publication) should be considered as potential targets of new cancer drugs.

DISCOVERY QUESTIONS

25. Hypothesize why gene clustering may be more challenging to use for breast cancer than for DLBCL.

26. What is the significance of tumor pairs (before and after chemotherapy) having gene profiles that are more similar to the person than the type of tumor? Do you believe each person's tumor is unique, or is there an alternative explanation?

27. Explain the rationale for using the intrinsic gene subset for clustering. Is this a valid approach? Support your opinion with data presented in this section.

28. Explain how three different studies could identify three different lists of genes that were all predictive for clinical outcome. Explain why the two genes on all three lists are not necessarily the best targets for new cancer drugs.

LINKS
Michael Wigler
METHODS
PCR

What Genomic Changes Occur in Cancer Cells?

In the breast cancer and the yeast aneuploidy (Section 6.2) case studies, we learned that microarrays can detect chromosomal duplications and deletions if the investigators look for them. Robert Lucito and his colleagues, working in the lab of Michael Wigler at Cold Spring Harbor Laboratory, wanted to examine directly the extent of aneuploidy in human cancers and to determine if tumors have particular patterns of aneuploidy. To perform this work, they used a partial human genome DNA microarray. The human genome was not completely sequenced at the time the paper was written, but Wigler's team devised a method that did not require a fully sequenced genome.

Wigler's method is a clever variation on the microarray methods we have seen so far. In the standard method, the target DNA on the glass slide is composed of genes. Since the human genome is too big and too complex to work with easily, Wigler et al. printed **genomic representations** of the genome instead. A genomic representation is a subset (about 2.5% of the total) of the whole genome, composed of pieces about 1 kb in length. To create a representation, investigators digest total genomic DNA with a "six cutter" restriction enzyme that recognizes a particular combination of six nucleotides. They chose Bgl II, which cuts the sequence AGATCT, to create sticky ends.

```
5′...A-OH      PO₄-G A T C T...3′
3′...T C T A G-PO₄      OH-A...5′
```

When the digestion is complete, the entire human genome is converted into restriction fragments of varying lengths with the same sticky ends. Each genomic restriction fragment is ligated to a short stretch of DNA (a **linker**) that has the complementary sticky ends to pair with the genomic DNA. These fragments are amplified using PCR under conditions that favor the formation of short (≤1 kb) fragments. The final outcome was 1,658 PCR products that ranged from 200 to 1,000 bp in size. These were spotted in duplicate on glass slides for a total of 3,316 spots, often called **features**, on one microarray.

The production of colored probes also had to be rethought, since tumors are small and heterogenous. Also, unlike yeast samples, the supply of cells was limited. Therefore, the investigators wanted to get away from working with RNA. They produced genomic representations for each tumor sample to create Cy3- and Cy5-labeled PCR products for probes on their microarrays. Genomic representations eliminated the need to know what was a real gene and what was not, as is required with typical DNA microarrays. The intensity of the green probe (X-axis) was plotted as a function of the red-to-green expression ratios (Y-axis) for each feature on the microarray (Figure 7.8).

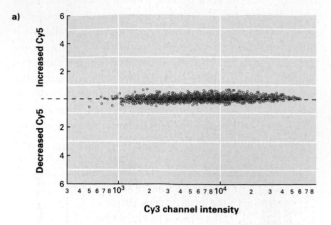

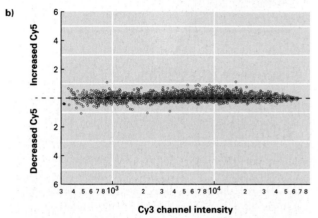

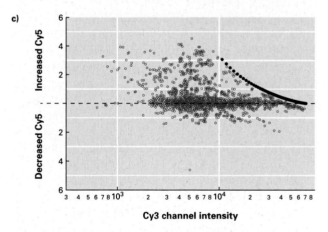

Figure 7.8 Representative genomic DNA microarrays to detect aneuploidy.

The X-axis (logarithmic scale) is the intensity of the Cy3 channel, and the Y-axis (linear scale) is the Cy5:Cy3 ratio. **a)** The same genomic representation was used to print the features and produce both Cy3 and Cy5 probes. **b)** The second experiment produced Cy3 and Cy5 probes from a single genome, but different from the genome used to print the microarray. **c)** The third experiment compared a breast cancer tumor to normal cells from a single person with the features derived from a second person. The filled black circles forming a diagonal indicate that the scanner's light detectors were saturated and could not measure any more light.

Genomic representation microarrays are new, so let's carefully describe how each experiment was performed. To do this, we will label different human genomes with different letters. Genome A is from person A; genomes B and B′ are from different tissues in person B. In the first experiment (Figure 7.8a), the features were genomic representations from genome A, and the Cy3 and Cy5 probes were created from the same genomic representation of genome A. In the second experiment (Figure 7.8b), the features were genomic representations from genome A, but the Cy3 and Cy5 probes were genomic representations from genome B. The features on the third experiment (Figure 7.8c) were produced from the same genomic representations from genome A. The Cy3 probe was a genomic representation produced from genome B taken from *wt* cells. However, the Cy5 probe was a genomic representation from cancerous cells of person B (genome B′). Therefore, we are comparing the genome from a single person in noncancerous vs. cancerous tissue.

Whenever you invent a new method, you have to demonstrate that it works as advertised. First, Wigler compared the representative genome to itself (all genome A), which resulted in a series of dots, one for each feature on the microarray (Figure 7.8a). For features that had low levels of dye (left side) or high levels of dye (right side), the ratio of the two colors was near 1:1, as you would expect. When a sample from a different person (genome B) was used as a source for the representative probe and compared with itself (B vs. B), the resulting distribution was still closely clustered around the 1:1 ratio (Figure 7.8b), though not as uniformly as before. Finally, a breast cancer biopsy was taken from a patient, the tumor was separated into individual cells, and sorted by fluorescence-activated cell sorting (FACS) into either cancerous (genome B′) or *wt* cells (genome B). These two cell types were used as the source for two different probes, and it is clear they do not have equivalent genomes, even though they were taken from a single person (Figure 7.8c). Keep in mind that the probes were derived from genomic DNA and not mRNA.

DISCOVERY QUESTIONS

29. Do you think genomic representation DNA microarrays is a valid method? Support your answer with the data from Figure 7.8.
30. Why was there more variation in the data when a second genome was compared to itself (Figure 7.8b), versus the original genome compared to itself (Figure 7.8a)?
31. What kinds of changes in genomic DNA do you see in the tumor (Figure 7.8c)?
32. How could you use this information to figure out which genes might be deleted or amplified?

Most of the DNA in Figure 7.8c that did not appear near the ratio of 1:1 appears to be duplicated, though there

is one feature that is greatly reduced in this particular tumor. These features represent the entire genome, not just ORFs. To determine whether a gene is duplicated, the particular feature must be sequenced and mapped to the genome. You could use PCR and **sequence-tagged site** (STS) primers in the chromosomal region of the deleted feature to determine the size of the deletion. Resolving the size of the aneuploidy would require some tumor DNA as PCR template, so hopefully they did not use all of the DNA in the earlier experiments.

If genomic representations worked once, they should work a second time. When two different breast cancer cell lines were compared in duplicate experiments, the results were reproducible (Figure 7.9). A perfect correlation between the two experiments would have produced a diagonal line, and we do see a scatter of points that fall on a diagonal. However, we are not given the correlation coefficient, so it is difficult to compare the precision of this method with that of others.

DISCOVERY QUESTIONS

33. What do you learn from Figure 7.9?
34. Do genomic representation DNA microarrays do what is claimed: detect genomic duplication and deletion events?
35. What additional data would you like to see?

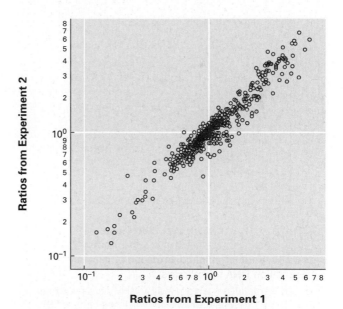

Figure 7.9 Comparison of duplicate experiments.
DNA from two breast cancer cell lines (MDA-MB-415 and SKBR-3) were used to generate representative probes and hybridized to an array of 938 genomic representation fragments, each spotted twice. The averaged ratios for each feature are plotted on the X-axis (experiment 1), and the averaged ratios for each feature are plotted on the Y-axis (experiment 2).

LINKS
Nazneen Rahman

METHODS
Southern blot

Can We Verify the Data?

Talk is cheap, and so is a new method that does not directly compare its data to a more familiar form. If Wigler's method can truly detect aneuploidy in tumors, it should compare well to a traditional method such as a Southern blot. The investigators compared some of the particular pieces of DNA on Southern blots using the same two breast cancer cell lines from Figure 7.9. Two types of Southern blots were performed; both genomic representations and total genomic DNA were blotted. These sources of DNA were probed with individual clones (identified by CHP numbers) and compared (Figure 7.10).

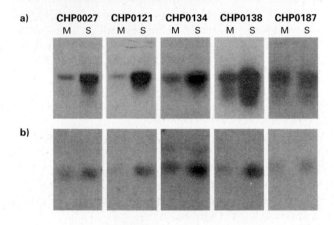

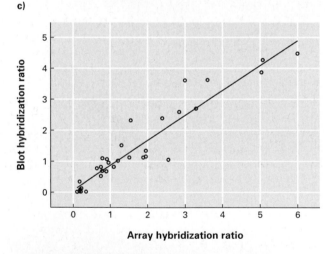

Figure 7.10 Southern blot confirmation of genomic representation DNA microarrays.

Five representatives (designated by CHP numbers) that proved to be present at different levels (see Figure 7.9) were used as probes for Southern blots using **a)** the 938 representative PCR products as target or **b)** total genomic DNA as target. CHP0187 was selected as an example of a feature that exhibited no differences between the two cell lines. The source of DNA in parallel experiments was either the MDA-MB-415 cell line (designated M) or the SKBR-3 cell line (designated S). **c)** A scatter plot comparing the ratios of DNA copy number obtained by microarray analysis (X-axis) vs. Southern blot hybridization (Y-axis).

DISCOVERY QUESTIONS

36. How well do the microarray data correlate with the blot data?
37. Would it be helpful to know the correlation coefficient in the comparison of microarrays vs. blots?

By eye, the representation microarray method looks as though it produces data comparable to Southern blots, but the data are neither quantitative nor analyzed by statistical means. Nevertheless, genomic representations are a clever use of microarrays that will improve with time. The reason so many people want to know which genes are duplicated or deleted is so that they can more accurately prescribe a medication that will work on a particular tumor. For example, if drug X works by binding to gene *Lmnop1*, then drug X will have no chance of helping a patient whose tumor has lost gene *Lmnop1*. The incentive to devise the best way to detect genomic aneuploidy is much more than an academic exercise in technology. Hopefully the people you know who have been diagnosed with cancer may benefit from DNA microarrays as a part of clinical assessment.

Since 1914, biologists have known that cancers often exhibit aneuploidy, but we have not known if the loss of genetic material was a cause or an effect. In the fall of 2004, British investigators led by **Nazneen Rahman** studied children with inherited mutations in both alleles of their *Bub1b* genes. Bub1b protein is known to help regulate chromosomal number in cells, and all these children had frequent aneuploidy in their cells. Some of the children developed cancer at very early ages; this indicates that aneuploidy can cause cancer, but this may not be true of all spontaneous cancers. Nevertheless, aneuploidy can be detected by microarrays and may be a good indicator of cancer risk as well as drug efficacy.

Summary 7.1

Like the mutated yeast genomes described in Section 6.2, human genomes that suffer from mutations are unstable. Uncontrolled cell growth is the consequence of an unregulated cell cycle, and genomes can suffer additional mutations, including aneuploidy. Breast cancer is a very complex compilation of cells and has proven difficult to categorize by microarrays. The degree of progression and the origin of the first cancerous cell make breast cancer particularly difficult to understand and combat. However, DLBCL is a more homogeneous disease that is easier to diagnose and categorize. Although it was once thought of as a single type of cancer with two distinct clinical outcomes, we have seen how DNA microarrays have reclassified DLBCL into two distinct clusters. Recognizing two subtypes of DLBCL may help us better understand why patients with the same diagnosis might respond very differently to the same chemotherapy treatment. Using DNA microarrays may help us personalize medical treatments to match each patient's genomic variations.

7.2 Improving Health Care with DNA Microarrays

Section 7.1 focused on disease; Section 7.2 focuses on treatments. The first case uses DNA microarrays to discover the evolution of the tuberculosis vaccine over the last 100 years. As the genome of an attenuated bacterium has evolved, the effectiveness of the vaccine has diminished. You will see how microarray data located the best source of a reinvigorated vaccine for tuberculosis. To analyze the effectiveness of over 100 drugs, we will examine the power of combining several databases with DNA microarray data in order to predict which medication might be the most effective for a particular cancer. The dream of many investigators is to be able to tailor medication to an individual patient's genome. Finally, you will examine obesity and its possible treatment with leptin. Is it possible to treat obesity with a simple injection of leptin? Or does leptin regulate several off-target integrated genomic circuits that are intimately connected?

Why Is the Tuberculosis Vaccine Less Effective Now?

DNA microarrays are versatile and powerful, but will they ever improve the development of medications? Whether the probe originates from RNA or genomic DNA will determine the type of information obtained. Using genomic DNA, we can compare entire genomes for ploidy differences in a single experiment. Cancer cells are not the only cells that add or delete large segments of DNA. Bacteria also experience duplications and/or deletions (see Chapters 2 and 3). If genome changes convey a selective advantage to the individual bacterium, the new mutation will become established in the population. Of course, this occurs every day in nature, but in the lab, we control the environment so prokaryotes rarely change their chromosomes and live to tell about it . . . at least, that is the commonly held assumption.

The bacterium *Mycobacterium tuberculosis* is the causative agent of the lethal and highly contagious respiratory disease **tuberculosis (TB)**. TB kills 2 million people in the world each year. A vaccine was first produced around 1900 when two French researchers, Albert Calmette and Camille Guérin, began a collaboration in Lille, France. They used *M. bovis*, a strain of bovine (cattle) tuberculosis, as a starting place (similar to using cowpox to vaccinate for smallpox). Over 13 years, they maintained the bacterium through a series of 231 subcultures before producing an attenuated strain of the bovine tubercle bacillus. The culture medium they used to grow their strain consisted of potato, glycerine, and bovine bile salts, and the vaccine was called *bacille de Calmette et Guérin* (BCG). From 1908 to 1921 (including World War I), the bacillus was maintained in culture until the clinicians demonstrated the vaccine was harmless to the tuberculosis-susceptible guinea pig. The first successful prophylactic use of BCG was carried out in 1921, when a physician persuaded Calmette to allow BCG to be used to protect a newborn child delivered by a mother who had TB. Nine years later, after many successful vaccinations, something went wrong.

In 1930, the public health director in Lübeck, Germany, asked Calmette to send a stock of BCG. The stock was grown for vaccine production and administered to 251 children. Within a year, 207 of the children had developed TB and 72 of them died. Skeptics of the BCG vaccine became outraged and very vocal in their criticism of Calmette and his vaccine. However, physicians who had successfully used different batches of BCG to protect people from TB rallied to support BCG and Calmette. An inquiry led by German health officials determined that the BCG stock had become contaminated in the Lübeck laboratory where the vaccine had been prepared. The BCG vaccine was determined to be safe and effective; the German physicians responsible for the contamination were punished with imprisonment.

Since its earliest use, the BCG vaccine has been amazingly successful. It was not until the 1960s (about 1,000 vaccine generations later) that the strain called BCG, which had been maintained in the Pasteur Institute, could be stored in a frozen state to minimize future changes. By the 1960s, BCG had been sent around the world, where different labs continued to grow their own stock supplies (only live bacteria produce robust immunity). From 1924 until it was frozen, some BCG strains occasionally were reported to be ineffective as a vaccine. This inconsistency led to the hypothesis that the locally prepared vaccine had changed because BCG had been successfully used elsewhere to prevent TB.

Marcel Behr and his colleagues at McGill University in Canada decided to examine the genome of *M. bovis* to see if it had changed over the years. To perform this study, Behr took advantage of the complete genomic sequence for the closely related species *M. tuberculosis*. He microarrayed 4,896 features that represented 3,902 ORFs, all but 22 of the pathogen's ORFs. Of these, 3,756 ORFs produced significant signals for analysis. Probes were made of total genomic DNA isolated from *M. bovis* and each of the 13 remaining vaccine-producing BCG strains Behr had collected from around the world. In each experiment, he used **random-priming** DNA polymerization reactions to produce red (Cy5-labeled) copies of *M. bovis* genomic DNA as a reference and green (Cy3-labeled) copies of the BCG

LINKS
Marcel Behr
M. bovis
METHODS
random-priming

vaccine strains as the experimental probe. Therefore, the genome of each BCG strain was compared to the genome of a recently isolated *wt* strain of *M. bovis* (Figure 7.11). Every feature bound by both genomes was yellow, and anywhere the vaccine had lost some genomic DNA was red.

Figure 7.11 *M. tuberculosis* DNA mircoarray probed with *M. bovis* (red) and BCG (green).

Go to www.GeneticsPlace.com to view this figure.

In *M. bovis* versus BCG genome comparisons, the investigators used software to highlight regions of the *M. bovis* genome that were missing from the BCG strains (Figure 7.12). Regions that contained two or more deleted ORFs that produced a different signal (red instead of yellow features) on the microarray were identified. Regions of the chromosome that contained these ORFs from the vaccine strains were sequenced to confirm the loss of genomic fragments.

DISCOVERY QUESTIONS

38. Using Behr's approach, would it be possible to detect genes that are not present in the *M. bovis* genome but are present in *M. tuberculosis*? What would the utility of this be?

39. Would it be possible to detect genes that are not present in the *M. tuberculosis* genome but are present in BCG vaccine genomes? Could this information have any practical consequence?

40. What types of mutations in the BCG strains would we not be able to detect by this method?

Based on the DNA microarray results and the sequenced *M. tuberculosis* genome, Behr's group determined that 16 regions of *M. bovis* had been deleted in the various BCG strains; they were designated RD1–RD16 (Regions Deleted). When examined at higher resolution, it was possible to estimate how many ORFs had been deleted in each region (Figure 7.13).

DISCOVERY QUESTIONS

41. Go to the BCG Genome web site and you will find a complete list of RD1–RD16. What are the smallest and largest deletions in terms of base pairs? In terms of ORFs?

42. At the BCG web site, you will find a list of all the deleted ORFs. Use the "Find" function of your browser to locate these three ORFs: Rv1985c, Rv1773c, and Rv3405c. (The "c" stands for Crick, one of the two strands of DNA.)
   a. What kind of molecules are these?
   b. Do you think their loss will have a small or large effect on the total protein content of the bacteria? Explain your answer.

In addition to finding particular deletions, it was possible to trace the evolution of the laboratory strains over the 40-odd years of growth in culture (Figure 7.14). The regions were lost at various times, with the earliest loss (RD1) occurring before the vaccine had been distributed around the world.

DISCOVERY QUESTIONS

43. How did Behr know that RD1 was lost before any of the strains were distributed?

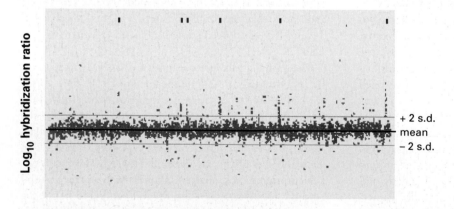

Figure 7.12 Genomic differences between *M. bovis* and BCG.
Computer plot of all features (X-axis) and the ratio of red to green signals. The mean ± two standard deviations (s.d.) is indicated. Purple bars at the top indicate regions where two or more features differed from the mean ratio by more than twice the standard deviation.

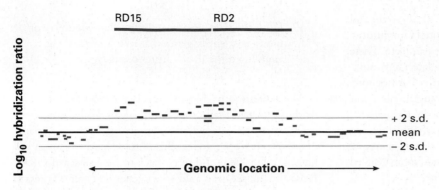

Figure 7.13 Examples of BCG genomic deletions.
Expanded view of RD15 and RD2 (purple bars), which are separated by 335 bp. Each deleted *M. bovis* ORF is indicated by a purple dash. The length of the purple dash indicates the relative size of the ORF.

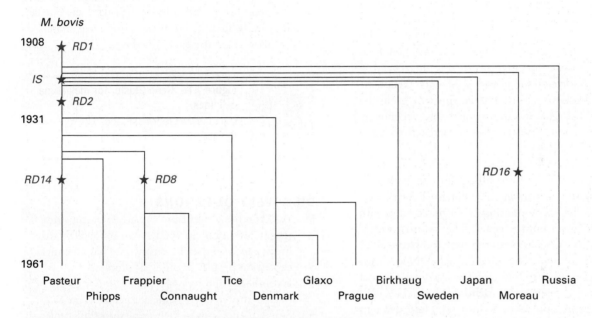

Figure 7.14 BCG genealogy deduced from microarray-detected deletions in the 13 currently available strains.
The vertical axis represents time, and the horizontal axis distributes different geographical locations of the strains used in this study. Six deletions are marked by the purple stars on branches of the dendrogram.

44. Which currently available strain(s) most closely resemble(s) the original strain of BCG? (The original strain was lost during World War I.)

45. Order the genomes from closest to most distant to *M. bovis*.

46. Why would evolution drive BCG to lose pieces of its genome?

Rarely do we get to see so clearly the consequences of genome instability. The BCG strain was grown for over 1,000 generations in laboratory conditions outside its normal habitat. Under these artificial conditions, it was subjected to new selection pressures. We saw the same evolution in yeast when it altered its genome in response to an investigator's selective deletion of a single gene (see pages 255–258). The bovine pathogen was grown in a cell-free environment with potatoes, sugar, and bile. Nature always produces genomic variation in a population, and in the lab environment, the old genome was no longer the most fit. As a result, laboratories around the world selected for strains that grew well in culture rather than worked best as vaccines. Behr's group studied the evolution of BCG and correlated it with efficacy of immunization. It should be possible to go back to the most "primitive" BCG strain and see if it produces the most robust immunity.

But the BCG story is not complete. These DNA microarrays could not detect any mutation less than 2 kb in length. Furthermore, they could not detect any chromosomal rearrangements, point mutations, deletions of duplicated genes, repetitive regions, or intergenic deletions. Would it be possible for the least deleted strain to produce the least protective vaccine? The genome of *M. bovis* is being sequenced, which will provide more detailed information to help us improve the TB vaccine.

Can We Choose the Most Effective Medication for Each Cancer?

The ultimate goal for many investigators using DNA microarrays is to improve medical care of patients. Diagnosis is one avenue of research, and drug development is a second. Pat Brown and David Botstein teamed up with John Weinstein from the **National Cancer Institute** (NCI) to determine how microarrays might be used to improve drug development and selection for individual patients. Weinstein's group was led by Uwe Scherf, and Pat Brown's group by Douglas Ross. Their goal was to use microarray technology to determine if cancers could be better matched with the most effective medications.

Ross and Scherf thought it would be prudent first to characterize the transcriptomes of the 60 cell lines used by NCI (called the NCI60) when testing new potential chemotherapy agents. They began by printing DNA microarrays with about 8,000 different human genes. The control cDNA, labeled green with Cy3, was produced from a mixture of 12 different human cell lines. The experimental cDNA, labeled red with Cy5, was produced from only one cell line at a time. The investigators focused their attention on 1,611 ORFs that produced expression ratios at least sevenfold different from control in at least 4 of the NCI60. Using these 1,611 ORFs, the 60 cell lines were clustered so that cell lines with similar expression profiles were placed next to each other (Figure 7.15). Two cell lines were grown in triplicate to test the reproducibility of the method.

Figure 7.15 NCI60 gene expression dendrogram.

Go to www.GeneticsPlace.com to view this figure.

DISCOVERY QUESTIONS

47. Do all cell lines cluster according to their tissue of origin?

48. Which tissues appear more difficult to cluster together than others? Which are easiest?

49. With the data already collected, can you think of a way to identify genes that might be better at unifying difficult-to-cluster tissue types?

The expression profiles of some tissue types are more easily distinguished than others, with breast tissue being especially hard to characterize, as we saw earlier (pages 268–269). If we wanted to cluster these hard-to-cluster cell lines, it might be useful to go back to the full set of ORFs. From the full set, we could select genes that were expressed maximally in one of these three tissue types (breast, prostate, or non–small cell lung) but not the other two tissue types, and add these new genes to the set of 1,611 genes for reclustering. In addition, many genes were not included on the original DNA microarrays, so there might be better ORFs that were not included in the first data set. The cell lines with clearly identifiable tissue types have subsets of genes that allow them to be identified easily (Figure 7.16).

Figure 7.16 Gene Cluster for melanoma cell lines.

Go to www.GeneticsPlace.com to view this figure.

DISCOVERY QUESTIONS

50. Based on the results from the breast cancer study (pages 268–269), do you think it more likely that breast cancer and melanoma would have identical expression profiles or different profiles?

51. Hypothesize about possible roles for the 16 ORFs that co-clustered in the melanoma set of genes.

52. Search GeneCards to determine the cellular roles of tyrosinase (tyrosinase is a member of the cluster in Figure 7.16). Type in "tyrosinase" and hit "Go." When the list of hits comes up, click on the "Display complete GeneCard" link to the left of the tyrosinase gene called "TYR." Are any phenotypes associated with the loss of tyrosinase?

53. Why does it make sense that melanomas would express high levels of tyrosinase?

Can We Predict Effectiveness of Chemotherapy?

Having clustered the NCI60 and identified genes that distinguish major cell types, Scherf and Ross had a pretty good handle on the normal transcriptomes for the NCI60. But this was only half of their project. They wanted to find a way to utilize the NCI60 and microarrays to predict the potential utility of a particular drug for a given tumor type.

Could microarrays be used to predict the success of one chemotherapy agent over another in a given patient?

To answer their question, Scherf and Ross integrated two data sets. The first was the NCI60 microarray data discussed previously. The second was a drug activity profile that measured the effectiveness of 118 commonly used chemotherapy medications tested by NCI on each of the NCI60. Scherf and Ross used 1,376 genes from their microarray data that provided the most robust signals and revealed the strongest differences among the NCI60. In the end, they produced a mammoth figure called a **clustered image map** (CIM). CIMs provide a visual way to interpret the intersection of two different data sets (Figure 7.17). On

Figure 7.17 Clustered image map.
Go to www.GeneticsPlace.com to view this figure.

the X-axis is the clustered data set of 1,376 genes based on their normal expression profiles in the NCI60. The Y-axis represents the clustering of 118 drugs based on their effectiveness in the NCI60.

DISCOVERY QUESTIONS

54. Do the two drugs in the insets in Figure 7.17 work better or worse when their respective gene targets are highly expressed?

55. Find a large region on this CIM where the drugs work best when genes are highly expressed.

56. How could this CIM be used to determine in advance whether a new drug is likely to work for a particular cancer?

57. Let's imagine that a new drug has been shown to be effective on some cancers. This drug is located at 15 on the Y-axis, and had a high negative correlation with genes around 780 on the X-axis. How could you use this information to screen cancer patients for the potential benefit of your new drug?

Scherf and Ross analyzed portions of the data, though everyone acknowledges that there are more data than any one group could analyze completely. In their paper, they focused on two interesting cases. Certain malignant tumors, such as acute lymphoblastic leukemia (ALL), lack asparagine synthetase (encoded by the gene *ASNS*). If a drug could be given to ALL patients to deplete extracellular asparagine, the cancer would starve from a lack of the amino acid asparagine. One drug that has been on the market for years is L-asparaginase which, as the name implies, is an enzyme that cleaves asparagine. The two ALL lines (MOLT-4 and CCRF-CEM) expressed the lowest levels of *ASNS* and

were the most sensitive to L-asparaginase. The chronic myelogenous leukemia cell line called K-562 expressed the highest levels of *ASNS* and was the least sensitive to L-asparaginase (Figure 7.18).

LINKS
CIM

The CIM revealed a negative correlation (−0.44) between expression of *ASNS* and the sensitivity of the NCI60 to L-asparaginase. There was a stronger correlation coefficient (−0.98) for the six leukemia cell lines.

DISCOVERY QUESTIONS

58. Why would a cell line (and by extrapolation a tumor) be more susceptible to L-asparaginase if it expresses less *ASNS*?

59. Does Figure 7.18 indicate that L-asparaginase is being used on the most responsive or the least responsive cancer types?

60. Ovarian cancer cell lines also had a negative correlation (−0.88) between *ASNS* expression and L-asparaginase sensitivity. What does this negative correlation suggest to you? Find a point not labeled in Figure 7.18 that represents a cell line likely to be responsive to the drug L-asparaginase.

61. *DPYD* encodes an enzyme that degrades uracil and thymidine and also the chemotherapy medication 5-FU. Would 5-FU work better in cells whose *DPYD* expression levels have a positive or negative correlation with the drug? Look back at the CIM data in Figure 7.17b. What does the CIM indicate is the correct answer to this question?

62. What gene numbers (from the X-axis in Figure 7.17) have a positive correlation with the drug 5-FU? What does this indicate for potential use of 5-FU in cancer patients?

This CIM project has provided new insight into drug discovery and ways to maximize the utility of any given drug. Of course, you would not want to treat a patient based on microarray data alone, but these data do provide a model for formulating predictions that are testable in clinical drug trials.

What Happens When You Accumulate Fat?

Fat homeostasis is complex and difficult to regulate by pharmacological intervention (see Chapter 5). The biggest problems are all the unknowns, because the protein **leptin** (encoded in mice by the gene *ob*) and the **lipostat** are connected to many "black boxes" including the brain, immune system, and sexual reproduction. Here are some reminders of the complexity involved:

• Untreated mice whose food intake is restricted to match the food intake of leptin-treated mice lose significantly

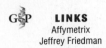

less weight than the leptin-treated mice. The food-restricted but non-leptin-treated mice are called **pair-fed** mice.

- Starvation causes a loss of both lean body mass and fat mass, while leptin treatment selectively reduces fat mass.
- Leptin treatment does not lead to the compensatory drop in energy expenditure seen in pair-fed mice.
- Therefore, leptin initiates metabolic circuitry distinct from the circuit induced by food restriction.

Can we understand the circuitry utilized by leptin signaling? Yeast won't be informative—there are no obese yeast. Neither flies nor worms have leptin genes. Grinding up people for research projects is frowned upon. What options are there? How can we learn more about the hidden circuitry behind the complex genomic responses initiated by leptin? Jeffrey Friedman, who first cloned *ob*, and his lab utilized a microarray approach to this problem, though they had some challenges to address before they could collect any data. First, the mouse genome had not been sequenced when they performed their research (1999). They used what was available: some mouse genes and ESTs. Rather than using cDNAs to spot on the

microarray, they printed **oligonucleotide microarrays** developed by the company Affymetrix. The principle is the same as spotting cDNAs, except that single-stranded oligonucleotides are attached to the glass slide. There are some other minor differences, but essentially it is a standard microarray experiment.

DISCOVERY QUESTIONS

63. What advantages are there to using oligonucleotides instead of cDNAs?

64. One benefit of oligonucleotides could not be fully utilized when making mouse microarrays in 1999. Why?

The partial mouse genome microarray represented about 6,500 ORFs. Using oligonucleotides instead of cDNAs has two advantages. First, you never have to denature the spotted DNA, since it was never double stranded; this saves time and one potential source of variation. Second, oligonucleotides can be chosen so that they do not bind to more than one gene's mRNA/cDNA; that is, areas of low sequence similarity can be chosen. Choosing gene-specific regions is not possible with full-length cDNAs produced by

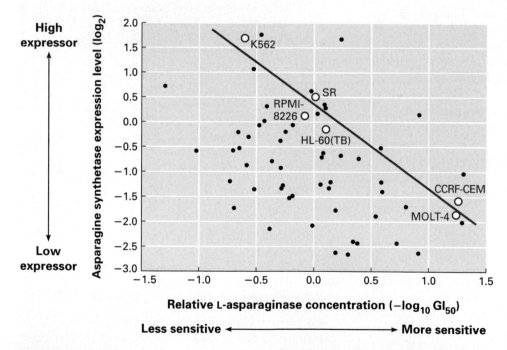

Figure 7.18 Correlation between *ASNS* expression and chemosensitivity to L-asparaginase for each of the NCI60.
The Y-axis represents the gene expression ratio (log$_2$) of *ASNS* in each cell line, relative to control. The X-axis is a relative scale of sensitivity to L-asparaginase, as measured by GI$_{50}$, meaning that growth was inhibited by 50%. A large value on the X-axis indicates that a low dosage was required to achieve GI$_{50}$. Large purple circles highlight six different leukemia cell lines with a best-fit line that had a correlation coefficient of −0.98, compared to a best-fit line correlation of −0.44 for all NCI60.

PCR. However, because the mouse genome had not been sequenced in 1999, it was impossible to be sure that the oligonucleotides would bind to only one gene.

The dye-labeled cDNAs used in this obesity research were produced from fat tissue of *wt* and *ob/ob* mice (homozygous mice lacking a functional copy of *ob*). Only 2,000 features were labeled at sufficient levels to be considered above background, and 77 of these were induced or repressed ≥ 3-fold when comparing *wt* to *ob* mice. As other groups had done previously, Friedman and his team wanted to make sure their microarray data could be confirmed by Northern blots. If anything, microarrays tend to underestimate the difference in expression ratios compared to the values obtained by Northern blots (Figure 7.19).

DISCOVERY QUESTIONS

65. How do you use the data provided by the cyclophilin blots? Is there a similar control for microarray data?
66. How could numerical values be obtained for Northern blots?
67. Which fold change data are better, Northern blots or microarrays? Answer this question both at the whole genome level and the individual gene level.

The first experiment was to examine the effect of leptin injections for *ob/ob* mice compared to *ob/ob* pair-fed mice (Figure 7.20). As previously reported, the leptin-treated *ob*

LINKS G&P
comparing *wt to ob*
METHODS
Northern blots

mice lost weight faster than the *ob* pair-fed mice. The control *ob* mice continued to gain weight during the 12 days of this experiment. Food intake for the leptin-treated and pair-fed mice was identical; the saline-injected *ob* mice got used to the injections and quickly regained their appetites.

DISCOVERY QUESTIONS

68. Summarize the data from both graphs in Figure 7.20.
69. Why were so many of the microarray time points taken early in the leptin treatment rather than evenly distributed over the 12 days?

The *ob/ob* mice in this study responded just as predicted and the control mice ate well and gained weight, which ruled out the influence of physical handling as a reason for any differences in the mice. These preliminary data allowed Friedman's group to proceed with the microarray analysis. Approximately 2,000 genes were clustered by expression profile in white fat tissue from leptin-treated and pair-fed *ob* mice. The investigators identified 16 clusters of genes, each of which showed different responses to leptin (Figure 7.21). The color scheme used by Friedman's lab is different from those we've seen before, but the principle is identical.

In all cases, expression ratios were calculated by comparing gene expression in untreated *wt* mouse white fat cells to

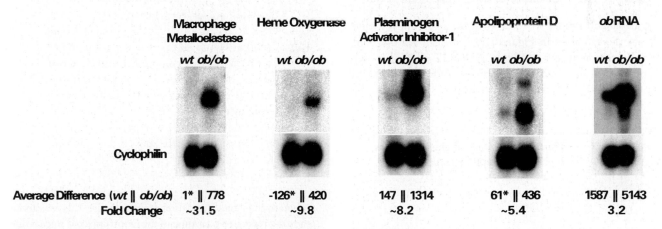

Figure 7.19 Comparison of microarray and Northern blot data.
Northern blots of RNA isolated from *wt* and *ob* mice were probed for the indicated genes. The fold change at the bottom was determined from the intensity of signals on Northern blots for *ob* vs. *wt* mice. When the *wt* signal was essentially zero, the ~ symbol was used to indicate an approximation. The average difference indicates the averaged numerical signal from the microarray data in *wt* vs. *ob* mice that was used to determine the difference in gene expression for ratios. An asterisk indicates no detectable signal. The gene cyclophilin was used as a control for the Northern blots.

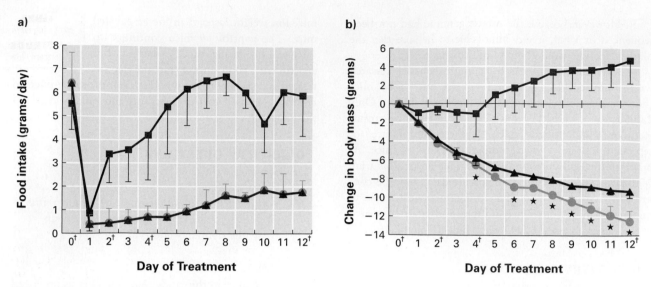

Figure 7.20 Effect of leptin on food intake and body mass for *ob/ob* mice injected with leptin, pair-fed, and injected with saline.
a) The amount of food eaten by *ob/ob* mice injected with saline (black squares), *ob/ob* mice injected with leptin (gray circles), and *ob/ob* pair-fed mice (purple triangles).
b) Body mass of the three groups of mice. Stars indicate significant differences ($p < 0.05$) comparing leptin-injected to pair-fed mice. The dagger (†) indicates time points at which microarray data were collected.

Figure 7.21 Clustered gene summaries of leptin-treatment (blue squares) and pair-feeding (green circles) on *ob/ob* white adipose tissue.
Go to www.GeneticsPlace.com to view this figure.

expression in similar samples from the experimental mice. From the examples in Figure 7.21, you can see that some genes were induced by leptin (red-yellow), others were repressed (from dark to light blue), and some remained unchanged (black). When the identities of the genes were examined, some intriguing findings surfaced (Figure 7.22).

Figure 7.22 Details of cluster L genes repressed by leptin injections in *ob/ob* mice.
Go to www.GeneticsPlace.com to view this figure.

In this cluster, many of the genes were involved with fatty acid biosynthesis, which you might have expected to be induced but instead were repressed. Included in this cluster was the gene "homologous to SREBP-1" (sterol response element binding protein), a cholesterol and fatty acid-sensitive transcription factor known to induce the expression of

Spot14. Interestingly, cytochrome P-450 IIE1, known to metabolize drugs, was also repressed.

DISCOVERY QUESTIONS

70. How do the individual gene data for cluster L in Figure 7.22 compare with the summary graphs? Can you compare the two methods for displaying the data? Explain your answer.

71. Look at the expression profile for cytochrome P-450 in Figure 7.22. Draw a quick graph to summarize its expression in leptin-injected and pair-fed mice. Would you have placed it fourth from the bottom in the cluster if you were clustering by eye?

72. Why would *SREBP-1* be repressed very little but *Spot14* be repressed the most of any gene in this cluster?

The overall trend for cluster L from Figure 7.21 was to be gradually repressed over time in the leptin-injected mice while only transiently repressed in the pair-fed mice. However, when we compare individual genes within cluster L (Figure 7.22), we see that some genes were only transiently repressed (e.g., cytochrome P-450), or slightly repressed (*SREBP-1*) in these leptin-treated mice. Note how subtle changes in a transcription factor (e.g., *SREBP-1*) can have major effects on the transcription of genes that require the same factor (e.g.,

Spot14). All of these data help explain why the response to leptin is very complex and not a simple circuit made up of only a few components. The leptin circuit complexity is made more apparent when we look at genes that were transiently induced by leptin treatment (Figure 7.23).

Figure 7.23 Detailed cluster of genes transiently induced by leptin injections in *ob/ob* mice.

Go to www.GeneticsPlace.com to view this figure.

DISCOVERY QUESTIONS

73. Look at the wide range of genes represented in Figure 7.23. What disease is associated with caveolin and calpain? (See Chapter 1.)

74. Search **OMIM** and determine the roles for these genes: integrin (click on alpha 7); *Rab* (click on 5a); synaptoporin; *Lim* (click on actin-binding Lim protein 1); nexin (click on sorting nexin 1); FK506 binding protein (click on 5); syntrophin (click on alpha-1).

75. Does leptin affect the expression of genes with known roles in the CNS and immune systems? Are any diseases associated with the short list of genes you found?

76. Pick one gene from Discovery Question 74 and graph its change in expression as revealed by leptin-treated and pair-fed mice.

What Effect Does Leptin Have on *wt* Adipose Tissue?

We would expect leptin injections to alter gene expression in mice that lacked the ability to produce leptin. What happens when *wt* mice are injected with leptin? Friedman knew that increasing either serum or cerebrospinal fluid concentrations of leptin in *wt* mice mimics the accumulation of excess fat. In response, a mouse will reduce its food intake and lose adipose mass but not muscle mass. How does leptin alter the transcriptome to accomplish these results in a *wt* mouse? Friedman's lab asked the same question and discovered that increasing leptin levels in a *wt* mouse also affects many genes, though not always in the same way as seen in *ob/ob* mice (Figure 7.24).

Figure 7.24 Cluster overviews of leptin treatment and pair-feeding in *wt* white adipose tissue.

Go to www.GeneticsPlace.com to view this figure.

LINKS
OMIM
METHODS
Western Blot

The first sign that *wt* mice respond differently from *ob* mice is that the same 2,000 genes were clustered into 13 groups instead of 16. In cluster A of Figure 7.24, you can see 58 out of 2,000 genes responded similarly in leptin-treated mice as pair-fed mice. In contrast, clusters B, K, D, G, and F reveal that some genes respond differently to reduced food intake as opposed to leptin treatment; cluster G contains the leptin gene. Cluster K contains *SREBP-1*, *Spot14*, and many of the fatty acid metabolizing genes that were repressed in *wt* mice just as they were in the *ob/ob* mice. The answer to our original question is that genes in a *wt* mouse respond to leptin in many different ways. About 20 transcription factors were affected by leptin treatment, which means that approximately 1% of genes with altered expression regulate the transcription of many other genes. The complexity of this genome-wide response reinforces the lesson that simple solutions to obesity are not going to be possible, since leptin interacts with many different circuits.

Although it is easy to measure the genomic response in the form of a transcriptome, measuring protein changes is more difficult. However, Friedman's group performed one last experiment to assess changes in the transcription factor *SREBP-1*. Since most cellular functions are carried out by proteins, people want to know what is happening to the proteome more than the transcriptome. (Chapter 8 focuses on **proteomics**, the study of all proteins present in a cell.) *SREBP-1* is translated into a protein that initially binds to the endoplasmic reticulum (ER) and thus cannot reach the nucleus. Upon stimulation by sterol molecules, a protease is activated and cleaves SREBP-1p to release a portion of the protein from the ER so it can enter the nucleus. This SREBP-1p fragment is the active form of the transcription factor. When the amount of active SREBP-1p was compared in *ob/ob* leptin-treated and pair-fed mice, the protein differences were substantial, even though the changes in transcription were minor (Figure 7.25). Comparing protein to mRNA levels of one gene made it clear that genomic control of integrated circuits is not restricted to transcription.

Figure 7.25 Comparison of transcriptional and post-transcriptional regulation.

Go to www.GeneticsPlace.com to view this figure.

DISCOVERY QUESTIONS

77. Graph the change in *SREBP-1* transcription for the *ob/ob* mice treated with leptin and *ob/ob* pair-fed mice. Then roughly estimate values for the Western blot and plot this on the transcription graph. Describe the differences you see.

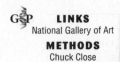

78. The change in SREBP-1p seen in the Western blot does not correlate **D602** well with *SREBP–1* transcription in *ob/ob* mice. Explain why.

79. Which form of data (protein or mRNA) gives you the better understanding of genomic responses to leptin? Support your position with data.

80. What impact does the Western blot have on your evaluation of microarray data?

This study of leptin-induced genomic response has opened more doors to understanding how leptin works, but it leaves many unanswered questions. It also illustrates both the strengths and weaknesses of microarray studies. Microarrays provide clues to the functions of unknown genes, illuminate new circuits, and show connections between circuits. However, the resolution is limited. An analogy for this limited resolution is the artistic expression perfected by Georges Seurat, **pointillism**. Go to the National Gallery of Art and search for one of Seurat's classic works. You can recognize patterns and trends in the overall picture, but when you look up close, you cannot recognize individual shapes as easily as you did from the global perspective. The same may be said about microarrays: they provide informative overviews, but gene-specific clarity is elusive. Nevertheless, as with the perspective provided by Seurat, microarrays are valuable and provide insights.

Summary 7.2

DNA microarrays are powerful tools for improving the quality of medicine. You saw how a well-established vaccine for tuberculosis might be reinvigorated now that we understand the genomic variations currently in use around the world. By compiling years of basic research, a clustered image map enabled you to predict which medication might be best for a particular person's cancer. However, the regulation of fat appears to be so deeply woven throughout our genome that the only effective treatment for obesity may be a low-fat diet and exercise. Simple cures may produce unacceptable consequences because of the integrated circuitry inherent in a genome.

Chapter 7 Conclusions

The central focus of Chapter 7 is the application of DNA microarrays for better understanding of the human condition. Substantial progress has been made toward the use of DNA chips to improve the treatment of cancer. Likewise, the development and understanding of new medications may be enhanced when DNA microarrays are incorporated into the process. What are the limitations of

DNA microarrays? The range of experiments is only as limited as the creativity and curiosity of the scientific community. Just as computers have become cheaper and faster, microarrays too will become increasingly accessible to more people, which will open the way for better experiments. Are DNA microarrays better at seeing the entire genome or individual genes? For another analogy from the artistic world, we might consider the work of Chuck Close and ask if we are seeing a comprehensive picture or just a bunch of dots.

If you choose a career path that leads you away from science, you will still encounter microarrays in your personal lives. Microarrays are being used to study the genomic rivalry between a host plant and its parasitic pathogen. Pathogen-host interaction could be studied in any organism (including humans) that can be attacked by parasites, bacteria, or viruses. Another research team examined the genomic diversity of the bacterium *H. pylori*, which is the causative agent of stomach ulcers. Cancer, acquired immunodeficiency syndrome (AIDS), TB, ebola—the list of potential uses for microarrays is as long as genomes are diverse.

References

Cancer Chips

Alizadeh, A. A., M. B. Eisen, et al. 2000. Distinct types of diffuse large B-cell lymphoma identified by gene expression profiling. *Nature.* 403: 503–511.

Dunn, S. 2002. Survival curves: Accrual and the Kaplan-Meier estimate. <http://www.cancerguide.org/scurve_km.html>. Accessed 22 April 2005.

Ein-Dor, L., I. Kela, et al. 2005. Outcome signature genes in breast cancer: Is there a unique set? *Bioinformatics.* 21(2): 171–178.

Golub, T. R., D. K. Slonim, et al. 1999. Molecular classification of cancer: Class discovery and class prediction by gene expression monitoring. *Science.* 286: 531–537.

Lossos, I. S., D. K. Czerwinski, et al. 2004. Prediction of survival in diffuse large-B-cell lymphoma based on the expression of six genes. *New England Journal of Medicine.* 350: 1828–1837.

Perou, C. M., T. Sørlie, et al. 2000. Molecular portraits of human breast tumors. *Nature.* 406: 747–752.

Ross, D. T., U. Scherf, et al. 2000. Systematic variation in gene expression patterns in human cancer cell lines. *Nature Genetics.* 24: 227–238.

Scherf, U., D. Ross, et al. 2000. A gene expression database for the molecular pharmacology of cancer. *Nature Genetics.* 24: 236–244.

Sørlie, T., R. Tibshirani, et al. 2003. Repeated observation of breast tumor subtypes in independent gene expression data sets. *PNAS USA.* 100: 8418–8423.

van't Veer, L. J., H. Dai, et al. 2002. Gene expression profiling predicts clinical outcome of breast cancer. *Nature.* 415: 530–536.

Weinstein, J. N., T. G. Myers, et al. 1997. An information-intensive approach to the molecular pharmacology of cancer. *Science.* 275: 343–349.

Aneuploidy Chips

Behr, M. A., M. A. Wilson, et al. 1999. Comparative genomics of BCG vaccines by whole-genome DNA microarray. *Science.* 284: 1520–1523.

Forozan, F., E. H. Mahlamaki, et al. 2000. Comparative genomic hybridization analysis of 38 breast cancer cell lines: A basis for interpreting complementary DNA microarray data. *Cancer Research*. 60: 4519–4528.

Grimm, D. 2004. Disease backs cancer origin theory. *Science*. 306: 389.

Hoslink. Pioneers in medical laboratory sciences. 2002. <www.hoslink.com/Pioneers.htm>. Accessed 7 June 2002.

Hughes, T. R., C. J. Roberts, et al. 2000. Widespread aneuploidy revealed by DNA microarray expression profiling. *Nature Genetics*. 25: 333–337.

Lucito, R., J. West, et al. 2000. Detecting gene copy number fluctuations in tumor cells by microarray analysis of genomic representations. *Genome Research*. 10: 1726–1736.

Rajagopalan, H., & C. Lengauer. 2004. Aneuploidy and cancer. *Nature*. 432: 338–341.

Leptin Chips

Belkin, L. 2000, December 24. The making of an 8-year-old woman. How do we understand puberty? Through the prism of our times. *New York Times Magazine*. 38–43.

Nohturfft, A., & R. Losick. 2002. Fats, flies and palmitate. *Science*. 296: 857–858.

Soukas, A., P. Cohen, et al. 2000. Leptin-specific patterns of gene expression in white adipose tissue. *Genes and Development*. 14: 963–980.

Clustering and Visualization of Microarray Data

Conklin, B. 2001. GenMapp: Gene MicroArray Pathway Profiler. <http://gladstone-genome.ucsf.edu/introduction.asp>. Accessed 8 February 2002.

D'haeseleer, P., S. Liang, & R. Somogyi. 2000. Genetic network inference: From co-expression clustering to reverse engineering. *Bioinformatics*. 16(8): 707–726.

Eisen, M. B., P. T. Spellman, et al. 1998. Cluster analysis and display of genome-wide expression patterns. *PNAS*. 95: 14863– 14868.

GCAT: The Genome Consortium for Active Teaching. 2005. <http://www.bio.davidson.edu/GCAT>. Accessed 22 April 2005.

Getz, G., E. Levine, & E. Domany. 2000. Coupled two-way clustering analysis of gene microarray data. *PNAS*. 97: 12079–12084.

Gilbert, D. R., M. Schroeder, & J. van Helden. 2000. Interactive visualization and exploration of relationships between biological objects. *Trends in Biotechnology*. 18: 487–494.

Heyer, L. J., D. Z. Moskowitz, et al. 2005. MAGIC Tool: Integrated microarray data analysis. *Bioinformatics*. 21: 2114–2115.

Kyoto Encyclopedia of Genes and Genomes (KEGG). 2001. <http://www.genome.ad.jp/kegg/>. Accessed 8 February 2002.

Lee, M.-L. T., F. C. Kuo, et al. 2000. Importance of replication in microarray gene expression studies: Statistical methods and evidence from repetitive cDNA hybridizations. *PNAS*. 97(18): 9834–9839.

Lemkin, P. F., G. C. Thornwall, et al. 2000. The microarray explorer tool for data mining of cDNA microarrays: Application for the mammary gland. *Nucleic Acids Research*. 28(22): 4452–4459. <http://www.lecb.ncifcrf.gov/MAExplorer/>.

Liao, B., W. Hale, et al. 2000. MAD: A suite of tools for microarray data management and processing. *Bioinformatics*. 16(10): 946–947. <http://pompous.swmed.edu/>.

Manduchi, E., G. R. Grant, et al. 2000. Generation of patterns from gene expression data by assigning confidence to differentially expressed genes. *Bioinformatics*. 16(8): 685–698.

National Center for Genome Resources. 2001. GeneX: A collaborative Internet database and toolset for gene expression data. <http://www.ncgr.org/research/genex/>. Accessed 8 February 2002.

Microarray Fabrication Methods

Brown, P. O., et al. 2001. The Mguide. <http://cmgm.stanford.edu/pbrown/mguide/>. Accessed 8 February 2002.

Kane, M. D., T. A. Jatkoe, et al. 2000. Assessment of the sensitivity and specificity of oligonucleotide (50mer) microarrays. *Nucleic Acids Research*. 28(22): 4552–4557.

Kumar, A., & Z. Liang. 2001. Chemical nanoprinting: A novel method for fabricating DNA microchips. *Nucleic Acids Research*. 29(2): e2.

Okamoto, T., T. Suzuki, & N. Yamamoto. 2000. Microarray fabrication with covalent attachment of DNA using bubble jet technology. *Nature Biotechnology*. 18: 438–441.

Proteomics

LINKS
Gene Ontology
Mike Snyder
METHODS
gene knockouts
transposons

Many researchers consider the 21st century the **postgenomic era**. By this, they are not implying that we should stop sequencing genomes or analyzing existing sequences. Postgenomic era simply means that the technical barriers to obtaining genomic information have been resolved. The next big challenge is to understand proteomes—all the proteins of an organism at a given time. **Proteomics** is a very difficult field because we lack methods for completely defining proteomes, partly because they present a moving target. Each of your cells has the same genome, the same DNA sequence you were born with and will have forever. However, each cell type in your body has a different proteome. In addition to cell-to-cell variation, any given cell will change its proteome over time. As you grow older, your proteomes change. When you have an infection, proteomes are altered to kill the pathogen. Medications, exercise, diet—your proteomes change in response to fluctuations in the intracellular and extracellular environments. This chapter uses a series of case studies to explore proteomic methods, research questions, and technical limitations.

8.1 Introduction

Biologists want to know what each protein does (molecular function), to which cellular circuits each protein contributes (biological process), and where in the cell each protein is located (cellular component). One way to discover these attributes is to knock out, or inactivate, a gene and look for new phenotypes. Deleting a single gene to discern phenotype is classical genetics and may not sound genomic at all. The postgenomic approach is to knock out all the genes, one at a time, and perform massive screens for particular phenotypes. This section examines three high-throughput approaches for deducing what each protein does in yeast cells.

What Do All These Proteins Do?

Until recently, the term "function" was used to describe what a protein does, but "function" is too vague. A consortium of genomic groups called Gene Ontology created three carefully chosen terms:

1. **Biological process** (why—why is this molecule being produced? e.g., movement of cell)
2. **Molecular function** (what—what kind of molecule is this? e.g., ATPase)
3. **Cellular component** (where—where is this molecule located? e.g., flagella, cytoplasm, etc.)

Using pre-genomic methods, biologists learned a lot about proteins, and these methods will continue to play an important role in the postgenomic era. For example, biologists have learned what happens to a cell that lacks a particular gene by using gene knockouts. Knocking out a gene allows the investigator to determine the consequences when a cell lacks a particular protein. Knockouts can be performed rather bluntly, so that every cell lacks the protein at all times, or more subtly, so that only certain cells in an organism at restricted times lack the protein. Either way, knockouts are not amenable to genome-wide analysis. What we need are methods that permit genome-wide analysis of biological processes, molecular functions, and cellular components.

How can we analyze every protein in three different ways? This daunting task was elegantly addressed by Petra Ross-Macdonald, Mike Snyder, and their collaborators at Yale University. Snyder's lab took advantage of transposons found in *Saccharomyces cerevisiae*. Transposons (also called jumping genes) are mobile pieces of DNA that can hop from one location in the genome to another. The location of the new insertion is random, as far as we know, which means a transposon can create random insertional mutations throughout the genome. But Snyder's group did not use an ordinary transposon.

Snyder's team created a very useful transposon derived from the wild-type version called Tn3. This multipurpose mini-transposon (called mTn) contained a number of genetic alterations to improve its utility in this genome-wide study of proteins (Figure 8.1). The inverted repeats (IR) on each of the ends are needed for mTn to insert into a chromosome. *lacZ* encodes the bacterial reporter enzyme β-galactosidase, which can turn a colorless substrate blue; when a cell turns blue, you know it has expressed β-galactosidase. However, mTn's version of *lacZ* did not contain a promoter or the first codon, so β-galactosidase was not expressed in most cells. You will notice there are two lox sites, which are binding sites for a viral protein that can excise all the DNA between two lox sites (more about this later). Finally, there is the site called 3xHA, which indicates the presence of three copies of the **hemagglutinin** (**HA**)

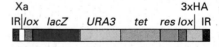

Figure 8.1 Diagram of mTn.

IR = inverted repeat; *lox* = a recombination site for excising intervening DNA; Xa = a restriction site; *lacZ* = a gene that encodes β-galactosidase; *URA3* = a gene required for uracil synthesis; *tet* = tetracycline resistance gene; *res* = required for transposition; 3xHA = three copies of the DNA-encoding hemagglutinin epitope tag. To create the mutations, a yeast genomic plasmid library in *E. coli* was randomly mutagenized by insertion of mTn. Individual plasmid clones were isolated and used for homologous recombination to replace the *wt* gene with the mTn-mutated one in diploid strains lacking *URA3*.

epitope tag encoding DNA. Hemagglutinin is a viral protein recognized by a commercially available monoclonal antibody. Epitope tags are short stretches of amino acids that are recognized by an antibody for detection on blots or by microscopy. The nine-amino-acid stretch of HA epitope tag has no functional role other than as an epitope.

DISCOVERY QUESTIONS

1. Why would you want to include a *wt URA3* gene in mTn?
2. Why would you want to use a version of *lacZ* that lacks a promoter and start codon?
3. Why would you want to put in lox sites that allow you to cut out all the intervening DNA?
4. When the investigators examined a subset of the mutagenized strains to verify that homologous recombination had occurred, they found that 47 out of 48 strains tested were the product of homologous recombination. What is the significance of their finding?

Using the mTn approach, Snyder's lab isolated 11,232 strains that were able to turn blue during vegetative growth. Since the *lacZ* gene lacked a promoter and start codon, the only way cells could turn blue would be if mTn inserted downstream of a promoter in the yeast genome. DNA sequencing was used to isolate the insertion point for 6,358 strains, with most of these landing within 200 bp of an annotated open reading frame (ORF). Due to the random nature of the mutagenesis, some ORFs had zero insertions while others were mutated multiple times. In total, 1,917 different annotated ORFs were mutated at least once. Interestingly, 328 **nonannotated ORFs** (**NORFs**) were also mutated. When the yeast genome was sequenced and annotated, only ORFs of more than 100 codons were considered real genes. Furthermore, if two ORFs of at least 100 codons overlapped, usually only the larger ORF was annotated. Therefore, it is likely that some functional yeast genes were not annotated. Only by random mutagenesis would NORFs have been subjected to experimental analysis.

No one would want to analyze 11,232 strains individually, so Snyder's group needed a high-throughput method to screen them. They developed phenotype *macro*arrays, using robots to spot each strain individually onto agar plates containing growth media and then determining which strains could grow under different growth conditions (Figure 8.2). YPD is normal growth medium for yeast, and you can see that there are some missing colonies in the macroarray of white colonies because some insertions were lethal (a total of 1,082 strains were nonviable). Testing for strains with lethal mutations must be performed in haploid cells, which is one of the benefits of working

with yeast. Benomyl is a drug that disrupts microtubules, and you can see a few resistant colonies. High doses (Calc^R) and low doses (Calc^S) of another drug revealed strains that are resistant or hypersensitive, respectively, to calcofluor, a wall-binding drug. Hygromycin affects cell wall biosynthesis, and there are some strains that are sensitive to it. When glycerol was used as the only carbon source, most cells were viable, but some were not. Using these and additional growth conditions, the investigators were able to rapidly screen thousands of strains, though obviously more than 20 growth conditions could have been analyzed.

LINKS
phenotype *macro*arrays
TRIPLES
METHODS
epitope tag
homologous
recombination

DISCOVERY QUESTIONS

5. How could Snyder's lab grow strains with mutated vital genes? If the strains are not viable, wouldn't they have died before they could be spotted on the macroarrays?
6. Explain how it is possible to screen for calcofluor-sensitive and -resistant cells. How is it possible to test for these two extremes?
7. How can a strain be sensitive to one drug that blocks cell wall formation but not another?
8. The full list of phenotypes is on the TRIPLES web site; click on "Disruption Phenotypes" and then "Search Phenotypic Data." Scroll through the list and design a new growth condition that was not tested by the group at Yale.

A clever idea in this experiment was to utilize the ability of yeast to turn red when metabolism by oxidative phosphorylation is functioning properly. All cells turned red

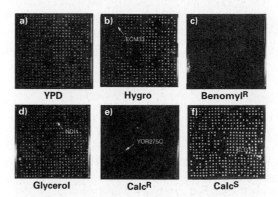

Figure 8.2 Phenotype macroarray analysis.
Examples of 21 x 21 cm macroarrays testing growth on different media: **a)** YPD, **b)** YPD with 46 μg/ml hygromycin, **c)** YPD supplemented with 20 μg/ml benomyl, **d)** YPGlycerol, **e)** YPD with 69.7 μg/ml calcofluor, and **f)** YPD supplemented with 12 μg/ml calcofluor. Arrows indicate strains mutated for genes functioning in cellular respiration (*NDI1*) or cell wall biogenesis (*ECM33*, *SLG1*, YOR275C).

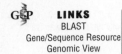

LINKS
BLAST
Gene/Sequence Resource
Genomic View

unless they lacked one of the genes responsible for oxidative phosphorylation, in which case they grew as white colonies (Figure 8.3). Another screen took advantage of the fact that all yeast cells contain the enzyme alkaline phosphatase in their vacuoles. Alkaline phosphatase modifies the colorless substrate BCIP into a blue product, in much the same way β-galactosidase can produce a blue product. When cells were placed on a weak detergent and BCIP, those with a weakened cell wall ruptured, spilling alkaline phosphatase, which created a blue halo around the cell wall mutants. Both of these phenotypes were easily scored, so that 11,232 strains could be screened in a couple of days.

After a few days of intense screening, Snyder's lab had collected a number of mutants that affected one or more of the 20 growth conditions they tested. A large number of mutants were identified, but some caveats are worth noting. First, not all the 11,232 strains were tested for all conditions; only 5,760 were screened for cycloheximide hypersensitivity, for example. Furthermore, some of the mutants had multiple insertions in the same gene, because the insertion site of mTn is random. Therefore, some of the mutant strains represent different alleles of the same mutated gene; those strains often, though not always, produced identical phenotypes. For example, there were 11 different strains with mTn inserted at different places in *IMP2*, a transcription factor involved in sugar utilization. Four of the insertions occurred in the 5′ and middle portions of the gene (at codons 46, 69, 230, and 270) and resulted in mutants unable to metabolize glycerol. However, one insertion, at codon 334 (11 codons prior to the

stop codon), created a mutant able to metabolize glycerol but unable to produce a cell wall. Insertions at codons 46, 230, 263, and 319 also produced the cell wall mutant phenotype. It appears *IMP2* needs its full length to stimulate the production of cell walls, but its carboxyl terminus is not necessary for glycerol metabolism.

Another useful component of the mTn method is the identification of NORFs. Snyder's team identified mTn insertions in 328 NORFs with sizes ranging from 50 to 247 amino acids long. Some of the disrupted NORFs exhibited phenotypes in the 20 growth conditions screened. For example, an insertion at base 198,816 of chromosome 12 resulted in hypersensitivity to the microtubule-destabilizing drug benomyl.

DISCOVERY QUESTIONS

9. Go to the *S. cerevisiae* **Genomic View** and click on chromosome 12 a little to the right of the centromere to see the range 100,000 to 200,000. Click on the blue "AAT2" box and the red "SNR30" box near base 198,816.
   a. What is the range in nucleotides of these two ORFs?
   b. What do these proteins do?
      Now go to the SGD **Gene/Sequence Resource** page. Enter nucleotides 198,416–199,215 and submit your search. This is an 800 bp window centered on the mTn insertion point at 198,816. The NORF described by Snyder's lab encodes 86 amino acids. Scroll down the results page a bit, and click "6-Frame Translation (with restriction map)." You will see the translation of all ORFs and NORFs in this region (Figure 8.4).

10. As you can see, two potential proteins are disrupted by an insertion in the colored box region—one on the top strand (row a) and another on the bottom strand (row d).
    a. On this web page, locate the 86-amino-acid NORF described by Macdonald.
    b. What are the first five amino acids?
       Go back to the SGD Gene/Sequence Resource page; enter the range of 198,728–198,988, select chromosome 12, and click on "Submit Form." In the new window, click on the protein translation with no header; copy that protein sequence. Perform a BLAST protein search with BLASTp. Paste in your protein sequence and BLASTp your NORF.

11. What have you learned about the NORF protein? Are there any orthologs? Any conserved domains?

12. How would you know if this NORF encodes a protein, or just looks like an ORF?

a)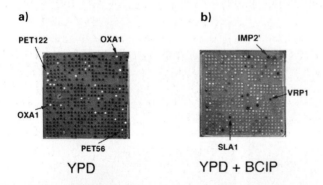

b)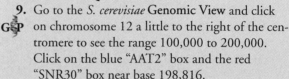

Figure 8.3 Macroarray analysis of metabolic pathways.

a) Metabolic mutants unable to carry out oxidative phosphorylation were identified as white (rather than red) colonies on YPD medium; all cells contained the *ade2* mutation, which leads to red metabolic pigment formation. Respiratory genes identified by this method are labeled with arrows. **b)** Genes functioning in cell wall maintenance were characterized through macroarray analysis of mutants grown on YPD overlaid with agar containing a mild detergent and BCIP. Dark colonies mark cells that lysed; arrows highlight genes known to be involved in cell wall maintenance.

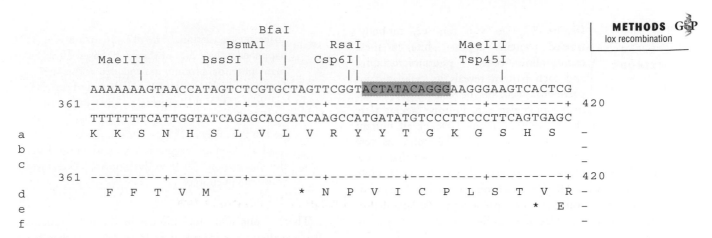

Figure 8.4 Region of interest in yeast genome, including putative NORF.
The site of mTn insertion is highlighted in purple. Above the double-stranded DNA sequence are restriction sites. Below the DNA are the six frames of translation (a, b, c top strand; d, e, f bottom strand) with a dashed line separating the three forward and reverse reading frames. Stars indicate stop codons.

13. How would you know whether disrupting the NORF or the overlapping gene "SNR30" caused benomyl sensitivity?

Snyder's team clustered the 20 different experimental conditions using software similar to that used for microarrays (Chapters 6 and 7). In the phenotype macroarrays, each growth condition is a column and each mutant strain is a row. The color key is different from what we have seen before, but the principle for clustering is identical: cluster the growth conditions based on the similarity of strains affected by each growth condition (Figure 8.5). When the clustering was performed, the investigators learned that some of the growth conditions affect more than one strain.

Figure 8.5 Clustered macroarray phenotype data.
Go to www.GeneticsPlace.com to view this figure.

DISCOVERY QUESTIONS

14. What advantage is there to clustering the phenotypes in this manner? Do you think it gives any new insight into gene function? Explain your answer.
15. Some of the genes identified in this analysis had no known function. How can clustering these data help us predict possible functions?

The predictive power of phenotype macroarrays goes beyond initial phenotype identification. The investigators used the phenotypes to cluster strains and experimental conditions. Clustering allows us to visualize which proteins may be involved in more than one process, or to identify a series of proteins that perform related functions in the same circuit. Proteins with previously unknown functions would also be clustered, allowing more detailed predictions about their possible functions.

You may have detected an old habit that is going to be hard to break: the use of "function" as a catchall term. We have said nothing about biological process, molecular function, or cellular component. With careful analysis of these data, at least one biological process performed by each protein could be predicted with a high degree of certainty. Molecular functions of individual proteins might become evident with additional experiments to refine the precise phenotypes. The only aspect missing is the cellular component.

Where Are These Proteins Located?

The Snyder lab had anticipated wanting to know the cellular location of the proteins when mTn was constructed. Remember the lox sites in Figure 8.1? All the DNA between the two lox sites could be excised (lox recombination) placing the epitope tag-encoding DNA near the 5′ end of the inserted DNA. If the mTn inserted within a coding region, then the lox-mediated excision could place the epitope tag onto the carboxyl end of the truncated protein. By tagging the protein with an epitope tag, the investigators could use the hemagglutinin monoclonal antibody to determine the cellular compartment that housed the tagged protein

G⚕P **LINKS**
NORF database
Ron Davis
METHODS
bar code
immunofluorescence

(Figure 8.6). As you can see, antibody-labeled proteins appear white in these immunofluorescence photomicrographs, and each pattern reveals its cellular compartment. Well, maybe you cannot see the differences, but yeast biologists can. Epitope tagging were performed on 1,340 strains (about 10% of all mutant strains) and the cellular localizations of these proteins were recorded. Interestingly, Snyder's team reported that a few NORFs were also epitope tagged. They cited one example located at base 1,236,754 on chromosome 4. When examined by immunofluorescence microscopy, the tagged protein was found to be localized to the nucleus.

DISCOVERY QUESTIONS

16. Go to *S. cerevisiae* Genomic View and find the location (base 1,236,754 of chromosome 4) of the NORF identified by this research. Find the annotated gene that overlaps with the insertion site.

17. Go to the SGD Gene/Sequence Resource page, retrieve the nucleotides on chromosome 4 ranging from 1,236,454–1,237,054, and click on

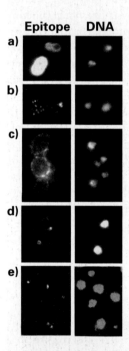

Figure 8.6 Localization of epitope-tagged proteins.
(Left column) Examples of immunofluorescence patterns in cells stained with monoclonal antibody against HA. (Right column) The same cells stained with a DNA-binding dye called DAPI. Numbers in parentheses indicate the number of strains with similar labeling patterns: **a)** Diffuse cytoplasmic labeling (189). **b)** A punctate pattern of cytoplasmic labeling (10). **c)** Localization to the plasma membrane. **d)** A ring in the mother-bud neck (2). **e)** The spindle body (5).

"6-Frame translation." The mTn insertion happened at base 300 of these 600 bases. How many potential proteins are encoded at this site?

18. Go to the TRIPLES NORF database and search for clone ID "V28B5," which is the name of the NORF from Discovery Question 17 that is localized to the nucleus. What does it say under the heading "Genome region"? What do you think of this description? Click on the blue V28B5 text to verify the phenotypes.

There is one more tool hidden in the mTn mutants. Because they were identified as blue colonies, they have landed downstream of a promoter. As you know, proteins are regulated in time and produced only when needed. Therefore, part of a protein's cellular role is determined by the timing of its expression. By subjecting the mTn-mutated cells to additional screens, it would be possible to determine when each protein is produced (During meiosis? During mating? At certain points during the cell cycle?).

The Snyder lab performed one of the largest functional analyses of proteins. In screening so many different mutants, though, a few mistakes were made in the analysis. By making their data publicly available, the Snyder lab has upheld the scientific standards that require reproducibility and peer review, and enabled you to make new discoveries with their data. The Snyder data are valuable tools that contribute to a larger understanding of the yeast proteome.

Though Snyder's was the largest screen, not all the data have been collected. Many of the mutants were not tested under all 20 growth conditions, and additional growth conditions could be tested on all 11,232 strains. Complete analysis will require more time than any single lab could provide. In keeping with the spirit of scientific collaboration and free access to reagents, the Snyder lab will share any strain of these mTn mutants with interested scientists, including students—maybe there is a thesis project waiting for you in proteomics! Before you hop on the Internet to order your yeast, there are other approaches you might find interesting.

Which Proteins Are Needed in Different Conditions?

Everyone wants to know what all these proteins do, and there are many different ways to find out. Elizabeth Winzeler, working with the lab of Ron Davis at Stanford University, was the first author on a paper with more than 50 coauthors in a multinational collaboration, including Ross-Macdonald and Snyder from Yale. In this mammoth project, 2,026 ORFs were individually deleted by homologous recombination and replaced with a selectable marker and 2 unique 20 bp sequences called UPTAGS and DOWNTAGS, which functioned like molecular bar codes

(4^{20} is more than 10^{12} possible bar codes). Flanking each unique UPTAG and DOWNTAG bar code was a pair of common **polymerase chain reaction** (**PCR**) primer sites, so that the unique bar codes for all UPTAGS and DOWN-TAGS in the different mutants could be amplified with two pairs of PCR primers. In conjunction with these mutants, DNA microarrays of 4,052 UPTAG and DOWNTAG bar code spots were constructed so that each spot contained the DNA from one bar code. Therefore, each mutant strain had two, and only two, spots to which its genomic DNA could bind (Figure 8.7). This bar code microarray method does not require the isolation of RNA for probe production. Instead, PCR products from genomic DNA template are labeled either red or green.

Figure 8.7 Bar Code microarray data.
Go to www.GeneticsPlace.com to view this figure.

Ordinarily, analyzing 2,026 different strains would have been tedious and unpleasant, but the team devised a high-throughput approach. They incubated 558 different strains in a single large flask. Immediately, an aliquot was taken and used as PCR template to produce a mixture of 558 red-colored pairs of UPTAG and DOWNTAG bar codes. Six hours later, another aliquot was removed, amplified by PCR, and labeled green. These two mixtures of PCR products were combined, denatured, and added to the microarray (see Figure 8.7). Bar code DNA chips allowed investigators to see which strains were unable to out-compete other strains under these growth conditions (Figure 8.7). Red spots indicate which strains grew slower compared to their flaskmates.

Most gene deletions resulted in yellow spots, because the loss did not affect cell growth under this culture condition. But a few spots exhibited shades of red, indicating the proportion of cells containing these bar codes decreased during the six hours. Since the loss of strains can be monitored over time and under many different growth conditions, the investigators had created a system to study evolution—changes in the gene pool over time due to selective advantages of certain genotypes. In this experiment, the flask contained a population of cells with genomic diversity competing for limited resources. Those with the best proteomes continued to replicate, while those with inferior proteomes gradually became less abundant.

DISCOVERY QUESTIONS

19. In the top and bottom left corners of Figure 8.7 there is a checkerboard pattern. Why would these control spots be necessary?

20. Some green spots are present on the microarray as well. Explain what must be happening in the flask to account for this.

21. Are any NORFs represented on this microarray? Explain your answer.

The bar code method showed which proteins provided a selective advantage in a mixed population of cells. By having two bar codes for each strain, the investigators established an internal control, since both bar codes within a particular sample should produce equal signals on the chip. This was verified by comparing the growth rates for each pair of bar codes at a given time point, and Figure 8.8a shows the correlation for cells grown in rich media. Bar code microarrays are the first time we have seen data from two spots of different sequences used to validate the results for individual genes. This type of validation is very helpful in identifying potential problems with the data before the data are used for interpretation of the role for any particular protein. Most of the data points are clustered in the upper right corner, which indicates *wt* growth rates. The data points that are lower on the diagonal line indicate cells that grew slower because of their mutations.

The growth profiles for the 10 fastest and 10 slowest strains illustrate the types of outcomes observed when a strain lacked a particular protein (Figure 8.8b). In this graph, the Y-axis shows the change in signal for each time point compared to the original signal. Since the same number of cells was sampled each time, strains growing at a normal exponential growth rate would have produced a signal at 1 (i.e., yellow spots on the microarray) for each time point. If a strain grew faster than typical cells, its fraction of DNA microarray signal would be greater than 1, as is seen for the 10 fastest strains. If a strain grew slower than typical cells, it would be less abundant in the sample of cells removed for probe production, and thus its fraction of signal would be below 1, as seen for the 10 slowest strains. If a particular bar code mutation caused a lethal phenotype, we would expect its representation in a sample from each time point to quickly become 0. Strains that failed to give any signal above background were excluded from consideration in Figure 8.8b.

When the bar code analysis was completed, the team classified by biological processes the role each protein played in vivo and tallied which biological processes were controlled by more essential genes than others (Figure 8.9). When the investigators examined the chromosomal distribution of essential and nonessential genes, they found that some essential genes were clustered near each other, but every chromosome contained essential genes. The bar code microarrays allowed the investigators to compare relative growth rates of mutant strains grown under any condition they chose. Therefore, they could measure the relative contribution of every protein in the proteome.

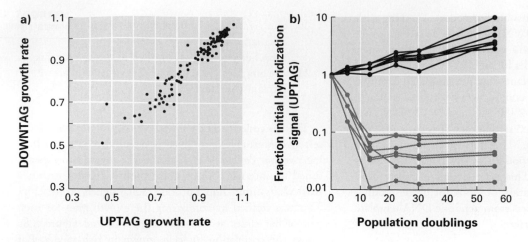

Figure 8.8 Analysis of bar code DNA microarrays.
a) Correlation of growth rate data obtained with both bar code (DOWNTAG and UPTAG) sequences for strains grown in rich medium. Data are shown for 331 strains that produced UPTAG and DOWNTAG hybridization signals that were both at least threefold over background at time zero. **b)** Normalized hybridization intensity data for the 10 slowest-growing (light purple) and 10 fastest-growing (dark purple) strains in rich medium. The data are presented as fraction of labeling intensity at time zero (log scale on Y-axis) over time in the form of population doubling (X-axis).

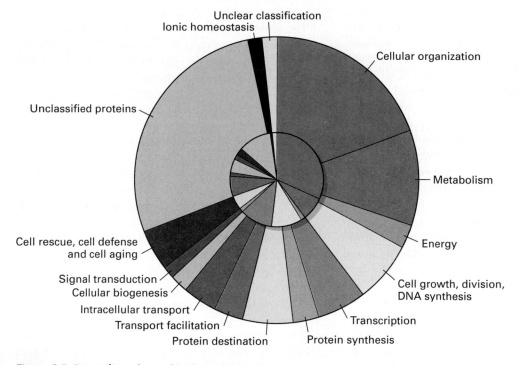

Figure 8.9 Bar code analysis of biological processes.
Distribution of functional classes of essential (inner circle) and nonessential (outer circle) genes using criteria from the Munich Information Center for Protein Sequences (MIPS).

Math Minute 8.1   **How Do You Know if You Have Sampled Enough Cells?**

A hidden assumption in the bar code study is that the proportion of each strain in a sample is the same as the proportion of the strain in the entire population at the time the sample is taken. If this assumption does not hold, the growth profiles of the 558 strains cannot be accurately assessed. The validity of the assumption depends on how large the sample is, compared to the size of the population. For example, suppose a population of 10^{10} cells contains an equal number of cells from each of 558 strains. If 10^9 cells (10%) are sampled from the population, it is more likely that the sample contains an approximately equal number of cells from each strain than if only 10^8 cells (1%) are sampled. If one strain were to constitute only 1% of the population, a very small sample might miss the strain completely, or get too few cells from the strain to be detected on a DNA microarray.

You can calculate the probability of a particular sampling outcome (i.e., getting a particular number of cells from each strain in a sample) using the multivariate **hypergeometric** frequency function, a formula involving the sample size and number of cells from each strain in the population. Specifically, suppose a population contains N cells from M different strains, with n_1 cells from strain 1, n_2 cells from strain 2, . . . and n_M cells from strain M. You can compute the probability that a random sample contains k_1 cells from strain 1, k_2 cells from strain 2, . . . and k_M cells from strain M, with the hypergeometric formula:

$$\frac{\binom{n_1}{k_1}\binom{n_2}{k_2}\binom{n_3}{k_3}\cdots\binom{n_M}{k_M}}{\binom{N}{K}}$$

where K is the sample size (i.e., $K = k_1 + k_2 + \ldots + k_M$). For example, if a population contains 100 cells from each of 3 different strains (a total of 300 cells), the probability that a sample of 60 cells contains exactly 20 cells from each strain is

$$\frac{\binom{100}{20}\binom{100}{20}\binom{100}{20}}{\binom{300}{60}}$$

In the hypergeometric formula, each pair of numbers in parentheses is a **binomial coefficient**

$$\binom{n}{k} = \frac{n!}{k!(n-k)!}$$

(read "n choose k"), which represents how many distinct sets of k cells of a particular strain can be chosen from the n cells of that strain in the population. For example, with $n_1 = 100$ and $k_1 = 20$,

$$\binom{n_1}{k_1} = \binom{100}{20} = \frac{100!}{20!(100-20)!} \approx 5.36 \times 10^{20}$$

Many calculators, as well as various mathematical and statistical software programs, have a binomial coefficient function that can save you a lot of computation.

In a sample of 60 cells from a population of 300 cells, the ideal sample would contain 20 cells from each strain, in perfect agreement with the population proportions of 1/3 for each strain. Evaluating the binomial coefficients in the preceding formula results in a probability of 0.017 of getting an ideal sample, that is, one that contains exactly 20 cells from each of the three strains. However, for practical purposes, we are satisfied with getting *close* to 20 cells from each strain. Suppose we are willing to accept a deviation of up to three cells from the ideal number (20 ± 3) for each strain, corresponding to a deviation of $3/60 = 1/20 = 0.05$ from the ideal proportion ($1/3 \pm .05$). There are 37 different sampling outcomes that satisfy this criterion. (Can you identify all 37?)

You can find the probability of getting one of these 37 outcomes by computing the probability of each outcome (using the hypergeometric formula) and adding the 37 probabilities. The result of these calculations—that is, the probability that the sample proportions are all within 0.05 of the population proportions—is approximately 0.47.

Table MM8.1 Probability that sample proportions are within a specified deviation from 1/3, for sample sizes of 60,120, and 180 cells.

Deviation	Probability		
	Sample 60 cells	Sample 120 cells	Sample 180 cells
0.025	0.113	0.345	0.502
0.05	0.470	0.766	0.954
0.075	0.649	0.954	0.998
0.1	0.887	0.995	0.999991
0.125	0.945	0.9997	≈1

In other words, we have a 53% chance of getting a sample that *differs* from the ideal by more than 3 cells in one or more strains.

To improve our chances of getting a good sample, we must sample more cells. Alternatively, we could relax our maximum deviation criterion, accepting a deviation of up to 5 cells (20 ± 5), for example. The probability that all strains are represented within a given deviation from 1/3 is shown in Table MM8.1 for three different sample sizes. The deviation is given as a proportion of the sample size; the deviation in number of cells is different for each sample size.

This table of probabilities shows that we should sample at least 180 cells from our population of 300 cells to be 95.4% certain that the sample proportions will be 1/3 ± 0.05 for all three strains. However, if we were willing to accept errors as large as 0.125 (i.e., sample proportions ranging from 0.208 to 0.458), a sample of 60 cells would probably do.

Now suppose we have a population containing 200 cells from strain 1; 97 cells from strain 2; and 3 cells from strain 3. In other words, the population is the same size as in our previous example (300 cells), but strain 3 constitutes only 1% of the population. What is the probability that a sample of 60 cells from this population contains at least one cell from strain 3? There are 61 sampling outcomes that have no cells from strain 3. By using the hypergeometric formula, you can sum the probabilities of these 61 outcomes to find that the probability of completely missing strain 3 is 0.511. Therefore, the probability that strain 3 is present in the sample of 60 cells is 1 − 0.511 = 0.489. Once again, sampling more cells improves our chances of getting a good sample—in this case, one in which a cell from strain 3 is present. Specifically, the probability that strain 3 is present is 0.785 when 120 cells are sampled, and 0.937 when 180 cells are sampled.

For very large samples, it is difficult to compute probabilities with the hypergeometric frequency function, and they are often approximated using the **normal distribution** (Math Minute 11.1). Whether exact or approximate, these probabilities show investigators that growth trends such as those in Figure 8.8 are not merely artifacts of the sampling process.

DISCOVERY QUESTIONS

22. Predict what might happen if only the slowest-growing strains were incubated together (see Figure 8.8b). Would their graphs be the same as when they were grown with all 558 strains?

23. Hypothesize why some of the lines in Figure 8.8b leveled off lower than others and why they never reach zero on the Y-axis.

24. In Figure 8.9, you can see that DNA synthesis has some nonessential genes. How can DNA synthesis be considered "nonessential" in a growing population? Explain this apparent contradiction.

25. What are the pros and cons of the bar code method compared to the mTn method? Is one better than the other?

Summary 8.1

Conceptually, the gene deletion approach is not genomic, but the capacity to analyze entire genomes does provide us with more comprehensive perspectives. The two yeast whole-proteome analysis methods could be applied to any organism, but, because of its small genome and its utility as a model eukaryote, yeast will continue to be a very popular system. Since the earliest days of *Drosophila* genetics, investigators have tried to learn what proteins do inside living cells. The first step is to characterize what each protein does (molecular function), why the cell requires this function (biological process), and where the function takes place (cellular component). However, to fully understand the proteome requires additional research tools and experimental designs, as you will see in the remaining sections of this chapter.

8.2 Protein 3D Structures

A major goal of biology is to describe how cells work and define the rules by which they live. All cells come from pre-existing cells. DNA is the heritable material. These rules have been validated repeatedly, and it is difficult to imagine a time when people doubted them. Another rule is that "form meets function," which means that if you know the shape of a molecule, you can understand how that molecule works. In this section, we examine several research efforts that focus on the 3D structure of proteins. Is there a high-throughput method for determining the 3D structure of entire proteomes? Can structure alone reveal how proteins work? Can we develop better medications once we know the 3D shape of a protein? Finally, can a change in shape be the difference between life and death, even when the amino acid sequence remains unchanged? We have provided several 3D Jmol tutorials for you to use; Jmol is freely available to anyone and works on all computers. You can explore any protein's structure by visiting FirstGlance.

Does a Protein's Shape Reveal Its Function?

We have sequences for many proteins, but amino acid sequence alone is insufficient to reveal what the proteins do, how they perform these roles, or with which proteins they interact. Many times, a protein's role is not fully understood until its 3D shape is known. Progress toward this goal has been sporadic, and driven by the perseverance of many individual laboratories around the world. Funding agencies have concentrated their resources on high-throughput efforts in **structural proteomics**.

To obtain detailed protein structure information at the atomic level, we need to purify the protein in large amounts (as much as 1 mg), crystallize the protein, and then use either X-ray crystallography or nuclear magnetic resonance (NMR) to determine the precise location of every atom in the protein. Structure determination is usually performed one protein at a time by individual labs. Structural proteomics demands faster methods, but progress has been slow. The first group to launch a proteome-wide 3D structure determination effort was the German Protein Structure Factory. About a year later, the National

Institutes of Health (NIH) initiated a $150 million structural genomics program to support research centers to expand the global effort of structural proteomics. The rate-limiting step for structural proteomics is the production of samples and crystals. Protein structural information is collected by the Protein Data Bank (PDB), an international repository for 3D structure files. The rate of 3D data collection has increased rapidly in recent years, but it will have to accelerate more if structural information is going to have a comprehensive impact on proteomics.

The importance of determining the 3D structure is recognized by most biologists around the world. A Canadian team led by Dinesh Christendat, Adelinda Yee, Aled Edwards, and Cheryl Arrowsmith wanted to determine the structure of all proteins in the archaeon *Methanobacterium thermoautotrophicum*. The genome of *M. thermoautotrophicum* contains 1,871 ORFs, which would present a formidable challenge to any group. As with any large project, the Canadian team used a rational approach to maximize success. First, they eliminated membrane proteins (about 30% of the proteome), because these proteins are notoriously difficult to crystallize. Next, they skipped any proteins that had obvious orthologs in the PDB (about 27% of the proteome), since these proteins would not provide as much new information. A total of 424 of the remaining 900 proteins were selected for structural analysis. The proteins were divided into small (<20 kDa) or large (>20 kDa) groups and carried through the requisite steps (Figure 8.10). As you can see, at each step, some proteins were excluded from the next phase, which illustrates the difficulty of the task.

LINKS G&P
3D data collection
Aled Edwards
Cheryl Arrowsmith
FirstGlance
German Protein Structure Factory
Jmol
PDB
structural genomics

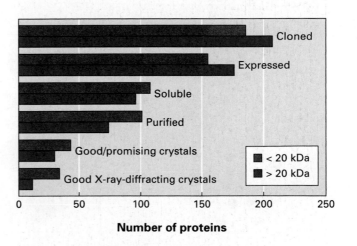

Figure 8.10 Structural proteomics of an archaeon. Histograms of the number of *M. thermoautotrophicum* proteins at the end of each step in cloning, expression, and sample preparation. Proteins were divided into two classes based on molecular weight.

STRUCTURES
aquaporin-1
ODCase

Structures for the first ten proteins were determined, leading to several new discoveries. Some folding patterns had never been seen before, including binding sites for coenzymes such as flavin mononucleotide (FMN) and nicotinamide adenine dinucleotide (NAD$^+$), as well as cation cofactors such as of calcium, nickel, and magnesium. Binding sites for these ligands could not have been predicted by amino acid sequences.

We will look at one protein as an example of what can be learned from structural proteomics. MTH129 is a known ortholog of Orotidine 5′ monophosphate DeCarboxylase (ODCase), which catalyzes the removal of one carbon dioxide in exchange for one hydrogen ion at carbon number six of uridine 5′ monophosphate (UMP). This exchange of CO_2 for H$^+$ occurs 10^{17} times faster than it would in the absence of ODCase. MTH129 is the first member of this enzyme family to have its structure determined. The key to its catalysis is the unusual localization of four amino acids of opposite charges (KDKD), which line the binding site. A quick series of electrostatic interactions permits the removal of the CO_2.

To further illustrate how function can be deduced from structural proteomics, let's look at an indispensable but often neglected protein, aquaporin-1. Introductory textbooks state that the cell membrane is "semipermeable." When you learned that membranes are made of phospholipids, you also learned that only hydrophobic molecules can pass through cell membranes. No charged or polar molecule can pass through, not even a proton. Yet you probably also learned that water, a polar molecule, can pass through a cell's membrane. This sets up a logical contradiction that is rarely discussed. In reality, water does not pass

through cell membranes; it travels through aquaporin and not the phospholipid bilayer. As the name implies, aquaporin is a protein channel that allows only water to pass. *Aquaporin-1* was the first water channel gene to be cloned, and thus the number one was added to its name. You can interact with aquaporin-1 online to see how its structure reveals its function.

The characterization of aquaporin answered many questions that had baffled biologists for years, but a big question remained: How can water (H_2O) pass through a channel when protons (H_3O^+) cannot? It would be a disaster if protons could sneak through with water, because membrane potentials and hydrogen gradients are vital for many cellular functions, including the production of ATP. Protons can move along a column of water by hydrogen bond exchange, which means that any stream of water rushing through aquaporin (at 3×10^9 water molecules per second) must be prevented from carrying H$^+$ ions.

A multinational group led by Kazuyoshi Murata determined the structure of perhaps the most important membrane protein. Aquaporin **monomers** are composed of 269 amino acids that form six membrane-spanning domains. The functional unit of aquaporin is a **homotetramer** (composed of four identical subunits) that allows water to pass through each monomer. The key to its function is the protein's hourglass-shaped pore, in which the narrowest constriction is only 3.0 Å wide (Figure 8.11). The size of this constriction is critical, because a water molecule is about 2.8 Å in diameter. In addition, several hydrophobic amino acids line the pore and help exclude other charged molecules. Although Murata's team could not be certain, the structure allowed them to predict that water does not form a continuous stream as it passes through the pore. Instead,

a) **b)** **c)**

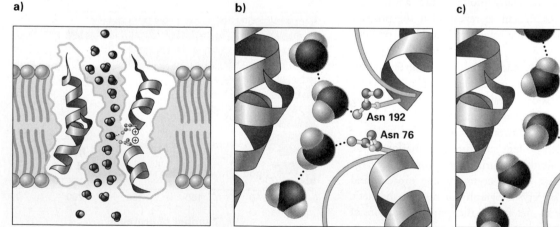

Figure 8.11 Structural basis for aquaporin function.
a) The charges from the helix control the orientation of the water molecules passing through the narrowest part of the channel. **b)** and **c)** The hydrogen bonding of a water molecule to asparagines 192 and 76, which extend their R groups to form the narrowest part of the channel.

as one water molecule reaches the constricted portion, the hydrogen bonds that had connected one water molecule to the next are temporarily transferred to two asparagine amino acids that form part of the constricted region (Figure 8.11b and c). Once the continuous column of hydrogen bonds is broken, protons cannot move along with water, and aquaporin has permitted water but not protons to pass through your membranes.

DISCOVERY QUESTIONS

26. How did the structure of ODCase lead to the discovery of a new functional motif (KDKD)? What is so unusual about this sequence of four amino acids?

27. Based on the 3D structure of aquaporin and its hypothesized mechanism of preventing protons from passing through the channel, predict two amino acids that should be conserved in every species.

Can We Use Structures to Develop Better Drugs?

By this time, you might be thinking that all this attention to structure is only for crystallographers and has no real-world applications. Let's look at one example of how the use of structures can lead to improved medical treatment. Vicki Nienaber and her colleagues at Abbott Laboratories in Illinois had a very simple but clever idea. If you soaked a crystallized protein in a solution of many different molecules, you might be able to co-crystallize compounds that bind in the active site (Figure 8.12). Using this approach, Nienaber's group worked to identify a new compound that could inhibit the enzyme urokinase. Slowing urokinase activity can slow the growth and metastasis of tumors without the unpleasant side effects of traditional chemotherapy. If the team could find an inhibitor that could be taken orally, this would reduce discomfort and expense for patients. Although the group did not find the ultimate drug they wanted, they did identify a "lead compound" that was able to enter the body after oral administration, and inhibited urokinase when present in the range of 2.5 μM. That new compound will be a good starting point for further drug development.

DISCOVERY QUESTIONS

28. Why is Nienaber's method better than current methods of drug discovery, in which many compounds are given to cell lines to see which drugs produce the desired effect?

29. Why is the co-crystallization method less appealing than the current method of drug discovery?

30. Go to the urokinase 3D tutorial. What similarities are there between the inhibitor in this 3D structure and the new one Nienaber discovered (shown in Figure 8.12)? What are the differences?

LINKS
Nobel Prize
Stanley Prusiner
Susan Lindquist
Vicki Nienaber
STRUCTURES
Prions
urokinase

Can One Protein Kill You?

There is another interesting story about shapes, but this one does not have a happy ending. Prions are proteins that can change their shapes, which is nothing new for proteins. But once a prion protein has changed its shape, it can make other prion molecules change too. As the number of converted proteins grows, a new phenotype becomes evident. Converted prions are the cause of scrapie in sheep, mad cow disease in cattle, and Creutzfeldt-Jakob disease (CJD) in humans. All three of these illnesses lead to a rapid, progressive dementia and eventually death. If you eat prion-"infected" meat, you can develop CJD. That humans become infected with the bovine (cattle) form of prions indicates a similarity in 3D structure between human and bovine prion proteins, which is evident when you compare their structures.

The existence of a contagious disease that lacks nucleic acid was unprecedented and considered impossible. Demonstrating this protein-only form of disease took many years, and eventually Stanley Prusiner's persistent efforts were rewarded with a Nobel Prize. The existence of highly conserved proteins whose only known function is to cause a lethal disease does not fit with our perception of evolution. Even more surprising has been the discovery of a prion-like protein in yeast, **Sup35p**. Why does yeast have a prion protein that cannot play any neurological role?

Sup35p is a translation termination factor with a twist. Its carboxyl end binds to the ribosomal complex to terminate translation. The other two parts (amino end and middle section) can be deleted, and the translation termination function remains intact. If Sup35p is converted to its infectious prion conformation, however, the shape change spreads throughout the cell and is passed on to all daughter cells. When in the prion conformation, Sup35p causes ribosomes to read through stop codons, extending the lengths of proteins and thus altering both their shapes and their functions. The loss of fidelity in translation does not sound evolutionarily adaptive, which makes us think prions should not have evolved—but they did evolve and are conserved in many eukaryotes. Could natural selection drive the ability of Sup35p to change shape and function?

Heather True and Susan Lindquist wanted to know why yeast would retain a protein with apparent detrimental translation capabilities. In their paper, they documented that Sup35p in the prion conformation conveys new phenotypes to cells (Figure 8.13a). As the translation fidelity is

a)

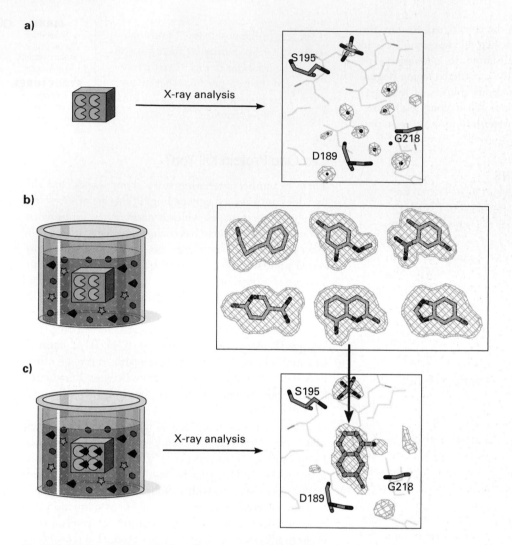

Figure 8.12 Structural method to discover new medications.
a) In an ideal situation, the active site is open to the solvent when the protein is crystallized. **b)** Different small molecules are mixed with the crystallized protein. **c)** Those with appropriate shapes will be incorporated into the crystal. In the structure web page of urokinase, you can see an inhibitor similar to aminopyrimidyl 2-aminoquinoline, which inhibits the enzyme at 0.37 μM.

reduced, extended proteins are formed, some of which provide the means to resist antibiotics.

Interestingly, the conversion of Sup35p to the prion conformation can occur spontaneously at a rate ranging from 10^{-5} to 10^{-7} (Figure 8.13b–d). The ability of converted Sup35p to produce new and heritable physiologies enables a single genotype to have multiple phenotypes. As cells grow in a changing environment, those with the prion form of Sup35p may be more fit and thus thrive while their genetically identical but phenotypically distinct sisters cannot survive. If this new phenotype were advantageous, subsequent genetic mutations might lead to the stabilization of the new phenotype (i.e., longer proteins), and Sup35p might subsequently revert to its nonprion conformation. Whether a

prion-induced advantage exists in mammals is unknown, but the work in yeast leads to interesting speculation on why mammals express potential prion proteins that are poised to create neurological diseases. Nevertheless, Sup35p illustrates rather bluntly why structural proteomics is needed to augment sequence information.

DISCOVERY QUESTIONS

31. In what way did the 3D shape change of Sup35p alter its function? Is the functional change related to the change in colony morphology? Explain your answer.

32. Is the change in shape of Sup35p due to any
G⚛P changes in DNA sequence? How could new phenotypes (see Figure 8.13a) become established in the gene pool if Sup35p reverted to its original shape and function? You can explore Sup35's structure using the PDB ID code 1R5N with FirstGlance.

Summary 8.2

The rule "form meets function" is reinforced by discoveries from structural proteomics. We can develop better medications once we know where an existing drug binds and the structure of its binding site. The ability to see the shape of aquaporin revealed how water molecules pass through a channel but tiny protons cannot. Although we still do not know why eukaryotes have prion proteins, the yeast prion-like Sup35p has offered us the first hint that a change in shape can lead to a beneficial change in function under certain circumstances. As we learned in Section 4.3, every adaptation comes with tradeoffs. Perhaps on an evolutionary time scale, prions have proven to be advantageous for some individuals and deleterious to others as selection pressures change. However, no protein acts in isolation, so to fully understand proteomes, we will need to define all the protein-protein interactions within cells.

8.3 Protein Interaction Networks

Which Proteins Interact with Each Other?

LINKS G⚛P
FirstGlance
METHODS
immunoprecipitation

To understand what a person does for a living, you need to know more than a simple job description. For example, every college and university has a development officer, but what does this person really do? If you knew with whom the development officer interacted, you would have a better understanding of this role in the life of your institution. Similarly, to fully understand what each protein does, we need to determine with which other proteins each individual protein interacts. In the pregenomic era, immunoprecipitation was the primary means of determining protein-protein interactions. In the postgenomic era, we need high-throughput methods to catalog protein interactions. As you will see, two methods have proven especially helpful for this purpose. A more vexing problem is how to display interaction networks so we can comprehend them, gain new insights, and make testable predictions.

At the risk of sounding like a conspiracy theorist, we know proteins rarely act alone. Most proteins need a partner to either allosterically or covalently modulate them to change their conformations, but not in a contagious way as prions do. If a protein has to be activated, often it must bind to another protein. When the cellular task is done,

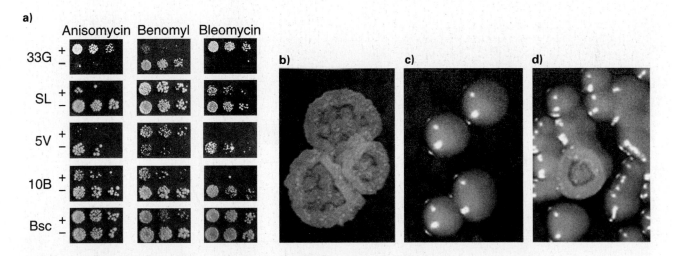

Figure 8.13 Functional consequences of protein conformation.
a) Sup35 prion conformation affects growth in different genetic backgrounds and under different conditions. Media contain the fungicides anisomycin, benomyl, or bleomycin. Five different strains are grown in the presence of a prion-shaped Sup35 (+) or the absence of the prion shape (−). **b)** Colony morphology is altered by prion conformation in Sup35. Cells were grown, diluted, and spotted onto plates containing potassium acetate and grown at 30°C. **c)** Cells lacking a prion form of Sup35 and **d)** the spontaneous appearance of a colony containing prion-shaped Sup35 amidst many colonies lacking prion-shaped Sup35.

LINKS
His3
mini-industry
Stan Fields
METHODS
comparative genomics
Y2H

proteins frequently have to interact with another protein to become inactivated. To fully describe a proteome, we must know each protein's interactions within a cell. Two proteomic methods currently amenable to high-throughput screening are DNA sequence analysis and yeast two-hybrid assays. We'll look at protein sequence analysis first and then turn our attention to the more prevalent two-hybrid method. Regardless of the method, the goal is to create protein interaction maps that enable us to visualize cellular functions more completely.

Can Sequence Analysis Uncover Interactions?

A series of papers published in 1999 described how protein-protein interactions could be deduced by comparative genomics (see Chapter 3). The rationale is simple. If species A has a protein with two domains (called 1 and 2), and species B has orthologous domains (contained in separate proteins 1' and 2'), then proteins 1' and 2' probably interact to fulfill the same function as found in species A. Using this approach, various groups have predicted over 100,000 protein-protein interactions in yeast, and thousands in *E. coli* and other prokaryotes. As complete genomes for more species are sequenced, this type of computer-based research and model building will continue.

A lot of useful proteomics information remains hidden in the published genome sequences. An in silico analysis of genomes provides us with reasonable models that can be tested in the lab, but is limited to proteins encoded by a single gene in one species and by two or more genes in another. Surely there must be protein interactions that do not fit these criteria, and we need a way to determine them as well. Sequence analysis is insufficient, so a more direct, high-throughput method is needed.

Can We Detect Protein Interactions?

In 1991, Stan Fields (University of Washington, Seattle) developed a new proteome-wide method to detect protein-protein interactions. It is worth noting that the method, called **yeast two-hybrid** (**Y2H**), was developed before the term proteomics was popular and thus was way ahead of its time. In short, the Y2H method uses a protein of interest as bait to discover proteins that physically interact with the bait protein; these proteins are called prey (Figure 8.14). Y2H experiments have been conducted in many labs to discover proteins that interact with each lab's favorite bait protein. In fact, Y2H has become a mini-industry, as illustrated by Fields's web pages dedicated to Y2H.

Here's how the Y2H method works. A single transcription factor is cut into two pieces called the DNA Binding Domain (DBD) and the Activation Domain (AD), which stimulates the RNA polymerase to begin transcription. Fused to the DBD is the bait protein of interest (B), which cannot initiate transcription on its own. Fused to the AD is the prey ORF, which can be any gene of known or unknown function. The prey protein of AD + ORF fused together cannot initiate transcription alone. When the bait and prey proteins are produced in the same cell, they might

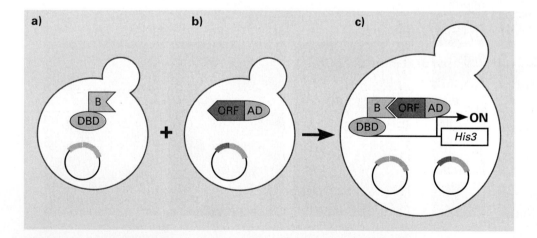

Figure 8.14 Yeast two-hybrid method.
a) The DNA binding domain (DBD) is fused onto the "bait" protein of interest, B. The bait is encoded by the plasmid shown in the same cell. **b)** The "prey" is encoded by its own plasmid and is composed of an ORF fused onto the activation domain (AD), which is capable of activating RNA polymerase. **c)** Both the bait and prey plasmids are inside the same cell, and if B and ORF proteins physically interact, RNA polymerase will be able to transcribe a reporter gene, in this case *His3*.

interact; if they do, transcription of the reporter gene is initiated. In Figure 8.14, the reporter gene is *His3*, which leads to the production of the amino acid histidine. Without transcription of *His3*, cells cannot grow unless histidine is added to the growth medium. Any ORF can be used as prey, which means a proteome-wide survey can be performed rapidly by transforming a cDNA library into cells that contain bait plasmids. In this way, every protein in a proteome can be tested individually for its potential to interact with the bait. You can watch a Quick-Time movie that illustrates a variation of Y2H in which interaction is signaled by the production of a blue color rather than histidine synthesis. The blue color is produced by the enzyme β-galactosidase, which is encoded in the *lacZ* reporter gene.

of a yeast cell if the prey ORF were the *His3* gene? Of course, it would grow regardless of interaction with the bait. As a control experiment, cells containing only prey must be tested on media lacking histidine. If prey-only cells grow, protein interaction was not necessary for growth, though protein interaction between bait and prey cannot be ruled out. Also, not all proteins work well inside the nucleus. The Y2H method can produce false positive or false negative results due to improper protein folding (i.e., 3D structure; Table 8.1). However, the greatest benefit of Y2H is that yeast cells can express genes from almost any species, which means this is a powerful proteomics method for *Drosophila, C. elegans, Arabidopsis,* zebra fish, mice, humans, and of course yeast, too.

LINKS GⓈP
Interaction Sequence Tags
Y2H results
METHODS
lacZ reporter

DISCOVERY QUESTIONS

33. The protein interactions detected by the Y2H method must take place inside the nucleus of the yeast cell. Why might this prove problematic?

34. What controls would you want to run to make sure the only positive cells are those with interacting proteins? In other words, how could you rule out that you had simply recloned the *His3* gene?

35. Since Y2H takes place inside yeast cells, is it restricted to yeast proteins only?

The Y2H method has been very popular and successful, but not perfect. For example, what would the phenotype be

DISCOVERY QUESTIONS

36. Go to the Y2H results web site from the Fields lab. Do a find function on your browser for RPC19. You will see that bait protein RPC19 bound to four prey proteins. Look at the next bait entry, called RPC40. What do you see as one of its prey? Compare RPC40 prey to the list of prey proteins when RPC19 was used as bait. Do you perceive any problems? What would you have expected to see? Hypothesize how this result could happen.

37. Go to Interaction Sequence Tags and you will see a small table of *C. elegans* proteins that have been tested by Y2H. Click on the following entries and

Table 8.1 The set of prey where RPC19 was used as the bait protein.
Interactions are ranked according to plausibility.

Gene	ORF	Description (from the YPD Database)
Rpc40	YPR110C	Shares subunit of RNA polymerases I and III (AC40)
	YLR266C	Protein with similarity to transcription factors, has Zn[2]-Cys[6] fungal-type binuclear cluster domain in the N-terminal region
Mtr2	YKL186C	Involved in mRNA transport, has similarity to *E. coli mbeA*
	YOL070C	Protein of unknown function
	YFR011C	Protein of unknown function
	YHL018W	Protein with weak similarity to human pterin-4-alpha-carbinolamine dehydratase
Gtr1	YML121W	GTP-binding protein involved in the function of the Pho84p phosphate transporter
Mas1	YLR163C	Beta (enchancing) subunit of mitochondrial processing peptidase

Color Codes

Previously known interacting protein (positive control)
Possible interaction partner
Significance unknown
Unlikely interaction partner

Source: http://depts.Washington.edu/%7Eyeastrc/th_8.htm (Stan Fields's web site).

G●P **LINKS**
Additional Y2H Results
CuraGen
David Eisenberg
DIP
MIPS search page
Peter Uetz

compare the results, but be aware that this database is a work in progress:

β-catenin

presenilin (associated with early onset of Alzheimer's)

MAP kinase

MAP kinase kinase

a. Why might some proteins make bad bait?

b. Are positive results always reproducible? What effect do the *C. elegans* data have on the Y2H method?

c. Based on the results, do you think MAP kinase or MAP kinase kinase is at the center of protein-protein interactions?

38. Fields has continued to push the envelope by systematically exploring the protein interactions in the yeast proteome. Go to the Y2H page that lists all the results, about 1,000 protein-protein interactions.

a. Search for *His3* and see what you find. Any surprises?

b. Now try the Additional Y2H Results web page. Search for *Swi*. The bait "SWI5" is in the first column, the prey gene is in the middle column, and the ORF name of this gene is in the third column.

Go to the MIPS search page and find out the function for Swi5 (search with YDR146c) and its

prey with a known function (Mei5; search with YPL121c). Does Swi5-Mei5 look like a probable interaction, or an artifact? Support your answer with information from the MIPS database.

In a rather large collaboration, Stan Fields and Peter Uetz at the University of Washington, Seattle, and CuraGen Corporation of New Haven, Connecticut, compiled a database of protein interactions. Many new interactions were detected which may lead to new insights about integrated circuits within the cell (Figure 8.15). When the analysis was completed, the investigators compiled the putative interactions (the interactome) into the cellular roles. Since many proteins have no assigned role and the number of protein-protein interactions is very small, we have a very limited understanding of how the proteome functions.

A group from UCLA, led by David Eisenberg, has created a very rich database they call the Database of Interacting Proteins (DIP). Using this database, they made some interesting predictions about the yeast prion-like protein Sup35p (Figure 8.16). Although Figure 8.16a is too crowded to show all the subtleties, it is clear that Sup35p is in the center of a complex interactome. What is especially intriguing is that in addition to translation, as you would expect for a translation termination factor such as Sup35p, its interactome includes many other processes. This predicted interaction network can be tested experimentally, which is one reason for producing complex interactomes and

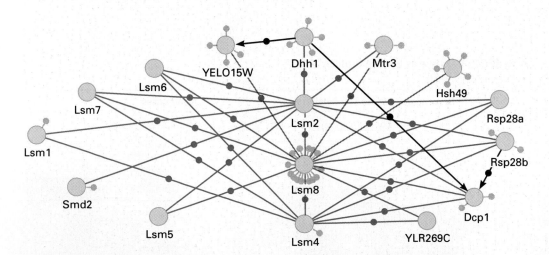

Figure 8.15 Interactome showing physical contacts of RNA splicing proteins.
The proteins are indicated by large purple nodes and their interactions with lines. The dots on the lines help you follow each line. Purple lines show interactions that were detected by traditional Y2H array screens, black from multiple high-throughput screens, and gray from literature and array screens. Black arrows point away from the protein used as bait in the screens. Small gray nodes indicate other protein-protein interactions not highlighted here.

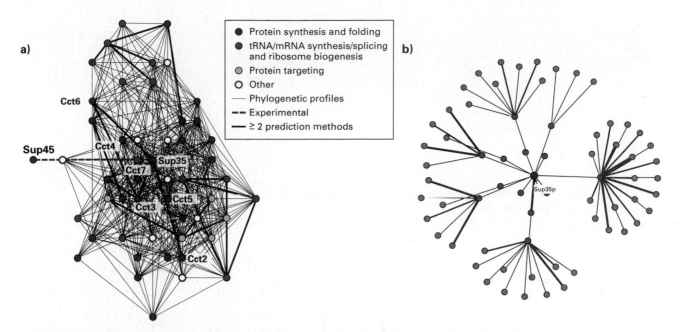

Figure 8.16 Network of protein interactions.
a) First-generation DIP image with higher (bold lines) and lower (thin lines) confidence. This network is centered on the yeast prion protein Sup35p. Many of the proteins in this network are involved in protein synthesis, including ribosomes, protein folding, sorting, modification, and targeting. **b)** Second-generation, interactive DIP image showing the query protein as a central purple node plus two layers of other proteins. The thickness of lines indicates confidence in the predicted interactions. An online legend explains additional details.

| Math Minute 8.2 | **Is Sup35 a Central Protein in the Network?** |

In Figure 8.16a, Sup35 appears to be central to a network of interacting proteins. However, looks can be deceiving, particularly in such a complex network. We would like to quantify whether Sup35 interacts with an unusually large number of proteins, compared to other proteins in the network.

As described in Math Minute 1.2, a mathematician would call the network of interacting proteins in Figure 8.16a an undirected **graph**. In graph theory, the number of edges touching a node is called the *degree* of the node. In graph theory terms, the question at hand is whether the Sup35 node has a significantly greater degree than is "typical" for such a graph. If so, investigators would be led to believe that Sup35 plays a central role in the function of the entire network of proteins.

One way to approach questions about graphs like that in Figure 8.16a is to build a probabilistic model of the graph, called a *random graph*. Under this model, an edge is drawn between each pair of nodes with a certain probability, each edge independent of the others. Advanced probability theory and graph theory provide precise ways to determine the probability that the maximum degree in an arbitrarily large graph exceeds a given number. However, under the simplifying assumption that the degree numbers are approximately normally distributed (see Math Minute 11.1), you can evaluate whether the degree of the Sup35 node is significantly greater than expected using the mean and standard deviation of the degree numbers. Since Figure 8.16a is too complex to illustrate this process, consider the smaller network in Figure MM8.1.

In Figure MM8.1, nodes 4, 5, and 7 have degree 1; nodes 2, 3, and 6 have degree 2; nodes 1 and 8 have degree 3; and node 9 has degree 5. The mean degree is the average

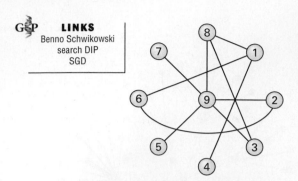

Figure MM8.1 Undirected graph with nine nodes representing a small gene network.
Edges indicate gene interactions.

of the nine degree numbers (2.22). The standard deviation of the nine degree numbers (1.23) is used to determine if the degree of node 9 is unusual. Specifically, if the degree of node 9 is more than two standard deviations larger than the mean degree, it can be considered unusually large. Because $5 > 4.68 = 2.22 + 2 \times 1.23$, we conclude that node 9 has an unusually large degree. This same procedure could be applied to the graph in Figure 8.16a to help quantify whether Sup35 plays a central role in this network of proteins.

making them freely accessible. The DIP results also advance the speculation that Sup35p, and its mammalian orthologs, may have additional functions that are much more integrated into cellular circuitry than we realized. As research continues, our understanding of protein interactions will become increasingly more complex and interconnected.

DISCOVERY QUESTIONS

39. Several proteins with the name Cct plus a number appear in Figure 8.16a. Search SGD, enter "Cct5," and find out what this protein does.

40. Now search DIP's database of protein interactions. Do a "Node" search for "Cct5" in the Node ID field, and click the "Query DIP" button. On the results page, click on the link under the heading "DIP Node" on the left side. Then click on the word "graph" in the top right corner and you will see Cct5 as the red node. Can you find Sup35p interacting with Cct5, as indicated in Figure 8.16a? (Click on a few nodes and read the new window that appears. Be sure to read the figure legend to help you interpret the interactome map.)

41. Perform a "Node" search for "Sup35". What do you find? Now try "Sup45" and see what role this protein plays. Click on the few nodes that directly interact with Sup45. Can you find Sup35 now? The thicker lines indicate interactions with more confidence. What problem do you see with the DIP database at this stage of its development?

Is It Possible to Understand Proteome-Wide Interactions?

By now, you must realize that proteomes contain very complex networks of more proteins than there are genes, and that there are many useful methods to help us understand functional roles for each interaction. The goal is to synthesize all this information into an integrated protein circuit diagram that shows which proteins interact with each other.

Not many people would care to spend countless hours sifting through all the different databases to synthesize a complete interactome map for yeast. But Fields and Uetz have led the way in collaboration with **Benno Schwikowski**, a computer scientist (now at the Pasteur Institute). The three investigators collated 2,709 interactions among 2,039 different proteins. When they integrated these protein connections, they synthesized integrated protein circuit diagrams containing 2,358 interactions among 1,548 proteins in the yeast proteome. The figures are too large to reproduce here, so we will look at the PDF files instead. Open the PDF file called "**Benno Figure 1**," which is from their paper. Here is the color key for the lines:

- Red: cellular role and subcellular localization of interacting proteins are identical
- Blue: localizations are identical but functions differ
- Green: cellular roles are identical but localizations differ
- Black: cellular roles and localizations differ
- Gray: localizations or cellular roles are unknown for one or both proteins

Here is the color code for the boxes:

- Blue: membrane fusion
- Gray: chromatin structure
- Green: cell structure
- Yellow: lipid metabolism
- Red: cytokinesis
- Light yellow: unknown role

When the PDF file opens, zoom to 800% magnification so you can read the names of the proteins. Use the find tool and enter Pex7. This protein is involved in lipid metabolism, as indicated by its yellow box. It is connected to several other proteins known to support lipid metabolism, including two other Pex proteins. Search SGD and determine the function of Pex7 and two other Pex proteins. Does SGD agree with the categorization of these proteins as lipid synthesis proteins? Does the PDF interactome diagram portray at least part of the circuitry involved in the process described in SGD?

The purpose of a good model is not to confirm what you already know but to provide new insights. In the same figure, just above Pex7 is a cluster of proteins with no known function (e.g., YLR095c). The model predicts that YLR095c is a central participant in a circuit that includes many proteins with known functions. Go to MIPS and determine the roles of Ami3 (YOL060c), Pup3 (YER094c), and Ilv6 (YCL009c). With this information, can you hypothesize the function of YLR095c and the other proteins in this circuit? What experiment could you design to test your hypothesis?

DISCOVERY QUESTIONS

42. In Benno Figure 1, locate Akr2. This protein is known to play a role in endocytosis. Can you tell which proteins are connected to Akr2 in this figure? You may need to zoom in further to see its connections.

DATA
aging.pdf
Benno Figure 1
Benno Figure 2
degradation.pdf
membrane.pdf

43. The interactome is so comprehensive that it is difficult to see individual connections. For example, Akr2 is connected to YHR105w, but you might not have been able to see this connection. The investigators produced a separate figure to show Akr2's connections to a subset of the full proteome circuitry:

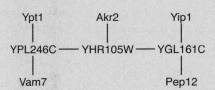

This version is readable and makes clear one weakness of the full interaction diagram. Describe at least one improvement you would like to see in the full protein interactome diagram.

44. Open aging.pdf, membrane.pdf, and degradation.pdf and use the find function to locate the prion protein "SUP35". In each file, see if Sup35p was given a particular function. Be sure to notice the color key in the upper left-hand corner of each figure before you zoom in 800 or 1,600%.

Go back to the aging file. Notice that Sup35 is hidden behind another protein called Kcs1. What is the cellular role of Kcs1, according to the aging.pdf network? Compare its "aging designation" with the role given in SGD. Its location on the network hints that it may interact with Sup35, but from this map we cannot be sure. Might Sup35p interact directly with Kcs1, based on their cellular roles?

As with microarrays, the power of a complex interactome may be more in its overview than in individual nodes. For example, Schwikowski tested the predicted function of proteins based on their interactions with other previously characterized proteins and found the deduced interactome was correct 72% of the time. Of course, this also means that the model was incorrect 28% of the time, which reinforces the need to use such maps only as general guides and starting places for formulating predictions. When the investigators examined interactions between functional roles rather than individual proteins, they found a few surprises (see Benno Figure 2). On this map, locate the RNA polymerase II cluster (pol II, which transcribes most genes). Prior to this study, you probably never would have predicted that transcriptional proteins interact with so many other cellular roles. It is worth noting that Sup35p was included in protein synthesis, which interacts (directly or indirectly) with the pol II group. From Discovery Question 44, you learned that Kcs1 was involved with pol II transcription and was located very near Sup35p in the interactome. Perhaps protein synthesis and pol II groups do interact directly. The Sup35p/Kcs1p example illustrates the power of a good model: it allows you to examine data in new ways and make predictions that can be tested experimentally. Data from your experiments can be used to refine the model and further your insights and understanding.

When Schwikowski redefined each protein by its cellular localization rather than its function (Figure 8.17), there were some more surprises. For example, there are 8 interacting pairs between the nucleus and the Golgi body and 27 interacting pairs between the nucleus and mitochondria.

DISCOVERY QUESTIONS

45. Does the cytoskeleton interact only with the plasma membrane, the nucleus, and the cytoplasm? Why does this map appear too simplistic, given what we know about vesicle and organelle movement?

46. There are three obvious examples of missing connections between two closely associated organelles. Identify as many missing interactions as you can.

Schwikowski's team undertook a gigantic task and produced a circuit diagram of many interactions that will help us better understand proteomes. Of course, their work is not complete; the yeast proteome contains more than

6,000 proteins, and this version of the interactome contains only 1,548 proteins. Nevertheless, the first phase of their work provides us with new perspectives. It is a huge step forward in understanding the yeast proteome and by extension the proteomes of many other eukaryotes.

At this point, you should be impressed with what we can learn about protein interactions, though they may never occur in a living cell. For example, although protein A and B can interact in a Y2H experiment, they may never be present at the same time inside a *wt* cell. This leaves us with a problem that must be addressed: What proteins are present at the same time?

Summary 8.3

As noted in the introduction to this chapter, the term "postgenomic era" does not imply that whole genome sequence analysis is completed. By reanalyzing genome sequences, we can predict which proteins are likely to interact with each other, and these predictions can be verified experimentally. Yeast two-hybrid experiments allow us to screen large numbers of proteins or protein fragments to determine which proteins bind to each other. However, the Y2H has its limitations, which **protein microarrays** might be able to offset, as you will see on pages 320–323. As we continue to accumulate new types of data in large quantities, it becomes more difficult to display them in a

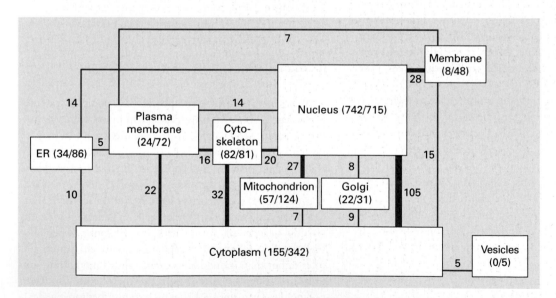

Figure 8.17 Protein interactions grouped by cellular compartments.
Numbers in parentheses indicate the number of interactions in the circuit diagram among proteins of this compartment/the total number of proteins in the same compartment. Lines connecting compartments indicate the number of protein interactions by the thickness of each line (numbers of connections are near each line). For example, there are 7 interactions between the 48 membrane proteins and the 72 plasma membrane proteins in PDF "Benno Figure 1."

meaningful and comprehensible way. Protein interaction circuits are beneficial but currently limited by the 2D constraints of a printed page. New visualization methods are being developed to address this limitation. The last piece of the proteomics puzzle, addressed in Section 8.4, is listing and quantifying all the members of a given proteome.

8.4 Measuring Proteins

It is important to understand what proteins can do, with what they interact, and their 3D shapes. However, this information is meaningless taken out of the context of a dynamic proteome. For example, let's imagine you knew the person who pumped up the basketballs used by Shaquille O'Neal. You would know what both people do, where they perform their jobs, and that they interact via the basketball. However, they do not play basketball together because they are never on the court at the same time. This simplistic example illustrates how important it is to distinguish between all possible interactions in all proteomes and actual interactions that occur because both proteins are expressed in one cell at the same time. Furthermore, it would be very instructive to know how many copies of each protein were present in a given proteome, just as it would to know if any basketball team had more than one Shaquille O'Neal.

This section examines the identification and quantification of proteins. DNA microarrays have proven so successful that many investigators are developing protein microarrays; we will read about three different approaches to creating them. Finally, we consider single-cell proteomics and experiments that study one molecule at a time; these single molecules can affect the metabolome, the entire metabolism of an organism.

Which Proteins Are Present?

An accurate list of proteins present in a proteome at various times is critical. What proteins do brain cells contain? Liver cells? Muscle cells? What about a yeast cell while it is switching from one food source to another? Completing each step of the cell cycle? Mating? These questions must be resolved if proteomics is going to have a significant impact. A primary objective of those working in proteomics is to identify and quantify every protein in a cell at a given time (i.e., define a proteome). Until recently, the best method for defining a proteome has been 2D electrophoresis.

What Are 2D Gels?

The proteomics workhorse since the mid-1970s has been **two-dimensional (2D) gel electrophoresis**. As shown in Figure 8.18, 2D gels separate proteins based on their isoelectric points (first dimension) and **molecular weights** (second dimension). For any protein, the net charge of the molecule depends upon the pH of the surrounding environment. As the pH changes, the net charge changes. When the net charge of a protein is zero and the protein stops moving, the pH of the local environment will be equivalent to the **isoelectric point** (**pI**). The isoelectric point of a protein depends on its amino acid sequence, a fact that allows us to separate a mixture of proteins based on charge. Once the proteins are sorted by isoelectric point, they are placed on a second gel and separated by molecular weight using standard SDS-PAGE. The end product is a pattern of spots and smears that defines each protein in a complex proteome. Unfortunately, some proteins do not resolve well by this method and thus they often escape detection. In particular, very basic proteins, very small peptides (Figure 8.18f), and proteins that are difficult to solubilize often escape detection by 2D gels. Although 2D gels separate many proteins, the limitations of this method frustrate investigators trying to define proteomes. Sample preparation is not a trivial component of the method, and depends in large part on which proteome you want to study. The most troublesome issues are (1) detection of spots, (2) quantification of each spot, and (3) identification of each spot.

Spot Detection As you can imagine, no single method of labeling proteins is going to work for every protein. Silver and Coomassie staining are not sensitive enough to detect every protein, though they work well for the most abundant ones. Radioactivity can be used for some samples, but is less successful with tissue biopsies taken from patients. Recently, fluorescent dyes have been developed that may bind to all proteins without altering protein mobility in 2D gels, though it is not clear yet how versatile the dyes will be. If you have an antibody to your favorite protein, then **Western blotting** of 2D gels can detect as few as 100 molecules per cell. However, Western blotting is labor intensive, limited by the availability of antibodies, and not suited for high-throughput proteomic analysis.

Quantification Figure 8.18 illustrates why quantification of each spot is difficult: some proteins smear, overlap other proteins, and are present in vastly different amounts than others. Fluorescent dyes hold the best hope for solving this problem as well.

Identification At the end of 2D gel analysis, there are several ways to identify each spot. Initially, investigators tried to create 2D "fingerprints" to compare one pattern against another, but this proved nearly impossible since 2D patterns were difficult to reproduce. The next approach was to cut out each spot and subject them separately to amino acid sequencing. This technique works better with the advent of improved methods for protein sequencing. Mass spectrometry (**MS**) is the preferred tool for determining the identity of the proteins.

METHODS G☯P

Mass spectrometry
SDS-PAGE
New visualization
Western blotting

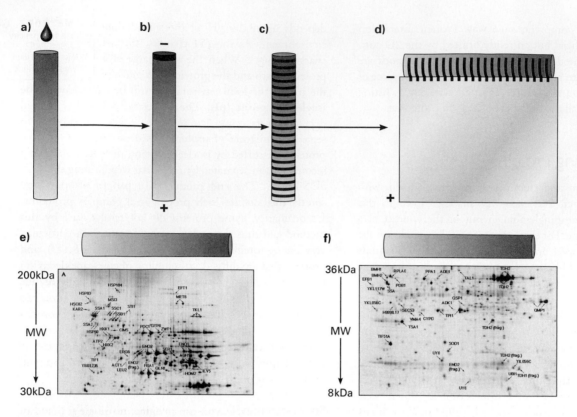

Figure 8.18 Two-dimensional (2D) gel electrophoresis.
Each column (a–c) with a gradation of gray shading represents the isoelectric focusing gel with a pH gradient. **a)** A mixture of proteins (purple drop) is applied to the isoelectric focusing gel and **b)** exposed to an electrical current. **c)** Proteins migrate to their isoelectric points and stop moving. **d)** This tubular gel is placed on top of a slab polyacrylamide gel that contains SDS and is subjected to electrophoresis (SDS-PAGE). Proteins migrate into the slab gel according to their molecular weights. Yeast cells were grown in rich media and subjected to 2D gel analysis. Using duplicate isoelectric focusing gels, large **e)** and small **f)** proteins were analyzed on separate gels. The same spots appear at the bottom of e) and the top of f). Molecular weights are resolved on the Y-axis and pIs on the X-axis. Panels e) and f) are from the Swiss 2D database at ExPASy.

Mass Spectrometric Identification of Proteins There are two key components to MS. The first is that all proteins can be sorted based on a **mass to charge ratio** (i.e., the molecular weight divided by the charge on this protein). The second is that proteins can be broken into peptide fragments, facilitating the identification of each protein. We will examine the four steps needed to identify each protein. Steps 2 and 4 utilize separate mass spectrometers and so the entire process is often referred to as **tandem MS**, or MS/MS.

Step 1: Ionization A protein sample is injected into a device that must ionize the proteins. Two methods are popular: matrix-assisted laser desorption/ionization (**MALDI**), and electrospray ionization (**ESI**). Both methods produce mixtures of proteins that have become charged, or ionized. In MALDI, a laser illuminates the protein mixture and the absorption of energy ionizes the proteins, which are sent on to a mass spectrometer (MS). In ESI, a strong electrical charge is imparted to the solvent containing the protein, after which the solvent evaporates, leaving the proteins ionized (Figure 8.19a). In both MALDI and ESI, ionized proteins enter an MS.

Step 2: Separation Once the proteins are ionized, their mass and charge must be determined (Figure 8.19b). This is a two-step process; first the ions are collected, and then the time it takes individual groups of identical proteins to move to the detector is measured. This timing determines mass (m) to charge (z) ratio (abbreviated as **m/z**). Detection of m/z is improving rapidly, but we do not need to focus on the details. Once the m/z ratios are determined for each protein, a computer allows identical proteins (defined by m/z ratios) from a single sample to be selected and sent to the activation chamber. This process is automated so proteins of different m/z ratios are sent to the activation chamber in quick succession.

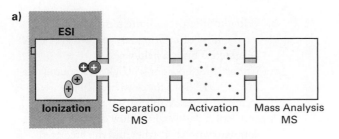

a)

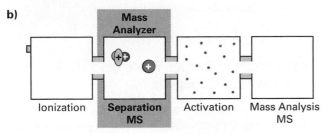

b)

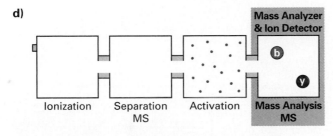

c)

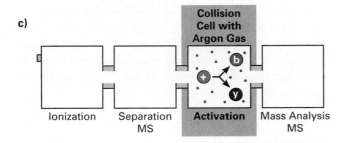

d)

Figure 8.19 Tandem mass spectrometry for protein identification.

a) ESI creates ionized proteins, represented by the colored shapes with positive charges. Each shape represents many copies of identical proteins. **b)** The ionized proteins are separated based on their mass to charge ratio (m/z) and sent one at a time into the activation chamber. Separation and selection take place in the first of the two MS devices. The solid purple protein has been selected for analysis; the other three are temporarily stored for later analysis. **c)** The group of m/z selected ionized proteins enters a collision cell that is filled with inert argon gas. The gas molecules collide with the proteins, which causes them to break into two peptide pieces (labeled b and y). **d)** Ionized peptide pieces are sent into the second MS device, which again measures the m/z ratio. A computer compares the spectrum of peptide pieces to a database of ideal spectra to identify the original group of identical proteins.

Step 3: Activation After being selected by m/z ratio, proteins must be broken into smaller fragments (Figure 8.19c). Fragmentation is accomplished by sending each protein into an argon-filled chamber. This inert gas collides with the ionized protein, and the vibrational energy of the gas causes the protein to break into two pieces. Each piece of the original protein, one containing the amino portion (labeled b) of the original protein and the other containing the carboxyl portion (labeled y), will carry a charge. Since each group of identical proteins can break in many places, more than one combination of peptide pieces will be produced. Bigger proteins can produce more combinations of peptide pieces, since they can be broken anywhere along the protein's length.

Step 4: Mass Determination This is the second MS of the two, and from it the term "**tandem MS**" was derived (Figure 8.19d). The final step is to determine the m/z ratios for the mixture of ionized protein pieces. Because there will be more than one fragment pair for each population of proteins (Figure 8.20a), the output of the second MS can be confusing (Figure 8.20b). However, a computer compares the spectrum of peptide pairs with ideal spectra for all known proteins in the database (see Math Minute 8.3). The ideal spectrum with the best match identifies the original protein. This process is repeated rapidly (within seconds) so that from one complex sample, hundreds of proteins can be identified.

Now it may be a bit more obvious why sample preparation is critical. If you took whole cell extracts and put them into an MS/MS system, there would be too many proteins to sort and identify. One solution is to separate proteins into functional domains (e.g., nucleus, plasma membrane, etc.). In addition, you could run a 2D gel and then cut out spots for MS/MS identification. In the end, 2D gels are labor intensive and cannot quantify how much of each identified protein was present in a particular sample. We will return to this quantification problem later.

DISCOVERY QUESTIONS

47. Go to the PROWL database, where you can search a large range of species.
    a. Look under Metazoa (multicellular organisms), Chordata (animals with backbones), Mammalia (mammals), Rodentia (rodents), and select "Mus musculus." In the "Enter Key Words" box, enter "leptin" and hit the "Search Keywords" button.
    b. Click on the link to the left of "leptin precursor—mouse."
    c. Click on the button called "Sequence Analysis," which will take you to a page with the mouse leptin protein sequence. What are the molecular

LINKS
ExPASy 2D
Fat 2D Gel
Michel Desjardins

weight (Mw) and pI of mouse leptin? Write down this information.

48. Go to ExPASy Mouse White Fat 2D Gel and you will see a 2D gel of white adipose tissue from mouse. Look in the area where you expect to see leptin. Click on some of the red plus signs in this area to see if leptin has been identified in this gel.

49. You have performed the steps investigators take to make an educated guess about which spot is their favorite protein. Based on your findings for Discovery Question 48, what should be done next to improve the PROWL database?

50. Search the ExPASy 2D database for your favorite protein by entering its name. As a starting place, type in "IDH" and hit "submit." You will see that IDH has been identified in mouse, *E. coli*, and *Staphylococcus aureus* (the "staph" infection pathogen). Click on one species and scroll down to the section called "2D PAGE maps for identified proteins." There are three links you should visit, in this order:

a. "Compute the theoretical pI/Mw"— At the bottom, click on "submit" and record the pI and Mw.

b. "How to interpret a protein map"— You will see the kind of information you will get in the next step. This is a help page.

c. Under "MapLocations:" click on a couple of the "SPOT" links to see where IDH is located. Compare this with the predicted pI and molecular weights. How did the 2D gels compare with each other and the predicted molecular weight and pI?

What Proteins Do Our White Blood Cells Need to Kill a Pathogen?

Our immune system has many ways to kill pathogens. One way is to engulf the invading pathogen by phagocytosis and destroy it. Phagocytosis is a part of our innate immune system that we share with other warm-blooded animals and even invertebrates such as insects and sea urchins. Therefore, the process of phagocytosis evolved millions of years ago, and the proteins involved must be well conserved. The challenge has been to identify all the proteins involved in phagocytosis.

A French and Canadian collaborative effort was able to define the proteome of the phagocytotic vesicles called **phagosomes**. Michel Desjardins led a group at the University

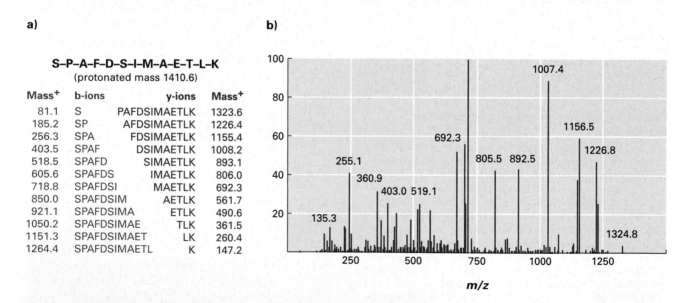

Figure 8.20 Protein identification through peptide fragment formation and separation.
When a group of identical proteins is broken into peptide pieces, more than one pair of b and y peptides will be formed. **a)** One protein sequence and its calculated mass on top, with the b peptides/masses (gray) and the y peptides/masses (purple) below. **b)** An experimentally determined mass/charge spectrum from the peptide in panel a). Some peaks are higher than others, which means that some b/y peptide pieces were more abundant than others. The spectrum is used to determine each peptide's amino acid sequence and protein identity.

MATH MINUTES G✆P
ExPASy 2D
ProFound
PROWL

| Math Minute 8.3 | **How Do You Identify Proteins on 2D Gels?** |

In Discovery Question 50, you found spots on 2D gels that contained IDH protein. Often, the protein of interest is located outside the expected pI and Mw range marked by colored boxes, while other spots are inside the expected range. Therefore, to identify a protein spot (e.g., IDH) on a 2D gel, the investigators excise the spot from the gel, analyze the excised protein with mass spectrometry, and process the measured masses with computer **algorithms**. In this Math Minute, we explore one method the investigators used, **protein mass fingerprinting** (**PMF**), using data from Discovery Question 50. PMF consists of three steps:

1. Break the protein into shorter peptides by digesting it with an enzyme (usually trypsin) that cleaves the protein at specific amino acids.
2. Use mass spectrometry to experimentally determine the mass of each peptide produced in step 1.
3. Search a database of proteins for those containing matching peptides. For a peptide to match, the string of amino acids should be flanked by two halves of an enzyme (e.g., trypsin) cleavage site and have approximately the same mass as one of the masses found in step 2.

Performing PMF is analogous to doing a BLAST search, since PMF requires sifting through a huge protein database to find amino acid sequences that match the experimentally measured query masses. However, unlike BLAST searches, the proteomics community does not have a universally accepted algorithm and probability model for PMF. We will use a program called ProFound, which is hosted at the PROWL site, but there are many other similar programs.

Go back to the ExPASy 2D database and search for "IDH." Click on the mouse hit, and scroll down to the two identified IDH spots on the liver map. Copy and paste the list of peptide masses from spot 2D-0013NW into the ProFound search form, inside the box labeled "monoisotopic masses." Under this box, change the charge state to MH^+. In the "Digestion" section, change the "Allow maximum" number of missed cleavages to 2. In the "Modifications" section, select "Iodoacetamide (Cys)" and check the box next to "Methionine oxidation." The search form now reflects the parameters described in the associated help page, reached by clicking on "Peptide Masses" from the ExPASy 2D page. Click on "Identify Protein" at the bottom of the search form to query the nr (nonredundant) protein database.

The resulting page lists proteins with matching peptides, ranked in order of the probability that the identified sequence is correct. The probability for each hit is the probability that the hit is the correct protein (K), given the experimental data (D) and background information (I). The mathematical notation for this probability is $P(K|DI)$. Experimental data are the peptide masses, and background information includes species, total predicted protein mass, and enzyme cleavage accuracy. You may have noticed that background information can be entered manually on the search form to narrow the search. The formula for computing $P(K|DI)$ is:

$$P(K|DI) = \frac{P(K|I)P(D|KI)}{P(D|I)}$$

The formula is based on **Bayes' Rule** (see Math Minute 4.4), adjusted for the fact that two variables (D and I) are given rather than one, as in the genetic testing example. The right side of the equation contains three probabilities that are calculated from a complex stochastic model of mass spectrometry measurements, the details of which we will not explore here. The important thing to know is that each $P(K|DI)$ probability can be used to reliably separate the true matching protein from proteins with peptides that match merely by chance (analogous to an **E-value** in BLAST searches). Using the calculated probabilities listed on the ProFound results page, you can evaluate the quality of the protein identifications and decide what threshold of probability is acceptable for you.

MATH MINUTE DISCOVERY QUESTIONS

1. In your ProFound results, click on the number in the "%" column of the top-ranked protein to see which peptides were identified as matching. Save this page. Repeat your ProFound search with the same masses and parameter settings, except select "Iodoacetic acid (Cys)" instead of "Iodoacetamide (Cys)." You should get the same highest-ranked protein, but this time one mass won't match a peptide in this protein sequence. Explain why this particular peptide no longer matches the protein.

2. To see many more proteins, click on the "+" next to the rank of the first protein. Why did you get so many matching proteins with equal rank of 1 and equally high probability? How could you select the best match from among these equally ranked proteins?

3. Go back to your original ProFound search, using "Iodoacetamide (Cys)" once again, and delete all but the first four peptide masses. Resubmit the search with your shortened list of masses. Does ProFound identify the protein correctly? How confident are you that the first hit listed is the correct one? Why did the probability go down for this search when you submitted four of the same masses from your first search?

---

of Montreal to define the proteome of the phagosome. The first challenge was sample preparation, since they did not want to define the complete proteome of the phagocyte. To simplify their work, they "fed" phagocytes latex beads that could be engulfed by cells. When the cells were lysed, the density of the latex beads allowed phagosomes to be separated from all the other cellular compartments. From this starting point, Desjardins's group used 2D gels to identify over 160 proteins associated with phagosomes (Figure 8.21). They identified most of the major and minor spots, though a few proteins were not visible on the 2D gel. They cut out each spot from the gel and identified the proteins using MALDI-MS/MS.

As we have discussed, some proteins are not easily resolved by 2D gels, especially membrane-associated proteins. Therefore, Desjardins performed additional analysis that used a detergent (Triton X-114) and SDS-PAGE. Using this combination, 31 additional proteins were identified that had not been resolved by the previous 2D gel. Altogether, the team amassed a long list of proteins associated with phagosomes; some of the proteins had not been associated with phagosomes before this research.

This accomplishment was substantial, but the group also wanted to determine which proteins were inside the phagosome (i.e., luminal side) and which were outside (i.e., cytoplasmic side). To determine which side each protein was on, the investigators incubated purified phagosomes with a **protease** that digests proteins. If the proteins were inside the phagosome, the protease would not digest them and the proteins would still be visible on a 2D gel. If they were

on the cytoplasmic side, the protease treatment would digest the proteins and their spots would be missing or greatly reduced. As you can see from Figure 8.22, this method worked very well and helped refine the cellular location of these proteins.

Having established which proteins are associated with phagosomes and which are on the inside or outside, Desjardins and his collaborators studied one more component of phagosomes: development. We have noted that proteomes can change over time. In this case, the proteome changes as the phagosome develops into a mature pathogen-killing organelle. The investigators were particularly interested in the accumulation of a family of digestive enzymes called cathepsins. From the data in Figure 8.23, you can see that some proteases accumulate over time (e.g., cathepsins D and S), others are reduced (cathepsin Z), and some are relatively constant (cathepsin A).

Among their findings were a few surprises. For example, proteins normally thought of as endoplasmic reticulum proteins were associated with phagosomes. Phagosome proteins normally associated with **apoptosis** and signal transduction also surprised the investigators. In their list of phagosome proteins are 17 with uncertain identities and 7 **expressed sequence tags** (**ESTs**), providing clues to the cellular roles of nonannotated genes. The group created a "virtual phagosome" to display all the parts of its proteome and which proteins might interact with each other. Many other labs will use this virtual phagosome information as a starting place to understand how some pathogens are killed by phagocytes and others escape destruction.

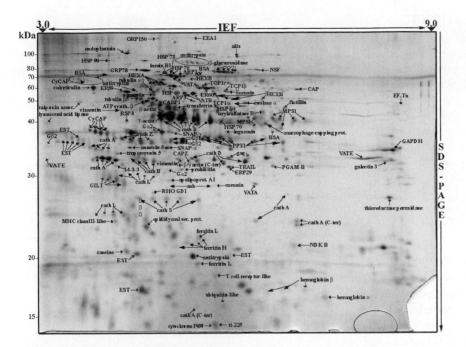

Figure 8.21 Phagosome 2D gel.
Phagosomes were isolated from macrophages and subjected to isoelectric focusing and SDS-PAGE. The pH gradient is presented above the gel and the molecular weight standards to the left. This gel was silver stained to locate the spots, which were excised and identified by tandem MS.

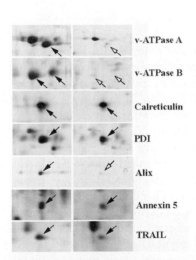

Figure 8.22 Localization of phagosome proteins.
Phagosomes were subjected to protease treatment to digest any proteins on the cytoplasmic side of the phagosome. Untreated (left) and protease-treated (right) phagosomes were compared on 2D gels. Proteins with substantial cytosolic portions are indicated with open arrows; those resistant to protease are indicated with solid arrows.

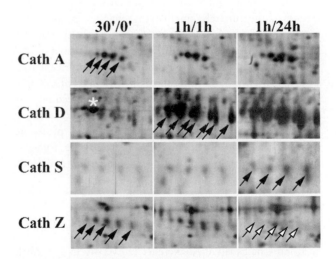

Figure 8.23 Cathepsin composition during phagosome maturation.
Phagosomes from macrophages were isolated after 30 minutes of exposure to latex beads (30'/0'), after 1 hour of incubation with latex beads followed by 1 hour of further development (1h/1h), or 1 hour of latex bead exposure followed by 24 hours of further development (1h/24h). The cathepsins are labeled on the left, with loss of signal indicated by the open arrows. Multiple arrows point to the multiple spots for each cathepsin. The star denotes a precursor form of cathepsin D.

DISCOVERY QUESTIONS

51. Perform a PubMed search using the G⬡P phrase "phagocytosis is a mechanism of entry into macrophages Desjardins." Read the abstract and see why some ER proteins were associated with the phagosome. Were the investigators unable to purify phagosomes, or did they discover a new aspect of immune systems?

52. If the phagosome proteome changes in 30 minutes (e.g., cathepsin D in Figure 8.23), do you think a DNA microarray would be able to detect these rapid changes in phagosomes? Explain your answer.

At this point, we have studied several ways to define a proteome. What proteins are present? What does each protein do? With which partners does each protein interact? However, two problems still exist that are critical to understanding a complete proteome. How much of each protein is present? Can we measure protein modifications such as phosphorylation?

How Much of Each Protein Is Present?

Defining the components of a proteome is a doable, though difficult, task. We can subdivide cells and tissues and gradually assemble the full contents of complex proteomes. Unfortunately, quantifying changes in abundance of each protein within a proteome is substantially more difficult. Furthermore, we would like to know when a protein has been modified, which may be more significant than how many copies of a protein a cell contains. For example, some proteins must be phosphorylated to become activated; measuring such posttranslational modifications is a critical aspect of understanding a dynamic proteome. Though the task is difficult, a few groups are trying to devise new technologies to address the quantity and modification issues.

Can We Quantify Proteomes in Cultured Cells?

In **Brian Chait**'s lab at Rockefeller University in New York, a team of biologists, chemists, and engineers devised a new approach to protein quantification and modification problems. At the heart of their method is the use of **stable isotopes** to label proteins as they are formed. Stable isotopes are nonradioactive versions of atomic elements that vary in atomic mass. A given molecule made with a heavy isotope will weigh more than the same molecule made with a lighter isotope. Using this principle, Chait's group developed a protocol (Figure 8.24) to compare the quantities of heavy and light proteins in each spot on a 2D gel, followed by MALDI-MS/MS for protein identification.

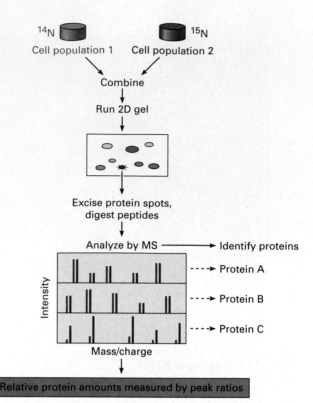

Figure 8.24 Quantifying differences in proteomes. Two cell pools are grown in the presence (purple = cell population 2) or absence (gray = cell population 1) of heavy nitrogen ^{15}N. The proteins are extracted, pooled, and subjected to 2D gel analysis; spots are excised; and proteins are identified and quantified by MS/MS. The relative areas under the pairs of heavy and light peptide peaks indicate relative abundance of each protein pair.

Two cell populations were grown under different conditions, similar to the way DNA microarray experiments are performed. One population of cells was grown in the presence of normal nitrogen (99.6% ^{14}N and 0.4% ^{15}N), while the second population was grown in a medium enriched with heavy nitrogen (>96% ^{15}N). When it was time to harvest the cells, the cultures were combined to ensure that all subsequent steps were performed equally on all cells, assuring that changes in relative amounts of each protein were not due to differences in handling of the samples. The mixed proteins were separated by 2D gels, from which spots were excised and subjected to sequence identification by MALDI-MS/MS. Heavy and light versions of a particular protein would co-migrate to the same spot on a 2D gel. However, unlike traditional 2D gel samples, each spot separated into two m/z peaks, because the ionized peptide fragments were separated into heavy (^{15}N) and light (^{14}N) versions of the same molecule. In Figure 8.24, proteins A and B were present in equal amounts in the two cell populations, but protein C was more abundant in cell population 2 than in cell population 1. Relative protein abundance is indicated by the height of the pairs of peaks (gray and purple) at different m/z ratios. If someone accidentally spilled half of the sample along the way, the relative amounts for all the proteins in the mixed sample would remain the same. Therefore, we cannot say exactly how many molecules of protein C were in either cell type, but we can say that protein C was five times more abundant in population 2 than in population 1. Using stable isotopes was a clever way to eliminate the problem of quantifying the intensity of spots in a 2D gel.

A slight variation of this method, illustrated in Figure 8.25, allowed the investigators to detect proteins that have been phosphorylated. Spots of interest were excised from 2D gels, and the ratios of unphosphorylated and phosphorylated proteins were measured. As shown in the bottom of Figure 8.25, the phosphorylation status of most peptides did not vary. The bar graph was of equal height above and below the horizontal line, and peptides A–D were represented by only one mass on the horizontal line. Peptide X, however, was differentially phosphorylated, as shown by the relative amounts of unphosphorylated (X) and phosphorylated (Xp) peptide. The total amount of protein X was constant in both cell populations; however, 30% of protein X in population 1 was phosphorylated compared to 70% in population 2.

With Chait's method, we can determine the variation each protein experienced in relative quantity and degree of phosphorylation. As a proof-of-concept experiment, Chait's group tested whether a particular yeast protein (Ste20) could be phosphorylated in two yeast strains. One strain lacked the protein Cln2, a cyclin that regulates a kinase during the G_1 to S transition in the cell cycle; the other strain expressed Cln2. Mutants lacking Cln2 were labeled with ^{14}N, and *wt* cells were labeled with ^{15}N. In the analysis of the Ste20 spot on the 2D gel, the investigators found that different peptides from Ste20 received different levels of phosphorylation (Figure 8.26). Panels a, b, c, and d represent different peptide fragments, phosphorylated very little (peptide b was near 0%), moderate amounts at one site (peptides c and d), or multiple times (peptide a could be phosphorylated on three different amino acids) in *wt* cells. Cln2 was required for Ste20 phosphorylation since cells that expressed Cln2 contained more phosphorylated Ste20.

DISCOVERY QUESTIONS

53. Using Chait's method, would it be possible to quantify the complete proteome of yeast? Explain your answer.
54. There is one extremely important limitation to this method. Can you figure out what it is? (Hint: What would you have to do to compare human biopsies using this method?)

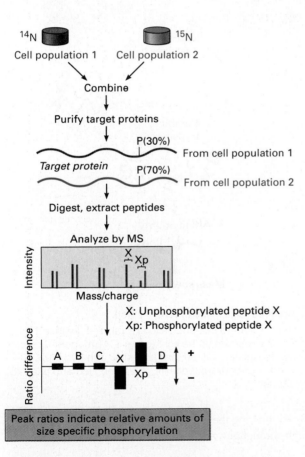

Figure 8.25 Quantifying relative levels of phosphorylation. Bar graphs A, B, C, and D indicate no difference for proteins A–D in the two populations. Compared to population 1, unphosphorylated protein X was reduced in population 2. Phosphorylated protein X (Xp) was increased in population 2.

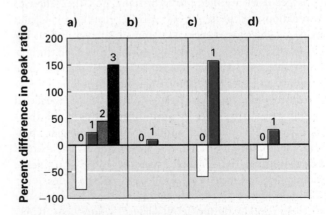

Figure 8.26 Phosphorylation of Ste20 requires Cln2. Four Ste20 peptides are phosphorylated in *wt* cells but not *Cln2*-deleted cells. Bar graphs show the loss of unphosphorylated Ste20 peptides (negative change) and the appearance of phosphorylated Ste20 peptides (positive change). The numbers above each bar indicate the number of phosphates added to each peptide.

LINKS
Ruedi Aebersold
METHODS
affinity chromatography
avidin
ICAT

Can We Quantify Proteins in Any Cell?

Chait's method was a huge improvement in quantification compared to traditional methods of measuring spot intensities on 2D gels. Using the stable isotope, we can compare relative amounts of each protein identified in two proteomes. However, there is a significant limitation: cells must be grown in a heavy isotope. Although this is not a problem for yeast, bacteria, or eukaryotic cells grown in tissue culture, it is much more difficult to accomplish with plants or large animals, and is not feasible to perform on humans. The use of isotopes was a good idea, but the requirement that protein synthesis incorporate the isotopes is problematic.

Ruedi Aebersold from the Institute for Systems Biology in Seattle developed a protocol that worked on any sample, even human biopsies: **isotope-coded affinity tags (ICAT)**. At the heart of the ICAT method is the labeling reagent Aebersold's group created (Figure 8.27). ICAT labeling reagents come in two forms, heavy and light. The heavy version has eight deuterium atoms, which increases its mass by eight atomic units over the light reagent, which contains eight hydrogen atoms. Other than this difference in mass, the two reagents are identical. On one end of the labeling reagents is a chemically reactive group that can bind to the amino acid cysteine. Any protein that contains a cysteine can be labeled with either the heavy or the light ICAT reagent. The other end of the reagent contains **biotin**, which binds very strongly to a protein called avidin. Avidin and biotin are like two halves of Velcro, except once they bind to each other, they almost never let go.

Anyone can take two protein populations and incubate one with the heavy ICAT reagent and one with the light ICAT reagent (Figure 8.28). All cysteine-containing proteins get labeled either heavy or light, and then the two populations of proteins are mixed, just as in Chait's method. However, Aebersold's team wanted to reduce the complexity of the proteins to be analyzed. First, all the proteins were cut with a protease called trypsin, and these trypsin-treated fragments were purified by avidin affinity chromatography, which retains only the ICAT-labeled trypsin fragments. Affinity purification reduced the complexity of the sample by about 90% without losing any protein identification or quantity information. The biotin/avidin-purified, ICAT-labeled peptides were sent into a tandem MS to ionize, sort, and identify the peptides.

In addition to protein identification, the relative amount of each peptide (heavy ICAT-labeled versus light ICAT-labeled peptides) was determined (Figure 8.29). In this example, cells in population 1 had only one copy of the protein, while cells in population 2 had three copies. Of course, this cartoon oversimplifies the situation because tandem MS cannot detect proteins in such small numbers, but it represents the relative quantification that ICAT can determine.

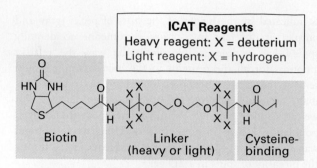

Figure 8.27 ICAT labeling reagent.
The reagent consists of three parts: a biotin affinity tag that binds irreversibly to avidin; a linker that contains eight stable isotopes; and a reactive group that will bind cysteines. The reagent exists in two forms: heavy contains eight deuterium atoms (purple), and light contains eight hydrogen atoms (gray).

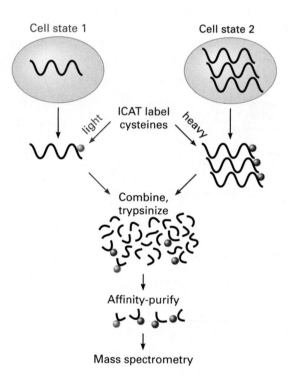

Figure 8.28 Schematic of ICAT method.
Proteins from two cell populations are labeled with either the heavy (purple) or the light (gray) ICAT reagent. The proteins are isolated, mixed, and digested with trypsin. ICAT-labeled peptide fragments are affinity-purified and analyzed by tandem MS.

By eliminating the 2D gel, proteome quantification became high-throughput, and all proteins could be detected.

Nice Idea, But Does ICAT Work?

Aebersold's group had invented a completely new approach to a vexing problem: relative quantification of any two proteomes. To test their new method, they created two protein

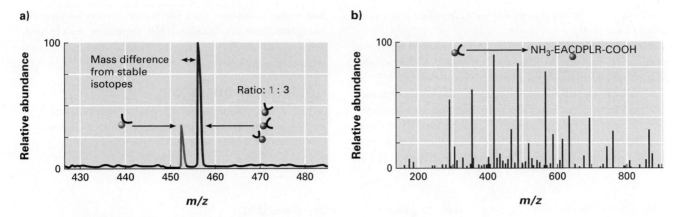

Figure 8.29 Quantification and identification of proteins by ICAT.
a) The heavy (purple) and light (gray) peptides are separated and quantified to produce a ratio for each protein from the two different cell populations. **b)** Each peptide bound to an ICAT reagent is subjected to MS analysis. Protein identification is performed by determining the amino acid sequence of the peptide.

Table 8.2 Sequence identification and quantification of six proteins in two mixtures.
Two or three peptide fragments were identified for each protein and gene names are according to Swiss-Prot nomenclature. ICAT-labeled cysteine residues are denoted by # signs. Ratios were calculated for each peptide as shown in Figure 8.29a. Expected ratios were calculated from the known amounts of proteins present in each mixture.

Gene Name	Peptide Sequence Identified	Observed Ratio (light/heavy)	Mean ± s.d.	Expected Ratio (light/heavy)	% Error
LCA_BOVIN	ALC#SEK	0.94	0.96 ± 0.06	1.00	4.2
	C#EVFR	1.03			
	FLDDLTDDIMC#VK	0.92			
OVAL_CHICK	ADHPFLFC#IK	1.88	1.92 ± 0.06	2.00	4.0
	YPILPEYLQC#VK	1.96			
BGAL_ECOLI	LTAAC#FDR	1.00	0.98 + 0.07	1.00	2.0
	IGLNC#QLAQVAER	0.91			
	#FDGVNSAFHLWC#NGR	1.04			
LACB_BOVIN	WENGEC#AQK	3.64	3.55 ± 0.13	4.00	11.3
	LSFNPTQLEEQC#HI	3.45			
G3P_RABIT	VPTPNVSVVDLTC#R	0.54	0.56 ± 0.02	0.50	12.0
	IVSNASC#TTNC#LAPLAK	0.57			
PHS2_RABIT	IC#GGWQMEEADDWLR	0.32	0.32 ± 0.03	0.33	3.1
	TC#AYTNHTVLPEALER	0.35			
	WLVLC#NPGLAEIIAER	0.30			

Source: Gygi, et al. 1999. *Nature Biotechnology.* 17: 996, Table 1.

mixtures composed of the same six proteins, but the amount of each protein was different in the two mixtures. These two mixtures of six proteins were labeled with either heavy or light ICAT reagents, pooled, and subjected to MS/MS analysis (Table 8.2). As you can see, the method successfully identified the proteins and quantified the relative abundance of all six proteins in the heavy and light protein mixtures. Because the investigators reported the

standard deviation, we can evaluate the scatter of the data, giving us more confidence in their methodology.

ICAT was working under controlled conditions, but how would it perform on something as complex as two complete proteomes? To find out, Aebersold's team compared the proteomes of yeast cells grown in the presence of either ethanol or galactose. In a one-hour time period, over 1,200 MS/MS scans were automatically recorded, and more than 800 different peptides were identified and quantified (Table 8.3).

How well did ICAT work? It performed well for those proteins it was able to identify. For example, the protein Pck1 had an ethanol/galactose expression ratio of 1.57 ± 0.15, which agreed with previously published reports and 2D gel electrophoresis analysis. In addition to identification and relative quantifications, ICAT performed analyses that other methods cannot. When the isozymes of alcohol dehydrogenase (*Adh1* and *Adh2*) were considered, ICAT

had better resolution than DNA microarrays because the two genes have very similar DNA sequences. Adh1p converts acetaldehyde to ethanol, while Adh2p converts ethanol to acetaldehyde (Figure 8.30).

DISCOVERY QUESTIONS

55. When ethanol is the only source of food, yeast cells rapidly convert it to acetaldehyde. Can you predict which form of ADH would be more

Table 8.3 Protein ratios from yeast grown on 2% galactose (Gal) or 2% ethanol (Eth). Gene names are according to the Yeast Proteome Database (YPD). Peptide sequences were identified by SEQUEST program. Protein expression ratios were calculated from the areas under the traces for each peptide as shown in Figure 8.29a. Data for only nine proteins are shown in this table.

Gene Name	Peptide Sequence Identified	Observed Ratio (Eth : Gal)
Ach1	KHNC#LHEPHMLK	>100 : 1
Adh1	YSGVC#HTDLHAWHGDWPLPVK	0.57 : 1
	C#C#SDVFNQVVK	0.48 : 1
Adh2	YSGVC#HTDLHAWHGDWPLPTK	>200 : 1
	C#SSDVFNHVVK	>200 : 1
Ald4	TFEVINPSTEEEIC#HIYEGR	>100 : 1
Bmh1	SEHQVELIC#SYR	0.95 : 1
Cdc19	YRPNC#PIILVTR	0.49 : 1
	NC#TPKPTSTTETVAASAVAAVFEQK	0.65 : 1
	AC#DDK	0.67 : 1
Fba1	SIAPAYGIPWLHSDHC#AK	0.60 : 1
	EQVGC#K	0.63 : 1
Gal1	LTGAGWGGC#TVHLVPGGPNGNIEK	1 : >200
Gal10	DC#VTLK	1 : >200
	HHIPFYEVDLC#DR	1 : >200

Source: Gygi, et al. 1999. *Nature Biotechnology.* 17:997, Table 2 (partial).

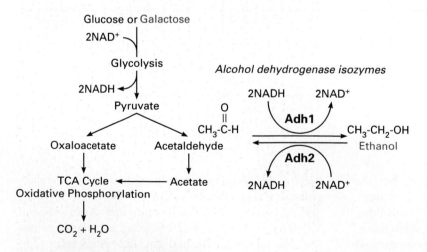

Figure 8.30 Reversible metabolic pathway for ethanol and galactose utilization.
The major proteins and products are shown. The two alcohol dehydrogenases are indicated as Adh1 and Adh2.

abundant when cells were grown on either galactose or ethanol? Support your answers using Figure 8.30.

56. Since the ICAT method relies on pairs of proteins (heavy and light), it does not measure any proteins that lack a partner peak. What is the problem with this limitation? If you were to use MS/MS to identify nonpaired proteins, what information would be lacking that ICAT is designed to produce?

ICAT was able to measure how yeast proteomes respond when the food source varies. When grown on galactose, Adh1 was present at twice the level as when grown on ethanol (see Table 8.3 for summary and Figure 8.31 for details). Conversely, Adh2 was present at more than 200 times the level of Adh1 when ethanol was the only source of food. This finding was particularly impressive from a technical standpoint; the two forms of Adh are 93% identical at the amino acid level and yet ICAT could distinguish between them. DNA microarrays often fail to distinguish the two *Adh* mRNAs, which are 88% identical in the coding region.

ICAT works very well in complex proteomes. It identified many of the proteins and determined relative amounts expressed in two different populations of cells. ICAT represents a more mature approach to proteomics that can examine dynamic cellular responses at the protein level as cells respond to changing environments. Using the ICAT method, it should be possible to examine variations among individuals of a species to understand the functional consequences of SNPs (see Chapter 4). We should be able to compare DNA microarray expression profiles with ICAT

proteomics data and see how well they correlate (Chapter 12). The utility of ICAT is impressive and will increase as it is refined by Aebersold's lab and others around the world.

The challenges of the proteome might appear solved except for a few details, but scientists always discover new questions to ask after data are collected. At this point, take a few minutes to dream where you would like to see the field of proteomics go. Think wildly and don't worry about how to do it experimentally.

Here's what we can do today:

- Use sequence information to predict functions and interacting proteins.
- Use yeast two-hybrid technology to determine protein-protein interactions.
- Identify many proteins in a proteome, though proteins expressed at very few copies per cell are still elusive.
- Use stable isotope methods to quantify relative amounts of proteins in two populations of cells.

What can't we do that you think would help us better understand what happens inside cells? Take a few minutes to write down your ideas.

Is the Last Unexplored Ecosystem on Earth Inside the Cell?

You may have written down some areas for exploration that are not included in this section. If so, keep thinking about your ideas; you might devise ways to substantially improve our understanding of cells. Although your list may have contained more than three new goals to pursue, we will briefly examine three that some labs have begun to

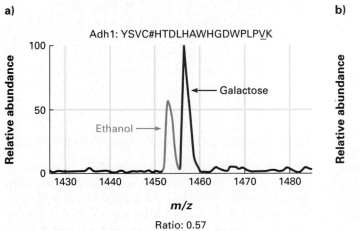

a)

Adh1: YSVC#HTDLHAWHGDWPLP<u>V</u>K

Ratio: 0.57

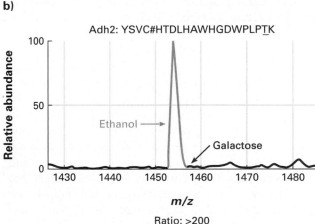

b)

Adh2: YSVC#HTDLHAWHGDWPLP<u>T</u>K

Ratio: >200

Figure 8.31 Reversible switch in metabolism.
ICAT detection of **a)** Adh1 and **b)** Adh2 from yeast cells grown in the presence of ethanol (gray) or galactose (purple). Relative amounts of these two proteins are indicated with ratios at the bottom.

LINKS
Antibody Resource
Gavin MacBeath
Pat Brown
protein chip paper
Stuart Schreiber

address: (1) protein microarrays, (2) single-cell analysis, and (3) metabolomics.

Can We Make Protein Microarrays?

DNA microarrays have been so popular and successful that many people are trying to create protein versions. So far, protein "chips" have been designed for three basic purposes: to identify and quantify proteins in a complex sample, to determine protein interactions, and to determine protein functions.

Identify and Quantify Pat Brown, who popularized DNA microarrays, has tried to do the same with protein chips. Brian Haab, Maitreya Dunham, and Pat Brown published their first protein chip paper in January 2001 in *Genome Biology*. (*Genome Biology* is an open-access journal, so you can freely download the PDF file for this paper.) Brown's team used monoclonal antibodies to detect specific proteins. For some protein microarrays, they spotted the antibodies and detected proteins in a mixed sample. On other chips, they spotted the proteins and detected antibodies in a mixed solution. Figure 8.32 shows the results of an **antibody microarray** that detected proteins labeled with the fluorescent dyes **Cy3** (green) and **Cy5** (red).

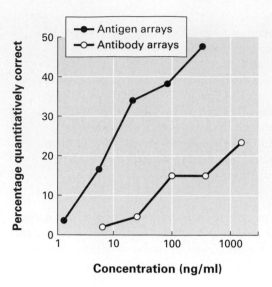

Figure 8.33 Comparison of two protein microarrays. Percentages of antibodies and antigens yielding quantitatively correct results when known protein mixtures were incubated on the microarrays.

Figure 8.32 Antibody arrays detected fluorescently labeled antigens.
Go to www.GeneticsPlace.com to view this figure.

For their proof-of-concept experiment, they worked with 115 antibody/**antigen** pairs (antigens are the proteins that antibodies recognize), creating 6 different cocktails of the 115 antigens, each with different concentrations of the various antigens. Each of these six cocktails was labeled red, and a single reference cocktail with a fixed concentration of the same 115 antigens was labeled green. Equal volumes of the experimental and reference proteins were mixed and then added to the antibody microarray and the ratio of the two colors was determined for each spot, or feature.

In an ideal protein microarray world, we could create arrays with monoclonal antibodies for each protein in the proteome and determine which proteins were present and in what amounts. Two protein samples would be applied to the microarray at the same time so we could compare two proteomes (e.g., healthy vs. diseased). Unfortunately, protein chips are not yet able to perform a two-proteome comparison. Furthermore, Brown's protein chip worked better when the antigens were spotted on the glass and used to identify which antibodies were present in a mixed solution (Figure 8.33). Antigen microarrays have limited utility

except in the study of immune responses to known antigens. Thus, this first attempt at using antibodies was interesting, but the search continues for better methods.

DISCOVERY QUESTIONS

57. What reagent is the limiting factor for the success of Brown's protein microarray method? Does this limitation make the method impractical?

58. Go to the Antibody Resource Page and choose a couple of links. How many monoclonal antibodies are commercially available? Do these sites change your evaluation of the practicality of Brown's antibody microarray method?

59. Of the two protein microarrays produced by Brown's lab, which would be more useful for defining a proteome? What is the limitation for this type of array?

Can Microarrays Detect Proteome Interactions?

Gavin MacBeath and Stuart Schreiber from Harvard University created useful protein microarrays for defining protein-protein interactions. They covalently attached proteins known to be one-half of interacting pairs of proteins to glass microscope slides. The spotted proteins created a microarray of in vitro "bait" analogous to the bait proteins used in yeast two-hybrids. In their first experiment, they demonstrated the specificity of the method, as three pairs of proteins were shown to bind to the appropriate partners but not the other bait proteins (Figure 8.34a). In a rather dramatic way, the investigators also showed that bait and prey proteins could discriminate even when only 1 feature out of 10,800 contained the appropriate bait protein (Figure 8.34b).

Figure 8.34 Detection of protein-protein interactions on protein microarrays.
Go to www.GeneticsPlace.com to view this figure.

In addition to determining protein-protein interactions, MacBeath and Schreiber wanted to measure the protein target for small molecules (i.e., ligands and metabolites). They incubated dye-labeled ligands with the appropriate receptor proteins that had been spotted onto a protein microarray (Figure 8.35). This experiment was a significant step forward into the next generation of "-omics"—metabolomics (see pages 325–327).

Figure 8.35 Detecting the targets of small molecules on protein arrays.
Go to www.GeneticsPlace.com to view this figure.

DISCOVERY QUESTIONS

60. In Figure 8.34b, only two bait proteins were spotted on the microarray. Think of a better experimental design for testing specificity than the one shown here. What negative control would you like to have seen in Figure 8.35?

61. In these protein microarray experiments, three different dyes were used. What properties would you expect these dyes to have so you could be sure the bottom panels of Figures 8.34 and 8.35 accurately represent the specificity of binding?

LINKS
microwell arrays
METHODS
kinase

Can Protein Microarrays Measure Kinase Activity?

Mike Snyder's lab (the team that created the mTn-tagged functional mutants described on pages 286–289) developed a high-throughput method to survey the functions of kinases. Their first innovation was the production of the protein array itself. Instead of spotting onto glass, Snyder's group developed an inexpensive way to produce silicon polymer microwell arrays, analogous to 96 well plates, only smaller (Figure 8.36). The microwells have rounded floors, are 300 mm deep, and hold approximately 300 nL of liquid. The microwell arrays are placed on microscope slides to facilitate handling, with two microwell arrays fitting on a single slide. For this work, the investigators pipetted all reagents by hand, but the technology is amenable to high-throughput, automated liquid-handling technology used in many labs.

The investigators produced 17 different microwell arrays, one for each kinase substrate (Figure 8.37). In the yeast genome, there are 122 kinase genes; Snyder's lab produced large amounts of 119 kinase enzymes and incubated each one in a microwell for each of the 17 microwell arrays. By using microwells, instead of traditional microarrays where the substrates would have been spotted onto a single glass microscope slide, the reaction conditions in each well could be manipulated independently to optimize the reactions.

For these kinase assays, radioactive ATP was used to label polypeptide substrates phosphorylated by a given kinase

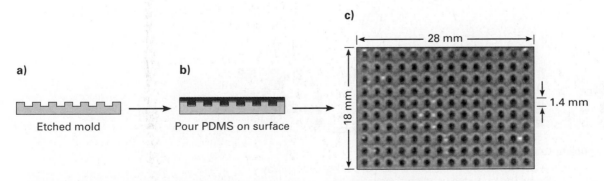

Figure 8.36 Microwell array manufacturing.
a) Molds were etched, and then **b)** monomeric elastomer was used to form a silicon-based polymer (purple). **c)** When removed from the mold, an array of microwells was produced and ready for the next step.

DATA
kinase tree
METHODS
unrooted phylogenetic
tree

(Figure 8.37c–f). In addition to testing each kinase with each substrate, the investigators also tested each kinase with itself to see if any were able to **autophosphorylate**; that is, phosphorylate themselves. For each microwell array, position I9 was a negative control. You can see that the kinase Mps1p (well B4) was able to phosphorylate all four substrates, including itself (Figure 8.37c).

DISCOVERY QUESTIONS

62. Microwell arrays worked very well with kinases; can you think of other proteome applications?

63. Could microwell arrays be used with a nonradioactive detection system?

64. Think about the correlation between the primary amino acid sequence, the 3D shape of each kinase, and the substrates that each kinase can phosphorylate. Snyder's group aligned the amino acid sequences for 107 kinases and created an unrooted phylogenetic tree. Look at the kinase tree PDF file that illustrates the relationship of the amino acid sequences for 107 kinases. Green indicates kinases that phosphorylate tyrosines in a

poly(tyrosine-glutamate) substrate; "blue kinases" phosphorylated the same substrate weakly; "yellow kinases" could not phosphorylate this substrate.

a. What correlation do you see between primary structure and substrate preference?

b. If you could see the 3D structures for all of these kinases, hypothesize what you would expect to see for all the "green kinases."

c. Hypothesize what you would see when you compared the 3D structure of a "green kinase" and a "blue" one, a "green kinase" and a "yellow" one.

It is striking to see so many green-coded kinases scattered all over in the phylogenetic tree. This means their amino acid sequences vary substantially, though subsets of them cluster on particular branches, indicating that kinases within a subset have similar amino acid sequences. Even though the green-coded kinase sequences vary, they all act on the same substrate, which indicates their active sites are conserved. This leads to the hypothesis that the differences in their sequences lie outside the active site. For those

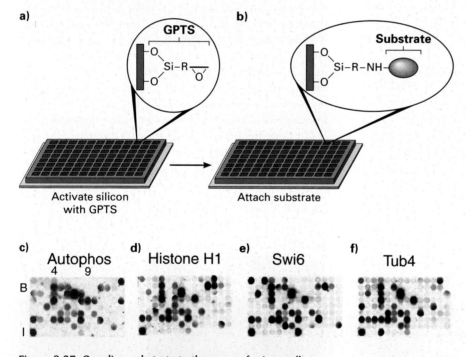

Figure 8.37 Coupling substrate to the array of microwells.

a) The silicon polymer was activated with a cross-linker called 3-glycidoxpropyl-trimethoxysilane (GPTS). **b)** The peptides used as substrates were covalently linked to the silicon. **c)** to **f)** Four of the 119 kinase microwell assays, with each microwell array containing one substrate (labeled at the top); a different kinase and radioactive ATP (^{33}Pγ-ATP) were added to each microwell. The microwell arrays were imaged with a sensitive device called a phosphoimager.

kinases that do not phosphorylate the same substrate yet cluster near the green-coded ones, we might expect to see a large degree of amino acid conservation overall but variations within the active site. This could be confirmed if we had access to all the 3D structures.

With the combination of all the different enzymes and all the different substrates in a proteome, Snyder's method will be very useful. Any enzyme that can produce a detectable product should be amenable to this approach. In addition to substrates, enzyme inhibitors, potential new medications, and even the effects of pesticides and herbicides could be measured by the microwell array method. Snyder's method is easy enough to fabricate in any lab and suitable for scaling up for large projects. There do not seem to be many limitations to this method, so you will probably hear more about it in the future.

Are All Cells Equal?

We have been talking about what happens inside a cell, but in reality all our data have been based on populations of cells. Now we have tools sensitive enough to detect individual molecules. Norm Dovichi (who moved from the University of Alberta to the University of Washington, Seattle) is an analytical chemist who "was trained to work with lasers" and has applied his skills to biological molecules. First, he helped invent the capillary electrophoresis technology that has driven high-throughput DNA sequencing. He chose DNA methods because, "compared to proteins, it

was easy." Now that the "easy" stuff is done, Dovichi says his lab group is working on the "really cool stuff."

LINKS G🧬P
Norm Dovichi
METHODS
capillary electrophoresis

The initial goal was to improve laser-based methods for detecting small amounts of proteins, which Dovichi's group accomplished, and in the process applied two new terms to proteins: **zeptomoles** (1 zmol = 10^{-21} moles = 602 molecules) and **yoctomoles** (1 ymol = 10^{-24} moles = 0.6 molecules). These terms had never been needed since no one could measure single molecules. Once Dovichi could detect a single molecule, he wanted to measure something of interest, so he turned to an enzyme called **alkaline phosphatase**. The first question was simple: If you isolate a single molecule of alkaline phosphatase and incubate it for varying amounts of time, will the reaction rate of that molecule behave the same way as a pool of molecules? A single molecule was incubated for one, two, four, and eight minutes, and the amount of product was measured (Figure 8.38). The individual enzyme created product at a linear rate, and the best-fit line has a correlation coefficient of 0.999. This result is not surprising, but is significant nonetheless. Dovichi has proven that his detection system can measure the output of a single enzyme molecule.

In science, we need to know that our results will be reproducible when the same experiment is performed a second time. When Dovichi repeated the time-course experiment, he noticed that each enzyme's rate of product formation was linear, but the individual enzymes did not form product at the same rate. His lab measured the

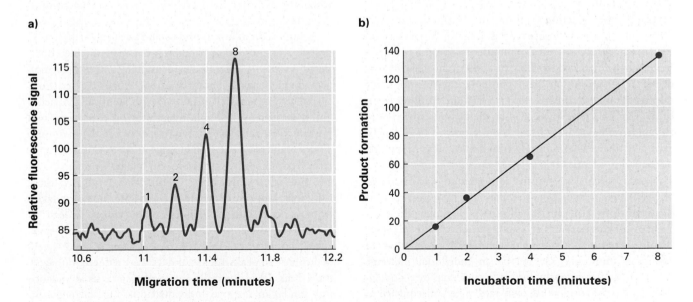

Figure 8.38 Measuring the rate of product formation from a single enzyme molecule.
a) Product formation generated by reacting a single enzyme for 1, 2, 4, and 8 minutes.
b) Kinetic plot of the data from panel a). The amounts of product for each incubation time are shown and the straight line is the least-squares fit to the data. Correlation coefficient is 0.999.

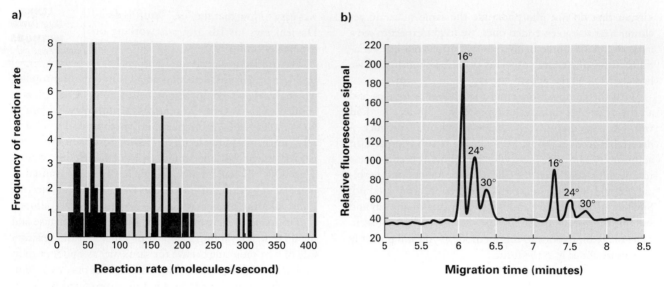

Figure 8.39 Variation between enzymes.

a) Histogram of reaction rates observed for 83 different alkaline phosphatase molecules assayed individually. Activity was determined from the incubation time and product formation. **b)** Electropherogram of alkaline phosphatase assay using two different molecules. Each molecule was incubated sequentially at 16°C, 24°C, and 30°C.

reaction rates (the slope of the line in Figure 8.38b) for 83 different molecules (Figure 8.39a). As you can see, not all molecules behave the same way under identical reaction conditions. The reaction rates for the 83 individual molecules of alkaline phosphatase varied over a tenfold range. Similarly, when reaction rates for individual enzymes were measured at different temperatures, the individual enzymes did not produce identical amounts of product (Figure 8.39b). Although he is not sure, Dovichi believes the degree of glycosylation on a particular enzyme molecule may be the determining factor for variation. Another possibility may be subtle variations in the tertiary structure of the folded enzyme. Whatever the reason, we have experimental evidence that multiple copies of a single protein do not function identically.

Next, Dovichi's lab wanted to determine what happens to individual molecules when they are subjected to **heat denaturation**. You can imagine two possible outcomes. Perhaps all the enzymes in a population would gradually become denatured, gradually decreasing the productivity of any given molecule. Or maybe an individual molecule would denature instantly, but each molecule in a population would denature at its own rate. The consequence of the second outcome would be that halfway through a denaturation experiment, individual enzymes would have either their normal level of activity or zero activity. Dovichi's team determined that alkaline phosphatase molecules denatured very rapidly, producing two kinds of

activities in a population: zero or normal. It is amazing that proteins behave digitally (toggling either on or off) rather than in a gradual (i.e., analog) fashion. Proteins acting as toggle switches are addressed further in Chapter 11, when we examine how proteins can make biochemical decisions.

As important as it is to understand what single molecules do, we ultimately want to know what the entire cell does. Based on Dovichi's results from studying single molecules, you might imagine that two mitotic clones derived from the same parental cell might not be identical. Dovichi's group decided to test this hypothesis by comparing two human colon adenocarcinoma cells grown in tissue culture. They conducted a one-dimensional analysis, utilizing capillary electrophoresis to separate and quantify proteins from two individual cells (Figure 8.40). They compared these two single-cell proteomes to a composite proteome taken from a pool of mitotic adenocarcinoma clones. At least 30 peaks were visible, and all three proteomes were very similar but not identical. In cell a, peaks 14 and 17 were barely detectible and peaks 12, 19, and 22 were reduced in magnitude. Peak 10 is especially interesting since its abundance was similar in the two individual cells but its migration time was altered. There are several plausible explanations, but Dovichi's group is especially intrigued by the possibility that altered migration time may represent differences in posttranslational modifications, such as the addition of a single phosphate group.

DISCOVERY QUESTIONS

65. How could you test the hypothesis that the peak 10 differences between cell a and cell b were due to phosphorylation?

66. With the ability to detect single molecules, what do you think the next step for Dovichi's group will be, as they develop tools to describe the proteome of single cells?

Genomic and proteomic approaches provide us with new tools, new types of data, and more importantly, new types of questions that reflect new ways of thinking about the cell. Dovichi's group has detected a difference in peak 10 that may represent differences in phosphorylation events as two genetically identical cells responded differently to their shared environment. We could use antibodies that are specific for phosphoproteins or Chait's stable isotope method to determine whether protein 10 was differentially phosphorylated. With time, Dovichi and others will address these technical challenges and further expand our understanding of cells.

For now, Dovichi's lab is busy developing the next generation of gel-free 2D analysis for proteins. His group has developed a capillary electrophoresis procedure to measure the molecular weight and then use the pI of proteins to produce 2D maps of proteomes from single cells. "What would be really cool," Dovichi mused, "would be to couple this single-cell 2D analysis with a method to identify each spot," such as MS/MS.

What Does a Proteome Produce?

LINKS G⊗P
Oliver Fiehn
METHODS
Arabidopsis

Imagine a day when we can define proteomes easily, determine which proteins interact, and measure their functions in vitro. What could be left? How about lipids? Sugars? Steroids? Ion concentrations? Metabolic intermediates? The list of cellular components is substantial, and so far we have discussed only proteins and mRNA. There is more to a cell than just transcription and translation. Although we cannot define complete proteomes yet, researchers around the world are already looking past this temporary limitation to defining **metabolomes**, or the entire metabolic state of cells.

Oliver Fiehn and his colleagues at the Max Planck Institute in Potsdam, Germany, are among the world leaders in profiling the metabolomes of model plants such as potatoes, tobacco, and *Arabidopsis*. *Arabidopsis* is a small flowering plant that goes from seed to flower in about two weeks, can be grown in culture, has been very important in understanding the biology of all flowering plants, and had its genome completely sequenced in 2000.

Fiehn's team wanted to compare the metabolomes of four different strains of *Arabidopsis*. Two strains were considered *wt*, though they differ from each other by carrying an undetermined number of allelic variations. The other two strains were mutants derived from these parental strains, each with a single gene deleted. *Col-2* is a *wt* strain from which the mutant strain *Dgd1* was derived; *Dgd1* lacked an enzyme used in the production of a particular lipid. Phenotypically, *Dgd1* did not photosynthesize well

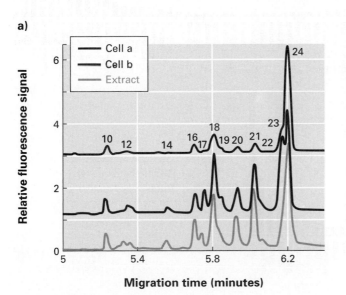
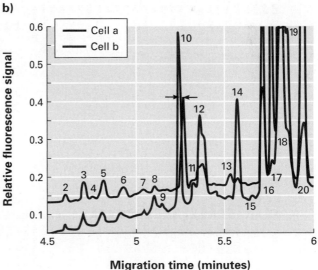

Figure 8.40 Capillary electrophoresis of single-cell proteomes.
a) Two individual human colon adenocarcinoma cells (purple and black; cell line HT29) were subjected to one-dimensional separation. The "extract" was produced from approximately 10⁶ cells (gray). Only the major peaks are labeled. **b)** Higher resolution of the central portion of panel a). The arrows mark differences in migration times between the two number 10 peaks.

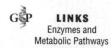
LINKS
Enzymes and
Metabolic Pathways

and was hypersensitive to light stress. *C24* is the other *wt* strain from which *Sdd1-1* was derived; *Sdd1-1* lacked a regulatory protein involved in controlling **stomata** development (the pores on the underside of leaves used for gas exchange). The only apparent phenotype in *Sdd1–1* was a two- to fourfold increase in the number of stomata per leaf.

Using **gas chromatography** to separate the cytoplasmic nonprotein compounds and then MS to identify them, Fiehn's group was able to determine the identity of 326 distinct compounds from the leaves of all four strains of *Arabidopsis*. In the *Dgd1* mutant, 153 of the 326 compounds showed significant concentration differences compared to the parental strain *Col-2* (Figure 8.41a). In contrast, in the *Sdd1–1* mutant strain, only 41 of the 326 compounds differed significantly from the *C24* parental strain (Figure 8.41b). It was surprising in both cases that although only one protein was lacking in each mutant proteome, the metabolome responded in complex and unexpected ways. Some metabolites were increased, while others were

reduced. In short, the proteome is complex, but the metabolome is exponentially more complex, perplexing, and challenging to study. To fully understand cells, we must measure all cellular components under many different conditions. Metabolomics will take a lot more effort and collaborations to produce the right technology and experiments.

DISCOVERY QUESTIONS

These two questions are designed to help you discover the wealth of metabolomic information currently available, and the complexity of defining metabolomes.

67. Go to the Enzymes and Metabolic Pathways database search page for an amazing project. In the "search terms" window, enter the word "orotidine" (this was one of the substrates for the ODCase we examined on page 296).

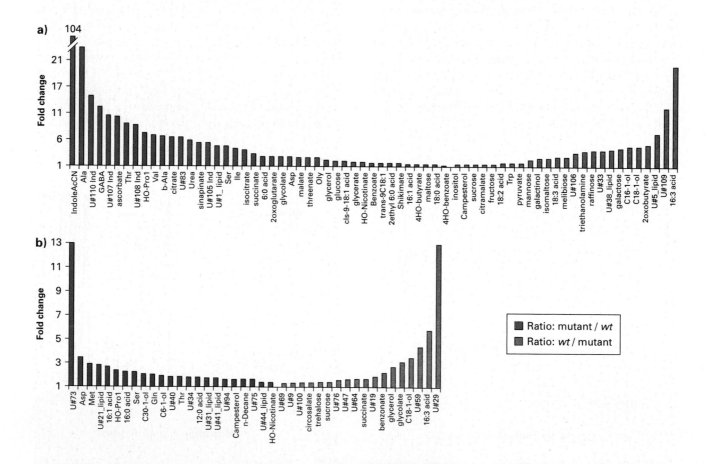

Figure 8.41 Significant changes in metabolomes when comparing *wt* and mutants.
a) Alteration in average metabolite levels (*t*-test *p*-value < 0.01) of *Dgd1* compared to its parental strain *Col-2*. Only 67 of the 153 significant differences detected are shown.
b) When compared to its parental strain (*C-24*), *Sdd1-1* had 41 significant differences in its metabolome; all 41 are shown.

a. Click on the "search" button. You will get several hits.

b. Click on "SVG" on the far right of hit "CO_2, NH_3—UMP, CO_2 anabolism (ATP, FAD^+) (cytosol, plasma membrane.)" You will see a diagram of the metabolic pathway to synthesize UMP. Mouse over any of the enzymes and compounds, and you will notice that they are all hyperlinks.

c. Click on UMP at the bottom of the list. You will get three options. In the left frame, click the button next to "UMP." In the right frame, you will see some text describing your choice of UMP. Click on the "depict" button. Soon you will see the structure of UMP.

d. Return to the pathway and click on "carbamoyl phosphate" and compare its structure with that of UMP.

e. Return to the pathway again. ODCase is the last enzyme (E.C. number 4.1.1.23) in this pathway to produce UMP. Notice that although carbon dioxide is removed during the final step, it is also consumed at the top of the pathway. Would this type of recycling present a problem if you wanted to define a metabolome?

68. You can explore the metabolic pathways of many different species at the Kyoto Encyclopedia of Genes and Genomes (KEGG) web site. Go to KEGG, click on any section of the metabolic overview, and gradually zoom in until you see individual enzymes and metabolites listed. This will give you a sense of how much work is involved in characterizing a single metabolome from one cell at one time point.

Summary 8.4

Proteomics is a very difficult field, and the methods have not yet produced the quality or quantity of data we want. 2D gels allow us to identify proteins in a proteome, but newer methods that utilize tandem mass spectroscopy enable us to identify more proteins. By incorporating stable isotopes, we can quantify as well as identify proteins in proteomes as they change over time. Protein chips are quickly becoming more sophisticated and no doubt will have a substantial effect on proteomics. As our tools become more precise, we can begin to study single proteins and discover new properties that were masked when we studied them in large pools of many molecules. From single-molecule functions to single-cell proteomics, we are beginning to recognize the heterogeneity and complex cellular circuitry of proteins and their metabolic products. When single proteins are deleted

from a proteome, the metabolome can be very different even if the phenotype appears unchanged. The proteome and metabolome may seem too daunting to understand—but in 1980, the human genome seemed too big to sequence. Time and new technologies can change the way we view new areas of research.

LINKS G⚕P
Human Proteome Initiative
KEGG

Chapter 8 Conclusions

The postgenomics era could be called the proteomics era, as the research community has begun to tackle this next huge challenge. The Human Proteome Initiative is gaining momentum and, more importantly, increasing its knowledge base. Proteomics has been more difficult than genomics because we need so much more information. We need to understand the many ways in which proteins work, their 3D structure, the protein interaction network, and which proteins and how much of each are present in any given proteome. In addition, we would like to understand more fully the stochastic nature of individual proteins and what effect this unpredictable behavior has on single-cell proteomes and metabolomes. You might have thought that the complete sequencing of the human genome meant all the challenging research had been completed, but hard work and great discoveries are still ahead. In addition to fully describing many different proteomes that change over time, we would like to integrate proteomics with all the other forms of information covered in the previous five chapters.

The area of proteomics is like the front line of an avalanche, where advances protrude from different places along the edge but each advancing edge is followed very quickly by the rest of a fast-paced progression. The methods of proteomics change quickly and investigators are making great progress. In a few years, our perspective on many topics will change. New perspectives will spawn new questions and insights into the complexity of cell webs. In Chapters 10 through 12, we will examine how proteomes, genomes, and metabolomes interact to produce cellular responses. This will culminate in the greatest challenge to date: synthesizing all the information into comprehensive models.

References
Functional Proteomics

Kumar, A., K.-H. Cheung, et al. 2000. TRIPLES: A database of gene function in *Saccharomyces cerevisiae*. *Nucleic Acids Research.* 28(1): 81–84.

Ren, B., F. Robert, et al. 2000. Genome-wide location and function of DNA binding proteins. *Science.* 290: 2306–2309.

Ross-Macdonald, P., P. S. R. Coelho, et al. 1999. Large-scale analysis of the yeast genome by transposon tagging and gene disruption. *Nature.* 402: 413–418.

Winzeler, E. A., D. D. Shoemaker, et al. 1999. Functional characterization of the *S. cerevisiae* genome by gene deletion and parallel analysis. *Science.* 285: 901–909.

Structural Proteomics

Abbott, A. 2000. Structure by numbers. *Nature.* 408: 130–132.

Christendat, D., A. Yee, et al. 2000. Structural proteomics of an archeaon. *Nature Structural Biology.* 7: 903–908.

Martz, E. 2001. Protein Explorer. <http://www.umass.edu/microbio/chime/explorer/>. Accessed 17 February 2002.

Murata, K., K. Mitsuoka, et al. 2000. Structural determinants of water permeation through aquaporin-1. *Nature.* 407: 599–605.

Nienaber, V. L., P. L. Richardson, et al. 2000. Discovering novel ligands for macromolecules using X-ray crystallographic screening. *Nature Biotechnology.* 18: 1105–1108.

Roepe, P. D. 2001. A peptide needle in a signaling haystack. *Nature Genetics.* 27: 6–8.

Sui, H., B.-G. Han, et al. 2001. Structural basis of water-specific transport through the AQP1 water channel. *Nature.* 414: 872–878.

True, H. L., & S. Lindquist. 2000. A yeast prion provides a mechanism for genetic variation and phenotypic diversity. *Nature.* 407: 477–483.

Wu, N., Y. Moi, et al. 2000. Electrostatic stress in catalysis: Structure and mechanism of the enzyme orotidine monophosphate decarboxylase. *PNAS.* 97: 2017–2022.

Quantitative Proteomics

Andersen, J. S., & M. Mann. 2000. Functional genomics by mass spectroscopy. *FEBS Letters.* 480: 25–31. [review article]

Celis, J., M. Kruhoffer, et al. 2000. Gene expression profiling: Monitoring transcription and translation products using DNA microarrays and proteomics. *FEBS Letters.* 480: 2–19.

Chalmers, M. J., & S. J. Gaskell. 2000. Advances in mass spectroscopy for proteome analysis. *Current Opinion in Biotechnology.* 11: 384–390.

Garin, J., R. Diez, et al. 2001. The phagosome proteome: Insight into phagosome functions. *Journal of Cell Biology.* 152(1): 1665–1680.

Gygi, S. P., B. Rist, & R. Aebersold. 2000. Measuring gene expression by quantitative proteome analysis. *Current Opinion in Biotechnology.* 11: 396–401.

Gygi, S. P., B. Rist, et al. 1999. Quantitative analysis of complex protein mixtures using isotope-coded affinity tags. *Nature Biotechnology.* 17: 994–999.

Mann, M. 1999. Quantitative proteomics? *Nature Biotechnology.* 17: 954–955.

Oda, Y., K. Huang, et al. 1999. Accurate quantitation of protein expression and site-specific phosphorylation. *PNAS.* 96: 6591–6599.

Patterson, S. D. 2000. Mass spectroscopy and proteomics. *Physiological Genomics.* 2: 59–65.

ProteomeMetrics. 2001. PROWL for Intranet. <http://prowl1.rockefeller.edu/>. Accessed 28 February 2005.

Interactions of the Proteome

Bartel, P. L., J. A. Roecklein, et al. 1999. A protein linkage map of *Escherichia coli* bacteriophage T7. *Nature Genetics.* 12(1): 72–77.

Butland, G., J. M. Peregrín-Alvarez, et al. 2005. Interaction network containing conserved and essential protein complexes in *Escherichia coli.* *Nature.* 433: 531– 537.

Enright, A. J., I. Iliopoulos, et al. 1999. Protein interaction maps for complete genomes based on gene fusion events. *Nature.* 402: 86–90.

Hybrigenics. 2001. PIMRider. <http://pim.hybrigenics.com/>. Accessed 24 February 2002.

Marcotte, E. M., M. Pellegrini, et al. 1999. A combined algorithm for genome-wide prediction of protein function. *Nature.* 402: 83–89.

Marcotte, E. M., M. Pellegrini, et al. 1999. Detecting protein function and protein-protein interactions from genome sequences. *Science.* 285: 751–753.

Mayer, M. L., & P. Hieter. 2000. Protein networks—Built by association. *Nature Biotechnology.* 18: 1242–1243.

Oliver, S. 2000. Guilt-by-association goes global. *Nature.* 403: 601–603.

Rain, J.-C., L. Selig, et al. 2001. The protein-protein interaction map of *Helicobacter pylori. Nature.* 409: 211–215.

Schwikowski, B., P. Uetz, & S. Fields. 2000. A network of protein-protein interactions in yeast. *Nature Biotechnology.* 18: 1257–1261.

Uetz, P., L. Grot, et al. 2000. A comprehensive analysis of protein-protein interactions in *Saccharomyces cerevisiae. Nature.* 403: 623–627.

Walhout, A. J., R. Sordella, et al. 2000. Protein interaction mapping in *C. elegans* using proteins involved in vulval development. *Science.* 287: 116–122.

Xenarios, I., E. Fernandez, et al. 2001. DIP: The Database of Interacting Proteins: 2001 update. *Nucleic Acids Research.* 29(1): 239–241.

Protein Arrays

Haab, B. B., M. J. Dunham, & P. O. Brown. 2001. Protein microarrays for highly parallel detection and quantitation of specific proteins and antibodies in complex solutions. *Genome Biology.* 2(2): 0004.1–0004.13.

MacBeath, G., & S. L. Schreiber. 2000. Printing proteins as microarrays for high-throughput function determination. *Science.* 289: 1760–1763.

Zhu, H., J. F. Klemic, et al. 2000. Analysis of yeast protein kinases using protein chips. *Nature Genetics.* 26: 283–289.

Zhu, H., & M. Snyder. 2001. Protein arrays and microarrays. *Current Opinion in Chemical Biology.* 5: 40–45.

Metabolomics

Fiehn, O., J. Kopka, et al. 2000. Metabolic profiling for plant functional genomics. *Nature Biotechnology.* 18: 1157–1161.

Glassbrook, N., C. Beecher, & J. Ryals. 2000. Metabolic profiling on the right path. *Nature Biotechnology.* 18: 1142–1143.

Single-Cell Proteomics

Chen, D., & N. J. Dovichi. 1998. Single-molecule detection in capillary electrophoresis: Molecular shot noise as a fundamental limit to chemical analysis. *Analytical Chemistry.* 68(4): 690–699.

Craig, D., E. A. Arriaga, et al. 1999. Studies on single alkaline phosphatase molecules: Reaction rate and activation energy of a reaction catalyzed by a single molecule and the effect of thermal denaturation—The death of an enzyme. *Journal of the American Chemical Society.* 118(22): 5245–5253.

Zhang, Z., S. Krylov, et al. 2000. One-dimensional protein analysis of an HT29 human colon adenocarcinoma cell. *Analytical Chemistry.* 72(2): 318–322.

Whole Genome Perspective

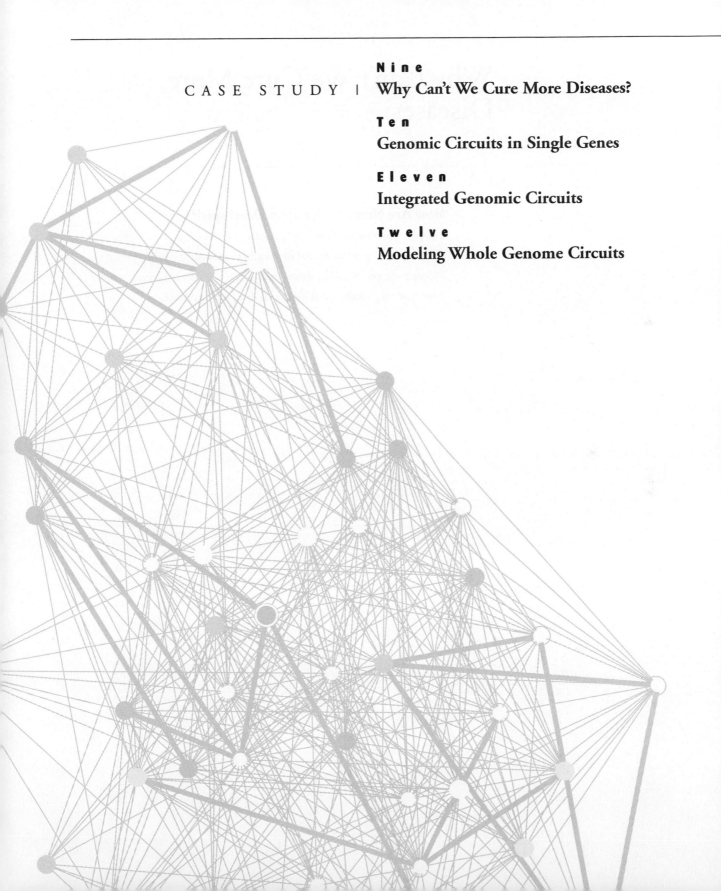

CASE STUDY

Why Can't We Cure More Diseases?

How Are New Medications Developed?

Evaluate Sources for New Medications

Understand requirements for developing a new medication.

Measure the potential for drugs to produce unexpected consequences.

Appreciate the challenge inherent in drug development.

The case studies in Chapters 1 and 5 addressed the complexity associated with muscular dystrophy and obesity. The purpose of these first two case studies was to illustrate a key point in the field of genomics: You need to think of cells the way an ecologist thinks of ecosystems. You have heard of the food web; now picture a **cell web**. Everything inside a cell is connected to something else. If you pluck one string of the cell web (e.g., mutate one gene), the entire cell reverberates and reacts to this perturbation. The pregenomic approach of reductionism is no longer the only one available, and traditional methods are still valuable for testing particular hypotheses.

We have learned a lot by dissecting the genome and understanding what a single gene does in isolation. The one-gene-at-a-time approach, though still very useful and beneficial, has its limitations. An ecologist would never study the food choice of a koala without also considering the animal's environment. It makes no sense for a mammal to eat only one type of leaf until you understand that koalas are surrounded by eucalyptus trees and have very few competitors for these leaves. Why should we be restricted to studying how one protein works in a microfuge tube? We must consider proteins in the context of their environment, which leads us to genomic medicine and potential treatments.

This chapter uses several case studies to examine why curing a disease by medical intervention sounds easier than it really is. We begin with a general overview of how to approach the task of developing a new medication, and then apply these lessons to specific examples. The examples range from making a better aspirin to curing muscular dystrophy by gene therapy. Although these two examples are extremes on a continuum of medical treatments, the difficulty of developing cures is universal. Can we introduce a drug into the cell web without perturbing the entire network?

How Are New Medications Developed?

Obesity and muscular dystrophies are genetically influenced conditions that have multiple causes and effects. Although the problems seem complex, surely we can devise some solutions. In this chapter, imagine yourself as a biomedical engineer who will design the "ultimate cure." You have unlimited resources (money, time, personnel, and a factory with unlimited space that generates nonpolluting electricity with wind and solar generators, etc.), so go crazy with creativity. Throughout this chapter we focus on different aspects of genomic medical therapy. For example, the first focus is on location: concentrate on where you want your medicine to go. Later, we focus our attention on other aspects, so keep tailoring your ideas to the appropriate focus.

Focus 1: Location, Location, Location

When you buy a home or rent an apartment, the most important consideration is location. Every town has some vacancies at any given moment. If inexpensive housing is an hour's drive from your workplace, you may consider this distance an unacceptable compromise in your pursuit of the right blend of cost and benefit. ("The price is right, but losing two hours of my day is too much.") Likewise, in genomic medicine, you have to address the location problem first. Where in the body do you need to send your "cure"? If the problem occurs in the brain, a skin ointment might not be a good idea. You might think that simple injections of the desired protein would work, but remember two things: (1) foreign proteins do not last long in your bloodstream, so injections would have to occur at least once a day, which is not the "ultimate cure" we were hoping for; and (2) proteins cannot cross the plasma membrane to enter cells. As you design a cure, think about the location of the problem in muscular dystrophy (Chapter 1) or obesity (Chapter 5). Where is the problem?

DISCOVERY QUESTIONS

1. Choose either obesity or muscular dystrophy, and define the location of the problem. Stick to the level of cells and organs for now. You probably will need to create an outline, with each Roman numeral being a different molecular cause for the disease.
2. Once you have the cellular location defined, list the proteins known to play a role in the disease.
3. For each protein you listed, describe its subcellular location (e.g., plasma membrane, endoplasmic reticulum, cytoplasm, bloodstream, etc.).
4. Finally, determine if you need to eliminate the defective protein or add a functional copy.

Focus 2: Delivery Vehicles

Once you have chosen a location, you have to select the right vehicle to deliver your cure. For example, if you delivered pizzas for a living and the caller's address was "blue boat moored 100 meters south of lighthouse," you would need a boat, not a car, to make the delivery. In the human body, there are three main delivery vehicles: viruses, protein carriers, and DNA.

Viral Vehicles Viruses contain nucleic acids (DNA or RNA) surrounded by a protein structure. When a virus binds to its target cell, it injects its genome into the new host cell and forces the host to make viral proteins. However, any given virus can infect only a narrow range of species and cell types. For example, the virus that gives you a cold binds to epithelial cells in your respiratory passages,

LINKS
Steven Dowdy
METHODS
liposome

but not to your skin or the epithelium that lines your blood vessels. Likewise, human immunodeficiency virus (HIV) infects only certain white blood cells. Viruses have evolved to be specific, and there is a virus for every cell type. If you could figure out which cells to target, you should be able to find a virus that infects this cell type.

There are a few common viruses currently in use for carrying medications. Adenoviruses, adeno-associated viruses, lentiviruses, herpesviruses, and retroviruses are among the most popular. Each of these has its advantages and disadvantages; it is unlikely nature has produced the perfect virus to be used as a vehicle for genomic cures. We are left with the same types of choices illustrated by the apartment that was inexpensive but located an hour away from your work.

DISCOVERY QUESTIONS

5. What viral structures determine the specificity of the viral infection? Could these structures be bioengineered to change binding specificity?
6. What are the negative consequences of infection with a virus? What does your immune system do to virally infected cells? What would this response do to your virally delivered cure?
7. Is there a way to create a virus that does not stimulate an immune response? What properties would such a virus have to exhibit in order to hide from the immune system?
8. Devise a method using a virus that could infect any cell and yet retain the specificity you wanted. You might want to incorporate surgery in your design.

Two strategies are currently in vogue due to the properties of the viruses. The first uses an **in vitro** (originally in glass petri dishes, but now everyone uses plastic) approach, which means the infection takes place in petri dishes. Cells are removed from the patient, infected in vitro, and then put back into the patient. This approach has the advantage of delivering high doses of a virus but not stimulating the immune system. However, it means the infected cells must be removable from the patient. Surgical removal would not be a good way to affect the hypothalamus or all skeletal muscles, but might work well on a short-term basis for heart, lung, or liver cells.

DISCOVERY QUESTIONS

9. Why would you want to use the patient's cells and not a cell line that has the cure gene already incorporated?
10. For some diseases, you might not have to infect the cells responsible for the disease. For example,

the inability to produce insulin causes diabetes. If you could get the cold virus to "convince" your lung cells to secrete insulin, you might be able to treat diabetes without having to infect pancreatic cells (the site of normal insulin production). Would muscular dystrophy or obesity be treatable via this indirect approach? Explain your answer.

The second approach leaves the diseased cells in the patient and infects them **in vivo** (in the living host). The in vivo approach requires the virus to be very specific, nonharmful, and invisible to the immune system. On the positive side, this approach is appropriate for tissues that are difficult to grow in vitro and successfully replace into the patient. The challenges of in vivo infection are addressed again near the end of this chapter.

Protein Carriers Viruses pose obvious problems as delivery vehicles, so you might wonder why they would ever be used. First, they can be produced very cheaply and on a large scale. Second, they are good at producing a lot of protein for a long time inside infected cells. So, economically, viruses are good solutions. However, there are alternatives. The traditional one is called a liposome, which consists of an artificial membrane (i.e., made in a microfuge tube and not by a living cell) stuffed with any cargo you want. A liposome can be used to deliver proteins or nucleic acids to a cell. However, liposomes are not specific, so they must be used in vitro on selected cells, in vivo on all cells, or by injection to a small localized set of cells (e.g., skeletal muscles).

The newest method is very cool but not well understood. Steven Dowdy found that some proteins can cross phospholipid bilayers unimpeded—another biological rule (proteins cannot cross membranes) with exceptions. These proteins all contain a protein-transduction domain (PTD), which means about 10 to 16 amino acids (a **domain**) are sufficient to move entire proteins across membranes (transduction). What is amazing is that even relatively large iron beads and large proteins can be imported into cells with PTDs (Figure 9.1). PTD-mediated delivery is a nonspecific method and would work on any cell type, which has advantages and disadvantages.

Nucleic Acids Surprisingly, striated muscle cells (cardiac and skeletal) can take up DNA without any special coating or delivery vehicle. So to treat such cells, you could simply inject your therapeutic DNA into muscles and they would take up the DNA. Even more surprising, the DNA can move to the nucleus and become expressed as protein. The physiology of skeletal muscles makes them especially amenable to DNA cures. Skeletal muscles are formed when individual cells fuse during development to form

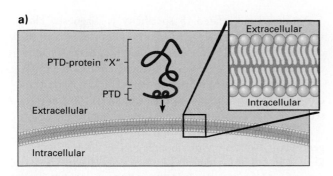

a)

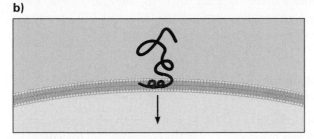

b)

c)

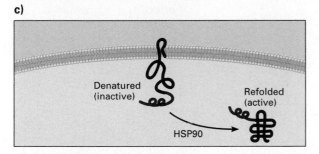

Figure 9.1 Cellular internalization of a large protein fused onto a protein-transduction domain.
a) The positively charged PTD makes contact with the negatively charged outer membrane. **b)** The protein translocates through the membrane in an unfolded state. **c)** Once inside the cell, members of the HSP90 protein family refold the larger protein into an active conformation.

multinucleated giant cells that can be as long as the distance between two bone joints. When therapeutic DNA enters a skeletal muscle, it has to enter only a small number of nuclei to affect a large muscle cell. The mechanism for DNA uptake is unknown and has not been observed in nonmuscle cells. DNA-mediated treatments could be combined with RNAi (see pages 169–172) to inhibit dominant disease alleles, something current gene therapy cannot accomplish.

DISCOVERY QUESTIONS

11. Choose either muscular dystrophy or obesity and design the vehicle you think would be best for delivering a miracle cure to the right cells and subcellular location.

12. Once you have designed your potential Nobel Prize–winning ultimate cure, you need to test it. What series of steps would you want to go through before you try it out on a real person? Outline your protocol leading up to the first human clinical trial.
13. Given that viral vehicles are often species-specific, what effect would this have on your protocol leading up to human clinical trials?

LINKS
genetic causes
Mary Relling
William Evans

Focus 3: Specificity

Now that you know how to deliver your ultimate cure to the right location, you need to think about specificity. Specificity has two meanings: (1) hit only the specific protein you want to target, and do not disturb other proteins that are working fine; (2) given that each person might have different alleles for a particular disease gene, ensure that your cure will work on that patient's specific alleles. The question of genomic variation is the heart of a new field called **pharmacogenomics**, which focuses on the genetic polymorphisms that translate into inherited differences in drug effects (see Chapter 4).

Since the advent of molecular methods, pharmacogenomics has tried to understand the molecular causes behind clinical observations first documented in the 1950s. All individuals with the same disease will not respond equally to the same treatment. Although there are many nongenetic reasons as well, sometimes the inability to be cured is seen within families and has genetic causes, as explained by William Evans and Mary Relling (see pages 196–198). Proteins that process the drugs into active forms, or destroy the drug, at a certain rate can lead to one form of drug-responsiveness polymorphism. The second polymorphism is when the drug binding site differs, so the active drug cannot bind properly. These two classes of polymorphisms can produce at least nine different genotypes in a population (Figure 9.2). Genetic diversity in the population is one reason why treatments such as "the patch" for smoking cessation are not effective for everyone.

As shown in Figure 9.2, different genotypes will respond differently to the same medication. The first column of graphs represents variation in drug metabolism and shows how some will modify the drug at different rates, which affects the amount of active drug in a person's circulation. The second column shows the effects of variation in the drug receptor when two alleles can produce three possible genotypes for this locus. Some alleles bind the medication better than others and thus can produce a stronger effective dose, even at identical drug concentrations.

LINKS
Elisabeth Dequeker
Jean-Jacques Cassiman
Minoxidil Part 1
Part 2

DISCOVERY QUESTIONS

14. Consider this example of genotypes affecting the outcome of a drug treatment. Go to Minoxidil Part 1 and Part 2. Propose a model to explain why only about half of the men (women's study not described here) who use Rogaine (its technical name is minoxidil) have more hair than they had initially.

15. What modification could you make to improve Rogaine's success rate?

Another issue related to specificity is diagnosis. How would you know which particular mutation has led to the disease? As we saw in muscular dystrophy and obesity, these diseases are not monogenic. You would want to determine which gene is mutated in each patient. This can be done by one of the many companies that employ small armies to sequence DNA, so you shouldn't worry about that . . . or should you? Figure 9.3 shows the results from three years of genotyping data for cystic fibrosis. In July 2000, Elisabeth Dequeker and Jean-Jacques Cassiman published the results from a multiyear study. In 1996, 1997, and 1998, these two investigators surveyed 136, 145, and 159 laboratories

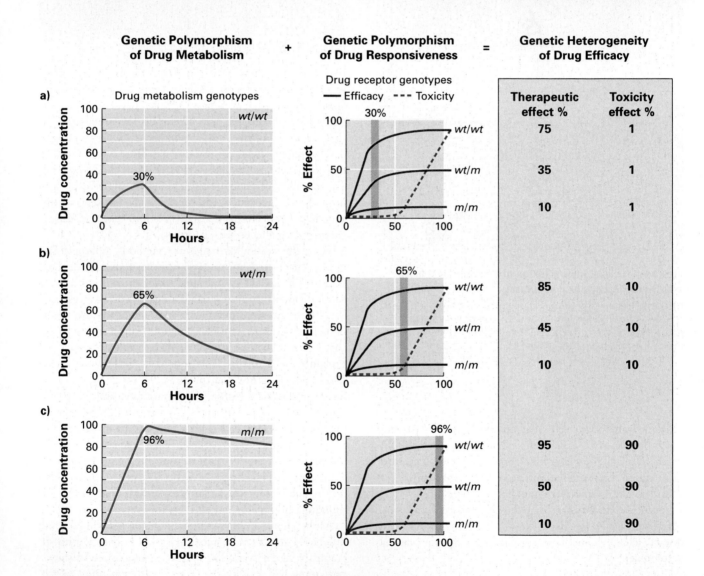

	Genetic Polymorphism of Drug Metabolism	Genetic Polymorphism of Drug Responsiveness	Genetic Heterogeneity of Drug Efficacy	
			Therapeutic effect %	Toxicity effect %
a)	wt/wt (30%)	wt/wt	75	1
		wt/m	35	1
		m/m	10	1
b)	wt/m (65%)	wt/wt	85	10
		wt/m	45	10
		m/m	10	10
c)	m/m (96%)	wt/wt	95	90
		wt/m	50	90
		m/m	10	90

Figure 9.2 Two loci that can affect a person's response to medications.

In the left column, three graphs illustrate the amount of active medication present in a person's circulation, depending on that person's genotype. The second column illustrates the effectiveness of three different concentrations of medication due to three genotypes for the drug receptor. Superimposed in the second column is a dotted line that shows the toxicity of the drug. The final table summarizes the consequences of medication for each of the nine genotypes.

Math Minute 9.1   **What's the Right Dose?**

One of the goals of pharmacogenomics is to find appropriate drugs and dosages for individuals of different genotypes. The hypothetical drug described in Figure 9.2 has minimal therapeutic effect on patients with *m/m* receptor genotypes, regardless of dose. However, patients with *wt/wt* or *wt/m* receptor genotypes can benefit from the drug, provided the dose is nontoxic. For concentrations greater than 50%, toxicity increases rapidly, while therapeutic effect increases very slowly. Therefore, drug concentration should probably be kept near 50% for all patients. Depending on factors such as disease severity, drug cost, and patient risk tolerance, lower concentrations might be desirable.

Using the data in Figure 9.2 and a mathematical model of drug metabolism, you can determine an appropriate dose for each drug metabolism genotype to achieve 50% peak concentration. A rough model of drug concentration in the bloodstream is based on what happens to the administered drug dose in an interval of time, say one hour. (A finer-scale model of drug concentration can be built using differential equations, in which the time interval is infinitesimal.)

For example, suppose 100 mg of the drug is administered orally to a patient with *wt/wt* metabolism genotype. The drug is absorbed into the bloodstream at a constant rate of 16 mg per hour, causing an increase in overall concentration over the first several hours. In what follows, we simplistically assume that the 16 mg enters the bloodstream all at once, at the beginning of the hour. We further assume that metabolizing proteins inactivate the drug at a rate proportional to the amount of active drug in the bloodstream at the beginning of the hour. Using an inactivation proportionality constant of 0.32, the following equations describe the amount of drug in the stomach at the end of hour n (S_n) and the amount of active drug in the bloodstream at the end of hour n (B_n):

$$S_{n+1} = \max(S_n - 16, 0)$$
$$B_{n+1} = (B_n + S_n - S_{n+1}) - 0.32 \times (B_n + S_n - S_{n+1}),$$

where the initial value of B is 0, and the initial value of S is 100. In the second equation, the quantity in parentheses is the amount of drug in the bloodstream at the beginning of hour $n + 1$, since B_n is the amount still active at the end of hour n and $S_n - S_{n+1}$ is the amount absorbed into the bloodstream at the beginning of hour $n + 1$. Thus, the second equation says that the amount of active drug in the bloodstream at the end of an hour is the amount at the beginning of the hour, minus 32% of that amount. The equation can be simplified algebraically by factoring out the quantity in parentheses. By entering these equations into a spreadsheet program or graphing calculator, you can verify that they generate the drug concentration curve for the *wt/wt* drug metabolism genotype (see Figure 9.2).

MATH MINUTE DISCOVERY QUESTIONS

1. Open the file drugmodel.xls in a spreadsheet program. The equations for the G⚕P *wt/wt* drug metabolism model are implemented in this file. Using the same 16 mg/hr absorption rate, find an inactivation proportionality constant that generates the drug concentration curve for the *wt/m* metabolism genotype (see Figure 9.2).

2. Using the inactivation proportionality constant you found in Math Minute G⚕P Discovery Question 1, adjust the initial dose down from 100 until the peak concentration is roughly 50%. This dosage would be more appropriate for patients with the *wt/m* metabolism genotype.

that perform genotyping services in Europe. They sent six known cystic fibrosis DNA clones carrying common mutations and asked that the samples be processed using routine procedures. Only 48% of the 114 labs that participated in all three years made zero mistakes. In addition, some of the kits used for detection have difficulty detecting particular mutations, such as the amino acid substitutions **551G → D** and 553R → X (where X represents any amino acid). Although the quality of genotyping is improving with time, it does raise the possibility that you might try to cure the wrong mutation!

How Many Drugs Does It Take to Cure a Disease?

We now have an interesting problem for drug development. If there are at least nine genotypes for each disease treatment (see Figure 9.2), does each genotype require a different medication? What effect will these different genotypes have on drug development and governmental approval? Each drug approved by the U.S. Food and Drug Administration has to be effective against the disease and safe to take. Does that mean that each drug must be tested in each of the nine genotypes? If so, that would require each of the nine drugs to be subjected to nine genotype-specific trials instead of the one genotype-blind trial currently used.

In our search for a simple cure, we have gone from developing one drug for one disease to be tested in one clinical trial to developing nine drugs, each of which has to be tested in nine clinical trials—an increase of 81-fold! Eighty-one is an underestimate since we have assumed only two

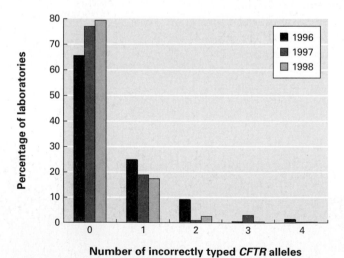

Figure 9.3 The percentage of laboratories that produced incorrect genotype determinations.
Interestingly, 10% of the mistakes were due to reporting mistakes, despite the correct sequence data. 1996 is purple; 1997 is dark gray; 1998 is light gray.

alleles at each of only two loci. What if more than one locus were required to process the drug? What if there were more than two alleles in the population? The potential expense for genotype-specific clinical testing may make pharmaceutical companies wish pharmacogenomics had stayed in the 1950s mentality: sometimes a drug works, sometimes it doesn't.

DISCOVERY QUESTION

16. Here is Discovery Question 12 again, though you might have a different answer this time. Once you have designed your potential Nobel Prize–winning ultimate cure, you will need to test it out. What series of steps would you want to go through before you tried it out on a real person? Outline your protocol leading up to the first human clinical trial.

What Type of Drug Works Best?

Now that you have thought about the location and specificity of your drug, it is time to decide what type of compound to make (not a small detail). Two types of drugs are available to you: chemical compounds and DNA molecules. Chemical medication can be represented by **aspirin**, a compound produced by pharmaceutical companies for over 100 years and used by people since **400 B.C.** This type of drug is not a protein but a small molecule with physical properties that are optimized for effect. For example, when you swallow a pill, the active ingredient should be resistant to destruction in the stomach, be readily taken up by cells, and have an appropriate half-life (not too long or too short).

Aspirin is a great example of a small and simple wonder drug. It is a very small molecule, which means it is easy and cheap to synthesize in vitro. Also, its simple structure means that it is stable at room temperature (good for shipping and storage) and it has a good half-life inside humans. What could be better? Well, that's not an easy question to answer, but a lot of chemists are busy trying to build a better aspirin.

Aspirin works because it binds to an enzyme called **cyclooxygenase** (**Cox**) and then adds an acetyl group $(CH_3-\overset{\overset{\text{O}}{\|}}{C}-)$ to one serine amino acid out of about 600 amino acids in the entire **cox enzyme**. When Cox is **acetylated** by aspirin, it is irreversibly inactivated because the modified serine is in the active site, which the normal substrate, arachidonic acid, can no longer reach. When functioning normally, Cox converts arachidonic acid into prostaglandins, which are lipid-signaling molecules that trigger the sensation of pain, inflammation, and fever. So if you want to stop the pain, inflammation, or fever, just take aspirin. But there is a complex twist to this simple story.

Humans have two genes that encode slightly different cyclooxygenases, called Cox-1 and Cox-2. Cox-1 is expressed in every cell in your body, with the highest levels in the stomach and kidneys. The prostaglandins produced here are needed to keep the cells working properly. Cox-2 is expressed primarily in the brain, lung, kidneys, and white blood cells when stimulated by cellular damage or infection. We want aspirin to block the activity of Cox-2 but leave Cox-1 unaffected. Unfortunately, aspirin has about 100 times higher affinity for Cox-1 than for Cox-2. This explains why some people have stomach problems if they take too much aspirin: their stomach-enriched Cox-1 is preferentially inhibited and the stomach cells lose the beneficial prostaglandins produced by Cox-1.

DISCOVERY QUESTIONS

17. Why does aspirin prefer to bind to Cox-1 rather than Cox-2? What is the structural reason for this difference? (These questions are meant to be general and not require sequence information.)

18. In general terms, how could you change the affinity of your "new and improved" aspirin so that it prefers Cox-2 over Cox-1?

19. What effect might allelic variations in Cox-1 and Cox-2 have on the effectiveness of new "super-aspirins"?

Modifying the structure of aspirin so that it preferentially binds and inhibits Cox-2 is like assembling a child's Christmas present late on December 24th. It sounds easy, but it's not. Many chemists and biologists have spent years designing the "perfect" super-aspirin that will cure your headache but protect your stomach. In 2004 and 2005, most of the super-aspirins (e.g., Celebrex and Vioxx) were removed from the market because the side effects might be lethal. In clinical trials, the incidence of heart attacks and strokes were doubled compared to the control populations. The drugs were withdrawn from the market even though many chronic pain sufferers had benefited from the drugs with no apparent negative consequences. The FDA later reversed itself, but the future of the drugs is uncertain (January 2006) since the manufacturers may have concealed clinical trial data indicating the side effects could be harmful. The potential benefits, financial implications, and the FDA approval process are interwoven, so the long-term resolution is uncertain. Could we have predicted problems would result from overly efficient Cox-2 inhibition?

Genetically engineered Cox-2 knockout mice (both alleles deleted) were produced that lacked any Cox-2 enzyme. Individuals lacking Cox-2 enzyme are biochemically equivalent to individuals subjected to long-term exposure to the new generation of super-aspirins. Sadly, these mice developed severe kidney problems and were susceptible to peritonitis (inflammation of the peritoneum, the membrane that lines the wall of the abdomen and covers the organs). Female mice showed reproductive problems in ovulation, fertilization, and implantation. Also, the area of highest Cox-2 activity is the brain, so there may be mental effects not detected in mice that would be unacceptable for humans. The good news was that loss of Cox-2 activity decreased the rate of colon cancer in these mice. The discovery of heart problems was made during a clinical trial testing the effectiveness of Vioxx for colon cancer.

LINKS
disease genes
Genentech
Immunex
METHODS
knockout

The bottom line in the aspirin case study is that what sounds simple at first may not be simple to implement. There are new forms of aspirin, but each will have new side effects that may outweigh the potential benefits of getting rid of a headache or reducing a fever. In case you think this is an isolated example, let's look at another situation that was even more difficult to detect.

A protein called TRAIL is being developed by two of the biotech giants, Genentech and Immunex. From 1996 to mid-2000, TRAIL was becoming very popular in cancer research labs because it appeared to kill many cancerous cells while leaving the noncancerous cells unharmed. Imagine a single drug that could treat many forms of cancer! A multicancer treatment is what we all want to see developed, but a good scientist should not jump to conclusions simply because it would be the best thing since . . . aspirin. It turns out that TRAIL does not harm healthy liver cells in mice or monkeys, but it kills human liver cells. Once again, we see the benefit of requiring that new drugs undergo long periods of testing. Given the diversity in the human gene pool, extreme care must be taken, since no drug can produce a uniform response in every person—not even simple drugs like aspirin.

Can Medication Do More Than Simply Mask Symptoms?

The sexiest solution to genetic diseases is **gene therapy** (replacing the defective gene with a functional one). When the human genome project was first proposed, many said it would change medicine as we know it. Although public health policies that guaranteed prenatal care and immunizations would be more cost-effective solutions to many health problems, such mundane solutions are not exciting enough to garner billions of dollars from the public and private sectors. Why not fix the problem at its source? Fix the mutated gene and cure the disease!

The first genetic disease to be understood was sickle cell disease, though the gene was not cloned until much later. The first "disease-causing gene" to be discovered was the dystrophin gene (see Chapter 1). The list of disease genes grows every day, but knowing the genetic cause of a problem does not mean that we can generate a genomic cure by simply replacing the inappropriate nucleotides. Gene therapy has been the dream of many investigators, but after many trials, there has been only limited success.

C A S E   S T U D Y

CASE STUDY

LINKS
Alain Fischer paper
Xiao Xiao

DISCOVERY QUESTIONS

20. Let's assume you could accurately diagnose which gene was mutated and caused the disease. How could you introduce a *wt* sequence into the nucleus of a cell? Can you insert the therapeutic gene anywhere in the genome, or does it have to go in a certain locus?

21. Do you need to replace every copy of the gene in all cells, or only a subset? Explain your answer.

22. What sort of regulation do you need over the transcription of the replacement gene?

23. If the patient never produced the protein before, how do you know the patient won't produce an immune response to the new protein (since it may be seen as foreign)?

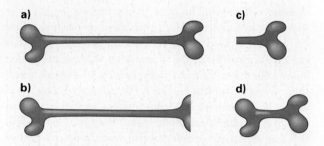

Figure 9.4 Three truncated versions of dystrophin.
a) Wild-type protein. **b)** Carboxyl-terminus truncation. **c)** Lacking most of the amino-terminus. **d)** Mini-dystrophin protein produced for gene therapy.

Do We Know the Answers?

Discovery Questions 20–23 are important and nearly impossible to answer. If these questions were easy, gene therapy would be routine. However, every disease requires a unique combination of answers. For example, there are dominant and recessive forms of limb-girdle muscular dystrophy (LGMD; see Chapter 1). Since a recessive disease is caused by the lack of a functional protein (only homozygous mutants have the disease) and a dominant disease is caused by one overactive protein (heterozygotes have the disease), inserting a single *wt* allele might cure a recessive disease but not a dominant disease. With obesity, the cause may be neuronal (located in a few nondividing cells) or due to the adipocytes (many mitotically active cells). Obesity may be due to a missing secreted ligand (leptin) or its receptor. Both of these produce the same symptoms but would require different strategies for gene therapy.

Since dystrophin was one of the earliest disease genes cloned, it was also one of the earliest candidates for gene therapy. The gene itself presented problems, though. First, it is a huge gene, approximately 3 million bp long. That is too big to fit inside any viral vector or liposome. In fact, it is too big to pipet! Unfortunately, the 14,000 bp cDNA was also too big for easy manipulation. In December 2000, Xiao Xiao from the University of Pittsburgh published a paper describing the use of a "minidystrophin gene" that was only about 4,000 bp long. The minigene was small enough to fit inside a replication-deficient adeno-associated virus and does not elicit an immune response. But here is the obvious question: How can a truncated version of dystrophin complement a mutant allele that is itself a truncated version of dystrophin? It all depends where the truncation occurs (Figure 9.4). The *wt* protein has two functional domains on either end and a long, rod-like middle section. Various mutant alleles encode for proteins that lack either one terminus or the other. The minigene created

in Xiao's lab contained both terminal domains and a shortened version of the center piece. When this minigene was given to *mdx* mice, the mice were cured. The *mdx* mice lack a dystrophin gene, so this minigene therapy worked for the most prevalent form of muscular dystrophy.

Kevin Campbell's group at the University of Iowa used gene therapy to treat mice that had LGMD2D. The Iowa group injected an adenovirus containing a *wt* allele of α-sarcoglycan into skeletal muscles and thereby corrected the mutant phenotype. In March 2005, a team from the University of Washington used lentivirus to deliver the dystrophin minigene to cultured muscle cells, hoping this virus would deliver the DNA to muscle cells more efficiently and stimulate a reduced immune response.

Sometimes gene therapy has been successful for children with severe combined immune deficiency (SCID, often called "the boy in the bubble disease"). SCID patients have almost no immune system because they lack a functional copy of an enzyme that metabolizes adenosine. Beginning in September 1990, a four-year-old girl with SCID was given the missing gene. Her immune system has improved, though she is also injected with the enzyme to bolster the amount produced by her therapeutic gene. In April 2000, a French group led by Alain Fischer treated several infants with SCID, and the children have been able to produce normal levels of white blood cells. In this case, a retrovirus was used to infect bone marrow stem cells for three days in vitro before being reinjected into the patients. Unfortunately, within a few years, several children developed leukemia, and one died, when the viruses inserted their genomes into oncogenes (two of the three had insertional mutations in the oncogene *LMO2*). Children with untreated SCID will eventually die from an infection and cannot enjoy normal childhoods. Because gene therapy has worked for some children, there is hope that a better delivery mechanism can be found.

Sometimes human errors are made that hamper research. In September 2000, Jesse Gelsinger, an 18-year-old man with a liver enzyme deficiency, was killed by his experimental gene therapy treatment. Investigators at the University of Pennsylvania's Institute for Human Gene Therapy (IHGT) were performing an experimental procedure when Gelsinger was given too many adenoviruses. He died four days after being injected with the potential cure—a tragic outcome for everyone involved, especially the Gelsinger family.

What should be done now? Many clinical gene therapy trials have been put on hold because of the unexpected leukemia produced in the French study. Should gene therapy be done only when no other therapy exists? Should we demand that more research be done in the area of traditional drugs, as illustrated with the aspirin story? Is gene therapy an inappropriate use of limited dollars available for health care? Should we insist that more cost-effective medical treatments be performed universally, before developing very expensive treatments for relatively rare genetic diseases? Should we insist that people who want these expensive treatments pay more of the expense? Should an experimental treatment be reserved for the rich who can afford to pay for it themselves? There are no easy answers, but one thing is certain. Difficult ethical questions should be discussed openly by people like you, who understand the complexity of the science and can communicate with those who do not. To address these ethical issues, we need a national dialogue, not small discussions restricted to academic institutions and insurance companies (see Chapter 3). Below is a series of questions without obvious answers. Consider these questions in light of what you know about developing new medications.

1. In October 2000, Tejvir Khurana from the University of Pennsylvania reported that in rats, 14 genes are expressed in skeletal muscles that control eye movement but not in skeletal muscles of the limb. Because patients with muscular dystrophy do not have any problems with the muscles that control eye movement, he speculated that one possible treatment for muscular dystrophy would be to "convert limb muscles to eye muscles by up-regulating the genes expressed in eye but not limb muscles."

    a. Based on what you know of muscular dystrophies, critique this proposal.

    b. Based on what you know about the complexity of cell webs, critique this proposal.

2. Gene therapy sounds like the ideal cure for genetic diseases, even though there are many technical difficulties. Assume that a given gene could be delivered to the desired cells. Hypothesize reasons why gene therapy might not work in some patients.

3. National health policies should not be based on individual cases, no matter how compelling they are. Health policy must be based on the greatest good for the most people. List the pros and cons of gene therapy as if you had been commissioned by the president

LINKS
IHGT
Tejvir Khurana

to study the issue. When you are done, decide whether to continue federal funding for gene therapy research.

4. There are evolutionary implications of gene therapy, too. For example, many people today wear glasses. However, 150,000 years ago, hominids did not need glasses because those with poor vision did not live long enough to reproduce. Today, the selection pressures have changed so that the gene pool is loaded with "bad-eyesight alleles." If gene therapy is conducted on people with genetic diseases, should we as a society require these people to be sterilized so they cannot pass on their disease alleles? If we do not, won't the gene pool become loaded with more disease alleles and thus require more and more gene therapy?

5. Since gene therapy may lead to an increased need for more gene therapy, should we require all gene therapy to be performed on fetuses, so that the resulting individual will not have the disease and will not pass it on to his or her children?

6. Complex traits, such as intelligence, beauty, and honesty, are influenced by our genomes. Should these traits be "treated" with gene therapy?

7. Now that we have raised many real-world ethical questions, step back into your hypothetical vacuum where you have unlimited resources. Choose either muscular dystrophy or obesity and design the ultimate cure. Of course, most of you will not be able to come up with a marketable idea, but consider the major issues, such as location, specificity, vehicle, and drug or gene therapy. Factor in ethical considerations while you design your cure.

Chapter 9 Conclusions

This chapter on genomic medicine was not very genomic, in the sense that it focused on only a few genes. However, genomics is more than just massive data sets and sequences of DNA. It is also a mindset, a new perspective. Genomics requires molecular biologists to think differently. We need to think of cells as integrated circuits with massive complexity and interconnections. Imagine a world in which every single person had a cell phone. How quickly would news spread around the world if we all had phones and unlimited free minutes? Unlimited free minutes is what each of your proteins has. Proteins talk to their neighbors all the time, who talk to their neighbors, etc. Some proteins communicate with proteins located in different cells, and so the message spreads from one cell to another, and soon the whole organism knows what has just happened in the little toe of your left foot.

The main purpose of the genomic medicine case studies (Chapters 1, 5, and 9) was to illustrate that Gregor Mendel was right and wrong at the same time. He correctly deduced how traits are passed on from one generation to the next. In the real world, outside the garden walls of an abbey, though, life is much more complicated. Cell webs, in which everything is connected to something else, are more complex than pea color and shape. Traits (e.g., muscular dystrophy, obesity, personality, intelligence, sexual orientation, etc.) have multiple inputs and connections. Now you are in the right frame of mind to think about systems biology, a holistic approach to genomics. Chapters 10 through 12 will expose you to a wide range of topics, methods, and data. Use the systems biology approaches and perspectives to discover hidden genomic information.

References

Allamand, V., K. M. Donahue, et al. 2000. Early adenovirus-mediated gene transfer effectively prevents muscular dystrophy in alpha-sarcoglycan-deficient mice. *Gene Therapy.* 16: 1385–1391.

Anderson, W. F. 2000. The best of times, the worst of times. *Science.* 288: 627.

Belkin, L. 2000, December 24. The making of an 8-year-old woman. *New York Times Magazine.* 38–43.

Black, H. 2000. Seeing a solution. *The Scientist.* 14(22): 18.

Campbell, K. 2000. Molecular studies of muscular dystrophy. <http://www.physiology.uiowa.edu/campbell/Research/Areas/researchareas.htm#MolecularStudies>. Accessed 10 May 2005.

Cavazzana-Calvo, M., S. Hacein-Bey, et al. 2000. Gene therapy of human severe combined immunodeficiency (SCID)-X1 disease. *Science.* 288: 669–672.

Dequeker, E. and J.-J. Cassiman. 2000. Genetic testing and quality control in diagnostic laboratories. *Nature Genetics.* 25: 259–260.

Evans, W. E. and M. V. Relling. 1999. Pharmacogenomics: Translating functional genomics into rational therapeutics. *Science.* 286: 487–491.

Feng, L., W. Sun, et al. 1993. Cloning two isoforms of rat cyclooxygenase: Differential regulation of their expression. *Archives of Biochemical Biophysics.* 307(2): 361–368.

Fischer, A., S. Hacein-Bey-Abina, and M. Cavazzana-Calvo. 2004. Gene therapy for immunodeficiency diseases. *Seminars in Hematology.* 41(4): 272–278.

Gura, T. 2000. Caution raised about possible new drug. *Science.* 288: 786–787.

Kaiser, J. 2005. Panel urges limits on X-SCID trials. *Science.* 307: 1544–1545.

Kalgutkar, A. S., B. C. Crews, et al. 1998. Aspirin-like molecules that covalently inactivate cyclooxygenase-2. *Science.* 280: 1268–1270.

Kay, M. A., J. C. Glorioso, & L. Naldini. 2001. Viral vectors for gene therapy: The art of turning infectious agents into vehicles of therapeutics. *Nature Medicine.* 7: 33–40.

Li, S., E. Kimura, et al. 2005. Stable transduction of myogenic cells with lentiviral vectors expressing a minidystrophin. *Gene Therapy.* 12(14): 1099–1108.

Loll, P. J., D. Picot, & R. M. Garavito. 1995. The structural basis of aspirin activity inferred from the crystal structure of inactivated prostaglandin H2 synthase. *Nature Structural Biology.* 2(8): 637–643.

Luong, C., A. Miller, et al. 1996. Flexibility of the NSAID binding site in the structure of human cyclooxygenase-2. *Nature Structural Biology.* 3(11): 927–933.

Marshall, E. 1999. Gene therapy death prompts review of adenovirus vector. *Science.* 286: 2244.

Morham, S. G., R. Langenbach, et al. 1995. Prostaglandin synthase 2 gene disruption causes severe renal pathology in the mouse. *Cell.* 83(3): 473–482.

Neimann, C. U., T. O. B. Krag, and T. S. Khurana. 2000. Identification of genes that are differentially expressed in extraocular and limb muscle. *Journal of the Neurological Sciences.* 179: 76–84.

OMIM: PROSTAGLANDIN-ENDOPEROXIDE SYNTHASE 2; PTGS2 2005. <http://www.ncbi.nlm.nih.gov:80/entrez/dispomim.cgi?id=600262>. Accessed 10 May 2005.

Pennisi, E. 1998. Building a better aspirin. *Science.* 280: 1191–1192.

Samad, T. A., K. A. Moore, et al. 2001. Interleukin-1β-mediated induction of Cox-2 in the CNS contributes to inflammatory pain hypersensitivity. *Nature.* 410: 471–475.

Service, R. 1996. Closing in on a stomach-sparing aspirin substitute. *Science.* 273: 1660.

Steppan, C. M., S. T. Bailey, et al. 2001. The hormone resistin links obesity to diabetes. *Nature.* 409: 307–312.

Wadia, J. S., & S. F. Dowdy. 2005. Transmembrane delivery of protein and peptide drugs by TAT-mediated transduction in the treatment of cancer. *Advanced Drug Delivery Reviews.* 57(4): 579–596.

Wang, B., J. Li, et al. 2000. Adeno-associated virus vector carrying human minidystrophin genes effectively ameliorates muscular dystrophy in *mdx* mouse model. *PNAS USA.* 97: 13714–13719.

Ye, X., K. Hama, et al. 2005. LPA₃-mediated lysophosphatidic acid signalling in embryo implantation and spacing. *Nature.* 435: 104–108.

Genomic Circuits in Single Genes

Embryonic development can be thought of in two halves: the early part is tightly controlled by genes, and the late part generates observable features that are unique to individuals, even identical twins. We are especially interested in the early part that is tightly controlled by the genome. Every cell in an organism contains the exact same DNA. Genetically, there is no difference between your liver cells, your brain cells, and your big-toe cells. How can our bodies produce these different tissues if the genes are all the same? The answer is that each cell only expresses a subset of its genes. The subset of genes expressed is tightly controlled during development, and this control is exerted over *location* (which cells do or do not express the gene?), *time* (when during development should the gene be turned on or off?), and *amount* (are only 2 molecules of the encoded protein needed per cell, or 20,000?).

Section 10.1 focuses on the **circuits** (interaction of proteins and DNA sequences in the promoter and enhancer) of a single gene that serves as a model for genomic control over the expression of genes. Perhaps better than any other gene, we understand gene regulation of **Endo16**, which is expressed in the developing gut of a sea urchin embryo. Over several years, Eric Davidson and his collaborators have slowly dissected *Endo16* to reveal an amazing circuitry that regulates its expression. To fully appreciate how a genome regulates the expression of individual genes, we will study how a sea urchin embryo regulates *Endo16*. The data are impressive, and the resulting model is elegant. In Section 10.2, we begin to coalesce what is known about the genomic regulation of genes into interactive models that are prototypes of what is needed to model entire genomes.

One final note: Chapter 10 is our window into the future. If you wanted to rewire your home, you would practice on a simple object first—a lamp, for example. Then you would gradually increase the level of complexity until you were ready to take on the fuse box and major appliances. Similarly, when we want to understand how genomes respond to their environment, we begin with a simple "wiring" diagram of a single gene. For the next 100 years, we will continue to analyze how genomes and proteomes are interconnected. Discovering how *Endo16* is regulated will help you understand Chapters 11 and 12, which expand the scale of gene and protein interactions.

10.1 Dissecting a Gene's Circuitry

Throughout your biology education, you have been told that promoters control transcription. Later you learned that individual proteins, called transcription factors, bind to particular sites within promoters to facilitate transcription. Repressor proteins bind to other sites to block transcription. For most biologists, this explanation has been

sufficient. Luckily, Eric Davidson wanted to understand exactly how genes are regulated during early development. He chose the sea urchin as his model organism, in part because it has been used for many years to study development. Of all the genes in sea urchin, why *Endo16*? Davidson explained that after years of studying *Endo16*, his team realized they had the molecular tools to analyze gene regulation at a finer resolution than ever before. In this section, we examine a lot of detailed information that Davidson used to produce a wiring diagram and a computer algorithm that model the genomic regulation of *Endo16*.

How Are Genes Regulated?

Every gene has to be controlled in three respects (location, time, amount), but we do not understand how this control is accomplished. Microarray data (Chapters 6 and 7) help us understand transcriptional output, and proteomics methods (Chapter 8) help us understand protein production. But the initial control mechanisms that regulate whether a gene will be active or silent are still hidden from us inside biological black boxes. Davidson describes components at the 5′ end of a gene that control transcription as **cis-regulatory elements**. Before we dissect the cis-regulatory elements of *Endo16*, we need to appreciate how tightly the expression of *Endo16* is controlled.

A gene's cis-regulatory elements are the DNA sequences upstream of the transcription start site that control transcription. Cis-regulatory elements are different from trans-regulatory elements, which are DNA sequences further away from the coding sequence, often located on a separate chromosome (Figure 10.1). The organization of cis-regulatory elements is modular. **Modular** in this context means the DNA can be divided by investigators into discrete functional units, each of which performs a particular job. Each cis-regulatory module is composed of DNA to which one or more proteins can bind, either to help initiate transcription (transcription factors) or to repress transcription (repressors).

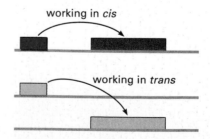

Figure 10.1 Spatial difference between cis- and trans-regulatory elements.
Cis-regulatory elements (dark purple) are adjacent to the coding sequences they regulate (dark gray). Trans-regulatory elements (light purple) regulate coding DNA located further away (light gray).

The sea urchin (***Strongylocentrotus purpuratus***) is a model organism that developmental biologists have studied for over 100 years. Sea urchins produce transparent embryos that are easy to produce and grow in the lab by in vitro fertilization. As with all model organisms, *S. purpuratus* is a useful tool for understanding the fundamentals shared by many species; in this case, we can learn a lot about animal development. As with all animals, the new zygote is formed by the fusion of egg and sperm. From this single diploid cell, all the different tissues must arise. How does this mixed bag of proteins, carbohydrates, lipids, and nucleic acids know to dedicate the correct number of cells to form each of the required organs?

In the case of *S. purpuratus*, we know a great deal about how a fertilized egg mitotically replicates and commits each subsequent daughter cell to a particular fate. From one cell come two cells, then four, etc. After six rounds of mitosis and cytokinesis, the sphere of cells is composed of 60 cells (Figure 10.2). Biologists have injected embryos with non-toxic dyes and followed the distribution of the dye in older tissues to produce a **fate map**, which indicates what type of tissue each cell will become. Fate maps have proven useful for understanding the significance of a gene being transcribed in a subset of cells at a particular time during development. Davidson's extensive analysis of sea urchin embryogenesis has contributed substantially to our understanding of animal development.

METHODS G&P
100 years

Developing embryos pass through two important stages: blastula and gastrula. **Blastula** refers to a hollow sphere of embryonic cells (Figure 10.2b) that is ready to undergo a process called gastrulation. **Gastrula** describes an embryo during the process of **gastrulation**, when a subset of blastula cells begin to invaginate, or move into the cavity of the blastula/gastrula (sea urchin web page has details and animations). These invaginating cells are genetically programmed to perform their movement, and without this internalization, the embryo will fail to develop normally. As the cells begin to invaginate, they repress some genes and activate others. The cells that form the elongating tube inside the gastrula will become the **endodermal** cells that line the gut of the future larva. The place where the cells begin to invaginate will become the mouth, and the place where this tube hits the other side of the embryo will become the anus. This elongated tube of cells is called the **archenteron**. As the archenteron continues to develop, it differentiates into foregut, midgut, and hindgut to denote the eventual fate of becoming the mouth, digestive tract, and anus, respectively. **Ectodermal** cells will form the exterior surface cells of the larva.

Molecular Dissection of Development

Eric Davidson and his colleagues have generated an amazing amount of high-quality data that illuminate the details hidden in what some people incorrectly referred to as "**junk DNA**." As we all know from yard sales, one person's junk is another person's treasure. Davidson's group has sifted through the cis-regulatory elements of the gene *Endo16* and found many hidden treasures. Although microarray technology (Chapters 6 and 7) has enabled us to simultaneously measure the activity of many genes, microarrays have not revealed the circuitry that controls individual gene transcription. We have to select one gene and then use a **reductionist** approach to take it apart. This approach seems obvious at first, but if you have ever taken apart an appliance, clock, or coaster brakes on a one-speed bike, you know that dissection is easier than reconstruction. Therefore, selecting which gene to take apart is the first important decision to be made. Davidson's group chose one gene that exhibits all three levels of transcriptional control—time, location, and amount. *Endo16* is as unlikely a hero as you could find anywhere in biology.

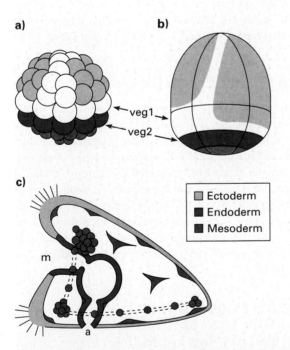

Figure 10.2 Sea urchin *S. purpuratus* embryogenesis.
a) 60-cell embryo undergoing rapid cell division. Cell fates have been determined for many of these cells, as shown by the color code for future ectoderm, mesoderm, and endoderm layers. Cells not yet committed to one of the three layers appear white. **b)** A 24-hour-old blastula composed of about 500 cells. **c)** Early larva 65 hours after fertilization, composed of about 1,500 cells. The mouth (m) and anus (a) are shown on the larva.

Ectoderm
Endoderm
Mesoderm

Expression of *Endo16*

Endo16 has been the subject of intense research mainly because it is an ideal (i.e., model) gene to use for investigating gene regulation during development (Figure 10.3).

G⚗P **METHODS**
CAT

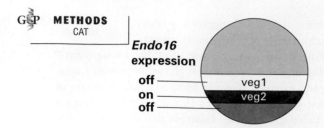

Figure 10.3 Location and timing of pre-blastula *Endo16* expression.
Prior to the blastula stage, *Endo16* is expressed in the veg2 cell layer. Only the veg2 layer will become endoderm, and the cell layers above and below veg2 do not express *Endo16*. Light gray indicates ectoderm, purple indicates endoderm, and dark gray indicates mesoderm; white is uncommitted.

Slightly before gastrulation, cells at the base of the blastula express *Endo16*, and later all cells of the invaginating archenteron express *Endo16* and become endodermal cells in the larva (Figure 10.4). This highly regulated expression pattern has made *Endo16* an excellent genetic marker to identify which cells will form the endoderm. Later in gastrulation, the cells of the foregut and hindgut repress *Endo16* so that only midgut cells still transcribe *Endo16*, which they do at an even higher rate than before.

The *Endo16* protein (Endo16p) is secreted by the cells and localized to the basal extracellular matrix (i.e., the cell side away from the hollow tube of the archenteron). Endo16p is a large glycoprotein that binds calcium. A similar gene has been identified in mice, and the two orthologs are similar in sequence to serum albumin, which is known to bind calcium. It appears that Endo16p and the mouse ortholog play active roles in cellular adhesion to the extracellular matrix, and possibly other cells too.

How Does a Gene Control Location, Timing, and Quantity of Transcription?

We study *Endo16* because its circuitry and logic are wonderful in their design and exquisite in their control. A vast majority of the work in this section was conducted by Chiou-Hwa Yuh, a senior research fellow at Caltech. The modules of *Endo16*'s cis-regulatory elements are within the 2,300 bp upstream of the coding DNA (Figure 10.5). Each module (G–A) has a function that can be studied individually because DNA-binding proteins recognize specific DNA sequences within each module, and these protein-DNA interactions regulate transcription.

DISCOVERY QUESTIONS

1. Why would a gene need so many different transcription factors to control its expression? Why not just use 5 to 10 DNA-binding proteins?

2. Of all the DNA-binding proteins shown in Figure 10.5, which would you want to study first? Why?

The term "watchmaker" can be used to describe anyone who successfully uses the reductionist approach to dissect an object and understand how it works. A related expression is that "for every job, you have to have the right tools." Davidson's group used molecular methods to build the right tools, and became expert watchmakers as they dissected the cis-regulatory element modules of *Endo16*. We do not need to understand every aspect of these tools, but we should recognize the strategy involved (Figure 10.6). The "tools" numbered 2–8 illustrate DNA constructs created to fuse each module individually onto the most basic promoter (Bp) that allows RNA polymerase to bind and begin transcription. Construct 1 is the *wt* cis-regulatory element, and construct 9 is the Bp alone placed upstream of the coding DNA. The investigators used a reporter gene rather than the coding portion of *Endo16*. As their reporter, they chose **chloramphenicol acetyl transferase (CAT)**. The CAT gene was cloned from antibiotic-resistant bacteria that neutralize the antibiotic chloramphenicol by adding an acetyl group. CAT is not normally found in sea urchin, and its output (both mRNA and protein) can be monitored easily. Each of the seven modules was studied individually, but later you will notice that modules D and C are often studied jointly.

Which Modules Control Location?

Constructs 2 through 8 can be divided into two categories: those that expressed CAT in the endoderm (where *Endo16* is normally expressed) and those that expressed CAT outside the endoderm. Table 10.1 summarizes these data, specific examples of which are provided in Figure 10.7. Notice that constructs 1, 2, 7, and 8 promote the production of CAT in endodermal cells. These constructs include modules G, B, and A. Constructs containing modules C, D, E, and F do not promote the production of CAT in endoderm cells, but do permit CAT production in **mesoderm** and ectoderm cells.

DISCOVERY QUESTIONS

3. Modules G, B, and A all promote about the same expression. Why would *Endo16* need three redundant modules that don't do much better in combination (GBA+Bp) than they do individually (see Table 10.1)?

4. Hypothesize how modules G, B, and A can be active in endodermal cells but not in other cell types. Refer to Figure 10.5 to formulate your hypothesis.

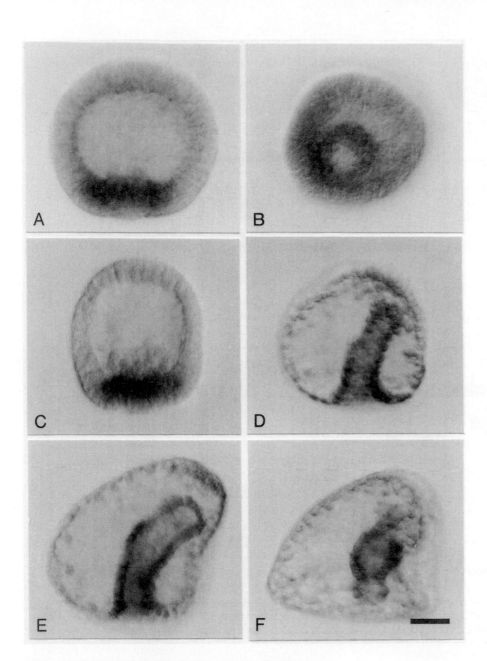

Figure 10.4 *Endo16* transcription in sea urchin development.
a) and **b)** Using in situ hybridization, *Endo16* mRNA is visible in 20-hour-old blastulas,
c) a 30-hour-old blastula at the beginning of gastrulation, **d)** a 48-hour-old gastrula,
e) a 60-hour-old embryo as it begins to take on the triangular shape of the larva, and
f) a 72-hour-old larva stained darkly in the bulbous midgut with the *Endo16* probe.
The lighter color outside the midgut is due to pigmentation in the embryo.

Why Do Modules F, E, and DC Promote Expression in the Wrong Cells?

From the data in Table 10.1, the roles of modules F–C are unclear; they do not promote the production of CAT in endodermal cells where *Endo16* is normally expressed, but they do promote its production in mesoderm and ectoderm. It seems counterproductive for portions of a gene's

cis-regulatory element to promote transcription in the wrong cell types. Therefore, new constructs were created that combined the best endoderm promoter (GBA+Bp) with each of the remaining modules (Figure 10.8). The idea was to test whether modules F–C had any influence on the transcription promoted by modules G, B, and A.

The new constructs exhibited different capacities to promote the formation of CAT (Table 10.2). The level of CAT

Modules:

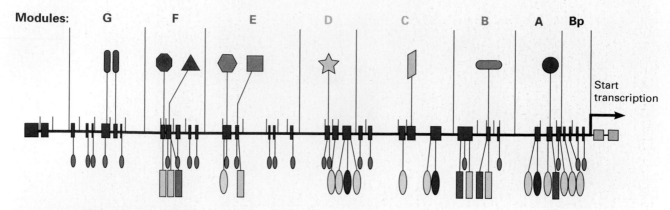

Figure 10.5 *Endo16* cis-regulatory element (horizontal black line) divided into the eight modules lettered G through A.

Within each module, DNA sequences are recognized by DNA-binding proteins (purple boxes on horizontal black line). The DNA-binding proteins are illustrated as colored shapes above and below the DNA. The larger shapes above the DNA bind to only one module, while those below bind to two or more modules. Bp is the "basic promoter" where the RNA polymerase begins transcription.

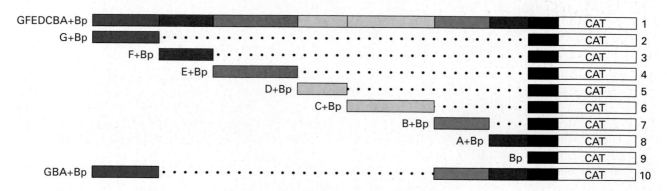

Figure 10.6 The first ten DNA constructs built by Davidson's lab to dissect *Endo16*'s cis-regulatory elements.

Each colored block represents a module (see Figure 10.5), and the modules are labeled G to A from left to right. Bp (black box) is located immediately upstream of the reporter gene CAT. Dotted lines indicate which portions of the *wt* cis-regulatory element have been deleted. For example, construct 2 (G+Bp) has module G fused directly onto the Bp, which was fused onto the CAT-coding DNA.

Table 10.1 Expression of CAT from different cis-regulatory modules as promoters (see Figure 10.6) in 48-hour embryos.
Numbers in the table indicate the percentage of embryos expressing CAT enzyme in the indicated cell types; some constructs expressed CAT in more than one cell type, so the total percentage in a column may exceed 100. Color-coding highlights the critical data for the three cell types (see Figure 10.2). "Total # embryos" indicates how many embryos exhibited a detectable signal. Numbers in parentheses indicate that no more than three cells per embryo expressed CAT, compared to the widespread expression for constructs 1, 2, 7, and 8.

	1	2	7	8	6	5	4	3	9
Construct	*wt*	**G+Bp**	**B+Bp**	**A+Bp**	**C+Bp**	**D+Bp**	**E+Bp**	**F+Bp**	**Bp**
Endoderm	97	94	89	95	0	0	0	0	0
Mesoderm	3.9	6	13	5.9	0	63	78	100	82
Ectoderm	5.8	12	19	12.3	0	41	28	4	27
Total # embryos	70	74	72	74	0	(11)	(41)	(45)	(13)

Source: Derived from Yuh and Davidson. 1996. *Development.* 122: 1074, Table 1 (partial).

production was essentially unchanged in endodermal cells, but the level of "inappropriate" CAT expression in mesoderm and ectoderm was altered. Modules DC (construct 13) reduced the capacity of the promoter to function in mesoderm cells, while modules F (construct 11) and E (construct 12) each reduced the expression of CAT in ectoderm cells. Therefore, modules F–C appear to function as cell-type specific repressors of transcription, which helps explain why *Endo16* is not expressed in ectoderm or mesoderm cells.

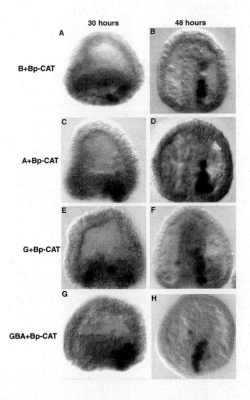

Figure 10.7 Photomicrographs showing the location of CAT mRNA.

In situ hybridization was used to detect CAT mRNA produced with different modules from the *Endo16* cis-regulatory element in embryos at 30 (left column) or 48 (right column) hours postfertilization. The constructs used in each case are described to the left, with the module controlling the expression written before the "Bp-CAT" core construct. The modular design of the constructs can be seen in Figure 10.6: **a)–b)**, construct 7 = B+Bp-CAT; **c)–d)**, construct 8 = A+Bp-CAT; **e)–f)**, construct 2 = G+Bp-CAT; and **g)–h)**, construct 10 = GBA+Bp-CAT.

Combining the data from Tables 10.1 and 10.2 with Figure 10.7, we can build our first comprehensive model for the function of each module. Modules G, B, and A strongly promote "appropriate" expression in the endoderm. Modules F and E repress expression in the ectoderm, while modules DC repress expression in the mesoderm. This level of promoter dissection is impressive but not unique. Many other research groups had cut and pasted modules before, but they often stopped at this point. Davidson's group wanted to go further, but they needed many more constructs than those we have already seen (Figure 10.9).

Davidson's team wanted to know which activating modules (G, B, or A) were specifically repressed by modules DC, E, and F. For this work, the investigators reported their data in a different format that requires a little explanation. Rather than simply reporting the amount of CAT enzyme activity in a large table, they used bar graphs to display the relative amount of activity compared to the lowest level of CAT activity when only the basal promoter (Bp) was placed upstream of the CAT-coding DNA. If there were no difference (e.g., Bp-CAT), no bar is visible, and you see only the vertical line at the number 1.00 along the logarithmic scale at the top of Figure 10.10. If a construct led to an increased production of CAT enzyme compared to Bp-CAT, then a bar extends to the right and displays fold induction. Constructs that expressed less CAT are shown by bar graphs extending to the left and displaying fold repression. Figure 10.10 illustrates six series of constructs. All bar graphs should be compared to the top one within each series.

Notice in the X-Bp series of constructs that the repressor modules can repress expression in some constructs but not others. When a repressor module is inserted between G and BA (GXBA+Bp series), each of the repressors reduces expression by about half. When a repressor was combined with module A alone (XA+Bp series), the repressor effects were even greater, especially when more than one repressor was combined. When repressor modules were combined with G, or B, or GB, the effects were not as uniform or strong compared to the XA+Bp series. Certain combinations of repressor modules did have stronger effects on either G or B, but not as strong as when module A was present. These data led the investigators to conclude that the repressors work primarily by affecting module A's ability to initiate transcription.

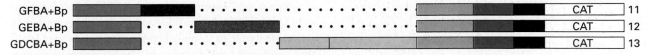

Figure 10.8 Three DNA constructs to elucidate the roles of modules F, E, and DC.

The colors and symbols are the same as used in Figure 10.6. The numbering of the constructs continues from Figure 10.6.

Table 10.2 Percentage of embryos expressing CAT enzyme when the best endoderm promoter was combined with the remaining four modules.
Location and frequency of CAT expression in 48-hour embryos containing constructs 10–13 (Figure 10.8). Some constructs expressed CAT in more than one cell type, and thus the total percentage in a column may exceed 100. Color-coding (see Figure 10.2) highlights the significant data.

Name	10 GBA+Bp	11 GFBA+Bp	12 GEBA+Bp	13 GDCBA+Bp
Endoderm	94	99	94	94
Mesoderm	9.4	10.4	30.4	1.4
Ectoderm	15	5.7	4.3	26

Source: Derived from Yuh and Davidson. 1996. *Development.* 122: 1075, Table 2.

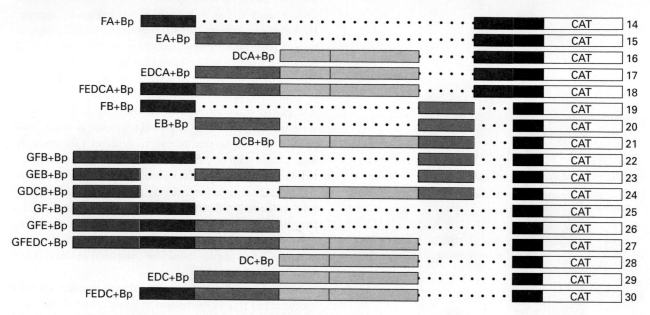

Figure 10.9 Additional DNA constructs to elucidate the roles of modules F, E, and DC.
The colors and symbols are the same as used in Figure 10.6. The numbering of the constructs continues from Figure 10.8.

DISCOVERY QUESTIONS

5. How do repressor modules exert their effect on *Endo16* expression?
6. Why does *Endo16* need two types of repressor modules (DC vs. F and E)?
7. Draw a picture that shows how repressor modules might be able to block transcription even when inducer modules are present. Review Figure 10.5 and include DNA-binding proteins in your model.

How Does Lithium Affect Transcription?

At this point, we know that modules G, B, and A enhance the basic promoter's ability to drive transcription. Modules DC act to repress expression in mesoderm, while modules E and F repress expression in ectoderm. Davidson and his lab wanted to understand another aspect of *Endo16*'s transcriptional control. It has been known for many years that lithium (Li) is a **teratogen**; it causes birth defects in embryos. The developmental effects on a fetus explain why physicians must know whether a woman is pregnant before prescribing Li to control her depression. If she were pregnant, the Li would lead to serious birth defects and probably a miscarriage.

Lithium has many effects inside sea urchin embryos, one of which is the conversion of cells from ectoderm fate into cells that will become endoderm. *Endo16* is expressed in endodermal cells, so Li-treated embryos express more *Endo16* than untreated control embryos. The investigators wanted to find out which modules were responsive to Li treatment. Their Li experiments used the same DNA constructs used in early studies (Figure 10.11). In this figure, the bar graphs

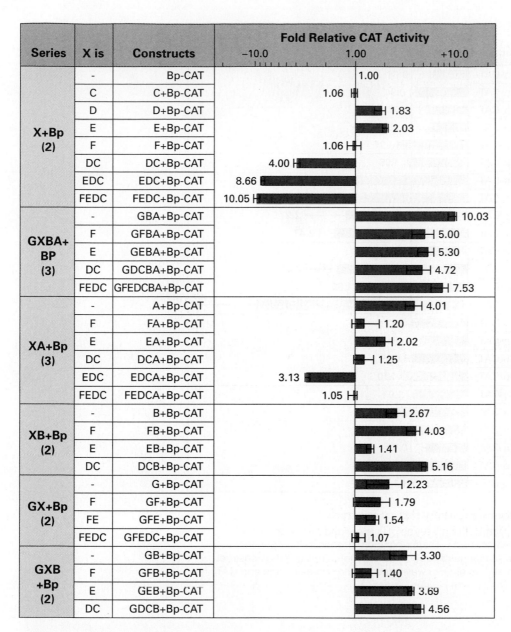

Series	X is	Constructs	Fold Relative CAT Activity
X+Bp (2)	-	Bp-CAT	1.00
	C	C+Bp-CAT	1.06
	D	D+Bp-CAT	1.83
	E	E+Bp-CAT	2.03
	F	F+Bp-CAT	1.06
	DC	DC+Bp-CAT	4.00
	EDC	EDC+Bp-CAT	8.66
	FEDC	FEDC+Bp-CAT	10.05
GXBA+ BP (3)	-	GBA+Bp-CAT	10.03
	F	GFBA+Bp-CAT	5.00
	E	GEBA+Bp-CAT	5.30
	DC	GDCBA+Bp-CAT	4.72
	FEDC	GFEDCBA+Bp-CAT	7.53
XA+Bp (3)	-	A+Bp-CAT	4.01
	F	FA+Bp-CAT	1.20
	E	EA+Bp-CAT	2.02
	DC	DCA+Bp-CAT	1.25
	EDC	EDCA+Bp-CAT	3.13
	FEDC	FEDCA+Bp-CAT	1.05
XB+Bp (2)	-	B+Bp-CAT	2.67
	F	FB+Bp-CAT	4.03
	E	EB+Bp-CAT	1.41
	DC	DCB+Bp-CAT	5.16
GX+Bp (2)	-	G+Bp-CAT	2.23
	F	GF+Bp-CAT	1.79
	FE	GFE+Bp-CAT	1.54
	FEDC	GFEDC+Bp-CAT	1.07
GXB +Bp (2)	-	GB+Bp-CAT	3.30
	F	GFB+Bp-CAT	1.40
	E	GEB+Bp-CAT	3.69
	DC	GDCB+Bp-CAT	4.56

Figure 10.10 Repressive functions of modules F, E, and DC on modules G, B, and A in 48-hour embryos.

The base level of CAT enzyme produced in the entire embryo promoted only by the Bp is set at 1.00. The ability of each construct to express more (bar graphs extending to the right) or less (bar graphs extending to the left) CAT enzyme is indicated. The number at the end of each bar graph denotes the fold change from the basal expression, which was 2.5×10^5 active CAT enzyme molecules per embryo. The scale is logarithmic, the error bars indicate standard error, and numbers in parentheses under the construct names denote how many separate experiments were performed.

show the ratio of CAT enzyme produced in Li-treated embryos to the amount of CAT produced in untreated embryos. For example, if Li treatment had no effect on a construct, the bar graph would be near 1, because the amount of CAT produced in the presence of Li would be equivalent to the amount of CAT produced in the absence of Li. If the ratio of CAT activity is below 1, Li treatment resulted in less CAT activity than in untreated embryos and the experimental construct was not sensitive to Li. However, if the bar graph extends beyond 1, the construct was sensitive to Li and produced more CAT in the presence of Li than in untreated embryos. In this manner, the

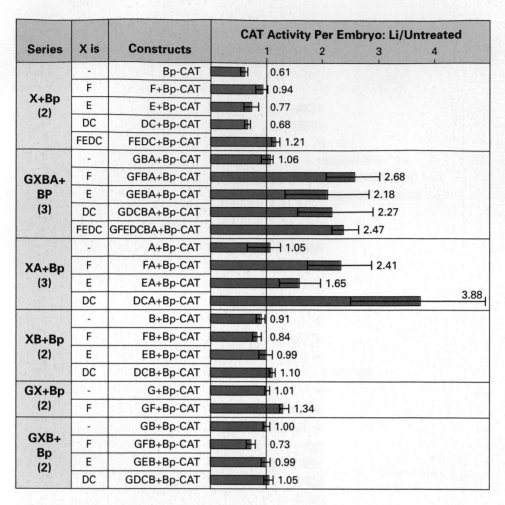

Series	X is	Constructs	CAT Activity Per Embryo: Li/Untreated			
			1	2	3	4
X+Bp (2)	-	Bp-CAT	0.61			
	F	F+Bp-CAT	0.94			
	E	E+Bp-CAT	0.77			
	DC	DC+Bp-CAT	0.68			
	FEDC	FEDC+Bp-CAT	1.21			
GXBA+ BP (3)	-	GBA+Bp-CAT	1.06			
	F	GFBA+Bp-CAT		2.68		
	E	GEBA+Bp-CAT		2.18		
	DC	GDCBA+Bp-CAT		2.27		
	FEDC	GFEDCBA+Bp-CAT		2.47		
XA+Bp (3)	-	A+Bp-CAT	1.05			
	F	FA+Bp-CAT		2.41		
	E	EA+Bp-CAT		1.65		
	DC	DCA+Bp-CAT				3.88
XB+Bp (2)	-	B+Bp-CAT	0.91			
	F	FB+Bp-CAT	0.84			
	E	EB+Bp-CAT	0.99			
	DC	DCB+Bp-CAT	1.10			
GX+Bp (2)	-	G+Bp-CAT	1.01			
	F	GF+Bp-CAT	1.34			
GXB+ Bp (2)	-	GB+Bp-CAT	1.00			
	F	GFB+Bp-CAT	0.73			
	E	GEB+Bp-CAT	0.99			
	DC	GDCB+Bp-CAT	1.05			

Figure 10.11 Effect of LiCl on the cis-regulatory element constructs.
The amount of CAT enzyme activity in the presence of Li is divided by the amount of CAT enzyme activity in untreated embryos. The further the bar goes to the right, the more CAT was produced in the presence of Li and thus the more sensitive the construct was to the endoderm-inducing effects of Li. Li was added at the two-cell stage and CAT enzyme assays were performed 48 hours postfertilization.

investigators could test each construct ± Li, and if the ratio was significantly larger than 1, the construct was sensitive to Li.

The results in Figure 10.11 are striking in their clarity. From the X+Bp series, we see that Li was unable to produce more CAT when each repressor module was tested individually. However, when module A was present (GXBA+Bp and XA+Bp series), Li treatment resulted in a substantial increase in CAT activity, which indicates these constructs were responsive to Li. When module A was combined with one of the repressors, the amount of CAT produced was increased. This experiment was well designed: there were constructs that contained module A but lacked repressor modules, and these constructs were not affected by Li. The bottom three series of constructs (all lacking module A) did not increase production of CAT

when exposed to Li, whether or not repressor modules were present. Therefore, Li does not activate modules G, B, or A of the cis-regulatory element; rather, it blocks the repression by modules F, E, and DC as long as module A is present.

What Controls the Timing of *Endo16* Transcription?

At this point, the investigators had established which modules promote and repress *Endo16* transcription in the appropriate location during embryogenesis (Figure 10.10 and Tables 10.1 and 10.2). Building on their success, they turned to the question of timing. Blastula formation occurs at about 24 hours and gastrula at about 48 hours. Sixty-five hours

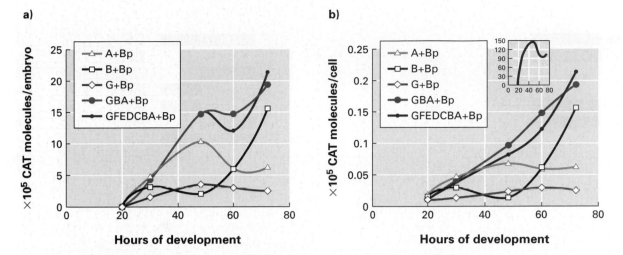

Figure 10.12 Measurement of CAT protein produced in 20- to 72-hour embryos.
a) The number of CAT molecules produced per embryo. The data points are averages of 100 embryos per time point; the smooth lines make it easier to see trends but are not intended to imply the level of CAT in times when samples were not taken. **b)** The number of CAT molecules produced per cell (the number of CAT molecules in the embryo divided by the number of cells that normally express endogenous *Endo16* at each time point). The inset graphs the number of cells (Y-axis) that normally express *Endo16* during development (X-axis).

after fertilization, the larva stage is achieved (see Figures 10.2 and 10.4, as well as the sea urchin web page, for details).

Data for the timing experiments are plotted as time courses for each DNA construct; CAT enzyme activity was measured at each time point (Figure 10.12). To provide a baseline for comparison, the investigators measured CAT activity using constructs 1, 2, 7, 8, and 10 (see Figure 10.6). From these data, we see that modules G, B, and A each promoted CAT transcription, and constructs that incorporated all three modules produced the most CAT. However, by measuring CAT production over time, we see that the three inducing modules exhibited different temporal profiles. Module A induced CAT production during the first 48 hours and then dropped off. Module B promoted CAT production primarily at the 60- and 72-hour time points (after module A had already peaked). Module G did not promote much CAT production by itself, though there was a marginal increase at around 48 hours. When modules GBA were combined, the expression level was almost equal to the *wt* cis-regulatory element containing all eight modules.

The investigators wanted to take into consideration the fact that the number of cells is changing over the same time period. They normalized their data by dividing the number of CAT molecules produced by the number of cells that typically produce Endo16p in *wt* embryos. To perform this normalization, they had to determine the number of Endo16p-expressing cells from 20–72 hours of development in *wt* sea urchin embryos (Figure 10.12b inset). The ability to count embryonic cells—one advantage of working with the sea urchin embryo—allowed the investigators

to calculate the number of CAT molecules produced per endodermal cell (Figure 10.12b). After this normalization, the whole-embryo data interpretation did not change substantially. During the first 48 hours, *Endo16* transcription was primarily under the control of module A. Module B began to exert an influence at 60 hours, when B and A equally contributed to CAT expression. By 72 hours, module A contributed a smaller percentage of the total expression, and module B had taken over control of expression. In fact, module B alone was almost sufficient to supply the entire amount produced by the *wt* cis-regulatory element at 72 hours. The little bit that B alone was lacking could be accounted for by the amount induced by module A. As seen in Figure 10.12a, module G played a trivial role by itself. It is interesting that the shape of the normalized B+Bp time course was very similar to the time course for the full-length cis-regulatory element.

Does Module G Have a Function?

To determine the role of module G in *Endo16* transcription, the investigators built DNA constructs 31 through 38 (Figure 10.13a). For these constructs, they removed the *Endo16* Bp and replaced it with a weakened viral promoter (SVp). When each inducing module was placed individually upstream of SVp, each construct displayed a different temporal profile (Figure 10.13b), though similar in shape to the analogous ones in Figure 10.12a. These profiles indicated that modules G, B, and A exerted their temporal influence on transcription without any amplification by Bp.

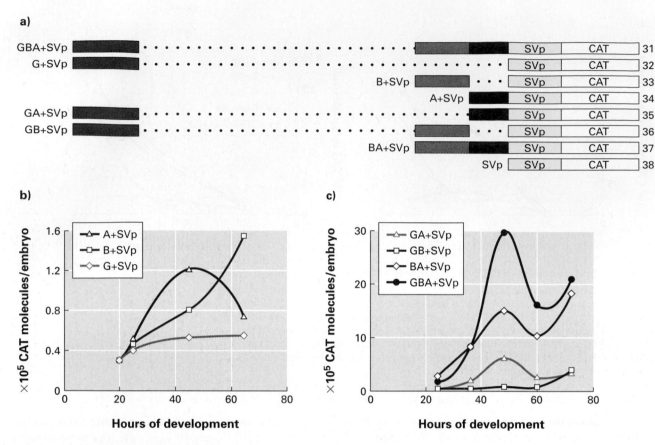

Figure 10.13 The role of module G.
a) For these constructs, the Bp was replaced with a basic promoter from the virus SV40 (SVp; same colors and symbols as in Figure 10.6). The numbering of the constructs continues from Figure 10.9. Measurement of CAT enzyme molecules produced per embryo from **b)** 20- to 72-hour embryos, or **c)** 22- to 72-hour embryos. Note the different Y-axis scales in panels b) and c), as well as the different time points sampled.

When the three inducing modules were analyzed in pairs or all together, the amplitudes and shapes of the time courses were altered. The amplitude of CAT production by module A was increased approximately fourfold when module G was added onto A+SVp, though the shape of the curve was essentially unchanged. Interestingly, module G did not alter the amplitude or shape of module B's ability to produce CAT. When modules B and A were combined, the shape and amplitude were very similar to those of GBA+Bp in Figure 10.12a. When module G was added onto BA+SVp, the amplitude was substantially increased at the 48-hour time point, which was when module A was exerting its maximum effect. Furthermore, the addition of module G (GBA+SVp) increased the output from module B at 60 and 72 hours, when module B becomes active. However, module G increased the output of module A (at 48 hours) to a greater extent than G increased the output of module B (at 60 and 72 hours). In short, module G acted as an amplifier for module A, and to a lesser extent module B, but only if B was combined with A.

DISCOVERY QUESTIONS

8. Summarize the major spatial, temporal, and amplitude aspects of modules G, B, and A.
9. Module G acts as an amplifier, but this might seem wasteful. Hypothesize why it might be beneficial to have a separate amplifier module rather than simply boosting the capacity of modules B and A directly.

Can We Draw a Transcription Circuit for *Endo16*?

Most biologists collect data until they can build a conceptual model that explains the process they study. A good model allows investigators to make predictions derived from the logical consequences of the model. These predictions can be tested in the lab or field, and one of two things will happen: The predicted behavior will be observed, or it will not. If the prediction is substantiated, the model has passed its first test. If not, the model must be refined to incorporate the new data. Predicting, experimental testing, and refining the model is a **reiterative** process that continues to improve the model. Once a model has been refined enough, it can provide understanding and insights that would not have been possible without it. These insights can lead to new areas of research and discoveries. In short, a good model is the end product of a lot of research and provides the starting place for additional research and understanding.

Circuit diagrams are one form of modeling. The use of circuit diagrams is relatively new, or, to use current jargon, not mature. We have been using forms of circuit diagrams throughout this book, and biologists have been drawing versions for a few years. However, we lack a standardized set of symbols. The issue of standardized circuit diagrams is discussed in Chapter 11, but for now, let's stick to the symbols used by Davidson and his group. Figure 10.14 shows us the "hard-wired circuitry" that has evolved in the sea urchin *Endo16* cis-regulatory elements.

The purpose of creating a circuit diagram is not to allow electrical engineers to create a copper version of *Endo16*. The real benefit of circuit diagrams such as this is their predictive power. When combined, modules B and A determine the shape of the transcription profile and module G amplifies the capacity of B and A. Early in development (up to 48 hours), regulation of *Endo16* is controlled by module A, but modules G and B can influence module A. Repressor modules F–C function by negatively affecting module A. Modules DC repress transcription in mesoderm, and modules F and E repress transcription in ectoderm. Li treatment induces "inappropriate" transcription of *Endo16* in nonendoderm cells by blocking the ability of the repressor modules to suppress transcription via module A. The circuit diagram visually summarizes the paragraph you just read.

DISCOVERY QUESTIONS

10. Use Figure 10.14 to predict whether module A would be active in ectoderm cells in the absence of modules F and E. Would module B be active in the absence of A?

11. Use Figure 10.14 to predict what proteins are present in, or absent from, endoderm cells that prevent the repression of *Endo16* by modules F–C.

12. Use Figure 10.14 to predict what would be needed for module B to become more productive later in development.

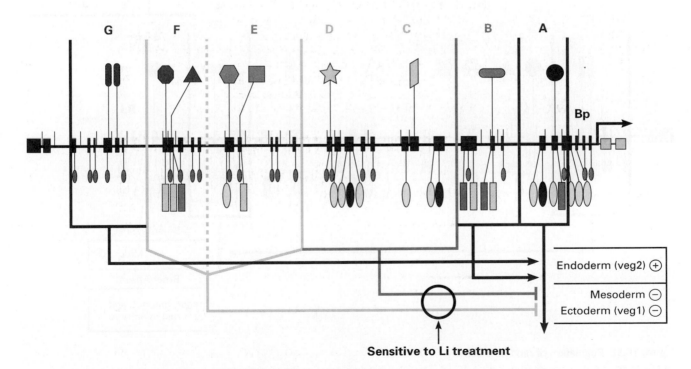

Figure 10.14 Circuit diagram of the *Endo16* cis-regulatory elements.
The modules of the *Endo16* cis-regulatory element and associated DNA-binding proteins are shown as in Figure 10.5. Modules that induce expression are denoted by purple lines; repressive modules are denoted by gray lines. The three cell types are listed to the right, and the net effect on each is shown: + indicates activation; − indicates repression.

So far, the circuit diagram explains the data we have covered, but that is to be expected since it was constructed from the same data. Does the model allow us to make any new predictions? Figure 10.15 shows the first prediction that could lead to new research and subsequent discoveries. Bmer is predicted to be one of the DNA-binding proteins that enables module B to control *Endo16*'s 60- to 72-hour location, timing, and amount specificity, with expression increased in the midgut but completely shut off in the foregut and hindgut. The gene encoding Bmer is predicted to have its own cis-regulatory elements that stimulate *Bmer* expression only in the midgut. The circuit diagram prediction also implies that there are transcription factors that bind to *Bmer*'s MG module and repressors that bind to its FG and HG modules. When produced in the appropriate cells, Bmer would activate module B, which can be amplified by module G, resulting in the production of *Endo16* at very high levels only in midgut cells beginning 60 hours after fertilization. All of these predictions are possible because of the more comprehensive understanding provided by the *Endo16* cis-regulatory element circuit diagram.

DISCOVERY QUESTION

13. This question is challenging. Use the predictive powers of the circuit diagrams in Figures 10.14 and 10.15a to predict which DNA-binding proteins are present in which cells and at what times during development. Only address the nine unique factors above the line in Figures 10.5, 10.14, and 10.15. Once you have made your prediction, write a brief summary of the research projects you would launch to test your predictions. You may use any method with which you are familiar.

Your answer to Discovery Question 13 could lead to new insights for gene regulation, animal development, and

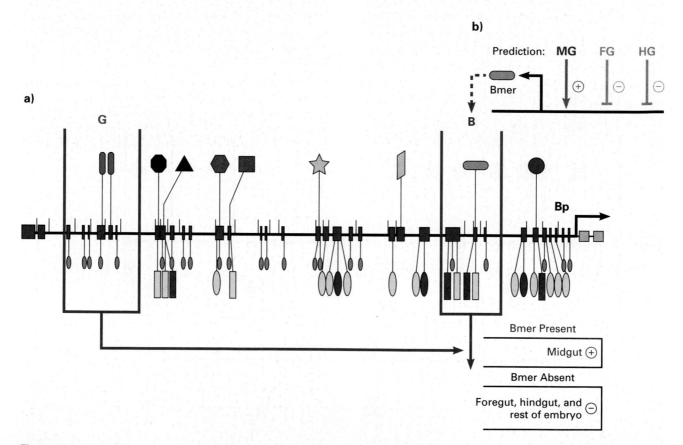

Figure 10.15 Regulation of module B activity.

a) Modified circuit diagram from Figure 10.14. The DNA-binding protein (Bmer) activates module B beginning about 60 hours after fertilization. Module B is amplified by module G to transcribe *Endo16* at a high rate only in midgut cells of the larva. **b)** Expression of Bmer is controlled by its cis-regulatory elements, which are composed of three modules: MG is an activator in midgut cells; FG is a repressor in foregut cells; HG is a repressor in hindgut cells. The dotted arrow indicates that Bmer would bind to module B of the *Endo16* cis-regulatory element to control the expression of *Endo16* in midgut cells.

genome control. The experiments you devise could be developed into research projects for publication, since you generated new hypotheses that have not been experimentally tested. You might be surprised that you could conduct genomic research with pencil, paper, and circuit diagrams, but one aspect of biology unaffected by genomics is that good research always starts at the desk and not the bench. Good ideas drive research. Good technology allows good ideas to be realized, but quiet, contemplative time is the catalyst behind scientific progress.

It seems that Yuh and Davidson and their colleagues must create time for contemplation, because their watchmaker-like dissection of *Endo16* continued. They looked at the prediction about the Bmer protein and decided to determine which DNA-binding proteins did what jobs. There are 15 different DNA-binding proteins that bind with high specificity to the cis-regulatory elements of *Endo16*. However, the investigators could not possibly analyze all 30-plus binding sites using over 15 DNA-binding proteins, so they decided to start with only one module. Which module would you choose?

Yuh and Davidson decided to start with module A (Figure 10.16). Why? As they explained in a paper, "Module A interacts with all of the other *Endo16* cis-regulatory elements and is either absolutely required for their operation or synergistically enhances their output. Moreover, it serves as a central switching unit, acting according to inputs from the other modules." They decided to start with the most critical module, where they were likely to see an effect when the system was perturbed.

What Makes Module A So Special?

Let's zoom in for a closer look at the DNA sequence and protein binding sites for module A (Figure 10.17). Some of the proteins that bind to and facilitate the actions of the eight modules (Bp, and G–A) have been identified. Of those given a name, some have been better characterized

than others. For example, protein "C" is a cAMP-responsive element-binding (CREB) protein family member, which has been shown to play a role in memory formation in flies and mammals, among other functions.

The functions of module A are mediated through interactions at eight different sites where at least four different proteins bind. Only two of these four genes have been cloned and sequenced. *Otx* is a member of the orthodenticle transcription family. *Otx* is a shorthand way of saying orthodenticle (*O*) transcription (*tx*). The other cloned gene is called *SpGCF1*, and its encoded protein binds many places in the eight modules. *Sp* stands for *S. purpuratus*, the Latin name for this species of sea urchin, and GCF1 represents the first factor to be cloned that binds to GC-rich DNA. SpGCF1p binds once in module A, twice in Bp, and appears to weakly stimulate transcription. Because SpGCF1p had been characterized before, Yuh and Davidson decided to study other proteins that bind to module A.

The researchers' challenge was to determine the roles for the remaining DNA-binding proteins in module A. As mentioned earlier, Otx protein binds to module A and nowhere else in *Endo16*'s cis-regulatory elements. The four binding sites labeled CG1, CG2, CG3, and CG4 (see Figure 10.17) are all bound by the same transcription factor, but the function of each binding site was unknown. Finally, two proteins called P and Z also bind only once in module A. Yuh and Davidson decided they needed more molecular tools for this job, so they created a new series of mutant cis-regulatory element DNA constructs and performed experiments similar to those described previously.

How Do Module A-Binding Proteins Work?

Like all good scientists, Yuh's group began their next steps where they ended previously. The investigators spliced together different portions of the cis-regulatory elements and placed them upstream of the Bp and the CAT reporter gene. These constructs were injected into developing

Module A functions:

Veg2 expression in early development:

Synergism with modules G and B enhancing endoderm expression later in development:

Repression in ectoderm (modules F and E) and mesoderm (module DC):

Modules F, E, and DC neutralized by Li treatment:

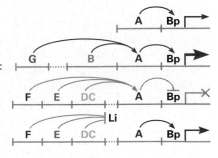

Figure 10.16 The roles of *Endo16* cis-regulatory element modules.
Module A communicates the output of all upstream modules to the Bp. Arrowhead denotes activation; ⊥ denotes repression; X denotes lack of transcription.

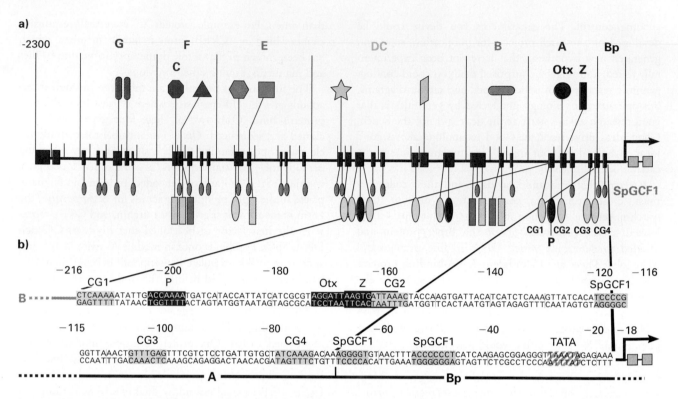

Figure 10.17 Module A and Bp showing protein binding sites.
a) Several of the module-specific DNA-binding proteins have been named. CG1–4 indicate that the same protein binds to four different sites. Note that all the small purple ovals below the line are copies of SpGCF1. **b)** The nucleotide sequence of module A and Bp. The sequence shown for module A starts at base −216, which is the 5′ edge of the CG1 binding site. Module A sequence continues onto the next line and terminates with the black vertical line at base −68 (5′ edge of a SpGCF1 binding site). Bp begins at base −67 and extends to base −1, though the sequence in this figure stops at base −18. The sequences to which DNA-binding proteins bind are enclosed in colored boxes.

embryos and CAT enzyme activity was measured (reported as the number of CAT enzyme molecules produced per embryo). When modules B and A work together, their output is a synergistic product of modules B and A assayed separately (Figure 10.18a). Interestingly, the output of B+A is exactly 4.2 times the output of module B alone! This numerical insight provided the investigators with a new perspective and new questions. First, why does the combined B+A output have the same shape but greater magnitude than the output from B alone? Why doesn't B+A equal the sum of individual parts B and A when measured in isolation? (Historical side note: This should sound familiar. "Why do wrinkled and smooth parental pea plants fail to produce mildly wrinkled peas?" Though Gregor Mendel's and Davidson's questions addressed very different aspects of genetics, it is interesting to see that the best questions never get too old to be asked again in new situations.) Second, what specific DNA-binding sites in module A are responsible for the 4.2-fold increase in the output of module B, as seen in Figure 10.18a?

To answer their questions, the investigators focused on the CG1 and P sites in module A (see Figure 10.17b). When either CG1 or P was mutated, the output dropped by about 50%, and the construct no longer had a 60- to 72-hour increase in output (Figure 10.18b). In fact, the BA constructs with either CG1 or P mutated looked very similar to module A alone (Figure 10.18c). Module A, however, was not affected by either mutation when compared directly (Figure 10.18d). What does all this mean? The investigators drew several conclusions from these data, as well as additional data not shown here:

1. Both CG1 and P sites are needed to provide the full 4.2X module B production of CAT (Figure 10.18b).
2. In module A, the CG1 and P sites have no other function since transcription from module A is unaffected by these mutations (Figure 10.18d).
3. In experiments not shown here, the investigators tested all other binding sites in module A and found that no other sites in module A had any effect on the output

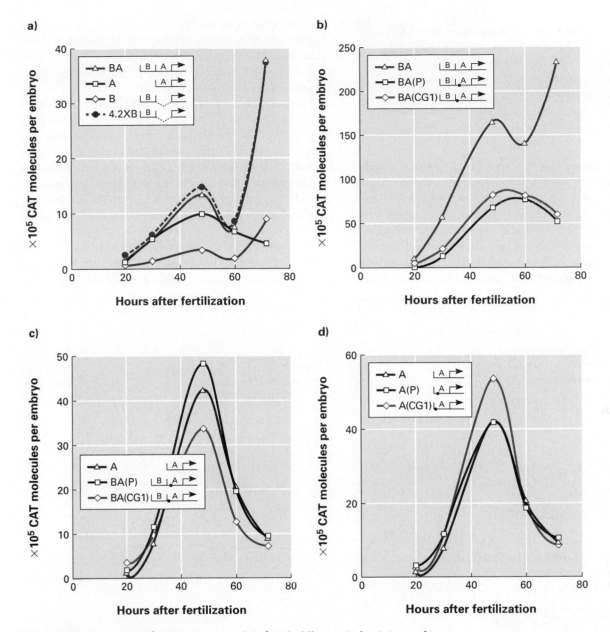

Figure 10.18 CAT output (molecules per embryo) with different *Endo16* cis-regulatory element DNA constructs.

In the panel legends, dots within a module indicate the location of mutated P or CG1 sites. The name of each construct contains the mutated site in parentheses. The dotted line in **a)** was produced by taking the time course of module B alone and multiplying it by 4.2; thus, it is a theoretical time course. All other time courses were experimentally derived.

of module A when studied in isolation (similar to Figure 10.18d).

4. Module B does not communicate directly with Bp, but must interact with CG1 and P of module A. This conclusion is based on the fact that none of the BA constructs with mutated CG1 or P have curves that rise late, the way module B does when studied in isolation (Figure 10.18a–c).

DISCOVERY QUESTIONS

14. In the preceding four-point summary, an assumption was made to formulate number 1. Design an experiment that would test directly whether there is a different effect when neither CG1 nor P sites are functional.

15. Look at Figure 10.17b and note where the CG1 and P binding sites are located. If both binding sites must be occupied to produce the 4.2X effect, propose a mechanism that explains why both transcription factors must be present. Be sure to consider the proximity of their binding sites.

In short, module A functions as a **bimodal toggle switch**. Either module A alone determines the output of *Endo16* (first 48 hours), or module B acts on the CG1 and P sites (beginning at 60 hours) to determine transcriptional output. In other words, module A toggles from complete control of gene productivity to complete submission of control to module B. When module B is activated (by the predicted Bmer protein; see Figure 10.15), module A acts

as a multiplier of module B's output. These interpretations beg the question: Which binding sites in module A determine its output in the absence of module B input (i.e., during the first 48 hours of development)? Can you design experiments to answer this question?

Many more constructs were needed, but only those involving Otx and Z sites are presented here. Notice from Figure 10.17 that the Otx and Z sites abut one another. For these experiments, the investigators produced various mutations in the Otx and Z DNA sequences to block the binding of the appropriate transcription factors. The different DNA constructs were placed into sea urchin embryos, and the amount of CAT enzyme produced was measured for each time point.

For comparison, the ability of module A alone to produce CAT was graphed (Figure 10.19a). When the Otx binding site was mutated, module A was unable to promote

a)

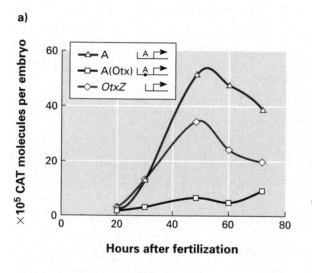

b)

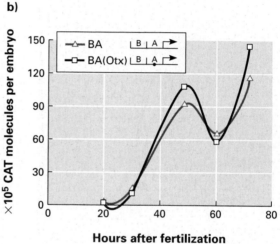

c)

Constructs	Endoderm
Endo16	97.0%
A	78.9%
A(Otx)	5.1%
OtxZ	56.5%
(Otx)Z	3.6%
Otx(Z)	72.0%

Figure 10.19 Role of Otx site in the output of module A.
a) Time-course expression of CAT protein using different versions of module A as the promoter. Panel legend uses dots to indicate the location of each mutated site, and the site inside the parentheses has been mutated. *OtxZ* (in italics) indicates that an 18 bp fragment of Otx and Z DNA was the only portion of module A upstream of Bp. **b)** Time-course expression of CAT protein using modules B+A or B + mutated A as the promoters. **c)** The percentage of embryos that exhibited CAT expression in the endoderm for each construct. Parentheses denote sites that were inactivated. To inactivate the Otx site, the sequence **ATTA** was changed to **GCCG**; Z was inactivated by changing **TGATTAA** to **CAGCCGG** (see Figure 10.17b). *OtxZ* indicates that an 18 bp fragment was used as in panel a).

CAT production. Interestingly, when only an 18 bp section of module A was used (*OtxZ*), transcription of CAT was recovered. It is surprising that a mere 18 bp fragment, which represents only 10% of module A, was able to produce a *wt* temporal profile with amplitude about half that of full-length module A. Using the results shown in Figures 10.18 and 10.19, with the circuit diagram in Figure 10.14, the investigators made a prediction and designed an experiment to test it. Prediction: If module B works through module A via the CG1 and P sites, then mutating Otx should not affect the ability of A to function as a conduit for module B.

When B was combined with *wt* A, or A with a mutated Otx, there was no difference in output (Figure 10.19b). These data confirm the prediction that module A uses only sites CG1 and P to amplify B; Otx does not play a role in this amplification. However, Otx was the critical site to determine the output of module A in the absence of module B (Figure 10.19a). The summary table in Figure 10.19c shows how important Otx is for module A to function properly. Especially noteworthy are the experiments performed on the *OtxZ* 18 bp fragment, which was able to produce about half as much CAT as full-length module A. When module A contained an inactivated Otx, CAT production dropped 15-fold. In contrast, when Z was inactivated, the productivity of the 18 bp fragment was enhanced so that it stimulated transcription better than the *OtxZ wt* fragment did.

DISCOVERY QUESTIONS

16. Look at Figure 10.17b and locate the Otx and Z binding sites. Formulate a hypothesis to explain why the *Otx(Z)* fragment promoted better than the *OtxZ* fragment.

17. Look at Figure 10.4 and predict the expression pattern for *Endo16* in an embryo that has a mutated Otx site. Remember that blastula are formed in 24 hours, gastrulation occurs at about 48–60 hours, and a larva is about 72 hours old.

Which Module A Sites Respond to Repression by Modules DC, E, and F?

Determining how a repressor interacts with module A required the use of Li again. In normal sea urchin embryos, Li causes the loss of repression in ectoderm and mesoderm cells that normally repress *Endo16*. A large set of experiments was needed to tease out the parts of module A important for repression. To minimize the number of variables, the investigators decided to use only module F as their test case for a repressor module; F normally represses *Endo16* in ectodermal cells. For these experiments, some new DNA constructs were built, and some constructs were used from earlier studies. For each construct tested, the

amount of CAT enzyme produced in Li-treated embryos was divided by the amount of CAT enzyme produced in untreated embryos. The ratios (Li-treated ÷ untreated) were displayed as bar graphs. If the Li-treated embryos produced the same amount of CAT enzyme as untreated embryos, then the ratio would be 1 (Figure 10.20). If the DNA construct was sensitive to Li, then the bar graph would exceed 1, because more CAT would have been produced per embryo when exposed to Li. Again, the goal was to determine which parts of module A interacted with the repressor modules. DNA constructs whose ratios exceeded 1 contained functional repressors, while those with ratios close to 1 did not.

The clarity of the results is impressive and should help you appreciate the quality of this research. The first constructs (GBA and GFBA) show how the system worked when repressor F was interacting normally with module A and Li was present. Note that the ratio of CAT expression increased dramatically when modules F and A were in the same construct. The next six constructs revealed that module F interacted with module A, and not modules G or B, to produce the Li response.

The final seven constructs in Figure 10.20 focused on smaller subregions within modules A and F. The

CAT Activity per Embryo: Li/Untreated

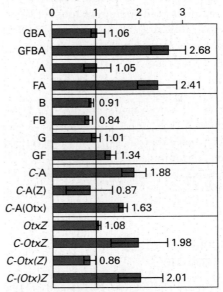

Figure 10.20 Repressor-sensitive portion of module A.
Relative amounts of CAT produced are displayed as bar graphs. Each bar graph represents a ratio of CAT enzyme produced in the presence of Li divided by CAT enzyme produced in the absence of Li. Bars that extend above 1 indicate that more CAT was produced in the presence of Li and therefore repressor module F was working properly. 48-hour-old embryos containing each construct were used to detect Li sensitivity. Otx and Z constructs use the same nomenclature as in Figure 10.19. *C* is a binding site in module F and stands for CREB, which is a DNA-binding protein.

nomenclature for the next set of constructs calls for a bit of clarification. The letter C stands for the CREB binding site within module F (see Figure 10.17a). When only the C portion was used upstream of Bp+CAT, an italic *C* was used in Figure 10.20. When *C* was paired with A (e.g., *C*-A), there was a strong Li effect (ratio larger than 1). When the Z site in module A was inactivated (*C*-A[Z]), the Li effect was abolished. When the Otx site in module A was inactivated (*C*-A[Otx]), the Li effect was retained. Even more impressive were the last four constructs, in which both the A and F modules were reduced to short DNA fragments and the rest of the modules were omitted. *OtxZ* alone was insensitive to Li treatment, as was the entire A module. *C-OtxZ* produced an effect similar to the much larger FA construct. When the Z site (*Otx[Z]*) was mutated, the Li effect was lost, but mutating the Otx site (*[Otx]Z*) did not reduce the Li effect at all.

It is worth noting the length of the DNA pieces used for these experiments. The combination of Otx and Z sites from module A consisted of 18 bp (`AGGATTAAGTGATTAAAC`). When Otx was mutated, the first `ATTA` was changed to `GCCG`, while the Z site was mutated from `TGATTAA` to `CAGCCGG` (see Figure 10.17b to locate Otx and Z sites). It's amazing how simple changes in "junk DNA" produce dramatic consequences. Z facilitates repression from module F (and perhaps other repressor modules) onto module A and is responsive to Li treatment. The Otx site facilitates transcription regulation from module A during the embryo's first 48 hours, but is unresponsive to Li since the teratagen affects *Endo16* by blocking repression.

DISCOVERY QUESTIONS

18. Summarize the data from Figure 10.20.
19. What assumptions were made by using only module F for these experiments? What were the practical reasons for these assumptions?
20. Are there any parts of module A that remain to be dissected?

How Does Module A Interact with the Basic Promoter?

If you compare the output in Figures 10.12a and 10.13a, you can see that the different basic promoters (Bp vs. SVp) did influence transcription. We know that module A is the conduit through which all others communicate with Bp. Therefore, Davidson's lab wanted to determine which portions of module A were responsible for communicating with Bp. In Figure 10.17, you can see that most of the module-A binding sites have been experimentally tested. The only three remaining are CG2, CG3, and CG4. Therefore, it is not surprising that the final experiments focused on these three sites.

The investigators conducted experiments in which different cis-regulatory element constructs were placed upstream of the CAT-coding DNA and injected into sea urchin embryos. The amount of CAT enzyme was measured at different time points. When modules B+A were used, we see the standard output. The deletion of the CG2 binding site, however, reduced the amplitude by about 50%, though the overall shape of the output was retained (Figure 10.21a). Thus, the CG2 site plays an important scaling role; that is, it influences amplitude of the output. When each of the other two CG sites were mutated and assayed in conjunction with module A but not module B, all constructs retained the same shape, but the amplitude was always reduced by half (Figure 10.21b). Binding sites CG2, CG3, and CG4 were important to module A's ability to multiply the output by a factor of 2. These final experiments completed the dissection of module A, since Davidson's lab had already found the roles played by binding sites Z, Otx, CG1, and P.

DISCOVERY QUESTIONS

21. Look carefully at Figure 10.17 and find the sequences altered in Figure 10.21. Describe fully the mutations generated in these constructs. Are any other binding sites altered in addition to the CG sites? If so, what effect should this have on the output of A+Bp constructs?
22. Do the locations of the CG binding sites relative to other binding sites in module A suggest any models for controlling transcription? Could these models lead to testable predictions?
23. Predict what would happen if the entire *wt Endo16* cis-regulatory element was placed upstream of the CAT sequence but CG2 was nonfunctional. Draw a graph (use Figure 10.12a as a template).
24. Create a table that summarizes the roles of binding sites Z, Otx, CG1, CG2-4, and P.

We have spent a significant amount of time and effort discovering how cis-regulatory elements work to control the expression of a single gene in the gut of a developing sea urchin. You might be wondering why. Remember, model organisms tell us a lot about all species when we study fundamental processes. Yeast taught us about cancer, and worms taught us about genomic tradeoffs in aging. The sea urchin has taught us about genomic regulation of development. Genes needed early in development are tightly regulated, and what appear to be redundant components (e.g., four CG sites) can play unique roles. We learned that *Endo16* contains three activator modules (G, B, and A) and three repressor modules (F, D, and C). Based on the experimental evidence, each DNA-binding site within a module

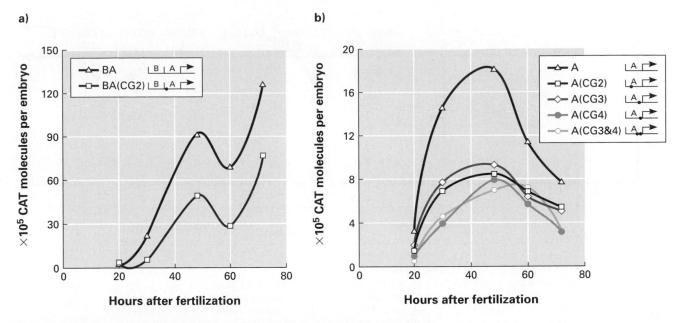

Figure 10.21 Role of CG2, CG3, and CG4 on interaction between module A and Bp.
To perform these experiments, **TCTAGA** was inserted in place of the
following nucleotides: CG2 = **TGATTAAACT**; CG3 = **TGTTTGAGTTT**;
CG4 = **ATCAAAGACAAAGG**. CAT activity was used to determine the number of
molecules per embryo. In the legend, dots denote the location of mutated sites, which
are shown within parentheses in the construct names.

has its own function. When the appropriate combinations of proteins bind to their appropriate sites, the downstream gene will be expressed when it is needed, where it is needed, and in the appropriate amount. Multiply this level of gene control by 22,287 annotated human genes and you begin to understand the level of control needed in each of your trillion cells to produce dynamic proteomes in each cell. Furthermore, cis-regulatory modules raise many evolutionary questions about their origins, and how they might be duplicated and used by other genes that require similar regulation. As more genomes become sequenced, we will be able to trace the evolution of individual cis-regulatory elements. Therefore, during the **postgenomic** era we will continue to mine genome sequences to understand genes and evolution.

Are Genes Hard-Wired?

The results presented in this chapter were the culmination of years of work performed on a scale never attempted by any other lab. It is worth remembering that these experiments used DNA constructs not found in sea urchin; the constructs were engineered and placed in *wt* embryos. The investigators assumed that the injected embryos expressed normal amounts of the DNA-binding proteins, and that the concentration of these proteins determined the regulation of gene expression. Given the need to perturb a system in order to understand it, all scientists must live with

similar assumptions, though it is worth remembering how the data were produced.

After thousands of experiments, we now have enough information to assign a role to each of the eight modules (G–A and Bp), as well as some specific binding sites. Module A is the central hub that performs a type of computational logic to determine where, when, and how much a gene should be transcribed (see Figure 10.16). Some functions of the genome are hard-wired into the DNA, with some of the instructions located in noncoding regions. Calling noncoding DNA "junk DNA" is equivalent to considering the keyboard and monitor the only important parts of a computer.

Throughout this book, we have found the ability to visualize data nearly as important as data collection. By using the term "hard-wired," Davidson created a mental image that everyone can understand. A vending machine that dispenses cold sodas is hard-wired; put the money in, push a button, and a soda appears. If you were to open the vending machine, you would see a series of wires connecting functional units (money counter, buttons, switching mechanism that chooses which soda to dispense, etc.). When a hard-wired machine is built, the electrical engineer draws a circuit diagram showing how each functional unit is connected to accomplish the desired output. Davidson wanted to use similar electrical engineering principles to capture all the information learned about the transcriptional control of *Endo16* (Figure 10.22).

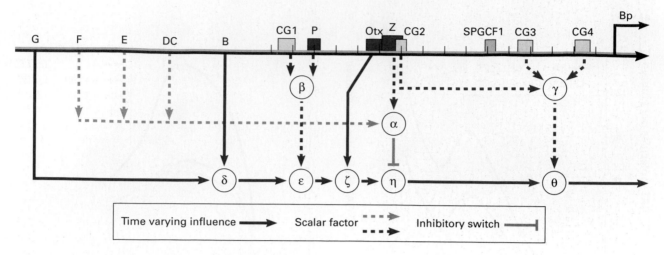

Figure 10.22 Circuit diagram for *Endo16* transcription.
Individual binding sites within module A are indicated by labeled boxes above the double line representing the DNA. The interactions of the upstream modules with elements of module A are indicated by circles and arrows beneath the DNA. Each labeled circle, or node, represents a specific regulatory interaction. Two types of regulatory influence are indicated: time-varying interactions (solid lines) that determine the temporal pattern of *Endo16* expression, and location/amount interactions that affect the level of *Endo16* transcription by constant scaling factors. Note that the scaling factor of the repressor modules determines which cells will repress *Endo16*.

To understand *Endo16*'s circuit diagram, think of each module or binding site as a source from which wires extend. When two wires meet at a node, a small transistor is located to perform a simple calculation. The overall flow of information in the circuit moves from left to right. The outcome of the cis-regulatory elements at the far right is analogous to a soda appearing after you have made your selection. For *Endo16*, the outcome of this circuit is the initiation of transcription at the right time, in the appropriate cells, to the correct level of RNA production.

Each node performs a calculation, and we need to understand each input and how it affects the output of each node. Node α represents the repressive action of modules F, E, and DC mediated through site Z. Node β receives input from binding sites P and CG1 to amplify the output of module B twofold. Node γ receives input from binding sites CG2–CG4 to produce a twofold amplification signal. Node δ receives positive input from modules B and G and produces a positive signal. The input from modules G and B are time dependent in that they become active later in development. Node ε receives input from nodes δ and β and produces the product of the two input nodes. Node ζ (zeta) receives input from Otx that is critical to the ability of module A to communicate with Bp early in development. Node ζ is also affected by node ε, which is regulated by the stage of development. Therefore, node ζ is capable of receiving information early (Otx) and late (G and B) in development to produce a signal that is a toggle switch. Node ζ will produce an outgoing signal that is driven either

by module A (via Otx) or by module B (and module G amplifies module B). The toggle switch of node ζ is controlled by the timing of protein production (e.g., the hypothesized "Bmer," which activates module B).

Once the developmental timing toggle switch ζ has determined whether module A or B is in control of *Endo16* transcription, a few more inputs are needed before the final level of transcription can be determined. One important input is whether a particular cell will transcribe *Endo16*. Node α receives its input of repression and sends a signal to node η (eta). Node η also receives a signal from node ζ, promoting transcription of *Endo16*. The decision made at node η is whether to pass along the positive signal, leading to transcription, or the negative signal, causing transcription to be blocked. Each cell must make its own determination as to whether to transcribe *Endo16*, and the decision is determined by the location of the cell. If the cell is located in the endoderm, then no repression signal reaches node η and *Endo16* is transcribed. If the cell is located in the mesoderm (modules DC) or ectoderm (modules F and E), then node α produces an inhibitory signal that negates the input from node ζ, the same way that pushing the money return button on a vending machine stops the transaction. Endodermal cells do not inhibit node η, and thus the original input from either module A or B is passed on to node θ (theta), which also receives input from node γ. If a positive signal is received from node η, the signal will be amplified by node γ. The final signal transmitted to the basic promoter (Bp) is determined by node θ. The final outcome from θ determines the timing, location, and amount of *Endo16* mRNA produced in each cell and thus helps shape the outcome of the developing embryo.

Models (e.g., circuit diagrams) are ways of displaying information so it can be more readily understood. Compare Figure 10.22 to the table you produced (Discovery Question 24) that lists the functions for the DNA binding sites. Your table contains the same information but is too limited to reveal all the intricacies of *Endo16* gene regulation. For example, what would happen if CG1 and P were nonfunctional? Your table probably indicates that the production of *Endo16* would be reduced by half, but from the circuit diagram we see that this is only true later in development, since node β does not influence node ζ if the signal from node δ is lacking. From the circuit diagram, we can better understand the interactions of the different modules and binding sites. However, the circuit-diagram visualization of *Endo16* regulation is not quantitative, so Davidson produced a different model to supplement the circuit diagram.

Do Genes Contain Miniature Computer Programs?

Most students of biology like to think of genes as small factories that produce RNA, but we often gloss over the issue of quantity. How does a gene know how much RNA to produce in each cell? How does the gene know when to increase its productivity when the cell needs it? As with any factory, there must be a way to regulate the rate of production.

However, the circuit diagram lacks quantitative information, so let's look at a computational model that allows you to make quantitative predictions (Table 10.3). Davidson wrote a simple computer program that summarized in seven steps what we know about *Endo16* gene regulation.

If you have ever written any computer code, you will be familiar with the style and logic of this quantitative model. If writing code is new to you, you might not recognize immediately the logic within the code. In fact, the two sentences you just read illustrate the logic in Davidson's model. Notice in Table 10.3 the pattern of several lines as "if . . . else . . ." conditional statements. These conditional statements indicate that two possible options exist. For this paragraph, either you have written computer code or you have not, and there is a consequence for each of the two possibilities (familiarity or no familiarity). As we walk through the *Endo16* computational model, we will examine several conditional statements and discover the consequences for each of the two possibilities. The model consists of seven steps and applies the following logic:

1. If any of the three repressor modules is activated (in mesoderm or ectoderm) and binding site Z is functional, then the consequence will be an activated ($\alpha = 1$) node and the transmission of a repressor signal to η, which will result in no input at θ (i.e., *Endo16* will be repressed). Alternatively, the node will be inactive ($\alpha = 0$), the inducing signal from ζ will be transmitted to η and θ, and *Endo16* will be transcribed.

Table 10.3 *Endo16* regulatory algorithm.

Variables correspond to nodes in the circuit diagram of Figure 10.22. The logic is in the left column and explanations are in the right column. Time, location, and amount of *Endo16* transcription are described in these seven steps.

Logic	Explanation
1. if (F = 1 or E = 1 or DC = 1) and (Z = 1) $\alpha = 1$ else $\alpha = 0$	Repression functions of modules F, E, and DC mediated by Z site
2. if (CG1 = 1 and P = 1) $\beta = 2$ else $\beta = 0$	Both CG1 and P are needed for synergistic link with module B
3. if (CG2 = 1 and CG3 = 1 and CG4 = 1) $\gamma = 2$ else $\gamma = 1$	Final step up of system output
4. $\delta(t) = B(t) + G(t)$ $\varepsilon(t) = \beta \times \delta(t)$	Positive input from modules B and G Synergistic amplification of module B output by CG1-P subsystem
5. if ($\varepsilon(t) = 0$) $\zeta(t) = Otx(t)$ else $\zeta(t) = \varepsilon(t)$	Toggle switch determining whether Otx site in module A or module B will control level of activity
6. if ($\alpha = 1$) $\eta(t) = 0$ else $\eta(t) = \zeta(t)$	Repression function inoperative in endoderm but active elsewhere
7. $\theta(t) = \gamma \times \eta(t)$	Final output communicated to Bp

Source: Yuh et al. 1998. *Science.* 279: 1901, Figure 6B only.

2. If both amplifiers CG1 and P are on, then β is 2, which represents its twofold enhancement of output from modules G+B. Alternatively, β is turned off if either CG1 or P is nonfunctional, and module B is unable to transmit its signal on to Bp. In this case, module A resumes its control over *Endo16* expression.

3. If the amplifiers CG2–CG4 are functioning, then γ is 2, which represents its twofold enhancement of transcription. Alternatively, γ is neutral and there is no additional enhancement of *Endo16* transcription.

4. Time of development controls module B activation. Node δ output depends on the time-dependent activation of modules G and B, which are regulated by upstream proteins such as Bmer described in Figure 10.15b. Module B assumes control during gastrulation about 60 hours after fertilization. Node ε is the product of δ and either 2 or 0 (see step 2) depending on the state of CG1 and P as described in the right column.

5. The toggle switch determines whether module A or module B governs the transcription of *Endo16*. If ε is not yet activated (i.e., β = 0 in step 4 because it is too early in development), then ζ equals the time-dependent output from the Otx binding site in module A. Alternatively, ζ equals the output from ε (i.e., β = 2 in step 4), which is governed by module B.

6. Transcription response of the entire gene is governed by the repressors described in step 1. If the repressors are activated (i.e., α = 1), then *Endo16* is not transcribed. Alternatively, the repressors are silent (i.e., α = 0) and the signal for transcription at node η equals the input from the toggle switch at node ζ (calculated in step 5).

7. At any given time, the instructions produced by θ and passed on to the basal promoter are the product of γ (either 2 or 1 from step 3) and η (calculated in step 6).

To predict the amount of *Endo16* transcribed, all we need to know is the experimental values for G+B in step 4 and the output of Otx in step 5. Steps 1–3 determine the amplitude and location of transcription; step 5 determines which module is governing transcription, depending on timing. Given these numerical values, the final outcome for *Endo16* is as certain as putting money in a vending machine and pushing the button.

Math Minute 10.1 **How Do You Make a Computer Understand Gene Regulation?**

In Table 10.3, the cis-regulatory element logic for determining *Endo16* transcription level [i.e., θ(*t*)] is stated in the language of everyday words. If you know the values of all the input variables [F, E, DC, Z, CG1, P, CG2, CG3, CG4, B(*t*), G(*t*) and ε(*t*)], you can follow the logic and calculate the resulting value of θ(*t*). However, because transcription of *Endo16* changes with time, you would have to calculate θ(*t*) for many different values of *t* to understand the dynamics of *Endo16* gene regulation. By expressing the logic in a programming language, you can let the computer do all the calculations, allowing you to focus on interpreting the results. This Math Minute explores how the Perl programming language can be used to express the logic in Table 10.3.

Open the Endo16 Perl script, and compare it to the logic in Table 10.3. Conceptually, they are almost identical, yet a computer could not understand Table 10.3. Let's walk through the details that let computers understand this Perl script. The computer keeps track of variables in a Perl script by assigning them names that begin with a dollar sign ($). Therefore, you might call a variable F in Table 10.3, but use $F in the Perl script. Greek letters are not allowed as variable names, though you can spell out the Greek letters to match the variable names listed in Table 10.3. Punctuation is an important part of a Perl script. For example, the parentheses and curly braces "{" and "}" must be in exactly the right place in "if . . . else" statements, and semicolons must be at the end of statements except those that end with a curly brace. Mathematical operations are represented in special ways in Perl. For example, you represent multiplication with an asterisk.

To assign a value to a variable, put the variable on the left side of an equal sign, and put the value on the right (e.g., $alpha = 1). A variable can also be on the right side of an equal sign, in which case the variable on the left is given the current value of the variable on the right. The variable on the right must be assigned earlier in the script. For example, the statement "$zeta = $epsilon" (line 22 of the Perl script; second half of number 5 in Table 10.3) assigns $zeta to have the same value that $epsilon has. The value of $epsilon was assigned in line 17 of the Perl script and the second half of number 4 in Table 10.3.

Variable assignment statements look like mathematical equations, but all they do is define variable values. In contrast, a double equal sign "$==$" is similar to a true/false question that is used to determine whether a variable has a particular value. (Note that "eq" is used in the same way as "$==$" for text variables such as $cg1.) For example, the phrase "$epsilon $==$ 0" is true if $epsilon has the value 0, and false otherwise (see line 19 in the Perl script and number 5 in Table 10.3). The "if . . . else" statement (lines 19–22 of the Perl script; number 5 of Table 10.3) uses the answer to the true/false question to decide whether to set $zeta to the value of $otx (if true) or to the value of $epsilon (if false).

A few more lines of Perl are needed to set the initial values of all the input variables in the Endo16 script. Perl runs on Windows, Mac OS X, and Unix workstations, so you could run the script on your own computer if you added lines to input the variable values. An alternative way to run a Perl script is through a web server. The Endo16 model web page is a form that sends inputs into the Endo16 Perl script. When you press "Submit," the web server executes the Perl script and returns the computed *Endo16* transcription level in the form of a web page.

By experimenting with different values for the input variables, investigators can test and refine the logic in Table 10.3 to create a more accurate model for *Endo16* transcription. A computer program such as the Perl script we have explored makes model refinement fast and free of calculation errors, enabling more efficient research into *Endo16* gene regulation.

LINKS   G‡P
Endo16 model
Susan Ernst

MATH MINUTE DISCOVERY QUESTIONS

1. What input variable values regulate *Endo16* transcription in ectoderm? What input variable values are irrelevant for modeling *Endo16* transcription in ectoderm or mesoderm?

2. Use the Endo16 model web page to determine *Endo16* transcription level after 24, 48, and 60 hours of development.

3. Use the Endo16 model web page to estimate the maximal *Endo16* transcription level while transcription is under module A control, and while transcription is under module B control. At approximately what times do these maximum values occur?

Davidson's lab has extensively dissected the cis-regulatory elements (2,300 bp) of *Endo16* that control when, where, and how much RNA will be transcribed. (The amount of mRNA is not controlled by the cis-regulatory elements; mRNA processing undergoes its own separate regulation.) You should appreciate that this 2.3 kb piece of DNA is but one part of one gene within one genome. There are cis-regulatory elements for every gene in every genome. The degree of transcriptional control in *Endo16* may appear to be sufficiently complex to explain all gene regulation, but we have not yet dissected the genome's full complexity. Susan Ernst and her colleagues at Tufts University reported that *Endo16* RNA can be alternatively spliced in three different ways to produce three different mRNAs and two different proteins. Two of the three mRNAs encode for the same amino acid sequence but have different 3′ untranslated regions. The longer form of the protein has additional calcium binding sites, which produce a different conformation

of the total protein and presumably a different function. All this *Endo16* variation means that understanding the circuitry of cis-regulatory elements is only the beginning to understanding the production of a protein.

DISCOVERY QUESTIONS

25. In Chapter 8, we discovered how proteins can behave stochastically. Is stochastic behavior possible in genes such as *Endo16*, or is its transcription identical in every endoderm cell?

26. If you isolated a mutant strain of sea urchin that lacked the *Bmer* gene, what effect would this have on *Endo16* transcription? What would the consequences be if a fertilized egg lacked a Z binding site? Refer to Figure 10.22 and Table 10.3 to help you answer these two questions.

Summary 10.1

To survive, an embryo must control when, where, and how much RNA a gene will produce. *Endo16* is regulated by the activity of its cis-regulatory elements, which work as if they were hard-wired like a vending machine. If you insert money and push the button, a soda of your choice appears. Your ability to control the vending machine is a consequence of its hard-wiring; every time you insert money you can predict the outcome. It is almost shocking to think of your genes, perhaps all genes, as being regulated by hard-wiring that can be diagrammed as electrical circuits or programmed like simple computer algorithms. We studied *Endo16* so you can begin to appreciate the complexity and predictability inherent in the transcriptional control of any genome (Chapters 11 and 12). Imagine what the circuit diagram would look like for 22,287 human genes. In Section 10.2, we will interact with two Java-based models that are prototypes for visualizing information, as an introduction to new ways of communicating genome regulation.

10.2 Integrating Single-Gene Circuits

Endo16 is an excellent model gene and has been very informative, but the complexity of whole genome regulation is too overwhelming to diagram as simple circuits. We need to revisit the problem encountered when considering protein-protein interaction maps. How can we best display complex circuits? It might seem a trivial problem, but imagine looking at a circuit diagram for 22,000 genes, or a computer program with enough lines of code to describe the development of a human embryo. Genomic information is accumulating faster than ever, and we need new tools to visualize all of it simultaneously. A group in Russia has produced some public-domain, interactive diagrams that we will use as a first step toward comprehensive models of genomic circuitry.

How Can We Describe to Others What We Know about a Genome Circuit?

We want to develop computer models that will enhance our understanding and lead to new discoveries. With a good computer simulation (see Math Minute 10.1), we could predict the consequences of particular changes to a system that might promote the maximum desired outcome and the fewest negative consequences. For example, what if you knew every gene that was activated during the progression of cancer (see Chapter 7)? You would like to develop a drug that could silence a critical gene or protein, but you don't know which one to choose. If you could predict the consequences of silencing a particular gene, you could choose a gene that killed the cancer cells but did the least

harm to healthy cells. That's what a good understanding of genome-wide circuits could provide.

People have been creating computer simulations for many years—ever since the vacuum tube was invented. In 1952, Hodgkin and Huxley used mathematics to accurately model the action potential of squid neurons, and their model still stands as a monument to experimental design, keen observation, and computational modeling. As computers improve, we can evaluate models faster. However, speed is not the limiting factor—biological understanding is. Nobel laureate Lee Hartwell and his colleagues explained, "the next generation of students should learn how to look for [biological] amplifiers and logic circuits, as well as to describe and look for molecules and genes." Hartwell was the principal investigator who helped discover proteins called cyclins, which control the cell cycle in every plant and animal with a circadian rhythm. Let's look at a simple interactive network for sea urchin. While viewing the model, think about whether it is better than paper diagrams to inform us about complex networks.

Does Interactivity Enhance Understanding?

Go to the gene circuits web page and let the Java applet load completely before you click on any buttons (see Figure 10.23). Green arrows and boxes are those that you have

Figure 10.23 Screen shots from a Java program.
Go to www.GeneticsPlace.com to view this figure.

not yet selected. Once selected, genes will appear either red or blue, according to this key:

Key to Symbols
(upstream genes)
Filled red arrows point to genes that become activated.
Hollow red arrows point to genes that become repressed.
(downstream targets)
Filled blue arrows point to genes that the highlighted gene activates.
Hollow blue arrows point to genes that the highlighted gene represses.

Technical Hints

1. For clarity, you can click once each on a series of genes while holding down the Shift key, turning the genes gray. Click on the "Pathway" button to eliminate all other genes from view.

2. You can find out a lot about any gene by holding down the Control button on your keyboard and clicking once on the gene. You will get a new window with information about the gene.

DISCOVERY QUESTIONS

27. Find and then click on "Endo16." What can you G⚛P conclude about the completeness of this model with regard to *Endo16* gene regulation?

28. Now click on "CyIIIa." What kind of control G⚛P is exerted on this gene? What type of protein does this gene encode? Use Technical Hint 2 to find out.

29. Compare this Russian interactive map to the G⚛P PDF maps produced by Schwikowski et al. (e.g., Benno Figure 1) and addressed by Discovery Question 42 in Section 8.3. Describe the benefits of an interactive map.

A Java applet to view protein-DNA circuits was created as an interactive means to show protein-protein interactions as well as protein-DNA interactions (Figure 10.24).

Figure 10.24 Screen shot of a Java web page that shows protein-protein and DNA-protein interactions.
Go to www.GeneticsPlace.com to view this figure.

An interaction is shown as a black node, with DNA drawn as boxes and proteins as ovals. Arrows leading toward a node indicate which two members are interacting. If a gene and a protein interact, the protein is binding to one of the gene's cis-regulatory elements. Arrows leading away from the nodes always point to the protein produced by the gene. If you hold down the Shift key and double-click on a node, it reveals the interaction with the regulatory protein and the target DNA. Choose a node of interaction (e.g., *Endo16* DNA and SpOtx protein) and look back at Figures 10.17 and 10.22 to compare the different types of models.

DISCOVERY QUESTIONS

30. Does this protein-DNA interaction applet help G⚛P you see the complex interactions better than the gene circuits applet?

31. Identify which gene has the largest number of G⚛P DNA-binding proteins governing its expression. List each of the proteins that regulate this gene.

32. Find which DNA-binding protein regulates the G⚛P most genes. List those genes.

We have considered two interactive ways to illustrate a number of the pathways related to *Endo16*. The field of creating gene and protein networks is new. Each group of genomic and proteomic investigators is creating their own set of tools to communicate their view of a network, and there are no universal standards. In October 2000, Isabelle Pirson, Jacques Dumont, and their colleagues at the Free University in Brussels, Belgium, proposed a simple set of symbols to illustrate on paper any network combination. However, because there are many competing visual languages in existence, it will take a lot of work and cooperation to produce a standardized way to communicate.

DATA G⚛P
Benno Figure 1
METHODS
protein-DNA circuits
visual languages

Summary 10.2

A common theme in genomics and proteomics is that simple unifying rules are usually too simplistic. Before we can fully understand how an embryo develops into an adult, we will need to dissect more genes to discover which mechanisms are common and which are unusual. As we accumulate data, we will build comprehensive models, but they must be comprehensible, too. Interactive models, 3D graphics, layered images—who knows what will work best? However, the two examples studied in this section illustrated why substantial improvements over paper diagrams are needed to facilitate our understanding of how genomes control gene expression.

Chapter 10 Conclusions

At first, it may have seemed that studying *Endo16* in depth was inappropriate for a genomics textbook. However, no other gene has been dissected as completely, so *Endo16* is the perfect model to help us understand how genomes are converted from silent code to dynamic cells. Genomics and proteomics are fields that can collect large amounts of data, and the two fields intersect when we study how an individual cell produces a dynamic proteome that is altered over time and in response to environmental changes. The best way to understand a machine is to take it apart and reassemble it. By taking apart *Endo16*, you can understand the logic behind one gene's wiring diagram and computer algorithm. The tools Davidson used need to be expanded to cover the entire genome, which raises a troubling question. How can we possibly understand the regulation of every gene in a genome if it took Davidson many years to model *Endo16*? To model entire genomes, we will need new tools and different types of questions. Chapter 11 will focus on genomic circuits composed of 2 to 50 genes. In Chapter 12, you will see what has been done to integrate many forms of data to create a systems approach to understanding complete genomes.

References

Bhalerao, J., P. Tylzanowski, et al. 1995. Molecular cloning, characterization, and genetic mapping of the cDNA coding for a novel secretory protein of mouse: Demonstration of alternative splicing in skin and cartilage. *Journal of Biological Chemistry*. 270(27): 16385–16394.

Boos, W., & A. Böhn. 2000. Learning new tricks from an old dog. MalT of *Escherichia coli* maltose system is part of a complex regulatory network. *Trends in Genetics*. 16: 404–409.

Davidson, E. H., R. A. Cameron, & A. Ransick. 1998. Specification of cell fate in the sea urchin embryo: Summary and some proposed mechanisms. *Development*. 125(17): 3269–3290.

Gladstone Institutes. 2005. Cell cycle and cell division. <http://www.genmapp.org/Yeast/Cell%20Cycle%20and%20Cell%20Division2.htm>. Accessed 11 May 2005.

Godin, R. E., L. A. Urry, & S. G. Ernst. 1996. Alternative splicing of the *Endo16* transcript produces differentially expressed mRNAs during sea urchin gastrulation. *Developmental Biology*. 179(1): 148–159.

Kirchhamer, C. V., C. H. Yuh, & E. H. Davidson. 1996. Modular cis-regulatory organization of developmentally expressed genes: Two genes transcribed territorially in the sea urchin embryo, and additional examples. *PNAS USA*. 93(18): 9322–9328.

Li, Q., S. Harju, & K. R. Peterson. 1999. Locus control regions coming of age at a decade plus. *Trends in Genetics*. 15: 403–408.

Ransick, A., S. Ernst, et al. 1993. Whole mount in situ hybridization shows *Endo16* to be a marker for the vegetal plate territory in sea urchin embryos. *Mechanisms of Development*. 42(3): 117–124.

Yuh, C.-H., H. Bolouri, & E. H. Davidson. 2001. Cis-regulatory logic in the *Endo16* gene: Switching from a specification to a differentiation mode of control. *Development*. 128: 617–629.

Yuh, C.-H., H. Bolouri, & E. H. Davidson. 1998. Genomic cis-regulatory logic: Experimental and computational analysis of a sea urchin gene. *Science*. 279: 1896–1902.

Yuh, C.-H., & E. Davidson. 1996. Modular cis-regulatory organization of *Endo16*, a gut-specific gene of the sea urchin embryo. *Development*. 122: 1069–1082.

Yuh, C.-H., E. R. Dorman, et al. 2004. An otx cis-regulatory module: A key node in the sea urchin endomesoderm gene regulatory network. *Developmental Biology*. 269(2): 536–551.

Yuh, C.-H., J. G. Moore, & E. H. Davidson. 1996. Quantitative functional interrelations within the cis-regulatory system of the *S. purpuratus Endo16* gene. *Development*. 122(12): 4045–4056.

Yuh, C.-H., A. Ransick, et al. 1994. Complexity and organization of DNA-protein interactions in the 5′-regulatory region of an endoderm-specific marker gene in the sea urchin embryo. *Mechanisms of Development*. 47(2): 165–186.

Review Literature

Bonifer, C. 2000. Developmental regulation of eukaryotic gene loci. *Trends in Genetics*. 16: 310–315.

Cho, R. J., & M. J. Campbell. 2000. Transcription, genomes, function. *Trends in Genetics*. 16: 409–415.

Clayton, D. F. 2000. The genomic action potential. *Neurobiology of Learning and Memory*. 74: 185–216.

Davidson, E. H. 2001. *Genomic regulatory systems: Development and evolution*. San Diego, CA: Academic Press.

Davidson, E. H. 1999. A view from the genome: Spatial control of transcription in sea urchin development. *Current Opinion in Genetics & Development*. 9: 530–541.

Endy, D., & R. Brent. 2001. Modelling cellular behavior. *Nature*. 409: 391–395.

Gladwell, M. 2000. *The tipping point: How little things can make a big difference*. Boston: Little, Brown.

Hartwell, L. H., J. J. Hopfield, et al. 1999. From molecular to modular cell biology. *Nature*. 402 Supplement: C47–C52.

Legrain, P., J.–L. Jestin, & V. Schächter. 2000. From the analysis of protein complexes to proteome-wide linkage maps. *Current Opinion in Biotechnology*. 11: 402–410.

Pirson, I., N. Fortemaison, et al. 2000. The visual display of regulatory information and networks. *Trends in Cell Biology*. 10: 404–408.

Wray, G. A. 1998. Promoter logic. *Science*. 279: 1871–1872.

Integrated Genomic Circuits

11.1 Natural Gene Circuits

Discover how genes can form toggle switches.

Understand how computer models of genomic circuits can lead to discoveries about learning.

Read genomic circuit diagrams to understand cancer.

Evaluate the influence of genome organization on the whole system.

11.2 Synthetic Biology

Utilize design principles to construct synthetic toggle switches.

Apply engineering principles to measure our understanding of genomes.

Integrate stochastic behavior of proteins and gene regulation.

LINKS
Adam Arkin
Harley McAdams

In biology, investigators must balance the utility of creating models against the danger of believing their models accurately represent a living system. Models of biological processes are never perfect, but they can help us make new discoveries. In Section 11.1, we examine gene regulation from a different level of control. Rather than determining exactly which DNA sequences control each aspect of a gene's overall productivity (Chapter 10), we will study gene regulation at the level of protein production. How can one gene influence another? Can genes work together to toggle between two alternative outcomes? Can we model genomic circuits to gain insights into how cells work? To answer these questions, we will explore a series of case studies that have led the way in modeling genomic responses on a small scale. Once we understand these types of **integrated circuits** (multigene interactions), can we calculate their reliability and ask why we are diploids and why we have apparently redundant genes? To understand how we learn new information, we will integrate a series of small circuits into a larger network. From this complex integrated circuit, we hope to discover new properties that were not apparent when each circuit was studied in isolation. Similar principles have been applied to understand cancer. Ultimately, we want to understand how organisms function by understanding how proteins work, both alone and as a part of integrated circuits.

11.1 Natural Gene Circuits

We know our genes are regulated to be activated in some cells and repressed in others (Chapter 10). We also know that proteomes are dynamic, changing in response to environmental influences and aging (Chapter 8). How does a cell know when to alter a particular gene's transcription? Cells need a mechanism to switch from on to off and vice versa. Genes need to sense their intracellular environment and respond accordingly. However, we don't want our cells to change so rapidly that genes are turned on and off every second of every minute. It would be a disaster for our brain cells to sense a drop in glucose and respond by converting themselves into liver cells that can store sugar. Therefore, our genes have to be tolerant of some cellular variations. Furthermore, cells need to have alternative means for accomplishing vital functions. Our genomes must be prepared for circumstances that might block one circuit from performing its cellular role. For example, human cells normally consume oxygen to produce adenosine triphosphate (ATP). Aerobic ATP production is a good strategy until you are being chased by a bear; then it is good to have an alternative (anaerobic) means to produce enough ATP to continue running. Knowledge of natural genomic circuits allows us to calculate the reliability of each component in

the circuit, which can further our understanding of genomes as they are regulated in living cells.

Can Genes Form Toggle Switches and Make Choices?

Let's look at one universal issue related to networks: **bistable toggle switches**. You know what a toggle switch is; it turns on your lights, computer, iPod, etc. A biological, bistable toggle switch will remain in one position (on or off) until the circuit determines the switch should be toggled to the other position. Bistable toggle switches are easy to understand in electrical engineering terms, but how can biological circuits determine when to flip a switch? In Chapter 10, we saw how transcription factors regulate whether a gene will be on or off, but what controls the transcription factors? And what controls the proteins that control them? Part of the answer is that an egg is not just an empty bag of water, but is filled (thanks to Mom) with many lipids, carbohydrates, nucleic acids (including mRNA ready for translation), and proteins (including transcription factors). Developmental biologists have discovered what causes an egg to enter mitosis and cytokinesis, and form a new organism. Nonembryonic cell division repeats itself according to some internal regulatory mechanism. Normally, our cells can control their cell-division toggle switch, but if they lose control of this switch, we develop cancer. How biological toggle switches exert control over gene expression is neither esoteric nor insignificant.

How Do Toggle Switches Work?

There are two ways to start answering this question: Start with data and build a model, or start with a model using engineering principles and improve the model with experimental data. Harley McAdams (Stanford University School of Medicine) and Adam Arkin (Physical Biosciences Division of the Lawrence Berkeley National Laboratory) combined the best of both approaches in an elegant analysis of genetic toggle switches. The first issue they had to address was the concept of noise.

Noise in a regulatory system such as a toggle switch means that, unlike your computer, genetic switches have to deal with a degree of uncertainty. We know that gene activation occurs when transcription factors bind to cis-regulatory elements. When a cell undergoes mitosis and cytokinesis (eukaryotes) or cell division (bacteria), the first source of noise is introduced: will both daughter cells receive the same number of transcription factors? Of course, if cells were as wise as Solomon, the pool of transcription factors would be split right down the middle, 50:50. However, cells are not "wise," and to some extent the partitioning process during cell division is random, or **stochastic**. For example, if a cell had 50 copies of the Otx transcription factor, 6% of the time a particular daughter cell might get 19 or fewer copies, while 6% of the time it might get at least 31 copies (Math Minute 11.1). That could have a profound effect on the subsequent regulation of *Endo16* expression.

Another component of genetic noise is the fact that few binding sites exist for each protein, and binding occurs at a slow rate. For example, Otx may be able to bind to only a few cis-regulatory elements in the entire genome, and it has to find these elements. Each cis-regulatory element must be found by a small number of DNA-binding proteins. The limited number of transcription factors and binding sites results in an increased range of times when all the transcription factors are in the right places for any given gene. Another example of slow reaction rate is that once the cis-regulatory element is fully occupied and ready to initiate transcription, the first RNA will be produced a variable amount of time later due to noise in the initiation of the transcription machinery. Transcription takes an average of several seconds to begin, but again this is an average, with a distribution of times both shorter and longer than the average.

What Effect Do Noise and Stochastic Behavior Have on a Cell?

In prokaryotes and eukaryotes, proteins are produced in bursts of translation of varying durations and with varying outputs. Therefore, the total number of proteins produced from any gene is not the same each time, but rather an average with a **normal distribution** (see Math Minute 11.1). By producing proteins in bursts rather than at a constant rate, the cell provides proteins a higher probability of forming a quaternary structure (e.g., a dimer) that may be required for full function. Most students learn that "gene Y is activated and produces X proteins per minute," but this summary statement is an oversimplification of a messy and mildly chaotic world inside each of your cells.

Genes are noisy, but what does this have to do with a genetic toggle switch? Everything. Let's imagine two proteins that each bind to different but overlapping binding sites, and that these sites have competing roles. For example, look back at the *Endo16* cis-regulatory element in Figure 10.17 and find the Z and CG2 binding sites. Here is a small segment of DNA that can accommodate two different proteins, but

---

Math Minute 11.1   **How Are Stochastic Models Applied to Cellular Processes?**

At first, it is hard to imagine that some cellular processes are random. But random doesn't necessarily mean chaotic; it is just a way of saying that the outcome is not exactly the same every time the process is repeated. Even sophisticated machinery designed to manufacture thousands of identical automobile parts produces parts that are nearly the same, but not 100% identical. The field of probability theory provides stochastic models for random processes. We have already seen one example in Math Minute 8.1: a model for sampling from a finite population using the hypergeometric frequency function. Now we will explore two stochastic models for cellular processes.

The Binomial Model

If 50 molecules of Otx (see Chapter 10) are floating around inside a nucleus prior to cell division, it seems likely that the two daughter cells will not always inherit exactly 25 molecules each. In this situation, randomness captures the idea that if a large number of identical cells divided, the outcome (i.e., the number of molecules inherited by each daughter) would vary. Some outcomes would occur quite often, while others would be rare. The fraction of the time that each possible outcome occurs in the long run (i.e., in a large number of cells) is an estimate of the probability of that outcome.

The standard stochastic model for situations like the allocation of Otx molecules between two daughter cells uses the *binomial frequency function*. This model assumes that a particular "experiment" is repeated n times, where each repetition, or trial, is independent of all the others. In the example of Otx, a trial consists of determining which daughter cell gets a particular molecule. We assume that the fate of each molecule is independent of the other 49, a reasonable assumption if each molecule of Otx has randomly selected a location inside the nucleus. Since all 50 molecules must wind up in one of the two daughter cells, there will be 50 trials ($n = 50$). Each trial results in a "success" with probability p. In this example, a trial is counted as a success if a particular daughter cell gets the Otx molecule in question. To keep track of how many molecules go to each daughter, it is helpful to distinguish the cells by their relative positions after cell division: daughter L (cell on the left) and daughter R (cell on the right). Let's arbitrarily pick daughter L as the one we follow. In other words, the number of successes in 50 trials is the number of Otx molecules that go to daughter L. Since daughter L is just as likely to get each molecule as is daughter R, $p = 0.5$.

Under the binomial model, you can compute the probability of achieving k successes out of n independent trials with the binomial formula:

$$\binom{n}{k} p^k (1 - p)^{n-k}$$

where $\binom{n}{k}$ is the binomial coefficient defined in Math Minute 8.1. Therefore, the probability that daughter L receives 25 molecules of Otx is

$$\binom{50}{25} (0.5)^{25} (1 - 0.5)^{50-25} \approx 0.112$$

You can find the probability that daughter L receives 19 or fewer molecules (meaning daughter R receives 31 or more molecules) by computing the probability of each outcome satisfying this criterion (there are 20 such outcomes), and adding the 20 probabilities to get 0.06. Similarly, you can determine the probability that daughter L receives 31 or more molecules (meaning daughter R receives 19 or fewer molecules) to be 0.06. Thus, with probability 0.12, each daughter will be 6 or more molecules away from the average value of 25.

The Normal Model

Many random factors influence the amount of protein produced by a gene at a particular time, including the number, location, and timing of all proteins needed to transcribe and translate the gene. In this situation, randomness means that if you measure the amount of protein produced by the same gene in thousands of identical cells (or in a single cell at thousands of time points), the outcome (i.e., number of protein molecules produced) will vary. Some outcomes will occur more frequently than others.

The standard stochastic model for a random quantity that represents the accumulation of many small random effects (e.g., protein production) is the normal distribution (also called the Gaussian distribution, or bell curve). The use of the normal distribution model is justified by one of the most powerful results in probability theory, the Central Limit Theorem.

Let X be the number of molecules of protein produced by a gene. You can compute the probability that the value of X is in a certain interval by finding the appropriate area under the curve given by the normal probability density function:

$$f(x) = \frac{1}{\sigma\sqrt{2\pi}} e^{-\frac{(x-\mu)^2}{2\sigma^2}}$$

In this function, μ is the mean of the distribution (the average, or expected, value of X) and σ is the standard deviation of the distribution (a measure of the variation in values of X). The values of μ and σ can be estimated by taking a random sample of measurements (i.e., measuring the quantity of protein produced at several randomly chosen times), and calculating the sample mean and sample standard deviation of these measurements.

For example, if $\mu = 300$ and $\sigma = 20$, the probability that X is between 310 and 330 is given by the shaded area in Figure MM11.1. You can look up the numerical value of this area (approximately 0.2417) in a table of normal probabilities, or you can use numerical integration to estimate the area. In addition, many mathematical, statistical, and spreadsheet programs provide a function for computing probabilities using the normal probability density function.

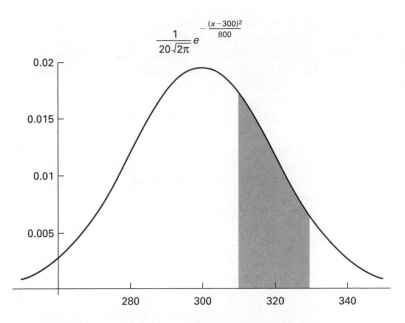

$$\frac{1}{20\sqrt{2\pi}}e^{-\frac{(x-300)^2}{800}}$$

Figure MM11.1 Normal probability density function with $\mu = 300$ and $\sigma = 20$.
The shaded area represents the probability that X is between 310 and 330.

A handy property of the normal probability distribution is that X is in the interval $\mu \pm \sigma$ 67% of the time; X is in the interval $\mu \pm 2\sigma$ 95% of the time; and X is in the interval $\mu \pm 3\sigma$ 99% of the time. For example, with $\mu = 300$ and $\sigma = 20$, we know that X is between 260 and 340 ($300 \pm 2 \times 20$) with probability 0.95. We used this property in Math Minute 8.2 to determine whether a particular node in a graph had an unusually large degree. Because the normal distribution is so often a reasonable approximation for random quantities, we can use this property any time we look at data with error bars to get a rough estimate of the probability that the measured quantity is within the interval denoted by the error bars.

Standard stochastic models are excellent starting points for understanding random cellular processes. However, these models rely on certain assumptions, which may or may not hold. Like all models, stochastic models can be refined after gathering experimental data.

---

only one at a time. Either Z can be occupied, or CG2, but not both. Both binding sites modify the output of module A; Z is responsible for repressing and CG2 for amplifying. Given the noise within the system, two genetically identical cells (descendants of the same fertilized sea urchin egg) may have exactly opposite developmental fates. Noise and stochastic genomic circuits help explain why even "identical" human twins have different fingerprints. As a result of noise, genetic toggle switches are affected by DNA-binding site competition and stochastic production of transcription factors. However, genetic toggle switches are too important to be determined by noise alone. Toggle switches need a feedback loop that reinforces what was initially a random "decision."

Let's understand Figure 11.1 (a theoretical switch) before we study a naturally occurring toggle switch. Protein A can bind to the cis-regulatory elements of genes b and c to initiate transcription for both genes. Protein B has three possible fates: it can be degraded by the cell; it can diffuse away and perform other functions; and, most importantly for us, it can repress the expression of gene c. Conversely, protein C has three fates, one of which is to repress gene b. Will protein A bind to b and indirectly suppress c? Or will A bind to c and suppress b? Either outcome is possible, because the determining factors are stochastic: the amount of A and its ability to find a limited number of binding sites upstream of b and c. Once the decision is made, a genetically identical population of cells can be split into two subtypes

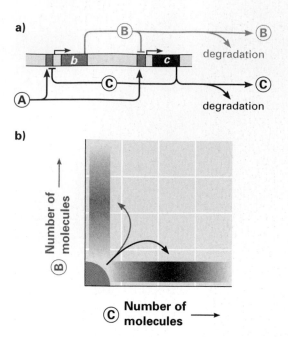

Figure 11.1 Toggle switch circuit.
a) Two promoters (small gray boxes) upstream of two genes (boxes with lowercase letters) are controlled by protein A (proteins are represented by capital letters inside circles). Protein B represses gene *c* and protein C represses gene *b*.
b) The shading displays the number of genetically identical cells containing different numbers of molecules (B or C). Initially, cells contain A but neither B nor C (mass of cells at the origin). Later, more cells are expressing C, as indicated by the darker shading gradation along the X-axis. Within each shaded gradation, the number of B or C molecules varies around a mean value, because protein production is stochastic. Arrows indicate the choice made by cells to express either B or C.

(Figure 11.1b). An individual cell will produce only B or C. No cells will make both, nor are there any genetic hardwiring instructions that allow us to predict which path any particular cell will choose.

DISCOVERY QUESTIONS

1. If two molecules of protein A were inside a single cell, would it be possible to produce equal amounts of proteins B and C in the same cell?
2. In Figure 11.1b, why did more cells produce protein C than protein B? Would you predict this same outcome if you repeated the experiment? Explain your answer.

Theory Is Nice, but Do Toggle Switches Really Exist?

Theoretical models help us comprehend general principles, but they are useful only if they approximate reality. Many pathogens evade our immune systems by changing their protein exteriors on a regular basis. How can genetically identical pathogens present different exteriors? They take advantage of noise and toggle switches. As your immune system learns to search and destroy, the pathogen changes its appearance. For the pathogen, it is easy to see that there are evolutionary forces at work to maintain noise and toggle switches, but mechanistically, how does it work? This section discusses several naturally evolved circuits and the toggle switches in each one; Section 11.2 highlights some recent research into synthetic circuits constructed by investigators but tested inside cells. Both types of research help us measure our understanding of noise, toggle switches, and biological circuits.

Let's take a closer look at a naturally evolved toggle switch that controls the behavior of the bacterial virus called λ phage. λ has two behaviors from which to "choose." It can either live quietly within its *Escherichia coli* host (**lysogenic** lifestyle), or it can replicate rapidly and blow up its host as the progeny are launched to infect new hosts (**lytic** lifestyle). The choice between peaceful coexistence and lethal parasitism is made by a single protein with the inconspicuous name of CII (pronounced C two).

The λ phage toggle switch (Figure 11.2) and the theoretical switch in Figure 11.1 are very similar. CII is equivalent to protein A. The amount of CII is the critical parameter, and one of two outcomes is possible. If CII finds the promoter P_{RE}, transcription will proceed toward the left of P_{RE} and lead to the transcription of *cI* (pronounced C one) further downstream. CI can bind to the promoter P_L upstream of *cIII* and lead to the production of CIII. CIII prevents the destruction of CII; thus, CIII indirectly reinforces its own production in a positive feedback loop. Dimerized CI reinforces its own production indirectly by binding to sites labeled O_{R1} and O_{R2} to repress the production of Cro protein (CI_2 acting as a repressor of *cro*). CI_2 binding to O_{R1} and O_{R2} also promotes its own production in a positive feedback loop by acting as a transcription factor for its own gene, *cI*. Once CII initiates this bistable toggle switch, λ is locked into peaceful lysogenic coexistence with its host *E. coli* unless new environmental forces perturb the system (e.g., UV light, change in nutrient availability). However, the toggle switch could have flipped the other way, depending on the noise and stochastic protein behaviors. CII protein could have been degraded if it took too long to find P_{RE}, because *E. coli* makes a protease that can destroy CII. If **Brownian motion** (random motion driven by kinetic energy) causes the protease to find CII before CII finds P_{RE}, the lytic lifestyle is chosen. In the absence of CII, the promoter labeled P_R is weakly active and begins transcribing to the right, resulting in the production of Cro protein. Cro_2 binds to O_{R3} and O_{R2}, which leads to repression of *cI* and increased transcription of *cro*. The positive feedback loop keeps the bistable toggle switch flipped toward *cro* transcription and a lytic lifestyle that eventually leads to the production of hundreds of fully mature viruses that swell and lyse the *E. coli* host cell.

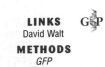
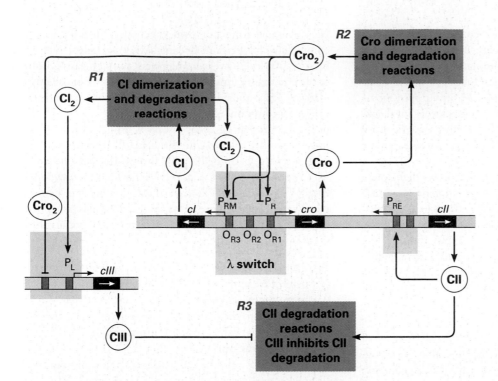

Figure 11.2 λ toggle switch that chooses between coexistence and murder.
DNA (light purple bands) and promoters (light gray boxes with arrows pointing to their genes) from λ phage. Genes are black boxes with white arrows indicating the direction RNA polymerase travels to transcribe the genes. Genes are induced (black arrows) or repressed ⊥ as indicated. Three regulatory regions (purple boxes labeled R1, R2, and R3) determine the lifestyle "decision" for λ phage. Named circles are proteins, with subscript 2 indicating dimerization. Arrows into and out of regulatory regions represent a flow of information.

There are several noisy factors in the choice made by λ phage, such as the limited number of proteins and binding sites as well as the variable amount of time it takes to initiate transcription. Another factor is the burst of protein production. Notice in Figure 11.2 that both Cro and CI must form homodimers to be functional. Dimerization is more likely to happen when proteins are produced in bursts than when the same number of proteins is made at a slow but steady rate. A final component worth noting is that environmental influences can skew this decision. For example, if the bacterium host happens to be growing in a nutrient-rich environment (e.g., in a flask with lots of glucose), the bacterium produces more protease, resulting in faster destruction of CII and the production of many new λ phage (lytic lifestyle). Conversely, if the bacterium happened to be in a nutrient-poor environment (e.g., on the bottom of your shoe), there are fewer (but not zero) protease molecules, so CII has a higher probability of finding its binding site on P$_{RE}$ before being destroyed. A longer half-life for CII leads to peaceful coexistence (lysogenic lifestyle), which makes good sense for the virus. Why should a virus reproduce rapidly if the environment is not conducive to making more potential hosts? Why not wait for the nutrients to arrive (e.g., when you step in something

yucky) so the bacteria can grow? When the nutrients arrive, bacteria will grow faster, proteases will be more numerous, CII will be destroyed more readily, more viruses will form, more bacteria will lyse, and viruses will infect more hosts. The selective advantage for a noise-tolerant toggle switch is impressive.

In recent studies, investigators have examined the amount of noise generated by different aspects of a genomic circuit. For example, graduate student Yina Kuang in David Walt's Chemistry Department lab at Tufts University led a team that studied gene expression in single *E. coli* cells. The investigators placed two different promoters (*recA* and *lacZ*) upstream of the reporter gene *GFP* and then measured the production of fluorescence in 200 individual cells when induced or under control conditions (Figure 11.3). The *recA* promoter is **constitutively** on (always activated) at a low

Figure 11.3 Monitoring *recA* and *lacZ* promoter activity in multiple individual cells.
Go to www.GeneticsPlace.com to view this figure.

level, and there is considerable variation (noise) among the 200 different cells. When induced, *recA* promoter stimulates large amounts of mRNA, as indicated by protein production, but the variation between cells is relatively low (see online *recA* movie for sample data). In contrast, *lacZ* exhibits a very low background level of transcription with little noise under control conditions, and induction does not produce as much increase over basal rate.

The behavior of *recA* and *lacZ* promoters might seem irrelevant until you consider the role each gene plays in a cell's life. RecAp is used to repair DNA damage. Cells need RecAp at all times, and thus cells tolerate a leaky and noisy *recA* promoter. When the cell senses DNA damage, the promoter requires only one step to switch to a higher expression rate with relatively less noise, because repairing DNA is a vital function that must be addressed before cell division can resume. In contrast, lacZp metabolizes lactose, and the gene is induced in the absence of glucose and the presence of lactose (or experimentally applied IPTG). Basal expression of *lacZ* is normally low because alternative sugars would be available. The toggle switch for *lacZ* induction requires several other proteins, and each of those proteins has its own level of noise. Therefore, *lacZ* induction is noisy because each step in the induction process brings its own level of noise to the combined process of *lacZ* transcription. It appears that the amount of noise in a toggle switch is related to each gene's function. These findings indicate noise may be more than just tolerated; rather, it appears to be a phenotype subject to selection pressure. Cells appear to benefit from some promoters with loose regulation, while others provide greater fitness when their transcription is very tightly regulated.

DISCOVERY QUESTIONS

3. What would be the consequences if CI degradation were more prevalent than CI dimerization? How does Cro$_2$ affect the ability of CII to switch λ from lytic to lysogenic?

4. If the P$_L$ promoter were inactivated, would this change the outcome of the toggle switch for lysogenic vs. lytic lifestyles? Explain your answer.

5. Which of the three regulatory regions (purple boxes) in Figure 11.2 would be subjected to the most noise? Hypothesize why tolerance of noise in this area of the λ life cycle may be advantageous.

How Can Multicellular Organisms Develop with Noisy Circuits?

The preceding examples may lead you to believe many genetic toggle switches are loaded with noise and impossible to coordinate—the genomic equivalent of herding cats.

But we know from our own experiences that life is not completely chaotic. You do not have brain cells trying to become liver cells. Every human went through gastrulation at the exact same time during gestation. How can cell populations with stochastic toggle switches work collectively toward a common goal? A team analogy may be useful, because coordinating genes in cells is similar to coordinating 11 football players on the field. Picture the offense with a quarterback (QB) who throws the ball, linemen who block defenders, and receivers who run downfield hoping to catch the ball, save the game, become heroes, etc. There are three keys to winning a football game, just as there are three keys to coordinating cell populations with noisy toggle switches.

1. Each player does not have to ensure that all the other players are in the right place. The QB and the two behind him can survey all other players and yell reminders to those who have lined up in the wrong place. This is called **cooperation through communication**.

2. At various times, the QB can consult a list of points to make sure everyone has made the right move. Watch how the QB will shout and sometimes raise and lower one leg to signal others to move a bit to the left or right. And what happens if everyone is confused? The QB can call a time-out to give the players a chance to get coordinated again. Each of these points prior to starting the play is called a **checkpoint**.

3. Any team that really wants to win has a contingency plan. Bill Cosby has a great comedy routine in which he relives a childhood football game where everyone is given very complex directions on where to go so the QB can throw the ball to someone. On real teams, the QB has two to four players running around, so if one is not a good target, the QB can look for other options. This duplication of options to accomplish a goal (winning the game) is beneficial **redundancy**.

Cells can use the same three keys to achieve coordination.

1. A subset of cells can secrete a product that will communicate a message to keep all cells synchronized.

2. Cellular proteins establish quality control at various checkpoints, such as DNA replication and the stages of mitosis. Checkpoints ensure the quality of the eventual outcome, but the exact timing for any given cell can vary due to noise and stochastic gene induction or protein function (e.g., regulation of *recA* and *lacZ* promoters).

3. Cells have redundant circuits to create fail-safe approaches to vital processes such as response to environmental signals. Some pathways have multiple ways of becoming activated and/or multiple ways of producing a cellular response. Redundancy also can be achieved by having isozymes that can perform

essentially the same job, though they may have slightly different tolerances to environmental perturbations.

Redundancy: Does Gene Duplication Really Increase Genome Reliability?

Measuring reliability is essentially an engineering question. Is it really necessary to have more than one way to stop your parked car from rolling down a hill? In this context, the answer seems obvious, but genomic redundancy is less intuitive. For many years, it has been argued that having more than one locus encoding a particular function imparts a selective advantage. The second copy might be a backup in case one gene is mutated and loses its function. The duplicated gene can mutate over time and produce new functions for the cell. Freshly evolved duplications can provide a wider range of tolerance to environmental conditions, to ensure the common function is accomplished (e.g., one may work better in cold temperatures and the other in hot). Biologists should not reinvent the wheel to measure the value of redundancy. Why not borrow from well-established engineering methods for reliability analysis to assess the likelihood that a particular function will be performed successfully (Figure 11.4)?

In equation (a) and Figure 11.4, we are interested in producing protein B. The reliability of any given protein being produced is arbitrarily set at 0.9 or 90% and is represented by the line segment with the gene labeled x. In this reliability analysis, we follow what happens when the DNA required to produce B is altered in different ways.

a.   Reliability (R) = P

= probability of B being produced
= 0.9 or 9,000 out of 10,000 success rate.

If production of B requires two genes (Figure 11.5; equation b), the reliability of this step drops from 0.9 to 0.81, because both x and y have to be functional and the rule of multiplication applies.

b. x and y must work:

$R = P^2$
= 0.9 × 0.9
= 0.81 or 8,100 out of 10,000 success rate.

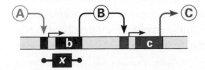

Figure 11.4 A three-gene pathway for the production of protein C.
The genetic unit represented by the line segment marked "x" indicates one or more genes that accomplish the task of producing B. In Figures 11.5–11.9, more than one gene/allele will be included in line segment x, but these are assumed to be unlinked.

If x were duplicated in the absence of y (Figure 11.6; equations c and d), the reliability of producing B substantially increases to 0.99. To determine this reliability, first we must calculate the probability of failure (Q) by taking the probability of either failure or success (total of 1) minus the probability of success (P). For a single gene x in a haploid genome (see Figure 11.4) the probability of failure is:

c.   Q = 1 − P
= 1 − 0.9
= 0.1.

For diploids (equation d), reliability equals all possible outcomes (1) minus the probability of failure (Q^2; Q × Q) because both the upper x and the lower x have to fail (multiplication rule again) for the production of B to be unsuccessful.

d. $R = 1 − Q^2$
= 1 − (0.1 × 0.1)
= 0.99 or 9,900 out of 10,000 success rate.

By evolving a diploid genome, we have increased reliability from 90% to 99%. This increase in reliability would also be true for haploids that duplicated a single locus. What is the reliability consequence if a diploid organism duplicates a locus (Figure 11.7; equation e)? Using the multiplication rule, the probability of failure for each gene (x and x') is multiplied, which produces another substantial increase in reliability.

Figure 11.5 A two-gene model to produce B.
These two genes are from unlinked loci in a haploid, though the genes have been diagrammed as adjacent for simplicity.

Figure 11.6 A diploid model to produce B.
The two alleles are on homologous chromosomes.

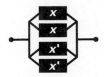

Figure 11.7 Diploid genome with a duplicated gene to produce B.
The two genes (x and x') are unlinked; alleles for a given gene are located on homologous chromosomes.

e. $R = 1 - (Q^2 \times Q^2)$
    $= 1 - Q^4$
    $= 1 - (0.1^4)$
    $= 1 - 0.0001$
    $= 0.9999$ or 9,999 out of 10,000 success rate.

We calculated that two different genes in a haploid genome were less reliable (equation b) than one gene (equation a). This makes sense because with two genes there are two ways to fail instead of just one. Redundancy should help ameliorate the weakness of two genes. Let's determine the reliability in diploids with duplicated genes when two genes are required to complete the function (Figure 11.8; equation f). To calculate this reliability, we need to combine equations b and e.

f. $R = (1 - Q^4)^2$
    $= (1 - 0.0001)^2$
    $= (0.9999)^2$
    $= (0.9998)$ or 9,998 out of 10,000 success rate.

Note that the reliability for two genes that have been duplicated in a diploid (equation f and Figure 11.8) was not quite as high as for one duplicated gene in a diploid (equation e and Figure 11.7). For half the number of alleles (4 instead of 8), the organism has a slightly higher reliability; if one allele is mutated, the function will still be completed with 3 remaining alleles. Even if the individual in Figure 11.8 carries a nonfunctional *x* allele and a nonfunctional *y* allele, the redundancy of the system maintains a high degree of reliability (Figure 11.9; equation g).

g. $R = $ modified equation f to take into account that only three viable alleles exist for each locus
    $= (1 - Q^3)^2$
    $= (1 - 0.001)^2$
    $= (0.999)^2$
    $= (0.998)$ or 9,980 out of 10,000 success rate.

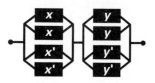

Figure 11.8 Diploid genome requiring two genes (*x* and *y*) to produce B in which both genes have been duplicated (*x'* and *y'*).
None of the genes (*x*, *x'*, *y*, and *y'*) are linked, though the pair of alleles for each locus is located on homologous chromosomes.

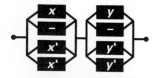

Figure 11.9 Mutant alleles affect the genome's reliability.
Genome from Figure 11.8, but one *x* allele and one *y* allele are nonfunctional, as indicated by "–".

Every time we model a biological circuit, we are trying to create a simple version of a complex system. From the lessons of simple circuits, is it possible to understand complex circuits? If we can understand complex circuits, will we discover new (emergent) properties that were not present in the dissected circuits? For the remaining two cases in Section 11.1, we will study complex genomic circuits that might reveal emergent properties that are undetectable when genes are studied one at a time. How do we convert environmental stimulation into memories? Is it possible to understand cancer formation by studying protein circuits? It is worth noting that the following two examples were investigated by mining data available in public databases and the literature. The investigators combined experimental data with in silico research to form new models and make predictions to refine the models with increasing accuracy.

Does Memory Formation Require Toggle Switches?

You may be familiar with the saying "easier said than done." In other words, to say that there is a way to coordinate all these genomic circuits sounds simple and is intuitively appealing—but where is the proof? By now, you should have an appreciation for the complexity of the problems ahead. Only with the benefit of genomic data analysis—sequences, variations, expression profiles, proteomics, biochemistry, computer science, mathematics, etc.—can we begin to piece together the necessary information to see coherent patterns and circuits. Upinder Bhalla (the National Center for Biological Sciences, Bangalore, India) and Ravi Iyengar (Mount Sinai School of Medicine) analyzed many years' worth of data, made some insightful discoveries, and set the pace for others to follow. They used an engineering approach to understand four cell-signaling circuits and discovered some interesting emergent properties. They used data in the public domain to create a complex computer model that accurately simulates the neurocircuitry necessary for learning.

LINKS G&P
GENESIS
STRUCTURES
cAMP

As we have seen before, to comprehend complexity, we need to simplify. That sounds like an oxymoron, but in fact, we do this all the time. No one says "compact disc read-only memory"; we just say CD-ROM. The five-letter acronym encapsulates a lot of information and facilitates comprehension. Bhalla and Iyengar decided to make a few simplifying assumptions first, rather than beginning with the most complex model possible. This simplification employs the principle known as **Occam's Razor**: start with the simplest possible explanation first, rather than more complex ones. As a model is compiled and analyzed, the need for specific experiments becomes apparent and the model may become more complex if necessary.

It is interesting to note how certain pathways are more popular than others. For example, every introductory biology textbook discusses how an adrenaline rush can stimulate the production of cyclic adenosine monophosphate (cAMP). Bhalla and Iyengar decided to model a different aspect of cAMP signaling and made a couple of simplifying assumptions. They assumed that the cytoplasm was a well-stirred bag of liquid in which all components have equal access to each other. Uniform distribution within the cytoplasm requires each component to have a mechanism for delivering its message only to the correct target molecule and only in one direction. If this were not the case, it would be like having uninsulated wires in your iPod. The electrical currents would be unorganized and multidirectional, and you would never hear any music. To generate a computer model of their uniformly mixed cell, the investigators needed to know the reaction rates of every enzyme and the concentration of every component in the four signaling circuits. However, easier said than done ... (Figure 11.10). The level of complexity in genomic circuits can grow to staggering proportions. A 3-molecule system requires only 7 measurements, but an 18-molecule system requires 162 measurements!

It became too difficult to use standard computational approaches for this level of complexity, so Bhalla and Iyengar utilized a neural network simulation program, called GENESIS, to analyze the four interacting pathways. Even with the help of very sophisticated computation, two simplifying assumptions were made that do not reflect reality. First, the investigators ignored **compartmentalization**. For example, some components may be embedded in phospholipid bilayers and thus not freely available to all other components. In fact, we know this is true of many components included in the analysis, but it was impossible to quantify this aspect. Second, they ignored **regional organization of components**. Real cells cluster some components near each other to significantly increase biochemical efficiency, rather than letting them drift by Brownian motion. The mitochondrion is an excellent example; metabolism is more efficient because of the clustering of molecules used in the electron transport pathways. Microorganisms also use gene clustering for the production of antibiotics.

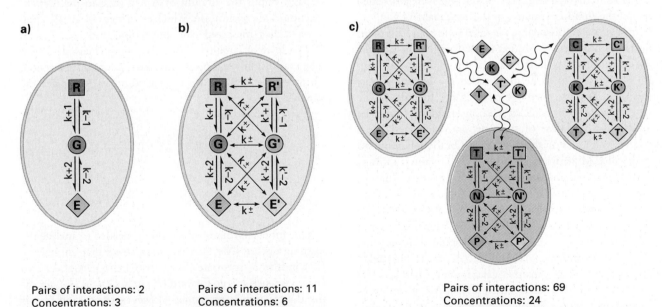

a)
Pairs of interactions: 2
Concentrations: 3
Rate constants: 4

b)
Pairs of interactions: 11
Concentrations: 6
Rate constants: 22

c)
Pairs of interactions: 69
Concentrations: 24
Rate constants: 138

Figure 11.10 Increased need for information as circuits become more complex.
a) to **c)** $k+x$ represents the rate constant for the forward reaction and $k-x$ the rate constant for the reverse reaction; 1 and 2 refer to pathways 1 and 2. In this simplified model, each component in a pathway can communicate only with its nearest neighbors. The effect of this simplifying assumption is most apparent in panel c). The symbols used are: R = receptor; G = G protein; E = effector; T = transcription factor; N = nucleic acids; P = proteins in the nucleus.

LINKS
long-term potentiation
METHODS
brain anatomy

Are Simple Models of Complex Circuits Worthwhile?

Most biologists assume that interconnected circuits work synergistically. The existence of synergy is why many in field biology dislike the reductionist approach used by molecular biologists. However, the reductionists' goal is not merely to disassemble a cell to look at the parts, but to understand the parts well enough to explain synergistic interactions and move beyond descriptions toward testable predictions. The particular case under investigation was one that philosophers and biologists have pondered for centuries: How do we learn?

Neurobiologists have given us great insights into the mechanisms utilized whenever an animal (e.g., flies or humans) learns something new. Intuitively, we know our neurons must undergo some sort of change in order for us to retain information. There is a genetic component to memory formation, so we know proteins are involved. Somehow, proteins must alter what a neuron does—become depolarized when stimulated and release neurotransmitters to relay this information. Sounds simple, right? A neuron's change in function is called long-term potentiation (**LTP**), which means the consequence of neuronal stimulation is maintained after the original stimulus is gone. To see a simple example of this, stare at a bright light and then turn away. Even though the light is no longer hitting your neurons, you still "see" it. Your neuron's function was changed so that it performed differently after the stimulus was removed. Bhalla and Iyengar decided to model the complex circuitry of the mammalian brain.

Before we begin dissecting, we need to learn a little brain anatomy to provide the context for learning. Where in the brain does learning taking place? Deep within your cerebrum is a collection of neurons called the **hippocampus**. For at least 30 years, memory research has focused on the hippocampus as the center of learning/memory. Inside the hippocampus are layers of neurons, and each layer has a name. We will focus on the CA1 layer. On the cell body and dendrite of CA1 neurons are bumps called spines. Embedded in these spines are integral membrane proteins, three of which are of particular interest to us. The neurotransmitter glutamate binds to its receptor, called mGluR (*mouse glut*amate *r*eceptor). As with all receptors, it facilitates **signal transduction**; that is, it transmits the extracellular signal across the plasma membrane. NMDAR (*N-m*ethyl-*D-as*partate *r*eceptor) is a voltage-sensitive calcium ion channel. The final plasma membrane component is AMPAR (α-amino-3-hydroxy-5-methyl-4-isoxazolepropionate receptor), another glutamate receptor that acts as an ion channel when stimulated by glutamate. LTP is initiated when mGluR and NMDAR are stimulated by a certain amount and frequency of stimuli. Experimentally, LTP can be induced in mouse neurons when stimulated with 3 mild electrical inputs of 100 Hz pulses, 1 second each and separated by 10 minutes. Bhalla and Iyengar set out to construct a computer model of the complex series of events from stimulation to LTP.

How Much Math Is Required to Model Memory?

Molecular biology is becoming mature enough to need the assistance of many other disciplines, especially mathematics. However, the math used in this study is quite simple. To start their in silico research, Bhalla and Iyengar considered only two types of connections: protein-protein interactions and **second messengers**. Two additional facts they considered were: (1) proteins **degrade**, that is, they are destroyed by the cell over time; and (2) enzymes have reaction rates—for example, there is an average amount of time it takes a kinase to consume ATP and add a phosphate onto its substrate. In the first reaction below, A and B join to form AB. This illustrates how two proteins, such as a kinase and its protein substrate, can bind to each other. The binding has a forward reaction rate (k_f) and a backward reaction rate (k_b). The second reaction shows the conversion of A and B into C and D. For example, we could measure the amount of Na^+ inside a cell (A) and the amount of K^+ outside a cell (B) as they are changed into Na^+ outside a cell (C) and K^+ inside a cell (D). The forward rate of this conversion (k_f) is the rate of the Na/K pump, and the backward rate (k_b) is the rate of ions passing through ion channels. This second reaction can be written as an equation, which says that the change (d) in the concentration of A ([A]) over a change in time ($/dt$) is equal to the production of A (the backward rate times the concentrations of C and D, which is written: k_b[C][D]) minus the amount of A lost (due to the forward reaction that consumes A, which is written: k_f[A][B]). You have studied these types of interactions and rates before, so this level of circuitry should be comprehensible so far. The math is only multiplication, division, and subtraction. To determine LTP, all we need is numbers to replace the variables.

$$A + B \underset{k_b}{\overset{k_f}{\rightleftharpoons}} AB \qquad A + B \underset{k_b}{\overset{k_f}{\rightleftharpoons}} C + D$$

$$d[A]/dt = k_b[C][D] - k_f[A][B]$$

One more enzymatic interaction is needed to analyze the learning circuit. The next equation states that an enzyme (E) binds to its substrate (S) to form a complex of the two (ES). This step can go forward (k_1) or backward (k_2), meaning the reaction is reversible. However, the second step is irreversible, as indicated by the forward arrow and only one rate constant (k_3). Therefore, the ES complex can be converted into the original enzyme (E; an enzyme is never consumed in a reaction) plus a new product (P). Each step occurs at a measurable rate, and these values are called **rate constants** (the k values). For example, the enzyme adenylyl cyclase produces cAMP from the substrate ATP. Adenylyl cyclase (E) binds to ATP (S) and forms (at rate k_1) a complex of the two (ES). ATP and adenylyl cyclase can fall apart (k_2) or proceed (k_3) toward an irreversible production of adenylyl cyclase plus cAMP (P).

$$E + S \underset{k_2}{\overset{k_1}{\rightleftharpoons}} ES \overset{k_3}{\rightarrow} E + P$$

That's all the math and chemistry you need to understand to grasp this very sophisticated in silico analysis of learning! We still have one problem, though. We don't have any real numbers to put into these equations. Luckily, biochemists and cell biologists have published all the values needed to initiate a GENESIS software analysis of a nerve's ability to learn.

How Do You Build Complex Models?

Fifteen well-studied circuits were needed to produce Bhalla and Iyengar's LTP simulation (Figure 11.11). They started by building computer models of each of the 15 circuits individually. At each step, they made sure that the models matched experimental data. This alone was an impressive task, but here is where the fun starts. They began to build integrated circuits by gradually combining each of the 15 individual circuits. Initially, they only allowed two types of connections between circuits: (1) secondary messengers arachidonic acid (AA) and diacylglycerol (DAG), and (2) an enzyme in one circuit bound to its substrate produced in another circuit.

Permitting these two types of interactions allowed the investigators to combine the circuits labeled a, b, e, f, h, k, and l in Figure 11.11 into one integrated circuit (Figure 11.12). Notice that epidermal growth factor (EGF) leads to the activation of two enzymes: mitogen-associated protein kinase (MAPK) (in circuit a) and PLC$_\gamma$ (in circuit f). MAPK has been studied intensively for many years. It is associated with substances that stimulate mitosis, and it phosphorylates proteins. Loss of control of MAPK can lead to the formation of cancerous cells. PLC$_\gamma$ is one isoform, or version, of phospholipase-C (there are several PLC genes in each person, and this version, or isoform, is called γ) that cuts the phospholipid called phosphotidyl inositol bisphosphate (PIP$_2$) into inositol trisphosphate (IP$_3$) and diacylglycerol (DAG). By connecting these seven circuits, a new layer of complexity became evident—feedback loops. A **feedback loop** occurs when the product has a stimulatory or inhibitory effect on one of the upstream components, such as an enzyme that leads to product formation.

Can a Transient Stimulus Produce Persistent Kinase Activation?

Do individual circuits behave synergistically when they are integrated into a larger pathway? Can a computer model accurately simulate LTP? Can we use this model to make predictions that lead to improved understanding of how we learn new information? We know that these individual components and circuits are involved in learning, and we

LINKS
15 circuits

know that LTP is the result of long-term activation of a kinase. But is it really possible to simulate something as complex as learning on a few megabytes of computer microcircuits?

How did the computer model compare with real data? In Figure 11.13a, you can see the simulation approximated the experimental data. In this graph, the MAPK activity was plotted as a function of time ($d[A]/dt$). The activation of MAPK was transient due to the normal degradation mechanisms in the cell. Equally impressive is Figure 11.13b, comparing simulated PLC$_\gamma$ activity (dashed lines) with experimental data (solid lines) in the presence (triangles) or absence (squares) of EGF over a 10,000-fold range of calcium concentrations. Given the close agreement between experimental data and the computer model, these two investigators had a good foundation on which to build more complex integrated circuits.

In Figure 11.12, you can see two areas that provide feedback: PKC and MAPK. PKC activates Raf, which activates another kinase MEK, which activates MAPK (two isoforms of MAPK were used in this simulation, numbers 1 and 2). MAPK activates cytosolic phospholipase A-2 (PLA2), to produce AA, which is half of the stimulus needed to activate PKC. What concentration and duration of stimulus are needed to produce a long-lasting activation of PKC and MAPK? Figure 11.14 is data intensive and worth the effort required to dissect it carefully. First, let's look at the PKC data (open symbols) in the figure. Neither 10 minutes at 5 nM EGF (open circle) nor 100 minutes at 2 nM EGF (open square) led to any significant activation of PKC. However, 100 minutes at 5 nM EGF (open triangle) was sufficient to lead to a protracted activation of PKC. Note how PKC was activated during the 100 minutes, and when the stimulation was removed, PKC activity remained elevated. A long-term change in function after removal of stimulus is a hallmark of LTP. MAPK activity (closed symbols) was a bit more complicated, but the final outcome was similar. Stimulation for 10 minutes at 5 nM produced transient activation, while 100 minutes at 2 nM produced a more prolonged activation, but at a lower amplitude, and the activation did not extend much beyond the 100 minutes of stimulation. However, 100 minutes of 5 nM stimulation produces a tenfold increase in activity (from 0.01 to 0.1 μM) as well as a prolonged activation (i.e., a bistable toggle switch). Notice how MAPK was activated during the 100 minutes and what happened at about 50 minutes: it jumped to a higher level. Something extraordinary (synergistic) was happening.

DISCOVERY QUESTION

9. Look at Figure 11.14 and hypothesize what may be the cause of the tenfold increase in MAPK activity between 30 and 50 minutes for the 100 minutes at 5 nM stimulation.

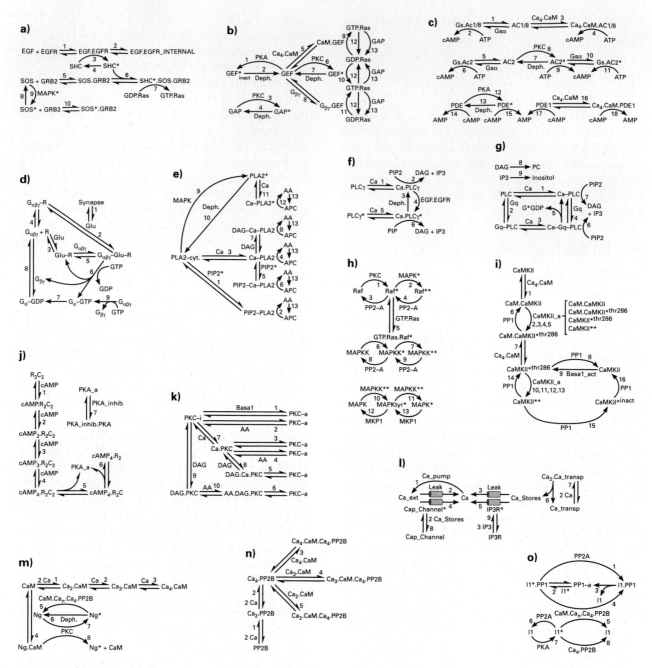

Figure 11.11 The 15 circuits that were modeled individually before being integrated into more complex networks.

Reversible reactions are indicated by double-headed arrows. Enzyme reactions are drawn as curved arrows, with the enzyme listed in the middle. For visual clarity, any component may be present more than once in a single circuit.

Why Does 100 Minutes of 5 nM EGF Achieve Long-Term Activation?

For the toggle switch of long-term enzyme activation, determining the importance of a particular combination of EGF concentration and duration of exposure is more complex than you might imagine at first. If MAPK could be transiently activated after 10 minutes at 5 nM or 100 minutes at 2 nM, then why were these levels of MAPK activation insufficient stimulus to create long-term activation? To determine the answer, we need to analyze Figure 11.15.

These two activity plots were generated by the concentration-effect curves for MAPK activation of PKC (purple line) and PKC activation of MAPK (black line), plotted on the same axes. The curves for PKC vs. MAPK

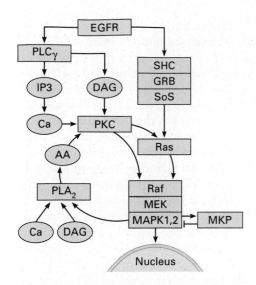

Figure 11.12 Circuit diagram of signaling pathway beginning with EGFR and ending with new gene activation inside the nucleus.

Rectangles represent enzymes, and circles represent messenger molecules. This integrated circuit utilized pathways a, b, e, f, h, k, and l from Figure 11.11.

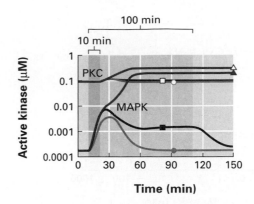

Figure 11.14 Activation of the feedback loop.
PKC (open symbols) and MAPK (closed symbols) activities were graphed to show the effect of a positive feedback loop. Three stimulus conditions are represented: 10 min at 5 nM EGF (circles), 100 min at 2 nM EGF (squares), and 100 min at 5 nM EGF (triangles). Dark and light gray shading in the graph represents the 10 and 100 minutes of EGF exposure, respectively.

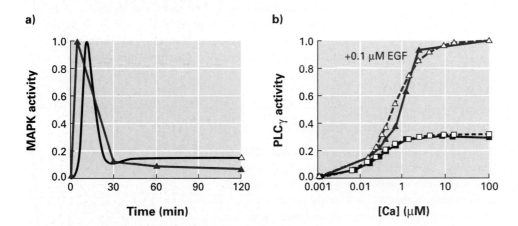

Figure 11.13 Computer model matches experimental data.
a) Measuring MAPK activity as a function of time. Simulation (open triangle) and real data (filled triangle) are very similar. The stimulus in both cases was a steady supply of 100 nM EGF. **b)** Measuring PLC$_\gamma$ activity as a function of calcium concentration, where dashed lines represent computer simulation and solid lines represent experimental data. Triangles indicate the presence of 100 nM EGF; squares indicate the absence of added EGF.

(purple) and MAPK vs. PKC (black) intersect at three points: A, B, and T. Point A represents high *a*ctivity for both PKC and MAPK, whereas point B represents *b*asal activity for both. A and B represent distinct steady-state levels of PKC and MAPK activities. A system with two distinct steady states is a **bistable circuit**. The bifurcation point T is important because it defines the *t*hreshold stim-ulation, which can be thought of as the stimulus to flip a toggle switch. If the initial stimulation of EGF (amplitude and duration) is sufficient to activate either PKC or MAPK

above T, then both enzymes would switch to steady state at point A. In contrast, if the initial stimulation is below T for both PKC and MAPK, then upon removal of the stimulus, both enzymes will relax to B, the basal level of activity.

We have returned to the concept of a bistable toggle switch, but this particular switch is produced by many more components than the simpler switch we studied earlier. In λ phage, the choice between lytic and lysogenic lifestyles was determined by the amount of CII protein present (see Figure 11.2). For EGF stimulation of simulated

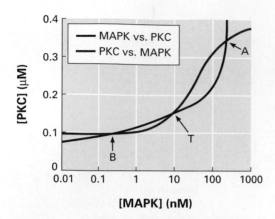

Figure 11.15 Bistability plot for feedback loop in Figure 11.12.
The PKC vs. MAPK activities plot (purple line) was constructed by holding the level of active MAPK constant, running the simulation until steady state, and reading the value for active PKC. This process was continued for a series of MAPK levels spanning the range of interest. A similar process was repeated for MAPK vs. PKC activities plot (black line) by holding the level of active PKC constant and calculating MAPK activity. Both plots were drawn with the concentrations of PKC on the Y-axis and MAPK on the X-axis. The curves intersect at three points: B (basal), T (threshold), and A (active). A and B are stable points, but T is the toggle switch point for the two steady-state "choices" of A and B.

LTP, a critical level of kinase activity in a positive feedback loop determines whether the stimulus leads to LTP (i.e., learning) or not.

Bhalla and Iyengar made an interesting analogy that was very appropriate given that they were simulating learning. They reminded us that a bistable (i.e., digital) system can store information the same way that RAM stores information on a computer. It is also worth noting that the simulated enzymatic bistable system was very reliable, due to the redundant mechanisms for achieving steady-state level A. High activity of either PKC or MAPK was sufficient to flip the system from off (steady-state B) to on (steady-state A). Earlier, we described biological systems in engineering terms, including "redundant," "reliable," and "fail-safe." Another engineering term, **robust**, means the circuit tolerates a wide range of environmental conditions. According to data not shown here, when the activities of five different enzyme concentrations were varied (PKC, MAPK, Raf, PLA2, and MAPKK, which phosphorylates MAPK to activate it), the simulated bistable circuit remained functional. An interesting consequence of this analysis was that most of the permutations were equally effective in achieving steady-state A, but PKC was the least tolerant to change, indicating it was the least robust enzyme of the five tested. The investigators hypothesized that PKC's lack of robustness was due to a limited number of PKC isoforms in their computer simulation (i.e., too little redundancy).

Math Minute 11.2   **Is It Possible to Predict Steady-State Behavior?**

Figure 11.15 depicts the steady-state MAPK concentration when [PKC] is held constant (black line), and the steady-state PKC concentration when [MAPK] is held constant (purple line). If we want to know the response of MAPK to a particular concentration of PKC, we begin at that value on the PKC axis, move horizontally until we hit the black line, move vertically until we hit the MAPK axis, and read the concentration at the point where we hit the MAPK axis. Conversely, if we want to know the response of PKC to a particular concentration of MAPK, we begin at that value on the MAPK axis, move vertically until we hit the purple line, move horizontally until we hit the PKC axis, and read the concentration at the point where we hit the PKC axis. Note that A, T, and B are stable points; if [PKC] or [MAPK] is at one of these three points, the response of the other kinase is at that same point.

You can analyze the stability of the feedback loop between MAPK and PKC by iteratively determining the response of [MAPK] to [PKC] and the response of [PKC] to [MAPK]. The assumption behind this iterative process is that [MAPK] at a particular time depends on [PKC] at an earlier time. Likewise, [PKC] at a particular time depends on [MAPK] at an earlier time. A graphical technique known as a cobweb diagram is helpful in following the iterative process.

Figure MM11.2 illustrates a cobweb diagram for an initial value of [MAPK] just above the threshold T, about halfway between 10 and 100 nM ([MAPK] $\approx 10^{1.5} \approx$ 32 nM). From this point on the MAPK axis, move vertically to the purple line and horizontally to the [PKC] axis. As described earlier, the resulting point (approximately 0.23 μM) is the response of PKC to the initial MAPK concentration. MAPK will now respond to this value (0.23 μM) of [PKC]. The resulting [MAPK] (approximately

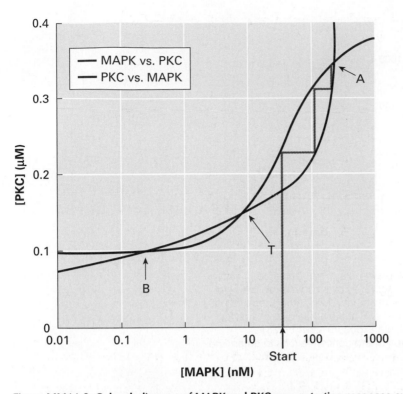

Figure MM11.2 Cobweb diagram of MAPK and PKC concentration response curves.
Start by determining the response of [PKC] to 32 nM [MAPK], and repeat this process with MAPK's response to the new PKC concentration, and so on, until converging to a point (e.g., point A).

100 nM) is found by moving horizontally from 0.23 μM on the PKC axis to the black line, and vertically to the MAPK axis. Repeat the process to find the following successive concentration responses: [PKC] = 0.31 μM; [MAPK] = 200 nM (point A); [PKC] = 0.35 μM (point A). The diagram, and thus the concentration of both kinases, has converged to A, as was observed in Figure 11.14.

You may have noticed that the lines from the curves to the axes were merely for keeping track of the concentrations of each kinase, and are always backtracked in the next kinase response determination. Therefore, the cobweb diagram can be constructed by beginning at [MAPK] = 32, moving vertically to the purple line, horizontally to the black line, vertically to the purple line, and so on, until the diagram converges to A.

You can construct cobweb diagrams starting at various values of [MAPK]. If you begin at [MAPK] > T, the diagram should converge to A, but if you begin at [MAPK] < T, it should converge to B. This shows that T is a threshold stimulation level. Given concentration-effect curves like those in Figure 11.15, cobweb diagrams allow us to predict the behavior of systems like this feedback loop.

DISCOVERY QUESTIONS

10. Design an experiment that allows you to test the prediction that robust LTP (i.e., learning) requires more than one isoform of PKC.

11. Describe how the variation in different isoforms of PKC relates to the equations we used to calculate reliability (see pages 377–378).

12. How do the terms "robust" and "reliability" relate to each other in Figure 11.12?

Can the Modeled Circuit Accommodate Learning and Forgetting?

Although we are oversimplifying what is required to form a memory in your brain, LTP is one critical step. If LTP requires the long-term activation of either PKC or MAPK, is it ever possible to inactivate these kinases and turn off LTP (i.e., forget a memory)? Bhalla and Iyengar added this aspect of learning to their already successful model. MAPK must be phosphorylated (by MAPKK) to be activated; to

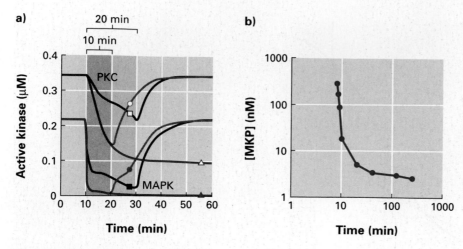

Figure 11.16 Inactivation of LTP by MKP.
a) Activities of PKC (open symbols) and MAPK (filled symbols) were simulated to show status of feedback. The feedback loop was initially activated by a stimulus above point T in Figure 11.15, and then one of three inhibitory inputs was applied: 10 minutes of 8 nM (circles), 20 minutes of 4 nM (squares), or 20 minutes of 8 nM (triangles) MKP activity. Dark and light gray shading denote 10- and 20-minute exposure times of MKP, respectively. **b)** MKP was applied for varying durations and concentrations to determine thresholds for inactivation of the feedback loop.

inactivate MAPK, there is a phosphatase (MAPK phosphatase [MKP]) that removes the activating phosphate. When MKP was added to their model, the investigators discovered more about the bistable circuit. Figure 11.16 is another data-rich figure worth careful dissection. PKC activity (Figure 11.16a, open symbols) was stimulated and then exposed to three levels of MKP activity: 10 minutes at 8 nM (open circle), 20 minutes at 4 nM (open square), or 20 minutes at 8 nM (open triangle). Only the greatest exposure to MKP was sufficient to fully inactivate PKC. Likewise, MAPK was treated with 10 minutes of 8 nM (closed circle), 20 minutes of 4 nM (closed square), or 20 minutes of 8 nM (closed triangle) MKP activity, achieving long-term inactivation (steady-state B in Figure 11.15) only with the greatest MKP exposure.

Only the most intense MKP activity was able to inactivate the feedback loop. The rebound in PKC and MAPK activities after the two lower exposures of MKP was due to a couple of factors: the persistence of AA, due to a relatively slow time course of degradation; and the time required to dephosphorylate the previously activated kinases in the MAPK circuit. The investigators simulated a wide range of MKP concentrations and durations of exposure to the bistable circuit and plotted the combinations needed to switch the circuit off (Figure 11.16b). At high MKP concentrations, inactivation occurs quickly, but there is a minimum threshold of nearly 10 minutes exposure. Conversely, when MKP was applied for very long times, at least 2 nM MKP was required to inactivate the feedback loop.

Bhalla and Iyengar created an integrated circuit simulation that exhibited feedback loop properties, a model that

has improved our understanding of LTP. Both the model and real neurons produce prolonged and elevated levels of activation that were initiated by an external signal (EGF binding to its receptor) even after the initial stimulation was removed. Their simulation has quantified what is required to turn off LTP by breaking the feedback loop.

DISCOVERY QUESTIONS

13. Look at Figures 11.12 and 11.16b and hypothesize which enzyme(s) were responsible for the required ten minutes of MKP exposure to inactivate the feedback loop even when the concentration of MKP was increased tenfold.

14. Explain what was happening to the level of phosphorylated MAPK when the amount of MKP was below 2 nM for 2 hours.

What Roles Do Other Integrated Circuits Play in LTP?

Having enjoyed this much success, the investigators decided to add more circuitry that interacts with the MAPK circuit we just studied. But first, we need to understand a little bit more about LTP (Figure 11.17). When calcium floods the stimulated neuron, these ions bind to a protein called calmodulin (CMD) to form a calcium/calmodulin (CaM) complex. CaM activates adenylyl cyclase isoforms 1 and 8 (AC1/8), calcineurin (CaN), and a kinase called Ca^{2+}/calmodulin-dependent protein kinase II (CaMKII). LTP is stimulated when CaMKII is activated for extended periods

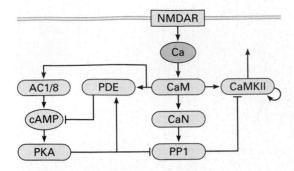

Figure 11.17 Circuit diagram examining the role of cAMP in LTP.
NMDAR is a voltage-gated calcium ion channel in the plasma membrane of neurons. AC1/8 represents two isoforms of adenylyl cyclase that produce cAMP. PDE is a phosphodiesterase that destroys cAMP. PP1 is a protein phosphatase that removes a phosphate (and thus inactivates) CaMKII. CaM is calmodulin that is activated when it binds calcium, and CaMKII is a kinase activated by CaM. CaMKII can autophosphorylate (small looping arrow). CaN, calcineurin, is a kinase that becomes activated when it binds CaM. PKA inhibits PP1 by adding a deactivating phosphate onto PP1.

of time. To be activated, CaMKII must be phosphorylated (by itself or another kinase). Bhalla and Iyengar hypothesized that protein kinase A (PKA) (a cAMP-dependent kinase) played a critical indirect role in the prolonged activation of CaMKII, and they wanted to use their computer simulation to test their prediction. They called this cAMP/PKA regulation of CaMKII a "gating" control, which can be thought of as another toggle switch. If enough PKA were activated, then CaMKII would become activated (a second bistable switch). The connection between CaM and CaMKII produced a new hard-wired connection that integrated smaller individual circuits (panels c, i, j, m, n, o in Figure 11.11). The new integrated circuit (Figure 11.17) would help the investigators determine whether interactions between CaMKII, cAMP, and CaN were sufficient to produce prolonged activation of CaMKII even after the amount of cytoplasmic Ca^{2+} inside the neuron returned to resting concentrations. The critical switch point in this circuit is CaM, which activates two competing signals that determine whether CaMKII will be activated or not. As we have seen before, stochastic behavior of proteins and noise in the switching mechanism will influence the outcome.

DISCOVERY QUESTIONS

15. How can one kinase (e.g., CaMKII) phosphorylate so many different proteins? In other words, predict how so many different substrates can bind to a single active site.

16. How can a kinase (e.g., PKA) activate one enzyme (PDE) and inhibit another (PP1) even though it adds a single phosphate in both cases?

17. How many pairs of proteins can you find in Figure 11.17 that are competing to produce opposite outcomes? What role would stochastic bursts of enzymatic activity play in the competitions you found in this figure?

When the new integrated circuit was stimulated with conditions that produce LTP in a real neuron, the simulated activities of four enzymes were graphed (Figure 11.18). When the concentration of cAMP was maintained at basal levels in the computer simulation, the external stimulation produced three transient increases in activities for CaMKII, CaN, and PP1, but had no effect on PKA. When the concentration of cAMP was allowed to rise as directed by the model, CaMKII, PKA, and PP1 exhibited significant changes in activities that were not present when cAMP concentration was fixed at resting levels, though CaN activity was unaltered. PKA needed elevated cAMP in order to become activated and PP1 was substantially reduced instead of slightly increased. Since we are particularly interested in the system's ability to initiate LTP via CaMKII, CaMKII's response to cAMP concentration changes was especially noteworthy. The amplitude of CaMKII activity was increased by about twofold, and the time it took for the activity to return to resting levels also doubled to about 20 minutes. The prolonged activation of CaMKII in the presence of elevated cAMP was due in large part to the significant inhibition of PP1, which otherwise would have inactivated CaMKII quickly. Nevertheless, CaMKII activity only increased transiently, which is not the expected behavior for a bistable toggle switch. Thus, the bistable toggle switch under these simulated conditions quickly reverted to the off position (equivalent to steady-state level B in Figure 11.15).

DISCOVERY QUESTIONS

18. Predict what is needed to stimulate CaMKII to become activated for the long term. Design an animal experiment to test your hypothesis.

19. Would cAMP levels rise or fall when given the "freedom" to vary after stimulation?

20. Which enzyme(s) exhibited the greatest duration of activity after the stimulus was removed?

Do They Need a More Complex Model to Match Reality?

Finally, Bhalla and Iyengar were ready to create a fully integrated circuit. This model integrates four individual signaling circuits (Figure 11.19). When electrical engineers view

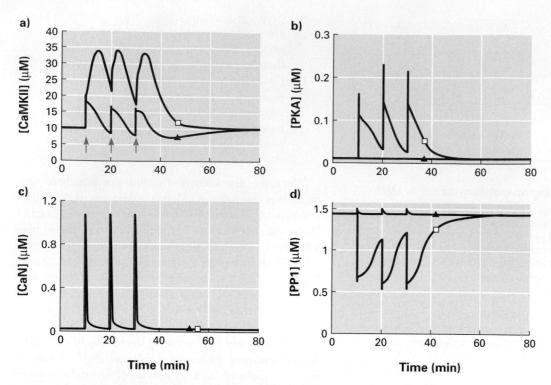

Figure 11.18 Temporary activation of four key enzymes for LTP.

Graphs show the simulated activities of **a)** CaMKII, **b)** PKA, **c)** CaN, and **d)** PP1 after stimulation with three 100 Hz pulses lasting 1 second each and separated by 10 minutes (which produces LTP in real neurons). Open squares indicate that cAMP concentrations were allowed to rise; filled triangles indicate that cAMP levels were artificially maintained at resting concentrations. The arrows in panel a) indicate when the three pulses were given to the system, at the 10-, 20-, and 30-minute time points.

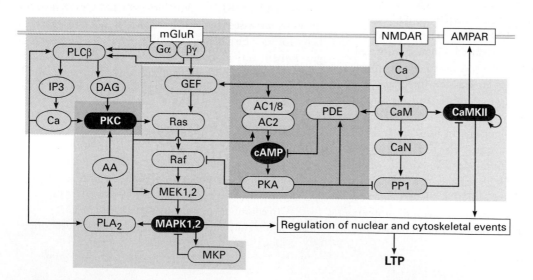

Figure 11.19 Full-circuit diagram with feedback loop, synaptic output, and CaMKII activity and regulation.

Two possible end points of the model are represented as AMPAR regulation and stimulation of nuclear/cytoskeletal events, which eventually leads to LTP. Note the feedback loop between PKC and MAPK with PLA$_2$ as the connecting enzyme. The four shaded regions highlight the four individual signaling circuits.

a circuit diagram such as this, they do not examine each piece individually. They look for functional units of circuitry and the connections between them. Figure 11.19 contains four functional units: PKC, MAPK, CaMKII, and the cAMP circuits. We have seen how PKC and MAPK circuits were connected (see Figure 11.12) and how the CaMKII and cAMP circuits were connected (see Figure 11.17). PKC was connected to the cAMP circuit via AC2. These connections enabled Bhalla and Iyengar to produce their final integrated circuit composed of four smaller circuits.

How powerful a research tool is the computer model that simulates integrated circuits? Can it elucidate critical features that result in LTP? Does the feedback loop between PKC and MAPK play a critical role in LTP? To build their integrated circuit shown in Figure 11.19, all the individual circuits shown in Figure 11.11 were incorporated. The in silico neuron was stimulated with glutamate and depolarized at the plasma membrane, which opens the NMDAR calcium ion channel. How did the full integrated circuit respond? Were there any unexpected activities?

The first pair of enzymes examined was PKC and MAPK (Figure 11.20), which are indirectly connected to each other. PKC exhibited a strong on/off response to the three electrical stimulations in the absence of the feedback loop (AA held at basal levels). When the feedback loop was allowed to function, notice how PKC activity substantially increased; also, the duration of the activation was maintained after the stimulation ceased. Compare the PKC activity to the activity of MAPK. In the absence of the feedback loop, MAPK activity was undetectable, but in the presence of the feedback loop, MAPK was strongly activated and the activation was sustained. Notice that MAPK activity began at about 30 minutes.

Four additional enzymes were examined in this simulation (Figure 11.21). PKA was activated +/− the feedback loop, as was CaMKII. But note that both of these enzymes exhibited a higher amplitude and sustained activity in the presence of feedback than in the absence of feedback. CaN was activated, but there was no difference +/− feedback loop. PP1 had its activity reduced, as one would expect, and its inhibition was of greater amplitude after the stimuli were removed. We are seeing the results of another bistable toggle switch that does not require any new transcription or translation. PP1 inhibits CaMKII, but PKA inhibits PP1 while also stimulating its own activity. In this toggle switch, CaN is not involved at all, so it is no surprise that CaN activity was not sustained when the feedback loop was functioning.

The most striking outcome we discovered was that complex circuits exhibit emergent properties without the need for new mRNA or protein production! What are the implications for microarray and proteomics research, when new biochemical properties emerge from a defined circuit that lacks additional transcription or translation? The emergent property is both good news and bad news for students who like to cram in material minutes before a test. The good news is that you can learn new information (i.e., LTP is produced)

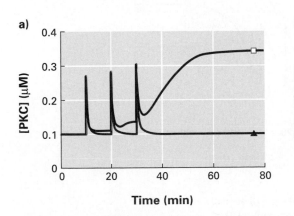
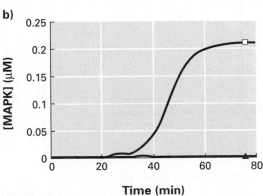

Figure 11.20 Activity profile of a) PKC and b) MAPK.
LTP is achieved with a fully functional integrated circuit (open squares) but not when the feedback loop is blocked by holding AA at resting concentrations (filled triangles). Electrical pulses of 100 Hz for 1 second were applied at 10, 20, and 30 minutes.

very quickly in the absence of new protein production, which would take about an hour to accumulate. The bad news is that LTP has two phases just like your memory: short-term and long-term. MKP can be thought of as the "off switch" to learning. MKP inactivates MAPK, which is needed to initiate transcription in the nucleus. MAPK indirectly controls the activity of CaMKII (via the feedback loop that includes PLA2, PKC, AC2, PKA, and PP1), which is also needed to produce LTP. Therefore, if your MKP activity exhibits a stochastic burst of activity, what you "learned" while cramming will be lost from your short-term memory and never reach your long-term memory.

MKP has the ability to determine which line (open squares vs. filled triangles in Figure 11.21) better represents your ability to remember something. The toggle switch of memory is controlled by the concentration of a few molecules to produce a system that is bistable. Therefore, MKP may act as a timer to bring to an end the early phase of LTP. Of course, you want to be able to retain information for a longer time, but occasionally you forget. MKP becomes the first checkpoint at which some people will not be able to commit a new memory to long-term storage. In Bhalla and Iyengar's full integrated circuit, MKP is not connected to

any other proteins, but this is unlikely to accurately reflect reality. What circuits control MKP activity? How do our neurons "decide" which way to flip the toggle switch for LTP?

Are LTP and Long-Term Memory Related?

It has been known for years that LTP is divided into two phases, slow and fast. You have just studied the fast phase, which was composed of four integrated circuits. In a real neuron, the slow phase of LTP changes the structure of synapses between neurons, requiring the production and intracellular transportation of new proteins (i.e., a dynamic proteome). The goal in learning is to change one synapse (to store a new memory) without altering other synapses (analogous to altering one computer file but not any others on your computer). MAPK and CaMKII lead to the slow production of new proteins that must be sent to the correct synapses. Bhalla and Iyengar hypothesize that the feedback loop described in their simulation controls the localization of new proteins via the cytoskeleton so they wind up at the correct synapse. What a huge conclusion to draw based on computer simulations! For the sake of science, it does not

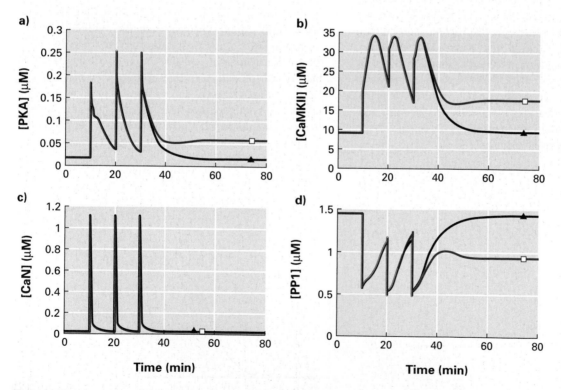

Figure 11.21 Activity profiles of PKA, CaMKII, CaN, and PP1.
The simulation was run with the feedback loop functioning (open squares) and with feedback blocked by holding AA fixed at resting concentrations (filled triangles). **a)** The Ca-stimulated PKA waveforms are almost identical, but the baseline rises when feedback is on, because PKC stimulates AC2 to produce additional cAMP. **b)** Activity profile of CaMKII. **c)** Activity profile of CaN is unaffected by the presence or absence of feedback. **d)** Activity profile of PP1 in the presence and absence of the feedback loop.

matter whether their hypothesis is right or wrong. Either way, the investigators have made a prediction based on their model that is testable in the lab. If they are correct, they may need to rent tuxedos and fly to Stockholm. If they are wrong, someone will experimentally demonstrate an inconsistency and advance the field by eliminating one incorrect possibility. Therefore, a completely computer-based research project has taught us a lot about integrated circuits, provided us with some new insights about LTP, and generated a new idea to explain long-term memory.

DISCOVERY QUESTIONS

23. What difference do you see when comparing the activation of CaMKII inFigure 11.21b and 11.18a? Propose a mechanism for this difference.
24. What is the consequence for LTP when PP1 is more inhibited in the presence of the feedback loop?
25. Explain to DNA microarray and proteomics experts why the LTP integrated circuit has profound implications for their research.
26. Hypothesize how a person who has a photographic memory might have a genotype that permits him or her never to lose fast LTP.
27. Hypothesize how an older person might retain his or her long-term memories but not be able to learn new things.

What Have We Learned (How Much LTP Have We Generated)?

Bhalla and Iyengar's paper had two major goals. Its primary goal was to understand how protein circuits work and to detect any synergistic properties. From this computer-based simulation, we now understand that integrated protein circuits can explain some old observations:

- It is possible to produce a prolonged signaling effect even after the original stimulus is removed.
- Protein circuits can activate feedback loops, which can provide the mechanism for synergistic properties.
- It is possible to understand the signaling threshold required to control a bistable toggle switch.
- A single stimulus can lead to multiple output pathways.
- Complex networks provide the mechanism for critical aspects of "biological design" that can provide the selective pressure for evolutionary steps. These design principles include redundancy, robustness, and fail-safe.
- All these features can be accomplished in the absence of new protein synthesis.
- Genetic variation in the population will result in certain genotypes that will respond differently to the same stimuli.

The second goal of their research was to gain a better understanding of the process of LTP, which leads to the formation of a new memory. In this area, the investigators predicted:

1. CaN does not play a major role in the LTP activation of CaMKII, which most believe is a critical protein in learning.
2. PKA is downstream of both PKC and MPAK and upstream of CaMKII.
3. PKC can be activated by more than one input, each with critical threshold concentrations to flip the bistable toggle switch to the on position.
4. The adenylyl cyclase isoforms play separate roles in LTP. AC1 is activated by CaM and allows the system to achieve a rapid inhibition of PP1 after Ca^{2+} influx. AC2 is activated by PKC and provides the sustained inhibition of PP1, which leads to the prolonged activation of CaMKII and thus LTP. AC2 provides the necessary step for the feedback loop, which is critical for the synergistic response.

We have spent a substantial amount of time and energy studying circuits. The work by Bhalla and Iyengar will lead to very focused and efficient research in the lab through computer-guided hypothesis formulation. These two investigators are quick to acknowledge that their model has limitations, such as compartmental constraints and changes in protein levels due to translation of new proteins. They wrote, "[M]odels such as these should not be considered as definitive descriptions of networks within the cell, but rather as one approach that allows us to understand the capabilities of complex systems and devise experiments to test these capabilities." They also leave us with an interesting question. Is the LTP integrated circuitry related to immunological "memory," in which memory white blood cells can "remember" pathogens years after initial contact? In silico circuit analysis may help us understand immunology as well as many other complex systems.

Can We Understand Cancer Better by Visualizing Its Circuitry?

The human genome project is not an end point, but rather a beginning. Knowing all the DNA content of humans, even if we knew every single nucleotide polymorphism (SNP) in the gene pool, would only give us the raw material to ask interesting questions. How do we learn? What drives our sexuality? What makes some people early risers and others night owls? Why do some people live to be over 100 years old?

We cannot answer these questions yet, because genomics is a work in progress. However, each year we have better tools and more data, allowing us to assemble more complex and comprehensive models that better describe biological

DATA
cell cycle
DNA repair
LINKS
Kohn's references
Kurt Kohn

processes. For example, Kurt Kohn from the **National Cancer Institute (NCI)** produced two huge circuit diagrams of cell cycle control and DNA repair. The first question most nonbiologists ask at this point is, "Why would the NCI want to study cell cycle and DNA repair?" Cancer results from the loss of cell cycle control and the inability to fix damaged DNA. A lot was already known about cell cycle and DNA repair (see Kohn's references), and Kohn summarized all this information in comprehensive circuit diagrams. But first he had to invent a language (Figure 11.22). Unlike Bhalla and Iyengar, Kohn permitted many more interactions to take place in his circuit diagrams, and therefore he needed a larger vocabulary. You should note the different types of interactions needed to complete Kohn's circuits. Some symbols are more intuitive than others, but each interaction is found inside your cells. Do not memorize the symbols, but make sure you understand what each symbol means in the vocabulary.

DISCOVERY QUESTIONS

28. Use Kohn's symbolic language to diagram how *Endo16* is regulated (see Figure 10.22).
29. Use Kohn's symbolic language to "translate" the PKC/MAPK feedback loop (see Figure 11.12).

Examine the web versions of the two circuit diagrams: the cell cycle and the DNA repair circuits. The first things you should notice are their sheer size and complexity. The circuit diagrams facilitate systems biology comprehension, which is appropriate because cancer is a malfunctioning system. The key to understanding complex circuit diagrams is to find functional units that are connected to other functional units. For example, look at the Myc functional unit that is magnified in Figure 11.23.

Myc is a known **oncogene**, and here is some information as it appeared in Kohn's annotations appendix:

C35: cdc25A may be transcriptionally activated by c-Myc; the Myc:Max heterodimer binds to elements in the cdc25A gene and activates its transcription.
M1: c-Myc and pRb compete for binding to AP2.
M2: AP2 and Max compete for binding to c-Myc. AP2 and Myc associate in vivo via their C-terminal domains.
M3: The E-cadherin promoter is regulated via AP2 recognition elements.
M4: c-Myc and pRb enhance transcription from the E-cadherin promoter in an AP2-dependent manner in epithelial cells (mechanism unknown). Activation by pRb and c-Myc is not additive, suggesting that they act upon the same site, thereby perhaps blocking the binding of an unidentified inhibitor. No c-Myc recognition element is required for activation of the

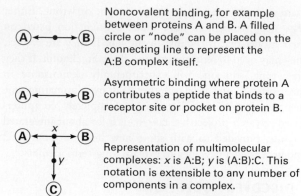

Noncovalent binding, for example between proteins A and B. A filled circle or "node" can be placed on the connecting line to represent the A:B complex itself.

Asymmetric binding where protein A contributes a peptide that binds to a receptor site or pocket on protein B.

Representation of multimolecular complexes: *x* is A:B; *y* is (A:B):C. This notation is extensible to any number of components in a complex.

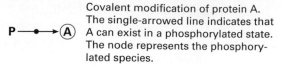

Covalent modification of protein A. The single-arrowed line indicates that A can exist in a phosphorylated state. The node represents the phosphorylated species.

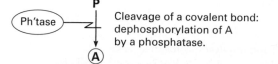

Cleavage of a covalent bond: dephosphorylation of A by a phosphatase.

Proteolytic cleavage at a specific site within a protein.

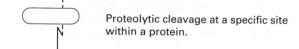

Stoichiometric conversion of A into B.

Transport of A from cytosol to nucleus. The filled circle represents A after it has been transported into the nucleus (the node functions like a ditto mark).

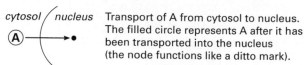

Formation of a homodimer. Filled circle on the right represents another copy of A. The filled circle on the binding line represent the homodimer A:A.

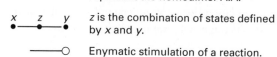

z is the combination of states defined by *x* and *y*.

Enzymatic stimulation of a reaction.

General symbol for stimulation.

General symbol for inhibition.

Shorthand symbol for transcriptional activation.

Shorthand symbol for transcriptional inhibition.

Degradation products.

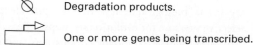

One or more genes being transcribed.

Figure 11.22 Symbolic language created for cell cycle and DNA repair circuit diagrams.

In the electronic version, black lines indicate binding interactions and stoichiometric conversions; red are covalent modifications and gene expression; green are enzyme actions; blue are stimulations and inhibitions.

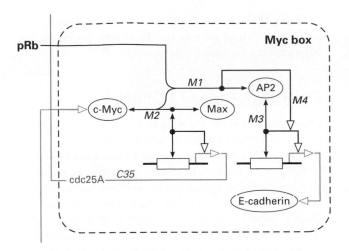

Figure 11.23 *Myc* **subsection from Kohn's cell cycle circuit.** The *Myc* box can be found in area D10, in the top right corner of the DNA repair circuit diagram. Symbols are defined in Figure 11.22.

E-cadherin promoter by c-Myc. Max blocks transcriptional activation from the E-cadherin promoter by c-Myc, presumably because it blocks the binding between c-Myc and AP2.

The proteins retinoblastoma (pRb, a **tumor suppressor**) and c-Myc compete to bind to the transcription factor AP2. The two possible combinations c-Myc/Max and pRb/AP2 can each bind to cis-regulatory elements to initiate transcription of E-cadherin. However, c-Myc can also bind to Max and lead to the transcription of cdc25A, a phosphatase that regulates the activity of a number of proteins in other functional domains. However, note that regulation of c-Myc in this functional unit is influenced by a different functional unit, as indicated by the arrow to the left of c-Myc.

From our examination of learning circuits, we know that feedback loops are very important. Let's look at one feedback loop in each of these circuit diagrams. Open the cell cycle file to locate cdc25C and cycB:cdk1 (in area H4).

1. cdc25C is activated by phosphorylation (blue line with green *C18* label).
2. cdc25C removes phosphates from cdk1 at amino acids T14 and Y15 (shown as a green line coming from cdc25C). Removal of the phosphates removes the inhibition from cdk1.
3. cdk1 interacts with cycA (black *C5* label), and this interaction stimulates the phosphorylation of cdc25C (green *C36*). Although this positive feedback loop is difficult to recognize at first glance, with practice and familiarity with the system, you would be able to see other examples.

The second feedback loop is taken from the DNA repair diagram. Locate **p53** in area E7–9 on the right side of the circuit. Immediately you can tell that p53 is a critical component, because there are so many interactions emanating from it. If you look below p53, you will see three black dots (black *P18*), which indicates that p53 (the protein) can form homotetramers. The homotetramer can initiate transcription (black *P17*) from seven different genes (one box with seven red lines emanating from it). One of the activated genes is Mdm2 (red *P39*), and the protein MDM2 can form a heterodimer with p53 (black *P28*), which inhibits transcription of the Mdm2 gene (blue *P29*). These components form a negative feedback loop. In the published paper accompanying this circuit diagram, Kohn reported that MDM2/p53 formation leads to the degradation of p53 (black *P30*), as indicated by the blue arrow labeled *P31*.

All these interactions were discernible from circuit diagrams with very few words, and they allow us to comprehend interactions and make predictions. Kohn's circuit diagrams are works in progress; don't make the mistake of assuming that if a component is shown in the diagram, all of its interactions are known. Our knowledge is limited, and studying circuit diagrams will focus our research more efficiently. We are beginning to make headway on larger-scale models, but we still have a lot to learn.

DISCOVERY QUESTIONS

30. In Kohn's circuit diagrams, try to locate one small functional subcircuit (similar to the Myc subcircuit), and describe in words what you see. Which components are linked to other functional units?

31. Find BRCA1, the breast cancer gene, located in area F10 of the DNA repair circuit. Describe in words how BRCA1 interacts with p53.

32. One of the blue lines below p53 is labeled *P31*. Notice that two different repressors can block the degradation of p53. Given what you know about circuits, is blocking this degradation an important process or not? What effect would two repressive interactions have on the reliability of blocking the degradation of p53? Can you find any other pathways that also block the degradation of p53?

33. If you became fluent in this symbolic language, do you think you should get a foreign language credit?

Summary 11.1

We have seen that genetic toggle switches can be formed with very few genes. Genomes contain hundreds or thousands of toggle switches so they can produce proteomes and metabolomes in response to environmental changes. The "choice" between two responses can be driven by particular factors (e.g., nutrient concentration for λ phage lysogenic

LINKS
Jim Collins
Charles Cantor
Timothy Gardner

versus lytic lifestyles) or by stochastic behavior of proteins (e.g., *recA* and *lacZ* promoters). Given our understanding of how bistable genetic switches operate, can we design and construct synthetic bistable toggle switches that function as predicted? If so, could these synthetic switches be used to produce useful devices? If you were designing a bistable switch for gene therapy, you would want to incorporate a level of redundancy that provided the patient with an acceptable degree of reliability. It might seem uncomfortable to treat genes like machines with reliability that can be calculated, but this approach to genome analysis is valuable. Applying engineering principles to genomics helps us understand why diploids evolved the apparent waste in complex circuits with redundancies. In Section 11.2, we study a series of synthetic genomic circuits that illustrate the cutting edge of our knowledge of how biological circuits work.

11.2 Synthetic Biology

Every time we build a model, we are trying to create a simple version of a complex system. Given the lessons in Section 11.1, is it possible to design and construct synthetic circuits? The new field of **synthetic biology** (analogous to synthetic chemistry) applies engineering principles to genomic circuits to construct small biological devices that should help us understand how naturally evolved circuits function. Perhaps someday we will be able to construct synthetic circuits that can perform beneficial tasks, but for the foreseeable future, we will use synthetic components, devices, and systems to elucidate how natural circuits function. In this section, we will consider four case studies that increase in complexity. Use these examples as starting places for synthetic circuits you might want to construct and test. Discovery Questions will allow you to explore the field of

synthetic biology, including an interactive construction web site which is part of the **BioBricks** project at the Massachusetts Institute of Technology (MIT).

Can Humans Engineer a Genetic Toggle Switch?

Often there is a division among biologists who like to argue about which experiments are "real science": those working with naturally produced organisms and molecules vs. those working with theoretical models to unify experimental observations. Both approaches are necessary for a complete understanding, but it is important to remember that predictions based on theoretical models must be tested with real experiments. In 2000, two research groups described genetic toggle switches that had been design and constructed. Both groups used *E. coli* as the host for their switches, but they made two different types of switches. One group built a toggle switch very similar to the one used by λ phage to choose its lifestyle. We'll examine this switch first. Later we examine a switch that oscillates like a circadian clock.

How to Build a Toggle Switch

Timothy Gardner and his colleagues Jim Collins and Charles Cantor at Boston University built a toggle switch that could choose between two states. However, rather than letting *E. coli* determine which direction to go, they constructed a toggle switch that could be regulated by the investigators (Figure 11.24). In this switch, they needed two constitutive promoters (colored black and gray) and two repressor genes (also colored black and gray). The black repressor protein silences the gray promoter, which drives production of the gray repressor protein. Conversely, the gray repressor protein silences the black promoter, which drives production of the black repressor protein.

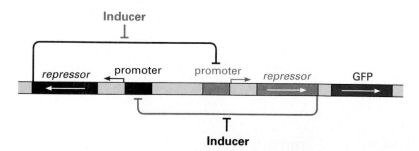

Figure 11.24 Theoretical two-gene bistable toggle switch.
The black gene "repressor" is transcribed from its black promoter. The black repressor protein binds to the gray promoter to block the production of the gray repressor protein. The gray repressor protein blocks the production of the black repressor protein when the gray repressor protein binds to the black promoter. To detect which state the toggle switch is in, GFP was placed downstream of the gray promoter so that the gray repressor and GFP are produced simultaneously.

Thus, if the black repressor protein were produced, the gray repressor protein could not be produced, and vice versa. This is an example of a synthetic bistable toggle switch similar to the theoretical one in Figure 11.1a. To control their toggle switch, the investigators utilized two inducer drugs, each of which incapacitates one of the repressor proteins.

The group from Boston University built and tested several different plasmids with a variety of promoters to see which would successfully produce a bistable toggle switch (Figure 11.25). To measure switching, the **reporter gene** GFP was added to the device to produce a glow-in-the-dark protein when the gray repressor protein was produced. As shown in Figure 11.25a, the investigators compared two plasmids, pIKE107 and pIKE105. The only difference between these two was the use of two different ribosomal binding sites that affected the efficiency of translation of the RNA into protein. Both 107 and 105 were capable of producing GFP (and thus the gray repressor as well), using an inducible promoter and a drug called IPTG. Note that 105 did not produce as much GFP as 107. When the inducer IPTG was removed, 105 was incapable of sustaining its output; therefore, it was not a bi*stable* switch. However, 107 was capable of sustaining the production of the gray repressor and GFP even after IPTG was removed; thus, it was stable. When 107 was exposed to the second inducer drug, called anhydrotetracycline (aTc), production of GFP (and thus the gray repressor) was eliminated within a couple of hours after induction of the black repressor. The experiment was well controlled (Figure 11.25b–c), since each half of the toggle switch was able to induce GFP production but neither of these "half-switches" was stable. Gardner's research demonstrated it is possible to take a theory (see Figure 11.1a) and construct a biologically functional switch in a test tube that works inside living cells.

LINKS
indicated plasmids
METHODS
inducible promoter

DISCOVERY QUESTIONS

34. Explain why pIKE105 was considered a failure and the control plasmid pTAK102 was deemed a success.

35. Based on the data in Figure 11.25, determine which drug (IPTG or aTc) was the gray inducer and which was the black inducer in Figure 11.24.

36. Design a bistable toggle switch that could be used in gene therapy to produce a protein on demand (e.g., insulin). Include in your design how the protein of interest could be turned on and off.

The success of Gardner's work is encouraging to those who want to understand genomic circuits. Synthetic circuits can be designed and validated experimentally. It also represents a success in "forward engineering," in which simple genetic circuits serve as models for more complex systems. In a more applied sense, the bistable toggle switch

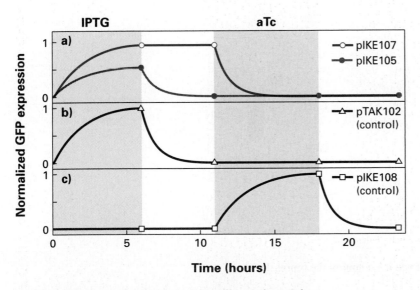

Figure 11.25 Experimental two-gene bistable toggle switch.
E. coli were transformed with the indicated plasmids and exposed to the inducer drugs as indicated by the shading. **a)** The only difference between pIKE107 and pIKE105 was different ribosomal binding sites that affected the rate at which the encoded proteins where translated. **b)** pTAK102 control plasmid contained only the IPTG-inducible promoter upstream of the GFP gene. **c)** pIKE108 control plasmid contained only the aTc-inducible promoter upstream of the GFP gene.

may prove useful in gene therapy and other biotechnology methods that require a silent gene to be activated at a certain point and then sustained. Finally, Gardner's team produced a bistable toggle switch, which is considered the first **genetic applet**, (a term derived from small computer programs called Java applets). Applets are self-contained programs, and Gardner's genetic applet is capable of being programmed (turned on or off, similar to the digital version of 1 or 0). Although the genetic applet is a long way from becoming a biological computer, the potential to store information in DNA is intriguing.

Can We Build a Synthetic Oscillating Clock?

One lesson from genomics is that interdisciplinary collaborations are the norm rather than the exception. Biologists need to collaborate with physicists, mathematicians, chemists, even graphic artists. While still a biology graduate student in Princeton University, Michael Elowitz teamed up with physicist Stan Leibler to construct a synthetic oscillating clock. To create a self-perpetuating cycling toggle switch required more genes than a two-gene genetic applet (Figure 11.26). First, you will notice that two plasmids were used. The larger **repressilator** plasmid controlled the cyclical nature of the output, while the smaller reporter plasmid

produced GFP. Three promoters and three protein repressors were used in the repressilator plasmid: the λ CI protein represses the production of LacI; the LacI protein represses the production of TetR protein; TetR represses the production of CI. You can see that each gene was repressed by one of the three repressors encoded on the repressilator plasmid. Therefore, the repressilator was a genetic closed circuit or triple negative feedback loop. Since it was impossible to directly observe the repressilator circuit in action, the reporter plasmid encoding GFP displayed the activity of the repressilator. GFP production was constitutive except when repressed by TetR that was part of the repressilator cycle; the *E. coli* cells lost their fluorescence every time the synthetic cycle produced TetR.

Figure 11.26 is a nice theoretical model, but does it actually work? Figure 11.27 dramatically illustrates how well the theory worked inside growing *E. coli*. The photos allow us to follow a single cell (Figure 11.27a–b) through its oscillations of GFP production (Figure 11.27c). The amount of GFP in one cell was measured, which revealed a fluorescence periodicity of about 150 minutes. What was so striking about the periodicity was that 150 minutes is longer than the *E. coli* cell cycle (about 65 minutes). The graph indicates cells divide more rapidly than the repressilator cycles; the investigators produced a cyclical circuit that outlived its cellular host.

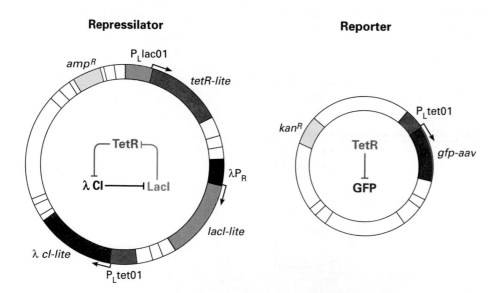

Figure 11.26 Two plasmids needed to make and see the output of the repressilator synthetic circuit.

The repressilator plasmid contains a cyclic negative feedback loop composed of three repressor genes and their corresponding promoters. $P_L$lac01 and $P_L$tet01 are strong promoters that can be tightly suppressed by LacI and tetracycline, respectively. The third promoter, λP_R, is repressed by CI (see Figure 11.2). The three repressor genes are appended with the suffix "lite" to indicate the encoded proteins degrade rapidly. The gfp-aav protein encoded on the reporter plasmid is also degraded rapidly, and its production is regulated by the $P_L$tet01 promoter.

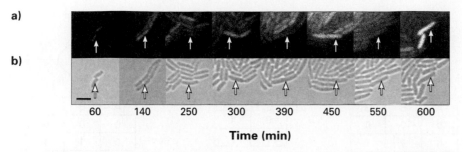

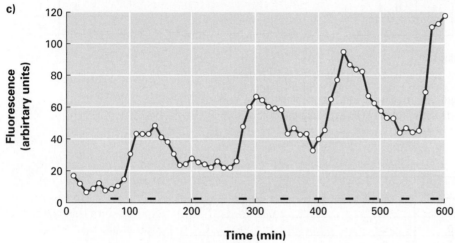

Figure 11.27 Cyclical toggle switching in live bacteria.

a) Fluorescence and **b)** phase contrast microscopy images of cells, revealing the time course of GFP expression and cell growth beginning with a single bacterium containing the repressilator and reporter plasmids (see Figure 11.26). Scale bar in **b)** indicates 4 μm in the photographs. **c)** The pictures in a) and b) correspond to peaks and troughs in the time course of GFP fluorescence intensity of the selected cell. Bars at the bottom of panel c) indicate the timing of cell division, as estimated from the phase contrast photomicrographs in panel b).

Math Minute 11.3 **How Can You Visualize Gene Regulation Logic?**

The schematic representation of the repressilator (Figure 11.26) contains information about how the device was constructed and how it works. However, schematic representations require you to follow the logic in your head to understand the behavior of the system. In this Math Minute, we explore how two other representations, truth tables and logic gates, are used to summarize gene regulation logic and behavior of systems like the repressilator.

Truth tables and logic gates describe a system of binary variables, each of which has the value 0 or 1. Gene regulation can be modeled with a binary system by representing gene expression as either "on" (= 1) or "off" (= 0) and regulatory elements as either "present" (= 1) or "absent" (= 0). For example, the simple binary function AND, which has the value 0 unless both input variables have the value 1, was used in statement 2 of Table 10.3 to model the toggle between control by module A and module B. The logic gate and truth table for the AND function are shown in Figure MM11.3a. The purpose of a truth table is to list the value of output variables for each possible combination of input variable values. A logic gate is a standardized symbol that encapsulates the same information for elementary functions; some of the most common logic gates and the corresponding truth tables are shown in Figure MM11.3.

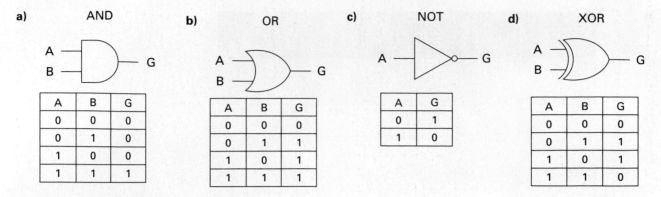

Figure MM11.3 Elementary logic functions.
Logic gate and truth table are given for each function. All gates, except NOT, receive two binary inputs (A and B) and return a single binary output (G). **a)** AND activates the output only if both inputs are present. **b)** OR activates the output if either (or both) of the inputs are present. **c)** NOT inverts the input. **d)** XOR activates the output if either (but not both) of the inputs is present.

Table M11.1 Repressilator truth table.

Input State				Output State			
λcI	LacI	TetR	GFP	λcI	LacI	TetR	GFP
0	0	0	0	1	1	1	1
0	0	1	1	0	1	1	0
0	1	1	0	0	1	0	0
0	1	0	0	1	1	0	1
1	1	0	1	1	0	0	1
1	0	0	1	1	0	1	1
1	0	1	1	0	0	1	0
1	1	1	1	0	0	0	0

A truth table can represent more complex functions that have many input and output variables. For example, the repressilator can be thought of as a function with four binary input and output variables: λcI, LacI, TetR, and GFP. The state of the repressilator at a particular time is represented by an ordered list of variable values; e.g., the state "1, 0, 0, 1" means that λcI is present, LacI is absent, TetR is absent, and GFP is present. Because the state of the system at the current time point determines the state of the system at the next time point, we can represent the behavior of the repressilator by describing the state of the system at regular time intervals.

In our example, the time interval is the time required to transcribe and translate a single copy of each gene. We also assume that all gene products are degraded in a single time interval. Thus, for example, if the input state is "1, 0, 0, 1," the output state will be "1, 0, 1, 1" because the input presence of λcI will prevent expression of LacI over the next time interval; the input absence of LacI will enable expression of TetR over the next time interval; and the input absence of TetR will enable expression of λcI and GFP over the next time interval. The truth table in Table MM11.1 summarizes input (current time point) and output (next time point) variable values for the repressilator.

To represent complex functions like the repressilator with logic gates, several gates must be combined into a circuit in which the output of one or more gates serves as input into other gates. A logic gate circuit model of the repressilator is shown in Figure MM11.4. Borrowing tools like truth tables and logic gates from the field of computer science gives us new ways to visualize complex information and helps us gain a deeper understanding of systems like the repressilator.

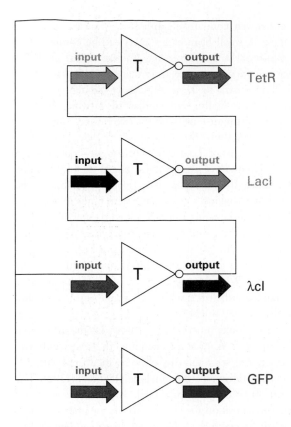

Figure MM11.4 Logic gate circuit model of repressilator.
The time interval for propagating values from one variable to another is represented by the "T" in each NOT gate. The input state of the system is indicated by the labeled arrows to the left of each NOT gate. The output state of the system is indicated by the labeled arrows to the right of each NOT gate.

MATH MINUTE DISCOVERY QUESTION

1. Why doesn't the repressilator get locked into a cycle of alternating between all 0's and all 1's?

The very regular periodicity of the repressilator helps you appreciate your own circadian rhythm and how it too can be controlled by a small number of genes that produce cyclical amounts of protein. However, there is a fly in this ointment of perfect timing. Noise and stochastic behavior of proteins are a fundamental property of gene regulation and protein production, so you might expect some problems with the regularity of the repressilator.

DISCOVERY QUESTIONS

37. How can a biological clock outlive its host cell?
38. Graph the production of CI, LacI, and TetR proteins on top of the graph in Figure 11.27c.

39. Why did the investigators choose proteins that are rapidly degraded by cells? What would have happened if the proteins were long-lived?
40. Predict what might happen to the repressilator periodicity inside sister cells after division.
41. What would be required for sister cells to maintain the exact same periodicity of GFP production? In other words, does the repressilator have any means for cooperation through communication or checkpoints (see pages 376)?

Given the clarity of the data in Figure 11.27c, you might think humans can construct a clock that is unaffected by

LINKS
Christina Smolke

the inherent noise of gene regulation. Still, a clock can only be as consistent as its component parts (Figure 11.28). We can follow one cell (purple trace in Figure 11.28c) and its sisters from two rounds of division (in gray and black) and see that although oscillation was retained in all three cells, timing and amplitude were not maintained through the generations. Interestingly, when cells stopped dividing due to nutritional limitations (**stationary phase**), the repressilator stopped working. Therefore, cell growth was required for proper functioning of the repressilator.

DISCOVERY QUESTIONS

42. Predict what would happen if IPTG were added to the repressilator (IPTG disrupts the function of LacI).

43. Hypothesize why some sibling cells altered periodicity and others altered amplitude.

Can Synthetic Devices Alter Gene Expression?

A chemical engineering graduate student in Christina Smolke's lab at Caltech, Travis Bayer, wanted to synthesize small devices that could predictably alter the activity of reporter genes. However, Bayer and Smolke wanted a faster response time than exhibited by the repressilator, so they chose to design RNA molecules that could respond quickly to environmental changes. They called their new molecules **antiswitches** because they were bistable toggle switches that performed part of their function by acting like antisense RNA. The term antiswitch should not be confused with an other new term, **riboswitch**, which describes the 5′ untranslated regions that regulate translation of the attached coding regions of some mRNAs.

An antiswitch contains an **aptamer**, a nucleic acid that can bind to a small ligand. Aptamers can be composed of either RNA or DNA. Bayer and Smolke found an RNA aptamer sequence in the literature that bound a drug called theophylline (used to treat respiratory ailments). Theophylline is structurally very similar to caffeine, but the aptamer can distinguish between the two related compounds (Figure 11.29a). Onto the aptamer, the investigators fused two different stems of bases. The short, single-stranded stem was called the aptamer stem and the longer, double-stranded one was called the antisense stem (Figure 11.29b). In the absence of theophylline, the antisense stem binds to itself and has no effect on the target mRNA. When theophylline binds to the aptamer, the entire molecule experiences a conformational change such that the antisense stem becomes single-stranded and the aptamer stem binds to a portion of the antisense stem. With this toggle switch of conformations, the single-stranded antiswitch stem is now able to bind to, and silence, the targeted mRNA.

The design looks functional, but how did the antiswitch perform in vivo? A plasmid encoding the antiswitch was transformed into yeast and the cells were incubated with different concentrations of theophylline (Figure 11.29c). The antiswitch functioned as predicted, with a discrete theophylline concentration for binding to the aptamer and time required for silencing the GFP signal. In subsequent experiments, Bayer and Smolke designed and built antiswitches that worked at different concentrations of theophylline, thereby demonstrating the flexibility of their devices.

Now that they could turn off GFP translation, the investigators wanted to test the specificity of their new device. Could they design and construct a system that would distinguish two different ligands to silence GFP and YFP (yellow fluorescent protein; Figure 11.30)? To test their constructs, they transformed yeast cells with two plasmids, each encoding a different antiswitch, and then exposed the cells to each

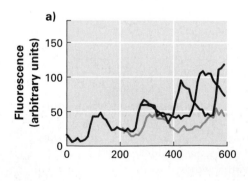
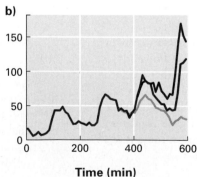
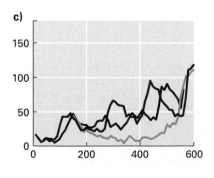

Time (min)

Figure 11.28 Examples of sister cells that maintain periodicity but not synchronicity.
a) to **c)** In each case, the fluorescence time course of the cell depicted in Figure 11.27 is redrawn in purple as a reference, and two of its siblings are shown in black and gray.
a) Siblings exhibiting delays in the phase after cell division, relative to the reference cell.
b) Phase is approximately maintained but amplitude varies significantly after division.
c) Reduced period (black) and long delay (gray).

ligand separately and then in combination. Once again, the synthetic devices worked exactly as designed.

Having mastered antiswitches that turn off active genes, the investigators designed and built a new antiswitch that repressed GFP production until the cells were incubated with ligand (Figure 11.31). They tested cells that contained one or the other of the two antiswitches and, as predicted, found that the two opposing antiswitches turned on or off GFP production at about 1 mM theophylline. Given this ability to turn a gene on or off, and the ability to distinguish different concentrations of ligand, it seems that Bayer and Smolke have designed and constructed very clever devices that should prove useful in future synthetic biology research.

DISCOVERY QUESTIONS

44. Explain why antiswitches might react faster than protein-based mechanisms for gene silencing.

45. Riboswitches, aptamers, and antiswitches are all new terms. Try a search on PubMed to see how many citations include these terms. It would be nice to know what aptamers are available. Try searching RNABase.org using the word "aptamer." Then try searching Aptamer Database to find particular ligands. Can you find the two used in the antiswitch case study? How could this database be improved?

LINKS
Aptamer Database
RNABase.org

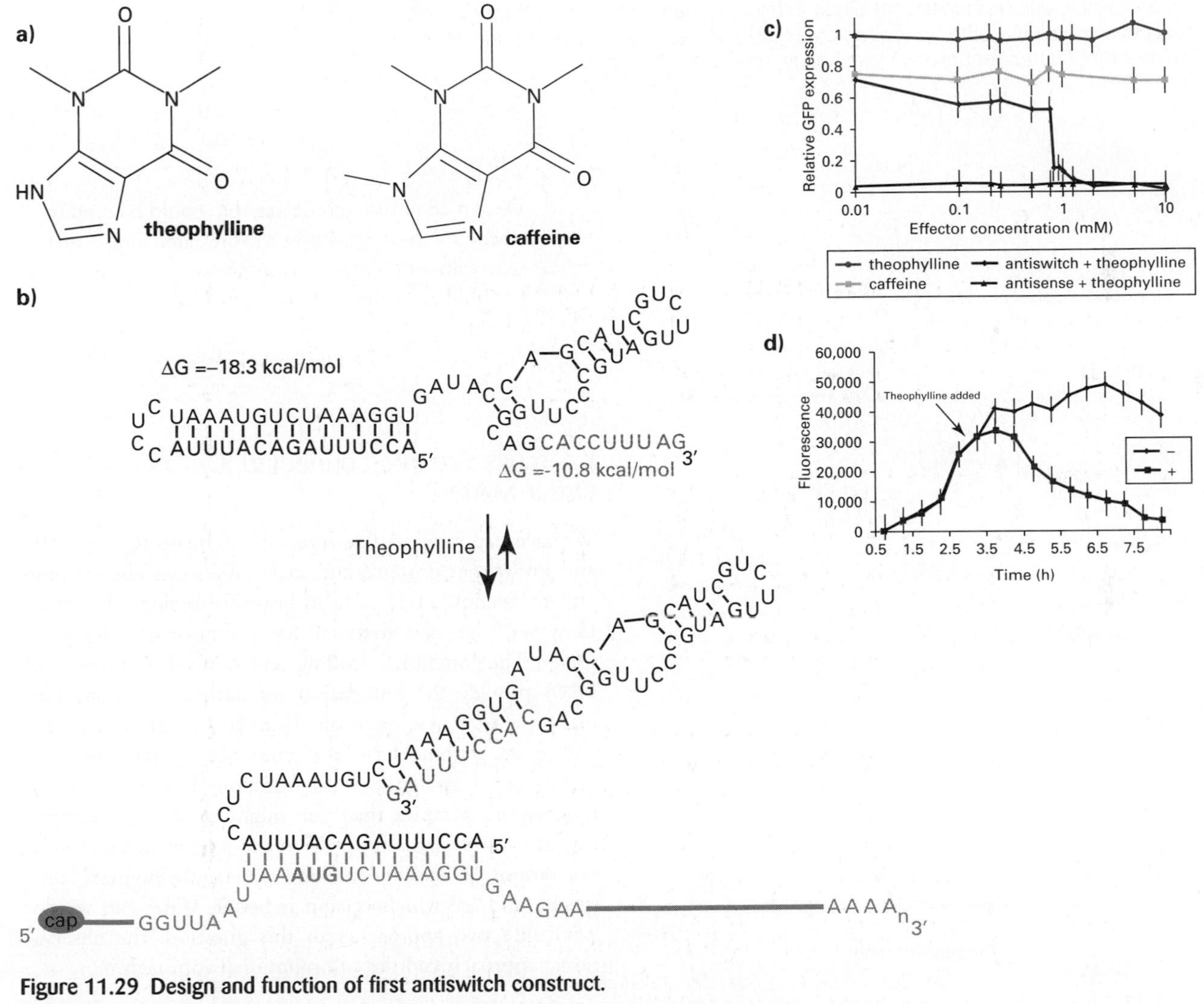

Figure 11.29 Design and function of first antiswitch construct.
a) Theophylline and its close relative, caffeine, which does not bind to the aptamer.
b) Sequence and conformational changes of antiswitch upon binding theophylline. Without ligand, the antiswitch antisense arm is closed and the aptamer arm is open. Theophylline binding causes the antiswitch arm to open and bind to the GFP mRNA (including start codon) while the aptamer arm closes. The stability of each arm when closed is shown as change of free energy (ΔG; larger negative number is more stable).
c) Three control experiments demonstrate the function meets the design expectations.
d) Two flasks of identical cells produce GFP, but one is incubated with theophylline to block further GFP production. All antiswitch experiments were performed in yeast cells.

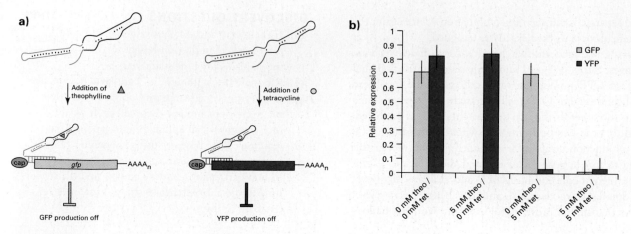

Figure 11.30 **Antiswitches respond specifically to different ligands and different mRNAs.**
a) New antiswitch was produced that binds to tetracycline and blocks YFP production.
b) Yeast cells containing both antiswitch plasmids and exposed to theophylline, tetra-cycline, or both; fluorescent protein levels measured with fluorometer.

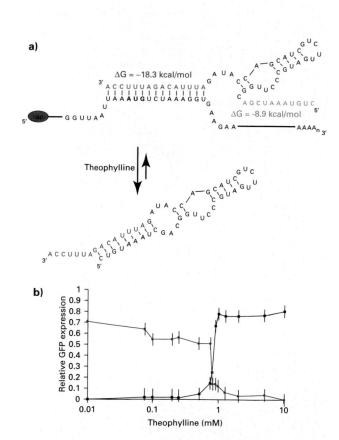

Figure 11.31 **Antiswitch responds to ligand and permits GFP production.**

a) Antisense and aptamer arm sequences are modified so that the greater stability of binding occurs between the anti-switch and mRNA in the absence of theophylline. **b)** Two different yeast cell populations expressing either the on-antiswitch (black) or the off-antiswitch (purple), both of which respond at the same ligand concentration.

46. Design an antiswitch device that could be used to treat a medical condition of your choosing. Don't worry about delivering your antiswitch to cells, just think creatively about ways to alter protein production.

If Circuits Are Interconnected, Does Gene Order Matter?

We have seen integrated cis-regulatory elements (Chapter 10), and know that proteins and metabolites can also be integrated (Section 11.1) to form beautiful biological circuits. However, the raw material for evolutionary change is DNA. The functional (coding and noncoding) aspects of DNA provide the foundation for natural selection. This brings us to a new question. Does the location of a gene within its genome have any influence on its role? This seems like a straightforward question, but it is actually much more complex than you might think. For example, how do you design an experiment to answer this question? We cannot go back in time and genetically engineer some species and see which version is better. What can we do? There are two approaches to this question: the observational approach and the computational approach.

Observational Approach

The first approach is the classic evolutionary biologist's approach of comparing what you find in nature. The rationale behind this is very sound. The best way to examine fitness is to see what has evolved over millions of years of selective pressure. Imagine some original cell on earth, some tiny robust prokaryote that was capable of surviving mutations to its own genome (RNA or DNA) and giving rise to a second species on earth. These two species continued to diversify as new genes evolved, became duplicated, modified, duplicated again, etc. All this genetic experimentation led to new arrangements of genes. For example, species 4 might have genes A, B, C, D, E, F, G, H, I, J, whereas species 5 evolved with a gene order of A, H, G, F, E, D, C, B, B', I, J. At some point during this time of species diversification, sex evolved and the order of genes became scrambled even more with chromosomal rearrangements.

DISCOVERY QUESTIONS

47. What kind of changes occurred to the DNA of species 5 compared to 4 (see preceding text)? What new opportunities does species 5 have by virtue of its two copies of gene B?

48. What would you predict about the degree of reliability (see pages 377–378) of the process encoded by genes B and B'?

49. The observational approach has limitations based on available genomes. Can you imagine a way to experimentally test the effects of altering gene order within a genome?

Evolutionary biologists assume that what works best must be what is still in existence today. Some look at morphology, some at behavior, and others at DNA sequences. As a result of genome comparisons, investigators have identified many gene complexes that are highly ordered. The most famous are the *Hox* genes, which are conserved not only in DNA sequence but also in gene order on chromosomes. DNA sequence and gene order are conserved from flies to humans, with genes arranged from head to tail along the length of the chromosomes in humans, mice, worms, and flies. To molecular biologists, however, there is something unsatisfying about this kind of research: it leads to correlations that have not been experimentally tested. Every molecular biologist accepts that the arrangement of *Hox* genes is evolutionarily advantageous, but it is impossible to test this by altering history.

Recent work by many researchers who analyze DNA sequences has produced some interesting observations. Elizabeth Williams and Laurence Hurst found that linked genes often evolve at similar rates. The coevolution of linked genes might seem obvious, but remember that after a few billion years of recombination and mutation, the probability that two genes would remain as neighbors and mutate at similar rates is very small. The caveat about this type of analysis is that you never can be sure if the

conserved gene order is due to chance or selective pressures. To address this concern, a group from the European Molecular Biology Laboratory (EMBL) in Germany analyzed the genomic sequences of nine different taxa of prokaryotes (proteobacteria, Gram-positive bacteria, and Archaea). They found, in many cases, that gene pairs were conserved in their order on chromosomes across diverse taxa. Clusters of two to seven genes were conserved in order and orientation for transcription. It is hard to imagine that chance alone could explain conserved gene order and orientation. Nevertheless, this observation is a correlation and not a causative analysis, and thus is not completely satisfying.

LINKS G◈P
Drew Endy
Elizabeth Williams and Laurence Hurst
EMBL

Computational Approach

There is a second approach that is more satisfying to the growing number of biologists exploring genomes, though this approach also has weaknesses. If we know all the sequences of the genes in a genome, and we know the roles for most of these genes, then why not build a computer simulation, make predictions, test them, refine the model, etc.? This computational approach to evaluating gene order for entire genomes is very new. One of the pioneers in this area is Drew Endy, currently working at MIT, and his whole genome approach to evolution is a new trend in biology.

Endy approached this experimentally difficult question of gene order by choosing a simple system (i.e., a model organism). Rather than selecting a species like yeast, which has about 6,200 genes, he chose to start smaller, with a viral genome of only 56 genes. Endy chose the virus T7 as his model organism (Figure 11.32). T7's double-stranded DNA genome is only 39,937 bp long. This is so small that it is hard to imagine the genome includes much complexity, but even this simple organism has some hidden secrets. Of the 59 proteins produced by T7's 56 genes, only 33 have known functions. That could present some problems with a simulated circuit, but computer models are not intended to be perfect; rather, they should incorporate current knowledge and allow predictions that can be tested experimentally to improve our understanding. Clearly, there is room for improvement when only 56% of the proteins have known functions.

Twenty-five minutes after T7 infects an *E. coli* bacterium, about 100 T7 progeny will be released when the host cell lyses. Unlike viruses that inject their genomes quickly, it takes about 10 minutes for the T7 genome to enter its host. The first 850 bp are inserted by the virus, and the rest is pulled into the host as it is being transcribed. Initially, the *E. coli* RNA polymerase binds to one or more of the 5 *E. coli*-recognized promoters in the T7 genome (spanning the first 15% of the genome in Figure 11.32). *E. coli* RNA polymerase pulls the T7 genome into the cell at about 45 bp

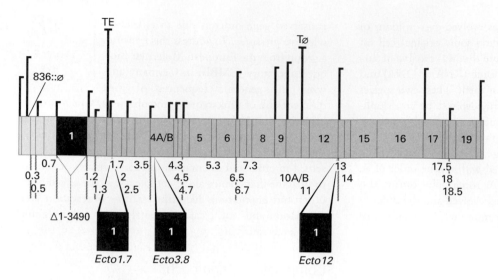

Figure 11.32 Wild-type T7 (T7⁺) genome contains 56 genes encoding 59 proteins.
Numbers represent coding regions (genes were numbered as space permitted). Vertical lines with half-bars represent *E. coli* (black) and T7 (purple) promoters, with bar height proportional to the strength of each promoter. *E. coli* RNAP (polymerase; TE) and T7 RNAP (Tø) terminators are shown as vertical lines with full bars. RNase III recognition sites are shown as vertical lines below the genome. All transcription moves from left to right. The positions of the *gene 1* (encoding the T7 RNA polymerase) in T7⁺ and in the three experimental strains constructed and characterized in the laboratory are shown below the genome and labeled *Ecto1.7*, *Ecto3.8*, and *Ecto12*; *wt gene 1* was deleted in the three Ecto-mutants.

per second. The first essential T7 gene to produce a protein is conveniently called *gene 1*. *gene 1* encodes the T7 RNA polymerase, which takes over the role of transcription and genome movement into *E. coli*. T7 RNA polymerase uses 17 different promoter sequences (in the remaining 85% of the genome) and pulls the DNA in at a rate of 200 bp per second—a fourfold increase in genome movement. Therefore, it is easy to see why T7 would evolve to have *gene 1* as the first coding sequence in its genome.

Does *Gene 1* Have to Be First?

T7 is a clear example that gene order might matter even for a small genome. How much flexibility is there for *gene 1* position within the genome? Could it be second? Third? If the promoters were moved around, would the change alter T7's evolutionary outcome? Endy started his computer simulation with the question about the optimum position for *gene 1* (Figure 11.32). Using in silico methods, Endy placed *gene 1* in every possible position (except as very first and very last) and calculated how long it would take T7 to double its numbers in a simulated flask of *E. coli*. He used this formula to determine the growth rate of T7:

$$\text{maximum doubling rate} = \mu_m = \max_t\{\log_2[Y(t)]/t\}$$

where $Y(t)$ is the computed number of intracellular progeny as a function of time (t).

The term μ_m defines the optimal time for phage-induced lysis in environments containing infinite uninfected hosts. For each of the 72 in silico *gene 1* positional mutants, μ_m was calculated and graphed. When he ran his simulation, Endy discovered some genotypes were better than the simulated *wt* strain (see Figure 11.33). The *wt* genotype (triangle) has a doubling rate of about 0.35 per minute, but there are several genotypes with higher rates—some as much as 30% higher. This simulated improvement on evolution flies in the face of evolutionary theory. Surely the T7⁺ genome is not suboptimal in its evolutionary fitness! But remember, this was a computer simulation. How did Endy's simulations compare with reality?

Three **ectopic** (not in their normal genomic location) strains, where copies of *gene 1* were placed into new locations, were generated in the lab. *Ecto1.7* had *gene 1* inserted into a nonessential gene called *1.7*. *Ecto3.8* had *gene 1* inserted into the coding region of the nonessential *gene 3.8*. In *Ecto12*, *gene 1* was inserted between a promoter and *gene 12*. Endy also constructed two control viruses which had a *gene 1*-sized piece of λ phage DNA inserted into genes *1.7* and *3.8*, and a third control strain in which a T7 late promoter was inserted at base 836.

DISCOVERY QUESTIONS

50. Think of reasons why the controls Endy used were not ideal. Design a different control for these experiments.

51. It is difficult to understand what would make a mutant strain more efficient, but easier to understand less efficient mutants. Explain why the mutations with *gene 1* positioned late in the genome would be the least efficient.

52. In the T7 computer simulation, G🧬P identical viruses were programmed to behave identically. In reality, would you expect each virus to perform exactly the same? Do you think stochastic events might occur in natural settings? If so, what effect might randomness have on the difference between computer growth curves and real growth curves?

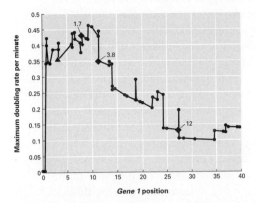

Figure 11.33 Predicted consequences of altered *gene 1* position.
Computed maximum phage doubling rate, μ_m, as *gene 1* was repositioned on the T7$^+$ genome. Black circles mark the 5′ end of the in silico ectopic (inserted) *gene 1* and the μ_m for 72 positional mutants. The purple diamonds indicate the computed μ_m for the ectopic *gene 1* strains that were constructed and characterized experimentally. The purple triangle indicates the μ_m computed for T7$^+$.

Endy compared the simulated viruses to his three genetically engineered mutant strains of T7. Each real strain was added to a separate population of *E. coli* in a flask, and the real doubling rates for these T7 mutants were experimentally measured (Figure 11.34). The simulated growth curve for T7$^+$ was four minutes faster than the experimentally determined growth curve for T7$^+$. Compare the slopes of computer growth rates and the real growth rates for T7, and you will see they are very similar. The simulated growth curve for *Ecto12* was ten minutes faster than the experimentally determined curve. Live *Ecto1.7* and *Ecto3.8* strains showed little agreement with predicted growth rates.

As part of his research, Endy simulated and measured plaque sizes, DNA replication, protein production, and lysis rates for his three real mutants. Go to the web site to view the T7 simulation (with a link to the 80-MB movie) that shows the rate of transcription and translation for every gene in the T7 genome. It also shows the *E. coli* and

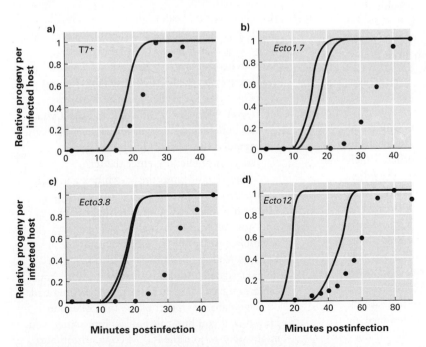

Figure 11.34 Computed (solid lines) and observed (black dots) intracellular one-step growth curves for T7$^+$(*wt*) and the ectopic *gene 1* strains.
Note the different time scale for *Ecto12*. **a)** *wt*, **b)** *Ecto1.7*, **c)** *Ecto3.8*, and **d)** *Ecto12* are shown (black), with T7$^+$ growth curve (purple line) for reference.

T7 RNA polymerases in action. All this work indicated that Endy was successful in the scientific sense. He constructed a model, made predictions, tested them, and now he will refine his model based on new experimental evidence.

DISCOVERY QUESTIONS

53. Which part of the real curve differs the most from the simulated curve (see Figure 11.34)? Explain why this part of each curve might be so different, given what you know about stochastic protein kinetics and noisy toggle switches.

54. Reevaluate the data in Figure 11.33 after analyzing the experimental data from Figure 11.34. What can you say about the optimum gene order for T7?

55. In the T7 simulation, which is produced in greater quantities, mRNA or proteins? By the end of infection, which genes are most highly transcribed? Which proteins are most abundant? How many copies are present inside the infected cell for the most abundant protein? At the end of the infection, the earliest T7 genes are no longer transcribed. Explain why this may be biologically adaptive and why this may indicate why the order of T7 genes is functionally significant.

56. Can the T7 RNA polymerase transcribe all the genes, or only the last 85%? In this simulation, what happens when a T7 RNA polymerase catches an *E. coli* RNA polymerase from behind? Do you think this is biologically accurate? Explain your answer.

57. Initially, what two proteins are the only ones present? What happens to each of them during the infection?

newest insights. Over several iterations, T7.*n* will lead to better predictions and a better understanding of genetic toggle switches and synthetic circuits.

DISCOVERY QUESTIONS

58. You may not realize it, but you can conduct synthetic biology research too. Go to the MIT BioBricks Registry to see a community-based effort to understand biological circuits and switches. The concept behind BioBricks is that DNA parts could be interchangeable, similar to Legos. If you could snap one part onto any other part, you could design and construct new devices and systems to test your understanding of how genomes perform some of their functions. Try these tasks to learn your way around.

   a. Click on "Parts, Devices & Systems" to learn the vocabulary, and "About Parts" in the left frame to learn what is in the BioBrick Registry.

   b. From the main page, click on the "reporter" icon to see a full list of reporter components. Find GFP and GFP-AAV. What is the difference between these two components?

   c. From the main page, enter "BBa_M0044" in the View Part search box on the bottom left side. What is AAV, and what information would you like to know that is missing?

   d. Click on the "Search" link on the left side, then click on the word "Example" above the middle search box. The system you have searched for is composed of four devices. Below it are the intermediate construction phases that are available in the BioBricks freezer. Mouse over the icons and see if you can understand the four devices, as well as the intended function of the overall system. To test your prediction, enter the term "BBa_I13907" into the search box and read the description.

59. Information in the Registry may be read by anyone. You can sign up as a guest member and assemble parts in the "sandbox" area of the Registry. Your part numbers have to be in the range designated for nonregistered visitors. You can create parts, but they will not be maintained on the database as a part of the full collection. However, if you find this process interesting, you might want to join the iGEM competition next year to conduct real research in synthetic biology. Although the field is cutting edge, the equipment requirements are modest.

Endy's research illustrates how difficult it will be to understand more complex genomes. Part of the problem with T7$^+$ is that some genes overlap each other, which makes it impossible to mutate one gene but not the other. To address this challenge, Leon Chan and Sriram Kosuri generated a synthetic T7 (called T7.1) in which every gene is completely separated from its neighbors. T7.1 sounds like a good idea that is bound to fail because synthetic biologists cannot fully understand the selection pressures that have driven maximization of fitness. However, T7.1 was viable and produced viable progeny. The plaque size and morphology were similar to T7$^+$, but the time to lysis was delayed for T7.1. As with computer models designed to approach biological reality, T7.1 is not perfect, but it does improve our previous understanding of the naturally evolved T7$^+$. With T7.1, we will be able to dissect each component of the genome to understand its function and later build an improved model of T7 that will test our

Summary 11.2

By applying engineering principles to toggle-switch design, we can construct biologically functional genetic switches and experimentally test their reliability. It might seem uncomfortable to treat genomes like machines with interchangeable parts, but by designing and building synthetic devices and systems, we can test our understanding more completely. Whether we assemble three-gene circuits, produce antiswitches to regulate the expression of proteins, or reengineer a model genome as a way of teasing apart the function of component parts, we are always trying to understand biology and evolution more completely. Eventually, synthetic biology may lead to new medical treatments, but for now, research using designed DNA parts is proving to be challenging and insightful.

Chapter 11 Conclusions

Understanding a genome requires more than just listing the component genes and defining their roles. We need to understand how the proteins interact, how genes regulate each other's transcription, and how positive and negative feedback loops produce synergistic properties that affect a cell's response to its changing needs. Toggle switches enable genomes to make "choices," and the inherent noise in toggle switches is a necessary part of how they function. Proteins exhibit stochastic behavior, so it may be impossible to produce a model that can predict with certainty how a particular cell will respond to a change in its environment. Complex circuits have unique properties, and as we construct increasingly complex circuit diagrams, we will discover more emergent properties that improve our understanding of cells and organisms. A cell web is composed of many interconnected components, and the ultimate challenge is to model the combined information of genome sequences, variations in the population, gene expression profiles, proteomics, metabolomics, and genomic circuits. These challenges are intimidating but exciting, too. In Chapter 12, we will use one case study to see what is possible now and where we want to go in the future.

References

Bayer, T. S., & C. D. Smolke. 2005. Programmable ligand-controlled riboregulators of eukaryotic gene expression. *Nature Biotechnology*. 23(3): 337–343.

Bhalla, U. S., & R. Iyengar. 1999. Emergent properties of networks of biological signaling pathways. *Science*. 283: 381–387.

Bhalla, U. S., P. T. Ram, & R. Iyengar. 2002. MAP kinase phosphatase as a locus of flexibility in a mitogen-activated protein kinase signaling network. *Science*. 297: 1018–1023.

Bonifer, C. 2000. Developmental regulation of eukaryotic gene loci. *Trends in Genetics*. 16: 310–315.

Chan, L. Y., Sriram, K., & D. Endy. 2005. Refactoring bacteriophage T7. *Molecular Systems Biology* 64–73 doi: 10.1038/msb4100025.

Cho, R. J., & M. J. Campbell. 2000. Transcription, genomes, function. *Trends in Genetics*. 16: 409–415.

Clayton, D. F. 2000. The genomic action potential. *Neurobiology of Learning and Memory*. 74: 185–216.

Dandekar, T., M. B. Snel, & P. Bork. 1991. Conservation of gene order: A fingerprint of proteins that physically interact. *Trends in Biochemical Sciences*. 23: 324–328.

Edwards, J. S., & B. O. Palsson. 2000. The *Escherichia coli* MG1655 in silico metabolic genotype: Its definition, characteristics, and capabilities. *PNAS*. 97: 5528–5533.

Edwards, J. S., & B. O. Palsson. 2000. Metabolic flux balance analysis and the in silico analysis of *Escherichia coli* K-12 gene deletions. *BioMed Central Bioinformatics*. 1(1): 1. <http://biomedcentral.com/1471-2105/1/1>

Elowitz, M. B., & S. Leibler. 2000. A synthetic oscillatory network of transcriptional regulators. *Nature*. 403: 335–338.

Endy, D., & R. Brent. 2001. Modelling cellular behavior. *Nature*. 409: 391–395.

Endy, D., L. You, et al. 2000. Computation, prediction, and experimental tests of fitness for bacteriophage T7 mutants with permuted genomes. *PNAS*. 97: 5375–5380.

Frankland, P. W., C. O'Brien, et al. 2001. α-CaMKII-dependent plasticity in the cortex is required for permanent memory. *Nature*. 411: 309–313.

Gardner, T. S., C. R. Cantor, & J. J. Collins. 2000. Construction of a genetic toggle switch in *Escherichia coli*. *Nature*. 403: 339–342.

Genoux, D., U. Haditsch, et al. 2002. Protein phosphatase 1 is a molecular constraint on learning and memory. *Nature*. 418: 970–975.

Gladwell, M. 2000. *The tipping point: How little things can make a big difference*. Boston: Little, Brown.

Hartwell, L. H., J. J. Hopfield, et al. 1999. From molecular to modular cell biology. *Nature*. 402 Supplement: C47–C52.

Hasty, J., J. Pradines, et al. 2000. Noise-based switches and amplifiers for gene expression. *PNAS*. 97: 2075–2080.

Hurst, L. D., Williams, E. J. B., & Pál, C. 2002 Natural selection promotes the conservation of linkage of co-expressed genes. *Trends in Genetics*. 18: 604–606.

Jeong, H. B., R. Tombor, et al. 2000. The large-scale organization of metabolic networks. *Nature*. 407: 651–654.

Kelley, B. P., R. Sharan, et al. 2003. Conserved pathways within bacteria and yeast as revealed by global protein network alignment. *PNAS*. 100(20): 11394–11399.

Kitami, T., & J. H. Nadeau. 2002. Biochemical networking contributes more to genetic buffering in human and mouse metabolic pathways than does gene duplication. *Nature Genetics*. 32: 191–194.

Kohn, K. 1999. Molecular interaction map of the mammalian cell cycle control and DNA repair systems. *Molecular Biology of the Cell*. 10: 2703–2734.

Kuang, Y., I. Biran, & D. R. Walt. 2004. Simultaneously Monitoring Gene Expression Kinetics and Genetic Noise in Single Cells by Optical Well Arrays. *Analytical Chemistry*. 76: 6282–6286.

Lee, J. F., J. R. Hesselberth, et al. 2004. Aptamer database. *Nucleic Acids Research*. 32(1): D95–100.

Legrain, P., J.-L. Jestin, & V. Schächter. 2000. From the analysis of protein complexes to proteome-wide linkage maps. *Current Opinion in Biotechnology*. 11: 402–407.

Lercher, M. J., J-V. Chamary, & L. D. Hurst. 2004. Genomic regionality in rates of evolution is not explained by clustering of genes of comparable expression profile. *Genome Research.* 14: 1002–1013.

McAdams, H., & A. Arkin. 1999. It's a noisy business! Genetic regulation at the nanomolar scale. *Trends in Genetics.* 15: 65–69.

McAdams, H., & A. Arkin. 1997. Stochastic mechanisms in gene expression. *PNAS.* 94: 814–819.

McAdams, H. H., & L. Shapiro. 1995. Circuit simulation of genetic networks. *Science.* 269: 650–656.

Miller, P., A. M. Zhabotinsky, et al. 2005. The stability of a stochastic CaMKII switch: Dependence on the number of enzyme molecules and protein turnover. *PLoS Biology.* 3(4): 0705–0717.

Milo, R., S. Shen-Orr, et al. 2002. Network motifs: Simple building blocks of complex networks. *Science.* 298: 824–827.

Murthy, V. L. 2005. The RNA structure database. <www.RNABase.org>. Accessed 20 May 2005.

Pirson, I., N. Fortemaison, et al. 2000. The visual display of regulatory information and networks. *Trends in Cell Biology.* 10: 404–408.

Ptashne, M. 1987. *A genetic switch: Gene control and phage λ.* Cambridge, MA: Cell Press & Blackwell Scientific Publications.

Rzhetsky, A., T. Koike, et al. 2000. A knowledge model for analysis and simulation of regulatory networks. *Bioinformatics.* 16(12): 1120–1128.

Suyama, M., & P. Bork. 2001. Evolution of prokaryotic gene order: genome rearrangements in closely related species. *Trends in Genetics.* 17(1): 10–13.

Weng, G., U. S. Bhalla, & R. Iyengar. 1999. Complexity in biological signaling systems. *Science.* 284: 92–96.

Wray, G. A. 1991. Promoter logic. *Science.* 279: 1871–1872.

Modeling Whole Genome Circuits

Up to this point, we have examined the complexity of and interactions within different subsets of genomes and proteomes. In this chapter, we begin to integrate those subsets into comprehensive and predictive models. The substantial difference between the field of genomics and traditional molecular biology is that genomics wants to understand all the molecules of life as they interact with each other. The methods developed over the last few years have provided us with new ways of looking within cells to see the connections made in biological circuits. On a weekly basis, leading scientific journals publish examples illustrating the rapidly evolving field of biology. Although numerous case studies exist, only one is discussed in this chapter: the modeling of the whole genome response in yeast to a change in energy sources. If you read *Science* or *Nature*, you know that the field of genomics is gathering momentum and that no textbook can compete with journals for current information. Therefore, use this chapter as an introduction to discovering the new perspective of genomics.

12.1 Is Genomics a New Perspective?

There is a movement afoot in biology. Some call it **systems biology**, but it has many synonyms: genomics, circuits, intentional, modular, etc. Here is how the Institute for Systems Biology (ISB) defines the term:

> Systems biology is a unique approach to the study of genes and proteins which has only recently been made possible by rapid advances in computer technology. Unlike traditional science which examines single genes or proteins, systems biology studies the complex interaction of all levels of biological information: genomic DNA, mRNA, proteins, functional proteins, informational pathways and informational networks to understand how they work together. Complex systems give rise to emergent or systems properties—for the brain, these are memory, consciousness, and the ability to learn. These systems properties cannot be understood by studying individual neurons; rather, the whole system or subsystems must be analyzed. For the past 30 years, biologists have tended to study individual genes or proteins. Systems biology requires global technologies both to define the elements of the system and to follow the elements' behavior as the system carries out its functions. For example, if one is to understand how a car functions, biologists would have studied the individual parts in isolation—the transmission; the ignition; the brakes, etc. The systems approach defines all of the elements in a system (discovery science) and then

studies how each behaves in relation to the others as the system is functioning. Ultimately, the systems approach requires a mathematical model which will both describe the nature of the system and its systems properties.

In his book *The Tipping Point*, Malcolm Gladwell dissects the components necessary for a small trend to spread to the rest of a culture. There are three critical components of this process: the people involved, the quality of the message, and the context within which the message is delivered. Gladwell is very clever in his exposition of the "tipping point" and gives many good examples. The question now is whether the newest fad in biology, systems biology, will reach some tipping point until "everyone is doing it." Let's look at the three critical components so you can decide if systems biology is the greatest thing since spliced genes.

The People Involved: Who Is Doing Systems Biology?

The list of people who are pushing for a systems approach reads like a who's who of biologists. Sydney Brenner, founder of the Molecular Science Institute (MSI), is responsible for the identification of mRNA (with François Jacob and Matthew Meselson); the idea that the genetic code consists of triplets (with Francis Crick); and the development of the nematode *Caenorhabditis elegans* as a model research organism (for which he shared a Nobel Prize). Most people would retire happily with that set of credentials, but Brenner is not resting on these laurels. MSI, located in Berkeley, California, includes on its board of trustees the 1999 Nobel laureate Günter Blobel, who works at The Rockefeller University in New York, and Drew Endy, who produced the online T7 program that evaluated gene order (Chapter 11). MSI describes its goal as follows:

> One consequence of this work will be to bring into being an intentional biology. By this we mean a rational and, as possible, controlled human interaction with the living world. Such a transition requires an ability to observe what is happening in biological systems, an ability to understand what we observe, and an ability to affect biological systems based upon our understanding. An intentional biology will allow humans to leverage the existing molecular infrastructure and cellular architecture to produce food, energy, and materials with greater efficiency than is currently possible.

Other notable biologists have moved to areas of research that are complex and require new approaches. Before his death, Francis Crick studied consciousness of the human mind/brain, while James Watson has been intrigued by "free will." To address the need for a new approach in biology, Cold Spring Harbor Laboratory (CSHL), where Watson is president, has created a new graduate school (Watson School of Biological Sciences) at a time when there is concern about producing too many PhDs for the market. Harvard University has launched a new Systems Biology Department to "explain how the higher level properties of complex systems materialize from the interactions among their parts." The Howard Hughes Medical Institute is creating a new Janelia Farm research community to focus on two large and complex topics: (1) the identification of general principles that govern how information is processed by neuronal circuits, and (2) the development of imaging technologies and computational methods for image analysis.

In the biography of *Seymour Benzer* entitled *Time, Love, Memory,* Benzer talks with Crick about the future of biology and the need for a new level of understanding. With genome sequences becoming huge parts lists for very diverse biological systems, many are looking toward new areas of research. Benzer, who pioneered the field of behavioral genetics, uses the model organism *Drosophila* to understand the brain and behavior. As Benzer continues to chip away at complex traits such as fly behavior, he reminds us many genes are involved. When it comes to human behavior, he is even more certain about the complexity. "So just having that [allele] doesn't mean you'll show that phenotype. Expression depends on a myriad of chemical reactions. And that's not generally understood. People think if you have the [allele], your fate is sealed." He is suggesting that we need to understand many genes, stochastic events, gene circuits, proteomics, single nucleotide polymorphisms—in short, systems biology.

The Quality of the Message: What Questions Do Systems Biologists Ask?

Nobel laureate Lee Hartwell is the director of the Fred Hutchinson Cancer Research Center and another proponent of the systems approach to examine the genetic variation in a population (see Chapter 4). In a 1999 commentary, Hartwell and his coauthors argued for a new approach to biological research:

> We have recently become intrigued by a very fundamental difference between the artificial way we work on problems like these model systems and the reality of human biology. In models, we study the role of genes in a single genetic background. All individuals are genetically identical. However, in the human population as well as with most organisms living in nature, the genetic background is diverse. Natural genetic variation in the breeding population is often such that two or more forms (polymorphisms) are present at each gene such that most individuals are genetically individual. How do organisms

accommodate this degree of variation? What variation contributes to the phenotypic diversity of the population? How do we think about genetic diversity in terms of molecular pathways? My lab is beginning to study the role of natural variation in the behavioral output of molecular pathways using cell division and cell mating as examples.

LINKS G$P
Janelia Farm
Lee Hartwell
Leroy Hood
Seymour Benzer
Systems Biology Department
Trey Ideker
Watson School
METHODS
respiration
STRUCTURES
Gal4p

12.2 Can We Model Entire Eukaryotes with a Systems Approach?

Given all the holes in our knowledge, will we ever be able to model a complete cell? While working on his PhD with Leroy Hood at the University of Washington in Seattle, Trey Ideker and others used a systems approach to understand yeast sugar metabolism. The ten-person team did not set out to understand every aspect of yeast's life; rather, they chose a well-defined metabolic pathway—galactose utilization. Yeast is normally grown in the lab with glucose as the energy source, but it can grow on galactose instead. Galactose is a six-carbon sugar with a very subtle but important distinction from glucose (Figure 12.1a). However, to metabolize galactose, the cell needs to induce some genes and repress others (Figure 12.1b). Galactose is given to the cells, but it is not able to enter the cytoplasm because it is too big and hydrophilic to cross the plasma membrane. The sugar must be imported by a galactose transport protein encoded by the gene *Gal2*. Genes *Gal1, Gal7, Gal10,* and *Gal5* encode metabolic enzymes that slowly convert galactose into glucose-6-phosphate, one of the basic molecules used in glycolysis and cellular respiration. Regulation of this galactose metabolism is controlled by a genome circuit with a toggle switch that is not bistable. *Gal4* encodes the transcription factor that activates all the genes previously listed as well as *Gal80* and *Gal3*. However, in the absence of galactose, the Gal4 protein (**Gal4p**) is bound by the Gal80 protein (Gal80p), preventing Gal4p from initiating transcription. When galactose is present in the cytoplasm, the sugar binds to Gal3p. When Gal3p has bound galactose, this complex binds to Gal80p, which must then let go of Gal4p. The inhibition of an inhibitor is analogous to a double negative in a sentence: "I will not not sleep through class" means you will sleep through class. Therefore, Gal3p-galactose inhibits Gal80p from inhibiting Gal4p (i.e., Gal4p becomes activated). When not inhibited by Gal80p, Gal4p is able to bind to the promoters of *Gal* genes *2, 1, 7, 10, 5, 80,* and *3* to activate them, which allows the cell to produce the proteins required to catabolize galactose. When galactose is totally consumed, the toggle switch flips back to the default pathway of metabolizing glucose.

METHODS
DNA microarrays
ICAT
knockout

Unlike investigators using a traditional approach, Ideker and his team combined four steps in their analysis that are the hallmark of a systems biology approach:

1. Define all the genes in the genome, and the subset of genes, proteins, and other molecules (mostly sugars) constituting the galactose pathway. Build a model based on this information to make predictions.

2. Perturb each pathway component through a series of genetic (i.e., knockout deletions) and environmental (e.g., sugar food source) perturbations.

3. Utilize both DNA microarrays and **isotope-coded affinity tags** (ICAT) data to create a more complete picture of a cell's response to experimentally introduced perturbations.

4. Refine the model when experimental data do not meet predictions.

With the complete genome sequence of *S. cerevisiae* available, as well as years of biochemical data, step 1 was easily completed using a computer. Step 2 was also simple, since deletion strains are available from other researchers

a)

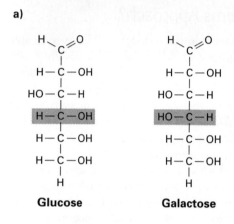

b)

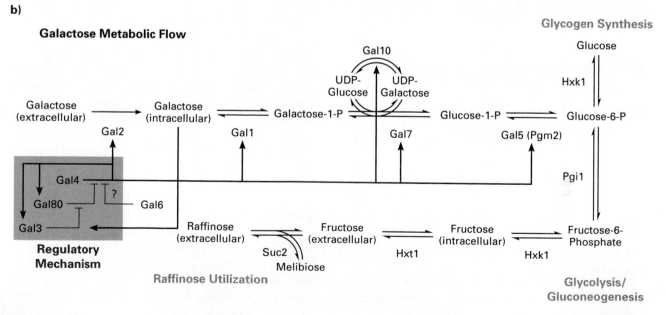

Figure 12.1 Galactose metabolism in yeast.
a) Line diagrams of glucose and galactose. Note the only difference is the position of one hydroxyl group. **b)** Metabolic circuit showing the galactose induction pathway. The toggle switch is controlled by the proteins located in the purple regulatory mechanism box. *Gal6* is a repressor of the galactose induction pathway, but its mechanism is uncertain, so it is shown with a question mark repressing *Gal4*.

and altering the sugar is as simple as taking a spoonful of medicine. The grunt work came at step 3.

In step 3, these investigators went to the lab to generate DNA microarray and ICAT data. They grew *wt* cells as well as knockout deletion strains (designated with the Greek letter delta, Δ) *gal2Δ*, *gal1Δ*, *gal5Δ*, *gal7Δ*, *gal10Δ*, *gal3Δ*, *gal4Δ*, *gal6Δ*, and *gal80Δ*. All strains were grown in the presence or absence (+/−) of galactose, and DNA microarrays were used to measure mRNA production. Each experiment was performed four times and the results averaged. The team compared some of their DNA microarray data with more traditional forms of mRNA analysis to ensure their data were accurate (Figure 12.2). The investigators wanted to get away from colorized microarray expression scales, so they created a black-and-white version. By eliminating colors, color-blind people are not disadvantaged, and it is much easier to reproduce black-and-white than color images. In this black-and-white scale, no change appears as

a neutral gray, which blends into the background of the figure so you do not see any spot if mRNA production is identical between the two conditions.

As with all figures, examine the control data first. The expression of actin was relatively stable, as shown by Northern blot and DNA microarrays. Remember that microarray data represent the average of four experiments, but the Northern blot data represent only one experiment. In the absence of galactose, *Gal80* mRNA is moderately reduced in three of the strains (average of $10^{-0.6}$ = fourfold repression; see Math Minute 6.1), but reduced 40-fold in *gal80Δ*. You would expect the *gal80Δ* strain to have a greater repression of *Gal80* mRNA than the other three strains, so this is a good control. However, when galactose was present, there is no change in the *wt* strain, because this is the reference level of expression (i.e., self vs. self). In the absence of galactose, *Gal1* is not expressed ($10^{-2.9}$ or

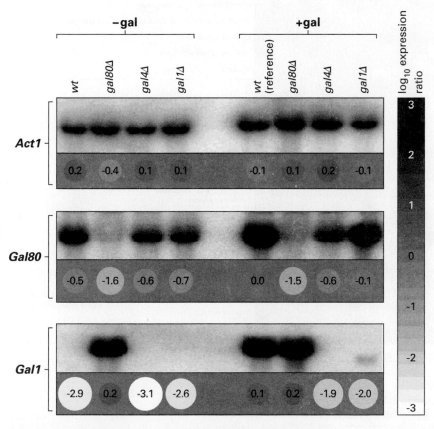

Figure 12.2 Comparison of Northern blot and microarray data.
Total RNA from yeast growing in each of eight conditions (column headings) was probed with radio-labeled *Gal1*, *Gal80* cDNA, or *Act1* (actin gene as a control probe). Corresponding changes in gene expression measured with the yeast DNA microarray are depicted graphically beneath each RNA band. Change in gene expression was relative to *wt* yeast growing in galactose (column 5), with medium-gray representing no change (see visual scale). Darker or lighter shades represent increasing or decreasing amounts of expression respectively, and spot size scales with the magnitude of change. The quantitative log_{10} change in expression level is annotated inside each spot.

794-fold repressed), as you would expect because *Gal1* requires galactose for its induction. In the *gal80Δ* strain minus galactose, *Gal1* was strongly induced, but when compared to *wt* cells in the presence of galactose, it showed only a 1.6-fold induction ($10^{0.2}$). In the presence of galactose, *wt* and *gal80Δ* produced about the same amount of *Gal1* mRNA, as you would expect because *wt* should inhibit Gal80p and the *gal80Δ* strain could never repress *Gal1*. However, in both *gal4Δ* and *gal1Δ* strains, *Gal1* mRNA is

not produced ($10^{-2.0} = 100$-fold repression), because the transcription factor (Gal4p) or the *Gal1* gene is absent.

Agreement between microarray and Northern blot data gave the investigators confidence to continue their microarray analysis (Figure 12.3). At this point, we will focus only on the top nine rows of data. It is always a good idea to begin with the control data (far left column), so first we will examine *wt* cells +/− galactose. In this column, the expression ratios for each gene were calculated as the amount of mRNA

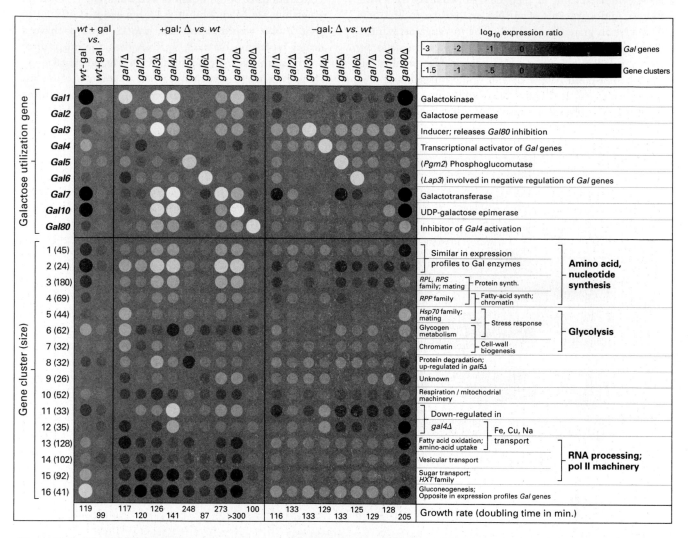

Figure 12.3 Galactose induction of yeast genes.

The top portion of the figure indicates the change in expression of a *Gal* gene (listed on the left side of each row) due to a particular perturbation (listed above each column). Medium-gray spots (the same color as the background) represent no change, darker spots represent increased mRNA production, and lighter spots indicate reduced mRNA production. The left half of the matrix shows expression changes for each deletion strain compared to *wt*, with both strains grown in the presence of galactose. The right half of the matrix compares mutants and *wt* cells in the absence of galactose. The far left two columns compare *wt* cells grown +/− galactose. The bottom portion of the figure illustrates average expression profiles for genes in each of 16 clusters. Clusters contain genes involved in various categories, as noted on the far right. Listed at the very bottom are the growth rates for each strain and growth condition. A larger number of genes were analyzed for Figure 12.3 than for the published version.

produced in the presence of galactose divided by the amount produced in the absence of galactose, $(wt + \text{gal}) \div (wt - \text{gal})$. In the presence of galactose, *wt* cells strongly induce the expression of *Gal1*, *Gal7*, and *Gal10*. Induced at lower levels were *Gal2*, *Gal80*, and *Gal3*. Only *Gal4* was repressed, but only by a little. Column 2 is a control self vs. self experiment, $(wt + \text{gal}) \div (wt + \text{gal})$, and you can see that all the genes showed no change in expression levels. The first two columns verified that the method was working as expected and therefore we can turn our attention to the experimental columns.

The next nine columns of data compare the effect of galactose on the *galΔ* mutant strains vs. *wt* cells ($+$gal; *Δ* vs. *wt*). As you would expect, there is a diagonal of white spots indicating each *Gal* gene was repressed in the strain from which it was deleted, when compared to *wt* cells. In the presence of galactose, *gal3Δ* and *gal4Δ* strains caused strong decreases of most, though not all, of the other *Gal* genes. This makes sense because Gal3p enables Gal4p to function by removing the inhibition caused by Gal80p (refer to Figure 12.1b). In the *gal80Δ* strain, the *Gal* genes (except the deleted *Gal80* gene) were expressed slightly more than in *wt* cells (shown as slightly darker gray spots), because the loss of the repressor Gal80p is functionally equivalent to galactose induction of the *Gal* genes. The only unexpected results in this panel were the column of lighter spots in the *gal7Δ* and *gal10Δ* strains. These two genes are downstream of the toggle switch, so it does not make sense that the absence of Gal7p or Gal10p would repress the other *Gal* genes. We will return to this issue.

In the absence of galactose (remaining nine columns), each *galΔ* mutant strain is compared to *wt* cells under identical growth conditions ($-$gal; *Δ* vs. *wt*). The most striking response is seen in the *gal80Δ* strain, which induced expression from many of the *Gal* genes because the repressor that normally inhibits their transcription had been deleted. Interestingly, the diagonal of white spots is not as obvious in the absence of galactose as it was in the presence of galactose, though *gal4Δ* is substantially repressed.

DISCOVERY QUESTIONS

1. Some of the data in Figure 12.3 do not fit with traditional expectations. Look at the diagonal of spots in the absence of galactose. Why did some genes not show signs of repression when they were deleted?

2. In the presence of galactose, when Gal80p is inactivated by Gal3p, *wt* cells still produce *Gal80* mRNA, as indicated by the white spot for the *gal80Δ* strain. Why would *wt* cells continue to produce *Gal80* mRNA when the cells do not want to repress the galactose pathway?

The bottom portion of Figure 12.3 shows the results of clustering 997 of the genes into 16 different clusters (see

Math Minute 6.3). These 997 are the only genes that showed significantly altered expression in the experiments. Each cluster represents genes that were regulated similarly to each other. The top two clusters contain the *Gal* genes; the bottom cluster contains genes involved in glycogen synthesis, which normally occurs only in the presence of excess glucose. The investigators chose a different way to display clustered gene expression profiles, illustrating yet again that visualization of data is an important component of the analysis. By clustering genes with known and unknown functions, we can formulate testable predictions about the unknown genes. Clustering is not a definitive test, but we need to characterize the 2,000 uncharacterized yeast genes, and combining a systems approach with clustering is a good beginning.

LINKS
Ruedi Aebersold

Does the Proteome Respond Like the Transcriptome?

Ideker's collaborators included Ruedi Aebersold, who invented ICAT, the quantitative proteomics method (see pages 316–319). As a part of the systems biology approach, the investigators wanted to integrate ICAT and DNA microarray data. They performed ICAT protein analysis on *wt* cells grown $+/-$ galactose. The proteomics investigators were able to identify and quantify 289 proteins from the two proteomes. They plotted the change in protein levels in the presence of galactose vs. the absence of galactose; a positive ratio indicated that the protein was more abundant in the presence of galactose (Figure 12.4). Changes in protein level are indicated by symbols further away from the horizontal line at zero. On the same graph, the investigators used the DNA microarray data to plot the expression ratios for the 289 genes that encoded the proteins measured by the ICAT method. Changes in mRNA production are indicated by symbols further away from the vertical line at zero. The diagonal dashed line indicates the location of genes/proteins whose protein and mRNA expression ratios were equal (i.e., both up or both down), as was the case for *Gal2* located in the top right quadrant.

Of the 289 quantified proteins, only 30 displayed substantial differences in abundance $+/-$ galactose (i.e., they appear further away from the horizontal line). Of these 30 proteins, 15 showed no difference in mRNA levels using DNA microarray analysis (very near the vertical line). In other words, there were significant changes in protein levels even though there were no significant changes in mRNA levels. This is a classic case of **posttranscriptional** control of the **cell web**.

A lack of correlation between mRNA and protein levels has been documented before in the study of single genes, but this was the first demonstration on a large scale, even though not all proteins could be measured. For the most part, metabolic and ribosomal components were up-regulated,

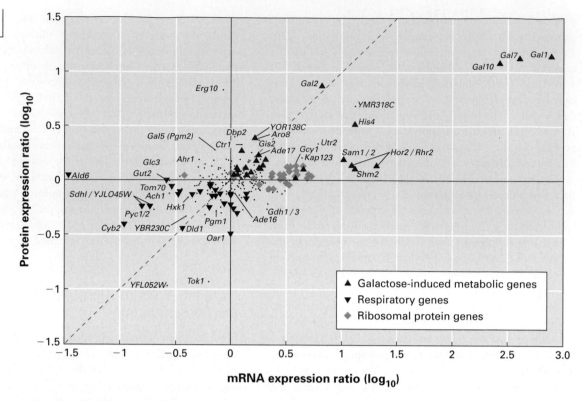

Figure 12.4 Scatter plot of protein vs. mRNA expression for the 289 proteins and genes identified by both ICAT and DNA microarray methods.

Purple triangles represent galactose-induced genes; gray diamonds represent ribosomal proteins; black inverted triangles denote components involved in cellular respiration. Dots represent other proteins and genes not in these three categories.

while respiration components were down-regulated. Look at the three *Gal* data points in the top right quadrant that are located far to the right of the diagonal line. Their location indicates that the genes were induced more than the proteins accumulated in the cells. Once again, we see that *Gal7* and *Gal10* behaved in similar but unpredicted ways given the model in Figure 12.1b.

DISCOVERY QUESTIONS

3. What effect does the difference between protein and mRNA levels have on your evaluation of all microarray data?

4. Find the protein in Figure 12.4 that increased its concentration about 8-fold but had its mRNA repressed about 1.3-fold ($10^{-0.1}$). How could the mRNA be less abundant but the protein more abundant?

Can We Build a Systems Model?

Throughout Chapters 10 and 11, we were reminded that the real benefit of a model is not building a perfect one, but using the best available model to make predictions, which

Figure 12.5 Circuit diagram of 348 nodes and 362 interactions overlaid with mRNA expression data.

Go to www.GeneticsPlace.com to view this figure.

can lead to new discoveries and improvements in the model. Ideker's group went back to the computer and used the research of Stan Fields (see pages 300–302) and his team to compile a list of 2,710 protein-protein interactions, as well as 317 protein-DNA interactions. From this list of proteins, the investigators isolated the 1,012 genes or proteins whose expression levels were altered in the presence of galactose. These 1,012 genes/proteins were analyzed by a computer program to overlay known interactions with expression changes for *gal4Δ* vs. *wt* + galactose. The final product was a large integrated circuit diagram (Figure 12.5). This figure contains a lot of information, but parts have been enlarged for us (panel b). Notice the central role played by Gal4p (protein in the center of the circuit). Also, notice how many of the genes pointed to by *Gal4* have reduced mRNA expression (white), as one would expect if the transcription factor Gal4p was missing.

The investigators acknowledged that their circuit diagram is not a complete model, as many of the proteins in the two proteomes (+/− galactose) were absent. They also pointed out predictions that were not supported by the data. The most striking example was that *gal7Δ* and *gal10Δ* strains cause down-regulation of many other genes. In fact, the *gal7Δ* and *gal10Δ* expression profiles in Figure 12.3 were very similar to *gal3Δ* and *gal4Δ* expression profiles. In Figure 12.1b, Gal7p and Gal10p have no apparent relationship to transcription, and yet the data indicate that they affect transcription. Are the data wrong, or is the model wrong?

Typically, if good data conflict with your model, trust your data. The investigators believed their data, so they reexamined their model. They recognized that what Gal7p and Gal10p had in common was **galactose-1-phosphate** (**Gal-1-P**). Perhaps accumulation of Gal-1-P is sensed by the cell and responded to by repressing the genes that help produce Gal-1-P. If this were true, then you would predict that the transcription of *Gal5* would not be affected in *gal7Δ* and *gal10Δ* strains, which is consistent with the data in Figure 12.3. However, correlation alone is not satisfying, so the investigators decided to test their revised model with two types of experiments.

For their first experiment, the investigators used a genetic approach. They grew a double-deletion mutant (*gal1Δ gal10Δ*), in the presence of galactose. For *gal10Δ* cells grown in galactose, the loss of Gal10p led to repression of *Gal 1, 2, 3, 7,* and *80* genes. However, in the double mutant, the loss of Gal1p would prevent the formation of Gal-1-P, and if their hypothesis was correct, then *Gal 2, 3, 7,* and *80* should not be repressed. Microarrays were used to measure mRNA expression in the double mutants compared to *wt* + galactose. The new prediction was correct; the other *Gal* genes were not repressed. Therefore, it seems possible that Gal-1-P can regulate the *Gal* genes in a new and unexpected way.

There is one problem with their conclusion, however. Maybe the double mutation perturbed an additional cellular circuit in addition to altering the production of Gal-1-P, and this different cellular circuit is the real cause, not

accumulation of Gal-1-P. To complement the double-mutant experiment, the investigators used a biochemical experiment. They grew two strains in parallel, *gal7Δ* and *gal10Δ* in the absence of galactose. In the absence of galactose, the *Gal* genes were repressed. Then galactose was added to both strains for 20 minutes, enough time for all the *Gal* genes to be transcribed but not for Gal-1-P to accumulate. When microarrays were used to measure mRNA from these two mutant strains compared to *wt* cells + galactose, there was no repression in the *Gal* genes. Therefore, even in single-mutation strains, the repression of *Gal* genes was not apparent within the first 20 minutes. This indicated that something other than mRNA (perhaps Gal-1-P) must accumulate before the *gal7Δ*- and *gal10Δ*-induced repression can occur.

Context of the Message: What Is the Impact of this Research?

Although the initial galactose metabolism model was unable to fully explain all the data, the experimental design was a success. The investigators used existing data to generate a model, made predictions, tested those predictions with DNA microarrays and proteomic methods, found where the data did and did not support the model, refined the model, made a new prediction, tested the new prediction with two additional experiments, and proposed a revised model. Some aspects of the data still cannot be explained, which means new and creative experiments must be designed to improve these weaknesses in the model. This reiterative approach will continue until the model gradually becomes better and better at predicting the ability of yeast to respond to changes in its environment.

Even with a lack of complete understanding of a relatively simple system, we gained some insight into emergent properties of an integrated biological circuit. First, it makes sense that a cell with too much Gal-1-P might want to slow

LINKS
Al Gilman
AFCS
Integrated Genomics
National Center for
Genome Research

down production of this sugar by repressing genes such as *Gal1*, *Gal10*, and *Gal7*. We saw in the learning circuitry (see pages 378–391) that feedback loops and subtle changes in the concentration of a few molecules are capable of regulating toggle switches. In the galactose circuit, there appears to be a toggle switch that can sense the accumulation of Gal-1-P and efficiently switch the circuit from on (make more Gal-1-P) to off (stop making Gal-1-P). When we examine the growth rate for the *gal7Δ* and *gal10Δ* strains + galactose, it appears that having too much Gal-1-P is not healthy (bottom of Figure 12.3): the doubling rates are very long (273 and >300 minutes, respectively) compared to the other strains. Is too much of a good thing (Gal-1-P) bad for the cells? The hypothesis of feedback inhibition by Gal-1-P is supported further when you examine the growth rate for the *gal80Δ* strain minus galactose. In these cells, the repression system is lost, so the cells transcribe *Gal* genes even in the absence of galactose, which also produces the slowest growth rate of any strain in this condition (205 minutes).

A second emergent property became evident with this study. Gal4p appears to regulate many other genes in addition to the *Gal* genes. Initially, the investigators looked for Gal4p-binding DNA sequences upstream of genes that were in the same clusters (see Figure 12.3). Specifically, they looked at the 270 genes in clusters 1 and 2 (the *Gal* gene clusters), clusters 15 and 16 (genes whose profiles were the inverse of clusters 1 and 2), and clusters 11 and 12 (genes that were down-regulated in *gal4Δ* +/− galactose). Out of the possible 270 genes, 22 of them (8%) contained Gal4p binding sites, a significantly higher percentage than in all of the remaining clusters. Included in these 22 genes were *Pcl10* (cluster 1), YMR318C (cluster 2), YJL045W (cluster 15), and YLR164W and *Icl1* (both of cluster 16). YMR318C and YJL045W are open reading frames (**ORFs**) with no known function. So, not only did we learn that Gal4p controls more genes than previously suspected, we also have good starting places to discern functions for two ORFs with unknown functions. Ideker's thesis research was a very good example of how a systems biology approach can lead to a deeper understanding than can be achieved when analyzing individual components in isolation. It is particularly interesting that Pcl10p is believed to play a role in repressing glycogen synthesis and that Gal4p activated the *Pcl10* gene. Thus, Gal4p indirectly represses glycogen synthesis, an emergent property previously undetected. Remember how electrical engineers look at integrated circuits? They look for functional units (such as galactose utilization and glycogen synthesis) and connections between functional units (such as Gal4p). With a systems approach, we hope to find as many connections as possible in order to build a more complete model of a cell.

12.3 Will Systems Biology Go Systemic?

We began this chapter by asking whether systems biology is a fad that will come and go, or a new perspective for biological research. According to *The Tipping Point*, one component needed to establish a new trend is the people involved. Early in this chapter we listed some of the elite of biology who are moving toward a systems approach. The most profound manifestation of this movement would be the creation of new research institutions dedicated to systems biology. Systems biologists include molecular cell biologists, chemists, computer scientists, physicists, engineers, and mathematicians who bring different skills, vocabularies, and perspectives to bear on complex problems. Systems biology is exemplified by Ideker's galactose research, where the investigators took a simple system and tried to understand it completely.

Nobel laureate Al Gilman, at the University of Texas Southwestern Medical Center in Dallas, created the Alliance for Cell Signaling (AFCS). Gilman's approach differs from ISB's in two major ways. First, rather than create a physical institution de novo, Gilman assembled a virtual community composed of people who retain their current jobs and institutional affiliations across the country. AFCS is a consortium of ~50 investigators working at ~20 different institutions rather than an assembly of ~50 people under one roof. Second, AFCS's goal is different; they want to create a virtual cell by understanding the relationships that vary both temporally and spatially between sets of inputs and outputs in signaling cells. They use multiple cell types and a wide range of methods to generate one integrated model. New models for research require new sources of funding. To launch AFCS, the National Institutes of Health awarded Gilman's team a "glue grant" of $25 million over 5 years to link the research efforts among different institutions. Gilman raised an additional $25 million from pharmaceutical companies and nonprofit research organizations.

Finally, there are a few other institutions whose missions are in line with a systems approach. The National Center for Genome Research is a nonprofit organization that develops computer tools that can be used by systems biologists. Integrated Genomics is a company that will create circuitry diagrams for the complete metabolism of your favorite organism. Of course, as a dot-com, it does this for profit, and thus it appears they expect the systems approach to be a growing business opportunity. Will Integrated Genomics be the biological equivalent of Amazon.com? Or will they be closer to eight-track-tapes.com, destined to fade into oblivion in a couple of years?

DISCOVERY QUESTIONS

10. What technical hurdles must systems biology overcome to become a major force in biology?

11. Pretend you have several billion dollars in your savings account. Instead of buying tons of snacks and T-shirts, you want to contribute to some institutions that will be successful. Would ISB or AFCS be a good place to donate money? Explain your answer.

12. The most important step in any plan is to clearly identify your goals. If you were Hood or Gilman, how would you define your goal so that you would be able to know you had successfully accomplished it? In other words, what would be a very good outcome for your new systems biology institution, and how would you know whether you had achieved it?

Chapter 12 Conclusions

We began with an open-ended question: Is systems biology a new perspective, or just the latest advance in the continuum of biology? Is systems biology merely a faddish term that will fade quickly from popular usage? Or do systems biology, whole genome research, biological circuits, etc., truly represent a paradigm change in furthering biological understanding? The case study in this chapter utilized many different forms of data discussed throughout the book. All aspects of life are important—sequence, variation, gene expression, proteomes, biological circuits. We can collect vast amounts of information, but can we process it? Can we build models that approach the complexity of a living yeast cell? Can the trillions of cells working cooperatively in your body be modeled in a comprehensive and realistic manner? Perhaps these goals will allow us to determine the success or shortcomings of systems biology. The pessimist may say humans will never know all there is to learn in biology, which is probably true. The optimist, however, may simply enjoy the process of attempting to learn everything and take pleasure from contributing to the cause.

References

Acar, M., A. Becskei, & A. van Oudenaarden. 2005. Enhancement of cellular memory by reducing stochastic transitions *Nature*. 435: 228–232.

Check, E. 2003. Harvard heralds fresh take on systems biology. *Science*. 425: 439.

Cho, R. J., & M. J. Campbell. 2000. Transcription, genomes, function. *Trends in Genetics*. 16: 409–415.

Clayton, D. F. 2000. The genomic action potential. *Neurobiology of Learning and Memory*. 74: 185–216.

Davidson, E. H. 2001. *Genomic regulatory systems: Development and evolution*. San Diego, CA: Academic Press.

Galitski, T. 2004. Molecular networks in model systems. *Annual Review in Genomics and Human Genetics*. 5: 177–187.

Gladwell, M. 2000. *The Tipping Point: How little things can make a big difference*. Boston: Little, Brown.

Hartwell, L. H., J. J. Hopfield, et al. 1999. From molecular to modular cell biology. *Nature*. 402 Supplement: C47–C52.

Hood, L., & D. Galas. 2003. The digital code of DNA. *Nature*. 421: 444–448.

Hood, L., J. R. Heath, et al. 2004. Systems biology and new technologies enable predictive and preventative medicine. *Science*. 306: 640–643.

Ideker, T., V. Thorsson, et al. 2001. Integrated genomic and proteomic analyses of a systematically perturbed metabolic network. *Science*. 292: 929–934.

Legrain, P., J.-L. Jestin, & V. Schächter. 2000. From the analysis of protein complexes to proteome-wide linkage maps. *Current Opinion in Biotechnology*. 11: 402–407.

Pirson, I., N. Fortemaison, et al. 2000. The visual display of regulatory information and networks. *Trends in Cell Biology*. 10: 404–408.

Shannon, P., A. Markiel, et al. 2003. Cytoscape: A software environment for integrated models of biomolecular interaction networks. *Genome Research*. 13: 2498–2504.

Glossary

Symbolic Notation

60-mer oligonucleotide 60 bases long. Any number can be used and the suffix -mer attached to indicate the length of the oligonucleotide.

123L→R a symbol using the pattern of **number:letter:letter** indicates that the first amino acid chime (L for leucine) at amino acid position 123 in the protein has been changed (to R for arginine).

Δtup1 the Greek letter delta (Δ) before or after a gene name is used to symbolize a deleted gene. In this case, the gene *Tup1* has been deleted. Italics is used in yeast to denote the gene is being described.

p53 a lower case "p" preceding a number indicates a protein of that molecular weight. For example, the **tumor suppressor** p53 has a molecular weight of 53,000 Daltons.

Tup1p when a gene name (e.g., *Tup1*) is followed by a lowercase "p," the *protein* encoded by *Tup1* is being discussed. This symbolic notation can be used for any gene/protein combination and, unlike the gene, the protein is not italicized.

A

accession number identification number given to every DNA and protein sequence submitted to NCBI or equivalent database. For example, the human leptin receptor SwissProt accession number is P48357.

algorithm step-by-step procedure for solving a problem (e.g., aligning two sequences) or computing a quantity (e.g. %GC). Typically written in Perl or other computer language.

alien genes genes found in a genome that appear to have come from another species via horizontal transfer.

aliquot can be used as a noun or verb. When a liquid volume is divided into portions and placed in microfuge tubes, these *aliquots* of liquid are *aliquotted* into different microfuge tubes.

alkaline phosphatase enzyme that cleaves phosphate from its substrate at alkaline pH. Commonly used as a reporter gene and to label probes for blots of all kinds.

ancient DNA DNA isolated from an ancient source such as animals trapped in amber, or bacteria found in archeological sites.

aneuploidy abnormal chromosomal number. For example, a loss of chromosome 4 and an extra copy of chromosome 21 could both be described as aneuploidy. Aneuploidy is also applied to portions of chromosomes.

annotated a gene is annotated when it has been recognized from a large segment of genome sequence and often we know something about its cellular role. Genomes can also be described as annotated once they have been analyzed for gene content.

antagonistic pleiotropy a theory of evolution that states for every beneficial gain in function, there is a compensatory loss in another function. For each genetic change, there are many effects (pleiotropy), some of which are beneficial while others are detrimental (antagonistic) in other environments.

anthrax (*Bacillus anthracis*) rod-shaped **Eubacteria** that can infect skin or lungs; may be used as a biological weapon.

antibody microarray one example of a **protein microarray** where antibodies are spotted onto glass to determine the nature and amount of **antigens** in a solution.

antigen any molecule that stimulates an immune response in the form of an antibody. Pollen and vaccines are both antigens.

antisense technology molecular method that uses a nucleic acid sequence complementary to an mRNA so that the two bind and the mRNA is effectively neutralized.

antiswitches RNA molecules that can activate or repress a specific mRNA sequence when aptamer portion is bound with a ligand.

apoptosis a normal function for many cells, apoptosis is a genetically encoded sequence of cellular actions that leads to the cell's death; often referred to as "programmed cell death."

aptamer nucleotide sequence that folds to a shape that can bind a small ligand such as caffeine or theophylline.

aquaporin protein that forms a channel (pore) across a membrane to allow the flow of water (*aqua*) molecules.

Arabidopsis a flowering plant about 15 cm tall that reproduces very quickly; sometimes referred to as the plant equivalent of the fruit fly. First plant to have its genome fully sequenced.

Archaea one of the three domains of life (along with **Eubacteria** and **Eukaryota**). This domain includes prokaryotes that typically live in extreme environments or utilize atypical sources of energy.

archenteron during **gastrulation**, a portion of the embryo **invaginates** to form a pocket inside the embryo. This pocket will become the gut and is called an archenteron prior to gut formation.

array an orderly pattern of objects. In genomic studies, there are *micro*arrays and *macro*arrays. Microarrays are small spots of DNA or protein, and the identity of the spotted material is known. Macroarrays are bacterial, yeast, or similar colonies on plates used to determine functional consequences of genomic manipulations.

autocatalytic an enzyme capable of stimulating its own (auto) activity (catalyst).

autophosphorylate an enzyme capable of phosphorylating itself.

avidin a protein found in egg whites and bacteria (e.g., *Streptomyces*) that binds with very high affinity to the vitamin **biotin** (also called vitamin H). In the lab, avidin is used to purify biotin-tagged molecules from complex mixtures as was done in the ICAT method.

B

bacterial artificial chromosome (BAC) cloning vector replicated in bacteria that can hold an insert of about 150,000 base pairs. BACs were used extensively in the public **HGP**.

bar codes analogous to the black-and-white striped bar codes on packages. In genomic studies, unique nucleotide sequences are used to replace each gene in the yeast genome. There are two halves to the bar codes: an **upstream** UPTAG and a **downstream** DOWNTAG.

Bayes rule formula for computing the probability of event A given event B, when the probability of event B given event A, the probability of event B given not event A, and the probability of event A are all known.

bimodal toggle switch a protein/DNA component that allows a circuit to accept input from two different sources to produce one of two outcomes. This switch is not stable.

binomial a probability model for counting the number of "successes" in an experiment with two possible outcomes; the same probability of success *p* holds for each trial of the experiment.

binomial coefficients the subject of Math Minutes 6.1 and 8.1; a counting formula, used to represent the number of ways a sample of k cells can be selected from a population of n cells.

BioBricks interchangeable DNA parts that can be assembled in multiple combinations to produce devices for research in synthetic biology; BioBrick registry <http://parts.mit.edu/>.

biofilms meshwork of prokaryotes living collectively and resistant to many chemical agents.

bioinformatics a field of study that extracts biological information from large data sets such as sequences, protein interactions, microarrays, etc. This field also includes the area of data visualization.

biological process coined by **Gene Ontology** to describe broad cellular outcomes, such as mitosis or energy production, that are accomplished by ordered assemblies of molecules.

biotin *See* **avidin**.

bistable circuit similar to **bistable toggle switch**, but describes the entire circuit and not just the switching mechanism.

bistable toggle switch a protein/DNA component of a circuit that chooses between two different outcomes. Once a choice is made, the switch maintains the outcome until the circuit receives new stimulus to change.

BLAST the protein and nucleic acid sequence search engine developed at **NCBI** that allows you to search sequence databases. BLASTn searches for nucleotide sequences; BLASTp searches for amino acid sequences; BLAST2 compares two sequences.

blastula an early embryonic stage of development where the embryo consists of a hollow ball of cells prior to **gastrulation**. It is from this stage that embryonic stem cells are isolated.

bootstrap analysis a statistical technique used to assess the reliability of the branching structure in phylogenetic trees.

Brownian motion random movement of particles or molecules that is powered by kinetic energy of the solution.

C

Caenorhabditis elegans nematode worm about the size of an eyelash and the first animal to have its genome sequenced completely. This model organism, pioneered by Sydney Brenner in the 1970s, is ideal for studying neurons; was instrumental in the discovery of **apoptosis**.

CAMP (Christie, Atkins, Munch-Peterson) factors secreted proteins that bind to antibodies (IgG and IgM) and can form pores in eukaryotic cell membranes. Distinct from cAMP; cyclic AMP.

capillary electrophoresis a method that allows small amounts of compounds to be separated for further analysis. This method is at the heart of automated DNA sequencing and proteomics.

CAT *See* **chloramphenicol acetyl transferase**.

caveolae indentations on the inner surface of plasma membranes that are used for endocytosis.

CD *See* **conserved domain**.

cDNA *See* **complementary DNA**.

CDS acronym for *co*ding *s*equences.

Celera for-profit company that competed with the human genome project to produce a human genome sequence; pioneered whole genome shotgun (WGS) sequencing method.

cell web a term coined in this book to describe the subcellular ecosystem of interacting components. Analogous to "food web," which indicates when one component of an interacting system is perturbed, the entire system is affected.

cellular component a term coined by **Gene Ontology** to describe subcellular structures, locations, and macromolecular complexes such as nucleus, telomere, and mitotic spindles.

checkpoint a pause during a vital and complex cell process to make sure events have progressed properly before proceeding.

chelate to bind and remove from solution. For example, EDTA chelates divalent cations such as Mg^{2+}. Receptors can chelate ligands as well.

chi-square test statistical method for determining whether two sets of ratios or frequencies are significantly different.

chloramphenicol acetyl transferase (CAT) a bacterial antibiotic resistance enzyme used as a reporter gene to assay the ability of cis-regulatory elements to promote transcription.

chromatogram (chromat) a four-colored graph produced from nonradioactive **dideoxy sequencing methods** including cycle sequencing.

CIM *See* **clustered image map**.

circuit a term borrowed from electrical engineering that indicates several interconnected proteins or genes. Circuits can describe connections needed to regulate a single gene or entire genomes.

circuit diagram a term borrowed from engineering that illustrates the interconnections among proteins or genes. Circuit diagrams allow us to visualize critical components and connections that are otherwise difficult to perceive.

cis-regulatory elements the portion of a gene that controls transcription of the same gene. The term describes the diverse DNA sequences that are bound by transcription factors and repressors and is inclusive of promoter and enhancer sequences.

clone noun or verb. A clone is any molecule/cell/organism present in more than one identical copy. To clone something means to produce more than one copy of the original molecule/cell/organism.

clustered image map (CIM) a complex integration of more than one data set. In this book, a CIM was used to integrate DNA microarray data with the effects of chemotherapy drugs on cells.

Clusters of Orthologous Groups (COG) NCBI compilation of evolutionarily related gene sequences from several microbial genomes. This site allows you to search by gene or cellular role and produces *dendrograms* to show sequence similarities.

cobweb diagram graphical analysis of stable and unstable equilibria of functions.

coding capacity the percentage of a genome that contains protein- or functional RNA-coding DNA.

coding sequence (CDS) abbreviation used at **NCBI** to indicate which bases constitute the **open reading frame**.

codon bias coding for a particular amino acid with a subset of codons more often than other codons.

COG *See* **Clusters of Orthologous Groups**.

compartmentalization a way to maximize efficiency of cellular processes that require multiple proteins to work in concert (e.g., organelles).

complement when one allele is able to encode a functional product and compensate for a nonfunctional gene.

complementary DNA (cDNA) created by investigators by incubating mRNA, reverse transcriptase, and **dNTPs**. Initially, the cDNA is single stranded and can be used as probes for **DNA microarrays**. Often a second strand of cDNA is produced to form **dsDNA**.

consanguineous synonym for incest; breeding among closely related individuals that can accumulate deleterious alleles in the offspring and thus more genetic diseases.

conserved domain (CD) a **domain** that has been retained during evolution presumably due to its essential role within the protein's structure. Conserved domain searches are a part of the **BLAST** search.

constitutive always active; constitutive promoters or genes are active at all times.

contiguous (contigs) overlapping DNA segments that as a collection form a longer and gapless segment of DNA.

controlled vocabulary a prescribed set of terms to standardize the description of characteristics as seen in Gene Ontology (GO).

cooperation through communication a principle that enhances efficiency of a complex cellular process. Proteins often signal each other as the process continues so the cell can achieve the intended outcome (e.g., **checkpoints**).

correlation coefficient a measure of the correspondence between two sets of data.

coverage based on the number of bases sequenced in a genome, the coverage represents how many times an average base was sequenced; finished genomes frequently have 8X coverage.

Cox *See* **cyclooxygenase**.

CpG islands patches of DNA bases of a C (cytosine) followed by a G (guanine) with an intervening phosphate (p).

Cy3 and **Cy5** fluorescent green and red dyes, respectively, that are commonly used for **microarray** experiments.

cycle a path in a graph that begins and ends at a particular node, without visiting any other node more than once.

cycle sequencing PCR-based method for sequencing DNA that uses fluorescent dyes and automated DNA sequencers.

cyclooxygenase (Cox) Cox converts arachidonic acid into prostaglandins, which are lipid-signaling molecules that trigger the sensation of pain, inflammation, and fever. Humans have two Cox genes called Cox-1 and Cox-2. Cox-1 is expressed in every cell, with the highest levels in the stomach and kidneys. Cox-2 is expressed primarily in the brain, lung, kidney, and white blood cells.

cytogenetic markers banding patterns on metaphase chromosomes used as physical landmarks along chromosomes, used to map the location of genes.

D

ddNTPs *See* **dideoxyribonucleotide triphosphates**.

degrade destroy a molecule such as a protein or mRNA. The implication is that the molecule is destroyed by a mechanism or protein and not simple entropy.

dendrogram a branching diagram that shows the relative sequence similarity between many different proteins or genes. Typically, horizontal lines indicate the degree of differences in sequences, but vertical lines are used for clarity to separate the branches. A scale bar should be included with each dendrogram.

deoxyribonucleotide triphosphates (dNTPs) these are the normal DNA monomers of dGTP, dCTP, dATP, and dTTP.

dicot a plant that emerges from its seed with two leaves and has branching veins (e.g., oak, soy bean, roses).

dideoxy [sequencing] method invented by Fred Sanger, a method for sequencing DNA that utilizes **ddNTPs**.

dideoxyribonucleotide triphosphates (ddNTPs) a modified **dNTP** that lacks the 3′ hydroxyl group. ddNTPs are used in the **dideoxy sequencing method** developed by Fred **Sanger**. There are four ddNTPs: ddGTP, ddCTP, ddATP, and ddTTP.

digenic a trait is described as digenic if two genes collectively encode the proteins that produce the trait. *Polygenic* is a related term that indicates more than two genes contributing to the phenotype. Hair and eye colors are examples of polygenic traits.

discovery science perhaps the oldest form of science, when a person performs an experiment to see "What if . . . ?" Discovery science is a departure from hypothesis testing, which has been the standard for many years in molecular biology.

DNA chips *See* **DNA microarrays**.

DNA microarrays or **DNA chips** synonyms for gene sequences spotted on glass slides used to measure simultaneously the level of transcription of many genes.

dNTPs *See* **deoxyribonucleotide triphosphates**.

domain (1) the highest level of taxonomic organization. All life is divided into three domains: **Archaea**, **Eubacteria**, and **Eukaryota**. (2) a region within a protein that has a particular shape and function (see **motif**).

dot plot graphical display comparing sequence conservation between two genomes with dots indicating strings of identical bases.

double-stranded DNA (dsDNA) formed when two DNA strands bind to each other.

double-stranded RNA (dsRNA) Formed when two complementary RNA sequences bind to each other. Transfer RNAs contain sections of dsRNA as do **external guide sequences**.

downstream a relative direction for DNA sequence. Since DNA is usually written with the 5′ end to the left, downstream would be to the right of a reference point. For example, the start codon is downstream of the promoter.

draft sequence a description of the degree of confidence in a DNA sequence. When the DNA has been sequenced only four times, it is described as a "draft sequence," compared to a "**finished sequence**," which has been sequenced about eight times. With a draft sequence, about 95% of the genes should be identifiable.

drug signature genes genes that are either **induced** or **repressed** when a particular drug is administered to cells and usually identified by **DNA microarrays.**

dsDNA *See* **double-stranded DNA**.

dsRNA *See* **double-stranded RNA**.

dystrophin gene/protein that causes muscular dystrophy when mutated. Dystrophin helps connect the contractile proteins inside muscles to the extracellular matrix. A mouse model of muscular dystrophy has a mutated dystrophin gene, called *mdx*.

E

EC numbers *See* **Enzyme Commission numbers**.

ectoderm one of the three layers of cells in a developing embryo. Ectoderm cells on the outer layer will form the epidermis and nervous system.

ectopic a gene inserted in an unnatural location. For example, in mutant strains of T7, ectopic copies of *gene 1* were inserted throughout the genome to test the positional effects of gene locations.

EGS *See* **External Guide Sequence**.

ELSI *See* **Ethical Legal and Social Issues**.

endoderm one of the three layers of cells in a developing embryo. Endoderm cells in the innermost layer will form the lining of the digestive system and connected organs such as liver and pancreas.

endosymbiotic symbionts are two organisms who live together so each benefits from the other. Endosymbionts live inside the host organism while both continue to benefit.

Endo16 sea urchin gene expressed early in development in endothelial cells that eventually become a part of the gut of the larva.

environmental stress response (ESR) coined as a result of **DNA microarray** experiments, ESR indicates a collection of genes regulated by sudden changes in the local environment (intracellular or extracellular).

Enzyme Commission (EC) numbers a systematic way to identify every enzyme regardless of species of origin or language used by investigator. For example, cytoplasmic IDH has an EC number of 1.1.1.42.

epigenetic regulation control of gene activity without altering DNA sequence. One example of epigenetic regulation is **imprinting**, which is affected in part by **methyltransferases** adding a methyl group ($-CH_3$) to cytosine bases in DNA.

epistatic Gene A is said to be epistatic to gene B if an allele of gene A masks the encoded effects of gene B.

epitope tag epitopes are portions of larger molecules to which an antibody binds. Epitope tags are added onto different molecules to provide an antibody binding site.

ESI *See* **mass spectrometry**.

ESR *See* **environmental stress response**.

ESTs *See* **expressed sequence tags**.

Ethical Legal and Social Issues (ELSI) a division of the **Human Genome Project**, ELSI was intended to fund research and educational efforts about the ethical impact of sequencing the human genome.

Eubacteria one of the three **domains** of life; includes all the gram-positive and gram-negative bacteria such as *E. coli*, *M. tuberculosis*, and *Salmonella*.

Euchromatin gene-rich, non-compacted portion of a chromosome that is transcriptionally active.

Eukaryotes one of the three **domains** of life; includes all plants, animals, fungi, and protists whose cells contain membrane-bound organelles such as nuclei, endoplasmic reticulum, Golgi body, etc.

E-value *See* **expect value**.

exons parts of RNA molecule spliced together to form mRNA.

expansionist a person who takes a systems approach to understanding complex, interconnected parts working as a whole; opposite of a reductionist.

expect value (E-value) when performing a **BLAST** search, you will obtain an E-value for each sequence that is retrieved. An E-value can be thought of as the probability that two sequences are similar to each other by chance. Therefore, E-values are best when they are small (e.g., 1×10^{-12}) compared to larger E-values (e.g., 0.06).

expressed sequence tags (ESTs) short DNA sequences obtained from either the 5′ or 3′ ends of **cDNAs**. An EST database has been established to help determine coding sequences within genomes, alternative splicing, and **single nucleotide polymorphisms**.

External Guide Sequence (EGS) short RNA sequences that bind to mRNAs to form a 3D shape that resembles tRNA precursors. RNaseP recognizes this pre-tRNA 3D shape and cleaves the mRNA, thus inactivating it.

extremophiles organisms that live in extreme environments, members of the domain **Archaea**. For example, Archaea can live in boiling water, extremely acidic water, or at the bottom of the ocean near hot vents.

F

FASTA simple text format for DNA or protein sequences.

fate map indicates the long-term outcome for each cell in an embryo. For example, a sea urchin fate map tells you what will happen to a particular cell once its development has locked it into a set pathway.

features in the context of **DNA microarrays**, each spot on the microarray can be called a feature. Spots and features are synonyms.

feedback loop biochemical term to describe how the products of a circuit pathway can either enhance (positive feedback) or repress (negative feedback) the circuit.

finished sequence DNA has been sequenced at least eight times and there are no gaps, and contains no more than 1 error in 10,000 base pairs (see **draft sequence**).

founder effect a population genetics term that explains a lack of genetic diversity in a population. For example, when a small number of adults land on a small island, all subsequent children typically contain no more variability than was present in the original inhabitants.

frozen genome description of the genome sequence from a particular date that does not change over time. It was necessary to create a frozen genome to annotate the draft human genome since it would be impossible to analyze the sequence if it changed every day as **draft sequence** became **finished sequence**.

function vague term that has been supplanted by Gene Ontology's controlled vocabulary of biological process, molecular function, and cellular component.

G

galactose 1-phosphate (Gal-1-P) an intermediate metabolite as galactose is converted into glucose for ATP production.

gap in sequencing, a segment of DNA that has not been sequenced but is flanked by sequenced DNA. Also used as a synonym for insertion or deletion (indel) in sequence alignment.

gas chromatography a method for separating and identifying many different types of molecules.

gastrula an embryo in the process of **gastrulation**.

gastrulation a process of embryonic development when part of a **blastula invaginates** and forms a second layer of embryonic cells.

GC content the percentage of DNA composed of G (guanine) or C (cytosine).

GCAT *See* **Genome Consortium for Active Teaching**.

GC-skew uneven distribution of guanine and cytosine bases between the two strands of DNA where GC base pairs occur.

GenBank developed and housed at **NCBI**, GenBank is the U.S. repository for all DNA and protein sequences.

Gene Ontology a collaborative effort of investigators to unify and standardize terms associated with the role a gene or protein plays in an organism. Represented model organisms include fruit fly (*Drosophila melanogaster*), yeast (*Saccharomyces cerevisiae* and *Schizosaccharomyces pombe*), mouse (*Mus musculus*), plant (*Arabidopsis thaliana*), worm (*C. elegans*), rat (*Rattus norvegicus*), and slime mold (*Dictyostelium discoideum*).

gene therapy correcting a defective gene by inserting a *wt* allele. Gene therapy can only work to correct recessive alleles unless the defective dominant allele is replaced by a knockout deletion.

genetic applet a term coined to indicate that simple toggle switches are capable of storing digital information (on or off).

genetic determinism the idea that all human traits are encoded in DNA. Examples of genetic determinism can be found in popular media stories that tout the discovery of a "smart gene" or "worry gene."

genetically modified organism (GMO) in contrast to **gene therapy**, GMO has a *transgene* inserted into the genomes in every cell in its body, as will all its offspring.

Genome Consortium for Active Teaching (GCAT) a nonprofit organization dedicated to bringing genomic methods into the undergraduate curriculum.

genomic representations a subset of the genome created by randomly choosing restriction fragments. Especially useful when the genome has not been sequenced but you want to survey genome-scale phenomena.

genomics a vague term that encompasses the study of **reference genome** sequences, variations within a species' genome, **DNA microarrays**, **circuits**, and **systems biology**. Some people include wider areas such as **proteomics**, **metabolomics**, etc., under the genomics umbrella. Due to its recent and changing definition, the term does not have a universally accepted meaning.

GMO *See* **genetically modified organism**.

Gram-negative bacteria a subset of **Eubacteria** that do not stain with a dye called Gram stain (e.g., *E. coli*).

graph mathematical term for a network in which objects (e.g. genes) are represented by nodes, and interactions between objects are represented by edges.

guilt by association an inference made from **DNA microarray** data. If Gene A is induced and repressed at the same time as Gene B, then guilt by association predicts Genes A and B perform similar functions.

H

haplotype (derived from haploid genotype) a collection of alleles in one individual that are located on one chromosome. Alleles within a haplotype often are inherited as a single unit from one generation to the next. In **SNP** studies, haplotypes refer to a group of genomic variations found repeatedly in many people within a population.

heat denaturation use of heat to deform the 3D shape of proteins or nucleic acids (e.g., turning dsDNA into ssDNA).

hemagglutinin (HA) a protein encoded by the human influenza virus and often used as an **epitope tag**.

heterochromatin gene-poor, compacted portion of a chromosome that is transcriptionally muted.

heterodimer protein made of two subunits. *Hetero* denotes different subunits, and *dimer* means it is composed of two subunits. There are also heterotrimers (three subunits), heterotetramers (four subunits), etc.

HGP *See* **Human Genome Project**.

hierarchical clustering a method for organizing large numbers of genes, tumors, or other objects into dendrograms.

high-throughput methods that produce large volumes of data and can process many samples quickly. Robots and computerized data collection are common themes in high-throughput methods.

hippocampus the location near the center of the brain where learning takes place and memories are stored. A critical layer of the hippocampus is the CA1 layer. See Brain Anatomy web page for details.

hits short-hand for sequences returned when searching a database such as NCBI.

hominid living or extinct human/human-like animals. Examples include the famous fossils called Lucy and Peking Man, as well as broader categories such as *Homo erectus, Neanderthal,* or *Homo sapiens.*

homologous recombination molecular method that targets a single gene for modification. Genes can be **knocked out** by this method, or more subtle changes can be created. Homologous recombination works better in some organisms (e.g., yeast) than others (e.g., humans).

homology a term with two different meanings that are often confused. Initially, homology referred to two sequences (DNA or amino acid) that were similar due to evolutionary relatedness. The newer and less specific meaning is simply two sequences that are similar. One other usage refers to homologous chromosomes, the pair of chromosomes in diploid organisms.

homotetramer proteins composed of four identical subunits. *Homo* denotes identical subunits, and *tetramer* means it is composed of four subunits. There are also homotrimers (three subunits), **homodimers** (two subunits), etc.

horizontal transfer the movement of DNA from one species to another without sexual transmission; mechanism uncertain, perhaps viral.

Human Genome Project (HGP) the multinational, public domain DNA sequencing consortium composed of academic labs in the U.S., Europe, and Asia. Funding for this work came from government agencies and private philanthropic organizations.

hybridize process of incubating a probe with its target as well as the process of the probe binding to its target. The probe can be a protein (e.g., an antibody) binding to another protein or a nucleic acid (e.g., **ssDNA**) binding to its complementary sequence.

hydropathy plot (Kyte-Doolittle) a computer-generated graph that uses the amino acid sequence to predict whether or not a protein will span a membrane.

hypergeometric a probability model for sampling from a population; under this model, sampled items are not returned to the population before sampling is repeated.

I

ICAT *See* **Isotope-Coded Affinity Tags**.

ideogram cartoon drawing of chromosomes.

IDH *See* **isocitrate dehydrogenase**.

immunoglobulin synonym for antibody.

immunoprecipitation method that uses an antibody to extract your favorite protein from a mixture of many proteins. Often, proteins that are bound to your favorite protein are co-purified along with your protein.

imprinting process through which mammalian paternal and maternal alleles can be treated differently, depending on which parent contributed them to the offspring.

in silico experimental process performed on a computer and not by bench research.

in vitro experimental process performed in a tube or petri dish and not in a living cell or organism. Literally translated as "in glass."

in vivo a term that means the experimental process was performed in live cells or organism, compared to **in vitro** or **in silico**.

indels collective noun that refers to insertions or deletions of bases in DNA sequences.

induced a gene with increased transcription. Typically refers to the switch from none to some transcription, but it could also refer to a switch from low to high transcription.

integrated circuits when a particular process requires more than one individual circuit of genes or proteins, it can be described as an integrated circuit. The distinction between a circuit and an integrated circuit is subjective and therefore indicates only relative complexity.

intergenic sequence DNA sequence between two genes, sometimes referred to as **"junk DNA."**

intrinsic gene subset coined as a part of a DNA microarray study of breast cancer, these genes are expressed in similar ways within a patient's tumor but not in tumors from other patients.

intron portion of the gene and initial RNA transcript that will be excised and not included in mRNA; usually begins with the sequence GT and end with AG.

invaginate when a mass of **blastula** cells migrate to form a pocket inside the larger embryo. Invaginating cells of a **gastrula** look as if someone is pushing a finger into a balloon.

isocitrate dehydrogenase (IDH) ubiquitous metabolic isozymes encoded by five genes in yeast and other eukaryotes. IDH converts isocitrate into α-ketoglutarate and requires a coenzyme (either NAD^+ or $NADP^+$) and a divalent cation (either Mn^{2+} or Mg^{2+}).

isoelectric point (pI) when the net charge of a protein is zero, the pH of the local environment will be equivalent to the isoelectric point of the protein.

isoforms or **isozymes** these closely related terms refer to two versions of highly similar proteins. Isozymes is specifically used for enzymes, while isoforms can be used for any protein.

Isotope-Coded Affinity Tags (ICAT) proteomic method that identifies proteins in a mixture and determines relative amounts for each protein from two protein mixtures.

isozymes *See* **isoforms**.

iteration when a process is repeated in an attempt to reach the ideal outcome, each repetition is called an iteration. Each iteration is slightly different from the previous one since we learn from the first and improve the second iteration.

J

jmol open source, java-based computer application that permits 3D viewing of molecules.

Joint Genome Institute (JGI) Department of Energy–funded organization that uses **high-throughput** methods and computational analysis to understand basic biology. Included in the JGI are the Human Genome Project, Microbial Genome Projects, fish, sea urchin, etc.

"junk DNA" an outdated term that indicates how little people knew about genomes. It was intended to recognize that about 98% of the human genome does not contain any genes and therefore had no function.

K

knockout a molecular method to target a single allele for deletion and replacement with DNA of your choice using **homologous recombination**. Through selective breeding, homozygous individuals can be produced.

knockout mouse a mouse that has had one or more (typically just one) gene deleted (**knocked out**).

Kyte-Doolittle plot *See* **hydropathy plot.**

L

leprosy a disfiguring disease caused by the bacterium *Mycobacterium leprae*. They infect white blood cells called **macrophages**, which normally engulf and kill pathogenic bacteria. Later, Schwann cells surrounding the nerves become infected, which leads to the loss of myelin, nerve damage, loss of sensation, and eventually loss of extremities due to reduced blood circulation.

leptin mammalian protein hormone (encoded by the *ob* gene) produced by fat cells that regulate fat homeostasis (the **lipostat**). If you produce too much leptin, you will lose fat, but if you don't produce enough, you will store more fat. In addition to fat homeostasis, leptin also influences sexual reproduction, immune and brain functions.

linkage when two *genes* are located near each other on the same chromosome; quantified by the frequency of recombination between two loci and measured in map units or centiMorgans, where 1 map unit equals 1% recombination frequency.

linkage disequilibrium linkage refers to *genes*, while linkage disequilibrium focuses on *alleles*. Linkage disequilibrium occurs when a set of alleles on one copy of a chromosome (a **haplotype**) stay associated with each other at a higher frequency than would be expected if recombination were completely random. When haplotypes are retained for many generations, it is assumed there is a selective advantage in not separating these particular alleles.

linker a short segment of **dsDNA** that can be ligated onto a second fragment of DNA to facilitate the cloning of that fragment. Linkers contain a restriction site, so they can be digested to produce the desired sticky ends for ligation.

liposomes spheres of lipids created **in vitro** to carry DNA or other biological reagent inside eukaryotic cells.

lipostat a term to describe the **integrated circuit** that uses **leptin** and other molecules to regulate fat homeostasis. Each person's lipostat is set for a percentage of body fat, and this set point can be increased by eating a high-fat diet or decreased by exercising.

log odds logarithm of a ratio of probabilities. Used to find motifs with the Position Weight Matrix method.

logic gates graphical representation of logic functions such as AND, OR, and NOT; may be combined to form complex logic circuits.

long-term potentiation (LTP) when a neuron is stimulated and the consequence of the stimulation is maintained after the original stimulus is gone. This is the central component to memory and learning, controlled in large part in the **hippocampus** area of the brain. See Brain Anatomy web page for details.

lysogenic bacterial viruses such as lambda (λ) insert their genomes into the chromosome of the host bacterium and remain dormant, or lysogenic.

lytic one of the two choices bacterial viruses such as lambda (λ) can make that result in the rupture of the infected bacterium and the release of many new viruses.

M

macroarrays individual spots in macroarrays are composed of colonies of cells such as yeast and permit **high-throughput** screening of whole cell phenotypes.

macrophage one type of white blood cell that engulfs **pathogens** and presents fragments of the destroyed pathogen to other white blood cells for subsequent immune response.

MALDI *See* **mass spectrometry.**

mass spectrometry (MS) a technique that allows investigators to separate proteins based on their **mass to charge ratio (m/z)**. The m/z for each protein allows them to be identified and quantified from complex mixtures. This **proteomics** tool is often used in pairs and called tandem mass spectrometry (MS/MS). Protein samples are first ionized and then inserted into MS/MS by either laser-based methods (MALDI and SELDI) or an electrospray method (ESI).

mass to charge ratio *See* **mass spectrometry.**

megabase (Mb) 1,000 kilobases or 1 million bases of DNA.

mesoderm one of the three layers of cells in a developing embryo. The middle layer of mesoderm cells will form many tissues such as heart, kidney, and reproductive organs as well as blood, bone, muscles, and tendons.

metabolic reconstruction using annotated genome to assemble the probable biochemical pathways within the species.

metabolome term coined to encompass the entire metabolic content of a cell or organism.

methylome term coined to refer to the methylation state of a genome (methyl-cytosine bases in the chromosome).

microarray *See* **DNA microarray**.

microsatellite a short segment of DNA (2 to 50 bases) repeated multiple times. Microsatellites vary in length and base composition, which makes them useful tools for distinguishing members of a population. For example, one allele may contain GCGCGC, while another might contain GCGCGCGCGC.

modular when a larger process or structure is composed of individual units that can perform functions, the larger process or structure can be described as modular or composed of modules.

molecular function coined by **Gene Ontology**, describes tasks performed by individual gene products such as transcription factor and calcium transportation.

molecular weights molecules of known mass used to determine the mass of unknowns. For example, protein molecular weight standards or markers are measured in Daltons (Da), and DNA/RNA are measured in bp or bases.

monocot a plant that emerges from its seed with one leaf and has parallel veins (e.g., grass, rice, corn).

monomers individual units that can be polymerized. For example, ATP is an RNA monomer.

most recent common ancestor (MRCA) a description of a prehistoric population or species that gave rise to two or more species. Parents are the MRCA for siblings; the MRCA for humans and gorillas lived about 1.2 million years ago.

motif a sequence of amino acids or nucleotides that performs a particular role and is often conserved in other species or molecules.

MRCA *See* **most recent common ancestor**.

MS *See* **mass spectrometry**.

multiplex when a series of reagents are mixed in a single tube so more than one outcome will be produced simultaneously; multiplex PCR produces several different-sized bands of DNA that can be detected.

mutation accumulation an evolutionary theory to explain why selective advantages in one aspect often are accompanied by loss of advantages in others. Random mutations accumulate gradually and can become fixed by genetic drift in genes that are not subject to selection pressure. Adaptation to one environment and loss of adaptation to another are caused by distinct and unrelated variations in the genome.

myelin the fatty layer of insulation that surrounds nerves outside the CNS. Myelin is produced by Schwann cells.

m/z *See* **mass spectrometry**.

N

Na/K ATPase enzyme found in the plasma membrane of all animal cells. It pumps three sodium ions out of the cell and two potassium ions into the cell. Moving these ions consumes one ATP and produces the membrane potential used in muscle and nerve depolarization.

National Cancer Institute (NCI) federally funded part of the U.S. **National Institutes of Health**, the NCI focuses on basic and applied research to treat and prevent cancer. Home for the Cancer Genome Anatomy Project, which intends to catalog the gene activity found in every type of cancer and compare this to the gene activity in *wt* tissues.

National Center for Biotechnology Information (NCBI) a federally funded part of the U.S. National Library of Medicine, NCBI is the home of **GenBank**, **BLAST**, **COG**, and many other genomic databases and computational tools.

National Institutes of Health (NIH) federally funded center for biomedical research and conduit for funding at institutions in the U.S. The NIH has several campuses located in many states, all of which are dedicated to understanding and improving human health.

NCBI *See* **National Center for Biotechnology Information**.

NCI *See* **National Cancer Institute**.

NCI60 a panel of 60 human cell lines that represent the major forms of cancer. New drugs being developed for chemotherapy are tested first on the NCI60 to determine their effects.

ncRNAs *See* **noncoding RNAs**.

NIH *See* **National Institutes of Health**.

noise in the context of genomic **circuits**, inputs and outputs that are not identical each time. A genomic circuit needs to tolerate a level of noise not typical of real electronic circuits.

nonannotated open reading frame (NORF) an open reading frame that was considered not to be a real gene when the genome was **annotated**.

noncoding RNAs (ncRNAs) DNA that is transcribed but not translated. The best known examples are rRNA and tRNA.

NORFs *See* **nonannotated open reading frame**.

normal distribution the subject of Math Minute 8.1, normal distribution indicates data have an average value around which all the individuals are clustered. A graph of normal distribution results in the classic "bell curve."

normalized data that has been corrected or standardized, for example, by subtracting the sample mean from each observation, and dividing the result by the sample standard deviation.

Northern blot named in reaction to **Southern blot**, RNA is separated according to size in an agarose gel, blotted onto a membrane, and then probed to detect specific sequences. Northern blots are used to determine which cells express the gene of interest, the size, and relative abundance of the mRNA.

O

ob *See* **leptin**.

Occam's Razor a guiding principle; when deciding which explanation to accept, always start with the simplest one.

oligonucleotide microarray *See* **DNA microarray**.

oligonucleotides (oligos) ssDNA polymers of unspecified length. The oligo sequence is determined by the investigator and synthesized **in vitro**. Oligos are used to probe blots, prime sequencing reactions, and PCR, as well as for spotting on **DNA microarrays**.

OMIM *See* **Online Mendelian Inheritance in Man**.

oncogenes mutant dominant alleles of vital genes (e.g., *Ras*). Oncogenes (acting like the accelerator on a car) force the cell cycle to speed up, while **tumor suppressors** are similar to brakes trying to slow down the cell cycle.

Online Mendelian Inheritance in Man (OMIM) a comprehensive web site that catalogs all the genetic and molecular information related to human diseases (not just male diseases).

open reading frame (ORF) a portion of a cDNA or gene that begins with the start codon and ends with the stop codon. Synonym for **coding sequence (CDS)** on **GenBank** results.

ORF *See* **open reading frame**.

orthologs two genes in different species that are evolutionarily related. For example, the mouse and human **leptin** genes are orthologous because they evolved from a common ancestral leptin gene.

overexpress when genes are bioengineered to produce excessive amounts of protein, they overexpress their encoded proteins.

P

p53 a protein of 53 kDa molecular weight and a very important **tumor suppressor**.

pair-fed when one mouse is treated with **leptin** injections and reduces its food intake, the amount of food consumed by a control mouse is restricted to equal that consumed by the leptin-injected mouse. The control mouse is called a pair-fed mouse since the amount it was fed was paired with a treated mouse.

paralogs two genes within the same species are called paralogs if one evolved from the other. For example, yeast has three *IDP* genes that encode different **isoforms** of **IDH**. These three could be called paralogous genes if two of them evolved in yeast from one ancestral gene.

pathogen bacterium that can harm its host.

PCR *See* **polymerase chain reaction**.

PDB *See* **Protein Data Bank**.

penetrance the percentage of individuals who develop the phenotype for a particular genotype. If ten people are homozygous at one locus but only eight of them exhibit the phenotype, then this gene has 80% penetrance.

Perl script a program written in Perl, which is optimized for manipulating and finding patterns in sequences.

phagosomes internal organelles of white blood cells that engulf and kill pathogens.

pharmacogenetics study of the relationship between particular genes or alleles and the effectiveness of medications.

pharmacogenomics very similar to **pharmacogenetics**, pharmacogenomics attempts to study genome-wide influences on the efficacy of medications.

phosphorylate (phosphoprotein) when an enzyme adds a phosphate to another protein, the enzyme is called a kinase and the substrate protein becomes a phosphoprotein.

phylogenetic tree graphic way to illustrate the evolutionary relatedness of genes, proteins, individuals, strains, or species.

phylotypes term intended to resolve the challenge of "species" when classifying prokaryotes using DNA sequence comparisons.

phytoplankton microscopic organisms that live in water and photosynthesize.

pI *See* **isoelectric point**.

plasmids natural or engineered circular DNA that is replicated inside prokaryotes but is not considered a chromosome.

pleiotropic one gene affects several seemingly unrelated phenotypes.

pointillism a school of art mastered by Georges Seurat in the 19th century and Chuck Close in the 20th. **DNA microarrays** are a collection of spots that can be examined as a whole or as a series of dots similar to the artwork of pointillists.

polymerase chain reaction (PCR) molecular method that allows you to mass-produce any segment of DNA as long as you have two **oligos** that **hybridize** to the two strands of target DNA with their 3' ends pointed toward each other.

postgenomic era describes the time in biology after entire genomes have become sequenced routinely. The postgenomic era began around the year 2000, though no single event signaled its beginning.

posttranscriptional control after a gene has been transcribed, the RNA can be modified and regulated, processes that constitute posttranscriptional control.

pRb *See* **retinoblastoma protein**.

prions proteins that have two shapes, one benign and the other contagious, which leads to the conversion of the benign shape to the contagious shape. Contagious prion proteins can spread from one organism to another and cause neurological diseases such as scrapie in sheep, mad cow disease, and Creutzfeldt-Jakob Disease (CJD) in humans.

prostaglandins produced when **cyclooxygenases** cleave **arachadonic acid**. Prostaglandins are lipid-signaling molecules that trigger the sensation of pain, inflammation, and fever.

protease enzyme **degrades** proteins into smaller pieces. Two examples are cathepsins found in **phagosomes** and trypsin produced by the pancreas to help digest food.

Protein Data Bank (PDB) database of every protein for which the 3D structure is known; it also contains a few nonprotein structures.

Protein Mass Fingerprinting (PMF) identification of proteins by matching the masses of peptides contained in the protein with those of known proteins in a database.

protein microarrays proteomic method similar to **DNA microarrays** in size and scale; proteins are spotted onto glass and are used to determine protein interaction or to identify and quantify molecules found in various solutions. One example is an **antibody microarray**.

proteome the complete collection of proteins in a cell/tissue/organism at a particular time. Unlike genomes, which are stable over the lifetime of the organism, proteomes change rapidly as each cell responds to its changing environment and produces new proteins and at different amounts. You have one body-wide proteome, about 200 tissue proteomes, and about a trillion individual cell proteomes.

proteomics the study of proteomes that includes determining the 3D shapes of proteins, their roles inside cells, the molecules with which they interact, and defining which proteins are present and how much of each is present at a given time.

pseudogenes segments of DNA that resemble genes by their sequence of bases but are nonfunctional. Pseudogenes often have **transposons** inserted in them, or they may have other mutations that led to their inability to encode a functional protein.

PubMed an extensive database of biomedical literature hosted by **NCBI** that is searchable. You can subscribe to PubCrawler and automatically search PubMed and receive email results on a schedule of your choosing.

***p*-value** probability associated with a statistical test of the difference between populations. Populations are considered significantly different if the associated *p*-value is small (typically 0.1 or smaller).

Q

qPCR quantitative PCR; synonym for real-time PCR.

quantitative trait loci (QTL) genes that encode phenotypes that can be measured on a scale, such as autism or skin tone.

quorum sensing a process by which cells detect localized crowding by many other cells.

R

rate constants a characteristic of enzymes that catalyze a reaction at a consistent pace. For example, the rate constant for **IDH** would indicate how many moles of isocitrate would be consumed per unit time.

real-time PCR (RT-PCR) a molecular method that is sometimes confused with reverse transcriptase PCR. Real-time PCR uses the specificity of PCR to measure the number of template molecules in your starting material and is being developed to detect biological weapons.

reductionist a person who dissects a complex system into increasingly smaller parts in order to understand it. For example, to understand how a watch works, you might take it apart to determine its components and deduce how they work together to form a watch.

redundant in the context of **genomic circuits**, redundant means that a critical process can be performed by more than one gene or individual pathway. For example, there are three redundant **IDH** genes that utilize $NADP^+$ and consume isocitrate to produce NADPH and α-ketoglutarate.

reference genome or **reference sequence** genome that was sequenced first for a species and thus represents a standard but not necessarily "normal" example. The term "reference" implies that variations exist within the population, but the reference is used as a common point for comparisons.

reference sequence *See* **reference genome**.

regional organization of components a principle that allows complex systems to behave more efficiently. When many proteins are required to perform a task, the task can be accomplished more rapidly if the necessary proteins are located near each other.

reiterative a process can be described as reiterative if you keep trying to improve the quality with each successive attempt. For example, models that describe how genes are regulated are reiterative because each version of the model is built upon more information so the model gradually approaches the truth.

reliability an engineering term used to measure the probability that the desired outcome will be accomplished. The reliability of a biological process is higher if **redundant** genes ensure there are multiple ways to accomplish the task.

reporter gene gene from another species that produces an mRNA or protein that can be detected easily. Reporter genes are often used to study the capacity of promoters or

cis-regulatory elements. **CAT**, GFP, and *lacZ* are three common examples.

repressed when the level of transcription is reduced, a gene has been repressed. Repressor proteins bind to **cis-regulatory elements** and block or hinder transcription.

repressilator synthetic biology device used to produce oscillating GFP signals; became standard for naming many devices by adding -ilator suffix.

retinoblastoma protein (pRb) a very important **tumor suppressor**.

reversal a synonym for an inversion used by computer scientists and bioinformaticists; used to measure differences between two genomes.

reverse genetics beginning with a gene sequence and deducing its function afterwards; the opposite of traditional genetics.

riboswitch 5′ portion of mRNA that can regulate translation of its own mRNA.

ribozymes RNA enzymes. Contrary to the initial rule "all enzymes are proteins," some enzymatic reactions are performed by RNA.

RIKEN Japanese Institute of Physical and Chemical Research; a genome center that hosts many databases and funds research.

RNA interference (**RNAi**) short **dsRNA** capable of inactivating genes by blocking the production of the encoded proteins via microRNA (miRNA) amd short inhibitory RNA (siRNA).

RNase P an RNA-cleaving enzyme that is required to produce mature tRNA molecules from precursors. **External guide sequences** form tRNA-like shapes when they bind to mRNAs, and **RNase P** cleaves the mRNA and thus prevents its translation.

robust an engineering term that indicates the ability to function in less than optimum conditions (i.e., **noise**). Methods such as **Southern blots** and **DNA sequencing** can be described as robust because they work properly even when investigators unintentionally alter the reaction conditions. Tolerant biological processes can also be described as robust.

RT-PCR *See* **real-time PCR**.

S

sample mean average of data points in a sample; denoted by $\bar{y}$.

Sanger method *See* **dideoxy [sequencing] method**.

satiety the satisfied feeling of no hunger.

scaffold a collection of contigs lumped together into one larger contig.

second messengers a vague term that refers to the way a cell relays information intracellularly. cAMP and calcium

ions are examples of second messengers since they are produced in response to a primary message such as the binding of a ligand to its receptor.

sensitivity probability that a diagnostic test will be positive, given the individual has the disease; i.e. probability of a true positive test result.

sequence-tagged site (STS) unique locus in a **genome** defined by PCR primers that amplify a single locus. STSs were used to map the human genome, but are still useful as markers defining chromosomal positions.

serotypes analogous to genotypes, serotypes describe pathogens based on the ability of antibodies to bind to different subsets. For example, *Neisseria meningitides* exists in several serotypes.

shotgun sequencing a strategy for sequencing whole genomes, it was pioneered by the for-profit company Celera. Genomes are cut into very small pieces, cloned into plasmids, sequenced, and then assembled into whole chromosomes or genomes. This method is faster than hierarchical shotgun sequencing but more prone to assembly errors.

signal sequence hydrophobic in nature, the first 20 amino acids of proteins synthesized that pause translation until the ribosome docks with the rough endoplasmic reticulum (ER).

signal transduction conveyance of information from the outside to the inside of a cell. When a ligand binds to its receptor, this information is conveyed to the rest of the cell through a complex pathway of signal transduction that involves **second messengers**.

signature genes often cited in **DNA microarray** experiments; a collection of genes that are characteristic for a particular sample. For example, you would expect all green leaves to express a set of signature genes necessary to conduct photosynthesis (see Math Minute 7.1 for details).

single nucleotide polymorphisms (SNPs) very similar to point mutations except SNPs are considered to represent the genetic variation present in *wt* genotypes. By definition, SNPs differ from the **reference sequence** of a species.

singleton any object that is not included in a collection of objects of its type, usually because it is not similar to any of the objects under consideration.

sliding window a computational sampling method that selects a set number of monomers within a polymer (e.g., 11 amino acids in a protein) and then shifts down by one monomer to form the next sample of the same size.

small nuclear RNAs (snRNAs) example of a **ncRNA**.

small nucleolar RNAs (snoRNAs) example of a ncRNA.

SNP minor alleles SNPs that are less common than most other SNPs at the same position.

SNPs *See* **single nucleotide polymorphisms**.

soluble receptor alternative form of a receptor that is not anchored to a membrane. Soluble receptors can be bio-engineered or naturally occurring (e.g., **leptin** receptor in pregnant women).

somatic hypermutation immunology term that in part refers to the ability of B cells to alter the DNA of antibody-encoding genes in order to produce antibodies with improved binding capacity. This takes place in the germinal centers of lymph nodes.

Southern blot named to honor its inventor, Dr. Ed Southern. A classic molecular method that allows the investigator to separate DNA by size in a gel, transfer the DNA to specialized paper (called a membrane), and hybridize the DNA with a particular probe.

specificity probability that a diagnostic test result will be negative, given that the individual does not have the disease; i.e., probability of a true negative test.

stable isotopes differing from radioactive isotopes such as ^{35}S and ^{32}P, stable isotopes are atoms that do not emit radiation but have heavier or lighter masses based on the number of neutrons they carry. They are used in **proteomic** methods such as **ICAT**.

standard deviation measure of variation from the mean; denoted by s or σ (sigma).

stationary phase a description of cells growing in culture that have slowed down their growth rate. Stationary cells are not increasing in number, but they are not dead.

stochastic not exactly the same as random, stochastic refers to genes or proteins that can produce widely variable outcomes.

stomata plural of stoma, the openings in leaves that permit gas exchange between the environment and cells within the leaves.

Strongylocentrotus purpuratus the Latin name for the sea urchin model organism; its genome is completely sequenced. Expresses the model gene *Endo16* during development.

structural proteomics a discipline within proteomics that focuses on the 3D shape of proteins.

STS *See* **sequence-tagged site**.

substitution matrix used to compute the raw score of an alignment; elements represent the evolutionary closeness of amino acids.

Sup35 a nontoxic yeast protein involved in termination of translation; has **prion**-like properties. It can assume two shapes, one of which is able to convert other Sup35 proteins to the contagious shape that no longer terminates translation.

Swiss-Prot a proteomics database based in Switzerland.

synergy the total output is greater than the sum of its parts. For example, sperm and egg synergistically work together to produce a new individual, not just a diploid cell.

synteny a term that has experienced a gradual change in its meaning, but current usage refers to multiple genetic loci from different species located on a chromosomal region of common evolutionary ancestry. Many mouse and human loci are syntenic.

synthetic biology the use of engineering principles to construct small biological devices by assembling new DNA circuits inside cells; term derived from analogy with synthetic chemistry.

systems biology coined to denote the new perspective for research in the **postgenomic era**. Systems biology studies whole cells/tissues/organisms not by a traditional **reductionist's** approach but by holistic means in a **reiterative** attempt to model the complete cell/tissue/organism.

T

T7 dsDNA virus used as a model system to understand larger biological principles.

tandem MS *See* **mass spectrometry**.

TATA box a portion of the eukaryote promoter 25 bases upstream of the start transcription site; contains the sequence TATA.

TB *See* **tuberculosis**.

teratogen a substance such as lithium or alcohol that causes developmental abnormalities.

The Institute for Genomics Research (TIGR) Maryland-based genomics and proteomics nonprofit organization that produces public domain genome sequences and analysis software.

The SNP Consortium (TSC) central repository of human SNP data hosted at Cold Spring Harbor Laboratories, NY.

threshold stimulation referring to a critical point that determines which direction a toggle switch will go. Similar to the action potential of neurons where depolarization above the critical point leads to one of two options (i.e., action potential or no response).

TIGR *See* **The Institute for Genomic Research**.

transcriptional phase variation frequency of G's (guanines) in DNA sequence that alters the initiation rate of transcription and thus produces diversity within clonal populations.

transcriptome coined to describe the complete RNA content of a cell/tissue/organism and often measured by **DNA microarrays**.

transmembrane domain the portion of a protein that spans a phospholipids bilayer, typically about 20 amino acids long and predominantly hydrophobic.

transposons sometimes referred to as "jumping genes," segments of DNA that can move from one place in a genome to another. Transposons have moved throughout the human genome and constitute a significant percentage of our genome. In yeast, the transposon Tn3 was bioengineered to be used as a genomics/proteomics research tool called mTn.

truth table description of the output values of a logic function (e.g., AND, OR, NOT), based on the values of input variables.

TSC *See* **The SNP Consortium.**

t-**test** statistical method for determining whether a mean, or the difference between two means, is significantly different from the hypothesized value.

tuberculosis (TB) the **pathogenic** bacterium *Mycobacterium tuberculosis* causes TB, which is a debilitating and potentially lethal respiratory infection that can be antibiotic resistant and thus difficult to treat.

tumor suppressor protein slows down the progression of the cell cycle and prevents cancers. Typically, tumor suppressors work at **checkpoints** to ensure the cell is functioning properly before permitting the cell cycle or mitosis to continue. **p53** and **pRB** are two examples.

Tup1p the yeast protein encoded by the gene *Tup1* that functions as a transcription repressor.

two-component systems paired prokaryotic protein environmental sensors where one component senses the outside world and the other component signals the cytoplasm.

two-dimensional (2D) gel electrophoresis proteomics method that separates proteins based on the **isoelectric point** (first dimension) and **molecular weight** (second dimension). Often spots from 2D gels are excised and sequenced to identify them. ExPASy (Expert Protein Analysis System) has a good database of 2D gels.

U

untranslated region (UTR) portion of mRNA that is not translated; found on 5' and 3' ends of mRNA.

upstream a relative direction for nucleic acids often used to describe the location of a promoter relative to the start transcription site. For example, the start codon is upstream of the stop codon.

V

virulent pathogen parasite that has the potential to do serious harm to its host.

W

Western blot proteins that have been separated by size using SDS-PAGE, transferred to a special type of paper (called a membrane), and probed with an antibody. Western blots are used to determine molecular weight, tissue distribution, and relative amount of the protein of interest.

whole genome shotgun (WGS) a genome sequencing strategy that skips the mapping stage and uses computers to reassemble huge numbers of independent sequencing runs that were generated from many random fragments from a genome.

wild-type (wt) allele, genotype, or phenotype that is considered to be the standard for a given strain or species. Wild-type alleles encode functional proteins and produce typical phenotypes. *Italics* is used if referring to genes or alleles.

Y

Y2H *See* **yeast two-hybrid.**

YAC *See* **yeast artificial chromosome.**

Yap1p a yeast transcription factor encoded by the *Yap1* gene.

yeast artificial chromosome (YAC) a cloning vector that replicates in yeast and can contain inserts about 1 Mb in size.

yeast two-hybrid (Y2H) proteomics method to detect protein-protein interactions. Variations of this method have been produced for mammalian and bacterial cells as well. The protein of interest is used as a *bait* to "fish out" proteins that bind to it (called *prey*).

yoctomoles 1 ymol = 10^{-24} moles = 0.6 molecules. A quantity recently coined due to increased sensitivity in proteomics.

Z

zeptomoles 1 zmol = 10^{-21} mol = 602 molecules. A quantity recently coined due to increased sensitivity in proteomics.

Credits

Chapter 1 Fig. 1.1a The University of Kansas Medical Center; **Fig. 1.1b** BP Healthcare *(BPHealthcare.com)*; **Fig. 1.2** Bakker et al. *Lancet.* 1985 Mar 23;1(8430):655–8, adapted from fig. 2. © Elsevier Science; **Fig. 1.3a–c** Uta Francke; **Fig. 1.3d–e, Fig. 1.4** Zatz et al. *J Med Genet.* 1981 Dec;18(6):442–7, adapted from figs 2, 3, and 4; **Fig. 1.5a–c** Crosbie et al. *J Cell Biol.* 1999 Apr 5;145(1):153–65, adapted from fig 4. © The Rockefeller University Press; **Fig. 1.6b** Protein Data Bank; **Fig. 1.7** Jim Ervasti; **Fig. 1.9** Jung et al. *J Biol Chem.* 1996 Dec 13;271(50):32321–9, adapted from fig 6; **Fig. 1.10** Holt et al. *J Biol Chem.* 1998 Dec 25;273(52):34667–70, fig 4c; **Fig. 1.11, 1.12a** Crosbie et al. *Hum Mol Genet.* 2000 Aug 12;9(13):2019–27, fig 2d; adapted from fig 10 and 4; **Fig. 1.13a** Chien et al. *Nature.* 2000 Sep 14;407(6801):227–32, adapted from fig 2. © Macmillan Publishers Ltd.; **Fig. 1.13b–c** Sweeney et al. *PNAS.* 2000 Dec 5;97(25):13464–6, adapted from fig 1. © National Academy of Sciences, U.S.A.; **Fig. 1.14a–b** Rybakova et al. *J Cell Biol.* 2000 Sep 4;150(5):1209–14, fig 4. © The Rockefeller University Press; **Fig. 1.16, 1.17** Hagiwara et al. *Hum Mol Genet.* 2000 Dec 12;9(20):3047–54, fig 3 and 5b; **Fig. 1.18a–d** Richard et al. *Am J Hum Genet.* 1999 Jun;64(6):1524–40, adapted from fig 1; **Fig. 1.19,1.20, 1.21, 1.22** Grady et al. *Nat Cell Biol.* 1999 Aug;1(4):215–20, adapted from figs 2d, 3, 5 and 7; **Table 1.2** Betto, et al. *Italian Journal of Neurological Sciences.* 1999. 20: 375, adapted from Table 2.

Chapter 2 Fig. 2.3 Salzberg et al. *PLoS Biol.* 2004 Sep;2(9):E285. Epub 2004 Sep 14, pg 1274; **Fig. 2.4** International Human Genome Sequencing Consortium. *Nature.* 2004 Oct 21;431(7011): pg 935. © Macmillan Publishers Ltd.; **Fig. 2.6** Banfi et al. *Science.* 2000 Jan 7;287(5450):138–42, adapted from fig 3b. © The American Association for the Advancement of Science. All rights reserved; **Fig. 2.8** Shirley Tilghman; **Fig. 2.9** www.niams.nig.gov/hi/topics/acne/acne.htm; **Figs. 2.10, 2.11, 2.12** Bruggemann et al. *Science.* 2004 Jul 30;305(5684):671–3, fig S1, S5 and S3. © The American Association for the Advancement of Science. All rights reserved; **Figs. 2.13, 2.14** Xu et al. *Science.* 2003 Mar 28;299(5615):2074–6, adapted from fig S1; fig S3. © The American Association for the Advancement of Science. All rights reserved; **Fig. 2.15** Fraser et al. *Science.* 1995 Oct 20;270(5235):397–403, adapted from fig 1. © The American Association for the Advancement of Science. All rights reserved; **Figs. 2.16, 2.17** Raoult et al. *Science.* 2004 Nov 19;306(5700):1344–50, fig 1 and 3. Epub 2004 Oct 14. © The American Association for the Advancement of Science. All rights reserved; **Fig. 2.18** Gift of Sallie M. Chisholm. Photo by Claire Ting, MIT Biology Dept.; **Figs. 2.19, 2.20, 2.21** Rocap et al. *Nature.* 2003 Aug 28;424(6952):1042–7, adapted from fig 1a; fig 2a and 3a. Epub 2003 Aug 13. © Macmillan Publishers Ltd.; **Fig. 2.22a** Wirth. *Nature.* 2002 Oct 3;419(6906):495–6, fig 1. © Macmillan Publishers Ltd.; **Fig. 2.22b** James D. Berger, Department of Zoology, University of British Columbia, Canada; **Fig. 2.23** Cowman et al. *Science.* 2002 Oct 4;298(5591):126–8, adapted from fig 1. © The American Association for the Advancement of Science. All rights reserved; **Fig. 2.24** Foth et al. *Science.* 2003 Jan 31;299(5607):705–8, adapted from fig 2. © The American Association for the Advancement of Science. All rights reserved; **Fig. 2.25a** The Genomics Crew **Fig. 2.25b** Gift from Thomas H. Giddings, Molecular, Cellular and Developmental Biology Dept., University of Colorado; **Fig. 2.26** NCBI (http://www.ncbi.nlm.nih.gov/mapview/map_search.cgi?taxid=4932); **Fig. 2.27** Dujon et al. *Nature.* 2004 Jul 1;430(6995):35–44, fig 6. © Macmillan Publishers Ltd.; **Figs. 2.28, 2.29, 2.30** Adams et al. *Science.* 2000 Mar 24;287(5461):2185–95, fig 1; adapted from fig 3; fig from online supplement. © The American Association for the Advancement of Science. All rights reserved; **Fig. 2.31** Rubin et al. *Science.* 2000 Mar 4;287(5461):2204–15, adapted from fig 1. © The American Association for the Advancement of Science. All rights reserved; **Fig. 2.32** Gill et al. *Genetics.* 2004 Oct;168(2):1087–96, fig 1; **Fig. 2.33 a–e, 2.34, 2.35** Yu et al. *Science.* 2002 Apr 5;296(5565):79–92, adapted from figs 3, 4 and 5; fig 8; fig 6. © The American Association for the Advancement of Science. All rights reserved; **Fig. 2.36** Goff et al. *Science.* 2002 Apr 5;296(5565):92–100, adapted from fig 5. © The American Association for the Advancement of Science. All rights reserved; **Fig. 2.37** Rice Chromosome 10 Sequencing Consortium. *Science.* 2003 Jun 6;300(5625):1566–9, adapted from fig 2. © The American Association for the Advancement of Science. All rights reserved; **Figs. 2.38, 2.39, 2.40. 2.41, 2.42** Jaillon et al. *Nature.* 2004 Oct 21;431(7011):946–57, fig 2 and 4b; adapted from fig 6; fig 9; adapted from fig 10. © Macmillan Publishers Ltd.; **Figs. 2.43, 2.44, 2.45, 2.46** Lander et al. *Nature.* 2001 Feb 15;409(6822):pg 878, 897, 896, 902. © Macmillan Publishers Ltd.; **Fig. 2.47** International Human Genome Sequencing Consortium. *Nature.* 2004 Oct 21; 431(7011):931–45, fig 6. © Macmillan Publishers Ltd.; **Fig. 2.48** ENCODE Project Consortium. *Science.* 2004 Oct 22;306(5696):636–40, figs 1 and 3. © The American Association for the Advancement of Science. All rights reserved; **Figs. 2B1, 2B2** She et al. *Nature.* 2004 Oct 21;431(7011):pg 929 and 930. © Macmillan Publishers Ltd.; **Fig. 2B3** Hoskins et al. *Science.* 2000 Mar 24;287(5461):2271–4, adapted from fig 2c. © The American Association for the Advancement of Science. All rights reserved; **Fig. 2B4** Benos et al. *Science.* 2000 Mar 24;287(5461):2220–2, adapted from fig 1. © The American Association for the Advancement of Science. All rights reserved; **Table 2B1** Hoskins et al. *Science.* 2000 Mar 24;287(5461):2271–4, table 1c. © The American Association for the Advancement of Science. All rights reserved; **Fig. MM2.1, MM2.2** Laurie Heyer; **Table 2.6** M. Campbell; **Table 2.8** Lander et al. *Nature.* 2001 Feb 15;409(6822):860–921, adapted from pg 896. © Macmillan Publishers Ltd.

Chapter 3 Fig. 3.1 Bentley and Parkhill. *Annu Rev Genet.* 2004;38:771–92, fig 1; **Figs. 3.2, 3.3** Venter et al. *Science.* 2004 Apr 2;304(5667):66–74, fig 2; adapted from fig 7. Epub 2004

Mar 4. © The American Association for the Advancement of Science. All rights reserved; **Fig. 3.4** Jordan et al. *Nature.* 2005 Feb 10; 433(7026):633–8, adapted from fig 1. Epub 2005 Jan 19. © Macmillan Publishers Ltd.; **Fig. 3.5** Gilad et al. *PLoS Biol.* 2004 Jan;2(1):E5, fig 2. Epub 2004 Jan 20; **Fig. 3.6a–b** Thomas et al. *Nature.* 2003 Aug 14;424(6950):788–93, fig 1. © Macmillan Publishers Ltd.; **Figs. 3.7, 3.8, 3.9** Andelfinger et al. *Genomics.* 2004 Jun;83(6):1053–62, fig 1, 2 and 3; **Fig. 3.10** Martin and Embley. *Nature.* 2004 Sep 9;431(7005):134–7, fig 2. © Macmillan Publishers Ltd.; **Fig. 3.11** M. Campbell; **Figs. 3.12, 3.13** Rivera and Lake. *Nature.* 2004 Sep 9;431(7005):152–5, adapted from fig 2, 3 and supplement. © Macmillan Publishers Ltd. Courtesy of James Lake; **Figs. 3.14, 3.15, 3. 16, 3.17** Horiike et al. *Nat Cell Biol.* 2001 Feb;3(2):210–4, fig 1; adapted from fig 2; **Fig. 3.19** Shigenobu et al. *Nature.* 2000 Sep 7;407(6800):81–6, fig 3. © Macmillan Publishers Ltd.; **Fig. 3.22** Cole et al. *Nature.* 2001 Feb 22;409(6823):1007–11, fig 3. © Macmillan Publishers Ltd.; **Fig. 3.23** Fuerst and Webb. *PNAS.* 1991 Sep 15; 88(18):8184–8, fig 2a. © National Academy of Sciences, U.S.A.; **Fig. 3.25** Madsen et al. *Nature.* 2001 Feb 1;409(6820):610–4, adapted from fig 1b. © Macmillan Publishers Ltd.; **Figs. 3.26, 3.27** Ingman et al. *Nature.* 2000 Dec 7;408(6813):708–13, adapted from fig 2; fig 1. © Macmillan Publishers Ltd.; **Fig. 3.28a** Ruiz-Pesini et al. *Science.* 2004 Jan 9;303(5655):223–6, fig S2. © The American Association for the Advancement of Science. All rights reserved; **Fig. 3.28b** KEGG, OxPhos; **Fig. 3.29a–b** U.S. Army Photos; **Fig. 3.30** DeMello. *Nature.* 2003 Mar 6;422(6927):28–9, adapted from fig 1. © Macmillan Publishers Ltd.; **Figs. 3.31, 3.32** Burns et al. *Science.* 1998 Oct 16;282(5388):484–7, adapted from figs 1 and 4. © The American Association for the Advancement of Science. All rights reserved; **Fig. 3.33** Belgrader et al. *Science.* 1999 Apr 16; 284(5413):449–50, fig 1. © The American Association for the Advancement of Science. All rights reserved; **Fig. 3.34** Read et al. *Science.* 2002 Jun 14;296(5575):2028–33, fig 1 and 2. Epub 2002 May 9. © The American Association for the Advancement of Science. All rights reserved; **Figs. 3.35, 3.36** Vreeland et al. *Nature.* 2000 Oct 19;407(6806):897–900, fig 1 and 2. © Macmillan Publishers Ltd.; **Figs. 3.37, 3.38** Salo et al. *PNAS.* 1994 Mar 15;91(6): 2091–4, fig 1, 2 and 4. © National Academy of Sciences, U.S.A.; **Fig. 3.39** Nerlich et al. *Lancet.* 1997 Nov 8;350(9088):1404. © Elsevier Science; **Fig. 3.40** Eisenach et al. *J Infect Dis.* 1990 May;161(5): 977–81, fig 3a. ©1990 University of Chicago Press; **Fig. 3.41** Laura Richman; **Fig. 3.42** Richman et al. *Science.* 1999 Feb 19;283(5405): 1171–6, fig 4b. © The American Association for the Advancement of Science. All rights reserved; Fig. 3.43 Robert Lanciotti; **Fig. 3.44** Lanciotti et al. *Science.* 1999 Dec 17;286(5448):2333–7, fig 2. © The American Association for the Advancement of Science. All rights reserved; **Fig. 3.45** Chinese SARS Molecular Epidemiology Consortium. *Science.* 2004 Mar 12;303(5664):1666–9. Epub 2004 Jan 29, fig 1. © The American Association for the Advancement of Science. All rights reserved; **Figs. 3.46, 3.47** Pizza et al. *Science.* 2000 Mar 10;287(5459):1816–20, fig 2 and 3. © The American Association for the Advancement of Science. All rights reserved; **Fig. 3.48ab** http://gsbs.utmb.edu/microbook/; **Fig. 3.48c** Pizza et al. *Science.* 2000 Mar 10;287(5459):1816–20, adapted from fig 2. © The American Association for the Advancement of Science. All rights reserved; **Fig. 3.49** Onishi et al. *Science.* 1996 Nov 8;274(5289):980–2, adapted from fig 3. © The American Association for the Advancement of Science. All rights reserved; **Fig. 3.50a** Ma et al. *Antisense & Nucleic Acid Drug Dev.* 1998 Oct;8(5):415–26, fig 1a; **Fig. 3.50b** Werner et al. *Antisense & Nucleic Acid Drug Dev.* 1999

Feb;9(1):81–8, fig 2a; **Fig. 3.51a** Ma et al. *Antisense & Nucleic Acid Drug Dev.* 1998 Oct;8(5):415–26, fig 1b; **Fig. 3.51b** Werner et al. *Antisense & Nucleic Acid Drug Dev.* 1999 Feb;9(1):81–8, fig 2b; **Fig. 3.52** Ma et al. *Antisense & Nucleic Acid Drug Dev.* 1998 Oct;8(5):415–26, fig 6; **Fig. 3.53** Jacque et al. *Nature.* 2002 Jul 25; 418(6896):435–8, adapted from fig 1. Epub 2002 Jun 26. © Macmillan Publishers Ltd.; **Figs. 3.54, 3.55a–c** Robertson et al. *PLoS Biol.* 2005 Jan;3(1):e5, adapted from fig 1 and 3b. Epub 2004 Dec 28; **DQ Fig. 67** Grenfell et al. *Science.* 2004 Jan 6;303(5656): 327–32, fig 1. © The American Association for the Advancement of Science. All rights reserved; **Table 3.2** Stewart Cole.

Chapter 4 Fig. 4.1 Kennie Comstock; **Fig. 4.4** Ginger Armbrust; **Fig. 4.5** Rynearson et al. *Limnology and Oceanography.* 2000; 45:1329–1340, adapted from fig 4; **Figs. 4.6, 4.7a** Rynearson et al. *Mol Ecol.* 2005 May;14(6):1631–40, figs 2b, 3 and 4; adapted from fig 2; **Fig. 4.7b** Ginger Armbrust; **Fig. 4.8** Daly et al. *Nat Genet.* 2001 Oct;29(2):229–32, fig 2; **Fig. 4.9** Rynearson et al. *Limnology and Oceanography.* 2000; 45:1329–1340, adapted from fig 6; **Fig. 4.10** Marx. *Science.* 2004 Apr 30;304(5671):658–9, pg 659. © The American Association for the Advancement of Science. All rights reserved; **Fig. 4.13** Evans et al. *Science.* 1999 Oct 15;286 (5439):487–91, fig 2. © The American Association for the Advancement of Science. All rights reserved; **Figs. 4.14, 4.15** Walker et al. *Nature.* 2000 May 18;405(6784):296–7, adapted from fig 1. © Macmillan Publishers Ltd.; **Table 4.1** Rynearson et al. *Limnology and Oceanography.* 2000; 45: pg 1334. Image supplied by Kennie Comstock; **Table 4.3** Sachidanandam et al. *Nature.* 2001 Feb 15; 409(6822):pg 929. © Macmillan Publishers Ltd.

Chapter 5 Figs. 5.3, 5.4, 5.5, 5.6 Zhang et al. *Nature.* 1994 Dec 1;372(6505):425–32, fig 2a, 3a, 4b, and 6a. ©1994 Macmillan Publishers Ltd.; **Fig. 5.7** Pelleymounter et al. *Science.* 1995 Jul 28; 269(5223):540–3, adapted from fig 2a–d. © The American Association for the Advancement of Science. All rights reserved; **Fig. 5.8** Friedman. *Nature.* 2000 Apr 6;404(6778):632–4, fig 1. © Macmillan Publishers Ltd.; **Fig. 5.11** Heymsfield et al. *JAMA.* 1999 Oct 27;282(16):1568–75, fig 2; **Fig. 5.12** Schwartz et al. *Nature.* 2000 Apr 6;404(6778):661–71, fig 21. © Macmillan Publishers Ltd.

Chapter 6 Figs. 6.5, 6.6, 6.9 Chu et al. *Science.* 1998 Oct 23; 282(5389):699–705, adapted from fig 3; fig 4. © The American Association for the Advancement of Science. All rights reserved; **Figs. 6.10, 6.11, 6.12, 6.13, 6.14, 6.15, 6.16, 6.17** DeRisi et al. - *Science.* 1997 Oct 24;278(5338):680–6, adapted from fig 1, 2, 3 and 5; fig 4. © The American Association for the Advancement of Science. All rights reserved; **Figs. 6.18, 6.19, 6.20, 6.21, 6.22, 6.23** Gasch et al. *Mol Biol Cell.* 2000 Dec;11(12):4241–57, adapted from fig 1, 3, 2, 5, 8a, and 6ab; **Figs. 6.24, 6.25, 6.26** Hughes et al. *Nat Genet.* 2000 Jul;25(3):333–7, adapted from fig 4; adapted from fig 3; **Figs. 6.27, 6.28, 6.29, 6.30, 6.31** Shoemaker et al. *Nature.* 2001 Feb 15;409 (6822):922–7, adapted from fig 2; fig 3; adapted from fig 4. © Macmillan Publishers Ltd.

Chapter 7 Figs. 7.1, 7.2, 7.3, 7.4, 7.5 Alizadeh et al. 2000. *Nature.* 403: 505, fig1, 2, 3, 5ab, and 5c. © Macmillan Publishers Ltd.; **Figs. 7.6, 7.7** Perou et al. 2000. *Nature.* 406: 748, fig 1 and 3. © Macmillan Publishers Ltd.; **Figs. 7.8, 7.9, 7.10** Lucito et al. 2000. *Genome Research.* 10: 1727, adapted from figs 1, 2, 3, and 4; **Figs. 7.11, 7.12, 7.14, 7.13** Behr et al. 1999. *Science.* 284: 1521, fig 1 and 2. © The American Association for the Advancement of Science. All rights reserved; **Figs. 7.15, 7.16** Ross et al. 2000. *Nature*

Index

Page references followed by *f* indicate an illustration; page references followed by *t* indicate a table.

abundance-based coverage estimators (ACE), 184
Acanthamoeba polyphaga, 66
α-actin, 20
γ-actinin, 20
adbn-/- mice, 25–26
adbn-/- muscle, pathology of, 25*f*
adipose tissue, effect of leptin on, 201–202
Advanced Nucleic Acid Analyzer (ANAA), 148, 148*f*
African descent, 141, 142*f*, 143
aging/life expectancy, 198–200
agouti-related protein (AgRP), 228–229
algorithm, 70
alien genes, 59
alkaline phosphatase, 323
alleles, 199
    comparison to mutations, 187
alternating cycle, 71–72
Altman, S., 166
alu sequences, 102, 124
amenorrhea, 226
amyloid precursor protein (APP), 106
ancient DNA, 149, 154
aneuploidy, 255, 270*f*
Angelman Syndrome (AS), 55–56
animal model system, need for, 9
annotation
    of genomes, 34, 41, 44–45, 54
    microarrays in improving, 259, 260*f*
    microarrays in validating, 259–260, 261*f*
antagonistic pleiotropy, 199
anthrax, as biological weapon, 145–150, 147*f*, 148*f*, 149*f*
antibiotics, 164–166
antibody microarrays, 320
antidystrophin antibody, 23
antigen pairs, 320
Antinori, S., 212
antisense technology, 166
antiswitches, 400, 401*f*, 402*f*
apicoplast, 75–76
apoptosis, 27, 88, 144, 312
appetite
    stimulation of, 229
    suppression of, 229

applied research with DNA microarrays, 264–282
    on cancer, 264–273
    on improving health care, 273–282
    signature genes in, 266
aptamer, 400
aquaporin-1, 296
Archaea, 126
archenteron, 343
arcs, 12
Arkin, A., 370
Armbrust, G., 178–186
Arntzen, Charles, 204
Arrowsmith, C., 295
authenticity criteria, 155
autocatalytic capacity, 23
automated sequencing, 36
autophosphorylate, 322
avian flu, 159–160
avidin, 316
AZT, 194

Bacillus subtilis, 65
bacteria
    evaluating, 62–63
    genomic tradeoffs, 201–202
    gram-negative, 165–167, 165*f*
    survival in salt crystal, 149–151, 151*f*
bacterial artificial chromosomes (BACs), 49–50, 86
bacterial genomes, size of, 66–67
bacteriorhodopsin, 120
Bacteroides thetaiotaomicron, 62–63
Bailey, L., 204
baker's yeast, 78–81, 79*f*
base compositions, 115, 115*t*
basic research with DNA microarrays, 234–261
    alternative uses in, 254–261
    gene clusters in, 242–243, 243*t*
    log-transformation of microarray data, 239–240
    measuring similarity between expression patterns, 240–241
Bates, J., 154–155
Bayer, T., 400–401
Bayes' Rule, 208, 311
BCG (*bacille de Calmette et Guérin*) vaccine, 273–276, 275*f*
Beckmann, J., 27, 29
Beckwith-Wiedemann Syndrome, 58

Behr, M., 273, 275
Belgrader, P., 149
Belkin, L., 231
Benzer, S., 411
Berkeley *Drosophila* Genome Project (BDGP), 51, 86
Bhalla, U., 378–381, 384, 386–387, 391
bimodal toggle switch, 358
binomial coefficient, 293
BioBricks projects, 394
biofilms, 62
biological plastics, 204
biological process of proteins, 49, 286, 292*f*
biological weapons, identification of, 146–150, 147*f*, 148*f*, 149*f*
biology
    synthetic, 65, 394–407
    systems, 213, 410–411, 416–419
biomedical genome research, 162–173
    antibiotics, 165–167, 165*f*, 166*f*
    External Guide Sequence, 167–168, 168*f*, 169*f*
    RNA inhibition (RNAi), 169–170, 170*f*
    targeting of RNA genomes, 170–171
    vaccines, 163–165, 164*f*
biotin, 316
bistable circuit, 383
bistable toggle switches, 370, 394*f*, 395*f*
bit score, 7
BLAST2, 45
BLAST hits, 44
BLASTp, 8, 9, 46–47
blastula, 343
Blattner, F., 114
Blobel, G., 410
BLOSUM62, 46–47, 46*t*, 47*t*
BMD, 10, 13
bootstrap analysis, 143
Borodovsky, M., 44
Botstein, D., 250, 251, 264, 268, 276
Boyles, W., 146
brain, role of, in weight homeostasis, 228–229
BRCA1, 193, 212
BRCA2, 212
breast cancer, categorization with microarrays, 268–272, 270*f*, 271*f*, 272*f*
Brenner, S., 410
Bridges, C., 83